T0255032

LONDON MATHEMATICAL SOCIETY STUDENT TEXTS

53 The prime number theorem, G. J. O. JAMESON
54 Topics in graph automorphisms and reconstruction, JOSEF LAURI & RAFFAELE SCAPELLATO
55 Elementary number theory, group theory and Ramanujan graphs, GIULIANA DAVIDOFF, PETER SARNAK & ALAIN VALETTE
56 Logic, induction and sets, THOMAS FORSTER
57 Introduction to Banach algebras, operators and harmonic analysis, GARTH DALES *et al*
58 Computational algebraic geometry, HAL SCHENCK
59 Frobenius algebras and 2-D topological quantum field theories, JOACHIM KOCK
60 Linear operators and linear systems, JONATHAN R. PARTINGTON
61 An introduction to noncommutative Noetherian rings: Second edition, K. R. GOODEARL & R. B. WARFIELD, JR
62 Topics from one-dimensional dynamics, KAREN M. BRUCKS & HENK BRUIN
63 Singular points of plane curves, C. T. C. WALL
64 A short course on Banach space theory, N. L. CAROTHERS
65 Elements of the representation theory of associative algebras I, IBRAHIM ASSEM, DANIEL SIMSON & ANDRZEJ SKOWROŃSKI
66 An introduction to sieve methods and their applications, ALINA CARMEN COJOCARU & M. RAM MURTY
67 Elliptic functions, J. V. ARMITAGE & W. F. EBERLEIN
68 Hyperbolic geometry from a local viewpoint, LINDA KEEN & NIKOLA LAKIC
69 Lectures on Kähler geometry, ANDREI MOROIANU
70 Dependence logic, JOUKU VÄÄNÄNEN
71 Elements of the representation theory of associative algebras II, DANIEL SIMSON & ANDRZEJ SKOWROŃSKI
72 Elements of the representation theory of associative algebras III, DANIEL SIMSON & ANDRZEJ SKOWROŃSKI
73 Groups, graphs and trees, JOHN MEIER
74 Representation theorems in Hardy spaces, JAVAD MASHREGHI
75 An introduction to the theory of graph spectra, DRAGOŠ CVETKOVIĆ, PETER ROWLINSON & SLOBODAN SIMIĆ
76 Number theory in the spirit of Liouville, KENNETH S. WILLIAMS
77 Lectures on profinite topics in group theory, BENJAMIN KLOPSCH, NIKOLAY NIKOLOV & CHRISTOPHER VOLL
78 Clifford algebras: An introduction, D. J. H. GARLING
79 Introduction to compact Riemann surfaces and dessins d'enfants, ERNESTO GIRONDO & GABINO GONZÁLEZ-DIEZ
80 The Riemann hypothesis for function fields, MACHIEL VAN FRANKENHUIJSEN
81 Number theory, Fourier analysis and geometric discrepancy, GIANCARLO TRAVAGLINI
82 Finite geometry and combinatorial applications, SIMEON BALL
83 The geometry of celestial mechanics, HANSJÖRG GEIGES
84 Random graphs, geometry and asymptotic structure, MICHAEL KRIVELEVICH *et al*
85 Fourier analysis: Part I - Theory, ADRIAN CONSTANTIN
86 Dispersive partial differential equations, M. BURAK ERDOĞAN & NIKOLAOS TZIRAKIS
87 Riemann surfaces and algebraic curves, R. CAVALIERI & E. MILES
88 Groups, languages and automata, DEREK F. HOLT, SARAH REES & CLAAS E. RÖVER
89 Analysis on Polish spaces and an introduction to optimal transportation, D. J. H. GARLING
90 The homotopy theory of $(\infty,1)$-categories, JULIA E. BERGNER
91 The block theory of finite group algebras I, M. LINCKELMANN
92 The block theory of finite group algebras II, M. LINCKELMANN

LONDON MATHEMATICAL SOCIETY STUDENT TEXTS

Managing Editor: Ian J. Leary

Mathematical Sciences, University of Southampton, UK

London Mathematical Society Student Texts 91

The Block Theory of Finite Group Algebras

Volume I

MARKUS LINCKELMANN
City, University of London

CAMBRIDGE
UNIVERSITY PRESS

University Printing House, Cambridge CB2 8BS, United Kingdom

One Liberty Plaza, 20th Floor, New York, NY 10006, USA

477 Williamstown Road, Port Melbourne, VIC 3207, Australia

314-321, 3rd Floor, Plot 3, Splendor Forum, Jasola District Centre, New Delhi - 110025, India

79 Anson Road, #06-04/06, Singapore 079906

Cambridge University Press is part of the University of Cambridge.

It furthers the University's mission by disseminating knowledge in the pursuit of education, learning and research at the highest international levels of excellence.

www.cambridge.org
Information on this title: www.cambridge.org/9781108441834
DOI: 10.1017/9781108349321

First published 2018

A catalogue record for this publication is available from the British Library

ISBN – 2 Volume Set 978-1-108-44190-2 Paperback
ISBN – Volume I 978-1-108-42591-9 Hardback
ISBN – Volume I 978-1-108-44183-4 Paperback
ISBN – Volume II 978-1-108-42590-2 Hardback
ISBN – Volume II 978-1-108-44180-3 Paperback

Contents

Volume I

Volume II

Introduction

Group representation theory investigates the structural connections between groups and mathematical objects admitting them as automorphism groups. Its most basic instance is the action of a group G on a set M, which is equivalent to a group homomorphism from G to the symmetric group S_M of all permutations of the set M, thus representing G as an automorphism group of the set M. Classical representation theory, developed during the last decade of the 19th century by Frobenius and Schur, investigates the representations of a finite group G as linear automorphism groups of complex vector spaces, or equivalently, modules over the complex group algebra $\mathbb{C}G$. Modular representation theory, initiated by Brauer in the 1930s, considers finite group actions on vector spaces over fields with positive characteristic, and more generally, on modules over complete discrete valuation rings as a link between different characteristics. Integral representation theory considers representations of groups over rings of algebraic integers, with applications in number theory. Topologists have extensively studied the automorphism groups of classifying spaces of groups in connection with K-theory and transformation groups. Methods from homotopy theory and homological algebra have shaped the area significantly.

Within modular representation theory, viewed as the theory of module categories of finite group algebras over complete discrete valuation rings, the starting point of block theory is the decomposition of finite group algebras into indecomposable direct algebra factors, called *block algebras*. The block algebras of a finite group algebra are investigated individually, bearing in mind that the module category of an algebra is the direct sum of the module categories of its blocks. Block theory seeks to gain insight into which way the structure theory of finite groups and the representation theory of block algebras inform each other.

Few algebras are expected to arise as block algebras of finite groups. Narrowing down the pool of possible block algebras with essentially representation theoretic methods has been very successful for blocks of finite and tame representation type, but remains a major challenge beyond those cases. Here is a sample list of properties that have to be satisfied by an algebra B which arises as a block algebra of a finite group algebra over a complete discrete valuation ring $\mathcal{O}$ with residue field k of prime characteristic p and field of fractions K of characteristic 0.

- B is symmetric.
- $K \otimes_{\mathcal{O}} B$ is semisimple.
- The canonical map $Z(B) \to Z(k \otimes_{\mathcal{O}} B)$ is surjective.
- B is separably equivalent to $\mathcal{O}P$ for some finite p-group P.
- The Cartan matrix of $k \otimes_{\mathcal{O}} B$ is positive definite, its determinant is a power of p, and its largest elementary divisor is the smallest power of p that annihilates all homomorphism spaces in the $\mathcal{O}$-stable category $\underline{\mathrm{mod}}(B)$.
- The decomposition map from the Grothendieck group of finitely generated $K \otimes_{\mathcal{O}} B$-modules to the Grothendieck group of finitely generated $k \otimes_{\mathcal{O}} B$-modules is surjective.
- B is defined over a finite extension of the p-adic integers $\mathbb{Z}_p$, and $Z(B)$ is defined over $\mathbb{Z}_{(p)}$.
- $k \otimes_{\mathcal{O}} B$ is defined over a finite field $\mathbb{F}_q$, where q is a power of p, and $Z(k \otimes_{\mathcal{O}} B)$ is defined over $\mathbb{F}_p$.

The dominant feature of block theory is the dichotomy of invariants associated with block algebras. Block algebras of finite groups have all the usual 'global' invariants associated with algebras – module categories, derived and stable categories, cohomological invariants including Hochschild cohomology, and numerical invariants such as the numbers of ordinary and modular irreducible characters. Due to their provenance from finite groups, block algebras have further 'local' invariants that cannot be, in general, associated with arbitrary algebras. The prominent conjectures that drive block theory revolve around the interplay between 'global' and 'local' invariants. Source algebras of blocks capture invariants from both worlds, and in an ideal scenario, the above mentioned conjectures would be obtained as a consequence of a classification of the source algebras of blocks with a fixed defect group. In this generality, this has been achieved in two cases, namely for blocks with cyclic and Klein four defect groups. The local structure of a block algebra B includes the following invariants.

- A defect group P of B.
- A fusion system $\mathcal{F}$ of B on P.

- A class $\alpha \in H^2(\mathcal{F}^c; k^\times)$ such that α restricts on $\mathrm{Aut}_{\mathcal{F}}(Q)$ to the Külshammer–Puig class α_Q, for any Q belonging to the category $\mathcal{F}^c$ of $\mathcal{F}$-centric subgroups in P.
- The number of weights of $(\mathcal{F}, \alpha)$.

A sample list for the global structure of B includes the following invariants of B as an algebra, as well as their relationship with the local invariants.

- The numbers $|\mathrm{Irr}_K(B)|$ and $|\mathrm{IBr}_k(B)|$ of isomorphism classes of simple $K \otimes_{\mathcal{O}} B$-modules and $k \otimes_{\mathcal{O}} B$-modules, respectively, with their heights.
- The $\mathcal{O}$-stable module category $\underline{\mathrm{mod}}(B)$ and its dimension as a triangulated category.
- The bounded derived category $D^b(B)$ and its dimension as a triangulated category.
- The module category $\mathrm{mod}(B)$ as an abelian category, structure and Loewy lengths of projective indecomposable B-modules.
- The generalised decomposition matrix of B.

All of the above local and global invariants can be calculated, at least in principle, from the source algebras of B, and hence methods to determine source algebras are a major theme in this book.

Volume I introduces the broader context and many of the methods that are fundamental to modular group representation theory. Chapter 1 provides background on algebras and introduces some of the main players in this book – group algebras, twisted group algebras as well as category algebras, for the sake of giving a broader picture. Chapter 2 switches the focus from algebras to module categories and functors. Chapter 3 develops the classical representation theory of finite groups – that is, representations over complex vector spaces – just far enough to prove Burnside's $p^a q^b$-Theorem and describe Brauer's characterisation of characters. Turning to modular representation theory, Chapter 4 handles the general theory of algebras over discrete valuation rings. Chapter 5 combines this material with group actions, leading to Green's theory of vertices and sources, Puig's notion of pointed groups, and further fundamental module theoretic results on special classes of modules, as well as Green's Indecomposability Theorem.

In Volume II, the core theme of this book takes centre stage. Chapter 6 develops in a systematic way block theory, including Brauer's three main theorems, some Clifford Theory, the work of Alperin and Broué on Brauer pairs and Puig's notion of source algebras. Chapter 7 describes modules over finite p-groups, with an emphasis on endopermutation modules. This is followed by another core chapter on local structure, containing in particular a brief introduction to fusion systems, connections between characters and local structure, the structure theory of nilpotent blocks and their extensions. Chapter 9 on isometries illustrates the interaction between the concepts introduced up to this point. Applications in subsequent chapters include the structure theory of blocks with cyclic or Klein four defect groups.

Along the way, some of the fundamental conjectures alluded to above will be described. Alperin's weight conjecture predicts that the number of isomorphism classes of simple modules of a block algebra should be determined by its local invariants. Of a more structural nature, Broué's abelian defect group conjecture would offer, if true, some explanation for these numerical coincidences at least in the case of blocks with abelian defect groups. The finiteness conjectures of Donovan, Feit and Puig predict that once a defect group is fixed, there are only 'finitely many blocks' with certain properties. These conjectures are known to hold for blocks of various classes of finite simple groups and their extensions. Complemented by rapidly evolving reduction techniques, this points to the possibility of proving parts of these conjectures by invoking the classification of finite simple groups. A lot more work seems to be needed to provide the understanding that would transform mystery into insight.

The representation theory of finite group algebras draws significantly on methods from areas including ring theory, category theory, and homological algebra. Rather than giving systematic introductions to those areas, we develop background material as we go along, trying not to lose sight of the actual topic.

Acknowledgements. The author wishes to thank David Craven, Charles Eaton, Adam Glesser, Radha Kessar, Justin Lynd, Michal Stolorz, Michael Livesey, Lleonard Rubio y Degrassi and Benjamin Sambale for reading, commenting on, and correcting parts of various preliminary versions. Special thanks go to Dave Benson, Morty Harris, Sejong Park, as well as to Robert Boltje and the members of his research seminar at UCSC, for long lists of very detailed comments and suggestions.

1
Algebras and Modules

Algebras tend to have far too many modules for explicit descriptions of their module categories. One way to circumvent this issue is to make comparative rather than descriptive statements, relating the module category of an algebra A to that of a 'simpler' algebra B. This leads to considering Morita, derived and stable equivalences. Another strategy is to focus directly on describing the algebra structure of A rather than its module category. Explicit descriptions of algebras include matrix algebras, polynomial algebras, group algebras, category algebras, their twisted versions by some 2-cocycles, as well as quotients, direct products and tensor products thereof. These constructions, which work over arbitrary commutative rings, are described in the first few sections of this chapter. We also review some properties of semisimple, projective and injective modules, as well as some basics from homological algebra. For much of this chapter, it is possible to ignore at a first reading the category theoretic terminology that appears as a background theme in order to prepare for the more systematic use of category theory in later chapters.

All rings are nonzero associative unital, and all ring homomorphisms are unital, unless stated otherwise. The *centre of a ring* A is the set, denoted $Z(A)$, consisting of all elements $z \in A$ satisfying $az = za$ for all $a \in A$. The centre $Z(A)$ of A is a subring of A and we have $Z(A) = A$ if and only if the ring A is commutative. Given any commutative ring k, a *k-algebra* is a pair consisting of a ring A and a ring homomorphism $\eta : k \to Z(A)$. If no confusion arises we write λa instead of $\eta(\lambda)a$, where $\lambda \in k$ and $a \in A$. For A, B two k-algebras, a *homomorphism of k-algebras* from A to B is a ring homomorphism $\sigma : A \to B$ satisfying in addition $\sigma(\lambda a) = \lambda\sigma(a)$ for all $a \in A, \lambda \in k$. Not every ring homomorphism between two k-algebras is necessarily a k-algebra homomorphism because of the extra compatibility condition on scalars. In particular, it is possible for two k-algebras to be isomorphic as rings but not as k-algebras.

Modules are a far reaching generalisation of the idea, describing an algebra as a subalgebra of a matrix algebra. A *left module* over a k-algebra A is an abelian group U together with a bilinear map $A \times U \to U$, sending (a, u) to an element in U denoted au, such that $1u = u$ and $a(bu) = (ab)u$ for all $u \in U$ and all $a, b \in A$. A *right A-module* is an abelian group V with a bilinear map $V \times A \to V$ with the analogous properties. Unless specified otherwise, modules over a k-algebra A will be left modules. Any A-module U inherits a k-module structure via the canonical map $k \to Z(A)$. In particular, A is a k-module via the canonical map $k \to Z(A)$, and the multiplication in A is k-bilinear. The *opposite algebra* A^{op} of a k-algebra is defined as follows: as a k-module we set $A^{\mathrm{op}} = A$, and the product in A^{op} is given by $a \cdot b = ba$, where $a, b \in A$ and where ab is the product of a and b in A. A right A-module V can be viewed as a left A^{op}-module. For U, U' two A-modules we denote by $\mathrm{Hom}_A(U, U')$ the k-module of A-homomorphisms from U to U'; we will frequently write $\mathrm{End}_A(U)$ instead of $\mathrm{Hom}_A(U, U)$. Composition of maps in $\mathrm{End}_A(U)$ induces a k-algebra structure on $\mathrm{End}_A(U)$, which is then a subalgebra of $\mathrm{End}_k(U)$. The map sending $a \in A$ to the k-endomorphism of U given by $u \mapsto au$, for any $u \in U$, is a k-algebra homomorphism $A \to \mathrm{End}_k(U)$. Similarly, for V, V' two right A-modules we denote by $\mathrm{Hom}_{A^{\mathrm{op}}}(V, V')$ the k-module of right A-module homomorphisms from V to V'. We will typically write $\mathrm{End}_{A^{\mathrm{op}}}(V)$ instead of $\mathrm{Hom}_{A^{\mathrm{op}}}(V, V)$, and the map sending $a \in A$ to the k-endomorphism $v \to va$ of V, for any $v \in V$, is a k-algebra homomorphism $A^{\mathrm{op}} \to \mathrm{End}_k(V)$. We refer to §A.1 in the appendix for details on tensor products of modules. We denote by $\mathrm{Mod}(A)$ the category of A-modules, with A-homomorphisms as morphisms, and, whenever useful, we identify $\mathrm{Mod}(A^{\mathrm{op}})$ with the category of right A-modules without further comment.

One of the main tools to describe functors between module categories are bimodules. Given two k-algebras A, B, an *A-B-bimodule* is an abelian group M that is a left A-module and a right B-module such that the two module structures commute; that is, such that $a(mb) = (am)b$ for all $a \in A$, $b \in B$ and $m \in M$; we write amb instead of $a(mb)$ or $(am)b$. In addition, we adopt the convention that the two k-module structures on M inherited from the left A-module structure and the right B-module structure coincide. Thus we may consider M as a module over the algebra $A \otimes_k B^{\mathrm{op}}$ via $(a \otimes b) \cdot m = amb$, where $a \in A$, $b \in B$ and $m \in M$. For M, M' two A-B-bimodules we denote by $\mathrm{Hom}_{A \otimes_k B^{\mathrm{op}}}(M, M')$ the k-module of A-B-bimodule homomorphisms from M to M'. The map sending $a \in A$ to the endomorphism $m \mapsto am$, for any $m \in M$, is a k-algebra homomorphism $A \to \mathrm{End}_{B^{\mathrm{op}}}(M)$, and the map sending $b \in B$ to the endomorphism $m \mapsto mb$ is a k-algebra homomorphism $B^{\mathrm{op}} \to \mathrm{End}_A(M)$. Given three k-algebras A, B, C, an A-B-bimodule M and a B-C-bimodule V, the tensor product $M \otimes_B V$ is an A-C-bimodule. Similarly, homomorphism spaces between bimodules can be used to construct bimodules. If N is an A-C-bimodule, then the homomorphism space $\mathrm{Hom}_A(M,N)$ from M to N viewed as left A-modules inherits a B-C-bimodule structure from the right actions of B on M and C on N, given explicitly by $(b.\mu.c)(m) = \mu(mb)c$ for any $\mu \in \mathrm{Hom}_A(M,N)$, $b \in B$, $c \in C$ and $m \in M$. Similarly, if M' is a B-A-bimodule and N' a C-A-bimodule, then $\mathrm{Hom}_{A^{\mathrm{op}}}(M',N')$ becomes a C-B-bimodule via $(c.\nu.b)(m') = c\nu(bm')$, for any $\nu \in \mathrm{Hom}_{A^{\mathrm{op}}}(M',N')$, $b \in B$, $c \in C$ and $m' \in M'$.

A *generating subset* of an A-module M is a subset S of M such that every $m \in M$ can be written as an A-linear combination $m = \sum_{s\in S} a_s s$ with coefficients $a_s \in A$ of which only finitely many are nonzero. An A-module M is called *finitely generated* if M has a finite generating subset. An A-module F is called *free* if F has a generating subset S such that every element in F can be written *uniquely* as an A-linear combination $\sum_{s\in S} a_s s$ with coefficients $a_s \in A$ of which only finitely many are nonzero. Any such subset S of F is then called a *basis of* F. The free A-module A with basis $\{1\}$ is called the *regular* A-module. Every A-module U is isomorphic to a quotient of a free A-module F; if M is finitely generated then F can be chosen to have a finite basis. The category of A-modules $\mathrm{Mod}(A)$ is a k-linear abelian category. This means essentially that for any A-modules U, V, W, the composition of A-homomorphisms yields k-bilinear maps $\mathrm{Hom}_A(V,W) \times \mathrm{Hom}_A(U,V) \to \mathrm{Hom}_A(U,W)$, and that the first isomorphism theorem holds; that is, $\mathrm{Im}(\varphi) \cong U/\ker(\varphi)$ for any A-module homomorphism $\varphi : U \to V$. The subcategory of finitely generated A-modules, denoted $\mathrm{mod}(A)$, is a k-linear subcategory but not necessarily an abelian subcategory of $\mathrm{Mod}(A)$ because a submodule of a finitely generated module need not be finitely generated.

A module M over an algebra A is called *Noetherian* if every ascending chain of submodules $U_0 \subseteq U_1 \subseteq \cdots$ of M is eventually constant; that is, if $U_i = U_{i+1}$ for all i large enough. An A-module M is Noetherian if and only if every submodule of M is finitely generated. In particular, any quotient and any submodule of a Noetherian module over an algebra is again Noetherian, and hence the category of Noetherian A-modules is an abelian subcategory of $\mathrm{Mod}(A)$. We say that A is left (resp. right) Noetherian, if A is Noetherian as a left (resp. right) A-module, and that A is Noetherian if it is left and right Noetherian. If A left Noetherian or commutative, then any free A-module F with a finite basis S having n elements has the property that any other basis of F has n elements. The integer n is then called the *rank of* F. The map sending the expression $\sum_{s\in S} a_s s$ with coefficients $a_s \in A$ to the n-tuple $(a_s)_{s\in S}$ in some fixed order is an isomorphism of A-modules $F \cong A^n$.

An A-module M is called *Artinian* if every descending chain $U_0 \supseteq U_1 \supseteq \cdots$ of submodules of M is eventually constant. We say that A is left (resp. right) Artinian, if A is Artinian as a left (resp. right) A-module, and that A is Artinian if it is left and right Artinian. If k is a field and A a finite-dimensional k-algebra, then every finitely generated A-module M has finite dimension over k, hence is Artinian and Noetherian. More generally, if k is Noetherian (resp. Artinian) and if A is a k-algebra that is finitely generated as a k-module, then A and $Z(A)$ are Noetherian (resp. Artinian), and hence every finitely generated left or right A-module is Noetherian (resp. Artinian).

If not stated otherwise, k is a unital commutative ring.

1.1 Group algebras

Let A be a k-algebra that is free as a k-module, and let $\{a_i | i \in I\}$ be a k-basis of A, where I is some indexing set. The multiplication in A can be specified in terms of the 3-dimensional matrix of multiplicative constants $\lambda(i, j, s) \in k$ satisfying $a_i a_j = \sum_{s \in I} \lambda(i, j, s) a_s$ for all $i, j \in I$ such that, for fixed i, j, only finitely many of the $\lambda(i, j, s)$ are nonzero as s runs over I. Any choice of a k-basis of A and a 3-dimensional matrix as above determines a k-bilinear – albeit not necessarily associative – map $A \times A \to A$. Group algebras are special cases of algebras with a basis that is closed under multiplication:

Definition 1.1.1 The *group algebra* kG of a group G over the commutative ring k is the algebra that is the free k-module having the set of elements of G as a k-basis, endowed with the unique k-bilinear multiplication induced by the group multiplication of G. More explicitly, the elements of kG are the formal sums $\sum_{x \in G} \lambda_x x$, where $\lambda_x \in k$ for all $x \in G$, with only finitely many of the coefficients λ_x nonzero. The sum in kG is given componentwise by the formula

$$\Big(\sum_{x \in G} \lambda_x x\Big) + \Big(\sum_{x \in G} \mu_x x\Big) = \sum_{x \in G} (\lambda_x + \mu_x) x,$$

the scalar multiplication in kG is given by

$$\lambda\Big(\sum_{x \in G} \lambda_x x\Big) = \sum_{x \in G} (\lambda \lambda_x) x,$$

and the product in kG is given by the formula

$$\Big(\sum_{x \in G} \lambda_x x\Big)\Big(\sum_{x \in G} \mu_x x\Big) = \sum_{x, y \in G} \lambda_x \mu_y xy = \sum_{z \in G}\Big(\sum_{x, y \in G, xy = z} \lambda_x \mu_y\Big) z,$$

where as before the coefficients $\lambda_x, \mu_x, \lambda$ are in k, with only finitely many of the λ_x and the μ_x nonzero.

The unit element 1_{kG} of the algebra kG is the image in kG of the unit element 1_G of the group G. The images in kG of the elements of the group G become invertible in the algebra kG in such a way that the image in kG of the inverse x^{-1} in G of an element $x \in G$ is the inverse of the image of x in kG. We tend not to notationally distinguish the elements of G from their images in kG unless this is needed to avoid confusion. The associativity of the product in G implies that the multiplication in kG is associative. The definition of kG makes sense more generally with G replaced by a semigroup, provided we allow possibly non-unital algebras. Semigroup algebras are exactly the non-unital algebras having a basis closed under multiplication.

An *idempotent* in a k-algebra A is a nonzero element $i \in A$ satisfying $i^2 = i$. Two idempotents $i, j \in A$ are said to be *orthogonal* if $ij = 0 = ji$. If i, j are orthogonal idempotents in A then $i + j$ is an idempotent in A. If i is an idempotent in A such that $i \neq 1$, then $1 - i$ is an idempotent in A, and the idempotents i, $1 - i$ are orthogonal. An idempotent $i \in A$ is called *primitive* if i cannot be written as a sum of two orthogonal idempotents. A *primitive decomposition of an idempotent e in a k-algebra A* is a finite set I of pairwise orthogonal primitive idempotents in A such that $\sum_{i \in I} i = e$. Detecting idempotents, invertible and nilpotent elements is fundamental for the structural description of an algebra. We record here one of the standard procedures used to construct idempotents in a group algebra.

Proposition 1.1.2 *Let G be a group and H a finite subgroup of G. Suppose that $|H|$ is invertible in k. Then the element $e_H = \frac{1}{|H|} \sum_{y \in H} y$ is an idempotent in kG. Moreover, if H is normal in G, then e_H is in the centre $Z(kG)$ of kG.*

Proof Left multiplication by $x \in H$ in H permutes the elements in H, hence $x \cdot \sum_{y \in H} y = \sum_{y \in H} y$. Taking the sum over all x in H yields $(\sum_{y \in H} y)^2 = |H| \cdot \sum_{y \in H} y$, hence dividing by $|H|^2$ shows that e_H is an idempotent in kG. If H is normal in G then conjugation by any element in G permutes the elements of H, hence fixes e_H. The result follows. □

Proposition 1.1.3 *Let G, H be groups and let $\varphi : G \to H$ be a group homomorphism. There is a unique k-algebra homomorphism $\alpha : kG \to kH$ such that $\alpha(x) = \varphi(x)$ for any $x \in G$.*

Proof Since the elements of G form a k-basis of kG, there is a unique k-linear map α sending $x \in G$ to $\varphi(x)$. Using that φ is a group homomorphism one sees that α is multiplicative. □

The previous proposition implies that the construction sending a group G to its group algebra kG over k is a *covariant functor from the category of groups to the category of k-algebras*. There is also a functor in the other direction: any k-algebra A gives rise to a group, namely the group $A^\times$ of invertible elements in A. Since an algebra homomorphism maps invertible elements to invertible elements this is indeed a covariant functor from the category of k-algebras to the category of groups. A k-algebra A is called a *division algebra* if $A^\times = A \setminus \{0\}$. A commutative division algebra is a field. All modules over a division algebra are free; this is proved in the same way as the fact that all vector spaces over a field have a basis. Given two k-algebras A, B, their tensor product $A \otimes_k B$ over k is again a k-algebra. The tensor product of two group algebras is again a group algebra:

Theorem 1.1.4 *Let G, H be groups. There is a unique algebra isomorphism $k(G \times H) \cong kG \otimes_k kH$ sending $(x, y) \in G \times H$ to $x \otimes y$.*

Proof The given map, extended bilinearly to $k(G \times H)$, is obviously an algebra homomorphism. As (x, y) runs over $G \times H$, the set $x \otimes y$ runs over a k-basis of $kG \otimes_k kH$. □

In contrast, the *direct* product of two group algebras need not be a group algebra. For instance, if k is a field of characteristic 2, the direct product $kC_2 \times kC_2$ of two copies of the group algebra of a cyclic group of order 2 cannot be isomorphic to a finite group algebra. For if it were, it would have to be the group algebra of a group of order 4, hence isomorphic to either kC_4 or the Klein four group algebra kV_4. However, $kC_2 \times kC_2$ has nontrivial idempotents $(1, 0)$ and $(0, 1)$ while it is easily checked directly that the unit element is the unique idempotent in kC_4 and kV_4 (this follows also from more general statements in 1.11.5 below). An immediate consequence of the basic properties of Noetherian and Artinian modules is the following observation:

Proposition 1.1.5 *Let G be a finite group. If the ring k is Noetherian (resp. Artinian), then the algebras kG and $Z(kG)$ are Noetherian (resp. Artinian).*

Thus, for any finite group G, the integral group ring $\mathbb{Z}G$ is Noetherian but not Artinian, while if k is a field, then the group algebra kG is both Noetherian and Artinian. The trivial kG-module is Noetherian or Artinian if and only if the ring k is Noetherian or Artinian, respectively.

The representation theory of a finite group G over the commutative ring k is essentially the module category $\mathrm{Mod}(kG)$ of its group algebra. It is sometimes convenient to use the equivalent concept of a *representation of G over k*:

Definition 1.1.6 Let G be a group. A *representation of G over k* is a pair (V, ρ) consisting of a k-module V and a group homomorphism $\rho : G \to \mathrm{GL}(V)$, where $\mathrm{GL}(V) = \mathrm{Aut}_k(V)$ is the group of k-linear automorphisms of V.

The representations of G over k form a category; more precisely, given two representations (V, ρ), (W, σ) of G over k, a morphism from (V, ρ) to (W, σ) is a k-linear map $\varphi : V \to W$ satisfying $\varphi \circ \rho(x) = \sigma(x) \circ \varphi$ for all $x \in G$.

If V is free of finite rank n over k, then $\mathrm{GL}(V)$ can be identified with the group of invertible $n \times n$-matrices over k, by choosing a basis of V, and hence a representation yields a group homomorphism from G to the group $\mathrm{GL}_n(k)$ of invertible $n \times n$-matrices with coefficients in k.

Any representation of a group G over k gives rise to a kG-module, and vice versa, as follows. Given a representation $\rho : G \to \mathrm{GL}(V)$ for some k-module V, we define a kG-module structure on V by setting $x \cdot v = \rho(x)(v)$ for all $x \in G$, $v \in V$, and by extending this linearly to a general element in kG in the obvious way, that is $(\sum_{x\in G} \lambda_x x) \cdot v = \sum_{x\in G} \lambda_x \rho(x)(v)$, where $\lambda_x \in k$ with only finitely many λ_x nonzero. Conversely, given a kG-module V, we can regard V as a representation of G over k by restricting the action of kG on V to G; that is, by setting $\rho(x)(v) = xv$ for all $x \in G$ and $v \in V$. In this way, the categories of kG-modules becomes equivalent to the category of representations of G over k. This is a particular case of an equivalence (which we will describe in 1.4.4 below) between functors over a category and modules over the corresponding category algebra, modulo viewing a representation $\rho : G \to \mathrm{GL}(V)$ as a functor from the category $\mathbf{G}$ with one object and endomorphism set G to $\mathrm{Mod}(k)$, sending the unique object of $\mathbf{G}$ to V and any $x \in G$ to $\rho(x)$. In addition to being abelian, the category $\mathrm{Mod}(kG)$ admits a tensor product. Given two kG-modules U, V, the tensor product $U \otimes_k V$ becomes a kG-module with $x \in G$ acting on $u \otimes v$ by $x \cdot (u \otimes v) = xu \otimes xv$, for all $u \in U$ and $v \in V$.

The construction principle of group algebras extends to sets endowed with a group action, yielding *permutation representations* and the corresponding *permutation modules*. A set M on which a group G acts is called a *G-set*.

Definition 1.1.7 Let G be a group, and let M be a G-set. We denote by kM the kG-module that is, as a k-module, free having the set M as basis, with kG-module structure obtained by extending bilinearly the action of elements of G on elements of M. More explicitly, the elements of kM are formal sums $\sum_{m\in M} \mu_m m$, with only finitely many μ_m nonzero, endowed with the action of kG given by

$$\Big(\sum_{x\in G} \lambda_x x\Big) \cdot \Big(\sum_{m\in M} \mu_m m\Big) = \sum_{x\in G}\sum_{m\in M} \lambda_x \mu_m xm$$

$$= \sum_{m\in M} \Big(\sum_{(x,n)\in G\times M, xn=m} \lambda_x \mu_n \Big) m.$$

A kG-module V is called a *permutation module* if $V \cong kM$ for some G-set M, or equivalently, if V is k-free and has a k-basis that is G-stable. If G acts transitively on a k-basis of V, then V is called a *transitive permutation module*.

Examples 1.1.8 Let G be a group.

(i) The trivial kG-module k is a permutation module obtained from the trivial action of G on a set with one element.
(ii) The *regular left kG-module* is the permutation module obtained from the regular action of G on itself by left multiplication.
(iii) For any positive integer n, the symmetric group S_n acts on the set $\Omega = \{1, 2, \ldots, n\}$, and the associated kG-module $k\Omega$ is called the *natural permutation module of S_n*.

Proposition 1.1.9 *Let G be a group, and let U, V be permutation kG-modules. Then $U \oplus V$ and $U \otimes_k V$ are permutation kG-modules.*

Proof Let M be a k-basis of U and N be a k-basis of V that are permuted by the actions of G on U and V, respectively. The disjoint union of M and N yields a k-basis of $U \oplus V$ that is permuted by G, hence $U \oplus V$ is a permutation kG-module. The image $\{m \otimes n | (m, n) \in M \times N\}$ of the direct product of $M \times N$ in $U \otimes_k V$ is a k-basis of $U \otimes_k V$ that is permuted by G, whence the result. □

Proposition 1.1.10 *Let G be a finite group, let M be a transitive G-set, let $m \in M$ and let $H = \{x \in G | xm = m\}$ be the stabiliser of m in G. The map sending $x \in G$ to xm induces an isomorphism of G-sets $G/H \cong M$, and induces an isomorphism of kG-modules $kG/H \cong kM$.*

Proof Let $x, y \in G$. We have $xm = ym$ if and only if $y^{-1}xm = m$, that is, if and only if $y^{-1}x \in H$, which in turn is equivalent to $xH = yH$. Thus the map sending xH to xm is well-defined and an isomorphism of G-sets. The rest is clear. □

Given a k-algebra A, an A-module U is called *indecomposable* if U is nonzero and cannot be written as a direct sum $U = V \oplus V'$ with nonzero A-submodules V, V'. Any idempotent $\pi \in \mathrm{End}_A(U)$ gives rise to a direct sum decomposition $U = \pi(U) \oplus (\mathrm{Id}_U - \pi)(U)$. In particular, U is indecomposable if and only if the identity map Id_U on U is a primitive idempotent in the endomorphism algebra $\mathrm{End}_A(U)$. An A-module S is called *simple* if S is nonzero but has no nonzero proper submodule. A simple module is always indecomposable, but an indecomposable module need not be simple. A transitive permutation kG-module need not be indecomposable, in general. Thus a direct summand of a permutation kG-module is not necessarily a permutation kG-module. We will see later that if G is a finite p-group for some prime p and k is a field of characteristic p, then every transitive permutation kG-module is indecomposable, and hence direct summands of permutation modules over p-groups in characteristic p are again permutation modules. Nontrivial permutation modules are not simple.

Proposition 1.1.11 *Let G be a finite group, and let M be a finite G-set. The permutation kG-module kM has a trivial submodule, namely $k\sum_{m\in M} m$. In particular, if $|M| \geq 2$, then kM is not a simple kG-module.*

Proof Since the action of $x \in G$ on M permutes the elements of M, we have $x(\sum_{m\in M} m) = \sum_{m\in M} m$. The result follows. □

Exercise 1.1.12 Suppose that k is a field of prime characteristic p and let G be a cyclic group of order p^a for some positive integer a. Let y be a generator of G. Using that the binomial coefficients $\binom{p^a}{m}$ are divisible by p, for $1 \leq m \leq p^a - 1$, show that $(y-1)^{p^a} = 0$ in kG and deduce from this that the map sending an indeterminate x to $y - 1$ induces an algebra isomorphism $k[x]/(x^{p^a}) \cong kG$.

Exercise 1.1.13 Let G be a finite cyclic group of order n and suppose that k is a field containing a primitive n-th root of unity ζ. Let y be a generator of G. Show that there is a unique k-algebra isomorphism $kG \cong k^n$ sending y to $(1, \zeta, \zeta^2, \ldots, \zeta^{n-1})$.

Exercise 1.1.14 Let G be a finite group and x a nontrivial element in G. Show that $1 - x$ is a zero divisor in kG; that is, there exists a nonzero element $z \in kG$ such that $(1 - x)z = 0$.

Exercise 1.1.15 Let $C_3 = \{1, x, x^2\}$ be a cyclic group of order three with generator x. Denote by $\mathbb{F}_2$ and $\mathbb{F}_4$ fields with two and four elements, respectively.

(1) Show that we have an isomorphism of $\mathbb{F}_2$-algebras $\mathbb{F}_2C_3 \cong \mathbb{F}_2 \times \mathbb{F}_4$ mapping x to $(1, \gamma)$ for some generator γ of $\mathbb{F}_4{}^\times$. Show that under this isomorphism, $1 + x + x^2$ corresponds to the unit element of $\mathbb{F}_2$ and $x + x^2$ corresponds to the unit element of $\mathbb{F}_4$.
(2) Show that we have an isomorphism of $\mathbb{F}_4$-algebras $\mathbb{F}_4C_3 \cong \mathbb{F}_4 \times \mathbb{F}_4 \times \mathbb{F}_4$ mapping x to $(1, \gamma, \gamma^2)$. In the terminology developed in 1.14.2 below this means that $\mathbb{F}_4$ is a splitting field for C_3 but $\mathbb{F}_2$ is not.

Exercise 1.1.16 Let A be a k-algebra and i an idempotent in A. Show that i is primitive in A if and only if i is the only idempotent in the algebra iAi.

Exercise 1.1.17 Suppose that k is a field, and let n be a positive integer. For $1 \leq i \leq n$ denote by e_i the matrix in $M_n(k)$ whose diagonal entry (i, i) is equal to 1 and all of whose other entries are 0. Show that the set $\{e_i\}_{1 \leq i \leq n}$ is a primitive decomposition of the unit element in $M_n(k)$.

Exercise 1.1.18 Let S_3 be the symmetric group on three letters and k a field of characteristic 2. Show that the set

$$\{(1) + (123) + (132), (1) + (12) + (13) + (123), (1) + (12) + (13) + (132)\}$$

is a primitive decomposition of 1 in kS_3. Show that the first element of this set is in the centre of kS_3, and that the second and third elements are conjugate via the transposition (23). Use this to show that there is an algebra isomorphism $kS_3 \cong kC_2 \times M_2(k)$.

Exercise 1.1.19 Let G be a finite group and M a G-set of order 2 on which G acts transitively. Let k be a field. Show that the permutation kG-module kM is indecomposable if and only if $\text{char}(k) = 2$.

Exercise 1.1.20 Let G be a group acting on an abelian group A. Show that the action of G on A extends uniquely to a $\mathbb{Z}G$-module structure on A. Show that this is an isomorphism between the category of abelian groups endowed with a G-action and the category of $\mathbb{Z}G$-modules.

1.2 Twisted group algebras

One of the classic themes in homological algebra is the parametrisation of extensions of a group G by an abelian group Z endowed with a G-action in terms of elements in the second cohomology group $H^2(G; Z)$. If G acts trivially on Z, this parametrises central extensions of G by Z. We review this as far as needed later. Given a group G and a map $\alpha : G \times G \to k^\times$ we can endow the group algebra kG with a 'twisted' k-bilinear multiplication by setting $x \cdot y = \alpha(x, y)xy$ for any two $x, y \in G$, where xy is the usual product of x and y in G. In other words, the group basis G of kG is no longer closed under multiplication – the product of two group elements yields a group element multiplied by some scalar in $k^\times$. Not every choice of α will yield an associative multiplication.

Theorem 1.2.1 *Let G be a group and let $\alpha : G \times G \to k^\times$ be a map. The unique k-bilinear map from $kG \times kG$ to kG defined by $x \cdot y = \alpha(x, y)xy$ for all $x, y \in G$ is associative if and only if $\alpha(xy, z)\alpha(x, y) = \alpha(x, yz)\alpha(y, z)$ for all $x, y, z \in G$.*

Proof Let $x, y, z \in G$. We have $(x \cdot y) \cdot z = (\alpha(x, y)xy) \cdot z = \alpha(xy, z)\alpha(x, y)xyz$ and $x \cdot (y \cdot z) = (\alpha(y, z)x) \cdot (yz) = \alpha(x, yz)\alpha(y, z)xyz$. These two expressions are equal if and only if $\alpha(xy, z)\alpha(x, y) = \alpha(x, yz)\alpha(y, z)$. □

We introduce the following terminology for the maps α arising in the above theorem. We do this in slightly greater generality than needed for twisted group algebras.

Definition 1.2.2 Let G be a group and let Z be an abelian group, written multiplicatively. Suppose that G acts on Z from the left, written ${}^{x}a$, where $x \in G$ and $a \in Z$. A *1-cocycle of G with coefficients in Z* is a map $\gamma : G \to Z$ satisfying $\gamma(xy) = \gamma(x)({}^{x}\gamma(y))$ for all $x, y \in G$. We denote by $Z^1(G; Z)$ the set of all 1-cocycles of G with coefficients in Z. A *2-cocycle of G with coefficients in Z* is a map $\alpha : G \times G \to Z$ satisfying the *2-cocycle identity*

$$\alpha(xy, z)\alpha(x, y) = \alpha(x, yz)({}^{x}\alpha(y, z))$$

for all $x, y, z \in G$. We denote by $Z^2(G; Z)$ the set of all *2-cocycles of G with coefficients in Z.*

If G acts trivially on Z, then the 2-cocycle identity is exactly the identity satisfied by the map α in Theorem 1.2.1. An easy verification shows that the sets $Z^1(G; Z)$ and $Z^2(G; Z)$ are abelian groups with group structure induced by that in Z; that is, for $\gamma, \gamma' \in Z^1(G; Z)$ we define $\gamma\gamma'$ by $(\gamma\gamma')(x) = \gamma(x)\gamma'(x)$, and for $\alpha, \alpha' \in Z^2(G; Z)$ we define $\alpha\alpha'$ by $(\alpha\alpha')(x, y) = \alpha(x, y)\alpha'(x, y)$. The identity element of $Z^1(G; Z)$ is the constant map sending $x \in G$ to the identity element 1_Z of the group Z, and the identity element of $Z^2(G; Z)$ is the constant map sending $(x, y) \in G \times G$ to 1_Z.

Proposition 1.2.3 *Let G be a group and Z an abelian group on which G acts. Let $\gamma \in Z^1(G; Z)$ and $\alpha \in Z^2(G; Z)$. We have $\gamma(1) = 1$, and for any $x \in G$ we have $\alpha(1, x) = \alpha(1, 1)$ and $\alpha(x, 1) = {}^{x}\alpha(1, 1)$.*

Proof Let $x, y \in G$. Applying the equation $\gamma(xy) = \gamma(x)({}^{x}\gamma(y))$ with $x = 1$ yields $\gamma(y) = \gamma(1)\gamma(y)$, hence $\gamma(1) = 1$. The 2-cocycle identity applied with $1, 1, x$ yields $\alpha(1, x)\alpha(1, 1) = \alpha(1, x)\alpha(1, x)$, hence $\alpha(1, 1) = \alpha(1, x)$. The second identity follows similarly from the 2-cocycle identity applied to $x, 1, 1$. □

Definition 1.2.4 Let G be a group and $\alpha \in Z^2(G; k^\times)$, where $k^\times$ is considered with the trivial action of G. The *twisted group algebra of G by α* is the k-algebra, denoted $k_\alpha G$, which is equal to kG as k-module, endowed with the unique k-bilinear product $kG \times kG \to kG$ defined by $x \cdot y = \alpha(x, y)xy$ for all $x, y \in G$.

Here $x \cdot y$ denotes the product to be defined on $k_\alpha G$, while xy is the product in G of x and y. When performing calculations in twisted group algebras there are a few subtleties one needs to be aware of. Let G be a group and $\alpha \in Z^2(G; k^\times)$. In the group algebra kG, the inverse x^{-1} of an element x in the group G is also the multiplicative inverse of x in kG because the unit element of the algebra kG is the unit element of the group 1. In the twisted group algebra $k_\alpha G$ we still have a unit element, but it need not be equal to the unit element 1 of the group G because, for any $x \in G$, we have $1 \cdot x = \alpha(1, x)x$. The value $\alpha(1, x)$ need not be equal to the unit element of $k^\times$, but it is independent of the element x, which in turn yields a unit element in $k_\alpha G$ as follows.

Proposition 1.2.5 *Let G be a group. Consider $k^\times$ with the trivial action of G. Let $\alpha \in Z^2(G; k^\times)$.*

(i) For any $x \in G$ we have $\alpha(1, x) = \alpha(1, 1) = \alpha(x, 1)$.
(ii) For any $x \in G$ we have $\alpha(x, x^{-1}) = \alpha(x^{-1}, x)$.
(iii) The unit element of $k_\alpha G$ is equal to $\alpha(1, 1)^{-1}1_G$.
(iv) For any $x \in G$ the multiplicative inverse of x in $k_\alpha G$ is equal to $\alpha(1, 1)^{-1}\alpha(x, x^{-1})^{-1}x^{-1}$.

Proof Let $x \in G$. Statement (i) follows from 1.2.3. The 2-cocycle identity applied to x, x^{-1}, x yields $\alpha(1, x)\alpha(x, x^{-1}) = \alpha(x, 1)\alpha(x^{-1}, x)$. By (i) we have $\alpha(1, x) = \alpha(x, 1)$, whence (ii). We have $\alpha(1, 1)^{-1}1 \cdot x = \alpha(1, 1)^{-1}\alpha(1, x)x = x$, where the last equality follows from (i). Similarly, $\alpha(1, 1)^{-1}x \cdot 1 = x$, whence (iii). Finally, $\alpha(1, 1)^{-1}\alpha(x, x^{-1})^{-1}x \cdot x^{-1} = \alpha(1, 1)^{-1}\alpha(x, x^{-1})^{-1}\alpha(x, x^{-1})xx^{-1} = \alpha(1, 1)^{-1}1$, which is the unit element by (iii). The result follows. □

Proposition 1.2.6 *Let G be a group and $\alpha, \beta \in Z^2(G; k^\times)$. There is a k-algebra isomorphism $k_\alpha G \cong k_\beta G$ mapping $x \in G$ to $\gamma(x)x$ for some scalar $\gamma(x) \in k^\times$ if and only if $\alpha(x, y) = \beta(x, y)\gamma(x)\gamma(y)\gamma(xy)^{-1}$ for all $x, y \in G$. In other words, there is a G-graded algebra isomorphism $k_\alpha G \cong k_\beta G$ if and only if the classes of α and of β in $H^2(G; k^\times)$ are equal.*

Proof Let $x, y \in G$. Denote by $x \cdot y$ the product of x and y in $k_\alpha G$. Denote by $x * y$ the product of x and y in $k_\beta G$. The image of $x \cdot y = \alpha(x, y)xy$ in $k_\beta G$ as stated is $\alpha(x, y)\gamma(xy)xy$. The product in $k_\beta G$ of the images of x and y is $\gamma(x)\gamma(y)x * y = \gamma(x)\gamma(y)\beta(x, y)xy$. The two elements coincide if and only if $\alpha(x, y) = \beta(x, y)\gamma(x)\gamma(y)\gamma(xy)^{-1}$. □

Let G be a finite group and Z an abelian group (written multiplicatively) on which G acts. One checks that if $z \in Z$, then the map $\gamma : G \to Z$ defined by $\gamma(x) = {}^x z z^{-1}$ for all $x \in G$ is a 1-cocycle, and the set of 1-cocycles arising in this way form a subgroup of $Z^1(G; Z)$. Similarly, one checks that for any map $\gamma : G \to Z$ the map $\alpha : G \times G \to Z$ defined by $\alpha(x, y) = \gamma(x)({}^x\gamma(y))\gamma(xy)^{-1}$ for all $x, y \in G$ belongs to the group of 2-cocycles $Z^2(G; Z)$. Moreover, the set of 2-cocycles that arise in this way form a subgroup of $Z^2(G; Z)$.

Definition 1.2.7 Let G be a group and let Z be an abelian group, written multiplicatively, on which G acts on the left. We denote by $B^1(G; Z)$ the set of maps $\gamma : G \to Z$ for which there exists an element $z \in Z$ such that $\gamma(x) = ({}^xz)z^{-1}$ for all $x \in G$. The elements in $B^1(G; Z)$ are called called *1-coboundaries of G with coefficients in Z*. The set $B^1(G; Z)$ is a subgroup of $Z^1(G; Z)$. The quotient group

$$H^1(G; Z) = Z^1(G; Z)/B^1(G; Z)$$

is the *first cohomology group of G with coefficients in Z*; the elements in $H^1(G; Z)$ are called the *first cohomology classes of G with coefficients in Z*. We denote by $B^2(G; Z)$ the set of all maps $\alpha : G \times G \to Z$ for which there exists a map $\gamma : G \to Z$ such that $\alpha(x, y) = \gamma(x)({}^x\gamma(y))\gamma(xy)^{-1}$ for all $x, y \in G$. The elements of $B^2(G; Z)$ are called *2-coboundaries of G with coefficients in Z*. The set $B^2(G; Z)$ is a subgroup of $Z^2(G; Z)$. The quotient group

$$H^2(G; Z) = Z^2(G; Z)/B^2(G; Z)$$

is the *second cohomology group of G with coefficients in Z*; the elements in $H^2(G; Z)$ are called the *second cohomology classes of G with coefficients in Z*.

If G acts trivially on Z, then $H^1(G; Z) = \text{Hom}(G, Z)$ is the set of group homomorphisms from G to Z. In particular, if k is a field, and $k^\times$ considered with G acting trivially, then $H^1(G; k^\times) = \text{Hom}(G, k^\times)$ is a finite group of order dividing $|G|$. If in addition k has prime characteristic p and P is a finite p-group, then $H^1(P; k^\times)$ is trivial. Proposition 1.2.6 implies that if α represents the trivial class in $H^2(G; k^\times)$, then there is a k-algebra isomorphism $kG \cong k_\alpha G$ sending $x \in G$ to a nonzero scalar multiple of x, and moreover, this isomorphism is unique up to a group homomorphism $G \to k^\times$. In other words, if there is a k-algebra isomorphism $kG \cong k_\alpha G$ sending $x \in G$ to a nonzero scalar multiple of x, then all such isomorphisms are parametrised by the group $H^1(G; k^\times) = \text{Hom}(G; k^\times)$. This is a classic homological algebra scenario: if an obstruction in some cohomology group H^n to a certain construction vanishes, then all possibilities for that construction are parametrised by H^{n-1}, where typically $n = 2$ or $n = 3$. We will encounter this scenario in the context of extending modules from normal subgroups. The following observation is a criterion for when this situation arises.

Proposition 1.2.8 *Let G be a finite group and $\alpha \in Z^2(G; k^\times)$. The class of α in $H^2(G; k^\times)$ is trivial if and only if $k_\alpha G$ has a module that is isomorphic to k as a k-module.*

Proof If the class of α is trivial, then $k_\alpha G \cong kG$, and the trivial kG-module is obviously isomorphic to k as a k-module. Conversely, if $k_\alpha G$ has a module isomorphic to k, then the structural homomorphism of that module yields an algebra homomorphism $\varphi : k_\alpha G \to k$. For this to be an algebra homomorphism, for any $x, y \in G$, we have $\alpha(x, y)\varphi(xy) = \varphi(x \cdot y) = \varphi(x)\varphi(y)$. Thus $\alpha(x, y) = \varphi(x)\varphi(y)\varphi(xy)^{-1}$, showing that α is a 2-coboundary. □

The cohomology groups group $H^1(G; Z)$ and $H^2(G; Z)$ are *covariantly* functorial in Z and *contravariantly* functorial in G. That is, any group homomorphism of abelian groups $\zeta : Z \to Z'$ that commutes with the G-actions on Z and Z' determines group homomorphisms $H^1(G; Z) \to H^1(G; Z')$ and $H^2(G; Z) \to H^2(G; Z')$ induced by the map sending $\gamma \in Z^1(G; Z)$ and $\alpha \in Z^2(G; Z)$ to $\zeta \circ \gamma$ and $\zeta \circ \alpha$, respectively. Any group homomorphism $\varphi : G \to G'$ determines group homomorphism $H^1(G'; Z) \to H^1(G; Z)$ and $H^2(G'; Z) \to H^2(G; Z)$ induced by the maps sending γ and α to the cocycles obtained by precomposing with φ. In particular, if H is a subgroup of G, then the inclusion of H into G induces obvious *restriction maps* $\mathrm{res}^G_H : Z^1(G; Z) \to Z^1(H; Z)$ and $\mathrm{res}^G_H : Z^2(G; Z) \to Z^2(H; Z)$ The restriction maps sends $B^1(G; Z)$ to $B^1(H; Z)$, and $B^2(G; Z)$ to $B^2(H; Z)$, and hence they induce restriction maps, still denoted $\mathrm{res}^G_H : H^1(G; Z) \to H^1(H; Z)$ and $\mathrm{res}^G_H : H^2(G; Z) \to H^2(H; Z)$.

Proposition 1.2.9 *Let G be a finite group.*

(i) Let Z be a multiplicatively written abelian group on which G acts. Then every element in $H^1(G; Z)$ and in $H^2(G; Z)$ has finite order dividing $|G|$. If Z is finite, then $H^1(G; Z)$ and $H^2(G; Z)$ are finite, and the order of every element in $H^1(G; Z)$ and $H^2(G; Z)$ divides both $|G|$ and $|Z|$. In particular, if the orders of G and Z are coprime, then $H^1(G; Z) = H^2(G; Z) = \{0\}$.

(ii) Suppose that k is an algebraically closed field; consider $k^\times$ with the trivial action of G. Let Z be the group of $|G|$-th roots of unity in $k^\times$. The inclusion $Z \to k^\times$ induces a surjective group homomorphism $H^2(G; Z) \to H^2(G; k^\times)$; in particular, the abelian group $H^2(G; k^\times)$ is finite.

(iii) Suppose that k is a perfect field of prime characteristic p. Let P be a finite p-group. Then $H^2(P; k^\times)$ is trivial.

Proof Let $\gamma \in Z^1(G; Z)$. Let $x, y \in G$. Then $\gamma(xy) = \gamma(x)({}^x\gamma(y))$. Fixing x and taking the product over all $y \in G$ yields $\prod_{y \in G} \gamma(y) = \gamma(x)^{|G|} \prod_{y \in G} {}^x\gamma(y)$, hence $\gamma(x)^{|G|} = \prod_{y \in G} \gamma(y)({}^x(\gamma(y)^{-1}))$. This shows that $\gamma^{|G|}$ is a 1-coboundary, and hence that every element in $H^1(G; Z)$ has finite order dividing $|G|$. If Z is finite, then so is the set of maps from G to Z, and hence in particular $H^1(G; Z)$ is finite. Since the group structure of $H^1(G; Z)$ is induced by that in Z, it follows that $|Z|$ annihilates $H^1(G; Z)$. Let $\alpha \in Z^2(G; Z)$. For $x \in G$ set $\mu(x) = \prod_{y \in G} \alpha(x, y)$. For $x, y, z \in G$, consider the 2-cocycle identity $\alpha(x, yz)({}^x\alpha(y, z)) = \alpha(xy, z)\alpha(x, y)$. Fixing x, y and taking the product over all z yields $\mu(x)({}^x\mu(y)) = \mu(xy)\alpha(x, y)^{|G|}$, which shows that $\alpha^{|G|} \in B^2(G; Z)$. Thus the order of the image of α in $H^2(G; Z)$ divides $|G|$. If Z is finite there are only finitely many maps from $G \times G$ to Z, implying the finiteness of $H^2(G; Z)$. Since the group structure of $H^2(G; Z)$ is induced by that of Z, it follows that $|Z|$ annihilates $H^2(G; Z)$, whence (i). Suppose now that Z is the group of $|G|$-th roots in $k^\times$, where k is an algebraically closed field, and where G acts trivially on $k^\times$. With $\alpha \in Z^2(G; k^\times)$ and μ as before, since k is algebraically closed, for any $x \in G$ there is $\nu(x) \in k^\times$ such that $\nu(x)^{|G|} = \mu(x)$. Set $\beta(x, y) = \alpha(x, y)\nu(x)^{-1}\nu(y)^{-1}\nu(xy)$. Then β represents the same class as α in $H^2(G; k^\times)$, and we have $\beta(x, y)^{|G|} = \alpha(x, y)^{|G|}\mu(x)^{-1}\mu(y)^{-1}\mu(xy) = 1$, which shows that β takes values in the finite subgroup Z of $k^\times$, whence (ii). We use a variation of the same argument to prove (iii). With the notation as before and using P instead of G, if k has prime characteristic p, then the group Z of $|P|$-th roots of unity in $k^\times$ is trivial, and if in addition k is perfect, then for any of the values $\mu(x) \in k^\times$ as before, there is an element $\nu(x) \in k^\times$ satisfying $\nu(x)^{|P|} = \mu(x)$. Thus the 2-cocycle β as defined before is constant equal to 1. □

Proposition 1.2.10 *Let G be a finite cyclic group and k an algebraically closed field. Consider $k^\times$ with the trivial action of G. The group $H^2(G; k^\times)$ is trivial.*

Proof Let x be a generator of G and let $\alpha \in Z^2(G; k^\times)$. Denote by n the order of G and by $\hat{x}$ the image of x in the twisted group algebra $k_\alpha G$. Since $x^n = 1$ we have $\hat{x}^n = \mu 1$ for some $\mu \in k^\times$. Since k is algebraically closed, there is $\nu \in k^\times$ such that $\nu^n = \mu\alpha(1, 1)$. Set $\tilde{x} = \nu^{-1}\hat{x}$. Then $\tilde{x}^n = \alpha(1, 1)^{-1}1$, which is the unit element of $k_\alpha G$. This shows that the map sending the powers of x to the corresponding powers of $\tilde{x}$ induces an algebra isomorphism $kG \cong k_\alpha G$, and hence α represents the trivial class by 1.2.6. □

Remark 1.2.11 There is a more structural interpretation of $H^2(G; Z)$ in terms of invariants of $\mathbb{Z}G$-modules, which is the reason why we tend to write the abelian group $H^2(G, Z)$ additively even if G and Z are written multiplicatively: we have $H^2(G; Z) \cong \mathrm{Ext}^2_{\mathbb{Z}G}(\mathbb{Z}, Z)$, where Z is viewed as a $\mathbb{Z}G$-module with the prescribed action of G on Z.

We briefly review the parametrisation of certain group extensions in terms of second cohomology groups. Let

$$1 \longrightarrow Z \xrightarrow{\iota} \hat{G} \xrightarrow{\pi} G \longrightarrow 1$$

be a group extension with Z abelian; that is, ι is injective, π is surjective, and $\ker(\pi) = \mathrm{Im}(\iota)$. If $\iota(Z)$ is contained in $Z(\hat{G})$, then such an extension is called a *central extension of G by Z*. In general, an extension of G by an abelian group Z induces an action of G on Z as follows (we identify Z to its image in $\hat{G}$): if $x \in G$ and $a \in Z$, define ${}^x a = \hat{x}a\hat{x}^{-1}$, where $\hat{x} \in \hat{G}$ is any inverse image of x. Any two different inverse images of x in $\hat{G}$ differ by an element in Z, and since Z is abelian, this definition does not depend on the choice of $\hat{x}$, and yields therefore a well-defined action of G on Z. This action is trivial precisely if $\hat{G}$ is a central extension of G by Z.

Any extension of G by Z as above defines an element $\alpha \in Z^2(G; Z)$ as follows. For any $x \in G$ choose $\hat{x} \in \hat{G}$ such that $\pi(\hat{x}) = x$. Then, for $x, y \in G$, we have $\pi(\widehat{xy}) = xy = \pi(\hat{x}\hat{y})$. Thus $\widehat{xy}$ and $\hat{x}\hat{y}$ differ by a unique element in Z; say

$$\hat{x}\hat{y} = \alpha(x, y)\widehat{xy}$$

for a uniquely determined element $\alpha(x, y) \in Z$. One verifies that the associativity of the group multiplication in $\hat{G}$ is equivalent to the 2-cocycle identity of α, so $\alpha \in Z^2(G; Z)$. The 2-cocycle depends on the choice of the elements $\hat{x}$ in $\hat{G}$, but we will see below that the class of α in $H^2(G; Z)$ is independent of this choice. The 2-cocycle α represents the trivial class in $H^2(G; Z)$ if and only if the above central extension is split. Indeed, if there is a section $\sigma : G \to \hat{G}$ satisfying $\pi \circ \sigma = \mathrm{Id}_G$, then choosing $\hat{x} = \sigma(x)$ for $x \in G$ yields the constant 2-cocycle 1. Conversely, if for some choice of elements $\hat{x}$ we have $\alpha \in B^2(G; Z)$, then there is a map $\mu : G \to Z$ satisfying $\alpha(x, y) = \mu(x)({}^x\mu(y))\mu(xy)^{-1}$ for x, $y \in G$. An easy calculation shows that the map σ defined by $\sigma(x) = \mu(x)^{-1}\hat{x}$ for $x \in G$ is a section of π. Any $\alpha \in Z^2(G; Z)$ occurs as a 2-cocycle of an extension of G by Z. Given $\alpha \in Z^2(G; Z)$ we construct an extension of G by Z as follows. As set, we take $\hat{G} = Z \times G$. We consider $\hat{G}$ endowed with the product defined by

$$(\lambda, x)(\mu, y) = (\alpha(x, y)\lambda\mu, xy)$$

for $x, y \in G$ and $\lambda, \mu \in Z$. The same calculations as in the proof of 1.2.5 show that $\hat{G}$ is indeed a group with unit element $(\alpha(1, 1)^{-1}, 1)$. We define $\pi : \hat{G} \to G$ as the projection $\pi(\lambda, x) = x$ for $(\lambda, x) \in \hat{G}$. This is a surjective group homomorphism with kernel $\{(\lambda, 1) | \lambda \in Z\} \cong Z$. By setting $\hat{x} = (1, x)$ for any $x \in G$ we get $\hat{x}\hat{y} = (\alpha(x, y), xy) = \alpha(x, y)\widehat{xy}$, which shows that α is determined by this extension.

Theorem 1.2.12 *Let G be a group and Z a multiplicatively written abelian group. Let*

$$1 \longrightarrow Z \longrightarrow \hat{G} \xrightarrow{\pi} G \longrightarrow 1$$

$$1 \longrightarrow Z \longrightarrow \check{G} \xrightarrow{\tau} G \longrightarrow 1$$

be two extensions of G by Z. Suppose that the two extensions induce the same action of G on Z. For any $x \in G$ choose $\hat{x} \in \hat{G}$ such that $\pi(\hat{x}) = x$ and $\check{x} \in \check{G}$ such that $\tau(\check{x}) = x$. Denote by α, β the 2-cocycles in $Z^2(G; Z)$ satisfying $\hat{x}\hat{y} = \alpha(x, y)\widehat{xy}$ and $\check{x}\check{y} = \beta(x, y)\check{xy}$ for all $x, y \in G$. The classes of α and β in $H^2(G; Z)$ are equal if and only if there exists a group isomorphism $\varphi : \hat{G} \cong \check{G}$ making the following diagram commutative:

$$\begin{array}{ccccccccc}
1 & \longrightarrow & Z & \longrightarrow & \hat{G} & \xrightarrow{\pi} & G & \longrightarrow & 1 \\
 & & \| & & \downarrow{\scriptstyle\varphi} & & \| & & \\
1 & \longrightarrow & Z & \longrightarrow & \check{G} & \xrightarrow{\tau} & G & \longrightarrow & 1
\end{array}$$

Proof Suppose that φ exists making the last diagram commutative. Then for any $x \in G$, we have $\tau(\varphi(\hat{x})) = \pi(\hat{x}) = x = \tau(\check{x})$. Thus there is $\gamma(x) \in Z$ such that $\varphi(\hat{x}) = \gamma(x)\check{x}$. It follows that

$$\varphi(\hat{x}\hat{y}) = \varphi(\alpha(x, y)\widehat{xy}) = \alpha(x, y)\varphi(\widehat{xy}) = \alpha(x, y)\gamma(xy)\check{xy}$$

$$\varphi(\hat{x})\varphi(\hat{y}) = \gamma(x)\check{x}\gamma(y)\check{y} = \gamma(x)({}^{x}\gamma(y))\beta(x, y)\check{xy}.$$

These two expressions coincide if and only if $\alpha(x, y) = \beta(x, y)\gamma(x)({}^{x}\gamma(y)) \gamma(xy)^{-1}$, in which case the images of α and β in $H^2(G; Z)$ are equal. Conversely, if the images of α and β in $H^2(G; Z)$ are equal there exists a map $\gamma : G \to Z$ such that $\alpha(x, y) = \beta(x, y)\gamma(x)({}^{x}\gamma(y))\gamma(xy)^{-1}$, in which case we define φ by the formula $\varphi(z\hat{x}) = \gamma(x)z\check{x}$ for any $x \in G$ and $z \in Z$. □

We noted in 1.2.5 that the image of 1_G in $k_\alpha G$ need not be the unit element of this algebra. What we will see now is that one can always achieve this by making a suitable choice for the 2-cocycle α.

Definition 1.2.13 Let G be a group and Z a multiplicatively written abelian group on which G acts. A 2-cocycle $\alpha \in Z^2(G; Z)$ is called *normalised* if $\alpha(1, x) = 1$ for all $x \in G$.

With the notation of Theorem 1.2.12, the 2-cocycle α is normalised if and only if $\hat{1}_G = 1_{\hat{G}}$.

Proposition 1.2.14 *Let G be a group and Z a multiplicatively written abelian group on which G acts. Any class in $H^2(G; Z)$ can be represented by a normalised 2-cocycle.*

Proof Let $\alpha \in Z^2(G; Z)$. By 1.2.3 we have $\alpha(1, 1) = \alpha(1, x)$, for all $x \in G$. For any $x \in G$ set $\mu(x) = \alpha(1, 1)$. For any $x, y \in G$ set $\beta(x, y) = \alpha(x, y)\mu(x)^{-1}({}^x\mu(y))^{-1}\mu(xy) = \alpha(x, y)({}^x\alpha(1, 1))^{-1}$. Then $\beta \in Z^2(G; Z)$ represents the same class as α and we have $\beta(1, x) = \beta(1, 1) = 1$ as required. □

If the restriction of $\alpha \in H^2(G; k^\times)$ to a subgroup H of G represents the trivial class in $H^2(H; k^\times)$, then the inclusion $H \to G$ and a suitable map $\gamma : H \to k^\times$ yield an injective algebra homomorphism $kH \to k_\alpha G$ sending $y \in H$ to $\gamma(y)y$ in $k_\alpha G$. In this way, $k_\alpha G$ becomes a kH-kH-bimodule with kH as a direct summand. Using a generalisation of the first argument in the proof of 1.2.9 it follows that in this situation, the order of α divides $|G : H|$.

Proposition 1.2.15 *Let G be a group, H a subgroup of finite index in G, and let Z be an abelian group, written multiplicatively, on which G acts. Let $\alpha \in H^2(G; Z)$. If $\mathrm{res}^G_H(\alpha)$ is the trivial class in $H^2(H; Z)$, then $\alpha^{|G:H|}$ is the trivial class in $H^2(G; Z)$.*

Proof Consider a group extension

$$1 \longrightarrow Z \longrightarrow \hat{G} \xrightarrow{\pi} G \longrightarrow 1$$

and elements $\hat{x} \in \hat{G}$ such that $\pi(\hat{x}) = x$ for $x \in G$ and such that $\hat{x}\hat{y} = \alpha(x, y)\widehat{xy}$, where we abusively denote again by α a 2-cocycle representing the class α. By the assumptions, α restricts to an element in $B^2(H; Z)$, and hence, after possibly replacing α by a different 2-cocycle in the same class, we may assume that α is constant 1 on $H \times H$. Thus for $y, z \in H$ we have $\hat{y}\hat{z} = \widehat{yz}$. Let $\mathcal{R}$ be a set of representatives of G/H in G. Thus every element in G can be written uniquely in the form xh for some $x \in \mathcal{R}$ and some $h \in H$. We modify the choice of the $\hat{x}$ as follows: if $x \in \mathcal{R}$ and $h \in H$, we keep our previous choices of $\hat{x}$ and $\hat{h}$, and we set $\widehat{xh} = \hat{x}\hat{h}$. With this choice of inverse images of elements of G, we now have $\hat{x}\hat{h} = \widehat{xh}$ for any $x \in G$ and any $h \in H$, hence

$\alpha(x, h) = 1$ for any $x \in G$ and any $h \in H$. Together with the 2-cocycle identity $\alpha(x, yh)({}^x\alpha(y, h)) = \alpha(xy, h)\alpha(x, y)$ this implies that $\alpha(x, yh) = \alpha(x, y)$, for any $x, y \in G$ and $h \in H$. In other words, the value $\alpha(x, y)$ depends only on the H-coset of y. Thus the expression $\mu(x) = \prod_{y\in\mathcal{R}} \alpha(x, y)$ does not depend on the choice of $\mathcal{R}$. For $x, y, z \in G$, consider the 2-cocycle identity $\alpha(x, yz)({}^x\alpha(y, z)) = \alpha(xy, z)\alpha(x, y)$. Fixing x, y, and taking the product over all $z \in \mathcal{R}$ yields $\mu(x)({}^x\mu(y)) = \mu(xy)\alpha(x, y)^{|G:H|}$, and hence $\alpha^{|G:H|} \in B^2(G; Z)$ as required. □

Corollary 1.2.16 *Let G be a finite group, p a prime, P a Sylow p-subgroup of G, and Z an abelian p-group on which G acts. The restriction map $H^2(G; Z) \to H^2(P; Z)$ is injective.*

Proof Set $m = |G : P|$. This is a positive integer prime to p. If $\alpha \in H^2(G; Z)$ such that $\mathrm{res}^G_P(\alpha) = 0$, then $\alpha^m = 0$ by 1.2.15, and hence $\alpha = 0$ since taking m-th powers is an automorphism of Z, as Z is an abelian p-group. □

There is a shorter proof of 1.2.15 using properties of transfer maps that we have not introduced at this stage: the composition $\mathrm{tr}^G_H \circ \mathrm{res}^G_H$ is equal to multiplication by $|G : H|$ on $H^2(G; Z)$. Thus if $\mathrm{res}^G_H(\alpha)$ is trivial, then so is $\mathrm{tr}^G_H(\mathrm{res}^G_H(\alpha)) = \alpha^{|G:H|}$. Moreover, one can describe the image of the restriction map in terms of *stable elements* in cohomology.

A twisted group algebra $k_\alpha G$ does not have, in general, a basis that is closed under multiplication, but it is a quotient of the group algebra $k\hat{G}$ of the central extension $\hat{G}$ of G by $k^\times$ determined by α.

Proposition 1.2.17 *Let Z be an abelian group, written multiplicatively, and let*

$$1 \longrightarrow Z \longrightarrow \hat{G} \xrightarrow{\pi} G \longrightarrow 1$$

be a central extension of a group G by Z. For any $x \in G$ choose $\hat{x} \in \hat{G}$ such that $\pi(\hat{x}) = x$ and denote by β the 2-cocycle in $Z^2(G; Z)$ determined by $\hat{x}\hat{y} = \beta(x, y)\widehat{xy}$ for $x, y \in G$. Let $\mu : Z \to k^\times$ be a group homomorphism. Then $\alpha = \mu \circ \beta$ is a 2-cocycle in $Z^2(G; k^\times)$ and the map sending $z\hat{x}$ to $\mu(z)x$, for any $x \in G$ and any $z \in Z$, induces a surjective algebra homomorphism

$$k\hat{G} \to k_\alpha G$$

Proof Since μ is a group homomorphism, applying μ to the 2-cocycle identity for β shows that α satisfies the 2-cocycle identity. The last statement is an easy verification. □

When dealing with the group algebra over k of a central extension $\hat{G}$ of a group G by $k^\times$, it is important to not get confused between the roles of elements in $k^\times$ as either group elements in $\hat{G}$ or as scalars in $k\hat{G}$. For instance, for $\lambda, \mu \in k^\times$ the expression $\lambda\mu$ could mean either the product in the subgroup $k^\times$ of $\hat{G}$, or the element in the group algebra $k\hat{G}$ obtained by multiplying the group element μ by the scalar λ. The extreme example for this potential confusion is the group algebra $kk^\times$ of the group $k^\times$ over k. One way around this notational conflict consists of giving a name to the canonical map $\iota : k^\times \to \hat{G}$ and then denoting by $\lambda\iota(\mu)$ the scalar multiple of the group element $\iota(\mu) \in \hat{G}$ in the group algebra $k\hat{G}$. For instance, in the next proposition, we consider a finite subgroup Z of $k^\times$. This subgroup gives rise to an idempotent $e_Z = \frac{1}{|Z|}\sum_{z\in Z} z$, and here the sum of the elements $z \in Z \subseteq k^\times$ is *not* the sum in k but the formal sum in the group algebra $k\hat{G}$, with Z viewed as a subgroup of $\hat{G}$. By dividing Z and $\hat{G}$ in the above proposition by $\ker(\mu)$ one can always adjust the situation so that $\mu : Z \to k^\times$ is injective. In that case one can be more precise. The fact that second cohomology classes with values in the multiplicative group of an algebraically closed field can be represented by 2-cocycles with values in a finite subgroup allows for identifying a twisted group algebra with a direct factor of the group algebra of a finite group.

Proposition 1.2.18 *Suppose that k is an algebraically closed field. Let G be a finite group, acting trivially on $k^\times$, let $\alpha \in H^2(G; k^\times)$, and let*

$$1 \longrightarrow k^\times \longrightarrow \hat{G} \longrightarrow G \longrightarrow 1$$

be a central extension of G by $k^\times$ representing α. There is a finite subgroup G' of $\hat{G}$ with the following properties.

(i) We have $\hat{G} = k^\times \cdot G'$ and $Z = k^\times \cap G'$ is equal to the subgroup of $|G|$-th roots of unity in $k^\times$. In particular, $|G'| = |Z| \cdot |G|$ and the exponent of G' divides $|G|^2$.

(ii) The inclusion $G' \to \hat{G}$ induces an isomorphism of k-algebras $kG' \cdot e_Z \cong k_\alpha G$, where e_Z is the idempotent in $Z(kG')$ defined by $e_Z = \frac{1}{|Z|}\sum_{z\in Z} z$.

Proof For $x \in G$ denote by $\hat{x}$ an inverse image of x in $\hat{G}$. Thus α is represented by the 2-cocycle, abusively still denoted α, satisfying $\hat{x} \cdot \hat{y} = \alpha(x, y)\widehat{xy}$, for all $x, y \in G$. By 1.2.9 there is a map $\mu : G \to k^\times$ such that the 2-cocycle β defined by $\beta(x, y) = \alpha(x, y)\mu(x)\mu(y)\mu(xy)^{-1}$ has values in the subgroup Z of all $|G|$-th roots of unity of $k^\times$. We may choose β to be normalised. Set $\tilde{x} = \mu(x)\hat{x}$. Then, for $x, y \in G$, a short calculation shows that we have $\tilde{x} \cdot \tilde{y} = \beta(x, y)\widetilde{xy}$. Since β is normalised we have $\tilde{1}_G = 1_{\hat{G}}$. Thus $G' = \{\zeta\tilde{x} | \zeta \in Z, x \in G\}$ has the properties stated in (i). Note that $|Z|$ is invertible in $k^\times$. Thus the element e_Z is an idempotent in $Z(kG')$ by 1.1.2. The inclusion from G' to $\hat{G}$ induces an algebra homomorphism $kG' \to k\hat{G}$. Composing this with the map $k\hat{G} \to k_\alpha G$ from 1.2.17 sending $\tilde{x}$ to x, where $x \in G$ is here viewed as an element in $k_\alpha G$, yields an algebra homomorphism $kG' \to k_\alpha G$. Since the canonical map $\hat{G} \to G$ sends G' onto G, this algebra homomorphism is surjective. By construction of G', this homomorphism sends every element $z \in Z$ viewed as an element of G' to 1, hence sends e_Z to 1, and hence induces a surjective algebra homomorphism $kG' \cdot e_Z \to k_\alpha G$. Since $|G'| = |G| \cdot |Z|$, both algebras have the same dimension, hence are isomorphic. □

This proposition is an important tool for generalising properties of finite group algebras to twisted group algebras.

Remark 1.2.19 In three of the above statements, we have used the hypothesis that k is algebraically closed, namely 1.2.9 (ii), 1.2.10, and 1.2.18. This hypothesis will therefore reappear when quoting these results, but it is worth noting that one can work around this hypothesis as follows. If k is not algebraically closed in 1.2.9 (ii), then the proof shows that for any class $\alpha \in H^2(G; k^\times)$ there is a finite field extension k'/k such that the image of α in $H^2(G; (k')^\times)$ is in the image of the canonical map $H^2(G; Z) \to H^2(G; (k')^\times)$. Similarly, in 1.2.10, if $\alpha \in H^2(G; k^\times)$, then there is a finite extension k'/k such that the image of $\alpha \in H^2(G; (k')^\times)$ is zero, and the conclusions in 1.2.18 hold with k replaced by a finite extension. In the context of finite group algebras, the point of this remark is that it is always possible to work with finite fields, if this is convenient.

Exercise 1.2.20 Let V be a Klein four group; that is, V is elementary abelian of order 4. Let k be a field of characteristic different from 2, and set $Z = \{1, -1\} \subseteq k^\times$. Show the following statements.

(1) There is a normalised 2-cocycle $\alpha \in Z^2(V; Z)$ corresponding to a central extension

$$1 \longrightarrow Z \longrightarrow Q_8 \xrightarrow{\pi} V \longrightarrow 1$$

where Q_8 is a quaternion group of order 8.

(2) There is a surjective algebra homomorphism $kQ_8 \to k_\alpha V$ in which the non identity element z of $Z(Q_8)$ gets identified with -1, and there is a surjective algebra homomorphism $kQ_8 \to kV$ in which the non identity element z of $Z(Q_8)$ gets identified with 1.

(3) We have algebra isomorphisms $kQ_8(1+z) \cong kV \cong k \times k \times k \times k$, $kQ_8(1-z) \cong k_\alpha V$, and $kQ_8 \cong kV \times k_\alpha V$. If k contains a primitive 4-th root of unity, then $k_\alpha V \cong M_2(k)$.

Proving the statements in the above exercise with 'bare hands' is somewhat tedious; they follow more easily from Wedderburn's Theorem 1.13.3 below. As we will see later, an 'untwisted' finite group algebra of a nontrivial group can never be a matrix algebra over a field. The above exercise illustrates how group algebras of central group extensions involve twisted group algebras, and that twisted group algebras may behave very differently from group algebras. Twisted group algebras that are matrix algebras are closely related to *groups of central type*.

1.3 G-algebras

Let A be a k-algebra. We denote by $\mathrm{Aut}(A)$ the group of all k-algebra automorphisms of A. If $u \in A^\times$, then the map $a \mapsto uau^{-1}$ given by conjugation with u is an algebra automorphism of A. Any algebra automorphism of A arising in this way is called an *inner automorphism* of A. The set $\mathrm{Inn}(A)$ of inner automorphisms of A is a normal subgroup of $\mathrm{Aut}(A)$, and the quotient $\mathrm{Out}(A) = \mathrm{Aut}(A)/\mathrm{Inn}(A)$ is called the *outer automorphism group of* A. Note that $\mathrm{Aut}(A) = \mathrm{Aut}(A^{\mathrm{op}})$.

Definition 1.3.1 Let G be a finite group. A *G-algebra over k* is a k-algebra A endowed with an action $G \times A \to A$ of G on A, written $(x, a) \mapsto {}^x a$ for any $a \in A$ and $x \in G$, such that the map sending $a \in A$ to ${}^x a$ is a k-algebra automorphism of A for all $x \in G$. If A is a G-algebra over k, we denote for every subgroup H of G by A^H the subalgebra of all H-fixed points in A; that is, $A^H = \{a \in A \mid {}^y a = a \text{ for all } y \in H\}$. An *interior G-algebra over k* is a k-algebra A endowed with a group homomorphism $\sigma : G \to A^\times$, called *the structural homomorphism of the interior G-algebra A*.

Equivalently, a G-algebra over k is a k-algebra A endowed with a group homomorphism $\tau : G \to \mathrm{Aut}(A)$. The action of G on A is then given by ${}^x a = \tau(x)(a)$, where $x \in G$ and $a \in A$. Every interior G-algebra A with structural homomorphism $\sigma : G \to A^\times$ is in particular a G-algebra, with the group homomorphism $\tau : G \to \mathrm{Aut}(A)$ defined by $\tau(x)(a) = {}^x a = \sigma(x)a\sigma(x^{-1})$; that is $\tau(x)$ is equal to conjugation with $\sigma(x)$. In particular, $\mathrm{Im}(\tau)$ is contained in the subgroup $\mathrm{Inn}(A)$ of inner k-algebra automorphisms of A. In other words, an interior G-algebra A has the property that every $x \in G$ acts as an inner automorphism on A. However, even if G acts by inner automorphism on an algebra A, this does not automatically imply that the G-algebra structure 'lifts' to an interior G-algebra structure on A. We have obvious notions of homomorphisms of (interior) G-algebras:

Definition 1.3.2 Let G be a group. If A, B are G-algebras over k, then a *homomorphism of G-algebras from A to B* is a k-algebra homomorphism $f : A \to B$ satisfying $f({}^x a) = {}^x f(a)$ for all $a \in A$ and all $x \in G$. If A, B are interior G-algebras over k with structural homomorphisms $\sigma : G \to A^\times$ and $\tau : G \to B^\times$, then a *homomorphism of interior G-algebras from A to B is a k-algebra homomorphism* $f : A \to B$ satisfying $f(\sigma(x)) = \tau(x)$ for all $a \in A$ and all $x \in G$.

A homomorphism of G-algebras $f : A \to B$ induces algebra homomorphisms $A^H \to B^H$ for any subgroup H of G; that is, taking fixed points is functorial.

Examples 1.3.3 Let G be a finite group.

(1) The group algebra kG is an interior G-algebra with structural homomorphism $G \to (kG)^\times$ mapping $x \in G$ to its canonical image in kG.
(2) Let N be a normal subgroup of G. Conjugation by $x \in G$ induces an algebra automorphism on the subalgebra kN of kG. Thus kN is a G-algebra, but need not be an interior G-algebra.
(3) Let U be a kG−module. The k-algebra $\mathrm{End}_k(U)$ becomes an interior G-algebra with structural homomorphism $G \to \mathrm{End}_k(U)^\times = GL_k(U)$ mapping $x \in G$ to the k-linear automorphism of U sending $u \in U$ to xu. In particular, $\mathrm{End}_k(U)$ is a G-algebra, and the action of $x \in G$ on $\varphi \in \mathrm{End}_k(U)$ is given by $({}^x\varphi)(u) = x\varphi(x^{-1}u)$ for any $u \in U$. This formula shows that for any subgroup H of G we have $(\mathrm{End}_k(U))^H = \mathrm{End}_{kH}(U)$, the endomorphism algebra of U considered as a kH-module.
(4) Let $\alpha \in Z^2(G; k^\times)$. The twisted group algebra $k_\alpha G$ is a G-algebra, with $x \in G$ acting on $k_\alpha G$ as the conjugation by the image of x in $k_\alpha G$. This action does not lift to an interior G-algebra structure, unless α represents the trivial class in $H^2(G; k^\times)$, in which case this is just the usual group algebra kG.
(5) If L/K is a Galois extension with Galois group G, then L is a G-algebra over K via the canonical action of G on L. If we denote by $\mathcal{O}_K$, $\mathcal{O}_L$ the rings of algebraic integers in K, L, respectively, then $\mathcal{O}_L$ is a G-algebra over $\mathcal{O}_K$.

Remark 1.3.4 Let G, G' be finite groups, let A be an interior G-algebra with structural homomorphism $\sigma : G \to A^\times$, and let A' be an interior G'-algebra with structural homomorphism $\sigma' : G' \to (A')^\times$.

(1) For any subgroup H of G and any idempotent $i \in A^H$, the algebra iAi becomes an interior H-algebra with structural homomorphism $H \to (iAi)^\times$ mapping $x \in H$ to $\sigma(x)i = i\sigma(x)$.
(2) The algebra $A \otimes_k A'$ is an interior $G \times G'$-algebra with structural homomorphism mapping $(x, x') \in G \times G'$ to $\sigma(x) \otimes \sigma'(x')$. If moreover $G = G'$, we can consider $A \otimes_k A'$ again as an interior G–algebra through the *diagonal embedding* $\Delta_G : G \to G \times G$ mapping $x \in G$ to (x, x).

The following construction, which generalises that of group algebras, turns a G-algebra into an interior G-algebra.

Definition 1.3.5 Let G be a group and A a G-algebra over k. The *skew group algebra of A and G* is the interior G-algebra, denoted AG, which as an A-module is free, having the elements of G as basis, with multiplication induced by $(ax)(by) = (a({}^x b))(xy)$, for all $a, b \in A$ and $x, y \in G$, and with structural homomorphism $G \to (AG)^\times$ sending $x \in G$ to $1_A x$.

For $A = k$ this construction yields the group algebra kG. If A is an interior G-algebra, then the skew group algebra can be described as follows:

Proposition 1.3.6 *Let G be a group and A an interior G-algebra with structural homomorphism $\sigma : G \to A$. Then we have an isomorphism of interior G-algebras $AG \cong A \otimes_k kG$ sending ax to $a\sigma(x) \otimes x$, where $a \in A$ and $x \in G$.*

Proof One verifies that the map sending ax to $a\sigma(x) \otimes x$ yields an algebra homomorphism, and that it has an inverse sending $a \otimes x$ to $(a\sigma(x)^{-1})x$. □

Skew group algebras are special cases of more general concepts.

Definition 1.3.7 Let G be a group. A k-algebra B is called *G-graded* if, as a k-module, B is a direct sum $B = \oplus_{x \in G} B_x$ satisfying $B_x B_y \subseteq B_{xy}$ for all $x, y \in G$.

Note that if $B = \oplus_{x \in G} B_x$ is G-graded, then the subspace B_1 indexed by the unit element of G is a subalgebra of B. If A is a G-algebra, then the skew group algebra $AG = \oplus_{x \in G} Ax$ is G-graded in the obvious way.

Definition 1.3.8 Let G be a group and A a k-algebra. A *crossed product of A and G* is a G-graded k-algebra $B = \oplus_{x \in G} B_x$ such that $A = B_1$ and such that B_x contains an invertible element in B, for all $x \in G$.

A skew group algebra AG with the obvious G-grading is a crossed product since for any $x \in G$ we have $x \in Ax \cap (AG)^\times$. Let $B = \oplus_{x \in G} B_x$ be a crossed product of the algebra $A = B_1$ and the group G. For $x \in G$ choose $u_x \in B_x \cap B^\times$. Note that $u_x^{-1} \in B_{x^{-1}}$. For any $b \in B_x$ we have $b = (bu_x^{-1})u_x \in Au_x$, and hence $B_x = Au_x$. Conjugation by u_x induces an algebra automorphism $\iota(u_x)$ on A, sending $a \in A$ to $u_x a u_x^{-1}$. For any other choice $u'_x \in B_x \cap B^\times$ we have $u'_x = vu_x$ for some $v \in A^\times$, and hence $\iota(u'_x)$ and $\iota(u_x)$ differ by an inner automorphism of A. Thus the map sending $x \in G$ to $\omega(x) = \iota(u_x)\mathrm{Inn}(A) \in \mathrm{Aut}(A)/\mathrm{Inn}(A)$ is a group homomorphism. See [64, §1.3.A] for a more detailed discussion of the obstructions for a group homomorphism $G \to \mathrm{Aut}(A)/\mathrm{Inn}(A)$ to give rise to a crossed product of A and G.

Example 1.3.9 Let G be a group and N a normal subgroup of G. Then $kG = \oplus_{x\in[G/N]} kNx$ is a crossed product of kN and G/N.

1.4 Category algebras

Although category algebras will not be needed until much later, we introduce them already at this point, since they are a straightforward generalisation of group algebras and monoid algebras. Given a category $\mathcal{C}$, we denote by $\mathrm{Ob}(\mathcal{C})$ the class of objects of $\mathcal{C}$ and by $\mathrm{Mor}(\mathcal{C})$ the class of all morphisms in $\mathcal{C}$. We denote by $\mathrm{Hom}_{\mathcal{C}}(X, Y)$ the set of morphisms in $\mathcal{C}$ from X to Y, where $X, Y \in \mathrm{Ob}(\mathcal{C})$. A category $\mathcal{C}$ is called *small* if the classes $\mathrm{Ob}(\mathcal{C})$ and $\mathrm{Mor}(\mathcal{C})$ are sets, and $\mathcal{C}$ is called *finite* if $\mathrm{Ob}(\mathcal{C})$ and $\mathrm{Mor}(\mathcal{C})$ are finite sets. Note that if $\mathrm{Mor}(\mathcal{C})$ is a (finite) set, then so is $\mathrm{Ob}(\mathcal{C})$, since $\mathrm{Ob}(\mathcal{C})$ corresponds to the identity morphisms in $\mathrm{Mor}(\mathcal{C})$.

Definition 1.4.1 Let $\mathcal{C}$ be a small category. The *category algebra $k\mathcal{C}$ of $\mathcal{C}$ over k* is the free k-module with basis $\mathrm{Mor}(\mathcal{C})$ endowed with the unique k-bilinear multiplication defined for all $\varphi, \psi \in \mathrm{Mor}(\mathcal{C})$ by $\varphi\psi = \varphi \circ \psi$ whenever the composition $\varphi \circ \psi$ of morphisms in $\mathcal{C}$ is defined, and $\varphi\psi = 0$ otherwise.

Equivalently, $k\mathcal{C}$ consists of all functions $f : \mathrm{Mor}(\mathcal{C}) \to k$ that are zero on all but finitely many morphisms. Given two such functions f, g, the product $g \cdot f$ in $k\mathcal{C}$ is the function sending a morphism $\alpha \in \mathcal{C}$ to $\sum_{(\psi,\varphi)} g(\psi)f(\varphi)$, with (ψ, φ) running over the set of all pairs of morphisms in $\mathcal{C}$ satisfying $\psi \circ \varphi = \alpha$; the above sum is finite by the assumptions on f and g, with the usual convention that it is zero if the indexing set of pairs is empty. The multiplication in $k\mathcal{C}$ is associative because composition of morphisms in $\mathcal{C}$ is associative. For any object X in $\mathcal{C}$, the identity morphism Id_X is an idempotent in $k\mathcal{C}$ and for any two different objects X, Y in $\mathcal{C}$ the idempotents Id_X and Id_Y are orthogonal in $k\mathcal{C}$. If the object set $\mathrm{Ob}(\mathcal{C})$ is finite, then $k\mathcal{C}$ is a unitary algebra with unit element

$$1_{k\mathcal{C}} = \sum_{X\in\mathrm{Ob}(\mathcal{C})} \mathrm{Id}_X.$$

Since category algebras need not be unital, we divert in this section from our convention and do not require algebra homomorphisms to be unital. The following list of examples shows that group algebras and matrix algebras are particular cases of category algebras.

Examples 1.4.2 Group algebras, monoid algebras, and matrix algebras are category algebras:

(1) Let G be a group or monoid. Denote by $\mathbf{G}$ the category with a single object E whose endomorphism set is equal to G such that composition of two endomorphisms is given by multiplication in G. The map sending $x \in G$ to the endomorphism $x \in \mathrm{End}_{\mathbf{G}}(E)$ induces an isomorphism of k-algebras $kG \cong k\mathbf{G}$.

(2) Let n be a positive integer. Denote by $\mathbf{M}_n$ a category whose object set has n elements $X_1, X_2, \ldots, X_n$ such that there is exactly one morphism $\varphi_{i,j} : X_i \to X_j$ in $\mathbf{M}_n$, for $1 \leq i, j \leq n$. Denote by $E_{i,j}$ the matrix in $M_n(k)$ whose (i, j)-entry is 1 and all of whose other entries are zero, for $1 \leq i, j \leq n$. The map sending $E_{i,j}$ to $\varphi_{j,i}$ induces an isomorphism of k-algebras $M_n(k) \cong k\mathbf{M}_n$.

(3) Let $(\mathcal{P}, \leq)$ be a partially ordered set. Then $\mathcal{P}$ can be viewed as a category, with exactly one morphism $x \to y$ for any $x, y \in \mathcal{P}$ such that $x \leq y$. The corresponding category algebra $k\mathcal{P}$ is called the *incidence algebra* of $\mathcal{P}$.

Every k-algebra A can be obtained as a quotient of a monoid algebra: just take the multiplicative monoid $(A, \cdot)$ and the obvious map $k[A] \to A$. For a finite-dimensional algebra A over an algebraically closed field k there is a canonical choice for a category $\mathcal{C}$ with finitely many objects such that $A \cong k\mathcal{C}/I$ for some ideal I (which cannot be chosen canonically, in general); this leads to the notion of the *quiver* of an algebra, discussed in §4.9. A category algebra is never far off a semigroup algebra: given a category $\mathcal{C}$ with at least two objects we define a semigroup $\mathcal{S}$ by adjoining a zero element; that is, we set $\mathcal{S} = \mathrm{Mor}(\mathcal{C}) \cup \{\nu\}$, where the product in $\mathcal{S}$ is given by $\psi\varphi = \psi \circ \varphi$ if ψ, $\varphi \in \mathrm{Mor}(\mathcal{C})$ and $\psi \circ \varphi$ is defined; in all other cases the product of two elements in $\mathcal{S}$ is ν. Then $k\nu$ is an ideal in $k\mathcal{S}$, and we have an isomorphism of k-algebras $k\mathcal{C} \cong k\mathcal{S}/k\nu$.

A covariant functor between small categories $\Phi : \mathcal{C} \to \mathcal{D}$ maps $\mathrm{Mor}(\mathcal{C})$ to $\mathrm{Mor}(\mathcal{D})$ and hence induces a unique k-linear map $\varphi : k\mathcal{C} \to k\mathcal{D}$. This map, however, need not be an algebra homomorphism: with the notation of the previous example, all objects in $\mathbf{M}_n$ are isomorphic, and hence there is a unique equivalence of categories $\mathbf{M}_n \to \mathbf{M}_1$ sending every object of $\mathbf{M}_n$ to the unique object in $\mathbf{M}_1$. The linear map $M_n(k) \to M_1(k) = k$ induced by this equivalence is however not multiplicative, unless $n = 1$.

Proposition 1.4.3 ([95, 2.2.3]) *Let $\mathcal{C}$, $\mathcal{D}$ be small categories and $\Phi : \mathcal{C} \to \mathcal{D}$ a covariant functor. Denote by $\alpha : k\mathcal{C} \to k\mathcal{D}$ the unique k-linear map sending $\varphi \in \mathrm{Mor}(\mathcal{C})$ to $\Phi(\varphi)$. Then α is a homomorphism of k-algebras if and only if the map from $\mathrm{Ob}(\mathcal{C})$ to $\mathrm{Ob}(\mathcal{D})$ induced by Φ is injective. Moreover, if the object sets of $\mathcal{C}$, $\mathcal{D}$ are finite then α is a unital homomorphism of k-algebras if and only if Φ induces a bijection between $\mathrm{Ob}(\mathcal{C})$ and $\mathrm{Ob}(\mathcal{D})$.*

Proof Suppose Φ is not injective on objects. Let X, X' be two different objects in $\mathcal{C}$ such that $\Phi(X) = \Phi(X') = Y$. Then $\mathrm{Id}_X \mathrm{Id}_{X'} = 0$ in $k\mathcal{C}$, hence $\Phi(\mathrm{Id}_X \mathrm{Id}_{X'}) = 0$, but $\Phi(\mathrm{Id}_X)\Phi(\mathrm{Id}_{X'}) = \mathrm{Id}_Y \mathrm{Id}_Y \neq 0$, so α is not an algebra homomorphism. Conversely, if Φ is injective on objects, then for any two morphisms φ, ψ in $\mathrm{Mor}(\mathcal{C})$ such that $\Phi(\varphi)$, $\Phi(\psi)$ are composable in $\mathcal{D}$, the morphisms φ, ψ are composable as well, and $\phi(\varphi \circ \psi) = \Phi(\varphi \circ \psi)$, hence α is multiplicative. The last statement is an obvious consequence of the fact that the unit elements of $k\mathcal{C}$ and $k\mathcal{D}$ are equal to $\sum_{X \in \mathrm{Ob}(\mathcal{C})} \mathrm{Id}_X$ and $\sum_{Y \in \mathrm{Ob}(\mathcal{D})} \mathrm{Id}_Y$, respectively. □

Modules over a category algebra $k\mathcal{C}$ are closely related to functors from $\mathcal{C}$ to the category of k-modules $\mathrm{Mod}(k)$. In fact, if the object set $\mathrm{Ob}(\mathcal{C})$ is finite, then the corresponding module and functor categories are equivalent. For $\mathcal{C}$ a small category and k a commutative ring, we denote by $\mathrm{Mod}(k)^{\mathcal{C}}$ the category whose objects are the covariant functors $\mathcal{C} \to \mathrm{Mod}(k)$ and whose morphisms are natural transformations. As before, we denote by $\mathrm{Mod}(k\mathcal{C})$ the category of left $k\mathcal{C}$-modules.

Theorem 1.4.4 ([66]) *Let $\mathcal{C}$ be a small category with finite object set $\mathrm{Ob}(\mathcal{C})$. There is a k-linear equivalence $\mathrm{Mod}(k\mathcal{C}) \cong \mathrm{Mod}(k)^{\mathcal{C}}$.*

Proof Let M be a left $k\mathcal{C}$-module. We define a functor $\mathcal{F} : \mathcal{C} \to \mathrm{Mod}(k)$ as follows. For each object X in $\mathcal{C}$ set $\mathcal{F}(X) = \mathrm{Id}_X \cdot M$. For each morphism $\varphi : X \to Y$ in $\mathcal{C}$ we have, in $k\mathcal{C}$, the obvious equalities $\varphi = \mathrm{Id}_Y \circ \varphi = \varphi \circ \mathrm{Id}_X$, and hence the action of φ on M induces a unique k-linear map $\mathcal{F}(\varphi) : \mathrm{Id}_X \cdot M \to \mathrm{Id}_Y \cdot M$, through which $\mathcal{F}$ becomes a contravariant functor. The assignment $M \mapsto \mathcal{F}$ defines a k-linear functor $\rho : \mathrm{Mod}(k\mathcal{C}) \to \mathrm{Mod}(k)^{\mathcal{C}}$. Conversely, given a functor $\mathcal{F} : \mathcal{C} \to \mathrm{Mod}(k)$ we define a $k\mathcal{C}$-module M by setting $M = \oplus_{X \in \mathrm{Ob}(\mathcal{C})} \mathcal{F}(X)$, and for any morphism $\varphi : X \to Y$ in $\mathcal{C}$ and any $m \in \mathcal{F}(Z)$ we define the $k\mathcal{C}$-module structure on M by setting $\varphi \cdot m = \mathcal{F}(\varphi)(m)$ provided that $Z = X$ and $\varphi \cdot m = 0$ if $Z \neq X$. The assignment $\mathcal{F} \mapsto M$ defines a k-linear functor $\sigma : \mathrm{Mod}(k)^{\mathcal{C}} \to \mathrm{Mod}(k\mathcal{C})$. It is easy to check that $\rho \circ \sigma$ is the identity functor on $\mathrm{Mod}(k)^{\mathcal{C}}$. Since $\mathrm{Ob}(\mathcal{C})$ is finite, it follows that $M = \oplus_{X \in \mathrm{Ob}(\mathcal{C})} \mathrm{Id}_X \cdot M$ as k-modules, for any $k\mathcal{C}$-module M, which implies that also $\sigma \circ \rho$ is the identity on $\mathrm{Mod}(k\mathcal{C})$. □

There is a twisted version of category algebras that is analogous to twisted group algebras. The following definition is a particular case of low degree functor cohomology; when specialised to the category **G** of a group G this yields the corresponding concepts described in 1.2.2 and 1.2.7.

Definition 1.4.5 Let $\mathcal{C}$ be a small category and M an abelian group, written multiplicatively. A *2-cocycle of $\mathcal{C}$ with coefficients in M* is a map α sending any two morphisms φ, ψ in $\mathcal{C}$ for which the composition $\psi \circ \varphi$ is defined to an element $\alpha(\psi, \varphi) \in M$ such that for any three morphisms φ, ψ, τ in $\mathcal{C}$ for which the compositions $\psi \circ \varphi$ and $\tau \circ \psi$ are defined we have the *2-cocycle identity*

$$\alpha(\tau, \psi \circ \varphi)\alpha(\psi, \varphi) = \alpha(\tau \circ \psi, \varphi)\alpha(\tau, \psi)$$

in M. The set of 2-cocycles of $\mathcal{C}$ with coefficients in M is denoted by $Z^2(\mathcal{C}; M)$; this is an abelian group with product induced by that in M. A *2-coboundary of $\mathcal{C}$ with coefficients in M* is a map β sending any two morphisms φ, ψ in $\mathcal{C}$ for which $\psi \circ \varphi$ is defined to an element $\beta(\psi, \varphi) \in M$ such that there exists a map $\gamma : \text{Mor}(\mathcal{C}) \to M$ satisfying

$$\beta(\psi, \varphi) = \gamma(\varphi)\gamma(\psi)\gamma(\psi \circ \varphi)^{-1}$$

for any two morphisms φ, ψ for which the composition $\psi \circ \varphi$ is defined. The set $B^2(\mathcal{C}; M)$ of all 2-coboundaries of $\mathcal{C}$ with coefficients in M is a subgroup of $Z^2(\mathcal{C}; M)$ and the quotient group

$$H^2(\mathcal{C}; M) = Z^2(\mathcal{C}; M)/B^2(\mathcal{C}; M)$$

is called the *second cohomology of $\mathcal{C}$ with coefficients in M*.

Definition 1.4.6 Let $\mathcal{C}$ be a small category and $\alpha \in Z^2(\mathcal{C}; k^\times)$. The *twisted category algebra of $\mathcal{C}$ by α* is the k-algebra, denoted $k_\alpha\mathcal{C}$, which is equal to $k\mathcal{C}$ as k-module, endowed with the unique k-bilinear product $k_\alpha\mathcal{C} \times k_\alpha\mathcal{C} \to k_\alpha\mathcal{C}$ defined, for all $\varphi, \psi \in \text{Mor}(\mathcal{C})$ by $\psi\varphi = \alpha(\psi, \varphi)\psi \circ \varphi$ if the composition $\psi \circ \varphi$ is defined, and $\psi\varphi = 0$ otherwise.

Theorem 1.4.7 *Let $\mathcal{C}$ be a small category and $\alpha, \beta \in Z^2(\mathcal{C}; k^\times)$. The following hold.*

(i) The k-algebra $k_\alpha\mathcal{C}$ is associative.
(ii) For any morphism $\varphi : X \to Y$ in $\mathcal{C}$ we have $\alpha(\text{Id}_Y, \text{Id}_Y) = \alpha(\text{Id}_Y, \varphi)$ and $\alpha(\varphi, \text{Id}_X) = \alpha(\text{Id}_X, \text{Id}_X)$.
(iii) There is an isomorphism $k_\alpha\mathcal{C} \cong k_\beta\mathcal{C}$ sending any morphism φ in $\mathcal{C}$ to $\gamma(\varphi)\varphi$ for some scalar $\gamma(\varphi) \in k^\times$ if and only if the images of α and β in $H^2(\mathcal{C}; k^\times)$ are equal.

Proof For statement (i) one verifies exactly as in 1.2.1 that the associativity of the multiplication in $k_\alpha \mathcal{C}$ is actually equivalent to the 2-cocycle identity. Statement (ii) is proved as 1.2.5 (i), and the proof of (iii) follows the lines of the proof of 1.2.6. □

As in the case of groups, second cohomology groups of categories are related to extensions of categories – see for instance [92] for details and further references.

Remark 1.4.8 If $\mathcal{C}$ is a category such that the sets $\mathrm{Hom}_{\mathcal{C}}(X, Y)$ have already a k-module structure (and such that the composition in $\mathcal{C}$ is k-bilinear) then there is no need to linearise the homomorphism sets again; one can define an algebra by taking the direct sum $\oplus_{X,Y \in \mathrm{Ob}(\mathcal{C})} \mathrm{Hom}_{\mathcal{C}}(X, Y)$ with product induced by composition of morphisms. Theorem 1.4.4 carries over to this situation if $\mathrm{Ob}(\mathcal{C})$ is finite: a k-linear functor $\mathcal{F} : \mathcal{C} \to \mathrm{mod}(k)$ defines a module $\oplus_{X \in \mathrm{Ob}(\mathcal{C})} \mathcal{F}(X)$.

Exercise 1.4.9 Let k be a field and let M be a finite monoid. Let $x \in M$. Show that if the image of x in kM is invertible in the algebra kM, then x is invertible in M.

Exercise 1.4.10 Let k be a field and $\mathcal{C}$ a finite category. Show that if $\mathcal{C}$ has at least two morphisms, then $k\mathcal{C}$ is not a division ring; that is, $k\mathcal{C}$ has a nonzero noninvertible element.

Exercise 1.4.11 Let k'/k be a finite field extension of degree at least 2. Show that k', regarded as a k-algebra, is not isomorphic to the algebra of any finite category.

Exercise 1.4.12 Let k be a commutative ring. Find a category whose category algebra over k is isomorphic to the polynomial algebra $k[x]$.

Exercise 1.4.13 Let $\mathcal{P}$ be a finite partially ordered set, viewed as a category with exactly one morphism $x \to y$ whenever $x \leq y$. Let ι be the sum in $k\mathcal{P}$, of all morphisms in $\mathcal{P}$. Show that ι has a multiplicative inverse μ in $k\mathcal{P}$. Consider μ as a function on morphisms in $\mathcal{P}$ to k, and show that $\mu(x \to x) = 1$ for all $x \in \mathcal{P}$ and $\sum_z \mu(x \to z) = 0$ for $x, y \in \mathcal{P}$ satisfying $x < y$, where in the sum z runs over all elements in $\mathcal{P}$ such that $x \leq z \leq y$. This function is known as the *Möbius function* of $\mathcal{P}$, due to Rota.

Exercise 1.4.14 Let n be a positive integer and denote by $[n]$ the totally ordered set $\{1, 2, 3, \ldots, n\}$, viewed as a category. Show that the category algebra $k[n]$ is isomorphic to the algebra of upper triangular matrices in $M_n(k)$.

1.5 Centre and commutator subspaces

If G is a finite group the centre of the group algebra kG admits a canonical k-basis, namely its conjugacy class sums:

Theorem 1.5.1 *Let G be a finite group. Let $\mathcal{K}$ be the set of conjugacy classes in G, and for any conjugacy class C in $\mathcal{K}$ denote by $\overline{C}$ the conjugacy class sum $\overline{C} = \sum_{y \in C} y$ in kG. The set $\{\overline{C} | C \in \mathcal{K}\}$ is a k-basis of $Z(kG)$, and for any two C, $C' \in \mathcal{K}$ there are nonnegative integers a_D for all $D \in \mathcal{K}$ such that $\overline{C} \cdot \overline{C'} = \sum_{D \in \mathcal{K}} a_D \overline{D}$, where we identify the integers a_D with their images in k. In particular, $Z(kG)$ is free as a k-module with rank equal to the number of conjugacy classes in G.*

Proof An element $\sum_{x \in G} \lambda_x x$ of the group algebra kG lies in the centre $Z(kG)$ if and only if it commutes with all elements in G, or equivalently, if and only if $\lambda_x = \lambda_{yxy^{-1}}$ for all $x, y \in G$. This means that the coefficients λ_x depend only on the conjugacy class of x. Thus the conjugacy class sums form a k-basis of $Z(kG)$ by the preceding remarks. Let C, $C' \in \mathcal{K}$ and write $\overline{C} \cdot \overline{C'} = \sum_{D \in \mathcal{K}} a_D \overline{D}$ for some $a_D \in k$. Let $D \in \mathcal{K}$ and choose an element $z \in D$. The coefficient a_D is equal to the number of pairs $(x, y) \in C \times C'$ satisfying $xy = z$. This is, in particular, a nonnegative integer. The result follows. □

Proposition 1.5.2 *Let G, H be finite groups. There is a unique algebra isomorphism $Z(k(G \times H)) \cong Z(kG) \otimes_k Z(kH)$ sending the conjugacy class sum of $(x, y) \in G \times H$ in $k(G \times H)$ to $C_x \otimes C_y$, where C_x, C_y are the conjugacy class sums of x and y in kG and kH, respectively.*

Proof The given map is an algebra homomorphism, induced by the algebra isomorphism in 1.1.4, restricted to the centre of $k(G \times H)$. The explicit description of the centre of a finite group algebra in terms of conjugacy class sums 1.5.1 implies that this induces an isomorphism as stated. □

The fact that taking centres commutes with taking tensor products of algebras as in the previous proposition holds in greater generality.

Theorem 1.5.3 *Let A, B be k-algebras. Suppose that A and $Z(B)$ are free as k-modules and that $Z(B)$ is a direct summand of B as a k-module. Then the centraliser of the subalgebra $1 \otimes B$ of $A \otimes_k B$ is $A \otimes_k Z(B)$, and we have $Z(A \otimes_k B) = Z(A) \otimes_k Z(B)$.*

Proof Since $Z(B)$ is a direct summand of B as k-module, we can identify $A \otimes_k Z(B)$ to its canonical image in $A \otimes_k B$, and this is a direct summand of $A \otimes_k B$ as k-module. Similarly, we can identify $Z(A) \otimes_k Z(B)$ with its image in $A \otimes_k B$.

Clearly $A \otimes_k Z(B)$ centralises $1 \otimes B$, and $Z(A) \otimes_k Z(B)$ is contained in $Z(A \otimes_k B)$. We need to show the reverse inclusions. Let $\{a_i\}_{i \in I}$ be a k-basis of A and let $\{z_j\}_{j \in J}$ be a k-basis of $Z(B)$, for some indexing sets I and J. Then the set $\{a_i \otimes 1_B\}_{i \in I}$ is a B-basis of $A \otimes_k B$ as a right B-module, and the set $\{1_A \otimes z_j\}_{j \in J}$ is an A-basis of $A \otimes_k Z(B)$ as a left A-module. Let $z \in A \otimes_k B$. By the above, z can be written in the form

$$z = \sum_{i \in I} a_i \otimes b_i$$

with uniquely determined elements $b_i \in B$ of which only finitely many are nonzero. Suppose that z centralises $1 \otimes B$. This is equivalent to $(1_A \otimes b)z = z(1_A \otimes b)$ for all $b \in B$, which is equivalent to $\sum_{i \in I} a_i \otimes bb_i = \sum_{i \in I} a_i \otimes b_i b$ for all $b \in B$. The uniqueness of expressions of this form implies that this is equivalent to $bb_i = b_i b$ for all $i \in I$ and all $b \in B$. Thus $b_i \in Z(B)$ for $i \in I$, and hence $z \in A \otimes_k Z(B)$. This shows the first statement. Suppose that $z \in Z(A \otimes_k B)$. By the first statement, we have $z \in A \otimes_k Z(B)$. Thus z can be written in the form

$$z = \sum_{j \in J} c_j \otimes z_j$$

with uniquely determined elements $c_j \in A$ of which at most finitely many are non zero. Then the equality $(a \otimes 1_B)z = z(a \otimes 1_B)$ for all $a \in A$ is equivalent to $ac_j = c_j a$ for all $a \in A$, hence $c_j \in Z(A)$ for all $j \in J$, which proves that $z \in Z(A) \otimes_k Z(B)$. The result follows. □

The centre of a twisted group algebra $k_\alpha G$ does not in general admit a basis as in 1.5.1, essentially because the multiplicative inverse in $k_\alpha G$ of a group element x need not be equal to the inverse x^{-1} of x as an element in the group G. Another measure for how far a k-algebra A is from being commutative is the *commutator space* $[A, A]$; this is the k-submodule of A generated by all *additive commutators* $[a, b] = ab - ba$, where $a, b \in A$. We have $[A, A] = \{0\}$ if and only if A is commutative. Note that $[A, A]$ is a $Z(A)$-submodule of A, and hence $A/[A, A]$ is a $Z(A)$-module. Indeed, for $a, b \in A$ and $z \in Z(A)$ we have $z[a, b] = z(ab - ba) = (za)b - b(za) = [za, b]$. Commutator subspaces of group algebras admit the following description.

Proposition 1.5.4 *Let G be a finite group.*

(i) The commutator space $[kG, kG]$ is spanned, as a k-module, by the set of commutators $[x, y] = xy - yx$, with $x, y \in G$.
(ii) If $x, x' \in G$ are conjugate, then $x - x' \in [kG, kG]$.

(iii) *An element $\sum_{x\in G}\alpha_x x$ of kG belongs to the commutator space $[kG,kG]$ if and only if $\sum_{x\in C}\alpha_x = 0$ for any conjugacy class C in G.*
(iv) *Let $\mathcal{K}$ be the set of conjugacy classes, and for $C\in\mathcal{K}$, let $x_C\in C$. The k-module $\sum_{C\in\mathcal{K}} kx_C$ is a complement of $[kG,kG]$ in kG.*

Proof Since any element in kG is a k-linear combination of group elements, it follows trivially that any element in $[kG,kG]$ is a linear combination of elements of the form $xy-yx$, where $x,y\in G$. This proves (i). Let x, x' be conjugate elements in G. Thus there is $y\in G$ such that $x'=yxy^{-1}$. Then $x-x'=x-yxy^{-1}=[xy^{-1},y]$, whence (ii). Since xy and yx are conjugate it follows that the elements of the form $xy-yx$ in kG have the property as stated in (iii). Conversely, let C be a conjugacy class of G and for $x\in C$ let $\alpha_x\in k$ such that $\sum_{x\in C}\alpha_x=0$. Fix an element $x_C\in C$. Using (ii) it follows that $\sum_{x\in C}\alpha_x x=\sum_{x\in C}\alpha_x(x-x_C)$ belongs to $[kG,kG]$. This proves (iii). An arbitrary element in kG can be written as a sum, over the set of conjugacy classes C, of elements of the form $\sum_{x\in C}\lambda_x x$. Setting $\mu=\sum_{x\in C}\lambda_x$, it follows from (iii) that $(\sum_{x\in C}\lambda_x x)-\mu x_C$ belongs to $[kG,kG]$. Thus every element in kG is contained in $[kG,kG]+\sum_{C\in\mathcal{K}} kx_C$. No nonzero element in the k-module $\sum_{C\in\mathcal{K}} kx_C$ has the property described in (iii), and hence the previous sum is direct. □

For a group algebra kG of a finite group G, we have a close connection between the k-dual of the centre and the commutator subspace. Since the centre $Z(kG)$ has as canonical k-basis the set of conjugacy class sums, the k-dual $Z(kG)^*=\mathrm{Hom}_k(Z(kG),k)$ therefore also has a canonical basis indexed by the set $\mathcal{K}$ of conjugacy classes of G, namely the *dual basis* $\{\sigma_C | C\in\mathcal{K}\}$, where σ_C sends C to 1 and $C'\neq C$ to 0, for $C,C'\in\mathcal{K}$. The dual of the centre and the commutator space of a finite group algebra are related as follows (see e.g. [76, §5 (2)]).

Proposition 1.5.5 *Let G be a finite group.*

(i) *We have an isomorphism of k-modules $kG/[kG,kG]\cong Z(kG)^*$ sending $x+[kG,kG]$ to the unique element $\sigma\in Z(kG)^*$ which maps the conjugacy class sum of x^{-1} to 1 and any other conjugacy class sum to zero. In particular, the k-module $kG/[kG,kG]$ is free of rank equal to the number of conjugacy classes of G and the image in $kG/[kG,kG]$ of a system of representatives of the conjugacy classes of G is a k-basis of $kG/[kG,kG]$.*
(ii) *We have an isomorphism of k-modules $(kG/[kG,kG])^*\cong Z(kG)$ sending $\varphi\in(kG/[kG,kG])^*$ to $\sum_{x\in G}\varphi(x^{-1})x$, where we regard φ as a map from kG to k via the canonical surjection $kG\to kG/[kG,kG]$.*

Proof It follows from 1.5.4 that $kG/[kG, kG]$ is free of rank equal to the number of conjugacy classes of G, and that the image in $kG/[kG, kG]$ of a set $\mathcal{R}$ of representatives of the conjugacy classes is a basis that is independent of the choice of $\mathcal{R}$ because conjugate group elements have the same image in $kG/[kG, kG]$. Since both $Z(kG)$ and its k-dual have bases indexed by the conjugacy classes of G, statement (i) follows. One can prove (ii) by observing that this is the dual map of the isomorphism in (i), but one can do this also directly. A linear map $\varphi : kG/[kG, kG] \to k$ inflated to a map $kG \to k$ is, by the above, constant on conjugacy classes, and hence the element $\sum_{x\in G} \varphi(x^{-1})x$ belongs to $Z(kG)$. Conversely, if $\sum_{x\in G} \lambda_x x$ is in $Z(kG)$, then the map $\varphi : kG \to k$ defined by $\varphi(x) = \lambda_{x^{-1}}$ is constant on G-conjugacy classes, hence induces a map $kG/[kG, kG] \to k$. These two assignments are clearly inverse to each other. □

The maps in the above Proposition are in fact $Z(kG)$-module isomorphisms. One can verify this directly, but this follows also from more general properties of finite group algebras as *symmetric algebras*; see Theorem 2.11.2 below.

Proposition 1.5.6 *Let A, B be k-algebras. We have $[A \otimes_k B, A \otimes_k B] = A \otimes [B, B] + [A, A] \otimes B$ as k-submodules of $A \otimes_k B$.*

Proof Let a, $a' \in A$ and b, $b' \in B$. Then $[a \otimes b, a' \otimes b'] = [a, a'] \otimes bb' + a'a \otimes [b, b']$, which shows that the left side is contained in the right side. The other inclusion follows from $a \otimes [b, b'] = [a \otimes b, 1 \otimes b']$ and $[a, a'] \otimes b = [a \otimes b, a' \otimes 1]$. □

Given a positive integer n, the *trace* of a square matrix $M = (m_{ij})_{1\leq i,j\leq n}$ in $M_n(k)$ is the sum of its diagonal elements $\mathrm{tr}(M) = \sum_{1\leq i\leq n} m_{ii}$. Taking traces defines a k-linear map $\mathrm{tr} : M_n(k) \to k$. This map is surjective as it maps a matrix with exactly one diagonal entry 1 (and all other entries 0) to 1. An elementary verification shows that the trace is *symmetric*; that is, $\mathrm{tr}(MN) = \mathrm{tr}(NM)$ for any two matrices $M, N \in M_n(k)$. The commutator space of a matrix algebra consists of all matrices whose trace is zero.

Theorem 1.5.7 *Let n be a positive integer. We have $[M_n(k), M_n(k)] = \ker(\mathrm{tr})$, as a k-module, $\ker(\mathrm{tr})$ is free of rank $n^2 - 1$, and the k-module $M_n(k)/[M_n(k), M_n(k)]$ is free of rank 1. Moreover, $M_n(k)$ is equal to the k-linear span of the group $\mathrm{GL}_n(k) = M_n(k)^\times$ of invertible elements in $M_n(k)$.*

Proof Since $\mathrm{tr}(MN) = \mathrm{tr}(NM)$ it follows that for any $M, N \in M_n(k)$ we have $[M_n(k), M_n(k)] \subseteq \ker(\mathrm{tr})$. For $1 \leq i, j \leq n$ denote by E_{ij} the matrix with entry (i, j) equal to 1 and all other entries zero. Elementary verifications show that

$E_{ij}E_{rs} = E_{is}$ if $j = r$ and zero otherwise. If $i \neq j$, then $\mathrm{tr}(E_{ij}) = 0$ and $E_{ij} = [E_{ii}, E_{ij}] \in [M_n(k), M_n(k)]$. For $2 \leq j \leq n$ denote by D_j the matrix with entry $(1, 1)$ equal to 1, entry (j, j) equal to -1 and zero everywhere else; that is, $D_j = E_{11} - E_{jj}$. Again $\mathrm{tr}(D_j) = 0$ and $D_j = [E_{1j}, E_{j1}] \in [M_n(k), M_n(k)]$. The set of matrices E_{ij} with $i \neq j$ and D_j with $2 \leq j \leq n$ is linearly independent, contained in ker(tr) and has $n^2 - 1$ elements. This set is a k-basis of ker(tr) because together with E_{11} it is a k-basis of $M_n(k)$. This shows that $\ker(\mathrm{tr}) \subseteq [M_n(k), M_n(k)]$, whence the first equality as stated. The last statement is trivial if $n = 1$. Assume that $n \geq 2$. It suffices to show that the matrices E_{ij} are in the k-span of $\mathrm{GL}_n(k)$. If $i \neq j$, then $\mathrm{Id} + E_{ij}$ is invertible, hence $E_{ij} = (\mathrm{Id} + E_{ij}) - \mathrm{Id}$ is in the span of $\mathrm{GL}_n(k)$. The matrix E_{ii} is conjugate to E_{11} via a permutation matrix, so it suffices to show that E_{11} is in the span of $\mathrm{GL}_n(k)$. Let J be the matrix with antidiagonal entries 1 and 0 elsewhere. Then J and $J + E_{11}$ are invertible, hence $E_{11} = (J + E_{11}) - J$ is in the span of $\mathrm{GL}_n(k)$, completing the proof. □

A k-algebra homomorphism $\alpha : A \to B$ need not send $Z(A)$ to $Z(B)$. But α sends any commutator $[a, b]$ in $[A, A]$ to the commutator $[\alpha(a), \alpha(b)]$ in $[B, B]$ and hence induces a k-linear map $A/[A, A] \to B/[B, B]$. Thus the quotient $A/[A, A]$ has better functoriality properties than the centre $Z(A)$. If A, B are group algebras then by taking k-duals one gets from 1.5.5 at least a k-linear (but not necessarily multiplicative) map $Z(B) \to Z(A)$, so taking centres becomes contravariantly functorial. The fact that the centre and the quotient by the commutator subspace are dual to each other holds more generally for symmetric algebras; see Section 2.11. There are bimodule versions of centres and commutator subspaces.

Definition 1.5.8 Let A be a k-algebra, and let M be an A-A-bimodule. We set $M^A = \{m \in M \mid am = ma \text{ for all } a \in A\}$. We denote by $[A, M]$ the k-submodule of M generated by the set of elements of the form $am - ma$ in M, where $a \in A$ and $m \in M$.

Remark 1.5.9 Let A be a k-algebra and let M, N be A-A-bimodules. Some of the earlier properties of centres and commutator spaces extend to fixed points and commutator spaces in bimodules. The sets M^A and $[A, M]$ are both left and right $Z(A)$-submodules of M, and we have $A^A = Z(A)$. A bimodule homomorphism $\theta : M \to N$ restricts to maps $M^A \to N^A$ and $[A, M] \to [A, N]$. If θ is injective, them both of these restrictions are injective as well. If θ is surjective, then the induced map $[A, M] \to [A, N]$ is surjective, but the induced map $M^A \to N^A$ need not be surjective. That is, taking A-fixed points on bimodules, viewed as a functor from the category of A-A-bimodules to the category of k-modules, is left exact but not right exact.

Remark 1.5.10 Let A be a k-algebra and U an A-module. Then $\mathrm{End}_k(U)$ is an A-A-bimodule through composing and precomposing linear endomorphisms of U by the action of elements in A on U. More precisely, for $\varphi \in \mathrm{End}_k(U)$ and $a, b \in A$ the k-endomorphism $a \cdot \varphi \cdot b$ of U is defined by $(a \cdot \varphi \cdot b)(u) = a\varphi(bu)$ for all $u \in U$. We have $(\mathrm{End}_k(U))^A = \mathrm{End}_A(U)$; indeed, the equality $a \cdot \varphi = \varphi \cdot a$ is equivalent to $a\varphi(u) = \varphi(au)$ for all $u \in U$. Note that this is a generalisation of Example 1.3.3 (3).

One can describe fixed points on bimodules as follows.

Proposition 1.5.11 *Let A be a k-algebra and M an A-A-bimodule. The map sending an A-A-bimodule homomorphism $\varphi : A \to M$ to its evaluation $\varphi(1) \in M$ induces an isomorphism of $Z(A)$-modules $\mathrm{Hom}_{A\otimes_k A^{\mathrm{op}}}(A, M) \cong M^A$. The inverse of this isomorphism sends an element $m \in M^A$ to the homomorphism $\psi : A \to M$ defined by $\psi(a) = am$ for all $a \in A$.*

Proof For $\varphi : A \to M$ an A-A-bimodule homomorphism, we have $a\varphi(1) = \varphi(a) = \varphi(1)a$ for all $a \in A$, which shows that $\varphi(1) \in M^A$, so the map in the statement is well-defined, and clearly this map is a $Z(A)$-module homomorphism. The map ψ, as defined in the statement, is indeed an A-A-bimodule homomorphism, since we have $am = ma$ for all $a \in A$ as $m \in M^A$. A trivial verification shows that the two maps in the statement are inverse to each other. □

Corollary 1.5.12 *Let A be a k-algebra. The map sending an A-A-bimodule endomorphism φ of A to $\varphi(1)$ is a k-algebra isomorphism $\mathrm{End}_{A\otimes_k A^{\mathrm{op}}}(A) \cong Z(A)$. Its inverse sends $z \in Z(A)$ to the endomorphism ψ of A defined by $\psi(a) = az$ for all $a \in A$.*

Proof It follows from 1.5.11 that the given maps are inverse to each other. An easy verification shows that these maps are k-algebra homomorphisms. □

Corollary 1.5.12 implies that idempotents in $Z(A)$ correspond bijectively to projections of A onto bimodule summands of A. This is the very first step towards block theory; we will elaborate on this in more detail in §1.7.

Exercise 1.5.13 Use Proposition 1.5.5 and Theorem 1.5.7 to show that if G is a nontrivial finite group, then the group algebra kG cannot be isomorphic to a matrix algebra $M_n(k)$, where n is a positive integer.

Exercise 1.5.14 Let $\mathcal{C}$ be a finite category, and let X, Y be two different objects in $\mathcal{C}$. Show that $k\mathrm{Hom}_{\mathcal{C}}(X, Y)$ is contained in $[k\mathcal{C}, k\mathcal{C}]$.

Exercise 1.5.15 Let A be a k-algebra. Show that the ideal I in A generated by the subspace $[A, A]$ is the unique minimal ideal in A such that the k-algebra A/I is commutative.

Exercise 1.5.16 Let n be a positive integer. Show that $M_n(k)$ has a k-basis consisting of invertible matrices. (If k is a field, this follows from the last statement of 1.5.7.)

1.6 The Hopf algebra structure of group algebras

The unit element 1_A of a k-algebra A determines a k-algebra homomorphism $k \to A$ sending $\lambda \in k$ to $\lambda \cdot 1_A$. Not any k-algebra A will admit however a unital algebra homomorphism $A \to k$. For instance, there is no unital algebra homomorphism $M_2(k) \to k$. If G is a group, then the group algebra kG does admit such an algebra homomorphism, namely the algebra homomorphism $kG \to k$ induced by the trivial group homomorphism $G \to \{1\}$.

Definition 1.6.1 Let G be a group. The *augmentation homomorphism of* kG is the unique k-algebra homomorphism $\epsilon : kG \to k$ sending every element $x \in G$ to 1_k; that is, $\epsilon(\sum_{x\in G} \lambda_x x) = \sum_{x\in G} \lambda_x$, where $\lambda_x \in k$ for $x \in G$ with only finitely many λ_x nonzero. We set

$$I(kG) = \ker(\epsilon) = \left\{ \sum_{x\in G} \lambda_x x \,\middle|\, \sum_{x\in G} \lambda_x = 0 \right\}$$

and call $I(kG)$ the *augmentation ideal of* kG.

Through the augmentation homomorphism, the ring k itself becomes a kG-module with the property that every $x \in G$ acts as identity on k; this module is called the *trivial* kG*-module*. An element $\sum_{x\in G} \lambda_x x \in kG$ acts on the trivial kG-module k by multiplication with the scalar $\epsilon(\sum_{x\in G} \lambda_x x) = \sum_{x\in G} \lambda_x$, where $\epsilon : kG \to k$ is the augmentation homomorphism in 1.6.1. Thus the augmentation ideal $I(kG)$ is the annihilator of the trivial kG-module. If k is a field, then the trivial kG-module is simple because it has dimension 1.

Proposition 1.6.2 *Let G be a group. The augmentation ideal $I(kG)$ is free as a k-module and the set $\{x - 1 | x \in G \setminus \{1\}\}$ is a k-basis of $I(kG)$. In particular, if G is finite, then $I(kG)$ is a free k-module of rank $|G| - 1$.*

Proof Let $\sum_{x\in G} \lambda_x x \in I(kG)$; that is, $\sum_{x\in G} \lambda_x = 0$. Then $\sum_{x\in G} \lambda_x x = \sum_{x\in G} \lambda_x(x - 1) = \sum_{x\in G, x\neq 1} \lambda_x(x - 1)$ which shows that the set $\{x - 1 | x \in G \setminus \{1\}\}$ generates $I(kG)$ as a k-module. Since G is a k-basis of kG one easily sees that this set is k-linearly independent, hence a k-basis of $I(kG)$. □

This proof involves a certain amount of notational abuse which we will adopt without further comment: the element $x - 1$ in kG means more precisely $1_k \cdot x - 1_k \cdot 1_G$. The augmentation ideal of a cyclic group over some commutative ring is a principal ideal:

Proposition 1.6.3 *Let G be a finite cyclic group, and let y be a generator of G. Then $I(kG) = (y-1)kG = kG(y-1)$. In particular, $I(kG)$ is a principal ideal in kG.*

Proof Since G is cyclic, kG is commutative and thus $(y-1)kG = kG(y-1)$. Thus $(y-1)kG$ is an ideal in kG. Since $y - 1 \in I(kG)$ and $I(kG)$ is an ideal we get $(y-1)kG \subseteq I(kG)$. For the reverse inclusion, let $\sum_{x\in G} \lambda_x x \in I(kG)$; that is, $\sum_{x\in G} \lambda_x = 0$. Therefore $\sum_{x\in G} \lambda_x x = \sum_{x\in G} \lambda_x(x-1)$. Since G is cyclic with generator y, for any $x \in G$ there is a positive integer n such that $x = y^n$. Then $x - 1 = y^n - 1 = (y-1)(1 + y + y^2 + \cdots + y^{n-1})$ belongs to $(y-1)kG$, which implies the result. □

The kernel of an algebra homomorphism α between group algebras induced by a group homomorphism φ has as kernel the ideal generated by the augmentation ideal of $\ker(\varphi)$.

Proposition 1.6.4 *Let G, H be finite groups. Let $\varphi : G \to H$ be a group homomorphism and let $\alpha : kG \to kH$ be the induced algebra homomorphism. Set $N = \ker(\varphi)$. We have $\ker(\alpha) = kG \cdot I(kN) = I(kN) \cdot kG$.*

Proof Let $\sum_{y\in N} \mu_y y \in I(kN)$; that is, $\sum_{y\in N} \mu_y = 0$. Since N is normal in G we have $\sum_{y\in N} \mu_y xyx^{-1} \in I(kN)$, hence $xI(kN) = I(kN)x$ for all $x \in G$, from which we get the equality $kG \cdot I(kN) = I(kN) \cdot kG$. This equality shows that $I(kN) \cdot kG$ is a 2-sided ideal. Since φ maps all elements in N to 1 we get $I(kN) \subseteq \ker(\alpha)$. As α is an algebra homomorphism, its kernel is an ideal and thus contains the ideal $I(kN) \cdot kG$ generated by $I(kN)$ in kG. In order to show the reverse inclusion, let $\sum_{x\in G} \lambda_x x \in \ker(\alpha)$. Denote by $[G/N]$ a system of representatives of the cosets G/N in G. Note that φ maps N to 1, and hence that $\varphi(xy) = \varphi(x)$ for all $x \in G$ and $y \in N$. Thus $\alpha(\sum_{x\in G} \lambda_x x) = \sum_{x\in G} \lambda_x \varphi(x) = \sum_{x\in[G/H]} \sum_{y\in N} \lambda_{xy}\varphi(xy) = \sum_{x\in[G/H]} \sum_{y\in N} \lambda_{xy}\varphi(x) = 0$ if and only if $\sum_{y\in N} \lambda_{xy} = 0$ for any $x \in G$. This means that $\sum_{y\in N} \lambda_{xy} y \in I(kN)$ for any $x \in G$, and hence $\sum_{x\in G} \lambda_x x = \sum_{x\in[G/H]} \sum_{y\in N} \lambda_{xy} xy = \sum_{x\in[G/H]} x(\sum_{y\in N} \lambda_{xy} y)$ belongs to $kG \cdot I(kN)$. □

The augmentation ideal is the smallest left or right ideal I of kG with the property that all elements of G act trivially on kG/I as a left or right module, respectively. We observe next that dually, the largest submodule of kG as left or right kG-module on which all elements of G act as identity is free of rank one over k.

Proposition 1.6.5 *Let G be a finite group. Every submodule of the regular kG-module kG that is isomorphic to the trivial kG-module is contained in $k\sum_{x\in G} x$.*

Proof Let $a = \sum_{y\in G} \lambda_y y$ be an element in kG. Suppose that $xa = a$ for all $x \in G$. This means that $\sum_{y\in G} \lambda_y y = \sum_{y\in G} \lambda_y xy = \sum_{y\in G} \lambda_{x^{-1}y} y$, hence $\lambda_y = \lambda_{x^{-1}y}$ for all $x, y \in G$. Thus the coefficients λ_y are all equal, say $\lambda = \lambda_y$ for all $y \in G$, and hence $a = \lambda \sum_{x\in G} x$ as stated. □

If n is an integer such that $n \geq 2$, then there is no unitary algebra homomorphism $M_n(k) \to k$. Any such homomorphism would have to send the commutator subspace $[M_n(k), M_n(k)]$ to zero because k is commutative, but the ideal generated by the commutator subspace is easily seen to be all of $M_n(k)$. We observed further in 1.2.8 that twisted group algebras do not have an augmentation in general.

The multiplication in a k-algebra A defines a unique k-bilinear map $\mu : A \otimes_k A \to A$ satisfying $\mu(a \otimes b) = ab$ for all $a, b \in A$. The map μ is a homomorphism of A-A-bimodules but not a homomorphism of k-algebras, in general. It is yet another special feature of group algebras that there is always a "diagonal" map $kG \to kG \otimes_k kG \cong k(G \times G)$ induced by the diagonal group homomorphism $\Delta : G \to G \times G$ sending $x \in G$ to (x, x).

Theorem 1.6.6 *Let G be a group. There is a unique k-algebra homomorphism $\Delta : kG \to k(G \times G)$ mapping $x \in G$ to (x, x).*

Proof The algebra homomorphism Δ is induced by the group homomorphism $G \to G \times G$ sending $x \in G$ to (x, x). □

One important consequence of the existence of the algebra homomorphism Δ has been noted earlier: the tensor product over the base ring k of two kG-modules U, V is again a kG-module, with $x \in G$ acting 'diagonally' on $u \otimes v \in U \otimes_k V$ by $x \cdot (u \otimes v) = xu \otimes xv$ for all $u \in U$ and $v \in V$. If not specified otherwise, we will always assume that the tensor product of two kG-modules over k is endowed with this kG-module structure. Group algebras are isomorphic to their opposites:

Theorem 1.6.7 *Let G be a group. There is a canonical isomorphism of k-algebras $\iota : kG \cong (kG)^{\mathrm{op}}$ mapping $x \in G$ to x^{-1}, and ι has the property $\iota \circ \iota = \mathrm{Id}_{kG}$.*

Proof For any $x, y \in G$ we have $(xy)^{-1} = y^{-1}x^{-1}$. This implies that for any two elements $a, b \in kG$ we have $\iota(ab) = \iota(b)\iota(a)$, the product taken in kG. Clearly

ι applied twice is the identity on kG. Thus ι is an algebra homomorphism from kG to $(kG)^{\mathrm{op}}$ that is its own inverse. □

The fact that a group algebra kG is isomorphic to its opposite algebra allows us to move freely between left and right modules over group algebras: a left kG-module U can be viewed as a right kG-module with $u \cdot x = x^{-1}u$, where the left hand side of this equation defines the right action of $x \in G$ on $u \in U$ and the right hand side is the given action of x^{-1} on u through the left module structure. Similarly, a right kG-module V can be viewed as a left kG-module with $x \cdot v = vx^{-1}$ for $v \in V$ and $x \in G$. In a category theoretic language this implies that the categories of left and right kG-modules are equivalent. Given two groups G, H, combining 1.1.4 and 1.6.7 yields an algebra isomorphism $k(G \times H) \cong kG \otimes_k kH \cong kG \otimes_k (kH)^{\mathrm{op}}$ sending $(x, y) \in G \times H$ to $x \otimes y^{-1}$. In this way, every $k(G \times H)$-module can be viewed as a kG-kH-bimodule and vice versa. More explicitly, if M is a kG-kH-bimodule, we can view M as $k(G \times H)$-module by setting $(x, y) \cdot m = xmy^{-1}$, where $(x, y) \in G \times H$ and $m \in M$. In particular, the regular kG-kG-bimodule kG can be viewed as $k(G \times G)$-module with $(x, y) \in G \times G$ acting on $a \in kG$ by $(x, y) \cdot a = xay^{-1}$. This bimodule has an important symmetry property which we describe further in 2.11. Summarising the above, the group algebra kG of a group G over k admits the following structural maps: a *unit* $\eta : k \to kG$, an *augmentation map*, also called *counit* $\epsilon : kG \to k$, a *multiplication* $\mu : kG \otimes_k kG \to kG$, a *comultiplication* $\Delta : kG \to kG \otimes_k kG$, and an *antipode* $\iota : kG \to kG$, which is here viewed as an *anti-automorphism* of kG rather than as an isomorphism from kG to its opposite. A trivial verification shows the following identities (where we implicitly identify $kG \otimes_k k = kG = k \otimes_k kG$ as appropriate):

Theorem 1.6.8 *Let G be a group. The unit ϵ, augmentation η, multiplication μ, comultiplication Δ and antipode ι have the following properties:*

(i) $\mu \circ (\mathrm{Id} \otimes \eta) = \mathrm{Id} = \mu \circ (\eta \otimes \mathrm{Id})$.
(ii) $\mu \circ (\mu \otimes \mathrm{Id}) = \mu \circ (\mathrm{Id} \otimes \mu) : kG \otimes_k kG \otimes_k kG \to kG$.
(iii) $(\mathrm{Id} \otimes \epsilon) \circ \Delta = \mathrm{Id} = (\epsilon \otimes \mathrm{Id}) \circ \Delta$.
(iv) $(\Delta \otimes \mathrm{Id}) \circ \Delta = (\mathrm{Id} \otimes \Delta) \circ \Delta : kG \to kG \otimes_k kG \otimes_k kG$.
(v) $\mu \circ (\mathrm{Id} \otimes \iota) \circ \Delta = \eta \circ \epsilon = \mu \circ (\iota \otimes \mathrm{Id}) \circ \Delta : kG \to kG$.

A k-algebra A with structural maps ϵ, η, μ, Δ, ι satisfying the properties as in 1.6.8 is called a *Hopf algebra*. The statement 1.6.8(ii) is the associativity of μ. The statements (iii), (iv) correspond to A being a *coalgebra*, with (iv) being the *coassociativity of A*. The counit and comultiplication in kG are algebra homomorphisms, which corresponds to kG being a *bialgebra*. Statement (v) on the relation with the antipode is what is required for a bialgebra to be a

Hopf algebra. The antipode ensures that the categories of left and right modules over a Hopf algebra A are equivalent. The comultiplication ensures that a tensor product over k of two A-modules is again an A-module. Group algebras are particular cases of Hopf algebras, and a number of statements in what follows could be made in greater generality for Hopf algebras.

As mentioned above, twisted group algebras are not Hopf algebras, in general, since they need not have an augmentation. Although a fixed twisted group algebra need not have an antipode or a comultiplication either, it is possible to extend these notions to maps between different twisted group algebras in the following way.

Proposition 1.6.9 *Let G be a finite group and α a normalised 2-cocycle in $Z^2(G; k^\times)$. The map sending $x \in G$ to $\alpha(x, x^{-1})x^{-1}$ induces a k-algebra isomorphism $(k_\alpha G)^{\mathrm{op}} \cong k_{\alpha^{-1}}G$.*

Proof Denote by $\Phi : (k_\alpha G)^{\mathrm{op}} \cong k_{\alpha^{-1}}G$ the unique k-linear isomorphism sending $x \in G$ to $\alpha(x, x^{-1})x^{-1}$. Let $x, y \in G$. Denote by $x \cdot y = \alpha(x, y)xy$ the product of x and y in $k_\alpha G$. Thus the product of x and y in $(k_\alpha G)^{\mathrm{op}}$ is $y \cdot x = \alpha(y, x)yx$, and hence

$$\Phi(y \cdot x) = \alpha(y, x)\alpha(yx, x^{-1}y^{-1})(yx)^{-1}.$$

The product of $\Phi(x) = \alpha(x, x^{-1})x^{-1}$ and of $\Phi(y) = \alpha(y, y^{-1})y^{-1}$ in $k_{\alpha^{-1}}G$ is equal to

$$\alpha(x, x^{-1})\alpha(y, y^{-1})\alpha(x^{-1}, y^{-1})^{-1}(yx)^{-1}.$$

Thus, in order to show that Φ is an algebra homomorphism, it suffices to show that

$$\alpha(y, x)\alpha(yx, x^{-1}y^{-1}) = \alpha(x, x^{-1})\alpha(y, y^{-1})\alpha(x^{-1}, y^{-1})^{-1}.$$

By the 2-cocycle identity applied to $y, x, x^{-1}y^{-1}$, the left side in this equation is equal to

$$\alpha(y, y^{-1})\alpha(x, x^{-1}y^{-1}).$$

Therefore, after replacing the left side by this expression and cancelling $\alpha(y, y^{-1})$, it suffices to show that

$$\alpha(x, x^{-1}y^{-1}) = \alpha(x, x^{-1})\alpha(x^{-1}, y^{-1})^{-1}.$$

Bringing the rightmost term to the left side yields an equation which is the 2-cocycle identity applied to x, x^{-1}, y^{-1}, where we use that $\alpha(1, y^{-1}) = 1$ as α is normalised. $\square$

Proposition 1.6.10 *Let G be a group, let $\alpha, \beta \in Z^2(G; k^\times)$, and set $\gamma = \alpha\beta$. The map sending $x \in G$ to $x \otimes x$ induces an injective k-algebra homomorphism $k_\gamma G \to k_\alpha G \otimes_k k_\beta G$. Moreover, $k_\alpha G \otimes_k k_\beta G$ is free as a left or right module over $k_\gamma G$ through this homomorphism, and $k_\gamma G$ is isomorphic to a direct summand of $k_\alpha G \otimes_k k_\beta G$ as a $k_\gamma G$-$k_\gamma G$-bimodule.*

Proof Let $x, y \in G$. The product of $x \otimes x$ and of $y \otimes y$ in $k_\alpha G \otimes_k k_\beta G$ is equal to $\alpha(x, y)xy \otimes \beta(x, y)xy = \gamma(x, y)(xy \otimes xy)$, which implies that the given map is an injective algebra homomorphism. The images of $1 \otimes G$ and of $G \otimes 1$ in $k_\alpha G \otimes_k k_\beta G$ are bases as a left $k_\gamma G$-module and as a right $k_\gamma G$-module. The complement of the diagonal $\Delta G = \{(x, x) | x \in G\}$ in $G \times G$ is a k-basis of a bimodule complement of the image of $k_\gamma G$ in $k_\alpha G \otimes_k k_\beta G$, whence the result. □

Corollary 1.6.11 *Let G be a group and $\alpha \in Z^2(G; k^\times)$. The map sending $x \in G$ to $x \otimes x$ induces an injective k-algebra homomorphism $kG \to k_\alpha G \otimes_k k_{\alpha^{-1}} G$ that is split injective as a kG-kG-bimodule homomorphism. Moreover, the right side is free as a left and right kG-module.*

Proof This is the special case $\beta = \alpha^{-1}$ of 1.6.10. □

Exercise 1.6.12 Let G, H be finite groups. Show that the canonical k-algebra isomorphism $k(G \times H) \cong kG \otimes_k kH$ from 1.1.4 sends $I(k(G \times H))$ to the image of $I(kG) \otimes kH + kG \otimes I(kH)$ in $kG \otimes_k kH$.

Exercise 1.6.13 Show that the polynomial algebra $k[x]$ is a Hopf algebra, with comultiplication defined by $\Delta(x) = x \otimes 1 + 1 \otimes x$, counit η sending a polynomial to its constant coefficient, and antipode sending x to $-x$.

1.7 Blocks and idempotents

Given two k-algebras B, C, the Cartesian product $B \times C$, with componentwise sum and multiplication, together with the canonical projections onto B and C, is a direct product of B and C in the category of k-algebras. A *direct factor* of a k-algebra A is a pair consisting of a k-algebra B and an algebra homomorphism $\tau : A \to B$ which is split surjective as an A-A-bimodule homomorphism. Theorem 1.7.1 below implies that this is equivalent to requiring that there is an algebra isomorphism $A \cong B \times C$ for some algebra C (possibly zero) with the property that the composition of this isomorphism with the canonical projection onto B is equal to τ. If τ is clear from the context, we simply say that B is a direct factor of A.

Let A be a k-algebra and let b be an idempotent in $Z(A)$. Then $Ab = \{ab | a \in A\}$ is a k-algebra with unit element b, and Ab is a quotient algebra of A via the algebra homomorphism sending $a \in A$ to ab. As a bimodule homomorphism, this is split surjective with section the inclusion map $Ab \subseteq A$. If $b \neq 1$, then $1 - b$ is an idempotent in $Z(A)$, and the algebra A decomposes as a direct product $A = Ab \times A(1 - b)$ in the category of k-algebras. This same decomposition is a direct sum $A = Ab \oplus A(1 - b)$ in the category of A-A-bimodules. We have encountered an example for this construction in 1.2.18. The following theorem spells out the details of the correspondence between idempotents in $Z(A)$ and projections onto bimodule summands of A determined by the algebra isomorphism $Z(A) \cong \text{End}_{A\otimes_k A^{\text{op}}}(A)$ from 1.5.12.

Theorem 1.7.1 *Let A be a k-algebra. Let B be a k-submodule of A. The following are equivalent.*

(i) B is a direct summand of A as an A-A-bimodule.
(ii) B is a direct factor of A as a k-algebra.
(iii) $B = Ab$ for some idempotent b in $Z(A)$.

Proof Suppose that (i) holds. Write $A = B \oplus N$ for some A-A-bimodule N. We have $BN \subseteq B \cap N = \{0\}$. Similarly, we have $NB = \{0\}$. Write $1_A = b + n$ for some $b \in B$ and $n \in N$. Then $b = b1_A = b^2 + bn$, hence $b^2 = b$ and $bn = 0$. This shows that b is an idempotent. The same argument shows that n is an idempotent, and by the above, the idempotents b, n are orthogonal. Let $a \in A$. Write $a = c + r$ for some $c \in B$ and $r \in N$. Then $a = a1_A = ab + an$ and $a = 1_A a = ba + na$, hence $b, n \in Z(A)$. Thus b, n are orthogonal idempotents in $Z(A)$, and we have $Ab = B$ and $An = N$. Thus $A = Ab \times An$ is a direct product of k-algebras. This shows that (i) implies (ii) and (iii). Suppose that (ii) holds. Write $A = B \times C$ for some k-algebra C. Then $1_A = (1_B, 1_C) = (1_B, 0) + (0, 1_C)$. It follows that $(1_A, 0)$ and $(0, 1_C)$ are orthogonal idempotents in $Z(A)$. Setting $b = (1_B, 0)$ we get that $B = Ab$, hence (ii) implies (iii). If (iii) holds, then $1 - b$ is an idempotent in $Z(A)$ which is orthogonal to b, and we have $A = Ab \oplus A(1 - b)$ as A-A-bimodules, and $A \cong Ab \times A(1 - b)$ as k-algebras. Thus (iii) implies (i) and (ii). □

Corollary 1.7.2 *Let A be a k algebra. The following are equivalent:*

(i) A is indecomposable as a k-algebra.
(ii) A is indecomposable as an A-A-bimodule.
(iii) The idempotent 1_A is primitive in $Z(A)$.

A k-algebra A is indecomposable as an algebra if and only if A is indecomposable as an A-A-bimodule. Thus a primitive decomposition of 1_A in $Z(A)$, if there is any, is equivalent to a decomposition of A as a direct product of indecomposable algebras.

Definition 1.7.3 Let A be a k-algebra. A *block of A* is a primitive idempotent b in $Z(A)$; the algebra Ab is called a *block algebra of A*.

A simple but effective technical ingredient is that any algebra can be viewed as the endomorphism algebra of a suitable module. If i is an idempotent in a k-algebra A, then iAi is again a k-algebra, with unit element i. For any two idempotents i and j in A, the k-module iAj is an iAi-jAj-bimodule via multiplication in A. The left ideal Ai generated by an idempotent i is a submodule of A as a left A-module. The following result implies in particular that Ai is in fact a direct summand of A, and that every direct summand of A as a left A-module has this form.

Proposition 1.7.4 *Let A be a k-algebra.*

(i) Let i, j be orthogonal idempotents in A. Then $A(i+j) = Ai \oplus Aj$ as A-modules. In particular, we have $A = Ai \oplus A(1-i)$.

(ii) Let e be an idempotent in A and let U, V be submodules of Ae such that $Ae = U \oplus V$. There are unique idempotents i, j in A such that $U = Ai$, $V = Aj$ and $e = i + j$. Moreover, the idempotents i and j are then orthogonal.

(iii) Let i be an idempotent in A. The left A-module Ai is indecomposable if and only if i is primitive.

(iv) Any direct summand of A as a left A-module is equal to Ai for some idempotent i in A.

(v) Let i be an idempotent in A and U an A-module. We have a k-linear isomorphism $\mathrm{Hom}_A(Ai, U) \cong iU$ *mapping* $\varphi \in \mathrm{Hom}_A(Ai, U)$ *to* $\varphi(i)$.

(vi) Let i, j be idempotents in A. We have a k-linear isomorphism $\mathrm{Hom}_A(Ai, Aj) \cong iAj$ *mapping* $\varphi \in \mathrm{Hom}_A(Ai, Aj)$ *to* $\varphi(i)$.

(vii) Let i be an idempotent in A. We have a k-algebra isomorphism $\mathrm{End}_A(Ai) \cong (iAi)^{\mathrm{op}}$ *mapping* $\varphi \in \mathrm{End}_A(Ai)$ *to* $\varphi(i)$.

(viii) We have a k-algebra isomorphism $\mathrm{End}_A(A) \cong A^{\mathrm{op}}$ *mapping* $\varphi \in \mathrm{End}_A(A)$ *to* $\varphi(1_A)$.

Proof With the notation in (i), we have $A(i+j) \subseteq Ai + Aj$. Since i, j are orthogonal we have $i = i(i+j)$, hence $Ai \subseteq A(i+j)$. Similarly $Aj \subseteq A(i+j)$. Thus $A(i+j) = Ai + Aj$. If $a \in Ai \cap Aj$ then $a = ai$ and $a = aj$, hence $a = ai = aij = 0$, so the sum $Ai + Aj$ is direct. This proves (i). For (ii), write $e = i + j$ for a unique $i \in U$ and a unique $j \in V$. Then $i = i^2 + ij$. Since

$i \in U$ and $j \in V$ we have $i^2 \in U$ and $ij \in V$, hence $ij = 0$ and $i^2 = i$. Similarly $j^2 = j$ and $ji = 0$. In other words, i and j are orthogonal idempotents. Thus $Ae = Ai \oplus Aj$ by (i). Since $Ai \subseteq U$ and $Aj \subseteq V$ this implies $Ai = U$ and $Aj = V$. This proves (ii). If Ai is indecomposable then i is primitive by (i). Conversely, if i is primitive then Ai is indecomposable by (ii), whence (iii). Statement (iv) follows from (iii). The given map in (v) has as inverse the map sending $c \in iU$ to the homomorphism $Ai \to U$ sending ai to aic, that is, the homomorphism given by right multiplication with c on Ai. Statement (vi) is the special case of (v) where $U = Aj$. By (v), the given map in (vii) is an k-linear isomorphism. If $\varphi, \psi \in \mathrm{End}_A(Ai)$ then $(\psi \circ \varphi)(i) = \psi(\varphi(i)) = \psi(\varphi(i)i) = \varphi(i)\psi(i)$, which show that this map is an algebra isomorphism as claimed. Statement (viii) is the special case $i = 1_A$ of (vii). □

Corollary 1.7.5 *Let A be a k-algebra, e an idempotent in A, let I be a finite set, and suppose that $Ae = \oplus_{i \in I} U_i$ for some A-submodules U_i of Ae, where $i \in I$. Then there is a unique set $\{e_i\}_{i \in I}$ of pairwise orthogonal idempotents e_i in A satisfying $e = \sum_{i \in I} e_i$ and $U_i = Ae_i$ for $i \in I$.*

Proof Apply 1.7.4 (ii) inductively. □

The existence of a decomposition of a k-algebra as a direct product of its block algebras is equivalent to the existence of a primitive decomposition of the unit element in the centre of the algebra.

Proposition 1.7.6 *Let A be a k-algebra. If A is Noetherian or Artinian, then every idempotent in A has a primitive decomposition.*

Proof Arguing by contradiction, let e be an idempotent in A that does not have a primitive decomposition. We are going to construct a sequence of idempotents $e = e_0, e_1, e_2, \ldots$ such that $e_{i+1} = e_{i+1}e_i = e_ie_{i+1} \neq e_i$ and such that none of the e_i has a primitive decomposition. By the assumptions, $e_0 = e$ has no primitive decomposition. Suppose we have constructed $e_0 = e, e_1, \ldots, e_m$ for some $m \geq 0$ such that $e_{i+1} = e_{i+1}e_i = e_ie_{i+1} \neq e_i$ for $0 \leq i \leq m-1$ and such that none of the e_i has a primitive decomposition, $0 \leq i \leq m$. In particular, e_m has no primitive decomposition, hence is not primitive, and we have $e_m = e_{m+1} + e'_{m+1}$ for some orthogonal idempotents e_{m+1}, e'_{m+1}. Since e_m has no primitive decomposition, at least one of e_{m+1}, e'_{m+1} has no primitive decomposition. Choose notation such that e_{m+1} has no primitive decomposition. By construction, we have $e_{m+1} = e_me_{m+1} = e_{m+1}e_m$ and $e_m - e_{m+1} = e'_{m+1} \neq 0$. This shows the existence of a sequence of idempotents e_i with the above properties. But then the sequence of A-modules $\{Ae_i\}_{i \geq 0}$ is an infinite descending sequence and the sequence of A-modules $\{A(e - e_i)\}_{i \geq 0}$ is an infinite ascending sequence, none of which become constant. Thus A is neither Artinian nor Noetherian. □

Theorem 1.7.7 *Suppose that k is Noetherian. Let A be a k-algebra such that Z(A) is finitely generated as a k-module. Then Z(A) is Noetherian, and the following hold.*

(i) The set of blocks $\mathcal{B}$ of A is the unique primitive decomposition of 1_A in Z(A). In particular, $\mathcal{B}$ is finite, and any two different blocks of A are orthogonal.

(ii) $A = \prod_{b \in \mathcal{B}} Ab$ is the unique decomposition of A as a direct product of indecomposable k-algebras.

(iii) $A = \bigoplus_{b \in \mathcal{B}} Ab$ is the unique decomposition of A as direct sum of indecomposable A-A-bimodules.

Proof Since k is Noetherian and $Z(A)$ is finitely generated as a k-module, it follows that $Z(A)$ is Noetherian. It follows further from 1.7.6 that 1_A has a primitive decomposition $\mathcal{B}$ in $Z(A)$. Let $b \in \mathcal{B}$ and let c be any other primitive idempotent in $Z(A)$. Suppose that $bc \neq 0$. Then $b = bc + b(1 - c)$, and both bc, $b(1 - c)$ are either zero or idempotents in $Z(A)$. Clearly their product is zero. Since b is primitive and $bc \neq 0$ we get that $b(1 - c) = 0$, hence $b = bc$. Exchanging the roles of b and c yields $c = bc$, hence $b = c$. This shows the uniqueness of $\mathcal{B}$, whence (i). The rest follows from 1.7.1. □

In an additive category, finite direct sums (or equivalently, coproducts) and finite direct products coincide. The category of k-algebras has obvious direct products, but this category is not an additive category because the sum of two algebra homomorphisms is not an algebra homomorphism. Thus the direct product of two algebras is not a coproduct. Coproducts can be constructed in a way that is similar to the construction of free products of groups. In the category of commutative k-algebras, the coproduct coincides with the tensor product. The category of A-A-bimodules is, of course, additive, and therefore has finite direct sums. This is why the – from a set theoretic point of view identical – decompositions of A into its blocks yield a direct *product* in 1.7.7 (ii) and a direct *sum* in 1.7.7 (iii).

Definition 1.7.8 Let A be a k-algebra and b be a block of A. We say that an A-module M *belongs to the block* b if $M = bM$.

Proposition 1.7.9 *Suppose that k is Noetherian. Let A be a k-algebra such that A is finitely generated as a k-module. Let $\mathcal{B}$ be the set of blocks of A.*

(i) *For any A-module M we have $M = \oplus_{b \in B} bM$ as a direct sum of A-modules.*
(ii) *For any indecomposable A-module M there is a unique block b of A such that $M = bM$, or equivalently, such that M belongs to b.*
(iii) *For any two A-modules belonging to two different blocks b, c of A, respectively, we have* $\mathrm{Hom}_A(M, N) = \{0\}$.

Proof (i) Since $\sum_{b \in B} b = 1$ it suffices to show that the sum $M = bM + (1 - b)M$ is a direct sum of A-modules. If $m \in bM$ then $bm = m$, and hence, for $a \in A$, we have $am = abm = bam \in bM$, so bM is an A-submodule of M; similarly, $(1 - b)M$ is an A-submodule of M. Since $1 = b + (1 - b)$ we have $M = bM + (1 - b)M$, and if $m \in bM \cap (1 - b)M$ then $m = bm = (1 - b)m = (1 - b)bm = 0$, so this sum is direct. Statement (ii) follows from (i). If the A-modules M, N belong to different blocks b, c of A then, for $m \in M$ we have $m = bm$ and for $n \in N$ we have $cn = n$. Thus, for any $\varphi \in \mathrm{Hom}_A(M, N)$, we have $\varphi(m) = c\varphi(bm) = cb\varphi(m) = 0$ for all $m \in M$, whence the result. □

An A-module M that belongs to a block b of A has the property that b acts as identity on M because b is an idempotent. Thus any A-module belonging to a block b of A can be viewed as a module for the block algebra Ab. Conversely, an Ab-module can be viewed as an A-module which is annihilated by $A(1 - b)$. We identify tacitly the module category $\mathrm{Mod}(Ab)$ of a block algebra with the full subcategory of the module category $\mathrm{Mod}(A)$ consisting of all A-modules on which b acts as identity. With this identification, the module category $\mathrm{Mod}(A)$ is the direct sum of the module categories $\mathrm{Mod}(Ab)$, with b running over the set of blocks of A. If S is a simple Ab-module, then S remains simple as an A-module. Thus the set of isomorphism classes of simple A-modules can be identified with the union of the sets of isomorphism classes of simple Ab-modules, with b running over the blocks of A. If A is a Hopf algebra, the comultiplication need not be compatible with the block decomposition of A.

Exercise 1.7.10 Suppose that k is Noetherian. Let A be a k-algebra such that A is finitely generated as a k-module, and let i be a primitive idempotent in A. Show that there is a unique block b of A such that $ib \neq 0$. Show that this is the unique block such that the A-module Ai belongs to b, and also the unique block such that $i \in Ab$. We say in that case that *i determines the block b of A*, or that *b is determined by i*. Show that if j is a primitive idempotent in A that is conjugate to i by an element in $A^\times$, then i and j determine the same block b.

Exercise 1.7.11 Let k be a field. Show that if $\mathrm{char}(k) = 2$, then kS_3 has two blocks, if $\mathrm{char}(k) = 3$, then kS_3 has a single block, and if $\mathrm{char}(k)$ is different from 2 and from 3, then kS_3 has three blocks. Determine the block idempotents in all cases.

Exercise 1.7.12 Let k be a field and $\mathcal{P}$ a finite partially ordered set. Show that the blocks of $k\mathcal{P}$ correspond bijectively to the connected components of $\mathcal{P}$.

Exercise 1.7.13 Let A be a k-algebra, I an ideal in A, and let i be an idempotent in $A \setminus I$. Denote by j the image of i in A/I. Show that j is an idempotent in A/I, that $iAi \cap I$ is an ideal in iAi, and that there is a canonical algebra isomorphism $jA/Ij \cong iAi/iAi \cap I$.

1.8 Composition series and Grothendieck groups

Definition 1.8.1 Let A be a k-algebra. A *composition series of an A-module* M is a finite chain $M = M_0 \supset M_1 \supset \cdots \supset M_n = \{0\}$ of submodules M_i in M such that M_{i+1} is maximal in M_i for $0 \leq i \leq n-1$. The simple factors M_i/M_{i+1} arising this way are called the *composition factors* of this series and the non-negative integer n is called its *length*. Two composition series of A-modules M and M', respectively, are called *equivalent* if they have the same length n and if there is a bijection between the sets of composition factors of each of these series such that corresponding composition factors are isomorphic.

Lemma 1.8.2 *Let A be a k-algebra, let M be an A-module, let U and N be submodules of M and let V be a maximal submodule of U. We have $U + N = V + N$ if and only if $V \cap N \subsetneq U \cap N$. If this is the case then $U/V \cong (U \cap N)/(V \cap N)$. Otherwise, $U/V \cong (U + N)/(V + N)$.*

Proof We have $V \subseteq (U \cap N) + V \subseteq U$. Since V is maximal in U either $V = (U \cap N) + V$ or $(U \cap N) + V = U$. In the first case, $U \cap N = V \cap N$ and $(U + N)/(V + N) \cong U/(U \cap (V + N)) = U/V$. In the second case, $V + N = U + N$ and $U/V = ((U \cap N) + V)/V \cong (U \cap N)/(V \cap N)$. □

Theorem 1.8.3 (Jordan–Hölder) *Let A be a k-algebra, and let M be an A-module. Then M has a composition series if and only if M is both Artinian and Noetherian. In that case, any two composition series of M are equivalent.*

Proof If M has a composition series of length $n \geq 1$, then M has a maximal submodule M_1 which has a composition series of length $n - 1$. Arguing by induction over n it follows that both M_1 and the simple module M/M_1 are Artinian and Noetherian, hence M is, too. Conversely, if M is Artinian and Noetherian, we construct a composition series as follows. Since M is Noetherian, every nonzero submodule of M has a maximal submodule by 1.10.3. Since M is also Artinian, any chain constructed inductively by $M_0 = M$ and M_{i+1} maximal in M_i if M_i is nonzero, eventually becomes constant, which forces $M_i = \{0\}$ for i large enough. Let $M = M_0 \supset M_1 \supset \cdots \supset M_n = \{0\}$ and $M = N_0 \supset N_1 \supset \cdots \supset N_k = \{0\}$ be two composition series of M. If $n \leq 1$, then either M is zero or simple, so we are done. Suppose that $n > 1$. Set $N = N_1$; note that the N_j, with $1 \leq j \leq k$ form a composition series of N of length $k-1$. Consider the chain

$$M = M_0 + N \supset M_1 + N \supset \cdots \supset M_n + N = N$$

$$= M_0 \cap N \supset M_1 \cap N \supset \cdots \supset M_n \cap N = \{0\}.$$

Since N is maximal in M there is exactly one index i, $0 \leq i \leq n-1$, such that

$$M = M_0 + N = \cdots = M_i + N \supset M_{i+1} + N = \cdots = M_n + N = N$$

and by 1.8.2 this is also the unique index i for which $M_i \cap N = M_{i+1} \cap N$. It follows that we have a composition series

$$M = M_i + N \supset M_{i+1} + N = N = M_0 \cap N \supset \cdots \supset M_i \cap N$$

$$= M_{i+1} \cap N \supset \cdots \supset M_n \cap N = \{0\}.$$

Deleting the first term in this series yields a series of N of length $n-1$. Thus, by induction, this series of N is equivalent to the series of the N_j, $1 \leq j \leq k$ which is of length $k-1$. This means that we have $k = n$ and up to a permutation, the composition factors $(M_j \cap N)/(M_{j+1} \cap N) \cong M_j/M_{j+1}$ for $0 \leq j \leq n-1$ and $j \neq i$ are isomorphic to the factors N_j/N_{j+1} for $1 \leq j \leq n-1$. The remaining factor M_i/M_{i+1} is, by 1.8.2, isomorphic to $(M_i + N)/(M_{i+1} + N) \cong M/N_1$, which completes the proof. □

Proposition 1.8.4 *Let A be a k-algebra, and let M be an A-module that has a composition series. Let U be a submodule of M. Then U and M/U have composition series, and there is a composition series $M = M_0 \supset M_1 \supset \cdots \supset M_n = \{0\}$ of M such that $U = M_i$ for some i, $0 \leq i \leq n$.*

Proof The property of being Noetherian and Artinian passes to submodules and quotients; thus U and M/U have composition series by 1.8.3. The inverse image in M of a composition series of M/U together with a composition series of U yield a composition series of M in which U appears. □

Proposition 1.8.5 *Let A be a k-algebra, and let U, V be A-modules having composition series. Suppose that no composition factor of U is isomorphic to a composition factor of V. Then every submodule of $U \oplus V$ is equal to $U' \oplus V'$ for some submodule U' of U and some submodule V' of V.*

Proof Let W be a submodule of $U \oplus V$. Denote by $\iota : W \to U$ and $\kappa : W \to V$ the components of the inclusion map $W \subseteq U \oplus V$. Every composition factor of

$W/(\ker(\iota) + \ker(\kappa))$ is a composition factor of both U, V. Since U and V have no isomorphic composition factors, it follows that $W = \ker(\iota) + \ker(\kappa)$. Since the sum of ι and κ is the inclusion map, hence injective, it follows that $\ker(\iota) \cap \ker(\kappa) = \{0\}$. Thus $W = \ker(\iota) \oplus \ker(\kappa)$. Thus $U' = \iota(W) = \iota(\ker(\kappa)) \subseteq W$ and $V' = \kappa(W) = \kappa(\ker(\iota)) \subseteq W$, which implies that $W = U' \oplus V'$ as stated. □

This is not true in general if U, V have a common composition factor. For instance, $U \oplus U$ has 'diagonal' submodules of the form $\{(u, \lambda u) | u \in U\}$, where λ is a fixed element in k. Identifying classes of modules with equivalent composition series leads to considering *Grothendieck groups*.

Definition 1.8.6 Let A be a a k-algebra. The *Grothendieck group of A* is the abelian group denoted $R(A)$ which is generated by the set of isomorphism classes $[U]$ of finitely generated A-modules U, subject to the relations $[U] + [W] = [V]$ whenever there is a short exact sequence of A-modules of the form

$$0 \longrightarrow U \longrightarrow V \longrightarrow W \longrightarrow 0.$$

We denote abusively by $[U]$ the image of U in $R(A)$, for any finitely generated A-module U. The Grothendieck group of a finite-dimensional algebra over a field has a particularly simple structure:

Proposition 1.8.7 *Suppose that k is a field, and let A be a finite-dimensional k-algebra. The Grothendieck group $R(A)$ is a free abelian group having as a basis the set of isomorphism classes of simple A-modules; in particular, $R(A)$ has finite rank equal to the number $\ell(A)$ of isomorphism classes of simple A-modules. More precisely, for any finitely generated A-module U we have* $[U] = \sum_S d(U, S) \cdot [S]$, *where S runs over a set of representatives of the isomorphism classes of simple A-modules and where $d(U, S)$ is the number of composition factors isomorphic to S in a composition series of U. In particular, the images in $R(A)$ of two finitely generated A-modules coincide if and only if they have equivalent composition series.*

Proof Let $0 \longrightarrow U \longrightarrow V \longrightarrow W \longrightarrow 0$ be a short exact sequence of finitely generated A-modules. We first note that for any simple A-module S we have $d(V, S) = d(U, S) + d(W, S)$. Indeed, this follows from the fact that the image in V of a composition series of U and the inverse image in V of a composition series of W yield together a composition series of V. Denote by $R'(A)$ the free abelian group $R'(A)$ having as basis a set of symbols (S), where S runs over a set of representatives of the isomorphism classes of simple A-modules. By the above, there is a surjective group homomorphism $R(A) \to R'(A)$ sending the image of $[U]$ in $R(A)$ to $\sum_S d(U,S)(S)$. In order to show that this is an isomorphism, it suffices to show that the formula $[U] = \sum_S d(U,S) \cdot [S]$ holds in $R(A)$ as well. This is done by induction over the length of a composition series of U. If $U = S$ is simple, there is nothing to prove. Otherwise, let V be a maximal submodule of U. Then $S = U/V$ is simple, and we have $[U] = [S] + [V]$. Induction applied to V yields the formula, whence the result. □

In other words, the elements in the Grothendieck group $R(A)$ of a finite-dimensional algebra over a field are formal $\mathbb{Z}$-linear combinations of the isomorphism classes of simple A-modules. We will later consider an important variation of the definition of Grothendieck groups in which a relation $[U] + [W] = [V]$ is introduced only for all *split* exact sequences of A-modules

$$0 \longrightarrow U \longrightarrow V \longrightarrow W \longrightarrow 0.$$

The resulting abelian group for finite-dimensional algebras will be shown to be again free (but not, in general, of finite rank), having as a basis a set of representatives of the isomorphism classes of indecomposable modules. This will require the Krull–Schmidt Theorem 4.6.7 on the uniqueness of decompositions of modules into direct sums of indecomposable modules.

Exercise 1.8.8 Let A be a finite-dimensional algebra over an infinite field k, and let S be a simple A-module. Show that $S \oplus S$ has infinitely many composition series.

Exercise 1.8.9 Let A be a finite-dimensional algebra over a field k, and let U be an indecomposable A-module such that a composition series of U has length 2. Show that U has a unique composition series.

A module with a unique composition series is called *uniserial.* A uniserial module has in particular a unique maximal submodule. Uniserial modules will be considered more systematically in 7.1.1 and §11.3 below. They play an important role in the context of blocks with cyclic defect groups.

1.9 Semisimple modules

Definition 1.9.1 Let A be a k-algebra. An A-module M is called *semisimple* if M is the sum of its simple submodules.

Every simple A-module S is a quotient of the regular A-module A. Indeed, if s is a nonzero element in S, then the map sending $a \in A$ to $as \in S$ is a nonzero A-homomorphism, hence surjective because S is simple.

Theorem 1.9.2 (Schur's Lemma) *Let A be a k-algebra. For any simple A-module S the k-algebra* $\mathrm{End}_A(S)$ *is a division algebra. For any two nonisomorphic simple A-modules S, T we have* $\mathrm{Hom}_A(S, T) = \{0\}$.

Proof Let S, T be simple A-modules, and suppose there is a nonzero A-homomorphism $\varphi : S \to T$. Then $\ker(\varphi) \neq S$, hence $\ker(\varphi) = \{0\}$ because S is simple. Thus φ is injective. In particular, $\mathrm{Im}(\varphi) \neq \{0\}$. Since T is simple this implies $\mathrm{Im}(\varphi) = T$. Thus φ is an isomorphism. The result follows. □

If A is a finite-dimensional algebra over a field k, then every simple A-module S is finite-dimensional. In that case $\mathrm{End}_A(S)$ is a finite-dimensional division algebra over k and its centre $Z(\mathrm{End}_A(S))$ is a finite-dimensional extension field of k. If k is algebraically closed, then $\mathrm{End}_A(S) \cong k$, which forces $\mathrm{End}_A(S) = \{\lambda \mathrm{Id}_S \mid \lambda \in k\}$; that is, the only endomorphisms of S are the scalar multiples of the identity map. We state this as a corollary to Schur's Lemma:

Corollary 1.9.3 *Suppose that k is an algebraically closed field, and let A be a k-algebra. For any simple A-module S of finite dimension over k we have* $\mathrm{End}_A(S) = \{\lambda \mathrm{Id}_S \mid \lambda \in k\}$.

Proof Let $\phi : S \to S$ be an A-endomorphism of S such that $\phi \neq 0$. Since k is algebraically closed, the characteristic polynomial of f has a root (in fact, all of its roots) in k, and hence f has an eigenvalue $\lambda \in k$. If v is an eigenvector for λ, we have $f(v) = \lambda v$, which is equivalent to $v \in \ker(f - \lambda \mathrm{Id}_S)$. Thus $f - \lambda \mathrm{Id}_S \in \mathrm{End}_A(S)$ is not injective and hence is zero by Schur's Lemma 1.9.2, and so $f = \lambda \mathrm{Id}_S$. □

Lemma 1.9.4 *Suppose that k is an algebraically closed field. Let A be a k-algebra, and let U, V be finite-dimensional A-modules. Any 2-dimensional subspace of* $\mathrm{Hom}_A(U, V)$ *contains a nonzero nonisomorphism.*

Proof Suppose that all nonzero elements in a 2-dimensional subspace T of $\mathrm{Hom}_A(U, V)$ are isomorphisms. Let α, $\beta \in T$ be linearly independent. Then $\beta^{-1} \circ \alpha$ is an automorphism of U. Since k is algebraically closed, this automorphism has an eigenvalue, say λ. Thus there is a nonzero $u \in U$ such that $\beta^{-1}(\alpha(u)) = \lambda u$, hence such that $\alpha(u) = \lambda\beta(u)$. Thus $\alpha - \lambda\beta$ has u in its kernel, hence is an element in T which is not an isomorphism, contradicting the assumptions. □

The proofs of 1.9.3 and 1.9.4 indicate that it may not be necessary to assume k to be algebraically closed, so long as the field k is 'large enough' for the characteristic polynomials under consideration to have a root in k.

The next theorem characterises semisimple modules. The proof involves Zorn's Lemma. Since we will need this theorem only in the context of finite-dimensional algebras and finitely generated modules, we point out that one can read the proof below without Zorn's Lemma: whenever we choose a maximal submodule with certain properties and refer to Zorn's Lemma for the existence of such a maximal submodule, we can eliminate that reference by simply choosing a submodule of maximal dimension with the required property.

Theorem 1.9.5 *Let A be a k-algebra and let M be an A-module. If M is semisimple, so is every quotient and every submodule of M. Moreover, the following are equivalent:*

(i) M is semisimple.
(ii) M is a direct sum of simple modules.
(iii) Every submodule U of M has a complement; that is, there is a submodule V of M such that $M = U \oplus V$ (or equivalently, such that $M = U + V$ and $U \cap V = \{0\}$).

Proof If M is the sum of its simple submodules, this is true for every quotient module, because the image of a simple module in any quotient is either zero or again simple. To see that this property passes down to submodules, we show that property (iii) passes down to submodules. Suppose that M has property (iii) and let U be a submodule of M. Let V be a submodule of U. We have to find a complement of V in U. By the assumption we know that there is a complement W of V in M; that is, $M = V \oplus W$. We show that $W \cap U$ is a complement of V in U. That is, we need to show that $V \cap (W \cap U) = \{0\}$ and that $V + (W \cap U) = U$. The first equality is clear because $V \cap (W \cap U) \subseteq V \cap W = \{0\}$, since W was chosen to be a complement of V in M. For the second equality, let $u \in U$. Since $M = V \oplus W$ there are unique elements $v \in V$ and $w \in W$ such that $u = v + w$. Then $w = u - v \in U$, hence $w \in W \cap U$, which proves the second equality. We turn now towards the equivalence of the three statements above. Assume first that (i) holds. Let $\mathcal{S}$ be a maximal set of simple submodules of M such that the sum $U = \sum_{S \in \mathcal{S}} S$ is a direct sum. Such a set $\mathcal{S}$ exists by Zorn's Lemma. If $U \neq M$ there is a simple submodule T of M such that $T \not\subseteq U$, hence such that $U \cap T = \{0\}$ because T is simple. But then the sum $U + T$ is a direct sum, contradicting the maximality of $\mathcal{S}$. This contradiction implies that $U = M$, hence (i) implies (ii). Suppose next that (ii) holds. Let U be a submodule of M. Let V be a maximal submodule of M such that $U \cap V = \{0\}$. Such a submodule exists again by Zorn's Lemma. If $U + V \neq M$ there exists a simple submodule S of M such that $(U + V) \cap S = \{0\}$. Then $U \cap (V + S) = \{0\}$. Indeed, if $u = v + s$ for some $u \in U$, $v \in V$, $s \in S$, then $s = u - v \in (U \oplus V) \cap S = \{0\}$, so $s = 0$, hence $u = v \in U \cap V = \{0\}$, and thus $u = 0$. This, however, contradicts the maximality of V. Thus (ii) implies (iii). Finally, suppose that (iii) holds. Let U be the sum of all simple submodules of M. Let V be a complement of U in M. If $V = \{0\}$ we are done. If $V \neq \{0\}$, let $v \in V \setminus \{0\}$. Let W be a maximal submodule of M such that $v \notin W$, which exists by Zorn's Lemma. Let S be a complement of W in V (this exists, because we showed at the beginning that hypothesis (iii) passes down to the submodule V). Thus $M = U \oplus V = U \oplus S \oplus W$. Since U was chosen to be the sum of all simple submodules of M, the submodule S cannot be simple. Thus S has a nonzero proper submodule S_1. This in turn has a complement S_2 in S; thus $S = S_1 \oplus S_2$. We show that this implies $W = (W \oplus S_1) \cap (W \oplus S_2)$. Clearly the left side is contained in the right side. If $w_1 + s_1 = w_2 + s_2$ for some $w_1, w_2 \in W$ and $s_1 \in S_1, s_2 \in S_2$, then $w_1 = w_2 + (s_2 - s_1)$. Since the sum $W \oplus S$ direct, this forces $w_1 = w_2$ and $s_2 - s_1 = 0$. Since the sum $S_1 \oplus S_2$ is direct, this in turn forces $s_1 = 0 = s_2$. But then, since $v \notin W$, either $v \notin W \oplus S_1$ or $v \notin W \oplus S_2$, contradicting the maximality of W with this property. Thus (iii) implies (i). □

Corollary 1.9.6 *Let A be a k-algebra. The following are equivalent.*

(i) A is semisimple as a left A-module.
(ii) Every finitely generated A-module is semisimple.
(iii) Every A-module is semisimple.

Proof We show that (i) implies (iii). Suppose that A is semisimple as a left A-module. Since a free A-module is a direct sum of a (possibly infinite) family of copies of A it follows that every free A-module is semisimple. Let M be an A-module. For any $m \in M$ we have an obvious A-homomorphism $A \to M$ sending $a \in A$ to $am \in M$. Taking the sum of these homomorphisms over all elements $m \in M$ yields a surjective A-homomorphism from a free A-module onto M, and thus, by 1.9.5, M is semisimple. The implications (iii)$\Rightarrow$ (ii) $\Rightarrow$ (i) are trivial. □

We will show in 1.10.19 that if A is Artinian and if all left A-modules are semisimple, then also all right A-modules are semisimple, and vice versa. For semisimple modules we obtain the following unique decomposition property:

Proposition 1.9.7 *Let A be a k-algebra, let M be an A-module, and let S_i, T_j be simple submodules of M for $1 \leq i \leq n$, $1 \leq j \leq k$, where n, k are positive* *integers. Suppose that*

$$M = \oplus_{1 \le i \le n} S_i = \oplus_{1 \le j \le k} T_j.$$

Then $n = k$ and there is a permutation π of the set $\{1, 2, \ldots, n\}$ such that $S_i \cong T_{\pi(i)}$ for all i, $1 \le i \le n$.

Proof By looking at partial sums $S_1 \oplus \cdots \oplus S_i$ we obtain a composition series with the S_i as composition factors; similarly with the T_j instead of S_i. The result follows from 1.8.3. □

In other words, decomposing a finitely generated semisimple module is unique up to an automorphism of that module. One can get a stronger uniqueness statement if one collects isomorphic simple summands:

Proposition 1.9.8 *Let A be a k-algebra, and let M be a semisimple A-module which is a sum of finitely many simple A-modules. There is a unique direct sum decomposition $M = \oplus_{S \in \Lambda} M_S$, where Λ is a finite set of pairwise nonisomorphic simple A-modules and where M_S is the sum of all submodules of M isomorphic to S, for each S in Λ. Moreover, for any two different S, T in Λ we have* $\mathrm{Hom}_A(M_S, M_T) = \{0\}$.

Proof Write $M = \oplus_{i=1}^n S_i$ for some simple submodules S_i of M. Let Λ be a set of representatives of the isomorphism classes of the simple A-modules that occur in this direct sum. For each $S \in \Lambda$, set $M_S = \oplus_j S_j$, where j runs over the indices between 1 and n for which $S_j \cong S$. Then clearly $M = \oplus_{S \in \Lambda} M_S$. Since M_S is a direct sum of simple A-modules isomorphic to S, for any simple A-module T not isomorphic to S we have $\mathrm{Hom}_A(T, M_S) = \{0\}$ by Schur's Lemma 1.9.2. Thus M_S is equal the sum of all submodules of M isomorphic to S. This shows the uniqueness of this decomposition, and the last statement is again a consequence of Schur's Lemma. □

The submodules M_S of M are called the *homogeneous* or *isotypic components* of M. Given an algebra A and a subalgebra B, the restriction of a simple A-module to B need no longer be simple, nor even semisimple. The following theorem, due to Clifford, describes a situation where simple modules do restrict to semisimple modules. For U a module over some algebra A and B a subalgebra we denote by $\mathrm{Res}^A_B(U)$ the B-module obtained from restricting the action of A on U to the subalgebra B. If $A = kG$ for some group G and $B = kH$ for some subgroup H of G we write $\mathrm{Res}^G_H(U)$ instead of $\mathrm{Res}^A_B(U)$.

Theorem 1.9.9 (Clifford) *Let G be a finite group, let N be a normal subgroup of G and suppose that k is a field. For any simple kG-module S the restriction* $\mathrm{Res}^G_N(S)$ *of S to kN is a semisimple kN-module. Moreover, if T, T' are simple*

kN-submodules of $\mathrm{Res}^G_N(S)$ *then there is an element $x \in G$ such that $T' \cong xT$. In other words, the isomorphism classes of simple kN-submodules of* $\mathrm{Res}^G_N(S)$ *are permuted transitively by the action of G.*

Proof Let T be a simple kN-submodule of S restricted to kN. Let $x \in G$. The subset xT is again a simple kN-submodule of S restricted to kN. Indeed, it is a submodule because for $n \in N$ we have $nxT = x(x^{-1}nx)T \subseteq xT$ as N is normal in G. It is also simple because if V is a kN-submodule of xT then $x^{-1}V$ is a kN-submodule of T. This shows that if we take the sum of all simple kN-submodules of S of the form xT, with $x \in G$, we get a kG-submodule of S. Since S is simple this implies that S is the sum of the xT. □

1.10 The Jacobson radical

Definition 1.10.1 The *Jacobson radical* $J(A)$ of a k-algebra A is the intersection of the annihilators of all simple left A-modules. More explicitly, $J(A)$ is equal to the set of all $a \in A$ satisfying $aS = \{0\}$ for every simple A-module S.

The set $J(A)$ is an ideal in A, and we will see that $J(A)$ coincides with the intersection of the annihilators of all simple right A-modules. If M is an A-module, then $J(A)M$ is a submodule of M, because $J(A)$ is an ideal. In fact, $J(A)M$ is contained in every maximal submodule of M:

Lemma 1.10.2 *Let A be a k-algebra, let M be an A-module and let U be a maximal submodule of M. Then $J(A)M \subseteq U$.*

Proof Since U is a maximal submodule of M, the quotient M/U is a simple A-module. Thus $J(A)M/U = \{0\}$, which is equivalent to $J(A)M \subseteq U$ as stated. □

Lemma 1.10.3 *Let A be a k-algebra and M be a finitely generated A-module. For every proper submodule U of M there is a maximal submodule V of M containing U. In particular, every nonzero submodule of a Noetherian module has a maximal submodule.*

Proof Let $\mathcal{V}$ be the set of all proper submodules of M containing U. This set contains U, hence is nonempty. We show that this set is inductively ordered; that is, we show that every totally ordered subset of $\mathcal{V}$ has an upper bound. The existence of a maximal element in this set follows then from Zorn's Lemma. Let $\mathcal{T}$ be a totally ordered subset of $\mathcal{V}$. Set $S = \cup_{T \in \mathcal{T}} T$. We need to show that S belongs to $\mathcal{V}$. Clearly S contains U, so we need to show that S is still a proper submodule of M. Suppose $S = M$. By assumption, there is a finite subset W of M such that $M = \sum_{w \in W} Aw$. In particular, $W \subseteq S$. Thus, for any $w \in W$ there is $T \in \mathcal{T}$ such that $w \in T$. Since $\mathcal{T}$ is totally ordered and W is finite, there is actually a $T \in \mathcal{T}$ such that $W \subseteq T$. But then $T = M$, a contradiction. The last statement follows from the first, since any submodule of a Noetherian module is finitely generated. □

Theorem 1.10.4 (Nakayama's Lemma) *Let A be a k-algebra. Let M be a finitely generated A-module. If N is a submodule of M such that $M = N + J(A)M$, then $M = N$. In particular, if $J(A)M = M$, then $M = \{0\}$.*

Proof Arguing by contradiction, suppose that N is a proper submodule of M. By 1.10.3, there is a maximal submodule V of M containing N. As observed before, $J(A)M$ is contained in V. But then $N + J(A)M \subseteq V$, contradicting the assumptions. The last statement follows from applying the first with $N = \{0\}$. □

An ideal I in an algebra A is called *nilpotent* if there is a positive integer n such that $I^n = \{0\}$, where I^n is the set of all finite sums of elements of the form $a_1 a_2 \cdots a_n$, where $a_i \in I$ for $1 \le i \le n$. Thus all elements $a \in I$ satisfy $a^n = 0$. In other words, all elements in a nilpotent ideal are nilpotent. It is, however possible, that a nonnilpotent ideal consists entirely of nilpotent elements.

Theorem 1.10.5 *Let A be a k algebra.*

(i) The set $1 + J(A)$ is a subgroup of $A^\times$.
(ii) The radical $J(A)$ contains every nilpotent ideal in A.
(iii) The radical $J(A)$ contains no idempotent.
(iv) For any ring automorphism α of A we have $\alpha(J(A)) = J(A)$. In particular, if a group G acts on A by ring automorphisms, then $J(A)$ is G-stable.

Proof Let $a \in J(A)$. Write $1 = a + (1 - a)$. Thus $A = Aa + A(1 - a) = J(A) + A(1 - a)$. As a left A-module, A is finitely generated (by the element 1), and hence Nakayama's Lemma 1.10.4 applied to the A-module A and the submodule $A(1 - a)$ implies that $A(1 - a) = A$. It follows that $1 - a$ has a left inverse. Write a left inverse of $1 - a$ in the form $1 - b$ for some $b \in A$. Thus $1 = (1 - b)(1 - a) = 1 - a - b + ba$, hence $b = ba - a \in J(A)$. But then $1 - b$ has a left inverse as well, and so $1 - b$ is invertible, which implies that $1 - a$ and $1 - b$ are inverse to each other. This shows (i). For statement (ii), let I be a nilpotent ideal in A. If S is a simple A-module then IS is a submodule of S; thus either $IS = S$ or $IS = \{0\}$ by the simplicity of S. Since I is nilpotent the equality $IS = S$ is impossible because this would imply $I^n S = S$ for all positive integers n. Thus $IS = \{0\}$, hence $I \subseteq J(A)$. Let e be an idempotent in A such that $e \in J(A)$. In particular, e is nonzero. By (i) the element $1 - e$ is invertible with an inverse of the form $1 - a$ for some $a \in J(A)$. Then $1 = (1 - e)(1 - a) = 1 - e - a + ea$, hence $a = ea - e$. Since e is an idempotent this implies $ea = ea - e$, hence $e = 0$. This contradiction completes the proof of (iii). Let α be a k-algebra automorphism of A, and let S be a simple A-module. We define an A-module ${}_\alpha S$ by setting ${}_\alpha S = S$ as an abelian group, such that $a \in A$ acts on ${}_\alpha S$ as $\alpha(a)$ on S. Then ${}_\alpha S$ is again a simple A-module, since any A-submodule of S is also an A-submodule of ${}_\alpha S$. Moreover, if $a \in A$, then $\alpha(a)$ annihilates S if and only of a annihilates ${}_\alpha S$. The correspondence sending S to ${}_\alpha S$ induces a bijection on the isomorphism classes of simple A-modules, because ${}_{\alpha^{-1}}({}_\alpha S) = S = {}_\alpha({}_{\alpha^{-1}}S)$. Thus an element $a \in A$ annihilates all simple A-modules if and only if $\alpha(a)$ annihilates all simple modules, proving (iv). □

The backdrop for the arguments in the proof of Theorem 1.10.5 (iv) are *Morita equivalences*; see 2.8.13 and 2.8.16 below. Note that we allowed ring automorphisms rather than only algebra automorphisms in the last statement; this will be useful in the context of extending Galois group actions on a field to group algebras over that field.

Theorem 1.10.6 *Let A be a k-algebra. The radical $J(A)$ is equal to any of the following:*

(i) the intersection of the annihilators of all right simple A-modules;
(ii) the intersection of all maximal left ideals in A; and
(iii) the intersection of all maximal right ideals in A.

Proof If M is a maximal left ideal in A then the left A-module A/M is simple. Thus $J(A)A/M = \{0\}$, or equivalently, $J(A) \subseteq M$. This shows that $J(A)$ is contained in all maximal left ideals in A. Let S be a simple A-module and let $s \in S \setminus \{0\}$. The map $A \to S$ sending $a \in A$ to As is a surjective A-homomorphism because S is simple. The kernel of this homomorphism is the annihilator $M_s = \{a \in A | as = 0\}$ of s. Thus $A/M_s \cong S$, which shows that M_s is a maximal left ideal in A as S is simple. Moreover, $I_S = \cap_{s \in S\setminus\{0\}} M_s$ is the annihilator of S in A; in particular, I_S is the intersection of the maximal left ideals containing I_S. Since $J(A)$ is the intersection of all I_S, as S varies over the simple A-modules, we get that $J(A)$ is the intersection of all maximal left ideals in A. If we denote by $J'(A)$ the intersection of the annihilators of all right simple A-modules, the same argument shows that $J'(A)$ is the intersection of all maximal right ideals in A. The right analogue of 1.10.5 (i) shows that $1 + J'(A) \subseteq A^\times$. Suppose now that $J'(A) \not\subseteq J(A)$. Since $J(A)$ is the intersection of all maximal left ideals this implies that there is a maximal left ideal M in A such that $J'(A) \not\subseteq M$. The maximality of M implies in turn that $A = J'(A) + M$. Write $1 = c + m$ for some $c \in J'(A)$ and some $m \in M$. Then $m = 1 - c$, so m is invertible. This however would force the contradiction $M = A$. Thus $J'(A) \subseteq J(A)$. The same argument shows $J(A) \subseteq J'(A)$, hence the equality $J(A) = J'(A)$. □

It is not always true that the radical $J(A)$ is equal to the intersection of all maximal two-sided ideals in A. One situation where this does hold is when $A/J(A)$ is Artinian; we prove this in the section on Wedderburn's Theorem below.

Theorem 1.10.7 *Let A be a k-algebra and let I be an ideal in A such that $I \subseteq J(A)$. We have $1 + I \subseteq A^\times$, and for any element $a \in A$ we have $a \in A^\times$ if and only if $a + I \in (A/I)^\times$. In particular, the canonical map $A \to A/I$ induces a short exact sequence of groups*

$$1 \longrightarrow 1 + I \longrightarrow A^\times \longrightarrow (A/I)^\times \longrightarrow 1.$$

Proof Since $I \subseteq J(A)$ we have $1 + I \subseteq A^\times$ by 1.10.5. If $a \in A^\times$ then clearly $a + I \in (A/I)^\times$. Conversely, if $a \in A$ such that $a + I \in (A/I)^\times$ then $A = Aa + I$. Since A is finitely generated as left A-module (with 1 as generator, for instance) and again since $I \subseteq J(A)$, Nakayama's Lemma 1.10.4 implies that $A = Aa$. This shows that $a \in A^\times$. Thus in particular the canonical group homomorphism $A^\times \to (A/I)^\times$ is surjective and has $1 + I$ as kernel. □

The radical of a finite-dimensional algebra over a field admits another useful characterisation.

Theorem 1.10.8 *Suppose that k is a field. Let A be a finite-dimensional k-algebra. Then $J(A)$ is the unique maximal nilpotent ideal in A. For any subalgebra B of A we have $J(A) \cap B \subseteq J(B)$, and we have $J(Z(A)) = J(A) \cap Z(A)$.*

Proof By 1.10.5 every nilpotent ideal in A is contained in $J(A)$. Consider the descending chain of ideals $J(A) \supseteq J(A)^2 \supseteq \cdots$. Since A is finite-dimensional, there is an integer $n > 0$ such that $J(A)^n = J(A)^{n+1}$. But then Nakayama's Lemma 1.10.4 implies that $J(A)^n = \{0\}$. Thus $J(A)$ is nilpotent and contains all nilpotent ideals in A. This proves the first statement. If B is a subalgebra of A, then $J(A) \cap B$ is a nilpotent ideal in B, hence contained in $J(B)$. Applying this to the subalgebra $Z(A)$ yields $J(A) \cap Z(A) \subseteq J(Z(A))$. For the other inclusion we note that $J(Z(A))A = AJ(Z(A))$ is also a nilpotent ideal in A, hence $J(Z(A)) \subseteq J(A) \cap Z(A)$. □

Corollary 1.10.9 *Suppose that k is a field. Let A be a finite-dimensional commutative k-algebra. Then $J(A)$ is equal to the set of nilpotent elements in A.*

Proof Every element in $J(A)$ is nilpotent by 1.10.8. Conversely, let x be a nilpotent element in A. Since A is commutative, we have $xA = Ax$. Thus xA is a nilpotent ideal, hence contained in $J(A)$ by 1.10.8. □

Theorem 1.10.10 *Let A be a k-algebra that is finitely generated as a k-module. We have $J(k)A \subseteq J(A)$.*

Proof Let S be a simple A-module. Then, as a k-module, S is finitely generated because A is so. Thus $J(k)S$ is a proper submodule of S as a k-module. But clearly $J(k)S$ is also an A-submodule of S. Thus $J(k)S = \{0\}$ as S is simple. It follows that $J(k)A$ annihilates all simple A-modules, and so $J(k)A \subseteq J(A)$. □

The inclusion $J(A) \cap B \subseteq J(B)$ holds in more general circumstances; see also 4.7.9 below.

Theorem 1.10.11 *Let A be a k-algebra and B a subalgebra of A. If $A \otimes_B T$ is nonzero for every simple B-module T, then $J(A) \cap B \subseteq J(B)$.*

Proof Let T be a simple B-module. Then T is generated, as a B-module, by any nonzero element y in T. Thus the A-module $A \otimes_B T$ is generated by a single element $1 \otimes y$. It is also nonzero, by the assumptions. Thus $A \otimes_B T$ has a maximal submodule V by 1.10.3. Thus $S = A \otimes_B T/V$ is a simple A-module. Denote by $\psi : A \otimes_B T \to S$ the canonical surjection. Then $\psi(1 \otimes y)$ generates S; in particular, $\psi(1 \otimes y)$ is nonzero. Thus the map $\varphi : T \to S$ sending $t \in T$ to $\psi(1 \otimes t)$ is a nonzero B-homomorphism. This shows that T is isomorphic to a submodule of the restriction to B of the simple A-module S. Since $J(A)$ annihilates S it follows that $J(A) \cap B$ annihilates T, whence the result. □

The hypothesis that $A \otimes_B T$ is nonzero for every simple B-module holds in particular if B is a direct summand of A as a right B-module (in that case we trivially have $A \otimes_B V \neq \{0\}$ for any nonzero B-module V). This hypothesis is equivalent to requiring that for any simple B-module T there exists an A-module U such that T is a submodule of $\mathrm{Res}^A_B(U)$. Indeed, if $T \subseteq \mathrm{Res}^A_B(U)$, then there is an A-homomorphism $A \otimes_B T \to U$ sending $a \otimes t$ to at. This map is nonzero as its image contains T, and hence $A \otimes_B T$ is nonzero. Conversely, if $A \otimes_B T$ is nonzero, then the argument of the proof above shows that T is a submodule of $\mathrm{Res}^A_B(S)$ for any simple quotient S of the A-module $A \otimes_B T$.

Theorem 1.10.12 *Let A be a k-algebra and let I be an ideal in A. We have $(J(A)+I)/I \subseteq J(A/I)$. If $I \subseteq J(A)$, then $J(A/I) = J(A)/I$.*

Proof Every simple A/I-module can be viewed as a simple A-module via the canonical map $A \to A/I$. Since $J(A)$ annihilates every simple A-module, it follows that its image $(J(A)+I)/I$ in A/I annihilates every simple A/I-module, whence the first inclusion. If $I \subseteq J(A)$, then I annihilates every simple A-module, and hence every simple A-module can also be regarded as a simple A/I-module. Thus $a \in A$ annihilates every simple A-module if and only if its image $a+I$ in A/I annihilates every simple A/I-module, whence the equality $J(A)/I = J(A/I)$. □

For A, B two k-algebras, we denote by $J(A) \otimes B$ the image of $J(A) \otimes_k B$ in $A \otimes_k B$. This is a quotient of $J(A) \otimes_k B$, but need not be isomorphic to $J(A) \otimes_k B$ because the map obtained from tensoring the inclusion $J(A) \subseteq A$ with $- \otimes_k B$ might no longer be injective. If B is *flat* as a k-module, or equivalently, if the functor $- \otimes_k B$ is exact, then $J(A) \otimes_k B$ can be identified with its image in $A \otimes_k B$. Similarly for $A \otimes J(B)$.

Theorem 1.10.13 *Let A and B be k-algebras. Suppose that A and B are finitely generated as k-modules. We have $J(A) \otimes B + A \otimes J(B) \subseteq J(A \otimes_k B)$.*

Proof Let X be a simple $A \otimes_k B$-module. Since A, B, and hence $A \otimes_k B$, are finitely generated as k-modules, it follows that X is finitely generated as a k-module, hence also as a module over A and over B via the canonical maps $A \to A \otimes_k B$ and $B \to A \otimes_k B$. Regard X as an A-module via this map. Nakayama's Lemma 1.10.4 implies that $J(A)X$ is a proper submodule of X. This is clearly an $A \otimes_k B$-submodule of X because the actions of A and B on X commute. Thus $J(A)X$ is zero as X is simple. This shows that $J(A) \otimes B$ annihilates X, hence is contained in $J(A \otimes_k B)$. A similar argument shows that $A \otimes J(B)$ is contained in $J(A \otimes_k B)$. □

See 1.16.15 below for a sufficient criterion for when the inclusion in the previous theorem is an equality.

Theorem 1.10.14 *Let A be a k-algebra and let e be an idempotent in A.*

(i) For any simple A-module S either $eS = \{0\}$ or eS is a simple eAe-module.
(ii) For any simple eAe-module T there is a simple A-module S such that $eS \cong T$.

Proof Let S be a simple A-module such that $eS \neq \{0\}$, and let V be a nonzero eAe-submodule of eS. Then, since S is simple and e is an idempotent we have $S = AV = AeV$, hence $eS = eAeV = V$, which proves (i). Let T be a simple eAe-module. Then the A-module $Ae \otimes_{eAe} T$ is finitely generated (namely by any element of the form $e \otimes t$ for t a nonzero element in T). It is also nonzero, as eAe is a direct summand of $Ae = eAe \oplus (1-e)Ae$ as a right eAe-module. It follows from 1.10.3 that $Ae \otimes_{eAe} T$ has a maximal submodule M. Thus $S = Ae \otimes_{eAe} T/M$ is a simple A-module. The canonical surjection $\pi : Ae \otimes_{eAe} T \to S$ is nonzero on the subspace $e \otimes T$ because this space generates $Ae \otimes_{eAe} T$ as an A-module. Thus multiplying π by e yields a nonzero eAe-homomorphism $T \to eS$. Since eS is simple by (i), this implies that $T \cong eS$, whence (ii). □

Corollary 1.10.15 *Let A be a k-algebra and let e be an idempotent in A. We have $J(eAe) = eJ(A)e = eAe \cap J(A)$.*

Proof Let $a \in J(A)$. Then $eae \in J(A)$ because $J(A)$ is an ideal; in particular, $eJ(A)e \subseteq eAe \cap J(A)$. Moreover, eae annihilates every simple A-module S, hence eS, and therefore eae annihilates every simple eAe-module by 1.10.14 (ii). This shows $eJ(A)e \subseteq J(eAe)$. Let $c \in J(eAe)$. Then $c = ce$ annihilates every simple A-module S by 1.10.14 (i), implying $J(eAe) \subseteq J(A)$. Since e is an idempotent, this implies $J(eAe) \subseteq eJ(A)e$ and $eAe \cap J(A) \subseteq eJ(A)e$, whence the result. □

Corollary 1.10.16 *Let A be a k-algebra and b an idempotent in $Z(A)$. We have $J(Ab) = J(A)b$. In particular, if A has a block decomposition, then $J(A)$ is the sum of radicals $J(Ab)$, with b running over the blocks of A.*

Proof This is a special case of 1.10.15. It can also be proved more directly, using the decomposition $A = Ab \times A(1-b)$ and the fact that a simple A-module is either a simple Ab-module or a simple $A(1-b)$-module. □

The following result, due to J. A. Green, makes 1.10.14 more precise: multiplying by an idempotent e in a k-algebra A induces a bijection between the isomorphism classes of simple A-modules not annihilated by e and the isomorphism classes of simple eAe-modules.

Theorem 1.10.17 ([34, §6.2]) *Let A be a k-algebra and let e be an idempotent in A. Let T be a simple eAe-module.*

(i) Let M be the largest submodule of the A-module $Ae \otimes_{eAe} T$ satisfying $eM = \{0\}$. Then M is the unique maximal submodule of $Ae \otimes_{eAe} T$; in particular, $S = Ae \otimes_{eAe} T/M$ is, up to isomorphism, the unique simple A-module satisfying $eS \cong T$.

(ii) Let N be the smallest submodule of the A-module $\mathrm{Hom}_{eAe}(eA, T)$ containing the subspace $\mathrm{Hom}_{eAe}(eAe, T)$. Then N is the unique simple submodule of $\mathrm{Hom}_{eAe}(eA, T)$; in particular, $S = N$ is, up to isomorphism, the unique simple A-module satisfying $eS \cong T$.

Proof Let M' be a proper submodule of $Ae \otimes_{eAe} T$. If $eM' = \{0\}$ then $M' \subseteq M$ by the definition of M, so we may assume that $eM' \neq \{0\}$. Then $eM' \cong T$ because T is simple, so $M' = e \otimes T$. Thus $M' \supseteq AeM' = Ae \otimes_{eAe} T$, a contradiction. Thus M contains every proper submodule of $Ae \otimes_{eAe} T$, hence M is maximal in $Ae \otimes_{eAe} T$. It follows that $S = A \otimes_{eAe} T/M$ is simple, and since e annihilates M, it follows further that $eS \cong T$. If S' is another simple A-module satisfying $eS' = T$, then the map $Ae \otimes_{eAe} T \to S'$ sending $ae \otimes t$ to aet is a nonzero A-homomorphism, hence surjective with kernel M, forcing $S' \cong S$. This proves (i). For (ii), observe first that the obvious decomposition $eA = eAe \oplus eA(1-e)$ as left eAe-modules yields a decomposition of homomorphism spaces $\mathrm{Hom}_{eAe}(eA, T) = \mathrm{Hom}_{eAe}(eAe, T) \oplus \mathrm{Hom}_{eAe}(eA(1-e), T)$. Moreover, we have $\mathrm{Hom}_{eAe}(eAe, T) \cong T$. Let N' be a nonzero submodule of $\mathrm{Hom}_{eAe}(eA, T)$. If $eN' \neq \{0\}$ then $eN' = \mathrm{Hom}_{eAe}(eAe, T) \cong T$ since T is simple, hence N' contains N and thus $eN \cong T$. If $eN' = \{0\}$ then $N' \subseteq \mathrm{Hom}_{eAe}(eA(1-e), T)$. Since N' is nonzero there is a nonzero eAe-homomorphism $\varphi : eA \to T$ belonging to N', hence satisfying $eAe \subseteq \ker(\varphi)$. Since N' is an A-module, we have $a \cdot \varphi \in N'$ for all $a \in A$, hence $eAe \subseteq \ker(a \cdot \varphi)$ for all $a \in A$. This means that for all $a \in A$ and all $u \in eAe$ we have $0 = (a \cdot \varphi)(u) = \varphi(ua)$, hence $eA \subseteq \ker(\varphi)$, or equivalently, $\varphi = 0$, hence $N' = \{0\}$, a contradiction. The uniqueness statement follows from that in (i). □

We observed earlier that for any module M over some algebra A, the submodule $J(A)M$ lies in the intersection of all maximal submodules of M. Here is a situation, where they are equal:

Theorem 1.10.18 *Let A be an Artinian k-algebra, M a finitely generated A-module and J a proper ideal in A.*

(i) M is semisimple if and only if $J(A)M = \{0\}$.
(ii) $J(A)M$ is the intersection of all maximal submodules of M.
(iii) $J(A/J) = (J(A) + J)/J$.

Proof If M is semisimple, then $J(A)M = \{0\}$ by the definition of $J(A)$ as the annihilator of all simple A-modules. For the converse we first show that $A/J(A)$ is semisimple as a left A-module. We know that $J(A)$ is the intersection of all maximal left ideals in A. Since A is Artinian, we can in fact find finitely many maximal left ideals M_i, $1 \leq i \leq m$ such that $J(A) = \bigcap_{1 \leq i \leq m} M_i$. Consider the canonical map $A \to \bigoplus_{1 \leq i \leq m} A/M_i$. The kernel of this map is $J(A)$. The right side is semisimple. Thus this map induces an injective A-homomorphism from $A/J(A)$ to a semisimple A-module, hence $A/J(A)$ is semisimple. Suppose now that $J(A)M = \{0\}$. Since M is finitely generated, it is a quotient of a free module A^n of finite rank n. Since $J(A)M = \{0\}$, the module M is in fact a quotient of $(A/J(A))^n$. As this module is semisimple, so is M. This shows (i). As pointed out earlier, if N is a maximal submodule of M then M/N is simple, hence $J(A)M/N = \{0\}$, or equivalently, $J(A)M \subseteq N$. Thus we may replace M by $M/J(A)M$, or equivalently, we may assume that $J(A)M = \{0\}$. We have to show that the intersection of all maximal submodules of M is then also $\{0\}$. Now M is semisimple, by (i). Since M is finitely generated, M is in fact a finite direct sum of simple A-modules, say $M = \oplus_{1\leq i\leq r} S_i$ for some simple A-modules S_i. Define submodules $N_j = \oplus_{1\leq i\leq r, i\neq j} S_i$. Then $M/N_j \cong S_j$, hence N_j is a maximal submodule of M for $1 \leq j \leq r$. The intersection $\cap_{1\leq j\leq r} N_j$ is obviously zero, which concludes the proof of (ii). Every simple A/J-module is a simple A-module via the canonical map $A \to A/J$. Thus the image $(J(A)+J)/J$ of $J(A)$ in A/J annihilates every simple A/J-module, hence is contained in $J(A/J)$. View the quotient $(A/J)/((J(A)+J)/J)$ as an A-module. This quotient is annihilated by $J(A)$, hence is semisimple as an A-module by (i). This quotient is also annihilated by J, hence remains a semisimple A/J-module. In particular, this quotient is annihilated by $J(A/J)$, which implies the inclusion $J(A/J) \subseteq (J(A)+J)/J$. This proves (iii). □

Corollary 1.10.19 *Let A be an Artinian k-algebra. The following are equivalent:*

(i) $J(A) = \{0\}$.
(ii) A is semisimple as a left A-module.
(iii) A is semisimple as a right A-module.
(iv) Every left or right A-module is semisimple.

Proof Clearly (iv) implies (iii), (ii), and any of (iii), (ii) implies (i). If (i) holds then every finitely generated left or right A-module is semisimple by 1.10.18 and its right analogue. An arbitrary A-module is the sum of its finitely generated submodules, hence (i) implies (iv). □

Corollary 1.10.20 *Suppose that k is a field. Let A, B be finite-dimensional k-algebras, and let X be an A-B-bimodule. Suppose that X is semisimple as an $A \otimes_k B^{\mathrm{op}}$-module. Then X is semisimple as a left A-module and as a right B-module.*

Proof By the assumptions, X is annihilated by $J(A \otimes_k B^{\mathrm{op}})$. By 1.10.13, $J(A \otimes_k B^{\mathrm{op}})$ contains $J(A) \otimes 1$ and $1 \otimes J(B^{\mathrm{op}})$. Thus X is annihilated by $J(A)$ on the left and by $J(B)$ on the right. The result follows from 1.10.18. □

Definition 1.10.21 Let A be a k-algebra and let M be an A-module. The *socle of M*, denoted by $\mathrm{soc}(M)$, is the sum of all simple submodules of M, with the convention that $\mathrm{soc}(M) = \{0\}$ if M has no simple submodule. The *radical of M*, denoted by $\mathrm{rad}(M)$, is the intersection of all maximal submodules of M, with the convention that $\mathrm{rad}(M) = M$ if M has no maximal submodule.

Equivalently, $\mathrm{soc}(M)$ is the largest semisimple submodule of M. If A is finite-dimensional over a field k and M is finitely generated, then $\mathrm{soc}(M)$ is also the largest submodule of M that is annihilated by $J(A)$, and $\mathrm{rad}(M) = J(A)M$ is the smallest submodule of M such that $M/\mathrm{rad}(M)$ is semisimple, by theorem 1.10.18. One way to analyse the structure of the module M is to consider the *radical series*

$$M \supseteq \mathrm{rad}(M) = \mathrm{rad}^1(M) \supseteq \mathrm{rad}^2(M) \supseteq \cdots$$

where $\mathrm{rad}^{n+1}(M) = \mathrm{rad}(\mathrm{rad}^n(M))$ for positive n, and the *socle series*

$$\mathrm{soc}(M) = \mathrm{soc}^1(M) \subseteq \mathrm{soc}^2(M) \subseteq \mathrm{soc}^3(M) \subseteq \cdots$$

where $\mathrm{soc}^{n+1}(M)$ is the inverse image in M of $\mathrm{soc}(M/\mathrm{soc}^n(M))$ for positive n. For notational convenience, we adopt the convention $\mathrm{rad}^0(M) = M$ and $\mathrm{soc}^0(M) = \{0\}$. If M has a composition series, then the radical series and the socle series "slice" the module M into layers of finitely generated semisimple modules, and while the two series need not be equal, they have the same length.

Theorem 1.10.22 *Suppose that k is a field. Let A be a finite-dimensional k-algebra and let M be a finitely generated nonzero A-module.*

(i) For any $i \geq 0$ we have $\mathrm{rad}^i(M) = J(A)^iM$.
(ii) For any $i \geq 0$ we have $\mathrm{soc}^i(M) = \{m \in M \mid J(A)^im = \{0\}\}$.
(iii) Let s be the smallest positive integer such that $J(A)^sM = \{0\}$, and let t be the smallest positive integer such that $\mathrm{soc}^t(M) = M$. Then $t = s$, and for $0 \leq i \leq s$ we have $\mathrm{rad}^iM \subseteq \mathrm{soc}^{s-i}(M)$.

Proof By 1.10.18 we have $\mathrm{rad}(M) = J(A)M$, and $\mathrm{soc}(M)$ is the annihilator of $J(A)$. Both (i) and (ii) follow inductively. Since $\mathrm{soc}^{i+1}(M)/\mathrm{soc}^i(M)$ is semisimple we have $J(A)\mathrm{soc}^{i+1}(M) \subseteq \mathrm{soc}^i(M)$ for $i \geq 0$. Hence $s \leq t$ and $J(A)^iM \subseteq \mathrm{soc}^{t-i}(M)$ for $0 \leq i \leq t$. In order to show the equality $s = t$ we proceed by induction over s. If $s = 1$ then $J(A)M = \{0\}$, hence M is semisimple and thus $M = \mathrm{soc}(M)$. Suppose that $s > 1$. The submodule $J(A)^{s-1}M$ of M is annihilated by $J(A)$, hence semisimple, and thus $J(A)^{s-1}M \subseteq \mathrm{soc}(M)$. It follows that $J(A)^{s-1}$ annihilates $M/\mathrm{soc}(M)$. Let r be the smallest positive integer such that $J(A)^r(M/\mathrm{soc}(M)) = \{0\}$. Since $r < s$ the induction hypothesis implies that r is also the smallest positive integer such that

$\operatorname{soc}^r(M/\operatorname{soc}(M)) = M$. Now $\operatorname{soc}^r(M/\operatorname{soc}(M)) = \operatorname{soc}^{r+1}(M)/\operatorname{soc}(M)$, and hence $\operatorname{soc}^{r+1}(M) = M$, which forces $r+1 \geq t$. As $r+1 \leq s \leq t$ we get the equality $s = t$. □

Definition 1.10.23 Suppose that k is a field, let A be a finite-dimensional k-algebra, and let M be a finitely generated nonzero A-module. The smallest positive integer s satisfying $J(A)^s M = \{0\}$ is called the *Loewy length* of the finitely generated A-module M, denoted $\ell\ell(M)$, and the successive quotients $J(A)^{i-1}M/J(A)^i M$, with $1 \leq i \leq s$, are called the *Loewy layers of* M. If M is zero, we set $\ell\ell(M) = 0$.

The isomorphism class of a module is not, in general, determined by its Loewy layers.

Exercise 1.10.24 Let A be a k-algebra and M a uniserial A-module. Show that M is finitely generated and isomorphic to a quotient of the regular A-module A.

Exercise 1.10.25 Let k be a field. Show that $J(k[x]) = \{0\}$.

Exercise 1.10.26 Let k be a field and n a positive integer. Let A be a subalgebra of $M_n(k)$ consisting of upper triangular matrices. Show that $J(A)$ is equal to the set of matrices in A whose diagonal is zero.

Exercise 1.10.27 Show that $J(\mathbb{Z}) = \{0\}$, and that $\mathbb{Z}$ is indecomposable but not simple as a $\mathbb{Z}$-module. Deduce that Theorem 1.10.18 and its Corollary 1.10.19 do not hold in general if A is not Artinian.

Exercise 1.10.28 Let A be a k-algebra, I, J ideals in A, and suppose that J is nilpotent. Show that there is a positive integer n such that $(I+J)^n \subseteq I$. Deduce that if both I, J are nilpotent, then so is $I+J$.

Exercise 1.10.29 Suppose that k is a field. Let A be a finite-dimensional k-algebra and M a finitely generated nonzero A-module. Show that for any quotient or submodule N of M we have $\ell\ell(N) \leq \ell\ell(M)$. Show that if X is a subset of M that generates M as an A-module, then $\ell\ell(M) = \max\{\ell\ell(Am) | m \in X\}$. Deduce that $\ell\ell(M) \leq \ell\ell(A)$, where A is considered as the regular A-module.

Exercise 1.10.30 Let $\mathcal{P}$ be a finite partially ordered set, viewed as a category with exactly one morphism $x \to y$ whenever $x \leq y$. Show that the morphisms $x \to y$ with $x < y$ in $\mathcal{P}$ span an ideal J in $\mathcal{P}$ which is contained in the radical $J(k\mathcal{P})$ of the algebra $k\mathcal{P}$. Use this, together with 1.10.5 (i), to give an alternative proof of the fact that the sum in $\mathcal{P}$ of all morphisms in $\mathcal{P}$ has a multiplicative inverse. Show that if k is a field, then $J = J(k\mathcal{P})$.

Exercise 1.10.31 Let G be a finite group. Suppose that k is a field. Set $z = \sum_{x \in G} x$. Show that $kGz = zkG = kz$ is a one-dimensional two-sided ideal in kG. Deduce that z is contained in the socle of kG as a left and as a right kG-module. (We will see later that the socles of kG as left and right module coincide, a property which holds in general for finite-dimensional selfinjective algebras.)

1.11 On the Jacobson radical of finite group algebras

The first result in this section shows that the Jacobson radical $J(kP)$ of a finite p-group algebra kP over a field k of prime characteristic p is equal to the augmentation ideal $I(kP)$.

Theorem 1.11.1 *Let p be a prime. Suppose that k is a field of characteristic p. Let P be a finite p-group. We have $I(kP)^{|P|} = \{0\}$. In particular, $J(kP) = I(kP)$, and the trivial kP-module k is, up to isomorphism, the unique simple kP-module.*

Proof The augmentation ideal $I(kP)$ is a maximal ideal in kP because the quotient $kP/I(kP) \cong k$ is one-dimensional. In particular, this quotient is simple as a kP-module. Moreover, the elements of P act as the identity on this quotient since $I(kP)$ contains all elements of the form $y - 1$, where $y \in P$. Thus all we have to show is that the ideal $I(kP)^{|P|}$ is zero. We proceed by induction on the order of P. For $P = \{1\}$ there is nothing to prove. Suppose $|P| > 1$. Then $Z(P)$ is nontrivial. Thus $Z(P)$ has an element z of order p. Let $Z = \langle z \rangle$ be the cyclic central subgroup of order p in P generated by z. Consider the canonical group homomorphism $P \to P/Z$. The kernel of the induced algebra homomorphism is $I(kZ)kP$ by 1.6.4, which is clearly contained in $I(kP)$. This algebra homomorphism sends $I(kP)$ to $I(kP/Z)$. By induction, $I(kP/Z)^{|P/Z|}$ is zero. This means that $I(kP)^{|P/Z|}$ lies in the kernel $I(kZ)kP$ of the algebra homomorphism $kP \to kP/Z$. It suffices therefore to show that the p-th power of this kernel is zero. Since k has characteristic p, we have $(z - 1)^p = z^p - 1^p = 0$ because z has order p. By 1.6.3 we have $I(kZ) = (z - 1)kZ$, and hence $I(kZ)^p = \{0\}$. Since $I(kZ)kP = kPI(kZ)$, we get that $(I(kZ)kP)^p = I(kZ)^p kP = \{0\}$. The result follows. □

We include two consequences of this theorem which we will revisit later in a more general context of selfinjective and symmetric algebras.

Corollary 1.11.2 *Let p be a prime. Suppose that k is a field of characteristic p. Let P be a finite p-group. The one-dimensional subspace $k \cdot (\sum_{y \in P} y)$ of kP is equal to the unique simple submodule of kP as a left kP-module, the unique* *simple submodule of kP as a right kP-module and the unique minimal two-sided ideal in kP. Equivalently,* $k \cdot (\sum_{y \in P} y)$ *is equal to the socle of kP as a left or right kP-module.*

Proof Every trivial left or right submodule of kP is equal to $k \cdot (\sum_{y \in P} y)$ by 1.6.5, and since kP has no nontrivial simple modules, this is the socle of kP as left or right module. This space is an ideal, and since any nontrivial ideal is a left and right submodule, it must contain the socle, whence the result. □

Corollary 1.11.3 *Let p be a prime. Suppose that k is a field of characteristic p. Let P be a finite p-group and U a kP-module that has no direct summand isomorphic to kP. Then* $(\sum_{y \in P} y)U = \{0\}$.

Proof Suppose there is $u \in U$ such that $\sum_{y \in P} yu \neq 0$. Then the kernel of the kP-homomorphism $\sigma : kP \to U$ sending $y \in P$ to yu does not contain the unique minimal ideal $k \cdot (\sum_{y \in P} y)$ of kP, so σ is injective. Therefore there is a k-linear map $\tau : U \to kP$ such that $\tau(u) = 1$ and $\tau(yu) = 0$ for $y \in P$, $y \neq 1$. An easy verification shows that the map $\rho : U \to kP$ defined by $\rho(v) = \sum_{y \in P} y\tau(y^{-1}v)$ for all $v \in U$ is a homomorphism of kP-modules. But then $\rho(u) = 1$, hence $\rho \circ \sigma$ is an endomorphism of kP sending 1 to 1, and thus the identity endomorphism of kP. This yields the contradiction that U has a direct summand isomorphic to kP. □

The two Corollaries above are special cases of Theorem 4.11.2 and Proposition 4.11.7 below. Another important consequence of Theorem 1.11.1 is that transitive permutation modules for finite p-groups are indecomposable:

Corollary 1.11.4 *Let p be a prime. Suppose that k is a field of characteristic p. Let P be a finite p-group. The regular kP-module kP is indecomposable, and for every subgroup Q of P the permutation module kP/Q is indecomposable.*

Proof Since $kP/J(kP) \cong k$ it follows from 1.11.1 that $J(kP)$ is the unique maximal submodule of kP viewed as a left kP-module. Thus kP is indecomposable as a kP-module. Let Q be a subgroup of P. There is a canonical kP-homomorphism $\pi : kP \to kP/Q$ mapping $u \in P$ to uQ. This homomorphism is surjective, hence nonzero, and so $\ker(\pi) \subseteq J(kP)$. But then the image $\pi(J(kP))$ is the unique maximal submodule of kP/Q, and hence kP/Q is indecomposable as well. □

Corollary 1.11.5 *Let p be a prime. Suppose that k is a field of characteristic p. Let P be a finite p-group. The unit element of kP is the unique idempotent in kP. In particular, kP has a unique block.*

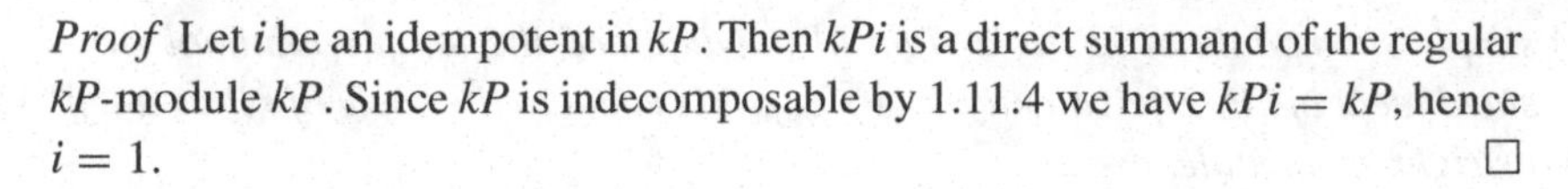

Proof Let i be an idempotent in kP. Then kPi is a direct summand of the regular kP-module kP. Since kP is indecomposable by 1.11.4 we have $kPi = kP$, hence $i = 1$. □

We mention without proof that as a consequence of a result of Jennings [43], the radical and socle series of finite p-group algebras coincide:

Theorem 1.11.6 *Suppose that k is a field of positive characteristic p. Let P be a finite p-group, and let s be the smallest positive integer such that $J(kP)^s = \{0\}$. Then, for $0 \leq i \leq s$, we have $J(kP)^i = \mathrm{soc}^{s-i}(kP)$. That is, the radical and socle series of kP as a left kP-module coincide.*

Proof See e.g. [5, 3.14.7] for a proof. □

Theorem 1.11.7 *Let G be a finite group and N a normal subgroup of G. Suppose that k is a field. We have $J(kG) \cap kN = J(kN)$ and $J(kN)kG$ is an ideal in kG contained in $J(kG)$.*

Proof By Clifford's Theorem 1.9.9 every simple kG-module S restricts to a semisimple kN-module, hence is annihilated by $J(kN)$. Thus $J(kN) \subseteq J(kG)$, and hence $J(kN) \subseteq J(kG) \cap kN$. This inclusion is an equality by 1.10.8. Since N is normal in G, conjugation by $x \in G$ induces an automorphism of kN, hence stabilises $J(kN)$, and therefore $J(kN)kG = kGJ(kN)$ is an ideal contained in $J(kG)$. □

Corollary 1.11.8 *Let G be a finite group, p a prime and let P be a normal p-subgroup in G. Suppose that k is a field of characteristic p. Then all elements in P act as the identity on every simple kG-module. We have $I(kP)kG \subseteq J(kG)$, or equivalently, the kernel of the canonical algebra homomorphism $kG \to kG/P$ is contained in $J(kG)$.*

Proof By 1.11.1 we have $J(kP) = I(kP)$, and hence 1.11.7 implies that $I(kP) \subseteq J(kG)$. In particular, elements of kG of the form $y - 1$ with $y \in P$ annihilate all simple modules, and thus y acts as identity on all simple kG-modules. The result follows, using 1.6.4 for the last statement. □

Theorem 1.11.9 *Let G be a finite group, H be a subgroup of G. Suppose that k is a field such that $|G : H|$ is invertible in k. Let M be a kG-module whose restriction to kH is semisimple as a kH-module. Then M is semisimple as a kG-module.*

Proof Let U be a kG-submodule of M. We have to show that U has a complement in M as a kG-module. Since M is semisimple as a kH-module, U has a complement V in M as a kH-module. Let $\pi : M \to U$ be the projection of M onto U with kernel V. Since V is a kH-submodule of M the map π is a kH-homomorphism, but not necessarily a kG-homomorphism. Define a map $\tau : M \to M$ by $\tau(m) = \frac{1}{|G:H|} \sum_{x \in [G/H]} x\pi(x^{-1}m)$ for all $m \in M$. Since π is a kH-homomorphism, this sum does not depend on the choice of $[G/H]$. Since $\pi(M) \subseteq U$ we also have $\tau(M) \subseteq U$. Moreover, for $u \in U$ we have $\pi(u) = u$, and hence $\pi(x^{-1}u) = x^{-1}u$ for all $x \in G$ because U is a kG-submodule of M. Thus $\tau(u) = u$. Finally, τ is a kG-homomorphism: if $y \in G$ and $m \in M$ then $y\tau(y^{-1}m) = \frac{1}{|G:H|} \sum_{x \in [G/H]} yx\pi(x^{-1}y^{-1}m) = \frac{1}{|G:H|} \sum_{x \in [G/H]} x\pi(x^{-1}m) = \tau(m)$, because if x runs over a set of representatives of the cosets G/H in G, then so does yx. It follows that τ is a projection of M onto U as a kG-module, and hence $\ker(\tau)$ is a complement of U in M. Thus M is semisimple by 1.9.5. □

The proof of 1.11.9 provides a recipe for how to construct a complement of a submodule in a semisimple module.

Theorem 1.11.10 *Let G be a finite group and let N be a normal subgroup of G. Suppose that k is a field such that either* $\mathrm{char}(k) = 0$ *or* $\mathrm{char}(k) = p$ *for a prime p that does not divide the order of G/N. Then $J(kG) = J(kN)kG = kGJ(kN)$.*

Proof By 1.11.7 we have $J(kN)kG \subseteq J(kG)$. In order to show that this inclusion is an equality it suffices to show that $kG/J(kN)kG$ is semisimple as a kG-module. Since the index $|G : N|$ is invertible in k it suffices, by 1.11.9, to show that $kG/J(kN)kG$ is semisimple as a kN-module. This, however, is clear since $J(kN)$ annihilates $kG/J(kN)kG$. □

Corollary 1.11.11 *Let G be a finite group and p be a prime. Suppose that k is a field of characteristic p and that G has a normal Sylow p-subgroup P. Then $J(kG) = I(kP)kG$.*

Proof By 1.11.10 we have $J(kG) = J(kP)kG$, and by 1.11.1 we have $J(kP) = I(kP)$, whence the result. □

Theorem 1.11.12 (Maschke's Theorem) *Let G be a finite group. Suppose that k is a field. Every kG-module is semisimple if and only if either* $\mathrm{char}(k) = 0$ *or* $\mathrm{char}(k) = p$ *does not divide the order of G. More precisely*

(i) If either $\mathrm{char}(k) = 0$ *or* $\mathrm{char}(k) = p$ *does not divide $|G|$ then $J(kG) = \{0\}$.*
(ii) If $\mathrm{char}(k) = p$ *divides the group order $|G|$ then $\sum_{x \in G} x \in J(kG)$; in particular, $J(kG) \neq \{0\}$.*

Proof By 1.11.10 applied to $N = \{1\}$, under the assumptions in (i), we have $J(kG) = \{0\}$. By 1.10.19, this is equivalent to stating that every kG-module is semisimple. Assume now that $\mathrm{char}(k) = p$ divides $|G|$; that is, the image of $|G|$ in k is zero. Set $z = \sum_{x \in G} x$; we clearly have $xz = z$ for any $x \in G$, and hence $z^2 = |G|z = 0$. Thus z is a nilpotent element in $Z(kG)$ and hence $zkG = kGz$ is a nilpotent ideal in kG, thus contained in $J(kG)$ by 1.10.8. □

Exercise 1.11.13 Show that the first statement of Maschke's Theorem extends to twisted group algebras: if G is a finite group, k a field of characteristic zero or of prime characteristic not dividing $|G|$, and if $\alpha \in Z^2(G; k^\times)$, then $k_\alpha G$ is semisimple. (*Hint:* show first that one may assume that k is algebraically closed, and then use 1.2.18 and Maschke's Theorem.)

1.12 Projective and injective modules

The direct sum of any two free A-modules is free: if F is a free A-module with basis S and F' is a free A-module with basis S', then $F \oplus F'$ is a free A-module with basis the disjoint union of S and S'. A direct summand of a free module, however, need not be free. This motivates the following definition.

Definition 1.12.1 Let A be a k-algebra. An A-module P is called *projective* if P is a direct summand of a free A-module; that is, if there is an A-module P' such that $P \oplus P'$ is free.

Examples 1.12.2

(a) Let A be a k-algebra and let i be an idempotent in A. Then the left A-module Ai is projective. Indeed, we have $A = Ai \oplus A(1 - i)$ as left A-modules, so Ai is a summand of the free A-module A of rank 1. Moreover, Ai is projective indecomposable if and only if the idempotent i is primitive (cf. 1.7.4). If k is a field, if A has finite dimension, and if $i \neq 1$, then Ai is projective but not free, since its dimension is strictly smaller than that of the free module A of rank 1.

(b) Let G be a finite group and let H be a subgroup of G. If the order $|H|$ of H is invertible in k then the permutation kG-module kG/H is projective. Indeed, by 1.1.2, $e_H = \frac{1}{|H|} \sum_{y \in H} y$ is an idempotent in kG, and one verifies that the map sending a coset xH to xe_H, where $x \in G$, induces an isomorphism of kG-modules $kG/H \cong kGe_H$. Thus kG/H is projective by the previous example.

(c) Let n be a positive integer. The set V of column vectors $(\lambda_1, \lambda_2, \ldots, \lambda_n)^T$ with coefficients $\lambda_i \in k$ is a left $M_n(k)$-module via multiplication of $n \times n$-matrices with column vectors. This is a projective $M_n(k)$-module. To see this, for $1 \leq i \leq n$, denote by V_i the subspace of $M_n(k)$ consisting of all matrices that are zero outside of the column i. Then

$$M_n(k) = V_1 \oplus V_2 \oplus \cdots \oplus V_n$$

is a decomposition of $M_n(k)$ as a left $M_n(k)$-module. Each summand V_i is isomorphic to V as an $M_n(k)$-module. In particular, V is isomorphic to a direct summand of $M_n(k)$ as a left $M_n(k)$-module and hence projective. An easy verification shows that if k is a field, then V is also simple. The analogous construction using row vectors and the rows of $M_n(k)$ yields a decomposition of $M_n(k)$ as a direct sum of projective (simple) right modules.

The following category theoretic characterisation of projective modules avoids any reference to free modules:

Theorem 1.12.3 *Let A be a k-algebra, and let P be an A-module. The following are equivalent:*

(i) The A-module P is projective.

(ii) Any surjective A-homomorphism $\pi : U \to P$ from some A-module U to P splits; that is, there is an A-homomorphism $\sigma : P \to U$ such that $\pi \circ \sigma = \mathrm{Id}_P$.

(iii) For any surjective A-homomorphism $\pi : U \to V$ and any A-homomorphism $\psi : P \to V$ there exists an A-homomorphism $\varphi : P \to U$ such that $\pi \circ \varphi = \psi$.

Proof Suppose that P is projective. Let $\pi : U \to V$ be a surjective A-homomorphism, and let $\psi : P \to V$ be an A-homomorphism. Let P' be an A-module such that $P \oplus P'$ is free and let S be a basis of $P \oplus P'$. Extend ψ to an A-homomorphism $P \oplus P' \to V$ in any which way (for instance, by sending P' to zero), still denoted by ψ. For $s \in S$ choose any element $u_s \in U$ such that $\pi(u_s) = \psi(s)$. Since $P \oplus P'$ is free with basis S there is a unique A-homomorphism $\varphi : P \oplus P' \to U$ such that $\varphi(s) = u_s$ for all $s \in S$. Thus $\pi \circ \varphi = \psi$ on $P \oplus P'$. Restricting φ to P yields the required lift of ψ on P. This shows that (i) implies (iii). Suppose that (iii) holds. Let $\pi : U \to P$ be a surjective A-homomorphism. Applying (iii) to π and to Id_P instead of ψ yields an A-homomorphism $\sigma : P \to U$ satisfying $\pi \circ \sigma = \mathrm{Id}_P$. Thus (iii) implies (ii). Suppose finally that (ii) holds. Let S be any generating set of P

as an A-module. Let F be the free A-module with basis S. Let $\pi : F \to P$ be the unique A-homomorphism sending s (viewed as a basis element of F) to s (viewed as an element in P). Since S generates P, it follows that the map π is surjective. Applying (ii) yields a map $\sigma : P \to F$ satisfying $\pi \circ \sigma = \mathrm{Id}_P$, and hence P is isomorphic to a direct summand of the free module F. Thus (ii) implies (i). □

The third characterisation in Theorem 1.12.3 has an interpretation in terms of functors. A functor from Mod(A) to Mod(k) is *exact* if it sends any exact sequence of A-modules to an exact sequence of k-modules. Given any A-homomorphism $\pi : U \to V$, composition with π induces a map $\mathrm{Hom}_A(P, U) \to \mathrm{Hom}_A(P, V)$ sending $\varphi \in \mathrm{Hom}_A(P, U)$ to $\pi \circ \varphi$. In this way, $\mathrm{Hom}_A(P, -)$ becomes a covariant functor from Mod(A) to Mod(k). Statement (iii) in Theorem 1.12.3 says that a projective module P is characterised by the property that if π is surjective, then so is the induced map $\mathrm{Hom}_A(P, U) \to \mathrm{Hom}_A(P, V)$. Using this, one easily checks the following statement:

Theorem 1.12.4 *Let A be a k-algebra, and let P be an A-module. Then P is projective if and only if the functor* $\mathrm{Hom}_A(P, -) : \mathrm{Mod}(A) \to \mathrm{Mod}(k)$ *is exact.*

A module M over an algebra A can always be written as a quotient of a free A-module, hence as a quotient of a projective A-module. Given a surjective A-homomorphism $\pi : P \to M$ from a projective A-module P onto M, the next result shows that $\ker(\pi)$ is determined by M 'up to projectives', regardless of the choice of (P, π). This will become important in the context of stable categories.

Theorem 1.12.5 (Schanuel's Lemma) *Let A be a k-algebra, let M be an A-module and let*

$$0 \longrightarrow U \longrightarrow P \xrightarrow{\pi} M \longrightarrow 0$$

$$0 \longrightarrow V \longrightarrow Q \xrightarrow{\tau} M \longrightarrow 0$$

be two short exact sequences of A-modules terminating in M such that P and Q are projective A-modules. Then $U \oplus Q \cong V \oplus P$.

Proof We consider the pullback T of π and τ; that is,

$$T = \{(p, q) \in P \oplus Q \mid \pi(p) = \tau(q)\}.$$

The projection $\sigma : T \to P$ sending $(p, q) \in T$ to p is surjective; indeed, since τ is surjective, for any $p \in P$ there is $q \in Q$ such that $\tau(q) = \pi(p)$, or equivalently, such that $(p, q) \in T$. Since P is projective, the map $T \to P$ splits, and hence $T \cong P \oplus \ker(\sigma)$. Now $\ker(\sigma) = \{(0, q) \in P \oplus Q | \tau(q) = 0\}$ and hence the map sending $(0, q)$ to q induces an isomorphism $\ker(\sigma) \cong \ker(\tau) = V$; together we get $T \cong V \oplus P$. The same argument applied to the second projection $T \to Q$ yields $T \cong U \oplus Q$, whence the result. □

Lemma 1.12.6 *Let A be a k-algebra, and let U, V be A-modules. There is a k-linear map $\mathrm{Hom}_A(U, A) \otimes_A V \to \mathrm{Hom}_A(U, V)$ sending $\varphi \otimes v$ to the map $u \mapsto \varphi(u)v$, where $u \in Y$, $v \in V$, and $\varphi \in \mathrm{Hom}_A(U, A)$. This map is natural in U and V.*

Proof The naturality of the map in the statement means that the family of maps $\mathrm{Hom}_A(U, A) \otimes_A V \to \mathrm{Hom}_A(U, V)$ as given is a natural transformation of bifunctors in the variables U and V. This is a trivial verification. □

We call the map in 1.12.6 the *canonical evaluation map*. Finitely generated projective modules admit further characterisations in terms of canonical evaluation maps.

Theorem 1.12.7 *Let A be a k-algebra and P an A-module. The following statements are equivalent.*

(i) The A-module P is finitely generated projective.
(ii) For any A-module U the canonical evaluation map $\mathrm{Hom}_A(P, A) \otimes_A U \to \mathrm{Hom}_A(P, U)$ is an isomorphism.
(iii) For any A-module U, the canonical evaluation map $\mathrm{Hom}_A(U, A) \otimes_A P \to \mathrm{Hom}_A(U, P)$ is an isomorphism.
(iv) The canonical evaluation map $\mathrm{Hom}_A(P, A) \otimes_A P \to \mathrm{Hom}_A(P, P)$ is an isomorphism.
(v) The canonical evaluation map $\mathrm{Hom}_A(P, A) \otimes_A P \to \mathrm{Hom}_A(P, P)$ is surjective.
(vi) There is a finite subset S in P and, for every $s \in S$, an A-homomorphism $\psi_s : P \to A$ such that $\sum_{s \in S} \psi_s(p)s = p$ for all $p \in P$.

Proof For any A-module U the k-linear map $\mathrm{Hom}_A(P, A) \otimes_A U \to \mathrm{Hom}_A(P, U)$ sending $\varphi \otimes u$ to $(q \mapsto \varphi(q)u)$, where $q \in P$, $u \in U$ and $\varphi \in \mathrm{Hom}_A(P, A)$, is in fact natural in P and U; that is, the family of maps obtained from fixing one of P or U and letting the other vary over all modules is a k-linear natural transformation of functors. For $P = A$ or $U = A$ this map is clearly an isomorphism. And because these isomorphisms are natural they commute with taking finite direct sums, so this map is an isomorphism if one P or U is free of finite rank. Again because of the naturality, this map also commutes with idempotent endomorphisms (that is, projections onto direct summands of P or U), and hence this map is an isomorphism if one of P or U is finitely generated projective. This shows that (i) implies (ii) and (iii). Taking $U = P$ shows that any of (ii) or (iii) implies (iv). Clearly (iv) implies (v). If (v) holds, then there is a finite subset S of P and, for any $s \in S$, an A-homomorphism $\psi_s \in \mathrm{Hom}_A(P, A)$ such that $\sum_{s\in S} \psi_s \otimes s$ maps to Id_P under the map in (v). But this means just that $\sum_{s\in S} \psi_s(p)s = p$ for any $p \in P$, hence (v) implies (vi). Suppose that (vi) holds. Consider a free A-module with a basis $\{e_s\}_{s\in S}$ indexed by the elements of S. The A-homomorphism $\oplus_{s\in S} Ae_s \to P$ sending ae_s to as for any $a \in A$, $s \in S$, has as a section the map sending $p \in P$ to $\sum_{s\in S} \psi_s(p)e_s$, which implies (i). □

The first part of the proof of Theorem 1.12.7 is the blueprint for many proofs in which the existence of an isomorphism is being shown 'by naturality'.

Proposition 1.12.8 *Let A be a k-algebra and let P be a projective right A-module. Then P is flat; that is, the functor $P \otimes_A -$ from* $\mathrm{Mod}(A)$ *to* $\mathrm{Mod}(k)$ *is exact.*

Proof If $P = A$ as right A-module then $P \otimes_A -$ is isomorphic to the forgetful functor sending an A-module U to U viewed as k-module, so in particular this functor is exact. Since the tensor product commutes with direct summands, this remains true for any free right A-module P, hence for any projective A-module. □

There are examples of flat modules that are not projective. For instance, $\mathbb{Q}$ is a flat $\mathbb{Z}$-module, but not a projective $\mathbb{Z}$-module. This is part of a more general statement on localisation being an exact functor. We mention a basic fact on the restriction to subgroups, which will be part of more general considerations in §2.1.

Proposition 1.12.9 *Let G be a finite group, and H a subgroup. As a left or right kH-module, kG is free. If U is a projective left or right kG-module, then its restriction* $\mathrm{Res}^G_H(U)$ *is a projective left or right kH-module.*

Proof It suffices to prove that the restriction of a free kG-module to kH is a free kH-module. For that it suffices to show that kG is free as a left kH-module. If $x \in G$, then right multiplication by x^{-1} yields an isomorphism $kHx \cong kH$ as a left kH-module. Thus if $\mathcal{R}$ is a set of representatives of the cosets Hx in G, then $\mathcal{R}$ is a basis of kG as a kH-module, hence free. A similar argument, using the cosets xH, proves this for right modules. □

Proposition 1.12.10 *Let A, B be k-algebras and let U, V be B-modules. Suppose that U is finitely generated projective as a B-module, or that A is flat as a k-module and B, U are finitely generated projective as k-modules. We have a* *natural k-linear isomorphism*

$$A^{\mathrm{op}} \otimes_k \mathrm{Hom}_B(U,V) \cong \mathrm{Hom}_{A\otimes_k B}(A \otimes_k U, A \otimes_k V)$$

sending $a \otimes \tau$ *to the map* $(c \otimes u) \mapsto (ca \otimes \tau(u))$, *where* $a,\ c \in A$, $\tau \in \mathrm{Hom}_B(U,V)$, *and* $u \in U$. *For* $U = V$ *this is an isomorphism of k-algebras* $A^{\mathrm{op}} \otimes_k \mathrm{End}_B(U) \cong \mathrm{End}_{A\otimes_k B}(A \otimes_k U)$.

Proof With the notation of the statement, the map sending $a \otimes \tau$ to the map $(c \otimes u) \mapsto (ca \otimes \tau(u))$ is a k-linear map

$$A^{\mathrm{op}} \otimes_k \mathrm{Hom}_B(U,V) \to \mathrm{Hom}_{A\otimes_k B}(A \otimes_k U, A \otimes_k V).$$

One verifies that this map is natural in U and V and that if $U = V$ this map is an algebra homomorphism. If $U = B$ this map is an isomorphism because both sides are canonically isomorphic to $A^{\mathrm{op}} \otimes_k V$ as k-modules. The naturality of this map implies that it is an isomorphism whenever U is a finitely generated projective B-module. Assume now that A is flat as a k-module and that B, U are finitely generated projective as k-modules. Then U is finitely generated as a B-module, and hence U is a quotient of a finitely generated projective module P_0. Since B is finitely generated projective as a k-module it follows that P_0 is finitely generated projective as a k-module. Any surjective B-homomorphism $P_0 \to U$ splits as a k-homomorphism since U is projective as a k-module. Thus the kernel of a surjective B-homomorphism $P_0 \to U$ is again finitely generated projective as a k-module, and hence, as a B-module, this kernel is a quotient of a finitely generated projective module P_1. Applying the two functors, for fixed V and variable U, occurring in the statement, as well as the natural transformation between them, to the exact sequence of B-modules of the form $P_1 \to P_0 \to U \to 0$ yields a commutative diagram of the form

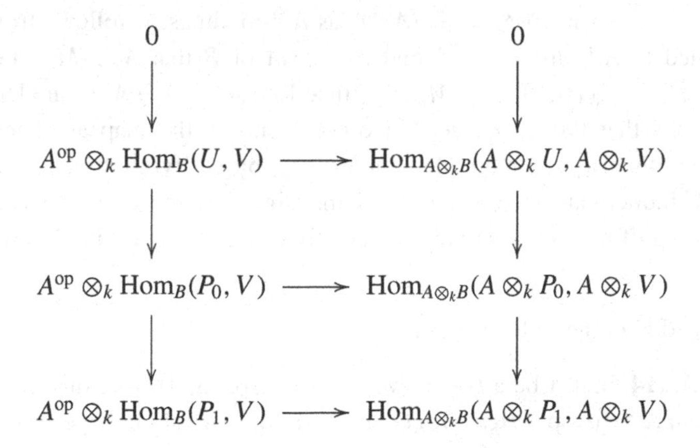

in which the columns are exact because of the exactness properties of homomorphism functors (cf. A.1.13) and the flatness assumption on A as a k-module. Moreover, the lower two horizontal maps are isomorphisms by the first part of the proof, because P_0, P_1 are finitely generated projective. An easy argument shows that the upper horizontal map is an isomorphism as well. □

Corollary 1.12.11 *Let A be a k-algebra and let U, V be A-modules. Let k' be a commutative flat k-algebra. Suppose that A and U are finitely generated projective as k-modules. We have a natural k'-linear isomorphism*

$$k' \otimes_k \mathrm{Hom}_A(U, V) \cong \mathrm{Hom}_{k' \otimes_k A}(k' \otimes_k U, k' \otimes_k V)$$

which for $U = V$ is an isomorphism of k'-algebras.

Proof The canonical map in the statement is obviously k'-linear. The rest follows from 1.12.10 applied to k, k', A instead of k, A, B, respectively. □

Corollary 1.12.12 *Let k' be a commutative k-algebra and U a finitely generated projective k-module. We have a natural k'-linear isomorphism $k' \otimes_k \mathrm{Hom}_k(U, k) \cong \mathrm{Hom}_{k'}(k' \otimes_k U, k')$.*

Proof This is the special case $A = k'$ and $B = k = V$ in 1.12.10. □

A frequently needed special case of this Corollary arises if k'/k is a field extension, A is a finite-dimensional k-algebra, and U a finitely generated A-module.

Corollary 1.12.13 *Let A be a k-algebra and n a positive integer. Let $a \in A$ and $M \in M_n(k)$. The map sending $a \otimes M$ to the matrix aM in $M_n(A)$ obtained from multiplying all entries of M by a induces a k-algebra isomorphism $A \otimes_k M_n(k) \cong M_n(A)$.*

Proof Let $U = k^n$. Then $A^{\mathrm{op}} \otimes_k U \cong (A^{\mathrm{op}})^n$ as A^{op}-modules. It follows from 1.12.10 applied to A^{op} instead of A and k instead of B that $A \otimes_k M_n(k) \cong A \otimes_k \mathrm{End}_k(U) \cong \mathrm{End}_{A^{\mathrm{op}}}((A^{\mathrm{op}})^n) \cong M_n(A)$, since $\mathrm{End}_{A^{\mathrm{op}}}(A^{\mathrm{op}}) \cong A$ as an algebra. One checks that this sequence of isomorphisms is the map as stated. One can show this also in a more elementary way, by verifying directly that $a \otimes M \mapsto aM$ induces an algebra isomorphism, using the canonical k-basis of $M_n(k)$ consisting of matrices that are 1 in exactly one entry and 0 in all other entries. □

We will need later the following notion.

Definition 1.12.14 Let A be a k-algebra. A left (resp. right) A-module M is called a *progenerator* (resp. a *right progenerator*) *of* A, if M is finitely generated projective and if A is isomorphic as a left (resp. right) A-module to a direct summand of M^n for some positive integer n.

Equivalently, M is a progenerator if the category $\mathrm{proj}(A)$ of finitely generated projective A-modules is equal to the smallest full subcategory, denoted $\mathrm{add}(M)$, of $\mathrm{Mod}(A)$ that contains M and that is closed under isomorphisms, direct summands and finite direct sums. Indeed, suppose that M is a progenerator. Then M belongs to $\mathrm{proj}(A)$, hence any summand of a finite direct sum of copies of M belongs to $\mathrm{proj}(A)$, which shows that $\mathrm{add}(M)$ is a subcategory of $\mathrm{proj}(A)$. For the reverse inclusion we use the fact that A is a direct summand of M^n for some positive integer n. Let U be in $\mathrm{proj}(A)$; that is, U is a summand of A^m for some positive integer m, and A^m is a direct summand of M^{mn}. Thus U is a direct summand of M^{nm}, hence belongs to $\mathrm{add}(M)$. This shows the equality $\mathrm{proj}(A) = \mathrm{add}(M)$ under the assumption that M is a progenerator. If M is not a progenerator, then at least one of the two defining properties of M fails: if M is not finitely generated projective, then $\mathrm{add}(M)$ is not contained in $\mathrm{proj}(A)$; if for any positive integer n the module A is not a summand of M^n, then A is not in $\mathrm{add}(M)$, and hence $\mathrm{proj}(A)$ is not contained in $\mathrm{add}(M)$. If M is a progenerator for A, then $\mathrm{Hom}_A(M, A)$ is a right progenerator for A. This follows from the fact that the A-duality functor $\mathrm{Hom}_A(-, A)$, considered in more detail in 2.9.12 below, sends finitely generated projective left A-modules to finitely generated projective right A-modules, and in particular it sends the regular left A-module A to the regular right A-module A.

Reversing the arrows in the characterisations of projective modules in abstract category theoretic terms gives rise to the dual concept of injective modules:

Definition 1.12.15 Let A be a k-algebra. An A-module I is called *injective* if for every injective A-homomorphism $\iota : U \to V$ and any A-homomorphism $\psi : U \to I$ there exists an A-homomorphism $\varphi : V \to I$ such that $\varphi \circ \iota = \psi$.

Dualising the corresponding proofs for projective modules yields immediately the following statements for injective modules:

Theorem 1.12.16 *Let A be a k-algebra and let I be an A-module. The following are equivalent:*

(i) The A-module I is injective.

(ii) Any injective A-homomorphism $\iota : I \to V$ from I to some A-module V splits; that is, there is an A-homomorphism $\kappa : V \to I$ such that $\kappa \circ \iota = \mathrm{Id}_I$.

(iii) The contravariant functor $\mathrm{Hom}_A(-, I) : \mathrm{Mod}(A) \to \mathrm{Mod}(k)$ *is exact.*

Theorem 1.12.17 *Let A be a k-algebra, let M be an A-module and let*

$$0 \longrightarrow M \longrightarrow I \xrightarrow{\pi} U \longrightarrow 0$$

$$0 \longrightarrow M \longrightarrow J \xrightarrow{\tau} V \longrightarrow 0$$

be two short exact sequences of A-modules starting in M such that I and J are injective A-modules. Then $U \oplus J \cong V \oplus I$.

Projective and injective modules are compatible with block decompositions of algebras in an obvious way:

Proposition 1.12.18 *Let A be a k-algebra and let b be an idempotent in $Z(A)$.*

(i) Let P be a projective (resp. injective) A-module. Then bP is a projective (resp. injective) Ab-module.
(ii) Let Q be a projective (resp. injective) Ab-module. Then Q remains projective (resp. injective) as an A-module.

Proof Obvious. □

Proposition 1.12.19 *Let A be a finite-dimensional algebra over a field. The following are equivalent.*

(i) We have $J(A) = \{0\}$.
(ii) Every A-module is semisimple.
(iii) Every A-module is projective.
(iv) Every A-module is injective.
(v) Every simple A-module is projective.
(vi) Every simple A-module is injective.

Proof The equivalence of (i) and (ii) has been observed in 1.10.19. Let V be an A-module. Suppose (ii) holds. Let $\pi : U \to V$ be a surjective A-homomorphism. Since U is semisimple, $\ker(\pi)$ has a complement U' in U, which shows that π is split surjective. Thus V is projective. This shows that (ii) implies (iii), and a similar argument applied to the image of an injective homomorphism shows that (ii) implies (iv). The implications (iii) $\Rightarrow$ (v) and (iv) $\Rightarrow$ (vi) are trivial. Suppose that (v) holds. Let U be an A-module, and M a maximal submodule. Then U/M is simple, hence projective, and thus the canonical map $U \to U/M$ splits. It follows that $U \cong M \oplus U/M$. Arguing by induction over $\dim_k(U)$ we get that any finite-dimensional A-module is semisimple. Thus A is semisimple, and hence any A-module is semisimple by 1.10.19 Thus (v) implies (ii). A similar argument, using simple submodules of U, shows that (vi) implies (ii). □

It follows from 1.9.6 that all statements in this theorem are equivalent to the corresponding statements for finitely generated A-modules instead of all A-modules.

The notions of surjective, injective homomorphisms and projective, injective modules can be generalised to arbitrary categories. Given a morphism $f : X \to Y$ in a category $\mathcal{C}$ we say that f is an *epimorphism* if for any two morphisms $g, g' : Y \to Z$ such that $g \circ f = g' \circ f$ we have $g = g'$. One verifies that epimorphism in module categories are surjective homomorphisms. Dually, f is called a *monomorphism* if for any two morphisms $g, g' : Z \to X$ such that $f \circ g = f \circ g'$ we have $g = g'$. Again, one verifies that monomorphisms in module categories are injective homomorphisms. An object P in $\mathcal{C}$ is called *projective* if for any epimorphism $f : X \to Y$ and any morphism $g : P \to Y$ there is a morphism $h : P \to X$ such that $f \circ h = g$. Dually, an object I in $\mathcal{C}$ is called *injective* if for any monomorphism $f : X \to Y$ and any morphism $g : X \to I$ there is a morphism $h : Y \to I$ such that $h \circ f = g$. Split epimorphisms or monomorphisms are special cases of split morphisms in a category:

Definition 1.12.20 A morphism $f : X \to Y$ in a category $\mathcal{C}$ is *split* if there is a morphism $s : Y \to X$ in $\mathcal{C}$ such that $f = f \circ s \circ f$.

Thus if $f : X \to Y$ is split and an epimorphism, then the equation $f = f \circ s \circ f$ implies that $\mathrm{Id}_Y = f \circ s$; equivalently, a split epimorphism is a morphism that has a right inverse. Similarly, if f is split and a monomorphism, then the equation $f = f \circ s \circ f$ implies that $\mathrm{Id}_X = s \circ f$; equivalently, a split monomorphism is a morphism that has a left inverse. The morphism s need not be split in this definition, but it can always be chosen to be split:

Proposition 1.12.21 *Let $\mathcal{C}$ be a category and $f : X \to Y$ a split morphism in $\mathcal{C}$. Then there is a split morphism $s : Y \to X$ such that $f = f \circ s \circ f$ and $s = s \circ f \circ s$.*

Proof Since f is split there is a morphism $t : Y \to X$ in $\mathcal{C}$ such that $f = f \circ t \circ f$. Set $s = t \circ f \circ t$. Then $f \circ s \circ f = f \circ (t \circ f \circ t) \circ f = (f \circ t \circ f) \circ t \circ f = f \circ t \circ f = f$, and $s \circ f \circ s = (t \circ f \circ t) \circ f \circ (t \circ f \circ t) = t \circ (f \circ t \circ f) \circ t \circ f \circ t = t \circ f \circ t \circ f \circ t = t \circ f \circ t = s$. □

Proposition 1.12.22 *Let A be a k-algebra, let $f : X \to Y$ be a nonzero split homomorphism of A-modules, and let $s : Y \to X$ be an A-homomorphism such that $f = f \circ s \circ f$ and $s = s \circ f \circ s$.*

(i) We have a canonical isomorphism $X \cong \ker(f) \oplus \mathrm{Im}(s \circ f)$.
(ii) We have a canonical isomorphism $Y \cong \mathrm{Im}(f) \oplus \ker(f \circ s)$.
(iii) If X *is indecomposable then* f *is a split monomorphism.*
(iv) If Y *is indecomposable then* f *is s split epimorphism.*

Proof Since $f = f \circ s \circ f$ we have $\ker(f) = \ker(f \circ s \circ f) \supseteq \ker(s \circ f) \supseteq \ker(f)$, hence all these inclusions are equalities. Similarly, $\mathrm{Im}(f) = \mathrm{Im}(f \circ s \circ f) \subseteq \mathrm{Im}(f \circ s) \subseteq \mathrm{Im}(f)$, hence again all these inclusions are equalities. Composing the equality $f = f \circ s \circ f$ on the left and on the right by s implies that $s \circ f$ and $f \circ s$ are idempotents in $\mathrm{End}_A(X)$ and $\mathrm{End}_A(Y)$, respectively. Thus $X = \ker(s \circ f) \oplus \mathrm{Im}(s \circ f) = \ker(f) \oplus \mathrm{Im}(s \circ f)$ and $Y = \mathrm{Im}(f \circ s) \oplus \ker(f \circ s) = \mathrm{Im}(f) \oplus \ker(f \circ s)$. This shows (i) and (ii); the statements (iii) and (iv) are obvious consequences of (i) and (ii), respectively. □

This proposition shows that split morphisms correspond to decompositions of modules; this holds more generally for any additive category in which idempotents of endomorphism rings correspond to direct summands.

Exercise 1.12.23 Let A be a k-algebra, U an A-module and V a submodule of U. Show the following.

(i) If U has no nonzero projective direct summand, then U/V has no nonzero projective direct summand.
(ii) If U has no nonzero injective direct summand, then V has no nonzero injective direct summand.

Exercise 1.12.24 Let A be a k-algebra, and let $f : U \to V$ and $f' : U' \to V'$ be homomorphisms of A-modules. Show that f and f' are split if and only if their direct sum $f \oplus f' : U \oplus U' \to V \oplus V'$ is split.

Exercise 1.12.25 Let A be a k-algebra. Let U, V be semisimple A-modules and $\alpha : U \to V$ an A-homomorphism. Show that α is split. Show that A is semisimple if and only if every morphism in $\mathrm{Mod}(A)$ is split.

Exercise 1.12.26 (Stolorz) Let $\mathcal{C}$ be a finite category and k a field. Show that if $k\mathcal{C}$ is semisimple, then every morphism in $\mathcal{C}$ is split.

1.13 Wedderburn's Theorem

If A is an algebra over a commutative ring k and $\lambda \in k$ such that the image $\lambda \cdot 1_A$ of λ in A is not invertible in A, then λA is a proper ideal in A. Thus if A is a simple algebra, then the image $k \cdot 1_A$ of k in $Z(A)$ is a field. We assume in this section that k is a field.

Definition 1.13.1 A k-algebra A is called *simple* if A is nonzero and has no ideals other than $\{0\}$ and A. A k-algebra A is called *semisimple* if A is semisimple as a left A-module, or equivalently, if the regular left A-module is a direct sum of simple A-modules.

The ideals in a k-algebra A are exactly the submodules of A when viewed as $A \otimes_k A^{\mathrm{op}}$-module via $(a \otimes b) \cdot c = acb$, where $a, b, c \in A$. Thus A is simple as a k-algebra if and only if A is simple as an $A \otimes_k A^{\mathrm{op}}$-module. With this terminology, Maschke's Theorem 1.11.12 states that the group algebra kG of a finite group G over a field k is semisimple if and only if $\mathrm{char}(k) = 0$ or $\mathrm{char}(k)$ does not divide the group order $|G|$.

Theorem 1.13.2 *A k-algebra A is simple Artinian if and only if $A \cong M_n(D)$ for some positive integer n and some division algebra D over k. In that case, A has, up to isomorphism, a unique simple module S, and moreover the following hold:*

(i) We have an isomorphism of left A-modules $A \cong S^n$; in particular, A is semisimple.
(ii) We have $D \cong \mathrm{End}_A(S)^{\mathrm{op}}$.
(iii) We have $A \cong \mathrm{End}_{D^{\mathrm{op}}}(S)$.

Proof Let n be a positive integer and let D be a division algebra over k. Let I be a nonzero ideal in $M_n(D)$. For any two integers i, j between 1 and n denote by E_{ij} the matrix with a 1 in position (i, j) and zero everywhere else. Let $C = (c_{ij})$ be a nonzero matrix in I. Fix i, j such that $c_{ij} \neq 0$. Then $E_{ki}CE_{jl} = c_{ij}E_{kl}$, where $1 \leq k, l \leq n$. Since c_{ij} is nonzero, it is invertible. Thus $E_{kl} \in I$, and since the E_{kl} span $M_n(D)$ it follows that $I = M_n(D)$. This shows that $M_n(D)$ is simple. As a D-module, $M_n(D)$ is free of rank n^2. Now D is Artinian because it is a division ring, hence so is $M_n(D)$ as a D-module. Since every $M_n(D)$-submodule of $M_n(D)$ is also a D-submodule it follows that $M_n(D)$ is Artinian. Conversely, suppose that A is simple and Artinian. Then A has a simple submodule S as a left A-module (otherwise we could easily construct an infinite descending chain of submodules). Let $a \in A$. As S is a left submodule of A, also Sa is a left submodule of A. The map $S \to Sa$ sending $s \in S$ to $sa \in Sa$ is a surjective A-homomorphism. Thus either $Sa = \{0\}$ or $Sa \cong S$. Set $I = \sum_{a \in A} Sa$. This is a two-sided ideal in A, hence equal to A. But I is also a sum of simple submodules, all isomorphic to S. It follows that A is a direct sum of copies of S as a left A-module. Since A is Artinian, A is in fact a finite direct sum of copies of S, say $A \cong S^n$ for some integer n. Since every simple A-module is a quotient of A this implies that every simple A-module is isomorphic to S. Set $D = \mathrm{End}_A(S)^{\mathrm{op}}$. We have $A \cong \mathrm{End}_A(A)^{\mathrm{op}} \cong \mathrm{End}_A(S^n)^{\mathrm{op}} \cong M_n(\mathrm{End}_A(S))^{\mathrm{op}} \cong M_n(\mathrm{End}_A(S)^{\mathrm{op}}) = M_n(D)$, where the penultimate isomorphism sends a matrix to its transpose. It remains to show (iii). By 1.12.13, we may identify $A = D \otimes_k M_n(k)$ and $S = D \otimes_k k^n$. In particular, S is free of rank n as a left D-module. Since A is simple, it follows that the structural map $A \to \mathrm{End}_{D^{\mathrm{op}}}(S)$ is injective. Both sides are free of rank n^2 as left D-modules, so this is an isomorphism. □

We will see in Section 2.8 that the statements (ii) and (iii) in Theorem 1.13.2 are a special case of a *Morita equivalence*; more precisely, the A-D-bimodule S induces a Morita equivalence between A and D. See 2.8.8 for details. Note that if A is simple Artinian, then the integer n and the division algebra D in the previous theorem are uniquely determined by the structure of A. Wedderburn's Theorem describes semisimple algebras as direct products of matrix algebras over division algebras:

Theorem 1.13.3 (Wedderburn) *Let A be a semisimple k-algebra. Then A has finitely many isomorphism classes of simple modules. Let $\{S_i | 1 \leq i \leq m\}$ be a set of representatives of the isomorphism classes of simple A-modules. Set $D_i = \mathrm{End}_A(S_i)$ and let n_i be the unique positive integer such that $A \cong \oplus_{1 \leq i \leq m}(S_i)^{n_i}$ as left A-modules. Then we have an isomorphism of k-algebras*

$$A \cong \prod_{1 \leq i \leq m} M_{n_i}(D_i^{\mathrm{op}}).$$

In particular, A is a finite direct product of simple k-algebras, and the simple algebra factors are parametrised by the isomorphism classes of simple A-modules.

Proof Since A is semisimple, we have $A = \oplus_{j \in J} T_j$, where J is a set and T_j is a simple left ideal for $j \in J$. Write $1_A = \sum_{j \in I} t_j$, where I is a finite subset of J and $t_j \in T_j$ for $j \in I$. Thus $A = A \cdot 1_A = \oplus_{j \in I} T_j$; that is, A is a finite direct sum of simple modules. Every simple A-module is a quotient of A, and hence A has finitely many isomorphism classes of simple modules. Thus $A \cong \oplus_{1 \leq i \leq m}(S_i)^{n_i}$ for some positive integers n_i. We have

$$\mathrm{End}_A(A) = \mathrm{End}_A(\oplus_{1 \leq i \leq m}(S_i)^{n_i}) \cong \prod_{1 \leq i \leq m} \mathrm{End}_A((S_i)^{n_i}) \cong \prod_{1 \leq i \leq m} M_{n_i}(D_i)$$

where we used Schur's Lemma 1.9.2 in the last two isomorphisms. Taking the transpose of matrices shows that the opposite of the matrix algebra $M_{n_i}(D_i)$ is isomorphic to $M_{n_i}(D_i^{\mathrm{op}})$, whence the result. □

Had we considered right A-modules, we could have avoided the step of taking opposite algebras. By formulating Wedderburn's Theorem using right modules, we find that the isomorphism classes of simple left and right A-modules correspond bijectively to each other, since they correspond to the matrix factors $M_{n_i}(D_i^{\mathrm{op}})$ in such a way that their endomorphism algebras are opposite division algebras. In particular, if all D_i are commutative in Theorem 1.13.3 then $A \cong A^{\mathrm{op}}$.

Corollary 1.13.4 *Let A be a k-algebra such that $A/J(A)$ is Artinian. The numbers of isomorphism classes of simple left A-modules and of simple right A-modules coincide. More precisely, there is a bijection between the sets of isomorphism classes of simple left and right A-modules such that if the class of the simple left A-module S corresponds to the class of the simple right A-module T, we have an algebra isomorphism* $\mathrm{End}_A(S) \cong (\mathrm{End}_{A^{\mathrm{op}}}(T))^{\mathrm{op}}$.

Proof By 1.10.12, we have $J(A/J(A)) = \{0\}$, and hence by 1.10.19, $A/J(A)$ is semisimple. The result follows from 1.13.3. □

Another application of Wedderburn's Theorem is the following theorem which makes previous statements on the radical more precise in the case where $A/J(A)$ is Artinian.

Theorem 1.13.5 *Let A be a k-algebra. Suppose that $A/J(A)$ is Artinian. The map sending a simple A-module S to its annihilator M_S in A induces a bijection between the set of isomorphism classes of simple A-modules and the set of maximal ideals in A. In particular, A has only finitely many maximal ideals, and the intersection of all maximal ideals is equal to $J(A)$. Moreover, $A/J(A)$ is semisimple of finite composition length as an $A \otimes_k A^{\mathrm{op}}$-module.*

Proof Wedderburn's Theorem 1.13.3 yields an algebra isomorphism $A/J(A) \cong E_1 \times E_2 \times \cdots \times E_r$ with simple algebras E_i, for $1 \leq i \leq r$ and some positive integer r. Each E_i is simple as an $A \otimes_k A$-module, whence the last statement. Set $M_i = E_1 \times E_2 \times \cdots \times E_{i-1} \times E_{i+1} \times \cdots \times E_r$. Then M_i is an ideal and A/M_i is isomorphic to the simple algebra E_i, hence M_i is a maximal ideal in $A/J(A)$. One checks that every maximal ideal in $A/J(A)$ is of this form. In particular, the intersection of all maximal ideals in $A/J(A)$ is zero, and the maximal ideals in $A/J(A)$ correspond bijectively to the isomorphism classes of simple $A/J(A)$-modules. By Nakayama's Lemma 1.10.4, every maximal ideal of A contains $J(A)$, and hence the inverse images of the M_i in A are precisely the maximal ideals in A. Since $J(A)$ annihilates every simple A-module, the isomorphism classes of simple modules over A and over $A/J(A)$ correspond bijectively to each other. The result follows. □

Theorem 1.13.6 *Let A, B be k-algebras, and let $f : A \to B$ be a surjective algebra homomorphism. Suppose that $A/J(A)$ is Artinian. Then $f(J(A)) = J(B)$ and $f(A^\times) = B^\times$. In particular, $B/J(B)$ is Artinian.*

Proof Any simple B-module restricts to a simple A-module via f and hence $f(J(A)) \subseteq J(B)$. Thus f induces a surjective algebra homomorphism $\bar{f}: A/J(A) \to B/J(B)$. Since $A/J(A)$ is Artinian, so is $B/J(B)$. Since $A/J(A)$ is also semisimple, Wedderburn's Theorem implies that $A/J(A) \cong E_1 \times E_2 \times \cdots \times E_r$ for some simple algebras E_i, for $1 \le i \le r$ and some positive integer r. Thus $\ker(\bar{f})$ is an ideal in this direct product of simple algebras, hence a product of some of these factors. Note that $f(J(A))$ is an ideal in B because f is surjective. But then the quotient $B/f(J(A))$ of $A/J(A)$ is also a product of certain of these E_i, and in particular, $J(B/f(J(A))) = \{0\}$. This shows that $f(J(A)) = J(B)$. Since $B/J(B)$ is a direct product of some of the factors E_i, it follows that $\bar{f}$ is a split surjective algebra homomorphism, hence induces a surjective map $(A/J(A))^\times \to (B/J(B))^\times$. Clearly $f(A^\times) \subseteq B^\times$. To show that this inclusion is an equality, let $v \in B^\times$. The image $\bar{v}$ of v in $B/J(B)$ is then invertible. By the previous argument, there is an invertible element $\bar{u}$ in $A/J(A)$ satisfying $\bar{f}(\bar{u}) = \bar{v}$. Let u be any inverse image of $\bar{u}$ in A. Then $u \in A^\times$, by 1.10.7. By construction we have $f(u) - v \in J(B)$. Thus there is $a \in J(A)$ such that $f(a) = f(u) - v$. Then $u - a \in A^\times$ and we have $f(u - a) = f(u) - f(u) + v = v$. Thus f maps $A^\times$ onto $B^\times$. □

Theorem 1.13.7 *Suppose that k is a field having at least three distinct elements. Let A be a finite-dimensional k-algebra. Then A has a k-basis consisting of invertible elements in A.*

Proof We show first that the property of having a basis consisting of invertible elements passes to direct products. Suppose that $A = B \times C$ for some finite-dimensional k-algebras B and C, such that X is a k-basis of B contained in $B^\times$ and Y a k-basis of C contained in $C^\times$. By the assumptions on k, there is an element $\lambda \in k \setminus \{0, 1\}$. Choose $b \in X$ and $c \in Y$. We will show that the set $Z = \{(x, \lambda c)\}_{x \in X} \cup \{(b, y)\}_{y \in Y}$ is a k-basis of A. Indeed, this set contains (b, c) and $(b, \lambda c)$, so its span contains $(0, c)$ and $(b, 0)$, where we use that λ is neither 1 nor 0. Since $(x, \lambda c) = (x, 0) + \lambda(0, c)$, it follows that the span of Z contains $X \times \{0\}$ and similarly, the span of Z contains $\{0\} \times Y$, which shows that Z is a basis of A. By construction, Z is contained in $A^\times$. Thus we may assume that A is indecomposable as an algebra. Since $1 + J(A) \subseteq A^\times$, it follows that the k-span of $A^\times$ contains $J(A)$, and therefore we may assume that $J(A) = \{0\}$. But then A is semisimple and indecomposable, hence isomorphic to $M_n(D) \cong M_n(k) \otimes_k D$ for some positive integer n and some finite-dimensional division algebra D. By 1.5.7 the algebra $M_n(k)$ has a basis contained in $M_n(k)^\times$, and any basis of D is contained in $D^\times$. Tensoring bases of $M_n(k)$ and D together yields therefore a basis of A contained in $A^\times$. □

Exercise 1.13.8 Let V be a vector space over a field k with a countably infinite basis. Set $E = \mathrm{End}_k(V)$. Show that E has exactly one proper nontrivial ideal I, consisting of all endomorphisms f of V with the property that $\mathrm{Im}(f)$ is finite-dimensional. Deduce that E/I is a simple algebra that is not Noetherian, not Artinian, and not semisimple.

1.14 Splitting fields

Let k be a field. If k is algebraically closed and A a finite-dimensional k-algebra, then 1.9.3 implies that $\mathrm{End}_A(S) \cong k$ for any simple A-module S. By Wedderburn's Theorem 1.13.3, this is equivalent to $A/J(A)$ being a direct product of matrix algebras over the field k. The following terminology describes this situation.

Definition 1.14.1 Let A be a finite-dimensional k-algebra. We say that A is *split* if $\mathrm{End}_A(S) \cong k$ for every simple A-module S, or equivalently, if $A/J(A)$ is a direct product of matrix algebras over k. An extension field k' of k is called a *splitting field for* A if the k'-algebra $k' \otimes_k A$ is split.

Wedderburn's Theorem implies that if n a positive integer, then the matrix algebra $M_n(k)$ is a split k-algebra. Every finite-dimensional algebra over an algebraically closed field is split, and the algebraic closure of a field k is a splitting field for every finite-dimensional k-algebra. We extend the terminology to groups in the obvious way:

Definition 1.14.2 Let G be a finite group. We say that k is a *splitting field for* G if the k-algebra kG is split.

Examples 1.14.3 Let p be a prime. Suppose that k is a field of characteristic p. Then k is a splitting field for any finite p-group P because $kP/J(kP) \cong k$ by 1.11.1. The field $\mathbb{F}_4$ with four elements is a splitting field of the cyclic group C_3 of order 3, but the prime field $\mathbb{F}_2$ is not. As we will see later, the reason for this is that $\mathbb{F}_4$ contains a primitive third root of unity, but $\mathbb{F}_2$ does not.

Given a k-algebra A and an A-module M, there is a k-algebra homomorphism $A \to \mathrm{End}_k(M)$ sending $a \in A$ to the k-linear map $m \mapsto am$, where $m \in M$. This algebra homomorphism is called the *structural homomorphism of* M, because it is equivalent to the A-module structure on M. The structural homomorphism is in general neither injective nor surjective. We have the following criterion for the surjectivity:

Theorem 1.14.4 *Let A be a finite-dimensional k-algebra, and let S be an A-module. The structural algebra homomorphism $A \to \mathrm{End}_k(S)$ is surjective if and only if S is a simple A-module and $\mathrm{End}_A(S) \cong k$.*

Proof After replacing A by A/I, where I is the annihilator of S, we may assume that the structural map $A \to \mathrm{End}_k(S)$ is injective. Note that then A is simple, since the annihilator of a simple module is a maximal ideal by 1.13.5. If this map is also surjective, then A is isomorphic to a matrix algebra over k, hence split, and S is a simple A-module, hence satisfies $\mathrm{End}_k(S) \cong k$. Since we know already that A is simple, the converse follows from 1.13.2 and comparing dimensions. □

Theorem 1.14.5 *Let A be a split finite-dimensional k-algebra, and let k'/k be a field extension. Then $k' \otimes_k A$ is a split k'-algebra, we have $J(k' \otimes_k A) = k' \otimes_k J(A)$, for every simple A-module S, the $k' \otimes_k A$-module $S' = k' \otimes_k S$ is simple, and the correspondence sending S to S' induces a bijection between the isomorphism classes of simple A-modules and of simple $k' \otimes_k A$-modules.*

Proof Since $J(A)$ is a nilpotent ideal in A, it follows that $k' \otimes_k J(A)$ is a nilpotent ideal in $k' \otimes_k A$, hence contained in $J(k' \otimes_k A)$. Thus we may replace A by its quotient by $J(A)$; that is, we may assume that A is semisimple. We need to show that $k' \otimes_k A$ is also semisimple. Since A is split, it follows from Wedderburn's Theorem 1.13.3 that A is a direct product of matrix algebras over k. Thus $k' \otimes_k A$ is a direct product of matrix algebras over k'. In particular, $k' \otimes_k A$ is indeed semisimple, hence its radical is zero. Since the direct matrix factors of A and $k' \otimes_k A$ correspond bijectively to the isomorphism classes of simple modules over A and $k' \otimes_k A$, the remaining statements follow. □

A variation of the above arguments yields the following version of Wedderburn's Theorem for split algebras:

Theorem 1.14.6 *Let A be a finite-dimensional k-algebra. Let $\{S_i\}_{1 \leq i \leq h}$ be a set of representatives of the isomorphism classes of simple A-modules. Then A is split if and only if the product of the structural homomorphisms $A \to \mathrm{End}_k(S_i)$ induces an isomorphism of k-algebras*

$$A/J(A) \cong \prod_{1 \leq i \leq h} \mathrm{End}_k(S_i).$$

Proof Suppose that A is split. The product of the structural maps $A \to \mathrm{End}_k(S_i)$ has as kernel the Jacobson radical $J(A)$ and hence induces an injective algebra homomorphism $A/J(A) \to \prod_{1 \leq i \leq h} \mathrm{End}_A(S_i)$. The algebra $A/J(A)$ is split semisimple and hence, by Wedderburn's Theorem 1.13.3, both sides have the same dimension, whence the isomorphism as stated. The converse is trivial. □

Proposition 1.14.7 *Let A be a split finite-dimensional k-algebra. Let S be a simple A-module, and let b be the unique block idempotent in $Z(A)$ to which S belongs. The block algebra Ab is isomorphic to a matrix algebra over k if and only if S is projective and injective. In that case, we have $Ab \cong \mathrm{End}_k(S)$.*

Proof If Ab is isomorphic to a matrix algebra over k, then $J(Ab) = \{0\}$, and hence S is projective and injective as an Ab-module by 1.12.19. But then S remains projective and injective as an A-module by 1.12.18. Conversely, suppose that S is projective and injective. Any surjective A-homomorphism $A \to S$ splits because S is projective. Thus A has a submodule isomorphic to S. Let U be the sum of all submodules of A that are isomorphic to S. Then $U \cong S^n$ for some positive integer. Since S is injective, so is U, and we therefore have $A = U \oplus V$ for some submodule V of A which has no submodule isomorphic to S. Then V has also no quotient isomorphic to S, because, as before, any such quotient map would split and yield a submodule isomorphic to S. It follows that there are no nonzero A-homomorphisms between U and V in either direction. Thus $A \cong \mathrm{End}_A(A)^{\mathrm{op}} \cong \mathrm{End}_A(U \oplus V)^{\mathrm{op}} \cong \mathrm{End}_A(U)^{\mathrm{op}} \times \mathrm{End}_A(V)^{\mathrm{op}}$, as a direct product of k-algebras. Since A is split, we have $\mathrm{End}_A(S) \cong k$. Since $U \cong S^n$, it follows that $\mathrm{End}_A(U)^{\mathrm{op}} \cong M_n(k)$. This is an indecomposable algebra, hence the block algebra to which S belongs. The result follows. □

The following lemma is a special case of a more general theorem, due to Noether and Deuring.

Lemma 1.14.8 *Let k'/k be a field extension, let A be a k-algebra, and set $A' = k' \otimes_k A$. Let S, T be simple A-modules. There is a nonzero homomorphism of A'-modules $k' \otimes_k S \to k' \otimes_k T$ if and only if $S \cong T$.*

Proof Suppose there is a nonzero homomorphism $\varphi : k' \otimes_k S \to k' \otimes_k T$ of A'-modules. This is in particular a homomorphism of A-modules, where we identify A with its image $1 \otimes A$ in A'. The homomorphism φ restricts to an A-homomorphism $\psi : S \to k' \otimes_k T$. Since φ is nonzero, it follows that ψ is nonzero because S generates $k' \otimes_k S$ as an A'-module. As an A-module, $k' \otimes_k T$ is isomorphic to a direct sum of copies of T (indexed by a basis of k' over k). Thus there is a nonzero A-homomorphism from S to T, and hence $S \cong T$ as S, T are simple. The converse is trivial. □

We use Lemma 1.14.8 to show that a subfield k of a splitting field k' of an algebra of the form $k' \otimes_k A$ remains a splitting field for A if and only if all simple $k' \otimes_k A$-modules can be realised over the k-algebra A.

Theorem 1.14.9 *Let k'/k be a field extension and let A be a finite-dimensional k-algebra such that the k'-algebra $A' = k' \otimes_k A$ is split. Then the k-algebra A is split if and only if for every simple A'-module S' there is a (necessarily simple) A-module S such that $k' \otimes_k S \cong S'$. In that case, the map sending S to $k' \otimes_k S$ induces a bijection between the sets of isomorphism classes of simple A-modules and simple A'-modules, and we have $J(A') = k' \otimes_k J(A)$.*

Proof Suppose that for every simple A'-module S' there is a simple A-module S such that $k' \otimes_k S \cong S'$. It follows from 1.14.8 that the map sending S' to S induces an injective map from the set of isomorphism classes of simple A'-modules to the set of isomorphism classes of simple A-modules. We show that this map is in fact a bijection on isomorphism classes of simple modules. Indeed, let S be a simple A-module. Since A is finite-dimensional over k, so is S, and hence $k' \otimes_k S$ is finite-dimensional over k'. Thus there is a simple A'-module W such that there is a surjective A'-homomorphism $k' \otimes_k S \to W$. By the assumptions, we have $W \cong k' \otimes_k T$ for some simple A-module T. By 1.14.8, we have $S \cong T$, and hence $k' \otimes_k S \cong k' \otimes_k T \cong W$, which is simple. This shows that there is a bijection between the isomorphism classes of simple A-modules and simple A'-modules as stated. The structural algebra homomorphism $A \to \prod_S \mathrm{End}_k(S)$, where S runs over a set of representatives of the isomorphism classes of simple A-modules, induces an injective algebra homomorphism $A/J(A) \to \prod_S \mathrm{End}_k(S)$. Since A' is split, and since every simple A'-module is by the above isomorphic to $k' \otimes_k S$ for a simple A-module S, it follows upon tensoring with k' that we get a surjective algebra homomorphism $k' \otimes_k A/J(A) \to \prod \mathrm{End}_{k'}(k' \otimes_k S)$ by 1.14.6. Comparing dimensions implies that these maps are isomorphisms, and hence A is split. Moreover, this shows that $k' \otimes_k A/J(A) \cong (k' \otimes_k A)/(k' \otimes_k J(A))$ is semisimple, and hence that $J(A') = k' \otimes_k J(A)$. The converse implication follows from 1.14.5. □

With the notation of Theorem 1.14.9, if A is not split, then the radical of $k' \otimes_k A$ could be bigger than $k' \otimes_k J(A)$ because in general the tensor product of two simple algebras need not be semisimple. Group algebras over a field have the property that some *finite* field extension will be a splitting field. Thus for any finite group G and any prime p there is a finite splitting field of characteristic p for G (this follows from the next theorem applied to the algebra $\mathbb{F}_p G$).

Theorem 1.14.10 *Let A be a finite-dimensional k-algebra. There is a finite field extension k'/k such that $k' \otimes_k A$ is split.*

Proof Let $\bar{k}$ be an algebraic closure of k and X a k-basis of A. Identify A to its image $1 \otimes A$ in $\bar{k} \otimes_k A$. The algebra $\bar{k} \otimes_k A$ is split since $\bar{k}$ is algebraically closed. Let S be a simple $\bar{k}G$-module. Choose a $\bar{k}$-basis $\{s_1, s_2, \ldots, s_t\}$ of S.

Then, for any $x \in X$, the action of x on S is described by a matrix $(\sigma_{i,j}) \in M_t(\bar{k})$ obtained from the equations $xs_i = \sum_{1\leq j\leq t} \sigma_{i,j}s_j$. Let k' be the subfield of $\bar{k}$ obtained by adjoining to k all elements $\sigma_{i,j}$ obtained from letting S vary over a set of representatives of the isomorphism classes of simple $\bar{k}G$-modules and by letting x vary over all basis elements in X. The extension k'/k is then finite because k' is obtained from adjoining finitely many algebraic elements to k. Going back to our initial notation, let $S' = \sum_{1\leq i\leq t} k's_i$ be the k'-vector space in S generated by the $\bar{k}$-basis $\{s_1, s_2, \ldots, s_t\}$ of S. This basis is also a k'-basis of S', and we have $S \cong \bar{k} \otimes_{k'} S'$. The point is that S' is in fact a $k' \otimes_k A$-module, thanks to the choice of k'. In other words, every simple $\bar{k} \otimes_k A$-module S is of the form $S \cong \bar{k} \otimes_{k'} S'$ for some $k' \otimes_k A$-module S'. By 1.14.9 the k'-algebra $k' \otimes_k A$ is split. □

Proposition 1.14.11 *Let A be a finite-dimensional k-algebra. If A is split, then Z(A) is split.*

Proof Suppose that A is split. By 1.10.8, the inclusion $Z(A) \subseteq A$ induces an injective k-algebra homomorphism from $Z(A)/J(Z(A))$ to $\prod_S \mathrm{End}_k(S)$, where S runs over a set of representatives of the isomorphism classes of the simple A-modules. The image of this map is contained in $\prod_S \mathrm{End}_A(S)$ precisely because the elements in $Z(A)$ commute with all elements in A. Since A is split, this is a product of copies of k. The algebra $Z(A)/J(Z(A))$ is a product of field extensions of k. Since there is no ring homomorphism from a proper field extension of k to k this is only possible if $Z(A)/J(Z(A))$ is also a product of copies of k. □

Proposition 1.14.12 *Let A be a finite-dimensional k-algebra, and let e be an idempotent in A. If A is split, then eAe is split.*

Proof Suppose that A is split. The image of e in $A/J(A)$ is nonzero, and by 1.10.15 we have $J(eAe) = eJ(A)e$. Therefore, by considering the images of e in the simple quotients of A, we may assume that A is isomorphic to a matrix algebra, hence equal to $\mathrm{End}_k(V)$ for some finite-dimensional k-vector space V. Then $eAe = e \circ \mathrm{End}_k(V) \circ e \cong \mathrm{End}_k(e(V))$, which is again isomorphic to a matrix algebra over k. Thus eAe is split. Alternatively, this follows also from 1.14.6. □

Schur's Lemma yields a nondegenerate scalar product on the Grothendieck group of finitely generated modules over a finite-dimensional split algebra.

Proposition 1.14.13 *Let A be a split finite-dimensional k-algebra. The Grothendieck group R(A) has a unique scalar product* $\langle -, - \rangle_A : R(A) \times$

$R(A) \to \mathbb{Z}$ satisfying $\langle [S], [S'] \rangle = \dim_k(\mathrm{Hom}_A(S, S'))$ for any two simple A-modules S, S'. This scalar product is nondegenerate and the set of isomorphism classes of simple A-modules is an orthonormal basis for $R(A)$.

Proof Since the set of isomorphism classes of simple A-modules is a $\mathbb{Z}$-basis of $R(A)$, there is a unique bilinear map $\langle -, - \rangle : R(A) \times R(A) \to \mathbb{Z}$ satisfying $\langle [S], [S'] \rangle = \dim_k(\mathrm{Hom}_A(S, S'))$ for any two simple A-modules S, S'. By Schur's Lemma 1.9.2 in conjunction with the fact that A is split we get that $\langle [S], [S'] \rangle_A = 1$ if $S \cong S'$, and zero otherwise. The result follows. □

It is sometimes useful to detect whether a particular simple module is defined over a field that is not a splitting field. Let k'/k be a finite Galois extensions with Galois group Γ. Let A be a k-algebra. Then Γ acts on the k'-algebra $A' = k' \otimes_k A$ by k-algebra automorphisms; more precisely, any $\gamma \in \Gamma$ induces a k-algebra automorphism of A' sending $\lambda \otimes a$ to $\gamma(\lambda) \otimes a$. Since $(k')^\Gamma = k$, it follows that $(A')^\Gamma = A$, where we identify A to its image $1 \otimes A$ in A'. Restriction along this automorphism induces an equivalence on $\mathrm{Mod}(A')$ as an abelian category. (If γ is not the identity, then the induced automorphism on A' is not a k'-algebra automorphism, and hence the induced equivalence on $\mathrm{Mod}(A')$ is not k'-linear.) More explicitly, the equivalence induced by γ sends an A'-module M to the A'-module ${}_\gamma M$, which is equal to M as a k-module, with $\lambda \otimes a$ acting as $\gamma(\lambda) \otimes a$ on M. An A'-module M is called Γ-*stable* if ${}_\gamma M \cong M$ for all $\gamma \in \Gamma$.

Proposition 1.14.14 *Let k'/k be a finite Galois extension with Galois group Γ. Let A be a finite-dimensional commutative k-algebra. Suppose that $A' = k' \otimes_k A$ is split semisimple. Let M be a Γ-stable simple A'-module. Then there is a simple A-module N such that $M \cong k' \otimes_k N$.*

Proof The k'-algebra A' is commutative and split semisimple, hence a direct product of copies of k', corresponding to the (necessarily central) primitive idempotents in A', each of which corresponds in turn to an isomorphism class of simple modules. Note that A is then semisimple as well, so a direct product of commutative division k-algebras. Let e be the primitive idempotent in A' such that $eM = M$. Thus $A' = k'e \times A'(1 - e)$. Since the isomorphism class of M is assumed to be Γ-stable, it follows that the action of Γ on A' fixes e. Thus e is contained in A (identified to $1 \otimes A$ in A'), and hence $A = Ae \times A(1 - e)$. Since $A' \cong k' \otimes_k Ae \times k' \otimes_k A(1 - e) \cong k' \times A'(1 - e)$, this forces $Ae = ke \cong k$. Consequently, the (up to isomorphism) unique simple A-module N satisfying $eN = N$ has dimension 1 and satisfies $k' \otimes_k N \cong M$ as required. □

The previous result remains true for noncommutative A, provided that the endomorphism algebras of the simple A-modules are commutative (this follows from the proof of Thévenaz [90, Lemma (51.11)]).

Definition 1.14.15 Let A be a k-algebra. We say that A *is defined over a subfield* k_0 *of* k if there exists a k_0-algebra A_0 such that $k \otimes_{k_0} A_0 \cong A$ as k-algebras.

The existence of A_0 is equivalent to the existence of a basis of A with multiplicative constants in k_0. As we will see later, blocks of finite group algebras over a field of positive characteristic are always defined over some finite field, thus limiting their possible algebra structures. We state here a crucial tool to address this type of question. The proof requires a theorem of Lang.

Proposition 1.14.16 ([46, Lemma 2.1]) *Suppose that k is an algebraically closed field of prime characteristic p. Let q be a power of p and let V be a k-vector space of finite dimension n. Let Φ be an additive automorphism of V and denote by V^Φ the subgroup of fixed points in V of Φ. Suppose that $\Phi(\lambda v) = \lambda^q \Phi(v)$ for all $v \in V$ and all $\lambda \in k$. Then V^Φ is an $\mathbb{F}_q$-subspace of V such that $V = k \otimes_{\mathbb{F}_q} V^\Phi$.*

Proof Let $v \in V^\Phi$ and $\lambda \in \mathbb{F}_q$. By the assumptions on Φ we have $\Phi(\lambda v) = \lambda^q \Phi(v) = \lambda v$, and hence V^Φ is an $\mathbb{F}_q$-subspace of V. Set $I = \{1, 2, \ldots, n\}$, and choose a k-basis $\{x_i\}_{i \in I}$ of V. Let $S = (s_{j,i}) \in M_n(k)$ be the matrix over k satisfying $\Phi(x_i) = \sum_{j \in I} s_{j,i} x_j$, for all $i \in I$. The matrix S is invertible. Indeed, the assumption $\Phi(\lambda v) = \lambda^q \Phi(v)$ for $\lambda \in k$ and $v \in V$ implies that $\ker(\Phi)$ is a k-subspace of V. Since Φ is a group automorphism of V, it follows that this kernel is trivial, and hence S is invertible. Denote by $F : k \to k$ the ring automorphism given by $F(\lambda) = \lambda^q$ for all $\lambda \in k$. By Lang's Theorem in [54] or [87, Theorem 10.1], there is an invertible matrix $T = (t_{i,j}) \in M_n(k)$ such that $T \cdot F(T)^{-1} = S$. For $i \in I$ set $y_i = \sum_{j \in I} t_{j,i} x_j$. Since T is invertible, the set $\{y_i\}_{i \in I}$ is a k-basis of V. Using the properties of Φ, we get that

$$\Phi(y_i) = \sum_{j \in I} t_{j,i}^q \Phi(x_j) = \sum_{u \in I} \Big(\sum_{j \in I} s_{u,j} t_{j,i}^q \Big) x_u.$$

Using $S \cdot F(T) = T$, the right side is equal to $\sum_{u \in I} t_{u,i} x_u = y_i$, and hence $y_i \in V^\Phi$. Since the y_i are a k-basis of V contained in V^Φ, it follows that a k-linear combination $\sum_{i \in I} \mu_i y_i$ is in V^Φ if and only if $\mu_i^q = \mu_i$ for $i \in I$, hence if and only if $\mu_i \in \mathbb{F}_q$. This shows that the set $\{y_i\}_{i \in I}$ is an $\mathbb{F}_q$-basis of V^Φ, completing the proof. □

Corollary 1.14.17 *Suppose that k is an algebraically closed field of prime characteristic p. Let q be a power of p and let A be a finite-dimensional k-algebra. Suppose that Φ is a ring automorphism of A satisfying $\Phi(\lambda a) = \lambda^q \Phi(a)$ for all $a \in A$ and all $\lambda \in k$. Then A^Φ is an $\mathbb{F}_q$-subalgebra of A satisfying $A = k \otimes_{\mathbb{F}_q} A^\Phi$.*

Proof By 1.14.16, A^Φ is an $\mathbb{F}_q$-subspace of A satisfying $A = k \otimes_{\mathbb{F}_q} A^\Phi$. Since Φ is a ring homomorphism, it follows that A^Φ is a subring of A, and the property $\Phi(\lambda a) = \lambda^q \Phi(a)$ for $a \in A$ and $\lambda \in k$ implies that A^Φ is an $\mathbb{F}_q$-subalgebra of A. □

An $\mathbb{F}_q$-basis of A^Φ is a k-basis of A. Thus the hypotheses in 1.14.17 imply that the 3-dimensional matrix of multiplicative constants of A can be chosen with coefficients in $\mathbb{F}_q$. There are at most q^{n^3} such choices, where $n = \dim_k(A)$.

Exercise 1.14.18 Let n be a positive integer. Show that the matrix algebra $M_n(k)$ is defined over any subfield of k.

Exercise 1.14.19 Let G be a group. Show that the group algebra kG is defined over any subfield of k. Use Theorem 1.5.1 to show that if G is finite, then $Z(kG)$ is defined over any subfield of k.

1.15 Simple modules of finite group algebras

Theorem 1.15.1 *Let G be a finite group and let k be a splitting field for G such that $|G|$ is invertible in k. The number of isomorphism classes of simple kG-modules is equal to the number of conjugacy classes in G.*

Proof Since $|G|$ is invertible in k, it follows from Maschke's Theorem 1.11.12 that the algebra kG is semisimple. As k is a splitting field the algebra kG is in fact isomorphic to a direct product of matrix algebras $\prod_{1\leq i\leq h} \mathrm{End}_k(S_i)$, where $\{S_i | 1 \leq i \leq h\}$ is a set of representatives of the isomorphism classes of simple kG-modules. Since $Z(\mathrm{End}_k(S_i)) = k \cdot \mathrm{Id}_{S_i}$ has dimension 1, it follows that $\dim_k(Z(kG))$ is equal to the number of isomorphism classes of simple kG-modules. By 1.5.1, $\dim_k(Z(kG))$ is also equal to the number of conjugacy classes of G, whence the result. □

Remark 1.15.2 In general there is no canonical bijection between the isomorphism classes of simple modules and conjugacy classes of G. Symmetric groups are notable exceptions to this rule: their isomorphism classes of simple modules in characteristic zero are labelled by partitions. Symmetric groups arise as Weyl groups of general linear groups; for treatments pursuing the representation theory of Weyl groups and finite groups of Lie type, see for instance [18], [31] and [63].

For fields of positive characteristic dividing $|G|$, counting the number of isomorphism classes of simple modules is more involved. Given a finite group

G and a prime p we call an element $x \in G$ a *p-element* if the order of x is a power of p, and we call x a *p'-element* if the order of x is not divisible by p.

Theorem 1.15.3 (Brauer) *Let G be a finite group and let k be a splitting field for G of prime characteristic p. The number of isomorphism classes of simple kG-modules is equal to the number of conjugacy classes of p'-elements in G.*

Brauer's proof of this theorem has a distinct Lie theoretic flavor. The p-restricted Lie algebra structure of kG, while not needed explicitly, provides the conceptual background for this proof. We start with a general result relating simple modules of a split algebra A to the additive commutator subspace $[A, A]$, generated as a k-vector space by the commutators $ab - ba$, where $a, b \in A$. This is followed by considerations of the p-power map on A and its effect on certain commutators in finite group algebras.

Theorem 1.15.4 *Let A be a finite-dimensional split algebra over a field k. The number of isomorphism classes of simple A-modules is equal to $\dim_k(A/(J(A) + [A, A]))$.*

Proof The algebra A and its semisimple quotient $A/J(A)$ have the same number of isomorphism classes of simple modules. The canonical map $A \to A/J(A)$ has kernel $J(A)$ and maps the commutator space $[A, A]$ onto the commutator space $[A/J(A), A/J(A)]$. Thus the inverse image of $[A/J(A), A/J(A)]$ in A is the space $J(A) + [A, A]$. Therefore, the canonical map $A \to A/J(A)$ induces an isomorphism

$$A/(J(A) + [A, A]) \cong A/J(A)/[A/J(A), A/J(A)].$$

It follows that we may replace A by its semisimple quotient $A/J(A)$, hence we may assume that A is split semisimple. In that case, A is a direct product of matrix algebras, one for each isomorphism class of simple modules. Therefore, we may in fact assume that A is a matrix algebra over k. But then A has a unique isomorphism class of simple modules, and by Theorem 1.5.7, we have $\dim_k(A/[A, A]) = 1$, whence the theorem. □

Corollary 1.15.5 *Let A be a finite-dimensional algebra over a field k. Then $[A, A]$ is a proper subspace of A.*

Proof Let k'/k be a field extension, and set $A' = k' \otimes_k A$. Then $[A', A'] = k' \otimes_k [A, A]$. Thus, by choosing for k' a splitting field of A, we may assume that A is split. The result follows from 1.15.4. □

Lemma 1.15.6 *Let G be a finite group and let k be a field of prime characteristic p. For any $x \in G$ there is a unique p-element u and a unique p'-element s*

such that $x = us = su$. Moreover, both u and s are elements of the cyclic group $\langle x \rangle$, and if the order of u is p^a then we have $(x - s)^{p^a} = 0$ in kG.

Proof Let $x \in G$ and denote by n the order of x. Write $n = p^a m$ with $(p, m) = 1$. Then there are $c, d \in \mathbb{Z}$ such that $cp^a + dm = 1$. Set $u = x^{dm}$ and $s = x^{cp^a}$. We have $u, s \in \langle x \rangle$, hence $us = su$. Furthermore, $x = x^1 = x^{cp^a + dm} = x^{cp^a} x^{dm} = us$. We have $u^{p^a} = x^{p^a dm} = x^{nd} = 1^d = 1$, so u is a p-element, and $s^m = x^{cp^a m} = x^{cn} = 1^c = 1$, so s is a p'-element. If $x = u's' = s'u'$ for any other p-element u' and p'-element s' then u' and s' commute with x, hence with any power of x and thus also with u and s. Thus $s = xu^{-1} = s'(u'u^{-1})$ is a p'-element, which forces $u'u^{-1} = 1$, hence $u = u'$ and $s = s'$. Since $\mathrm{char}(k) = p$ and since x and s commute, we have $(x - s)^{p^a} = x^{p^a} - s^{p^a} = u^{p^a} s^{p^a} - s^{p^a} = s^{p^a} - s^{p^a} = 0$ as claimed. □

With the notation of 1.15.6, we call u the *p-part of x* and s the *p'-part of x*.

Lemma 1.15.7 *Let A be a finite-dimensional algebra over a field k of prime characteristic p. For any $a, b \in A$ and any nonnegative integer n we have $(a + b)^{p^n} = a^{p^n} + b^{p^n} + c$ for some $c \in [A, A]$.*

Proof The statement holds clearly if a and b commute (with $c = 0$). Thus we may assume $a \neq b$. By developing the expression $(a + b)^{p^n}$ we get a sum of products of the form $x_1 x_2 \cdots x_{p^n}$, where each x_i is either a or b. For any such summand, the summand $x_{p^n} x_1 x_2 \cdots x_{p^n - 1}$ occurs as well. Therefore, by setting $\gamma(x_1 x_2 \cdots x_{p^n}) = x_{p^n} x_1 x_2 \cdots x_{p^n - 1}$ we get an action of of the cyclic group $\langle \gamma \rangle$ of order p^n with generator γ on the set of summands in the development of $(a + b)^p$. Note that $\gamma(x_1 x_2 \cdots x_{p^n}) - x_1 x_2 \cdots x_{p^n}$ is a commutator. Thus if $x_1 x_2 \cdots x_{p^n}$ belongs to a nontrivial $\langle \gamma \rangle$-orbit of length p^m the sum of all summands in this orbit is equal to $p^m \cdot (x_1 x_2 \cdots x_{p^n}) + c = c$ for some $c \in [A, A]$. If γ fixes a summand $x_1 x_2 \cdots x_{p^n}$ then either all x_i are equal to a, or all x_i are equal to b. Together we get that $(a + b)^{p^n} = a^{p^n} + b^{p^n} + c$ for some $c \in [A, A]$ as claimed. □

Lemma 1.15.8 *Let A be a finite-dimensional algebra over a field k of prime characteristic p. For any $a \in [A, A]$ and any nonnegative integer n we have $a^{p^n} \in [A, A]$.*

Proof Since $a \in [A, A]$ we can write a as a finite sum of commutators; it follows from 1.15.7 that we may assume $a = uv - vu$ for some $u, v \in A$. It follows from 1.15.7 that $(uv - vu)^{p^n} \in (uv)^{p^n} - (vu)^{p^n} + [A, A] = [u, (vu)^{p^n - 1} v] + [A, A] = [A, A]$, whence the result. □

Lemma 1.15.9 *Let A be a finite-dimensional split algebra over a field k of prime characteristic p. Let $a \in A$. We have $a \in J(A) + [A, A]$ if and only if $a^{p^n} \in [A, A]$ for some positive integer n.*

Proof Suppose that $a^{p^n} \in [A, A]$ for some positive integer n. Let $\bar{a} = a + J(A)$ be the image of A in the semisimple quotient $\bar{A} = A/J(A)$, which is a direct product of matrix algebras over k as A is split. Thus $\bar{a}^{p^n} \in [\bar{A}, \bar{A}]$ corresponds to a family of matrices of trace zero, and hence $\bar{a}$ corresponds to a family of matrices whose p^n-th power have trace zero. But over a field of characteristic p, the trace of the p^n-th power of a matrix is the p^n-th power of the trace of that matrix. (This is clear for an upper triangular matrix, and in general, this follows from the fact that upon extending k to an algebraic closure $\bar{k}$ of k, any matrix becomes conjugate to an upper triangular matrix.) Thus $\bar{a}$ itself corresponds to a family of matrices of trace zero and hence is contained in $[\bar{A}, \bar{A}]$. But then its inverse image a is contained in $J(A) + [A, A]$. Conversely, if $a \in [A, A]$ and $b \in J(A)$ we prove that $(a + b)^{p^n} \in [A, A]$ for some positive integer n. If $a \in [A, A]$ and $b \in J(A)$ there is a positive integer n such that $b^{p^n} = 0$. By 1.15.8, we have $a^{p^n} \in [A, A]$. Thus $(a + b)^{p^n} \in a^{p^n} + b^{p^n} + [A, A] = a^{p^n} + [A, A] = [A, A]$ as required. □

Proof of Theorem 1.15.3 Let R be a set of representatives of the conjugacy classes of G and let R' be a set of representatives of the conjugacy classes of p'-elements in G. By 1.15.4 it suffices to show that the image of R' in $kG/(J(kG) + [kG, kG])$ is a k-basis. It follows from 1.5.5 that the image of R generates this quotient as a k-module. For $x \in G$, write $x = us = su$, with u a p-element and s a p'-element in G. By 1.15.6 there is a positive integer n such that $(x - s)^{p^n} = 0$. By 1.15.9 this implies that $x - s \in J(kG) + [kG, kG]$. Thus the images of x and s in $kG/(J(kG) + [kG, kG])$ are equal. It follows that the image of the set R' generates $kG/(J(kG) + [kG, kG])$ as a k-module; we need to check linear independence. Assume that there are $\lambda_s \in k$ for $s \in R'$ such that $\sum_{s \in R'} \lambda_s s \in J(kG) + [kG, kG]$. Again by 1.15.9, for some positive integer n, we have $(\sum_{s \in R'} \lambda_s s)^{p^n} \in [kG, kG]$, and by 1.15.7 this implies that $\sum_{s \in R'} \lambda_s^{p^n} s^{p^n} \in [kG, kG]$. One checks easily that the set $\{s^{p^n}\}_{s \in R'}$ is just another set of representatives of the conjugacy classes of p'-elements in G. By 1.5.5 the image of this set in $kG/[kG, kG]$ is k-linearly independent. This forces $\lambda_s^{p^n} = 0$ for all $s \in R'$, and hence $\lambda_s = 0$. This completes the proof. □

Example 1.15.10 The set of permutations $\{(1), (1, 2)(3, 4), (1, 2, 3), (1, 2, 3, 4, 5), (1, 2, 3, 5, 4)\}$ is a set of representatives of the conjugacy classes of the alternating group A_5. Thus, if k is a splitting field for A_5, the number of isomorphism classes of simple kA_5-modules is equal to 4 if char(k) is 2 or 3, equal to 3 if char(k) is 5 and equal to 5 if char(k) is 0 or greater than 5.

Combining the Theorems 1.15.1 and 1.15.3 applied to centralisers of p-elements in a finite group yields a statement relating numbers of simple modules in different characteristics. For A a finite-dimensional algebra over a field of any characteristic, we denote by $\ell(A)$ the number of isomorphism classes of simple A-modules.

Theorem 1.15.11 (Brauer) *Let G be a finite group, K a splitting field for G of characteristic zero and k a splitting field for all subgroups of G of positive characteristic p. Denote by $\mathcal{U}$ a set of representatives of the conjugacy classes of p-elements in G. We have $\ell(KG) = \sum_{u \in \mathcal{U}} \ell(kC_G(u))$.*

Proof If $x, y \in G$ are conjugate then their p-parts are conjugate. If x, y have the same p-part u then their p'-parts are conjugate in $C_G(u)$. Thus, if u runs over $\mathcal{U}$ and, for any such u we let s run over a set of representatives of the p'-conjugacy classes in $C_G(u)$, then us runs over a set of representatives of the conjugacy classes of G. In particular, the number of conjugacy classes of G, which by 1.15.1 is $\ell(KG)$, is equal to the sum, taken over $u \in \mathcal{U}$, of the numbers of p'-conjugacy classes in $C_G(u)$. This sum is by 1.15.3 equal to the sum of the numbers $\ell(kC_G(u))$, with u running over $\mathcal{U}$, as claimed. □

Note that $1 \in \mathcal{U}$; since $C_G(1) = G$, the corresponding term in the sum in 1.15.11 is $\ell(kG)$. In particular, if p divides $|G|$ this sum has at least two terms, and thus $\ell(KG) > \ell(kG)$. Theorem 1.15.11 is part of core techniques, both in finite group theory and modular representation theory, relating 'global' invariants of a finite group (algebra) to 'p-local' invariants. Alperin's weight conjecture captures the intuition that we should expect close connections between these two worlds.

Conjecture 1.15.12 (Alperin's Weight Conjecture for finite groups) *Let G be a finite group and k an algebraically closed field of prime characteristic p. Then*

$$\ell(kG) = \sum_{Q} w(kN_G(Q)/Q),$$

where Q runs over a set of representatives of the G-conjugacy classes of p-subgroups of G, and where $w(kN_G(Q)/Q)$ denotes the number of isomorphism classes of simple projective $kN_G(Q)/Q$-modules.

The term $Q = 1$ in the right hand side of the statement of Alperin's weight conjecture yields in particular the number $w(kG)$ of isomorphism classes of simple projective kG-modules. If p does not divide $|G|$, then this is the only summand, and every simple kG-module is projective, so this conjecture holds trivially, as both sides are equal to the number of conjugacy classes of G. We will see later that Alperin's weight conjecture admits a refinement for blocks of finite group algebras. Alperin's weight conjecture is known for many finite groups, including for symmetric groups, general linear groups, p-solvable groups. See 6.10.2 below for more details.

Definition 1.15.13 Let G be a finite group and k a field of prime characteristic p. A projective simple $kN_G(Q)/Q$-module is called a *weight of* G, or a *p-weight of* G, if the ambient characteristic needs emphasising.

It is a strong restriction on a p-subgroup Q to give rise to a weight; we mention the following elementary fact. For G a finite group and p a prime, we denote by $O_p(G)$ the largest normal p-subgroup of G.

Proposition 1.15.14 *Let G be a finite group, k a field of characteristic p and S a simple projective kG-module. Then for any normal subgroup N of G we have $O_p(N) = \{1\}$.*

Proof If Q is a normal p-subgroup of G, then Q acts trivially on S by 1.11.8. But S is also projective as a kQ-module, hence Q is trivial. Let N be a normal subgroup of G. By 1.12.9, the restriction of S to kN is projective, and by 1.9.9 it is semisimple. Thus kN has a projective simple module. It follows from the first argument that N has no nontrivial normal p-subgroup. □

1.16 Central simple and separable algebras

A central simple algebra over a field k is a simple k-algebra that need not be split, but whose centre is isomorphic to the base field k. If A is a simple ring then $Z(A)$ is a field, because any non invertible element x in $Z(A)$ generates a two-sided ideal of the form xA. Thus, if A is a simple algebra over a field k then $Z(A)$ is an extension field of k, and hence if in addition $\dim_k(A)$ is finite then $Z(A)/k$ is a finite field extension.

Definition 1.16.1 An algebra A over a field k is called *central simple* if A is a simple k-algebra such that $Z(A) = k$.

Given a field k and a positive integer n, the matrix algebra $M_n(k)$ is a central simple k-algebra. More generally, if D is a division ring such that $k = Z(D)$, then $M_n(D)$ is a central simple k-algebra. Wedderburn's Theorem implies that any finite-dimensional central simple k-algebra is of that form. The tensor product of two simple algebras need not be simple in general; it is simple provided one of the two algebras is central simple.

Proposition 1.16.2 *Let A be a finite-dimensional central simple algebra over a field k and let B be a simple k-algebra. Then $A \otimes_k B$ is simple. Moreover, if B is central simple then $A \otimes_k B$ is central simple.*

Proof Let I be a non zero ideal in $A \otimes_k B$. Any $x \in I$ can be written in the form $x = \sum_{i=1}^{m} a_i \otimes b_i$ for some integer m and some $a_i \in A$, $b_i \in B$, for $1 \leq i \leq m$. Choose a nonzero $x \in I$ and an expression of the form above such that m is smallest possible. We show that then actually $m = 1$. Note that the minimality of m implies that the a_i, b_i are all nonzero, and that the set $\{b_1, b_2, \ldots, b_m\}$ is k-linearly independent. In particular, since A is simple, the ideal generated by any of the a_i is equal to A. Thus there is a finite sum of the form $\sum_{j=1}^{n} y_j a_1 z_j = 1$, for some integer n and some $y_j, z_j \in A$, where $1 \leq j \leq n$. The element $\sum_{j=1}^{n}(y_j \otimes 1)x(z_j \otimes 1) = \sum_{i=1}^{m}(\sum_{j=1}^{n}(y_j a_i z_j)) \otimes b_i$ belongs to the ideal I. We therefore may assume that $a_1 = 1$. Suppose that $m \geq 2$. Since a_1 and a_2 are linearly independent and since $Z(A) \cong k$ we have $a_2 \notin Z(A)$. Thus there is an element $u \in A$ such that $ua_2 - a_2 u \neq 0$. The linear independence of the b_i implies that $(u \otimes 1)x - x(u \otimes 1)$ is a nonzero element in I. Since $a_1 = 1$ this element is now a sum of $m - 1$ elementary tensors, a contradiction. Thus $m = 1$ and I contains an element of the form $1 \otimes b$ for some nonzero $b \in B$. Since B is simple, the two-sided ideal of $A \otimes_k B$ generated by $1 \otimes b$ is equal to $A \otimes_k B$. This shows that $A \otimes_k B$ is simple. If B is central simple, then by 1.5.3 we have $Z(A \otimes_k B) \cong Z(A) \otimes_k Z(B) \cong k$, and hence $A \otimes_k B$ is central simple. □

In order to illustrate that the hypothesis on A being central simple cannot be dropped in general, consider a quadratic field extension L/K such that $L = K(\alpha)$ for some root α of $x^2 + 1$. Then L can be viewed as a K-algebra, and an easy verification shows that in $L \otimes_K L$ we have $(1 \otimes 1 + \alpha \otimes \alpha)(1 \otimes 1 - \alpha \otimes \alpha) = 0$. Thus the two elements in this product are zero divisors and therefore generate proper non zero ideals in $L \otimes_K L$, which shows that $L \otimes_K L$ is not a simple K-algebra.

Corollary 1.16.3 *Let A be a central simple algebra over a field k. For any positive integer n the algebra $M_n(A)$ is central simple.*

Proof Since $M_n(A) \cong M_n(k) \otimes_k A$ and since $M_n(k)$ is central simple this follows from 1.16.2. □

Corollary 1.16.4 *Let D be a finite-dimensional central simple division algebra over a field k. Then $\dim_k(D) = n^2$ for some positive integer n.*

Proof Let $\bar{k}$ be an algebraic closure of k. Then in particular $\bar{k}$ is a simple k-algebra, and hence $\bar{k} \otimes_k D$ is simple. By 1.9.3 the algebra $\bar{k} \otimes_k D$ is also split as $\bar{k}$-algebra, and hence $\bar{k} \otimes_k D \cong M_n(\bar{k})$ for some positive integer. Therefore $\dim_k(D) = \dim_{\bar{k}}(\bar{k} \otimes_k D) = n^2$. □

Proposition 1.16.5 *Let A be a finite-dimensional algebra over a field k. Then A is a central-simple k-algebra if and only if the canonical algebra homomorphism $\Phi : A \otimes_k A^{\mathrm{op}} \to \mathrm{End}_k(A)$ defined by $\Phi(a \otimes b)(c) = acb$ for all $a, b, c \in A$ is an isomorphism.*

Proof Suppose that Φ is an isomorphism. Every ideal J in A yields an ideal $J \otimes A^{\mathrm{op}}$ in $A \otimes_k A^{\mathrm{op}}$. Since $\mathrm{End}_k(A)$ is a simple algebra this implies that A is simple. Moreover, on one hand we have $Z(\mathrm{End}_k(A)) \cong k$, and on the other hand we have $Z(A \otimes_k A^{\mathrm{op}}) = Z(A) \otimes_k Z(A^{\mathrm{op}})$ by 1.5.3. This forces $Z(A) \cong k$, and hence A is central simple. Suppose conversely that A is central simple. Then, again by 1.5.3, we have $Z(A \otimes_k A^{\mathrm{op}}) \cong k$, and by the previous proposition, $A \otimes_k A^{\mathrm{op}}$ is simple. Thus A is, up to isomorphism, the unique simple $A \otimes_k A^{\mathrm{op}}$-module, and the map Φ is injective. But then Φ is an isomorphism as both sides have the same dimension. □

Two finite-dimensional central simple algebras A, B over a field k are called *equivalent* if there are positive integers m, n such that $M_m(A) \cong M_n(B)$. We denote by $[A]$ the equivalence class of A. The equivalence class of the trivial algebra k consists of all matrix algebras $M_n(k)$, with n running over the positive integers. By 1.16.5, for any finite-dimensional central simple k-algebra we have $[A \otimes_k A^{\mathrm{op}}] = [k]$. One can use these observations to define the *Brauer group of a field k*. This is the set whose elements are equivalence classes of finite-dimensional central simple k-algebras endowed with the product

$$[A] \cdot [B] = [A \otimes_k B]$$

for any two finite-dimensional central simple k-algebras A, B. This product is compatible with the equivalence relation, associative, with unit element $[k]$ and inverses given by $[A]^{-1} = [A^{\mathrm{op}}]$ for any finite-dimensional central simple k-algebra A. The Brauer group of an algebraically closed field is trivial. By a theorem of Frobenius, the Brauer group of the real number field $\mathbb{R}$ has order 2; the nontrivial element is represented by the Hamiltonians.

Theorem 1.16.6 (Skolem–Noether) *Let A be a finite-dimensional central simple algebra over a field k. Any k-algebra automorphism of A is an inner automorphism.*

Proof Let V be a simple A-module, and set $D = \mathrm{End}_A(V)$. Since A is simple we have $A \cong \mathrm{End}_D(V) \cong M_n(D^{\mathrm{op}})$ for some positive integer n. Consider V as $A \otimes_k D$-module in the obvious way. By 1.16.2 the algebra $A \otimes_k D$ is simple, and hence V is up to isomorphism its unique simple module. Let α be an automorphism of A as k-algebra. Then the map sending $a \otimes d$ to $\alpha(a) \otimes d$, for $a \in A$, $d \in D$, is an automorphism of $A \otimes_k D$, which we are going to denote abusively by α again. Now restriction via α defines an $A \otimes D$-module structure on V with $a \otimes d$ acting on $v \in V$ as $\alpha(a) \otimes d$. The resulting $A \otimes_k D$-module, denoted ${}_\alpha V$ is again simple, hence isomorphic to V. Thus there is a k-linear isomorphism $\tau : V \cong V$ satisfying $\tau((a \otimes d) \cdot v) = (\alpha(a) \otimes d) \cdot \tau(v)$ for all $a \in A$, $d \in D$, $v \in V$. In particular, τ is an isomorphism of D-modules. Using the fact that $A \cong \mathrm{End}_D(V)$ it follows that there is an invertible element $x \in A^\times$ satisfying $\tau(v) = xv$ for all $v \in V$. In particular, applying this to $d = 1$ we get that $(xa) \cdot v = (\alpha(a)x) \cdot v$ for all $v \in V$, and hence $xa = \alpha(a)x$ for all $a \in A$, which shows that α is the inner automorphism given by conjugation with the element x. □

Corollary 1.16.7 *Let k be a field and n a positive integer. Any k-algebra automorphism of $M_n(k)$ is inner.*

The above theorem and its corollary do not hold for ring automorphisms that are not algebra automorphisms: no nontrivial automorphism of a field can be inner. Given a subalgebra B of an algebra A, we denote by $C_A(B)$ the set of all $a \in A$ satisfying $ab = ba$ for all $b \in B$ and call $C_A(B)$ the *centraliser of B in A*. Clearly $C_A(B)$ is a unital subalgebra of A.

Theorem 1.16.8 *Let A be a finite-dimensional central simple algebra over a field k and let B be a simple subalgebra of A. Then $C_A(B)$ is a simple subalgebra of A, we have $\dim_k(A) = \dim_k(B)\dim_k(C_A(B))$ and $C_A(C_A(B)) = B$. Moreover, if B is central simple, then multiplication in A induces an isomorphism of k-algebras $B \otimes_k C_A(B) \cong A$.*

Proof Let V be a simple A-module and set $D = \mathrm{End}_A(V)$. By 1.13.2 there is a positive integer n such that $A \cong M_n(D^{\mathrm{op}})$ as algebras and $A \cong V^n$ as left A-modules. In particular, $\dim_k(A) = n \cdot \dim_k(V) = n^2\dim_k(D)$, hence $\dim_k(V) = n \cdot \dim_k(D)$. Both A and $D = \mathrm{End}_A(V)$ act on V, and these actions commute. Thus we may consider V as $A \otimes_k D$-module in the obvious way. Note that D is central simple, and thus $A \otimes_k D$ is central simple by 1.16.2. Thus the structural algebra homomorphism $A \otimes_k D \to \mathrm{End}_k(V)$ is injective. Both sides have the same dimension $n^2 \cdot \dim_k(D)^2$, so this is an isomorphism $A \otimes_k D \cong \mathrm{End}_k(V)$. Taking centralisers with respect to the subalgebra $1 \otimes D$ and using $Z(D) \cong k$ yields, together with 1.5.3, an algebra isomorphism $A \cong \mathrm{End}_D(V)$, as already observed in 1.13.2. Taking centralisers again with respect to the subalgebra B yields an algebra isomorphism $C_A(B) \cong \mathrm{End}_{B\otimes_k D}(V)$. Since D is central simple, the algebra $B \otimes_k D$ is simple. Thus, if we denote by W a simple $B \otimes_k D$-module, then $\mathrm{Res}^{A\otimes_k D}_{B\otimes_k D}(V) \cong W^m$ for some positive integer m. By 1.13.2 we have $B \otimes_k D \cong \mathrm{End}_E(W)$, where $E = \mathrm{End}_{B\otimes_k D}(W)$. It follows that $C_A(B) \cong \mathrm{End}_{B\otimes_k D}(V) \cong \mathrm{End}_{B\otimes_k D}(W^m) \cong M_m(E)$, and this is a simple algebra as E is a division algebra. We have $\dim_k(B \otimes_k D) = \dim_E(W)^2 \dim_k(E)$ and $\dim_k(C_A(B)) = m^2 \dim_k(E)$. Their product is equal to $m^2 \dim_k(W)^2 = \dim_k(V)^2 = \dim_k(A \otimes_k D)$. This shows that $\dim_k(A) = \dim_k(B)\dim_k(C_A(B))$. Since $B \subseteq C_A(C_A(B))$, comparing dimensions yields $B = C_A(C_A(B))$. If B is central simple, then $B \otimes_k C_A(B)$ is simple by 1.16.2. Thus the map $B \otimes_k C_A(B) \to A$ induced by multiplication in A is injective. But then this map is an isomorphism as both sides have the same dimension. □

Corollary 1.16.9 *Let D be a finite-dimensional central simple division algebra over a field k and let n be the positive integer satisfying* $\dim_k(D) = n^2$. *Let K be a maximal subfield of D. Then* $\dim_k(K) = n$.

Proof The maximality of K implies that $K = C_D(K)$. Thus $\dim_k(D) = \dim_k(K)^2$ by 1.16.8. □

Definition 1.16.10 A algebra A over a commutative ring k is called *separable* if A is projective as an $A \otimes_k A^{\mathrm{op}}$-module, or equivalently, as an A-A-bimodule.

Multiplication in A defines a surjective A-A-bimodule homomorphism $A \otimes_k A \to A$, and hence if A is separable, then this homomorphism splits. Conversely, if this homomorphism splits then A is isomorphic to a direct summand of the projective A-A-bimodule $A \otimes_k A$, and hence A is separable. A trivial verification shows that a direct product $A \times B$ of two k-algebras A and B is separable if and only both A and B are separable. Matrix algebras are separable.

Proposition 1.16.11 *Let n be a positive integer. The matrix algebra $M_n(k)$ is separable.*

Proof For integers i, j such that $1 \le i, j \le n$, denote by e_{ij} the matrix that has (i, j)-entry 1 and all other entries zero. There is a unique bimodule homomorphism $M_n(k) \to M_n(k) \otimes_k M_n(k)$ sending the unit matrix to $z = \sum_{i=1}^{n} e_{i1} \otimes e_{1i}$. Indeed, an easy verification shows that for any integers s, t such that $1 \le s$, $t \le n$, we have $e_{st}z = e_{s1} \otimes e_{1t} = ze_{st}$, and hence the map sending $a \in M_n(k)$

to $az = za$ is a bimodule homomorphism. Multiplication in $M_n(k)$ sends z to the identity matrix, and hence this homomorphism is a section of the canonical map $M_n(k) \otimes_k M_n(k) \to M_n(k)$ given by matrix multiplication. □

A finite-dimensional separable algebra over a field is always semisimple, but the converse need not be true.

Proposition 1.16.12 *Let A be a finite-dimensional separable algebra over a field k. Then $J(A) = \{0\}$, or equivalently, every left or right A-module is semisimple.*

Proof Let U be an A-module. Since A is separable, the map $A \otimes_k A \to A$ induced by multiplication in A splits as a homomorphisms of A-A-bimodules. Tensoring this homomorphism and its section by $- \otimes_A U$ shows that the map $A \otimes_k U \to U$ sending $a \otimes u$ to au splits as a homomorphism of left A-modules, where we use the canonical isomorphism $A \otimes_A U \cong U$. This shows that U is isomorphic to a direct summand of $A \otimes_k U$. Since k is a field, U has a k-basis X. The image $1 \otimes X$ in $A \otimes U$ is then an A-basis of $A \otimes_k U$, hence $A \otimes_k U$ us a free A-module. Thus U is projective, hence semisimple by 1.12.19. The same argument shows that every right A-module is projective, hence semisimple. □

The earlier proof of Maschke's Theorem actually proved that finite group algebras over a field of characteristic zero are not just semisimple, but in fact separable. Slightly more generally, we have the following result.

Theorem 1.16.13 (Maschke's Theorem revisited) *Let G be a finite group. Suppose that that $|G|$ is invertible in the commutative ring k. Then the k-algebra kG is separable. In particular, if k is a field of characteristic zero or of prime characteristic not dividing the order of G, then every kG-module is semisimple.*

Proof We need to show that the canonical map $\pi : kG \otimes_k kG \to kG$ induced by multiplication in kG has a section as a bimodule homomorphism. We claim that there is a unique bimodule homomorphism σ sending 1 to $w = \frac{1}{|G|} \sum_{x \in G} x \otimes x^{-1}$. To see that this yields a bimodule homomorphism we need to check that $yw = wy$ for all $y \in G$. If x runs over all elements in G then so does yx, and hence $yw = \frac{1}{|G|} \sum_{x \in G} yx \otimes x^{-1}y^{-1}y = wy$ as required. Since $\pi(w) = 1$, the bimodule homomorphism σ is clearly a section of π. The second statement follows from 1.16.12. □

Proposition 1.16.14 *Let k be a field, A a finite-dimensional separable k-algebra, and B a finite-dimensional k-algebra. Then $J(A \otimes_k B) = A \otimes_k J(B)$. In particular, if B is semisimple, then $A \otimes_k B$ is semisimple.*

Proof Since $A \otimes_k J(B)$ is an ideal that is contained in $J(A \otimes_k B)$, we may replace $A \otimes_k B$ by $A \otimes_k B/J(B)$; that is, we may suppose that B is semisimple. Then every B-module is projective. Since A is separable, it is isomorphic to a direct summand of $A \otimes_k A$. Thus $A \otimes_k B$ is isomorphic to a direct summand of $(A \otimes_k B) \otimes_B (A \otimes_k B)$, where we identify B with its image $1 \otimes B$ in $A \otimes_k B$. Thus any $A \otimes_k B$-module is isomorphic to a direct summand of $(A \otimes_k B) \otimes_B V$ for some B-module V. Since V is projective, so is $(A \otimes_k B) \otimes_B V$. Thus any $A \otimes_k B$-module is projective. It follows from 1.12.19 that $A \otimes_k B$ is semisimple, or equivalently, that $J(A \otimes_k B) = \{0\}$. The result follows. □

Corollary 1.16.15 *Let A and B be finite-dimensional k-algebras such that $A/J(A)$ is separable. Then $J(A \otimes_k B) = J(A) \otimes_k B + A \otimes_k J(B)$.*

Proof The image of $J(A) \otimes_k B$ in $A \otimes_k B$ is a nilpotent ideal, hence contained in $J(A \otimes_k B)$. Thus we may divide by this ideal, which amounts to replacing A by $A/J(A)$. Since $A/J(A)$ is separable, it follows from 1.16.14 that $J(A/J(A) \otimes_k B) = A/J(A) \otimes_k J(B)$. Taking inverse images in $A \otimes_k B$ implies the result. □

Corollary 1.16.16 *Let A be a finite-dimensional k-algebra such that $A/J(A)$ is separable. Then $J(A)$ is equal to the radical of A as an $A \otimes_k A^{\mathrm{op}}$-module; in particular, $A/J(A)$ is semisimple as an $A \otimes_k A^{\mathrm{op}}$-module.*

Proof By 1.16.15 we have $J(A \otimes_k A^{\mathrm{op}}) = J(A) \otimes_k A^{\mathrm{op}} + A \otimes_k J(A^{\mathrm{op}})$. Note that $J(A^{\mathrm{op}}) = J(A)$ as subspaces of A. Thus $J(A \otimes_k A^{\mathrm{op}}) \cdot A = J(A)$, whence the result. □

Proposition 1.16.17 *Let k'/k be a field extension, and let A be a k-algebra. Then A is separable if and only if the k'-algebra $k' \otimes_k A$ is separable.*

Proof An easy verification shows that $(k' \otimes_k A) \otimes_{k'} (k' \otimes_k A) \cong k' \otimes_k (A \otimes_k A)$. It follows immediately that if A is separable, then so is $k' \otimes_k A$. If $k' \otimes_k A$ is separable, then the canonical map $k' \otimes_k (A \otimes_k A) \to k' \otimes_k A$ splits. Write the image of $1_{k' \otimes_k A}$ under a section in the form $\sum_\lambda \lambda \otimes z_\lambda$, with $\lambda \in k'$ running over a k-basis of k' containing 1, and where $z_\lambda \in A \otimes_k A$. One verifies that there is a unique section $A \to A \otimes_k A$ as bimodules that sends 1_A to z_1. □

Proposition 1.16.18 *Let k'/k be a finite field extension. Then k'/k is a separable field extension if and only if k' is a separable k-algebra.*

Proof Let $\alpha \in k'$, and let $f \in k[x]$ be the minimal polynomial of α over k. Let E be a splitting field for f. Write $f(x) = \prod_{i=1}^n (x - \alpha_i)^{m_i}$, where the α_i are the distinct roots of f in E, and where $m_i \geq 1$ is the multiplicity of the root α_i. Then $E \otimes_k k(\alpha) \cong E \otimes_k k[x]/fk[x] \cong E[x]/fE[x]$. By the Chinese Remainder Theorem, this is isomorphic to $\prod_{i=1}^{n} E[x]/(x-\alpha_i)^{m_i}E[x]$. If f is separable, then all m_i are 1, hence $E[x]/fE[x]$ is a direct product of fields, all isomorphic to E. In particular, the E-algebra $E \otimes_k k(\alpha) \cong E[x]/fE[x]$ is separable. If $m_i > 1$ for some i, then $(x-\alpha_i) + (x-\alpha_i)^{m_i}E[x]$ is a nonzero nilpotent element in the commutative algebra $E \otimes_k k(\alpha) \cong E[x]/(x-\alpha_i)^{m_i}E[x]$, and hence this algebra has a nonzero radical. Thus $E \otimes_k k'$ has a nonzero radical; in particular, $E \otimes_k k'$ is not separable. If k'/k is separable, then $k' = k(\alpha)$ for some $\alpha \in k'$ with a separable minimal polynomial f. By the preceding argument, if we denote by E a splitting field of $f \in k[x]$, then $E \otimes_k k'$ is a separable E-algebra. Thus k' is a separable k-algebra by 1.16.17. If k'/k is not a separable field extension, then k' contains an element with a minimal polynomial f having a root α_i with multiplicity $m_i \geq 2$. By the above, there is a field extension E of k such that $E \otimes_k k'$ is not separable, and hence neither is k', where we use again 1.16.17. □

Proposition 1.16.19 *Let A be a finite-dimensional algebra A over a field k. The following are equivalent.*

(i) The k-algebra A is separable.
(ii) For any field extension k'/k the k'-algebra $k' \otimes_k A$ is semisimple.
(iii) For any finite field extension k'/k the k'-algebra $k' \otimes_k A$ is semisimple.

Proof If A is separable, then so is $k' \otimes_k A$ for any field extension k'/k, hence semisimple by 1.16.12. Thus (i) implies (ii), and (ii) implies trivially (iii). Suppose that (iii) holds. By 1.14.10 there is a finite field extension k'/k such that k' is a splitting field for A. Then $k' \otimes_k A$ is split semisimple, hence a direct product of matrix algebras over k'. Thus $k' \otimes_k A$ is a separable k'-algebra by 1.16.11. It follows from 1.16.17 that A is a separable k-algebra. Thus (iii) implies (i), whence the result. □

Corollary 1.16.20 *Any finite-dimensional central simple algebra over a field is separable.*

Proof This follows from combining 1.16.19 and 1.16.2. □

One can show that the finite-dimensional separable algebras over a field k are precisely the finite direct products of simple algebras $M_n(D)$, where D is a division algebra over k such that $Z(D)/k$ is a separable field extension.

Proposition 1.16.21 *Let k be a field and G a finite group. Then $kG/J(kG)$ is a separable k-algebra.*

Proof If $\mathrm{char}(k) = 0$, then $J(kG) = \{0\}$ and kG is separable by 1.16.13. Suppose that $\mathrm{char}(k) = p > 0$. Denote by $\mathbb{F}_p$ the prime field of k. Every finite extension field E of $\mathbb{F}_p$ is separable over $\mathbb{F}_p$. It follows from 1.16.14 and 1.16.18 that $J(EG) = E \otimes_{\mathbb{F}_p} J(\mathbb{F}_pG)$. Thus $\mathbb{F}_pG/J(\mathbb{F}_pG)$ is a separable $\mathbb{F}_p$-algebra. But then $k' \otimes_{\mathbb{F}_p} \mathbb{F}_pG/J(\mathbb{F}_pG)$ is semisimple for any field extension k' of $\mathbb{F}_p$, by 1.16.19. Thus for any field extension k' of k, the algebra $k' \otimes_k kG/J(kG) \cong k' \otimes_k (k \otimes_{\mathbb{F}_p} \mathbb{F}_pG/J(\mathbb{F}_pG)) \cong k' \otimes_{\mathbb{F}_p} \mathbb{F}_pG/J(\mathbb{F}_pG))$ is semisimple. The result follows from 1.16.19. □

1.17 Complexes and homology

We review basic terminology from homological algebra. Although we have chosen to state definitions and results for additive categories whenever possible, in most applications the underlying categories are module categories.

Definition 1.17.1 A *graded object over a category* $\mathcal{C}$ is a family $X = (X_n)_{n\in\mathbb{Z}}$ of objects X_n in $\mathcal{C}$. Given two graded objects $X = (X_n)_{n\in\mathbb{Z}}$ and $Y = (Y_n)_{n\in\mathbb{Z}}$ in $\mathcal{C}$, a *graded morphism of degree m* is a family $f = (f_n)_{n\in\mathbb{Z}}$ of morphisms $f_n : X_n \to Y_{n+m}$ in $\mathcal{C}$. The category of graded objects over $\mathcal{C}$ with graded morphisms of degree zero is denoted $\mathrm{Gr}(\mathcal{C})$.

More generally, composition of morphisms in $\mathcal{C}$ induces a composition of graded morphisms between graded objects over $\mathcal{C}$ such that the degrees of the graded morphisms add up. The category $\mathrm{Gr}(\mathcal{C})$ admits a *shift automorphism*, denoted $[i]$, for any integer i, defined by $X[i]_n = X_{n-i}$, where X is a graded object over $\mathcal{C}$ and n runs over $\mathbb{Z}$. On morphisms this automorphism is defined similarly by $f[i]_n = f_{n-i}$.

Definition 1.17.2 A *chain complex* over an additive category $\mathcal{C}$ is a pair (X, δ) consisting of a graded object X in $\mathcal{C}$ and a graded endomorphism δ of degree -1, called the *differential of the complex*, satisfying $\delta \circ \delta = 0$. Explicitly, δ is a family of morphisms $\delta_n : X_n \to X_{n-1}$ satisfying $\delta_{n-1} \circ \delta_n = 0$. Dually, a *cochain complex* over a $\mathcal{C}$ is a pair (X, δ) consisting of a graded object $X = (X^n)_{n\in\mathbb{Z}}$ in $\mathcal{C}$ and a graded endomorphism $\delta = (\delta^n : X^n \to X^{n+1})_{n\in\mathbb{Z}}$, called *differential* of the cochain complex, of degree 1 satisfying $\delta \circ \delta = 0$, or equivalently, $\delta^{n+1} \circ \delta^n = 0$ for $n \in \mathbb{Z}$.

One can visualise a chain complex as a possibly infinite sequence of morphisms in which the composition of any two consecutive morphisms is zero. In order to distinguish between chain complexes and cochain complexes, the standard notational convention is to use subscripts for chain complexes and superscripts for cochain complexes. One can always switch from a chain complex to a cochain complex and vice versa by setting $X^n = X_{-n}$ and $\delta^n = \delta_{-n}$. Through this correspondence, any terminology in the context of chain complexes has an analogue for cochain complexes.

Definition 1.17.3 A *chain map* between two chain complexes (X, δ), (Y, ϵ) over an additive category $\mathcal{C}$ is a graded morphism of degree zero $f = (f_n : X_n \to Y_n)_{n\in\mathbb{Z}}$ satisfying $f \circ \delta = \epsilon \circ f$, or equivalently, $f_{n-1} \circ \delta_n = \epsilon_n \circ f_n$ for $n \in \mathbb{Z}$. Cochain maps are defined similarly. The chain complexes, together with chain maps, form the category $\mathrm{Ch}(\mathcal{C})$ of *chain complexes over* $\mathcal{C}$.

If the differential of a complex (X, δ) is clear from the context, we adopt the notational abuse of just calling X a chain complex. We have a forgetful functor $\mathrm{Ch}(\mathcal{C}) \to \mathrm{Gr}(\mathcal{C})$ mapping a chain complex (X, δ) to its underlying graded object X. We have a functor from $\mathrm{Gr}(\mathcal{C}) \to \mathrm{Ch}(\mathcal{C})$ sending a graded object X to the complex $(X, 0)$ with zero differential; when composed with the forgetful functor this yields the identity functor on $\mathrm{Gr}(\mathcal{C})$.

There is a *shift automorphism* of $\mathrm{Ch}(\mathcal{C})$ extending the shift on $\mathrm{Gr}(\mathcal{C})$. For (X, δ) a chain complex, we define a graded object as before by $X[i]_n = X_{n-i}$, together with differential $\delta[i]_n = (-1)^i\delta_{n-i}$, for $n, i \in \mathbb{Z}$. Note the sign convention here. The shift functor $[i]$ applied to a cochain complexe (X, δ) is defined by $X[i]^n = X^{n+i}$ and $\delta[i]^n = (-1)^i\delta^{n+i}$. A chain complex X is called *bounded above* if $X_n = 0$ for n large enough; we denote by $\mathrm{Ch}^+(\mathcal{C})$ the full subcategory of $\mathrm{Ch}(\mathcal{C})$ consisting of all bounded above chain complexes over $\mathcal{C}$. A chain complex X is called *bounded below* if $X_n = 0$ for n small enough; we denote by $\mathrm{Ch}^-(\mathcal{C})$ the full subcategory of $\mathrm{Ch}(\mathcal{C})$ consisting of all bounded below chain complexes over $\mathcal{C}$. A chain complex X is called *bounded* if $X_n = 0$ for all but finitely many i; we denote by $\mathrm{Ch}^b(\mathcal{C})$ the full subcategory of $\mathrm{Ch}(\mathcal{C})$ consisting of all bounded chain complexes over $\mathcal{C}$. If $\mathcal{C}$ is an abelian category then $\mathrm{Ch}(\mathcal{C})$ is again an abelian category. More precisely, for any chain map f from a complex (X, δ) to a complex (Y, ϵ), the differential δ restricts to a differential on the graded object $\ker(f) = (\ker(f_n : X_n \to Y_n))_{n\in\mathbb{Z}}$, and the resulting chain complex $(\ker(f), \delta|_{\ker(f)})$ is a kernel of f; similarly, ϵ induces a differential on the cokernel of f as a graded morphism, which yields a cokernel of f in the category $\mathrm{Ch}(\mathcal{C})$. In particular, f is a monomorphism (resp. epimorphism) in $\mathrm{Ch}(\mathcal{C})$ if and only if all f_i are monomorphisms (resp. epimorphisms) in $\mathcal{C}$. The categories $\mathrm{Ch}^+(\mathcal{C})$, $\mathrm{Ch}^-(\mathcal{C})$, $\mathrm{Ch}^b(\mathcal{C})$ are full abelian subcategories of $\mathrm{Ch}(\mathcal{C})$. For cochain complexes we adopt the convention, that if X is a cochain complex, then X is called *bounded below* (resp. *bounded above*) if the corresponding chain complex defined by $X_n = X^{-n}$ is bounded above (resp. below).

For a complex (X, δ) over an abelian category $\mathcal{C}$, the condition $\delta \circ \delta = 0$ means that there is a canonical monomorphism $\mathrm{Im}(\delta) \subseteq \ker(\delta)$. Its cokernel is defined to be the homology of X:

Definition 1.17.4 Let $\mathcal{C}$ be an abelian category. The *homology* of a chain complex (X, δ) over $\mathcal{C}$ is the graded object $H_*(X, \delta) = \ker(\delta)/\mathrm{Im}(\delta)$; more explicitly, $H_n(X, \delta) = \ker(\delta_n)/\mathrm{Im}(\delta_{n+1})$ for $i \in \mathbb{Z}$. If the differential δ is clear from the context we write $H_*(X)$ instead of $H_*(X, \delta)$. If $H_*(X) = \{0\}$ then X is called *exact* or *acyclic*. Similarly, the *cohomology* of a cochain complex (Y, ϵ) is the graded object $\ker(\epsilon)/\mathrm{Im}(\epsilon)$; explicitly, $H^i(Y, \epsilon) = \ker(\epsilon^n)/\mathrm{Im}(\epsilon^{n-1})$ for $n \in \mathbb{Z}$. If ϵ is clear from the context we write again simply $H^*(Y)$, and if $H^*(Y) = \{0\}$ then Y is called as before *exact* or *acyclic*.

Any short exact sequence in an abelian category can be viewed as bounded exact complex. Taking homology or cohomology is functorial: any chain map $f : (X, \delta) \to (Y, \epsilon)$ induces, upon restriction, chain maps $\ker(\delta) \to \ker(\epsilon)$ and $\mathrm{Im}(\delta) \to \mathrm{Im}(\epsilon)$, and hence a graded morphism $H_*(f) : H_*(X) \to H_*(Y)$. Given two composable chain maps f, g, one verifies easily that $H_*(g \circ f) = H_*(g) \circ H_*(f)$. In this way, taking homology defines a functor $\mathrm{Ch}(\mathcal{C}) \to \mathrm{Gr}(\mathcal{C})$; similarly for cohomology. The chain map f is called a *quasi-isomorphism* if the induced map on homology $H_*(f)$ is an isomorphism $H_*(X) \cong H_*(Y)$. Note that if $f : X \to Y$ is a quasi-isomorphism, there need not be a chain map $g : Y \to X$ inducing the inverse isomorphism $H_*(Y) \cong H_*(X)$. The relation of two complexes X, Y being *quasi-isomorphic* is the smallest equivalence relation $\sim$ satisfying $X \sim Y$ if there is a quasi-isomorphism $X \to Y$. Similarly for cochain maps between cochain complexes.

One of the fundamental features of complexes over an abelian category is that short exact sequences of complexes give rise to long exact (co-)homology sequences. We state and prove this for module categories, as this will be sufficient for the purpose of this book. For general abelian categories, one can either directly modify the proofs, or use Freyd's embedding theorem, saying that any abelian category can be fully embedded into a module category.

Theorem 1.17.5 *Let A be a k-algebra. Any short exact sequence of chain complexes of A-modules*

$$0 \longrightarrow X \xrightarrow{f} Y \xrightarrow{g} Z \longrightarrow 0$$

induces a long exact sequence

$$\cdots \longrightarrow H_n(X) \xrightarrow{H_n(f)} H_n(Y) \xrightarrow{H_n(g)} H_n(Z) \xrightarrow{d_n} H_{n-1}(X) \longrightarrow \cdots$$

depending functorially on the short exact sequence.

The functorial dependence in this theorem means that given a commutative diagram of chain complexes with exact rows

$$\begin{array}{ccccccccc} 0 & \longrightarrow & X & \xrightarrow{f} & Y & \xrightarrow{g} & Z & \longrightarrow & 0 \\ & & \downarrow a & & \downarrow b & & \downarrow c & & \\ 0 & \longrightarrow & X' & \xrightarrow[f']{} & Y' & \xrightarrow[g']{} & Z' & \longrightarrow & 0 \end{array}$$

we get a commutative "ladder" of long exact sequences:

$$\begin{array}{ccccccccccc} \cdots \to & H_n(X) & \xrightarrow{H_n(f)} & H_n(Y) & \xrightarrow{H_n(g)} & H_n(Z) & \xrightarrow{d_n} & H_{n-1}(X) & \xrightarrow{H_{n-1}(f)} & H_{n-1}(Y) & \to \cdots \\ & \downarrow H_n(a) & & \downarrow H_n(b) & & \downarrow H_n(c) & & \downarrow H_{n-1}(a) & & \downarrow H_{n-1}(b) & \\ \cdots \to & H_n(X') & \xrightarrow[H_n(f')]{} & H_n(Y') & \xrightarrow[H_n(g')]{} & H_n(Z') & \xrightarrow[d'_n]{} & H_{n-1}(X') & \xrightarrow[H_{n-1}(f')]{} & H_{n-1}(Y) & \to \cdots \end{array}$$

The morphism d_n is called a *connecting homomorphism*. This translates verbatim to cochain complexes, except that the connecting homomorphism $d^n : H^n(Z) \to H^{n+1}(C)$ is of degree 1.

Proof of Theorem 1.17.5 Denote by δ, ϵ, ζ the differentials of X, Y, Z, respectively. We define $d_n : H_n(Z) \to H_{n-1}(X)$ as follows. Any element in $H_n(Z)$ is of the form $z + \mathrm{Im}(\zeta_{n+1})$ for some $z \in \ker(\zeta_n)$. Then, since g is surjective in each degree, there is $y \in Y_n$ such that $g_n(y) = z$. Since $\zeta_n(z) = 0$ we get that $g_{n-1}(\epsilon_n(y)) = \zeta_n(g_n(y)) = \zeta_n(z) = 0$. Thus $\epsilon_n(y) \in \ker(g_{n-1}) = \mathrm{Im}(f_{n-1})$, by the exactness of the sequence in the statement. Thus there is $x \in X_{n-1}$ satisfying $f_{n-1}(x) = \epsilon_n(y)$. Moreover, $f_{n-2}(\delta_{n-1}(x)) = \epsilon_{n-1}(f_{n-1}(x)) = \epsilon_{n-1}(\epsilon_n(y)) = 0$. Since f_{n-2} is a monomorphism, this shows that $\delta_{n-1}(x) = 0$, and hence $x + \mathrm{Im}(\delta_n)$ is an element in $H_{n-1}(X)$. One verifies that if $z \in \mathrm{Im}(\zeta_{n+1})$ then $x \in \mathrm{Im}(\delta_n)$. This implies that the class $x + \mathrm{Im}(\delta_n)$ depends only on the class $z + \mathrm{Im}(\zeta_{n+1}$. Thus there is a well-defined map $d_n : H_n(Z) \to H_{n-1}(X)$ sending $z + \mathrm{Im}(\zeta_{n+1})$ to $x + \mathrm{Im}(\delta_n)$. An easy, but lengthy verification shows that the long homology sequence becomes exact in this way, and another equally tedious verification shows the functoriality. □

Corollary 1.17.6 *Let A be a k-algebra and let*

$$0 \longrightarrow X \xrightarrow{f} Y \xrightarrow{g} Z \longrightarrow 0$$

be a short exact sequence of chain complexes of A-modules.

(i) f is a quasi-isomorphism if and only if Z is acyclic.
(ii) g is a quasi-isomorphism if and only if X is acyclic.
(iii) If two of the complexes X, Y, Z are acyclic, so is the third.

Proof The long exact homology sequence shows that if $H_{n+1}(Z) = H_n(Z) = \{0\}$, then $H_n(f)$ is an isomorphism, and if $H_n(f)$, $H_{n-1}(f)$ are isomorphisms, then the maps $H_n(g)$, d_n are zero, hence $H_n(Z) = \{0\}$. This shows (i), and the rest follows similarly. □

The following observation is used to compare the long exact homology sequences via a commutative ladder as above:

Lemma 1.17.7 (The 5-Lemma) *Let A be a k-algebra and let*

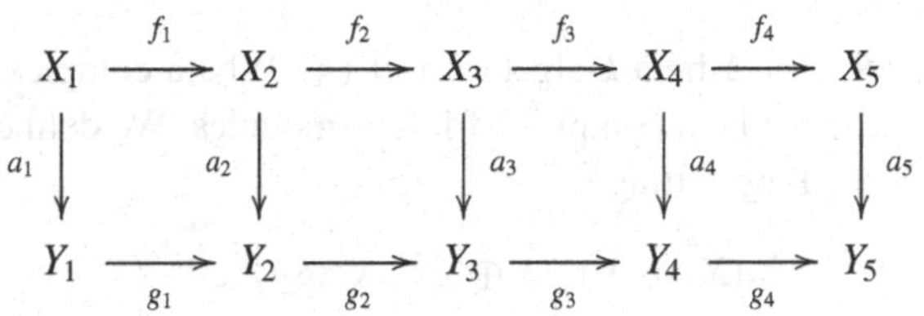

$$\begin{array}{ccccccccc} X_1 & \xrightarrow{f_1} & X_2 & \xrightarrow{f_2} & X_3 & \xrightarrow{f_3} & X_4 & \xrightarrow{f_4} & X_5 \\ \downarrow{\scriptstyle a_1} & & \downarrow{\scriptstyle a_2} & & \downarrow{\scriptstyle a_3} & & \downarrow{\scriptstyle a_4} & & \downarrow{\scriptstyle a_5} \\ Y_1 & \xrightarrow[g_1]{} & Y_2 & \xrightarrow[g_2]{} & Y_3 & \xrightarrow[g_3]{} & Y_4 & \xrightarrow[g_4]{} & Y_5 \end{array}$$

be a commutative diagram of A-modules with exact rows. If a_1, a_2, a_4, a_5 are isomorphisms then a_3 is an isomorphism.

Proof Let $x \in \ker(a_3)$. Then $a_4(f_3(x)) = g_3(a_3(x)) = 0$, hence $f_3(x) = 0$ as a_4 is an isomorphism. Thus $x \in \ker(f_3) = \mathrm{Im}(f_2)$, and so there is $y \in X_2$ such that $f_2(y) = x$. Then $g_2(a_2(y)) = a_3(f_2(y)) = a_3(x) = 0$, hence $a_2(y) \in \ker(g_2) = \mathrm{Im}(g_1)$, and so there is $z \in Y_1$ satisfying $g_1(z) = a_2(y)$. As a_1 is an isomorphism, there is $w \in X_1$ such that $a_1(w) = z$. Then $a_2(f_1(w)) = g_1(a_1(w)) = g_1(z) = a_2(y)$. Since a_2 is an isomorphism this implies that $f_1(w) = y$. But then $x = f_2(y) = f_2(f_1(w)) = 0$, and so a_3 is injective. A similar argument shows that a_3 is surjective. □

Corollary 1.17.8 *Let A be an algebra over a commutative ring k and let*

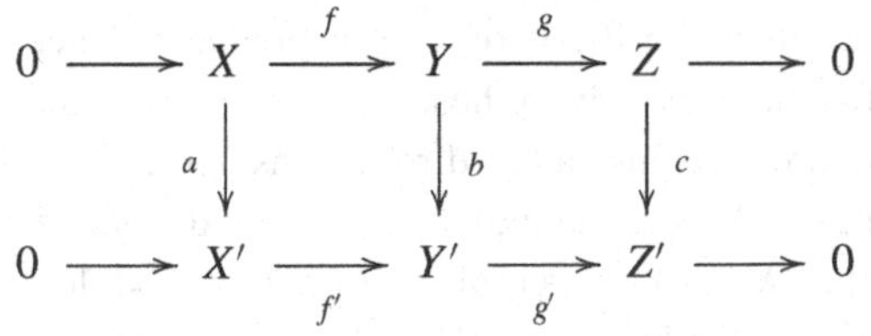

$$\begin{array}{ccccccccc} 0 & \longrightarrow & X & \xrightarrow{f} & Y & \xrightarrow{g} & Z & \longrightarrow & 0 \\ & & \downarrow{\scriptstyle a} & & \downarrow{\scriptstyle b} & & \downarrow{\scriptstyle c} & & \\ 0 & \longrightarrow & X' & \xrightarrow[f']{} & Y' & \xrightarrow[g']{} & Z' & \longrightarrow & 0 \end{array}$$

be a commutative diagram of chain complexes of A-modules with exact rows. If two of a, b, c are quasi-isomorphisms, so is the third.

Proof Apply the 5-Lemma to the five terms in the commutative ladder following Theorem 1.17.5. □

Tensor products of modules and homomorphism spaces between modules extend to complexes as follows. Let A be a k-algebra. If X is a right A-module, and (Y, ϵ) a complex of left A-modules, then applying the covariant functor $X \otimes_A -$ to the components and the differential of Y yields a complex $X \otimes_A Y$ of k-modules whose component in degree n is $X \otimes_A Y_n$, with differential $\mathrm{Id}_X \otimes \epsilon$ given by the family of maps $\mathrm{Id}_X \otimes \epsilon_n : X \otimes_A Y_n \to X \otimes_A Y_{n-1}$. Similarly, if (X, δ) is a complex of right A-modules and Y a left A-module, then applying the functor $- \otimes_A Y$ to X yields a complex $X \otimes_A Y$. If both X, Y are complexes, one can combine this to a definition of a tensor product of complexes, modulo introducing appropriate signs.

Definition 1.17.9 Let A be a k-algebra. Let (X, δ) be a complex of right A-modules, and let (Y, ϵ) be a complex of left A-modules. We define a complex of k-modules $X \otimes_A Y$ by setting

$$(X \otimes_A Y)_n = \oplus_{i+j=n} X_i \otimes_A Y_j,$$

the direct sum taken over all pairs of integers (i, j) satisfying $i + j = n$, with differential given by taking the direct sum of the maps $(-1)^i \mathrm{Id}_{X_i} \otimes \epsilon_j : X_i \otimes_A Y_j \to X_i \otimes_A Y_{j-1}$ and $\delta_i \otimes \mathrm{Id}_{Y_j} : X_i \otimes_A Y_j \to X_{i-1} \otimes_A Y_j$.

More explicitly, the above differential sends $x \otimes y \in X_i \otimes_A Y_j$ to the pair $(\delta_i(x) \otimes y, (-1)^i x \otimes \epsilon_j(y))$ in $X_{i-1} \otimes_A Y_j \oplus X_i \otimes_A Y_{j-1}$. The sign is required to ensure that this map is indeed a differential. This construction is bifunctorial. If $f : Y \to Z$ is a chain map of complexes of A-modules, then the direct sum of the maps $\mathrm{Id}_{X_i} \otimes f_j : X_i \otimes_A Y_j \to X_i \otimes_A Z_j$ yields a chain map, denoted $\mathrm{Id}_X \otimes f$. In this way $X \otimes_A -$ becomes a functor from $\mathrm{Ch}(\mathrm{Mod}(A))$ to $\mathrm{Ch}(\mathrm{Mod}(k))$. This functor need not preserve quasi-isomorphisms: see 1.17.11 for a sufficient criterion for $X \otimes_A -$ to preserve a quasi-isomorphism. Similar statements hold with the roles of X and Y exchanged. Note that the sum over all pairs (i, j) satisfying $i + j = n$ is finite if both X, Y are bounded below, or if both complexes are bounded above, or if one of X, Y is bounded. Since we will use the tensor product of complexes only in these cases, we ignore the analogous definition using direct products instead of direct sums. As in the case of the tensor product of modules, if X is a complex of B-A-bimodules and Y a complex of A-C-bimodules, then $X \otimes_A Y$ is a complex of B-C-bimodules. There is a similar construction using the bifunctor $\mathrm{Hom}_A(-, -)$ instead of $- \otimes_A -$, which besides sign conventions, requires some additional care because $\mathrm{Hom}_A(-, -)$ is contravariant in the first argument. If X is an A-module and (Y, ϵ) a complex of A-modules, then applying the covariant functor $\mathrm{Hom}_A(X, -)$ to Y yields a complex $\mathrm{Hom}_A(X, Y)$. If (X, δ) is a complex of A-modules and Y an A-module, then applying the contravariant functor $\mathrm{Hom}_A(-, Y)$ to X yields a cochain complex $\mathrm{Hom}_A(X, Y)$. In order to extend the construction $\mathrm{Hom}_A(X, Y)$ to the case where both X and Y are complexes, we need to reinterpret cochain complexes as complexes via the convention mentioned after 1.17.2.

Definition 1.17.10 Let A be a k-algebra, and let (X, δ) and (Y, ϵ) be complexes of A-modules. We define a complex $\mathrm{Hom}_A(X, Y)$ by setting

$$\mathrm{Hom}_A(X, Y)_n = \prod_{j-i=n} \mathrm{Hom}_A(X_i, Y_j),$$

the product taken over all pairs of integers (i, j) satisfying $j - i = n$, with differential sending $\alpha \in \mathrm{Hom}_A(X_i, Y_j)$ to the pair $((-1)^{n+1}\alpha \circ \delta_{i+1}, \epsilon_j \circ \alpha)$ in $\mathrm{Hom}_A(X_{i+1}, Y_j) \oplus \mathrm{Hom}_A(X_i, Y_{j-1})$.

Note that when using the notation $\mathrm{Hom}_A(X, V)$, where X is a complex and V a module, we will need to specify which convention we use, since this notation could either mean the cochain complex obtained from applying the contravariant functor $\mathrm{Hom}_A(-, V)$ to X, or the chain complex obtained from this with the sign convention in Definition 1.17.10. For instance, with the conventions of Definition 1.17.10, if (X, δ) is a complex of A-modules, then the k-dual $X^* = \mathrm{Hom}_A(X, k)$, viewed as a chain complex of right A-modules, has degree n term equal to $(X_{-n})^*$ and differential $(-1)^{n+1}(\delta_{-n+1})^* : (X_{-n})^* \to (X_{-n+1})^*$. As in the case of bimodules, if X is a complex of A-B-bimodules and Y a complex of A-C-bimodules, then $\mathrm{Hom}_A(X, Y)$ is a complex of B-C-bimodules.

Proposition 1.17.11 *Let A be a k-algebra, and let X be a complex of finitely generated projective right A-modules. Let Y, Z be complexes of A-modules, and let $f : Y \to Z$ be a quasi-isomorphism. Suppose that X is bounded, or that X, Y and Z are bounded below. Then the induced chain map $\mathrm{Id}_X \otimes f : X \otimes_A Y \to X \otimes_A Z$ is a quasi-isomorphism of complexes of k-modules.*

Proof In order to test whether a chain map is a quasi-isomorphism, we need to check that in any fixed degree, the induced map on homology is an isomorphism. Since the tensor product $X \otimes_A Y$ of bounded below complexes involves in any fixed degree only a finite number of terms of X and of Y, there is no loss of generality in assuming that X is bounded. We then proceed by induction over the length of X. If X is concentrated in a single degree, then $X \otimes_A -$ is an exact functor, hence commutes with taking homology, and therefore preserves quasi-isomorphisms. Suppose that X has at least two nonzero terms. Let m be the smallest integer satisfying $X_m \neq \{0\}$, and let n be the largest integer satisfying $X_n \neq \{0\}$. Write

$$X = \ 0 \longrightarrow X_n \longrightarrow \cdots \longrightarrow X_{m+1} \longrightarrow X_m \longrightarrow 0$$

and denote by X' the complex

$$X' = \ 0 \longrightarrow X_n \longrightarrow \cdots \longrightarrow X_{m+1} \longrightarrow 0 \longrightarrow 0$$

obtained from eliminating X_m from X. Consider X_m as a subcomplex of X, concentrated in degree m. Then we have an obvious exact sequence of complexes of right A-modules

$$0 \longrightarrow X_m \longrightarrow X \longrightarrow X' \longrightarrow 0.$$

In each fixed degree, this sequence is clearly a split exact sequence of right A-modules. Therefore, tensoring this sequence with either Y or Z yields exact sequences of complexes of k-modules (which are still split in each fixed degree). Thus we have a commutative diagram of complexes of k-modules

$$\begin{array}{ccccccccc}
0 & \longrightarrow & X_m \otimes_A Y & \longrightarrow & X \otimes_A Y & \longrightarrow & X' \otimes_A Y & \longrightarrow & 0 \\
 & & \downarrow{\scriptstyle \mathrm{Id}_{X_m} \otimes f} & & \downarrow{\scriptstyle \mathrm{Id}_X \otimes f} & & \downarrow{\scriptstyle \mathrm{Id}_{X'} \otimes f} & & \\
0 & \longrightarrow & X_m \otimes_A Z & \longrightarrow & X \otimes_A Z & \longrightarrow & X' \otimes_A Z & \longrightarrow & 0
\end{array}$$

with exact rows. The two vertical chain maps $\mathrm{Id}_{X_m} \otimes f$ and $\mathrm{Id}_{X'} \otimes f$ are quasi-isomorphisms by induction, and hence so is $\mathrm{Id}_X \otimes f$, by 1.17.8. □

The canonical evaluation maps from 1.12.6 can be assembled to yield chain maps, modulo taking care of signs.

Proposition 1.17.12 *Let A be a k-algebra, and let U, V be complexes of A-modules. For any two integers i and j denote by $\Phi_{i,j} : \mathrm{Hom}_A(U_{-i}, A) \otimes_A V_j \to \mathrm{Hom}_A(U_{-i}, V_j)$ the canonical evaluation map. Then the maps $(-1)^{ij}\Phi_{i,j}$ induce a chain map* $\mathrm{Hom}_A(U, A) \otimes_A V \to \mathrm{Hom}_A(U, V)$.

Proof For any integer i, the degree i term of $\mathrm{Hom}_A(U, A)$ is $\mathrm{Hom}_A(U_{-i}, A)$. Thus, for any integer n, the degree n term of $\mathrm{Hom}_A(U, A) \otimes_A V$ is the direct sum $\oplus_{i,j}\mathrm{Hom}_A(U_{-i}, A) \otimes_A V_j$, taken over all pairs of integers (i, j) such that $i + j = n$. Similarly, the degree n term of $\mathrm{Hom}_A(U, V)$ is the direct sum $\oplus_{i,j}\mathrm{Hom}_A(U_{-i}, V_j)$, taken over the same set of pairs of integers. It follows that the evaluation maps $\Phi_{i,j}$ induce a graded map from $\mathrm{Hom}_A(U, A) \otimes_A V$ to $\mathrm{Hom}_A(U, V)$, but this need not commute with the differentials. Using the sign conventions in the definitions of tensor product and homomorphism complexes one verifies that the maps $(-1)^{ij}\Phi_{i,j}$ commute with the differentials of the chain complexes $\mathrm{Hom}_A(U,A) \otimes_A V$ and $\mathrm{Hom}_A(U,V)$. □

If at least one of U, V has the property that all its terms are finitely generated projective, then by 1.12.7 the canonical evaluation maps are isomorphisms, and hence the chain map in 1.17.12 is an isomorphism of chain complexes.

1.18 Complexes and homotopy

Definition 1.18.1 Let $\mathcal{C}$ be an additive category and let (X, δ), (Y, ϵ) be complexes over $\mathcal{C}$. A *(chain) homotopy from X to Y* is a graded morphism $h : X \to Y$ of degree 1; that is, h is a family of morphisms $h_n : X_n \to Y_{n+1}$ in $\mathcal{C}$, for any $n \in \mathbb{Z}$. We do not require h to commute with the differentials. Two chain morphisms $f, f' : X \to Y$ are called *homotopic*, written $f \sim f'$, if there is a homotopy $h : X \to Y$ such that $f - f' = \epsilon \circ h + h \circ \delta$, or equivalently, if

$$f_n - f'_n = \epsilon_{n+1} \circ h_n + h_{n-1} \circ \delta_n$$

for any $n \in \mathbb{Z}$. In that case we say that *h is a homotopy from f to f'*. A chain map $f : X \to Y$ is a *homotopy equivalence* if there is a chain map $g : Y \to X$ such that $g \circ f \sim \mathrm{Id}_X$ and $f \circ g \sim \mathrm{Id}_Y$; in that case, g is called a *homotopy inverse of* f, and the complexes X, Y are said to be *homotopy equivalent*, written $X \simeq Y$. If $X \simeq 0$ (the zero complex), we say that X is *contractible*. For cochain complexes, we define analogously a *cochain homotopy* to be a graded morphism of degree -1.

Let $\mathcal{C}$ be an additive category. The *homotopy category of complexes over* $\mathcal{C}$ is the category $K(\mathcal{C})$ whose objects are the complexes over $\mathcal{C}$ and whose morphisms are the homotopy equivalence classes

$$\mathrm{Hom}_{K(\mathcal{C})}(X, Y) = \mathrm{Hom}_{\mathrm{Ch}(\mathcal{C})}(X, Y)/\sim$$

of chain maps, for any two complexes X, Y over $\mathcal{C}$. The composition of morphisms in $K(\mathcal{C})$ is induced by that in $\mathrm{Ch}(\mathcal{C})$. We denote by $K^+(\mathcal{C}), K^-(\mathcal{C}), K^b(\mathcal{C})$ the full subcategories of $K(\mathcal{C})$ consisting of bounded above, bounded below, bounded complexes over $\mathcal{C}$, respectively. If two chain maps $f, f' : X \to Y$ over an additive category $\mathcal{C}$ are homotopic via a homotopy $h : X \to Y$, then for any integer i, the "shifted" chain maps $f[i], f'[i] : X[i] \longrightarrow Y[i]$ are homotopic via the homotopy $h[i]$ given by $h[i]_n = h_{n-i}$ for any $n \in \mathbb{Z}$, and thus the automorphism $[i]$ of the additive category $\mathrm{Ch}(\mathcal{C})$ induces an automorphism, still denoted by $[i]$, of the homotopy category $K(\mathcal{C})$. This automorphism preserves any of the subcategories $K^+(\mathcal{C})$, $K^-(\mathcal{C})$, $K^b(\mathcal{C})$. The following observation is a trivial consequence of the above definitions:

Proposition 1.18.2 *For a chain complex X over an additive category $\mathcal{C}$ we have $X \simeq 0$ if and only if $\mathrm{Id}_X \sim 0$.*

Note that the first 0 in this statement means the zero complex and the second 0 means the zero chain map on X.

Proposition 1.18.3 *Let $\mathcal{C}$ be a abelian category, and let $f, f' : (X, \delta) \to (Y, \epsilon)$ be chain maps of complexes over $\mathcal{C}$.*

(i) For any homotopy $h : X \to Y$, the graded morphism $h \circ \delta + \epsilon \circ h : X \to Y$ is a chain map inducing the zero morphism from $H_*(X)$ to $H_*(Y)$.
(ii) If $f \sim f'$ then $H(f) = H(f') : H_*(X) \to H_*(Y)$.
(iii) If f is a homotopy equivalence, then f is a quasi-isomorphism.
(iv) If $X \simeq 0$ then X is acyclic.

Proof We have $\epsilon \circ (h \circ \delta + \epsilon \circ h) = \epsilon \circ h \circ \delta = (h \circ \delta + \epsilon \circ h) \circ \delta$, hence $h \circ \delta + \epsilon \circ h$ is a chain map from X to Y. Moreover, the map $\ker(\delta) \to \ker(\epsilon)$ induced by $h \circ \delta + \epsilon \circ h$ is equal to the map induced by $\epsilon \circ h$ and hence factors through the canonical monomorphism $\mathrm{Im}(\epsilon) \subset \ker(\epsilon)$, which shows in turn that it induces the zero map on homology, whence (i). If $f \sim f'$, then by (i), the difference $f - f'$ induces the zero map on homology and thus $H(f) = H(f')$, which proves (ii). Suppose f has a homotopy inverse g. Then, by (ii), we have $\mathrm{Id}_{H_*(X)} = H(g \circ f) = H(g) \circ H(f)$, and similarly $\mathrm{Id}_{H_*(Y)} = H(f) \circ H(g)$. Thus $H(g)$ and $H(f)$ are inverse to each other, proving (iii). If $X \simeq 0$, then X is quasi-isomorphic to zero by (iii), which is equivalent to $H_*(X) = 0$, whence (iv). □

For A an algebra, X a complex of A-modules, V an A-module, and n an integer, we denote by $V[n]$ the complex that is equal to V in degree n and zero in all other degrees, with the zero differential. We regard $\mathrm{Hom}_A(X, V)$ as a cochain complex, obtained from applying the contravariant functor $\mathrm{Hom}_A(-, V)$ to X. The following Proposition describes two instances in which the (co-)homology of a complex is reinterpreted as a space of homotopy classes of chain maps.

Proposition 1.18.4 *Let A be a k-algebra, V an A-module, and (X, δ) a complex of A-modules. Let n be an integer.*

(i) There is a natural isomorphism

$$H_n(\mathrm{Hom}_A(V, X)) \cong \mathrm{Hom}_{K(\mathrm{Mod}(A))}(V, X[n]).$$

In particular, there is a natural isomorphism

$$H_n(X) \cong \mathrm{Hom}_{K(\mathrm{Mod}(A))}(A, X[n]).$$

(ii) There is a natural isomorphism

$$H^n(\mathrm{Hom}_A(X, V)) \cong \mathrm{Hom}_{K(\mathrm{Mod}(A))}(X, V[n]).$$

Proof An element in degree n of $\mathrm{Hom}_A(V, X)$ is given by an A-homomorphism $\zeta : V \to X_n$. This belongs to the kernel of the differential of $\mathrm{Hom}_A(V, X)$ if and only if $\delta_n \circ \zeta = 0$. This is equivalent to asserting that ζ defines a chain map $V[n] \to X$. In that case, ζ belongs to the image of the differential of $\mathrm{Hom}_A(V, X)$ if and only if $\zeta = \delta_{n+1} \circ \eta$ for some A-homomorphism $\eta : V \to X_{n+1}$. This is equivalent to asserting that ζ, as a chain map, is homotopic to zero. This shows the first isomorphism in (i). The second isomorphism follows from this and the fact that $\mathrm{Hom}_A(A, -)$ is isomorphic to the identity functor on $\mathrm{Mod}(A)$. The naturality in X of the map thus constructed is obvious. This proves (i). Similarly, an element in degree n of the cochain complex $\mathrm{Hom}_A(X, V)$ is an A-homomorphism $\zeta : X_n \to V$. This belongs to the kernel of the differential of $\mathrm{Hom}_A(X, V)$ if and only if $\zeta \circ \delta_{n+1} = 0$. This is equivalent to asserting that ζ is a chain map $X \to V[n]$. In that case, ζ belongs to the image of the differential of $\mathrm{Hom}_A(X, V)$ if and only if $\zeta = \eta \circ \delta_n$ for some $\eta : X_{n-1} \to V$. This is equivalent to asserting that ζ is homotopic to zero, when regarded as a chain map $X \to V[n]$. Statement (ii) follows. □

Theorem 1.18.5 *Let $\mathcal{C}$ be an abelian category, let P be a complex of projective objects in $\mathcal{C}$, I a complex of injective objects in $\mathcal{C}$ and let*

$$0 \longrightarrow X \xrightarrow{f} Y \xrightarrow{g} Z \longrightarrow 0$$

be a short exact sequence of complexes over $\mathcal{C}$.

(i) Suppose that X is acyclic and that one of P, Y is bounded below. The map $\mathrm{Hom}_{\mathrm{Ch}(\mathcal{C})}(P, Y) \to \mathrm{Hom}_{\mathrm{Ch}(\mathcal{C})}(P, Z)$ given by composition with g is surjective and induces an isomorphism

$$\mathrm{Hom}_{K(\mathcal{C})}(P, Y) \cong \mathrm{Hom}_{K(\mathcal{C})}(P, Z).$$

(ii) Suppose that Z is acyclic and that one of Y, I is bounded above. The map $\mathrm{Hom}_{\mathrm{Ch}(\mathcal{C})}(Y, I) \to \mathrm{Hom}_{\mathrm{Ch}(\mathcal{C})}(X, I)$ given by precomposition with f is surjective and induces an isomorphism

$$\mathrm{Hom}_{K(\mathcal{C})}(Y, I) \cong \mathrm{Hom}_{K(\mathcal{C})}(X, I).$$

Proof (i) Denote by $\delta, \epsilon, \zeta, \pi$ the differentials of X, Y, Z, P, respectively. Given any chain map $q : P \to Z$ we construct inductively a chain map $p : P \to Y$ such that $q = g \circ p$. Since one of P, Y is bounded below, we have $q_i = 0$ for all sufficiently small integers i, so take $p_i = 0$ for i sufficiently small. Let

n be an integer. Suppose we have already constructed morphisms $p_i : P_i \to Y_i$ satisfying $g_i \circ p_i = q_i$ and $\epsilon_i \circ p_i = p_{i-1} \circ \pi_i$ for $i < n$. We construct p_n as follows. Since g_n is an epimorphism and P_n is projective, there is a morphism $p'_n : P_n \to Y_n$ such that $g_n \circ p'_n = q_n$. We have to adjust p'_n to make sure, that it is compatible with the differentials. We have $g_{n-1} \circ (\epsilon_n \circ p'_n - p_{n-1} \circ \pi_n) = \zeta_n \circ g_n \circ p'_n - g_{n-1} \circ p_{n-1} \circ \pi_n = \zeta_n \circ q_n - q_{n-1} \circ \pi_n = 0$ because q is a chain map. Thus we have a canonical monomorphism $\mathrm{Im}(\epsilon_n \circ p'_n - p_{n-1} \circ \pi_n) \subset \ker(g_{n-1}) = \mathrm{Im}(f_{n-1})$. Consequently, there is a morphism $\sigma : P_n \to X_{n-1}$ such that $f_{n-1} \circ \sigma = \epsilon_n \circ p'_n - p_{n-1} \circ \pi_n$. Moreover, $f_{n-2} \circ \delta_{n-1} \circ \sigma = \epsilon_{n-1} \circ f_{n-1} \circ \sigma = \epsilon_{n-1} \circ \epsilon_n \circ p'_n - \epsilon_{n-1} \circ p_{n-1} \circ \pi_n = -p_{n-2} \circ \pi_{n-1} \circ \pi_n = 0$, hence $\delta_{n-1} \circ \sigma = 0$ as f_{n-2} is a monomorphism. Therefore we have a canonical monomorphism $\mathrm{Im}(\sigma) \subset \ker(\delta_{n-1}) = \mathrm{Im}(\delta_n)$, where the last equality holds as X is acyclic. Thus there is a morphism $\rho : P_n \to X_n$ such that $\sigma = \delta_n \circ \rho$. Set $p_n = p'_n - f_n \circ \rho$. We still have

$$g_n \circ p_n = g_n \circ p'_n - g_n \circ f_n \circ \rho = g_n \circ p'_n = q_n,$$

and we now also have the compatibility with the differentials $\epsilon_n \circ p_n = \epsilon_n \circ p'_n - \epsilon_n \circ f_n \circ \rho = \epsilon_n \circ p'_n - f_{n-1} \circ \delta_n \circ \rho = \epsilon_n \circ p'_n - f_{n-1} \circ \sigma = \epsilon_n \circ p'_n - (\epsilon_n \circ p'_n - p_{n-1} \circ \pi_n) = p_{n-1} \circ \pi_n$ as required. This shows the surjectivity of the map given by composition with g. We need to show that $p \sim 0$ if and only if $q \sim 0$. If $p \sim 0$ there is a homotopy $h : P \to Y$ such that $p = \epsilon \circ h + h \circ \pi$. Composing with g yields $q = g \circ \epsilon \circ h + g \circ h \circ \pi = \zeta \circ g \circ h + g \circ h \circ \pi$, thus $q \sim 0$ via the homotopy $g \circ h : P \to Z$. In order to show the converse, observe first that since g_{n+1} is an epimorphism, any morphism $P_n \to Z_{n+1}$ lifts to a morphism $P_n \to Y_{n+1}$, and thus every homotopy $P \to Z$ lifts to some homotopy $P \to Y$. This means that if $q \sim 0$, then there is some chain map $p' : P \to Y$ such that $p' \sim 0$ and $g \circ p' = q$, but p' need not be equal to p. It suffices to show that $p - p' \sim 0$. Since $g \circ (p - p') = 0$, we may therefore assume that $q = 0$. Then $g \circ p = q = 0$, hence we have a canonical monomorphism $\mathrm{Im}(p) \subset \ker(g) = \mathrm{Im}(f)$. This implies that there is a chain map $u : P \to X$ such that $f \circ u = p$. It suffices to show that $u \sim 0$. This is again done inductively. Given an integer n, suppose that we have morphisms $h_i : P_i \to X_{i+1}$ satisfying $u_i = \delta_{i+1} \circ h_i + h_{i-1} \circ \pi_i$ for any $i < n$. Using this equality for $i = n-1$ we get $\delta_n \circ (u_n - h_{n-1} \circ \pi_n) = \delta_n \circ u_n - \delta_n \circ h_{n-1} \circ \pi_n = \delta_n \circ u_n - (u_{n-1} - h_{n-2} \circ \pi_{n-1}) \circ \pi_n = \delta_n \circ u_n - u_{n-1} \circ \pi_n = 0$, as u is a chain map. Again, we get a canonical monomorphism $\mathrm{Im}(u_n - h_{n-1} \circ \pi_n) \subset \ker(\delta_n) = \mathrm{Im}(\delta_{n+1})$. Therefore, as P_n is projective, there is $h_n : P_n \to X_{n+1}$ such that $\delta_{n+1} \circ h_n = u_n - h_{n-1} \circ \pi_n$ as required. This completes the proof of (i). The proof of (ii) is similar. □

Corollary 1.18.6 *Let $\mathcal{C}$ be an abelian category, let P be a complex of projective objects in $\mathcal{C}$, let I be a complex of injective objects in $\mathcal{C}$, and let X be an acyclic complex of objects in $\mathcal{C}$.*

(i) If one of X, P is bounded below, then $\mathrm{Hom}_{K(\mathcal{C})}(P, X) = \{0\}$.
(ii) If one of X, I is bounded above, then $\mathrm{Hom}_{K(\mathcal{C})}(X, I) = \{0\}$.

Proof Apply the above theorem to the short exact sequences $0 \to X \to X \to 0 \to 0$ and $0 \to 0 \to X \to X \to 0$. □

An abelian category $\mathcal{C}$ is said to have *enough projective objects* if for any object X in $\mathcal{C}$ there exists a projective object P in $\mathcal{C}$ and an epimorphism $\pi : P \to X$. Dually, $\mathcal{C}$ is said to have *enough injective objects* if for any object X in $\mathcal{C}$ there exists an injective object I and a monomorphism $\iota : X \to I$. If A is a k-algebra, then the category of A-modules $\mathrm{Mod}(A)$ has enough projective and injective objects. If A is Noetherian, then the category of finitely generated A-modules $\mathrm{mod}(A)$ has enough projective objects, but need not have enough injective objects.

Corollary 1.18.7 *Let $\mathcal{C}$ be an abelian category and X a complex over $\mathcal{C}$.*

(i) If $\mathcal{C}$ has enough projective objects, X is acyclic if and only if $\mathrm{Hom}_{K(\mathcal{C})}(P, X) = \{0\}$ *for any bounded below complex P of projective objects in $\mathcal{C}$.*
(ii) If $\mathcal{C}$ has enough injective objects, X is acyclic if and only if $\mathrm{Hom}_{K(\mathcal{C})}(X, I) = \{0\}$ *for any bounded above complex I of injective objects in $\mathcal{C}$.*

Proof (i) If X is acyclic, then $\mathrm{Hom}_{K(\mathcal{C})}(P, X) = \{0\}$ for any bounded below complex P of projective objects in $\mathcal{C}$ by 1.18.6. If X is not acyclic, then there is an integer n such that $H_n(X)$ is not zero, or equivalently, such that the canonical monomorphism $\mathrm{Im}(\delta_{n+1}) \subset \ker(\delta_n)$ is not an isomorphism, where δ is the differential of X. Let P be the complex that is zero in any degree other than n and that in degree n is a projective object P_n in $\mathcal{C}$ such that there is an epimorphism $\pi : P_n \to \ker(\delta_n)$; this is possible since $\mathcal{C}$ has enough projective objects. Then π defines a chain map from P to X that cannot be homotopic to zero, because π does not factor through δ_{n+1}. This shows (i). By dualising the above proof, one shows (ii). □

Corollary 1.18.8 *Let $\mathcal{C}$ be an abelian category.*

(i) If P is an acyclic bounded below complex of projective objects in $\mathcal{C}$, then $P \simeq 0$.

(ii) If I is an acyclic bounded above complex of injective objects in $\mathcal{C}$, then $I \simeq 0$.

Proof Apply 1.18.6 shows that $\mathrm{Id}_P \simeq 0$, hence $P \simeq 0$, whence (i). Similarly for (ii). □

Definition 1.18.9 Let $\mathcal{C}$ be an additive category. The *cone of a complex* (X, δ) *over* $\mathcal{C}$ is the complex $(C(X), \Delta)$ over $\mathcal{C}$ given by $C(X)_n = X_{n-1} \oplus X_n$ with differential

$$\Delta_n = \begin{pmatrix} -\delta_{n-1} & 0 \\ \mathrm{Id}_{X_{n-1}} & \delta_n \end{pmatrix} : X_{n-1} \oplus X_n \to X_{n-2} \oplus X_{n-1}$$

for all integers n.

The sign in the definition of Δ ensures that $\Delta \circ \Delta = 0$, so that $C(X)$ is indeed a complex. A chain map $f : X \to Y$ of complexes over an additive category $\mathcal{C}$ is called *degreewise split* if the morphism $f_n : X_n \to Y_n$ is split for all integers n; that is, if there are morphisms $s_n : Y_n \to X_n$ satisfying $f_n = f_n \circ s_n \circ f_n$ for all integers n. This does not necessarily imply that f is split as a chain map because the family $(s_n)_{n\in\mathbb{Z}}$ is not required be a chain map. The cone of X comes along with a canonical degreewise split monomorphism of complexes $i_X : X \to C(X)$ given by the canonical monomorphisms $X_n \hookrightarrow X_{n-1} \oplus X_n$ for any integer n, and with a degreewise split canonical epimorphism $p_X : C(X)[-1] \to X$ given by the canonical projections $X_n \oplus X_{n+1} \twoheadrightarrow X_n$ for any integer n (this matches with our sign convention: the differential of $C(X)[-1]$ is $-\Delta$ shifted by one degree). These chain maps yield a degreewise split short exact sequence of complexes

$$0 \longrightarrow X \xrightarrow{i_X} C(X) \xrightarrow{p_X[1]} X[1] \longrightarrow 0.$$

If is X is bounded below, bounded above or bounded, so is $C(X)$.

Proposition 1.18.10 *Let $\mathcal{C}$ be an additive category and let X be a complex over $\mathcal{C}$. We have $C(X) \simeq 0$.*

Proof Denote by δ and Δ the differentials of X and $C(X)$, respectively, and define a homotopy h on $C(X)$ by

$$h_n = \begin{pmatrix} 0 & \mathrm{Id}_{X_n} \\ 0 & 0 \end{pmatrix} : X_{n-1} \oplus X_n \longrightarrow X_n \oplus X_{n+1}$$

for any integer n. A straightforward matrix calculus shows that $h \circ \Delta + \Delta \circ h = \mathrm{Id}_{C(X)}$; thus $\mathrm{Id}_{C(X)} \sim 0$, or equivalently, $C(X) \simeq 0$. □

This proposition and the preceding short exact sequence highlight a formal similarity between the 'shift' functor [1] in a homotopy category of complexes and the injective Heller operator Σ defined in 2.14.1 below: both show up as the right hand side of a short exact sequence with a middle term whose image in the homotopy category and stable category, respectively, is zero.

Corollary 1.18.11 *Let A be a k-algebra, Y, Z complexes of A-modules, and let $g : Y \to Z$ be a chain map. The following are equivalent.*

(i) The chain map $g : Y \to Z$ is a quasi-isomorphism.
(ii) For any bounded below complex P of projective A-modules, composition with g induces an isomorphism $\mathrm{Hom}_{K(\mathrm{Mod}(A))}(P, Y) \cong \mathrm{Hom}_{K(\mathrm{Mod}(A))}(P, Z)$.
(iii) For any bounded above complex I of injective A-modules, precomposition with g induces an isomorphism $\mathrm{Hom}_{K(\mathrm{Mod}(A))}(Z, I) \cong \mathrm{Hom}_{K(\mathrm{Mod}(A))}(Y, I)$.

Proof Suppose that g is a quasi-isomorphism. By 1.18.10, the cone $C(Z)$ is contractible. Thus, by adding if necessary, to Y a copy of $C(Z)$ shifted by one degree, together with the canonical map $C(Z)[-1] \to Z$, we may assume that g is surjective. Then, by 1.17.5, the complex $X = \ker(g)$ is acyclic. It follows from 1.18.5 that (i) implies (ii). Suppose conversely that (ii) holds. Applying (ii) to the complex $A[n]$ which is equal to A in degree n and zero in all other degrees, for $n \in \mathbb{Z}$, in conjunction with 1.18.4 shows that g is a quasi-isomorphism. Using the fact that $\mathrm{Mod}(A)$ has enough injective objects, a variation of the above arguments shows the equivalence between (i) and (iii). Alternatively, this Corollary is in fact a formal consequence of the triangulated structure of homotopy categories; see A.4.9. □

The following Proposition has an interpretation in terms of relative projectivity; see 2.7.6 below.

Proposition 1.18.12 *Let $\mathcal{C}$ be an additive category and X a complex over $\mathcal{C}$. Denote by $\mathcal{F} : \mathrm{Ch}(\mathcal{C}) \to \mathrm{Gr}(\mathcal{C})$ the forgetful functor sending a complex (X, δ) to the underlying graded object X. The following are equivalent.*

(i) $X \simeq 0$.
(ii) For any chain map $f : Y \to Z$ of complexes over $\mathcal{C}$ and any chain map $u : X \to Z$ such that there is a graded morphism $v : X \to Y$ satisfying $u = f \circ v$ as graded morphisms, there is a chain map $w : X \to Y$ satisfying $u = f \circ w$.
(iii) For any chain map $f : Y \to Z$ of complexes over $\mathcal{C}$ and any chain map $u : Y \to X$ such that there is a graded morphism $v : Z \to X$ satisfying $u =$

$v \circ f$ as graded morphisms, there is a chain map $w : Z \to X$ satisfying $u = w \circ f$.

Proof Suppose that (i) holds. Let h be a homotopy on X such that $h \circ \delta + \delta \circ h = \mathrm{Id}_X$, where δ is the differential of X. Let $f : Y \to Z$, $u : X \to Z$ be chain maps and $v : X \to Y$ be a graded morphism such that $u = f \circ v$ as graded morphisms. Denote by ϵ, ζ the differentials of Y, Z, respectively. Set $w = v \circ h \circ \delta + \epsilon \circ v \circ h$. Clearly w is a chain map, and we have $f \circ w = f \circ v \circ h \circ \delta + f \circ \epsilon \circ v \circ = u \circ h \circ \delta + \zeta \circ f \circ v \circ h = u \circ h \circ \delta + \zeta \circ u \circ h = u \circ h \circ \delta + u \circ \delta \circ h = u \circ (h \circ \delta + \delta \circ h) = u$. Thus (ii) holds. A similar argument shows, that (i) implies (iii). Conversely, suppose that (ii) holds. The canonical projections $X_{n-1} \oplus X_n \to X_n$ define a graded morphism $v : C(X) \to X$ satisfying $v \circ i_X = \mathrm{Id}_X$. By the hypothesis, there is actually a chain map $w : C(X) \to X$ satisfying $w \circ i_X = \mathrm{Id}_X$. Moreover, since $C(X) \simeq 0$ there is a homotopy h on $C(X)$ such that $\Delta \circ h + h \circ \Delta = \mathrm{Id}_{C(X)}$, where Δ is the differential of $C(X)$. Composing with w on the left and i_X on the right yields $\delta \circ w \circ h \circ i_X + w \circ h \circ i_X \circ \delta = w \circ i_X = \mathrm{Id}_X$, and hence $X \simeq 0$ via the homotopy $w \circ h \circ i_X$ on X. Thus (i) holds. A similar argument, using the fact that p_X is a split epimorphism in each degree, shows that (iii) implies (i). □

Proposition 1.18.13 *Let $\mathcal{C}$ be an additive category and let $f : X \to Y$ be a chain map of complexes over $\mathcal{C}$. The following are equivalent.*

(i) $f \sim 0$.
(ii) f factors through some complex which is homotopy equivalent to zero.
(iii) f factors through the chain map $i_X : X \to C(X)$.
(iv) f factors through any degreewise split monomorphism $X \to Z$.
(v) f factors through the chain map $p_Y : C(Y)[-1] \to Y$.
(vi) f factors through any degreewise split epimorphism $Z \to Y$.

Proof Since $C(X) \simeq 0$, both (iii) and (iv) imply (ii), and (ii) implies clearly (i). Furthermore, (iv) implies (iii) and (vi) implies (v), because i_X is a degreewise split monomorphism and p_X is a degreewise split epimorphism. Conversely, since $C(X) \simeq 0$, it follows from 1.18.12, that (iii) implies (iv) and that (v) implies (iv). Suppose that (i) holds. It remains to show that both (iii) and (v) hold. Denote by δ, ϵ the differentials of X, Y, respectively, and let $h : X \to Y$ be a homotopy satisfying $f = h \circ \delta + \epsilon \circ h$. We define a graded morphism $r : C(X) \to Y$ by setting

$$r_n = \begin{pmatrix} h_{n-1} & f_n \end{pmatrix} : X_{n-1} \oplus X_n \to Y_n$$

for any integer n. This is easily seen to be a chain map satisfying $f = r \circ i_X$. Thus (iii) holds. Similarly, we define a graded morphism $s : X \to C(Y)[-1]$ by setting

$$s_n = \begin{pmatrix} f_n \\ h_n \end{pmatrix} : X_n \to Y_n \oplus Y_{n+1}$$

for any integer n. Again, this is a chain map satisfying $f = p_X \circ s$, and whence (v) holds. □

Definition 1.18.14 Let $\mathcal{C}$ be an additive category. A complex (X, δ) over $\mathcal{C}$ is called *split* if there is a graded endomorphism s of degree 1 of X such that $\delta \circ s \circ \delta = \delta$, or equivalently, if each δ_n is a split morphism in $\mathcal{C}$.

The graded morphism s in this definition need not commute with the differential. If (X, δ) is split, then 1.12.21 implies that s can be chosen such that $\delta \circ s \circ \delta = \delta$ and $s \circ \delta \circ s = s$.

Proposition 1.18.15 *Let A be a k-algebra and X a complex of A-modules.*

(i) The complex X is split if and only if $X \cong Y \oplus H_*(X)$ for some contractible complex Y, where $H_*(X)$ is considered as a complex with zero differential.
(ii) The complex X is contractible if and only if X is split acyclic.

Proof Denote by δ the differential of X. Suppose that X is split. Let s be a graded endomorphism of X of degree 1 satisfying $\delta \circ s \circ \delta = \delta$ and $s \circ \delta \circ s = s$. Then $\delta \circ s$ and $s \circ \delta$ are graded idempotent endomorphisms of degree 0 of X. Thus we have $X = \mathrm{Im}(s \circ \delta) \oplus \ker(s \circ \delta) = \mathrm{Im}(\delta \circ s) \oplus \ker(\delta \circ s)$ as graded A-modules. We have $\ker(\delta) \subseteq \ker(s \circ \delta) \subseteq \ker(\delta \circ s \circ \delta) = \ker(\delta)$, hence all inclusions are equalities. A similar argument yields $\mathrm{Im}(\delta) = \mathrm{Im}(\delta \circ s)$. Thus $X = \mathrm{Im}(s \circ \delta) \oplus \ker(\delta) = \ker(\delta \circ s) \oplus \mathrm{Im}(\delta)$ as graded A-modules. In particular, $\mathrm{Im}(\delta)$ and $\ker(\delta)$ are direct summands of X as graded A-modules. Since $\mathrm{Im}(\delta) \subseteq \ker(\delta)$ it follows that $\mathrm{Im}(\delta)$ is a direct summand of $\ker(\delta)$ as a graded A-module. Write $\ker(\delta) = \mathrm{Im}(\delta) \oplus H$ for some complement of $\mathrm{Im}(\delta)$ in $\ker(\delta)$ as a graded A-module. Thus $X = \mathrm{Im}(s \circ \delta) \oplus \mathrm{Im}(\delta) \oplus H$ as a graded A-module. Since $H \subseteq \ker(\delta)$, the graded submodule H of X is in fact a subcomplex of X with zero differential. By construction, $H \cong H_*(X)$. The graded submodule $Y = \mathrm{Im}(s \circ \delta) \oplus \mathrm{Im}(\delta)$ is a subcomplex of X because δ maps $\mathrm{Im}(\delta)$ to zero and $\mathrm{Im}(s \circ \delta)$ to $\mathrm{Im}(\delta \circ s \circ \delta) = \mathrm{Im}(\delta)$. We need to show that Y is contractible. Define a homotopy h on Y such that h is zero on the summand $\mathrm{Im}(s \circ \delta)$ and equal to s on the summand $\mathrm{Im}(\delta)$. Since $s \circ \delta$ is the identity on $\mathrm{Im}(s \circ \delta)$, a straightforward verification shows that Y is contractible with this homotopy. For the converse it suffices to note that a contractible complex (Y, ϵ) is split acyclic. Indeed, Y is acyclic by 1.18.3 (iii), and if h is a homotopy on Y such that $\epsilon \circ h + h \circ \epsilon = \mathrm{Id}_Y$, then composing by ϵ on the right yields $\epsilon \circ h \circ \epsilon = \epsilon$, hence Y is split. Both statements follow. □

Corollary 1.18.16 *Let A be a k algebra, and let X, Y be split complexes of A-modules. A chain map $f : X \to Y$ is a quasi-isomorphism if and only if f is a homotopy equivalence.*

Proof By 1.18.15 we may assume that X and Y have zero differentials. Then $X \cong H_*(X)$ and $Y \cong H_*(Y)$. It follows that if f is a quasi-isomorphism, then f is an isomorphism. The converse is clear by 1.18.3 (iii). □

Proposition 1.18.17 *Let A be a k-algebra, and let X be a bounded complex of projective A-modules such that $H_i(X)$ is projective for all integers i. Then X is split.*

Proof If X is zero or consists of a single nonzero term, then $X \cong H_0(X)$, so this is trivial. We argue by induction over the length of X. Suppose that X is nonzero, and let m be the smallest integer such that $X_m \neq \{0\}$. Denote by $\delta = (\delta_i)_{i \in \mathbb{Z}}$ the differential of X. We have $H_m(X) \cong X_m/(\mathrm{Im}(\delta_{m+1})$, which is projective by the assumptions. Thus the map $X_m \to X_m/(\mathrm{Im}(\delta_{m+1}))$ is split surjective. It follows that $X_m \cong \mathrm{Im}(\delta_{m+1}) \oplus H_m(M)$. Since X_m is projective, so is $\mathrm{Im}(\delta_{m+1})$, and hence the map $\delta_{m+1} : X_{m+1} \to \mathrm{Im}(\delta_{m+1})$ is split surjective. It follows that the complex X is the direct sum of $H_m(X)$, viewed as a complex concentrated in degree m, a contractible complex $\mathrm{Im}(\delta_{m+1}) \cong \mathrm{Im}(\delta_{m+1})$, where the two terms are in degree $m + 1$ and m, respectively, and a complex X' that coincides with X in all degrees, except in degree $m + 1$, where X'_{m+1} is the summand of X_{m+1} obtained from stripping off a summand isomorphic to $\mathrm{Im}(\delta_{m+1})$, and $X'_m = \{0\}$. All terms of X' and the homology of X' are still projective, and hence X' is split by induction. □

The previous proposition holds more generally for bounded below complexes.

Theorem 1.18.18 *Let C be an abelian category and let*

$$0 \longrightarrow X \xrightarrow{f} Y \xrightarrow{g} Z \longrightarrow 0$$

be a degreewise split short exact sequence of chain complexes over C.

(i) The chain map f is a homotopy equivalence if and only if $Z \simeq 0$. In that case, f is a split monomorphism.

(ii) The chain map g is a homotopy equivalence if and only if $X \simeq 0$. In that case, g is a split epimorphism.

Proof Suppose that f is a homotopy equivalence. Let $f' : Y \to X$ be a homotopy inverse of f. Then $f' \circ f \sim \mathrm{Id}_X$. By 1.18.13, the difference $\mathrm{Id}_X - f' \circ f$ factors through f since f is a degreewise split monomorphism. Let $t : Y \to X$ be a chain map such that $\mathrm{Id}_X - f' \circ f = t \circ f$. Then $\mathrm{Id}_X = (f' + t) \circ f$, hence f is split as a chain map with retraction $f' + t$. This shows that f and g induce an isomorphism $Y \cong X \oplus Z$. It follows that $Z \simeq 0$ as f is a homotopy equivalence. Conversely, suppose that $Z \simeq 0$. Since g is degreewise split surjective, it follows from 1.18.12 (ii) that g is split surjective as a chain map. Thus f and g induce an isomorphism $Y \cong X \oplus Z$, and since $Z \simeq 0$ this implies that f is a homotopy equivalence. This shows (i), and a similar argument proves (ii). □

Corollary 1.18.19 *Let X, Y be chain complexes over an abelian category $\mathcal{C}$. We have $X \simeq Y$ if and only if there are contractible complexes P, Q such that $X \oplus P \cong Y \oplus Q$ in $\mathrm{Ch}(\mathcal{C})$. In that case, if the complexes X, Y are both bounded above (resp. bounded below, bounded), then the complexes P, Q can be chosen to be bounded above (resp. bounded below, bounded), too.*

Proof Suppose that $f : X \to Y$ is a homotopy equivalence. Set $P = C(Y)[-1]$ and $p = p_Y$. Note that p is a degreewise split epimorphism. Thus the chain map $(f, p) : X \oplus P \to Y$ is a degreewise split epimorphism. Since $P \simeq 0$, the chain map (f, p) is still a homotopy equivalence. By 1.18.18, the complex $Q = \ker(f, p)$ satisfies $Q \simeq 0$ and $X \oplus P \cong Y \oplus Q$ in $\mathrm{Ch}(\mathcal{C})$. The converse is trivial. The last statement follows from the fact that if Y is bounded above (resp. bounded below, bounded), so is $C(Y)$. □

Corollary 1.18.20 *Let $\mathcal{A}$ be an abelian category.*

(i) Let P, Q be bounded below chain complexes consisting of projective objects in $\mathcal{A}$, and let $f : P \to Q$ be a chain map. Then f is a quasi-isomorphism if and only if f is a homotopy equivalence.
(ii) Let I, J be bounded below cochain complexes consisting of injective objects in $\mathcal{A}$, and let $g : I \to J$ be a cochain map. Then g is a quasi-isomorphism if and only if g is a homotopy equivalence.

Proof Suppose that f is a quasi-isomorphism. Since the cone $C(P)$ is contractible by 1.18.10, we may replace Q by $Q \oplus C(P)$ and f by $\binom{f}{i_X}$. Thus we may assume that f is degreewise split injective. Then the cokernel Z of f is an acyclic complex by 1.17.5. By construction, Z is also a bounded below complex of projective objects, and hence Z is contractible by 1.18.8. It follows from 1.18.18 that f is a homotopy equivalence. This shows (i); a similar argument proves (ii). □

Remark 1.18.21 Let A be a k-algebra, X a complex of right A-modules, and Y, Z complexes of left A-modules. If $h : Y \to Z$ is a homotopy, then $\mathrm{Id}_X \otimes h$ is a homotopy from $X \otimes_A Y$ to $X \otimes_A Z$. In particular, if $Y \simeq Z$, then $X \otimes_A Y \simeq X \otimes_A Z$. In other words, the functor $X \otimes_A -$ from $\mathrm{Ch}(\mathrm{Mod}(A))$ to $\mathrm{Ch}(\mathrm{Mod}(k))$ induces a functor from $K(\mathrm{Mod}(A))$ to $K(\mathrm{Mod}(k))$. Similar statements hold for functors of the form $\mathrm{Hom}_A(Y, -)$ and $\mathrm{Hom}_A(-, Z)$.

2

Functors Between Module Categories

Given two algebras A, B over a commutative ring k, an A-B-bimodule M gives rise to a plethora of functors between the module categories of A and B. The tensor products over A and B yield covariant right exact functors $M \otimes_B - : \mathrm{Mod}(B) \to \mathrm{Mod}(A)$ and $- \otimes_A M : \mathrm{Mod}(A^{\mathrm{op}}) \to \mathrm{Mod}(B^{\mathrm{op}})$. With the notational conventions described at the beginning of Chapter 1, taking homomorphism spaces yields covariant left exact functors $\mathrm{Hom}_A(M, -) : \mathrm{Mod}(A) \to \mathrm{Mod}(B)$ and $\mathrm{Hom}_{B^{\mathrm{op}}}(M, -) : \mathrm{Mod}(B^{\mathrm{op}}) \to \mathrm{Mod}(A^{\mathrm{op}})$ as well as contravariant functors $\mathrm{Hom}_A(-, M) : \mathrm{Mod}(A) \to \mathrm{Mod}(B)$ and $\mathrm{Hom}_{B^{\mathrm{op}}}(-, M) : \mathrm{Mod}(B^{\mathrm{op}}) \to \mathrm{Mod}(A^{\mathrm{op}})$. Moreover, $\mathrm{Hom}_A(M, A)$ and $\mathrm{Hom}_{B^{\mathrm{op}}}(M, B)$ become B-A-bimodules, and so we also have the analogous functors with A and B exchanged. Under suitable hypotheses, functors defined by tensoring and by taking homomorphism spaces are isomorphic: by Theorem 1.12.7, if M is finitely generated projective as a left A-module, then we have an isomorphism of functors $\mathrm{Hom}_A(M, -) \cong \mathrm{Hom}_A(M, A) \otimes_A -$, and if M is finitely generated projective as a right B-module, then we have an isomorphism of functors $\mathrm{Hom}_{B^{\mathrm{op}}}(M, -) \cong - \otimes_B \mathrm{Hom}_{B^{\mathrm{op}}}(M, B)$. The most basic instance for a functor that can be interpreted in both ways is the restriction functor sending a module over a group algebra to its restriction as a module over a subgroup algebra. Functors between module categories always induce functors between corresponding categories of complexes, preserving homotopies, and hence many of the formal properties we investigate in this chapter carry over – sometimes with minor adaptations – to complexes. Many of the concepts treated in this chapter have in common that they make sense over any commutative ring. This includes relative projectivity, relative trace maps, Frobenius reciprocity, Mackey's formula, Higman's criterion and the notion of symmetric algebras. If not stated otherwise, k denotes a commutative ring with unit, and k-algebras are associative and unital.

2.1 Induction and restriction

Given a k-algebra A, we denote by $\mathrm{Mod}(A)$ the k-linear abelian category of left A-modules and by $\mathrm{mod}(A)$ the full subcategory of finitely generated A-modules. We identify $\mathrm{Mod}(A^{\mathrm{op}})$ with the category of right A-modules.

Definition 2.1.1 Let G be a finite group and let H be a subgroup of G. We denote by

$$\mathrm{Res}^G_H : \mathrm{Mod}(kG) \to \mathrm{Mod}(kH)$$

the *restriction functor* sending a kG-module U to its restriction $\mathrm{Res}^G_H(U)$ to kH and sending a kG-homomorphism $\varphi : U \to U'$ to φ viewed as a kH-homomorphism. We denote by

$$\mathrm{Ind}^G_H : \mathrm{Mod}(kH) \to \mathrm{Mod}(kG)$$

the *induction functor* sending a kH-module V to the kG-module $\mathrm{Ind}^G_H(V) = kG \otimes_{kH} V$ and sending a kH-homomorphism $\psi : V \to V'$ to the kG-homomorphism $\mathrm{Id}_{kG} \otimes \psi : kG \otimes_{kH} V \to kG \otimes_{kH} V'$.

If V is a finitely generated kH-module then the kG-module $\mathrm{Ind}^G_H(V)$ is finitely generated; similarly, if U is a finitely generated kG-module then the kH-module $\mathrm{Res}^G_H(U)$ is finitely generated. Thus Ind^G_H and Res^G_H restrict to functors between $\mathrm{mod}(kG)$ and $\mathrm{mod}(kH)$. If $x, y \in G$ then the two right H-cosets xH, yH are either equal or disjoint, and hence the right H-cosets in G form a partition of G. We denote by $[G/H]$ a subset of G such that $G = \cup_{x\in[G/H]} xH$ is a disjoint union. The set $[G/H]$ is called a *set of representatives of the set of right H-cosets in G*. Similarly, we denote by $[H\backslash G]$ a set of representatives of the left H-cosets Hx in G, where $x \in G$. If L is another subgroup of G, again any two *double cosets* HxL, HyL are either equal or disjoint, and we denote by $[H\backslash G/L]$ a set of representatives in G of the set of H-L-double cosets in G. In particular, for any two subgroups H, L of G and $x \in G$, the left coset Hx gives rise to the left kH-module $k[Hx]$, the right coset xH gives rise to a right kH-module, and from the double coset HxL we obtain the kH-kL-bimodule $k[HxL]$. Right multiplication by x and x^{-1} on kH and $k[Hx]$, respectively, induce inverse isomorphisms $kH \cong k[Hx]$ of left kH-modules. Thus we have an isomorphism of left kH-modules $kG \cong (kH)^{|G:H|}$, which shows that kG is free of rank $|G : H|$ as a left kH-module. Similarly, we have an isomorphism of right kH-modules $kG \cong (kH)^{|G:H|}$, hence kG is free of rank $|G : H|$ as a right kH-module. The kH-kL-bimodule $k[HxL]$ is free as a left kH-module because HxL is a union of certain left H-cosets, and similarly it is also free as a right kL-module. Writing G as the disjoint union of H-L-double cosets yields a decomposition of kG as a direct sum of kH-kL-bimodules

$$kG = \oplus_{x\in[H\backslash G/L]}k[HxL].$$

Given any kH-module V, the induced module $\mathrm{Ind}_H^G(V)$ can be written as a direct sum of k-modules as follows:

$$\mathrm{Ind}_H^G(V) = kG \otimes_{kH} V = \oplus_{x\in[G/H]}k[xH] \otimes_{kH} V = \oplus_{x\in[G/H]}x \otimes V.$$

This is a decomposition of $\mathrm{Ind}_H^G(V)$ as a k-module, not as a kG-module. Every subspace $x \otimes V$ is isomorphic to V as a k-module, but the action of G permutes these summands. In other words, as a k-module, $\mathrm{Ind}_H^G(V)$ is the direct sum of $|G : H|$ copies of V which are permuted by the action of G, in a way that encodes the permutation action of G on G/H.

Any functor $\mathcal{F}$ between the module categories of two algebras A and B extends in an obvious way to a functor between the categories of complexes of modules over A and B, preserving homotopies, hence inducing a functor between the homotopy categories of complexes (which will again be denoted by $\mathcal{F}$ if no confusion arises). In particular, restriction and induction between a finite group algebra kG and a subgroup algebra kH yield functors Res_H^G from $\mathrm{Ch}(\mathrm{Mod}(kG))$ and $K(\mathrm{Mod}(kG))$ to $\mathrm{Ch}(\mathrm{Mod}(kH))$ and $K(\mathrm{Mod}(kH))$, respectively; similarly for Ind_H^G. Restriction and induction extend to bimodules: if B is a k-algebra, and if U is a kG-B-bimodule, then we use the same notation $\mathrm{Res}_H^G(U)$ for the kH-B-bimodule obtained from restricting the left kG-module structure to kH; this yields a functor

$$\mathrm{Res}_H^G : \mathrm{Mod}(kG \otimes_k B^{\mathrm{op}}) \to \mathrm{Mod}(kH \otimes_k B^{\mathrm{op}}).$$

Similarly, induction yields a functor

$$\mathrm{Ind}_H^G : \mathrm{Mod}(kH \otimes_k B^{\mathrm{op}}) \to \mathrm{Mod}(kG \otimes_k B^{\mathrm{op}})$$

sending a kH-B-bimodule V to the kG-B-bimodule $\mathrm{Ind}_H^G(V) = kG \otimes_{kH} V$. Induction and restriction satisfy the following obvious transitivity property.

Proposition 2.1.2 *Let G be a finite group and let H, L be subgroups of G such that $L \leq H \leq G$. For any kG-module or complex of kG-modules V we have a natural isomorphism $\mathrm{Res}_L^G(V) \cong \mathrm{Res}_L^H(\mathrm{Res}_H^G(V))$ and for any kL-module or complex of kL-modules W we have a natural isomorphism $\mathrm{Ind}_L^G(W) \cong \mathrm{Ind}_H^G(\mathrm{Ind}_L^H(W))$.*

Examples 2.1.3 Let G be a finite group and let k be a commutative ring.

(1) The regular kG-module kG is isomorphic to $\mathrm{Ind}_1^G(k) = kG \otimes_k k$. Thus every free kG-module is of the form $\mathrm{Ind}_1^G(V)$ for some free k-module V;

we use here the fact that the tensor product commutes with direct sums (cf. A.1.9). In particular, every projective kG-module is isomorphic to a direct summand of $\mathrm{Ind}_1^G(V)$ for some free k-module V.

(2) Let M be a transitive G-set. Let $m \in M$ and let $H = \{y \in G | ym = m\}$ be the stabiliser of m in G. The map sending a coset xH to xm is a bijection of G-sets $G/H \cong M$. Indeed, this map is well-defined because $xhm = xm$ for any $x \in G$ and any $h \in H$, this map is surjective because M is transitive, and this map is injective, because the equality $xm = ym$ is equivalent to $y^{-1}xm = m$, hence to $xH = yH$. We have an isomorphism of kG-modules $k[G/H] \cong \mathrm{Ind}_H^G(k)$ sending xH to $x \otimes 1_k$, where k is understood as the trivial kH-module. The inverse of this map sends $x \otimes \lambda$ to λxH, where $x \in G$ and $\lambda \in k$. Thus the permutation kG-module kM of an arbitrary finite G-set is a direct sum of modules of the form $\mathrm{Ind}_H^G(k)$, one summand for each G-orbit in M.

Let G be a finite group, H a subgroup of G and k a commutative ring. The basic properties of the tensor product imply that for any two kH-modules V, V', the map $\mathrm{Ind}_H^G : \mathrm{Hom}_{kH}(V, V') \to \mathrm{Hom}_{kG}(\mathrm{Ind}_H^G(V), \mathrm{Ind}_H^G(V'))$ is k-linear, and that the functor Ind_H^G maps any short exact sequence of kH-modules to a short exact sequence of kG-modules. Similar statements hold for the restriction functor Res_H^G. This can be expressed slightly more precisely as follows:

Theorem 2.1.4 *Let G be a finite group and let H be a subgroup of G. The covariant functors $\mathrm{Ind}_H^G : \mathrm{Mod}(kH) \to \mathrm{Mod}(kG)$ and $\mathrm{Res}_H^G : \mathrm{Mod}(kG) \to \mathrm{Mod}(kH)$ are k-linear exact. Moreover, both functors send (finitely generated) projective modules to (finitely generated) projective modules.*

Proof The exactness of a sequence of kG-modules $U \xrightarrow{\varphi} V \xrightarrow{\psi} W$ is equivalent to $\mathrm{Im}(\varphi) = \ker(\psi)$. For the purpose of checking the exactness of a sequence of kG-modules it suffices to check the exactness of the underlying sequence of k-modules (that is, we may ignore the action of G). The exactness of the functor Res_H^G is clear, because restricting the group action to a subgroup does not alter the underlying k-module structure. Tensoring a sequence of kH-modules by $kH \otimes_{kH} -$ is exact because $kH \otimes_{kH} -$ is isomorphic to the identity functor. Thus tensoring by a right kH-module which is a finite direct sum of copies of kH is exact. It follows that the functor Ind_H^G is exact because $kG = \oplus_{x \in [G/H]} xkH$ is free of finite rank as a right kH-module. Tensoring with a bimodule over two k-algebras is always a k-linear functor, so restriction and induction are k-linear functors. The functor Ind_H^G sends the free kH-module kH of rank 1 to the free kG-module of rank 1. Thus Ind_H^G sends a free kH-module to a free kG-module (of the same rank if that rank is finite), and thus Ind_H^G sends projective kH-modules (resp. finitely generated projective kH-modules) to projective kG-modules (resp. finitely generated projective kG-modules). The functor Res_H^G sends the free kG-module kG of rank 1 to a free kH-modules of rank $|G : H|$, and hence a similar argument as before shows that Res_H^G preserves the property of being (finitely generated) projective. □

As pointed out before, the covariant functors Ind_H^G and Res_H^G induce functors between the categories of chain complexes $\mathrm{Ch}(\mathrm{Mod}(kG))$ and $\mathrm{Ch}(\mathrm{Mod}(kH))$. Since they are exact, they 'commute' with taking homology (and the analogous results hold for the cohomology of cochain complexes):

Corollary 2.1.5 *Let G be a finite group and H a subgroup of G. Let X be a complex of kG-modules and Y a complex of kH-modules. Then $H_n(\mathrm{Res}_H^G(X)) \cong \mathrm{Res}_H^G(H_n(X))$ and $H_n(\mathrm{Ind}_H^G(Y)) \cong \mathrm{Ind}_H^G(H_n(Y))$ for any integer n.*

Proof Straightforward verification; in general, exact functors between abelian categories induce functors on the categories of (co-)chain complexes that commute with taking (co-)homology. □

Induction and restriction functors can be defined for any subalgebra B of a k-algebra A. Restricting an A-module U to U viewed as a module over a subalgebra B of A yields an exact functor Res_B^A from $\mathrm{Mod}(A)$ to $\mathrm{Mod}(B)$. The correspondence sending a B-module V to the module $\mathrm{Ind}_B^A(V) = A \otimes_B V$ yields a right exact functor Ind_B^A from $\mathrm{Mod}(B)$ to $\mathrm{Mod}(A)$. This functor sends a homomorphism of B-modules $\varphi : V \to V'$ to the homomorphism of A-modules $\mathrm{Id}_A \otimes \varphi : A \otimes_B V \to A \otimes_B V'$. Tensoring by a finitely generated projective module is an exact functor. Thus if A is finitely generated projective as a right B-module, then the functor Ind_B^A is exact. The identity functor on $\mathrm{Mod}(A)$ is isomorphic to the functors $A \otimes_A -$ and $\mathrm{Hom}_A(A, -)$. Thus the restriction functor Res_B^A is isomorphic to ${}_BA \otimes_A -$, where ${}_BA = A$ viewed as a B-A-bimodule, and also to $\mathrm{Hom}_A(A_B, -)$, where $A_B = A$ viewed as an A-B-bimodule. Restriction functors are always exact. By contrast, the induction functor $\mathrm{Ind}_B^A = A_B \otimes_B -$ need not be of the form $\mathrm{Hom}_B(M, -)$ for any A-B-bimodule M because it need not be exact. In the case of finite group algebras, the functor Ind_H^G from 2.1.4 is exact, and one can show that Ind_H^G is isomorphic to the functor $\mathrm{Hom}_{kH}(kG, -)$, with kG viewed as a kH-kG-bimodule. This is a special case of more general statements on bimodules over symmetric algebras; see 2.12.3 below.

In fact, induction and restriction make sense for any algebra homomorphism $\alpha : B \to A$. Given an A-module U, we denote by ${}_\alpha U$ the B-module that is equal to U as a k-module, endowed with the action of $b \in B$ as left multiplication by $\alpha(b)$. This defines an exact functor $\mathrm{Res}_\alpha : \mathrm{Mod}(A) \to \mathrm{Mod}(B)$. This functor is isomorphic to the functor ${}_\alpha A \otimes_A -$, where ${}_\alpha A$ is the B-A-bimodule that is equal to A as a right A-module, endowed as before with the left action of b by multiplication with $\alpha(b)$. Analogously, define A_α to be the A-B-bimodule that is equal to A as a left A-module, with right B-module structure given by the action of $b \in B$ by right multiplication with $\alpha(b)$. This bimodule induces a functor $\mathrm{Ind}_\alpha = A_\alpha \otimes_B - : \mathrm{Mod}(B) \to \mathrm{Mod}(A)$, which need not be exact, as A_α need not be projective as a right B-module. If α is an algebra isomorphism, then Res_α is in equivalence of categories $\mathrm{Mod}(A) \cong \mathrm{Mod}(B)$, with inverse $\mathrm{Res}_{\alpha^{-1}}$. This is a special case of a Morita equivalence; see 2.8.1.

Up to this point, the material of this section extends to twisted group algebras. Let G be a finite group and $\alpha \in Z^2(G; k^\times)$. For any subset U of G denote by $k_\alpha U$ the k-module generated by the image of U in $k_\alpha G$. If H is a subgroup of G then $k_\alpha H$ is a subalgebra of $k_\alpha G$, and if $x \in G$ then $k_\alpha[Hx] = k_\alpha H \cdot x$ is a left $k_\alpha H$-submodule of $k_\alpha G$ isomorphic to the regular module $k_\alpha H$ because the image of x in $k_\alpha G$ is invertible. By letting x run over a set of representatives of G/H in G, one sees that $k_\alpha G$ is free of rank $|G : H|$ as a left $k_\alpha H$-module. Similarly, $k_\alpha G$ is free of rank $|G : H|$ as a right $k_\alpha H$-module. If L is another subgroup of G, then $k_\alpha[HxL] = k_\alpha H \cdot x \cdot k_\alpha L$ is a $k_\alpha H$-$k_\alpha L$-subbimodule of $k_\alpha G$. The corresponding restriction and induction functors $\mathrm{Res}^{k_\alpha G}_{k_\alpha H}$ and $\mathrm{Ind}^{k_\alpha G}_{k_\alpha H} = k_\alpha G \otimes_{k_\alpha H} -$ are as in the untwisted case exact and transitive.

If H is a subgroup of a finite group G of index n and $\{x_i | 1 \le i \le n\}$ a set of representatives of G/H in G, and V a kH-module, then the action of an element $x \in G$ on an element in $\mathrm{Ind}_H^G(V) = \oplus_{i=1}^n x_i \otimes V$ combines the action of H on V and the permutation action of G on G/H. This action can be explicitly described by

$$x \cdot (x_i \otimes v) = x_{\pi^{-1}(i)} \otimes y_i v,$$

where π is the permutation in the symmetric group S_n satisfying $xx_i = x_{\pi^{-1}(i)} y_i$ for some $y_i \in H$. Tensor induction is obtained by replacing the direct sum by a tensor product over k.

Definition 2.1.6 Let G be a finite group, H a subgroup and V a kH-module. We define a kG-module $\mathrm{Ten}_H^G(V)$ as follows. Set $n = |G : H|$, and let $\{x_i | 1 \le i \le n\}$ be a set of representatives of G/H in G. Set $\mathrm{Ten}_H^G(V) = \otimes_{i=1}^n (x_i \otimes V)$, the tensor product over k of the k-submodules $x_i \otimes V$ of $kG \otimes_{kH} V$, endowed with the following action of G. Let $x \in G$, $\pi \in S_n$ and $y_i \in H$ such that $xx_i = x_{\pi^{-1}(i)} y_i$, and let $v_i \in V$, for $1 \le i \le n$. Set

$$x \cdot (\otimes_{i=1}^n (x_i \otimes v_i)) = \otimes_{i=1}^n (x_i \otimes y_{\pi(i)} v_{\pi(i)}).$$

One checks that this is a well-defined construction, and that up to isomorphism, $\mathrm{Ten}_H^G(V)$ does not depend on the choice of the set of representatives of G/H in G. If V is free of finite rank m as a k-module, then $\mathrm{Ten}_H^G(V)$ is free of rank m^n as a k-module. There is a more structural way to describe tensor induction, which also takes care of the issue of being well-defined. With the notation of 2.1.6, denote by H^n the direct product of n copies of H. The wreath product $H \wr S_n$ is equal to the semidirect product $H^n \rtimes S_n$, with S_n acting on H^n by permuting the components. Any choice of coset representatives x_i as in 2.1.6 defines an injective group homomorphism

$$\tau : G \to H \wr S_n,$$

sending $x \in G$ to $((y_i)_{1 \le i \le n}, \pi)$, where the $y_i \in H$ and $\pi \in S_n$ are defined as in 2.1.6. A different choice of coset representatives yields a group homomorphism that is conjugate to τ by an element in $H \wr S_n$. We use the superscript $\otimes n$ to indicate the tensor product over k of n copies of a k-module. Since $k(H^n) \cong (kH)^{\otimes n}$, the k-module $V^{\otimes n}$ becomes a module for kH^n, such that the i-th copy of H acts on the i-th copy of V. This kH^n-module structure extends in the obvious way to a $k(H \wr S_n)$-module structure, by letting S_n act on $V^{\otimes n}$ by permuting the components. One checks that

$$\mathrm{Ten}_H^G(V) \cong \mathrm{Res}_G^{H \wr S_n}(V^{\otimes n}),$$

where G is identified to a subgroup of $II \wr S_n$ via τ. By the above remarks, this identification is unique up to conjugacy, and hence the restriction to G does not depend, up to isomorphism, on the choice of τ. Tensor induction is not an additive functor, but it is transitive and compatible with tensor products of modules:

Proposition 2.1.7 *Let G be a finite group, and let H, L be subgroups of G. Let V, V' be kH-modules and let W be a kL-module.*

(i) If $L \le H$, then $\mathrm{Ten}_L^G(W) \cong \mathrm{Ten}_H^G(\mathrm{Ten}_L^H(W))$.
(ii) We have $\mathrm{Ten}_H^G(V \otimes_k V') \cong \mathrm{Ten}_H^G(V) \otimes_k \mathrm{Ten}_H^G(V')$.

Proof Straightforward verifications. □

As mentioned before, extending induction and restriction to chain complexes poses no problems. Extending tensor induction to chain complexes is also possible, but requires more care because of the signs in the differentials of tensor products of chain complexes. At the level of group cohomology, tensor induction yields the *Evens norm map*; see [6, §4.1] for more details.

2.2 Frobenius reciprocity

Theorem 2.2.1 (Frobenius reciprocity) *Let G be a finite group and H a subgroup of G. Let U be a kG-module and V a kH-module.*

(i) We have a natural k-linear isomorphism

$$\begin{cases} \mathrm{Hom}_{kG}(\mathrm{Ind}_H^G(V), U) & \cong & \mathrm{Hom}_{kH}(V, \mathrm{Res}_H^G(U)) \\ \varphi & \mapsto & (v \mapsto \varphi(1_G \otimes v)). \end{cases}$$

(ii) We have a natural k-linear isomorphism

$$\begin{cases} \mathrm{Hom}_{kH}(\mathrm{Res}_H^G(U), V) & \cong & \mathrm{Hom}_{kG}(U, \mathrm{Ind}_H^G(V)) \\ \psi & \mapsto & (u \mapsto \sum_{x \in [G/H]} x \otimes \psi(x^{-1}u)). \end{cases}$$

Proof One checks that the given map in the statement (i) is well-defined, and that the map sending $\psi \in \mathrm{Hom}_{kH}(V, \mathrm{Res}_H^G(U))$ to the unique linear map φ defined by $\varphi(x \otimes v) = x\psi(v)$ for $x \in G$ and $v \in V$ is its inverse. Similarly, one checks that the given map in (ii) is well-defined, and one constructs the inverse as follows: given $\varphi \in \mathrm{Hom}_{kG}(U, \mathrm{Ind}_H^G(V))$, and $u \in U$, write $\varphi(u) = \sum_{x \in [G/H]} x \otimes v_x$ for some $v_x \in V$. Then exactly one element $y \in [G/H]$ lies in H; we set $\psi(u) = yv_y$. □

The naturality of these isomorphisms in both arguments implies that we can replace U and V by a complex of modules over kG and kH, respectively; see Theorem 2.2.7 below. There is a tensor product version of Frobenius reciprocity.

Theorem 2.2.2 *Let G be a finite group and H a subgroup of G. Let U be a kG-module and V a kH-module. There is a natural isomorphism of kG-modules*

$$\begin{cases} U \otimes_k \mathrm{Ind}_H^G(V) & \cong & \mathrm{Ind}_H^G(\mathrm{Res}_H^G(U) \otimes_k V) \\ u \otimes (x \otimes v) & \mapsto & x \otimes (x^{-1}u \otimes v) \end{cases}$$

where $u \in U$, $x \in G$ and $v \in V$.

Proof The given k-linear map is a kG-homomorphism. The unique k-linear map sending $x \otimes (u \otimes v)$ to $xu \otimes (x \otimes v)$ is its inverse. □

The first statement in 2.2.1 carries over to the restriction and induction functors for an arbitrary subalgebra B of an algebra A.

Theorem 2.2.3 *Let A be a k-algebra and B a subalgebra of A. For any A-module U and any B-module V we have natural inverse isomorphisms of k-modules*

$$\begin{cases} \mathrm{Hom}_A(A \otimes_B V, U) & \cong & \mathrm{Hom}_B(V, \mathrm{Res}^A_B(U)) \\ \varphi & \mapsto & (v \mapsto \varphi(1 \otimes v)) \\ (a \otimes v \mapsto a\psi(v)) & \leftarrow & \psi \end{cases}.$$

This is easily verified directly, but follows also from a more general result which we will describe next. Given two k-algebras A, B and an A-B-bimodule M, we consider the following two functors. We have a functor

$$M \otimes_B - : \mathrm{Mod}(B) \to \mathrm{Mod}(A)$$

sending a B-module V to the A-module $M \otimes_B V$ and sending a B-homomorphism $\psi : V \to V'$ to the A-homomorphism $\mathrm{Id}_M \otimes \psi : M \otimes_B V \to M \otimes_B V'$, and we have a functor

$$\mathrm{Hom}_A(M, -) : \mathrm{Mod}(A) \to \mathrm{Mod}(B)$$

sending an A-module U to $\mathrm{Hom}_A(M, U)$, viewed as a B-module via $(b.\mu)(m) = \mu(mb)$ for all $\mu \in \mathrm{Hom}_A(M, U)$ and $b \in B$, and sending an A-homomorphism $\varphi : U \to U'$ to the B-homomorphism $\mathrm{Hom}_A(M, U) \to \mathrm{Hom}_A(M, U')$ which maps $\mu \in \mathrm{Hom}_A(M, U)$ to $\varphi \circ \mu$. If B is a subalgebra of A and $M = A$, viewed as an A-B-bimodule, then $M \otimes_B -$ is the induction functor Ind^A_B and $\mathrm{Hom}_A(M, -)$ is isomorphic to the restriction functor Res^A_B. Indeed, any A-homomorphism from A to an A-module U is determined by its value at 1, and for any $u \in U$ there is exactly one A-homomorphism $A \to U$ sending 1 to u (namely the homomorphism sending $a \in A$ to au) whence $\mathrm{Hom}_A(A, U) \cong U$. If A is viewed as an A-B-bimodule, then this is an isomorphism of B-modules. Thus the special case $M = A$ of the next result yields the previous result.

Theorem 2.2.4 *Let A, B be k-algebras and let M be an A-B-bimodule. For any A-module U and any B-module V we have natural inverse isomorphisms of k-modules*

$$\begin{cases} \mathrm{Hom}_A(M \otimes_B V, U) & \cong & \mathrm{Hom}_B(V, \mathrm{Hom}_A(M, U)) \\ \varphi & \mapsto & (v \mapsto (m \mapsto \varphi(m \otimes v))) \\ (m \otimes v \mapsto \psi(v)(m)) & \leftarrow & \psi \end{cases}.$$

Proof This is a series of straightforward verifications: one checks that

(1) the map $m \mapsto \varphi(m \otimes v)$ is an A-homomorphism from M to U;
(2) the map $v \mapsto (m \mapsto \varphi(m \otimes v))$ is a B-homomorphism from V to $\mathrm{Hom}_A(M, U)$;

(3) the map $\varphi \mapsto (v \mapsto (m \mapsto \varphi(m \otimes v)))$ is k-linear;
(4) the map $m \otimes v \mapsto \psi(v)(m)$ is well-defined (that is, one needs to check that $mb \otimes v$ and $m \otimes bv$ have the same image, for $b \in B$);
(5) the map $m \otimes v \mapsto \psi(v)(m)$ is an A-homomorphism from $M \otimes_B V$ to U;
(6) the map $\psi \mapsto (m \otimes v \mapsto \psi(v)(m))$ is inverse to the map $\varphi \mapsto (v \mapsto (m \mapsto \varphi(m \otimes v)))$. □

This theorem can be restated as saying that the functor $M \otimes_B -$ is *left adjoint* to the functor $\mathrm{Hom}_A(M, -)$. The general formal background on adjoint functors will be developed in the next section. Statement (i) in 2.2.1 is the particular case of 2.2.4 applied to $A = kG$, $B = kH$, $M = kG$ viewed as a kG-kH-bimodule in the obvious way. Indeed, $M \otimes_B V = \mathrm{Ind}_H^G(V)$ in that case, and $\mathrm{Hom}_A(M, U) \cong \mathrm{Res}_H^G(U)$, because a kG-homomorphism from $M = kG$ to U is determined by its value at 1, and so sending $\mu \in \mathrm{Hom}_{kG}(kG, U)$ to $\mu(1) \in U$ is a linear isomorphism, and one checks that this is also an isomorphism of kH-modules. Statement (ii) in 2.2.1 is however not just a direct application of 2.2.4; it involves a duality property which is specific to the class of *symmetric algebras*, introduced in §2.11 below. The naturality of the maps in 2.2.4 implies that these maps are compatible with right module structures on U and on V, yielding the following generalisation of 2.2.4:

Theorem 2.2.5 *Let A, B, C be k-algebras, let M be an A-B-bimodule, U an A-C-bimodule U and V a B-C-bimodule. We have a natural isomorphism of k-modules*

$$\mathrm{Hom}_{A\otimes_k C^{\mathrm{op}}}(M \otimes_B V, U) \cong \mathrm{Hom}_{B\otimes_k C^{\mathrm{op}}}(V, \mathrm{Hom}_A(M, U)).$$

Proof This is proved exactly as 2.2.4, taking into account the right C-module structures as appropriate. □

Natural transformations between functors of the form $\mathrm{Hom}_A(M, -)$ and $M \otimes_B -$ are easy to describe. Let M and M' be two A-B-bimodules. Any bimodule homomorphism $\psi : M' \to M$ gives rise to a natural transformation $\mathrm{Hom}_A(M, -) \to \mathrm{Hom}_A(M', -)$ given by the family of maps $\mathrm{Hom}_A(M, U) \to \mathrm{Hom}_A(M', U)$ sending $\varphi \in \mathrm{Hom}_A(M, U)$ to $\varphi \circ \psi$, with U running over the A-modules. Similarly, a bimodule homomorphism $\sigma : M \to M'$ gives rise to a natural transformation $(M \otimes_B -) \to (M' \otimes_B -)$ given by the family of maps $\sigma \otimes \mathrm{Id}_U : M \otimes_B U \to M' \otimes_B U$. The next result states that there are no other natural transformations between these pairs of functors:

Theorem 2.2.6 *Let A, B be k-algebras, and let M, M′ be A-B-bimodules.*

(i) There is a bijection $\mathrm{Nat}(\mathrm{Hom}_A(M, -), \mathrm{Hom}_A(M', -)) \cong \mathrm{Hom}_{A\otimes_k B^{\mathrm{op}}}(M', M)$ *sending a natural transformation* $\eta : \mathrm{Hom}_A(M, -) \to \mathrm{Hom}_A(M', -)$ *to* $\eta(M)(\mathrm{Id}_M)$. *The inverse of this bijection sends a bimodule*

homomorphism $\psi \in \mathrm{Hom}_{A\otimes_k B^{\mathrm{op}}}(M', M)$ *to the natural transformation* η_ψ *given by* $\eta_\psi(U)(\varphi) = \varphi \circ \psi$ *for all A-modules U and* $\varphi \in \mathrm{Hom}_A(M, U)$.

(ii) *There is a bijection* $\mathrm{Nat}((M \otimes_B -), (M' \otimes_B -)) \cong \mathrm{Hom}_{A\otimes_k B^{\mathrm{op}}}(M, M')$ *sending a natural transformation* $\eta : (M \otimes_B -) \to (M' \otimes_B -)$ *to the map* $\eta(B) : M = M \otimes_B B \to M' = M' \otimes_B B$. *The inverse of this bijection sends* $\psi \in \mathrm{Hom}_{A\otimes_k B^{\mathrm{op}}}(M, M')$ *to the natural transformation* η_ψ *given by* $\eta_\psi(U) = \psi \otimes \mathrm{Id}_U$ *for all A-modules U.*

Proof This is a straightforward exercise. One needs to make use of the naturality of η with respect to the right B-module structures on M, M' to show that in (i) the map $\eta(M)(\mathrm{Id}_M)$ is a bimodule homomorphism from M' to M, and that in (ii) the map $\eta(B)$ is a bimodule homomorphism from M to M'. □

Statement (i) in the preceding theorem is a special case of Yoneda's Lemma. We have ignored here some set theoretic issues concerning classes and sets. Modulo taking care of signs in tensor products of complexes, the results in this section – and in particular Theorem 2.2.4 – can be extended to complexes instead of modules.

Theorem 2.2.7 *Let A, B be k-algebras and let M be a bounded complex of A-B-bimodules. For any complex of A-modules U and any complex of B-modules V we have a natural isomorphism of complexes*

$$\mathrm{Hom}_A(M \otimes_B V, U) \cong \mathrm{Hom}_B(V, \mathrm{Hom}_A(M, U)).$$

Proof The fact that M is bounded ensures that in any fixed degree, the complexes $M \otimes_B V$ and $\mathrm{Hom}_A(M, U)$ are finite direct sums of terms of the form $M_i \otimes_B V_j$ and $\mathrm{Hom}_A(M_i, U_s)$, respectively. Let n be an integer. The term in degree n of the complex $\mathrm{Hom}_A(M \otimes_B V, U)$ is equal to the direct sum of the k-modules $\mathrm{Hom}_A(M_i \otimes_B V_j, U_s)$, with (i, j, s) running over all triples of integers such that $n = s - i - j$. Similarly, the term in degree n of the complex $\mathrm{Hom}_B(V, \mathrm{Hom}_A(M, U))$ is the direct sum of k-modules $\mathrm{Hom}_A(V_j, \mathrm{Hom}_A(M_i, U_s))$, with (i, j, s) running over the same set of triples. Theorem 2.2.4 yields for any such triple a natural isomorphism

$$\Phi_{i,j,s} : \mathrm{Hom}_A(M_i \otimes_B V_j, U_s) \cong \mathrm{Hom}_B(V_j, \mathrm{Hom}_A(M_i, U_s)).$$

The sum of these isomorphisms yields a graded isomorphism between $\mathrm{Hom}_A(M \otimes_B V, U)$ and $\mathrm{Hom}_B(V, \mathrm{Hom}_A(M, U))$, but this is not an isomorphism of complexes due to the signs in the definitions 1.17.9 and 1.17.10 of the differentials of the involved complexes. An easy verification shows that the family of maps $(-1)^{ij}\Phi_{i,j,s}$ yields an isomorphism of complexes as stated. □

2.3 Adjoint functors

Adjunction is one of the most important formal principles in category theory. We describe this, and observe that Frobenius reciprocity is a particular case of two biadjoint functors.

Definition 2.3.1 Let $\mathcal{C}$, $\mathcal{D}$ be categories and let $\mathcal{F} : \mathcal{C} \to \mathcal{D}$, $\mathcal{G} : \mathcal{D} \to \mathcal{C}$ be covariant functors. We say that $\mathcal{G}$ is *left adjoint to* $\mathcal{F}$ and that $\mathcal{F}$ is *right adjoint to* $\mathcal{G}$, if there is an isomorphism of bifunctors $\mathrm{Hom}_{\mathcal{C}}(\mathcal{G}(-), -) \cong \mathrm{Hom}_{\mathcal{D}}(-, \mathcal{F}(-))$. If $\mathcal{G}$ is left and right adjoint to $\mathcal{F}$ we say that $\mathcal{F}$ and $\mathcal{G}$ are *biadjoint*.

An isomorphism of bifunctors as in Definition 2.3.1 is a family of isomorphisms

$$\mathrm{Hom}_{\mathcal{C}}(\mathcal{G}(V), U) \cong \mathrm{Hom}_{\mathcal{D}}(V, \mathcal{F}(U)),$$

with U an object in $\mathcal{C}$ and V an object in $\mathcal{D}$, such that for fixed U we get an isomorphism of contravariant functors $\mathrm{Hom}_{\mathcal{C}}(\mathcal{G}(-), U) \cong \mathrm{Hom}_{\mathcal{D}}(-, \mathcal{F}(U))$, and for fixed V we get an isomorphism of covariant functors $\mathrm{Hom}_{\mathcal{C}}(\mathcal{G}(V), -) \cong \mathrm{Hom}_{\mathcal{D}}(V, \mathcal{F}(-))$. Such an isomorphism of bifunctors, if it exists, need not be unique. If $\mathcal{C}$, $\mathcal{D}$ are k-linear categories for some commutative ring k, we will always require such an isomorphism of bifunctors to be k-linear. Given an adjunction isomorphism $\Phi : \mathrm{Hom}_{\mathcal{C}}(\mathcal{G}(-), -) \cong \mathrm{Hom}_{\mathcal{D}}(-, \mathcal{F}(-))$, evaluating Φ at an object V in $\mathcal{D}$ and $\mathcal{G}(V)$ yields an isomorphism $\mathrm{Hom}_{\mathcal{D}}(V, \mathcal{F}(\mathcal{G}(V))) \cong \mathrm{Hom}_{\mathcal{C}}(\mathcal{G}(V), \mathcal{G}(V))$. We denote by $f(V) : V \to \mathcal{F}(\mathcal{G}(V))$ the morphism corresponding to $\mathrm{Id}_{\mathcal{G}(V)}$ through this isomorphism; that is, $f(V) = \Phi(V, \mathcal{G}(V))(\mathrm{Id}_{\mathcal{G}(V)})$. One checks that the family of morphisms $f(V)$ defined in this way is a natural transformation

$$f : \mathrm{Id}_{\mathcal{D}} \to \mathcal{F} \circ \mathcal{G}$$

called the *unit* of the adjunction isomorphism Φ, where $\mathrm{Id}_{\mathcal{D}}$ denotes the identity functor on $\mathcal{D}$ (sending every object and every morphism in $\mathcal{D}$ to itself). Similarly, evaluating Φ at an object U in $\mathcal{C}$ and at $\mathcal{F}(U)$ yields an isomorphism $\mathrm{Hom}_{\mathcal{C}}(\mathcal{G}(\mathcal{F}(U)), U) \cong \mathrm{Hom}_{\mathcal{D}}(\mathcal{F}(U), \mathcal{F}(U))$. We denote by $g(U) : \mathcal{G}(\mathcal{F}(U)) \to U$ the morphism corresponding to $\mathrm{Id}_{\mathcal{F}(U)}$ through the isomorphism $\mathrm{Hom}_{\mathcal{C}}(\mathcal{G}(\mathcal{F}(U)), U) \cong \mathrm{Hom}_{\mathcal{D}}(\mathcal{F}(U), \mathcal{F}(U))$; that is, $g(U) = \Phi(\mathcal{F}(U), U)^{-1}(\mathrm{Id}_{\mathcal{F}(U)})$. Again, this is a natural transformation

$$g : \mathcal{G} \circ \mathcal{F} \to \mathrm{Id}_{\mathcal{C}}$$

called the *counit* of the adjunction isomorphism Φ.

Example 2.3.2 Let G be a finite group and H a subgroup of G. The first statement in 2.2.1 says that the functor Ind_H^G is *left adjoint to* Res_H^G, and the second statement in 2.2.1 says that Ind_H^G is also *right adjoint to* Res_H^G.

Example 2.3.3 Let A, B be k-algebras and M an A-B-bimodule. Theorem 2.2.4 means that the functor $M \otimes_B - : \mathrm{Mod}(B) \to \mathrm{Mod}(A)$ is left adjoint to the functor $\mathrm{Hom}_A(M, -) : \mathrm{Mod}(A) \to \mathrm{Mod}(B)$. The adjunction isomorphism is given explicitly and can be used to calculate the unit and counit of this adjunction as follows. For any B-module V, the adjunction unit evaluated at V is the B-module homomorphism

$$V \longrightarrow \mathrm{Hom}_A(M, M \otimes_B V)$$

sending $v \in V$ to the map sending $m \in M$ to $m \otimes v$, and for any A-module U, the adjunction counit evaluated at U is the A-module homomorphism

$$M \otimes_B \mathrm{Hom}_A(M, U) \longrightarrow U$$

sending $m \otimes \mu$ to $\mu(m)$, where $m \in M$ and $\mu \in \mathrm{Hom}_A(M, U)$. Unlike in the case of finite group algebras it is not true in general that $M \otimes_B -$ is right adjoint to $\mathrm{Hom}_A(M, -)$. The reason why this does hold in the context of induction and restriction of finite group algebras, as described in the previous example, is that finite group algebras are *symmetric*. We will come back to this theme in 2.12.7.

Exercise 2.3.4 Show that the map sending a group G to its group algebra kG is a covariant functor from the category of groups to the category of k-algebras which has as a right adjoint the functor sending a k-algebra A to its group of invertible elements $A^\times$. Determine the unit and counit of this adjunction.

An adjunction isomorphism is uniquely determined by its unit and counit. To state this properly we need the following notation. Given two functors $\mathcal{F}$, $\mathcal{F}' : \mathcal{C} \to \mathcal{D}$ and a natural transformation $\varphi : \mathcal{F} \to \mathcal{F}'$, we denote for any functor $\mathcal{G} : \mathcal{D} \to \mathcal{E}$ by $\mathcal{G}\varphi : \mathcal{G} \circ \mathcal{F} \to \mathcal{G} \circ \mathcal{F}'$ the natural transformation given by $(\mathcal{G}\varphi)(U) = \mathcal{G}(\varphi(U)) : \mathcal{G}(\mathcal{F}(U)) \to \mathcal{G}(\mathcal{F}'(U))$ for any object U in $\mathcal{C}$. Similarly, for any functor $\mathcal{H} : \mathcal{E} \to \mathcal{C}$ we denote by $\varphi\mathcal{H} : \mathcal{F} \circ \mathcal{H} \to \mathcal{F}' \circ \mathcal{H}$ the natural transformation given by $\varphi(\mathcal{H}(W)) : \mathcal{F}(\mathcal{H}(W)) \to \mathcal{F}'(\mathcal{H}(W))$ for any object W in $\mathcal{E}$. We denote by $\mathrm{Id}_\mathcal{F}$ the identity natural transformation on $\mathcal{F}$, given by the family of identity morphisms $\mathrm{Id}_{\mathcal{F}(U)}$, with U running over the objects of $\mathcal{C}$.

Theorem 2.3.5 *Let $\mathcal{C}$, $\mathcal{D}$ be categories and let $\mathcal{F} : \mathcal{C} \to \mathcal{D}$, $\mathcal{G} : \mathcal{D} \to \mathcal{C}$ be covariant functors.*

(i) Suppose there is an adjunction isomorphism $\Phi : \mathrm{Hom}_\mathcal{C}(\mathcal{G}(-), -) \cong \mathrm{Hom}_\mathcal{D}(-, \mathcal{F}(-))$. The unit f and counit g of Φ satisfy $(\mathcal{F}g) \circ (f\mathcal{F}) = \mathrm{Id}_\mathcal{F}$ and $(g\mathcal{G}) \circ (\mathcal{G}f) = \mathrm{Id}_\mathcal{G}$.

(ii) Let $f : \mathrm{Id}_\mathcal{D} \to \mathcal{F} \circ \mathcal{G}$ and $g : \mathcal{G} \circ \mathcal{F} \to \mathrm{Id}_\mathcal{C}$ be two natural transformations satisfying $(\mathcal{F}g) \circ (f\mathcal{F}) = \mathrm{Id}_\mathcal{F}$ and $(g\mathcal{G}) \circ (\mathcal{G}f) = \mathrm{Id}_\mathcal{G}$. There is a unique adjunction isomorphism $\Phi : \mathrm{Hom}_\mathcal{C}(\mathcal{G}(-), -) \cong \mathrm{Hom}_\mathcal{D}(-, \mathcal{F}(-))$ such that f is the unit of Φ and g is the counit of Φ.

(iii) *Let* $\Phi : \mathrm{Hom}_{\mathcal{C}}(\mathcal{G}(-), -) \cong \mathrm{Hom}_{\mathcal{D}}(-, \mathcal{F}(-))$ *be an adjunction isomorphism with unit* f *and counit* g. *Then* $\Phi(V, U)(\varphi) = \mathcal{F}(\varphi) \circ f(V)$ *for any object* U *in* $\mathcal{C}$, *any object* V *in* $\mathcal{D}$ *and any morphism* $\varphi : \mathcal{G}(V) \to U$ *in* $\mathcal{C}$, *and* $\Phi(V, U)^{-1}(\psi) = g(U) \circ \mathcal{G}(\psi)$ *for any morphism* $\psi : V \to \mathcal{F}(U)$ *in* $\mathcal{D}$. *In particular, we have* $\varphi = g(U) \circ \mathcal{G}(\mathcal{F}(\varphi) \circ f(V))$ *and* $\psi = \mathcal{F}(g(U) \circ \mathcal{G}(\psi)) \circ f(V)$.

Proof Let U be an object in $\mathcal{C}$ and V an object in $\mathcal{D}$. Let $\varphi \in \mathrm{Hom}_{\mathcal{C}}(\mathcal{G}(V), U)$. Suppose we have an isomorphism of bifunctors $\Phi : \mathrm{Hom}_{\mathcal{C}}(\mathcal{G}(-), -) \cong \mathrm{Hom}_{\mathcal{D}}(-, \mathcal{F}(-))$. The functoriality in the second argument implies that we have a commutative diagram

$$\begin{array}{ccc} \mathrm{Hom}_{\mathcal{C}}(\mathcal{G}(V), \mathcal{G}(V)) & \xrightarrow{\mathrm{Hom}_{\mathcal{C}}(\mathcal{G}(V),\varphi)} & \mathrm{Hom}_{\mathcal{C}}(\mathcal{G}(V), U) \\ \downarrow & & \downarrow \\ \mathrm{Hom}_{\mathcal{D}}(V, \mathcal{F}(\mathcal{G}(V))) & \xrightarrow[\mathrm{Hom}_{\mathcal{D}}(V,\mathcal{F}(\varphi))]{} & \mathrm{Hom}_{\mathcal{D}}(V, \mathcal{F}(U)) \end{array}$$

where the vertical arrows are the natural bijections induced by Φ and the horizontal arrows are the maps induced by φ and $\mathcal{F}(\varphi)$. Chase now the element $\mathrm{Id}_{\mathcal{G}(V)}$ in this diagram. The image of $\mathrm{Id}_{\mathcal{G}(V)}$ under the top horizontal map in $\mathrm{Hom}_{\mathcal{C}}(\mathcal{G}(V), U)$ is φ. The image of $\mathrm{Id}_{\mathcal{G}(V)}$ under the left vertical map in $\mathrm{Hom}_{\mathcal{D}}(V, \mathcal{F}(\mathcal{G}(V))$ is $f(V)$, which is sent to $\mathcal{F}(\varphi) \circ f(V)$ under the lower horizontal map. Thus the right vertical map sends φ to $\mathcal{F}(\varphi) \circ f(V)$ in $\mathrm{Hom}_{\mathcal{D}}(V, \mathcal{F}(U))$. By considering a similar diagram with inverted roles one constructs a map sending ψ to $g(U) \circ \mathcal{G}(\psi)$ which is then an inverse of the preceding map. It follows that $\varphi = g(U) \circ \mathcal{G}(\mathcal{F}(\varphi) \circ f(V))$. This equality applied to $\mathrm{Id}_{\mathcal{G}(V)}$ shows that $\mathrm{Id}_{\mathcal{G}(V)} = g(\mathcal{G}(V)) \circ \mathcal{G}(\mathcal{F}(\mathrm{Id}_{\mathcal{G}(V)}) \circ f(V)) = (g(\mathcal{G}(V))) \circ (\mathcal{G}(f(V)))$, which implies that the composition $(g\mathcal{G}) \circ (\mathcal{G}f)$ is the identity transformation on $\mathcal{G}$. Similarly one shows that the other composition in the statement is the identity. This shows (i). Suppose that $f : \mathrm{Id}_{\mathcal{D}} \to \mathcal{F} \circ \mathcal{G}$ and $g : \mathcal{G} \circ \mathcal{F} \to \mathrm{Id}_{\mathcal{C}}$ are natural transformations satisfying $(\mathcal{F}g) \circ (f\mathcal{F}) = \mathrm{Id}_{\mathcal{F}}$ and $(g\mathcal{G}) \circ (\mathcal{G}f) = \mathrm{Id}_{\mathcal{G}}$. Consider the diagram:

$$\begin{array}{ccccc} \mathcal{G}(V) & \xrightarrow{\mathcal{G}(f(V))} & \mathcal{G}(\mathcal{F}(\mathcal{G}(V))) & \xrightarrow{g(\mathcal{G}(V))} & \mathcal{G}(V) \\ \downarrow{\scriptstyle \mathrm{Id}_{\mathcal{G}(V)}} & & \downarrow{\scriptstyle \mathcal{G}(\mathcal{F}(\varphi))} & & \downarrow{\scriptstyle \varphi} \\ \mathcal{G}(V) & \xrightarrow[\mathcal{G}(\mathcal{F}(\varphi)\circ f(V))]{} & \mathcal{G}(\mathcal{F}(U)) & \xrightarrow[g(U)]{} & U \end{array}$$

This diagram is commutative; indeed, the left rectangle commutes since $\mathcal{G}$ is a functor, and the right rectangle commutes since g is a natural transformation. Since, by the hypotheses on f and g we have $g(\mathcal{G}(V)) \circ \mathcal{G}(f(V)) = \mathrm{Id}_{\mathcal{G}(V)}$ it follows that $\varphi = g(U) \circ \mathcal{G}(\mathcal{F}(\varphi) \circ f(V))$. Similarly, we have a commutative diagram

$$\begin{array}{ccccc}
V & \xrightarrow{f(V)} & \mathcal{F}(\mathcal{G}(V)) & \xrightarrow{\mathcal{F}(g(U)\circ\mathcal{G}(\psi))} & \mathcal{F}(U) \\
\downarrow{\scriptstyle\psi} & & \downarrow{\scriptstyle\mathcal{F}(\mathcal{G}(\psi))} & & \downarrow{\scriptstyle\mathrm{Id}_{\mathcal{F}(U))}} \\
\mathcal{F}(U) & \xrightarrow[f(\mathcal{F}(U))]{} & \mathcal{F}(\mathcal{G}(\mathcal{F}(U))) & \xrightarrow[\mathcal{F}(g(U))]{} & \mathcal{F}(U)
\end{array}$$

from which we deduce that $\psi = \mathcal{F}(g(U) \circ \mathcal{G}(\psi)) \circ f(V)$. This shows that the maps sending φ to $\mathcal{F}(\varphi) \circ f(V)$ and ψ to $g(U) \circ \mathcal{G}(\psi)$ are inverse bijections. One easily checks that these bijections are natural, hence they define an isomorphism of bifunctors $\mathrm{Hom}_{\mathcal{C}}(\mathcal{G}(-), -) \cong \mathrm{Hom}_{\mathcal{D}}(-, \mathcal{F}(-))$. This proves (ii) and (iii). □

The restriction functor from the module category of an algebra A to the module category of a subalgebra B is trivially an exact functor. It has as a left adjoint the induction functor $A \otimes_B -$. This left adjoint sends the free B-module B of rank 1 to the free A-module $A \cong A \otimes_B B$ of rank 1, hence preserves projectives. This is a special case of more general principle: for functors between abelian categories, having a left or right adjoint has consequences for the exactness properties of that functor, and exactness properties of adjoint functors have in turn implications for preserving projective or injective objects.

Theorem 2.3.6 *Let $\mathcal{C}$, $\mathcal{D}$ be abelian categories. Let $\mathcal{F} : \mathcal{C} \to \mathcal{D}$ and $\mathcal{G} : \mathcal{D} \to \mathcal{C}$ be additive covariant functors. Suppose that $\mathcal{G}$ is left adjoint to $\mathcal{F}$.*

(i) $\mathcal{G}$ is right exact and $\mathcal{F}$ is left exact.
(ii) If $\mathcal{F}$ is exact then $\mathcal{G}$ preserves projectives.
(iii) If $\mathcal{G}$ is exact then $\mathcal{F}$ preserves injectives.

Proof Let $U \to V \to W \to 0$ be an exact sequence in $\mathcal{D}$. Let X be an object in $\mathcal{C}$. One checks that the induced sequence

$$0 \longrightarrow \mathrm{Hom}_{\mathcal{D}}(W, \mathcal{F}(X)) \longrightarrow \mathrm{Hom}_{\mathcal{D}}(V, \mathcal{F}(X)) \longrightarrow \mathrm{Hom}_{\mathcal{D}}(U, \mathcal{F}(X))$$

is exact. Thus, using the adjunction, we get an exact sequence

$$0 \longrightarrow \mathrm{Hom}_{\mathcal{C}}(\mathcal{G}(W), X) \longrightarrow \mathrm{Hom}_{\mathcal{C}}(\mathcal{G}(V), X) \longrightarrow \mathrm{Hom}_{\mathcal{C}}(\mathcal{G}(U), X).$$

We use this to show that the sequence

$$\mathcal{G}(U) \longrightarrow \mathcal{G}(V) \longrightarrow \mathcal{G}(W) \longrightarrow 0$$

in $\mathcal{C}$ is exact. If $\mathcal{G}(W) \to X$ is a cokernel of the morphism $\mathcal{G}(V) \to \mathcal{G}(W)$, then the composition $\mathcal{G}(V) \to \mathcal{G}(W) \to X$ is zero, hence the cokernel $\mathcal{G}(W) \to X$ is in the kernel of the injective map $\mathrm{Hom}_{\mathcal{C}}(\mathcal{G}(W), X) \to \mathrm{Hom}_{\mathcal{C}}(\mathcal{G}(V), X)$, so this cokernel is zero. This proves that the morphism $\mathcal{G}(V) \to \mathcal{G}(W)$ is an epimorphism. A similar argument shows that a cokernel $\mathcal{G}(V) \to Y$ of the morphism $\mathcal{G}(U) \to \mathcal{G}(V)$ factors through $\mathcal{G}(W)$, from which one deduces the exactness at $\mathcal{G}(V)$. Thus $\mathcal{G}$ is right exact. A similar argument shows that $\mathcal{F}$ is left exact, whence (i). Suppose that $\mathcal{F}$ is exact. Let P be a projective object in $\mathcal{D}$. Let $V \to W \to 0$ be an exact sequence in $\mathcal{C}$. Then $\mathcal{F}(V) \to \mathcal{F}(W) \to 0$ is exact in $\mathcal{D}$. Since P is projective, the induced sequence $\mathrm{Hom}_{\mathcal{D}}(P, \mathcal{F}(V)) \to \mathrm{Hom}_{\mathcal{D}}(P, \mathcal{F}(W)) \to 0$ is exact. Using the adjunction we get an exact sequence $\mathrm{Hom}_{\mathcal{C}}(\mathcal{G}(P), V) \to \mathrm{Hom}_{\mathcal{C}}(\mathcal{G}(P), W) \to 0$, and hence $\mathcal{G}(P)$ is a projective object in $\mathcal{C}$. This proves (ii), and a similar argument shows (iii). □

Left and right adjoints, if they exist, are unique up to isomorphism of functors:

Theorem 2.3.7 *Let $\mathcal{C}$, $\mathcal{D}$ be categories, $\mathcal{F}$ and $\mathcal{F}'$ covariant functors from $\mathcal{C}$ to $\mathcal{D}$. For any natural transformation of bifunctors $\Psi : \mathrm{Hom}_{\mathcal{D}}(-, \mathcal{F}(-)) \to \mathrm{Hom}_{\mathcal{D}}(-, \mathcal{F}'(-))$ there is a unique natural transformation of functors $\psi : \mathcal{F} \longrightarrow \mathcal{F}'$ such that $\psi(U) = \Psi(\mathrm{Id}_{\mathcal{F}(U)})$, and then Ψ is an isomorphism of bifunctors if and only if ψ is an isomorphism of functors. In particular, if $\mathcal{F}$ and $\mathcal{F}'$ are both right adjoint to a covariant functor $\mathcal{G} : \mathcal{D} \to \mathcal{C}$ then $\mathcal{F} \cong \mathcal{F}'$.*

Proof The first statement is a straightforward formal verification. If $\mathcal{F}$ and $\mathcal{F}'$ are both right adjoint to $\mathcal{G}$, we have isomorphisms of bifunctors

$$\mathrm{Hom}_{\mathcal{D}}(-, \mathcal{F}(-)) \cong \mathrm{Hom}_{\mathcal{C}}(\mathcal{G}(-), -) \cong \mathrm{Hom}_{\mathcal{D}}(-, \mathcal{F}'(-)),$$

and then the first statement implies that $\mathcal{F} \cong \mathcal{F}'$. □

Proposition 2.3.8 *Let A be a k-algebra and I a proper ideal in A. Let U be an A-module. The map sending $u \in U$ to $1_{A/I} \otimes u$ in $A/I \otimes_A U$ induces an isomorphism of A/I-modules $U/IU \cong A/I \otimes_A U$ which is natural in U. In particular, the canonical functor* $\mathrm{Mod}(A) \to \mathrm{Mod}(A/I)$ *sending U to U/IU is left adjoint to the canonical functor* $\mathrm{Mod}(A/I) \to \mathrm{Mod}(A)$ *induced by the surjection $A \to A/I$.*

Proof The map $U \to A/I \otimes_A U$ sending u to $1_{A/I} \otimes u$ has IU in its kernel, hence induces a surjective A/I-module homomorphism $U/IU \to A/I \otimes_A U$. The naturality in U is obvious. We construct an inverse as follows. Let $a \in A$ and $u \in U$. The assignment sending $(a+I, u)$ to $au + IU$ is easily seen to be a well-defined and A-balanced map, hence induces a unique map $A/I \otimes_A U \to U/IU$ sending $(a+I) \otimes u$ to $au + IU$. The two maps between U/IU and $A/I \otimes_A U$ constructed as above are clearly inverse to each other. The last statement follows from this isomorphism and the fact that by 2.2.4, the functor $A/I \otimes_A -$ on Mod(A) is left adjoint to the functor $\mathrm{Hom}_A(A/I, -)$ on Mod(A/I). The functor $\mathrm{Hom}_A(A/I, -)$ is in turn easily seen to be isomorphic to the functor induced by the surjection $A \to A/I$ (see also the next exercise for a slightly more general statement). □

Exercise 2.3.9 Let A, B be k-algebras and $\alpha : B \to A$ a homomorphism of k-algebras. Consider A as a left or right B-module via α. Show that the restriction functor Res_α : Mod(A) $\to$ Mod(B) induced by α has as a left adjoint the functor $A \otimes_B -$ and as a right adjoint the functor $\mathrm{Hom}_B(A, -)$.

Exercise 2.3.10 Give an alternative proof of Proposition 2.3.8 by showing directly that the canonical functor $U \mapsto U/IU$ from Mod(A) to Mod(A/I) is left adjoint to the functor from Mod(A/I) to Mod(A) induced by the surjection $A \to A/I$, and then use the uniqueness of adjoint functors from Theorem 2.3.7. Give a third proof of Proposition 2.3.8 by applying the functor $A/I \otimes_A -$ to the short exact sequence $0 \to IU \to U \to U/IU \to 0$ and using the fact that tensor product functors are right exact.

2.4 Mackey's formula

Mackey's formula describes the composition of the induction functor Ind_L^G from a subgroup L of a finite group G followed by the restriction functor Res_H^G to some possibly different subgroup H of G. This amounts essentially to interpreting the double coset decomposition $G = \cup_{x \in [H\backslash G/L]} HxL$ in terms of induction and restriction functors. For any subgroup H of a finite group G and any $x \in G$, the right kH-module $k[xH]$ is also a left k^xH-module because $xH = xHx^{-1}x = {}^xHx$ is also a left ${}^xH$-coset. Thus $k[xH]$ is in fact a k^xH-kH-bimodule, and for any kH-module V we have a k^xH-module ${}^xV = k[xH] \otimes_{kH} V$, obtained from V through the group isomorphism ${}^xH \cong H$ mapping ${}^xh$ to h for any $h \in H$. Equivalently, we could have defined ${}^xV$ by setting ${}^xV = V$ as a k-module, with $h \in {}^xH$ acting on $v \in {}^xV$ as $x^{-1}hx$ on $v \in V$. Mackey's formula is traditionally stated for modules, but extends verbatim to complexes.

Theorem 2.4.1 (Mackey's formula) *Let G be a finite group and let H, L be subgroups of G.*

(i) We have $kG = \oplus_{x\in[H\backslash G/L]}k[HxL]$ as kH-kL-bimodules.
(ii) For any $x \in G$ we have an isomorphism of kH-kL-bimodules

$$k[HxL] \cong kH \otimes_{k(H\cap{}^xL)} k[xL]$$

mapping yxz to $y \otimes xz$, where $y \in H$ and $z \in L$.
(iii) Let B be a k-algebra, and let W be a kL-B-bimodule or a complex of kL-B-bimodules. There is a natural isomorphism of (complexes of) kH-B-bimodules

$$\mathrm{Res}^G_H\mathrm{Ind}^G_L(W) \cong \oplus_{x\in[H\backslash G/L]} \mathrm{Ind}^H_{H\cap{}^xL}\mathrm{Res}^{{}^xL}_{H\cap{}^xL}({}^xW).$$

Proof Statement (i) follows from the partition of G into H-L-double coset. For statement (ii), we first check that the assignment sending yxz to $y \otimes xz$ is well-defined. Let $y, y' \in H$ and $z, z' \in L$ such that $yxz = y'xz'$. We need to show that $y \otimes xz$ $y' \otimes xz'$ are equal. Multiplying by y^{-1} on the left and by $(xz')^{-1}$ on the right yields $y^{-1}y' = xz(z')^{-1}x^{-1} \in H \cap {}^xL$. Thus, setting $w = y^{-1}y'$, we have $w^{-1} = xz'z^{-1}x^{-1}$. Since the tensor product on the right side in (ii) is taken over $k(H \cap {}^xL)$, it follows that $y \otimes xz = yw \otimes w^{-1}xz = y' \otimes xz'$ as required. A similar argument shows that there is a well-defined map sending $y \otimes xz$ to yxz. The two maps thus constructed are kH-kL-bimodule homomorphisms which are inverse to each other, proving (ii). Statement (iii) follows by tensoring the formulae in (i), (ii) with $- \otimes_{kL} W$. □

Corollary 2.4.2 *Let G be a finite group, H a subgroup of G and V a kH-module or a complex of kH-modules. Then V is isomorphic to a direct summand of $\mathrm{Res}^G_H(\mathrm{Ind}^G_H(V))$.*

Proof We have $kG = kH \oplus k[G \setminus H]$ as kH-kH-bimodules, hence $\mathrm{Res}^G_H\mathrm{Ind}^G_H(V) = kG \otimes_{kH} V = (kH \otimes_{kH} V) \oplus (k[G \setminus H] \otimes_{kH} V)$, and $kH \otimes_{kH} V \cong V$. □

Corollary 2.4.3 *Let G be a finite group, N a normal subgroup of G, and V a kN-module or complex of kN-modules. We have a natural isomorphism of (complexes of) kN-modules*

$$\mathrm{Res}^G_N(\mathrm{Ind}^G_N(V)) = \oplus_{x\in[G/N]}{}^xV.$$

Proof Since N is normal in G we have $xN = Nx = NxN$ and $N = {}^xN = N \cap {}^xN$ for all $x \in G$. Thus $\mathrm{Ind}^N_{N\cap{}^xN}\mathrm{Res}^{{}^xN}_{N\cap{}^xN}({}^xV) = {}^xV$, and now the statement follows from Mackey's formula in 2.4.1. Again, one could have seen this directly by observing that the k-module decomposition $kG = \oplus_{x\in[G/N]}k[xN]$ is a decomposition of kG as a kN-kN-bimodule because N is normal in G. Thus $\mathrm{Res}^G_N\mathrm{Ind}^G_N(V) = \oplus_{x\in[G/N]}k[xN] \otimes_{kN} V = \oplus_{x\in[G/N]}{}^xV$. □

Corollary 2.4.4 *Let G be a finite group and let H and L be subgroups of G such that $G = HL$. Let V be a kH-module or a complex of kH-modules. We have* $\mathrm{Res}^G_L(\mathrm{Ind}^G_H(V)) \cong \mathrm{Ind}^L_{H\cap L}(\mathrm{Res}^H_{H\cap L}(V))$.

Proof This is the particular case of Mackey's formula in which G has exactly one H-L-double coset. □

By making use of our standing convention, regarding kH-kL-bimodules as $k(H \times L)$-modules, statement (ii) in 2.4.1 can be reformulated as follows:

Corollary 2.4.5 *Let G be a finite group, let H, L be a subgroups of G and $x \in G$. Set $R = \{(h, x^{-1}hx)|h \in H \cap {}^xL\}$. Then R is a subgroup of $H \times L$, and we have an isomorphism of $k(H \times L)$-modules*

$$k[HxL] \cong \mathrm{Ind}^{H\times L}_R(k)$$

sending yxz to $(y, z^{-1}) \otimes 1$, where $y \in H$ and $z \in L$. In particular, we have an isomorphism of $k(G \times G)$-modules

$$kG \cong \mathrm{Ind}^{G\times G}_{\Delta G}(k),$$

where $\Delta G = \{(y, y)|y \in G\}$.

Proof The first isomorphism is a straightforward verification. The second isomorphism follows from the first applied to $G = H = L$ and $x = 1$. □

Corollary 2.4.6 *Let G be a finite group and H a subgroup of G. For any $y \in G$ set $\Delta_yH = \{(h, y^{-1}hy)|h \in H \cap {}^yH\}$. We have an isomorphism of $k(H \times H)$-bimodules*

$$kG \cong kH \oplus (\oplus_y \mathrm{Ind}^{H\times H}_{\Delta_yH}(k))$$

where $y \in G \setminus H$ runs over a set of representatives in G of the H-H-double cosets in G different from H.

Proof This isomorphism follows from 2.4.5, applied to the case $H = L$, where the first summand corresponds to the trivial H-H-double coset H in G. □

Combining Mackey's formula and Frobenius reciprocity yields the following theorem which we will use later to prove the simplicity of certain two-dimensional modules of dihedral groups.

Theorem 2.4.7 *Let G be a finite group, N a normal subgroup of G. Suppose that k is a splitting field for both G and N, and that $|G : N|$ is invertible in k.*

Let S be a simple kN-module. Then $\mathrm{Ind}_N^G(S)$ is simple if and only if $S \not\cong {}^xS$ for all $x \in G \setminus N$.

Proof Note that ${}^xS$ is a simple kN-module for any $x \in G$. It follows from 2.4.3 that $\mathrm{Res}_N^G\mathrm{Ind}_N^G(S) = \oplus_{x\in[G/N]}{}^xS$ is semisimple. Since $|G:N|$ is assumed to be invertible in k it follows from 1.11.9 that $\mathrm{Ind}_N^G(S)$ is semisimple. Since k is a splitting field for G, the module $\mathrm{Ind}_N^G(S)$ is simple if and only if $\mathrm{End}_{kG}(\mathrm{Ind}_N^G(S))$ has k-dimension 1. Using Frobenius reciprocity 2.2.1, we have $\mathrm{End}_{kG}(\mathrm{Ind}_N^G(S)) = \mathrm{Hom}_{kG}(\mathrm{Ind}_N^G(S), \mathrm{Ind}_N^G(S)) \cong \mathrm{Hom}_{kN}(S, \mathrm{Res}_N^G\mathrm{Ind}_N^G(S)) = \oplus_{x\in[G/N]}\mathrm{Hom}_{kN}(S, {}^xS)$. By Schur's Lemma, the right side is 1-dimensional if and only if $S \not\cong {}^xS$ for all $x \in G \setminus N$. □

Another useful application of Mackey's formula arises as a consequence of the following reciprocity property of tensor products:

Proposition 2.4.8 *Let G be a finite group and H a subgroup of G. For any kG-module U and any kH-module V we have a natural isomorphism of kG-modules*

$$\mathrm{Ind}_H^G(V) \otimes_k U \cong \mathrm{Ind}_H^G(V \otimes_k \mathrm{Res}_H^G(U))$$

sending $(x \otimes v) \otimes u$ to $x \otimes (v \otimes (x^{-1}u))$, where $x \in G$, $u \in U$ and $v \in V$.

Proof One checks that the given map is a homomorphism of kG-modules, and that the map sending $x \otimes (v \otimes u)$ to $(x \otimes v) \otimes (xu)$ is its inverse. □

Modulo keeping track of signs in tensor products of complexes, this proposition can be extended to bounded complexes – we leave the details to the reader.

Corollary 2.4.9 *Let G be a finite group, and let H, L be subgroups of G. For any kH-module V and any kL-module W we have a natural isomorphism of kG-modules*

$$\mathrm{Ind}_H^G(V) \otimes_k \mathrm{Ind}_L^G(W) \cong \oplus_{x\in[H\backslash G/L]}\mathrm{Ind}_{H\cap {}^xL}^G(\mathrm{Res}_{H\cap {}^xL}^H(V) \otimes_k \mathrm{Res}_{H\cap {}^xL}^{{}^xL}({}^xW)).$$

Proof Applying 2.4.8 yields $\mathrm{Ind}_H^G(V) \otimes_k \mathrm{Ind}_L^G(W) \cong \mathrm{Ind}_H^G(V \otimes_k \mathrm{Res}_H^G(\mathrm{Ind}_L^G(W)))$. Mackey's formula implies that this is isomorphic to the direct sum $\oplus_{x\in[H\backslash G/L]}\mathrm{Ind}_H^G(V \otimes_k \mathrm{Ind}_{H\cap {}^xL}^H(\mathrm{Res}_{H\cap {}^xL}^{{}^xL}({}^xW)))$. Applying 2.4.8 again (with H instead of G) and transitivity of induction yields the formula; all isomorphisms are functorial in V and W. □

Corollary 2.4.10 *Let G be a finite group, and let U, V be kG-modules that are finitely generated projective as k-modules. If U is projective, so is $U \otimes_k V$.*

Proof If U is finitely generated projective as a kG-module, then U is isomorphic to a direct summand of $\mathrm{Ind}_1^G(W)$ for some finitely generated projective k-module W. Thus $U \otimes_k V$ is isomorphic to a direct summand of $\mathrm{Ind}_1^G(\mathrm{Res}_1^G(U) \otimes_k W)$, by 2.4.8 applied to $H = 1$. By the assumptions, $\mathrm{Res}_1^G(U) \otimes_k W$ is finitely generated projective as a k-module, whence the result. □

By 1.1.9, the tensor product over k of two permutation kG-modules is again a permutation kG-module. Specialising 2.4.9 to trivial modules makes this more precise:

Corollary 2.4.11 *Let G be a finite group, and let H, L be subgroups of G. We have*

$$\mathrm{Ind}_H^G(k) \otimes_k \mathrm{Ind}_L^G(k) \cong \oplus_{x \in [H\backslash G/L]} \mathrm{Ind}_{H \cap {}^xL}^G(k).$$

There are bimodule versions of some of the above statements:

Proposition 2.4.12 *Let G be a finite group, H a subgroup of G and A a k-algebra. Let M be a kH-A-bimodule and V a kH-module. Consider $V \otimes_k M$ as a kH-A-bimodule with G acting diagonally on the left, consider V as a module for $k\Delta H$ via the canonical isomorphism $\Delta H \cong H$ and consider $\mathrm{Ind}_{\Delta H}^{G\times H}(V)$ as a kG-kH-bimodule. We have a natural isomorphism of kG-A-bimodules*

$$\mathrm{Ind}_H^G(V \otimes_k M) \cong \mathrm{Ind}_{\Delta H}^{G\times H}(V) \otimes_{kH} M$$

sending $x \otimes (v \otimes m)$ to $((x, 1) \otimes v) \otimes m$, where $v \in V$ and $m \in M$.

Proof One verifies that the given map is a homomorphism of kG-A-bimodules from $kG \otimes_{kH} (V \otimes_k M)$ to $(k(G \times H) \otimes_{k\Delta H} V) \otimes_{kH} M$. One checks then that this map is an isomorphism, with inverse sending $((x, y) \otimes v) \otimes m$ to $x \otimes (v \otimes y^{-1}m)$. The naturality in M and V is straightforward. □

In most applications of Proposition 2.4.12, the algebra A will be a finite group algebra kL, and in that case, the kG-kL-bimodule $\mathrm{Ind}_H^G(V \otimes_{\mathcal{O}} M)$ is the same as $\mathrm{Ind}_{H\times L}^{G\times L}(V \otimes_{\mathcal{O}} M)$, with the usual identifications.

Corollary 2.4.13 *Let G be a finite group, let U and V be $k\Delta G$-modules. We have a natural isomorphism of kG-kG-bimodules*

$$\mathrm{Ind}_{\Delta G}^{G\times G}(U) \otimes_{kG} \mathrm{Ind}_{\Delta G}^{G\times G}(V) \cong \mathrm{Ind}_{\Delta G}^{G\times G}(U \otimes_k V).$$

Proof Applying 2.4.12 with $H = G$ yields an isomorphism of $k(G \times G)$-modules $\mathrm{Ind}_{\Delta G}^{G\times G}(U) \otimes_{kG} \mathrm{Ind}_{\Delta G}^{G\times G}(V) \cong U \otimes_k \mathrm{Ind}_{\Delta G}^{G\times G}(V)$, where the left action of G is given diagonally and the right action is given by the right action of G on the second module. We can regard U as a $k(G \times G)$-action with the trivial right action. In this way, the expression $U \otimes_k \mathrm{Ind}_{\Delta G}^{G\times G}(V)$ becomes a tensor product of two $k(G \times G)$-modules. Applying 2.4.8 to this tensor product yields the result. One can also check directly that there is a unique $k(G \times G)$-module isomorphism sending $((1, 1) \otimes u) \otimes ((1, 1) \otimes v)$ to $(1, 1) \otimes (u \otimes v)$, where $u \in U$ and $v \in V$. □

Proposition 2.4.14 *Let G be a finite group and H a subgroup of G. Let V be a $k\Delta G$-module. We have a natural isomorphism of kG-kH-bimodules*

$$\mathrm{Res}^{G \times G}_{G \times H}(\mathrm{Ind}^{G \times G}_{\Delta G}(V)) \cong \mathrm{Ind}^{G \times H}_{\Delta H}(\mathrm{Res}^{\Delta G}_{\Delta H}(V)).$$

Proof We have $G \times G = (G \times H)\Delta G$ and $G \times H \cap \Delta G = \Delta H$. Thus the statement is a special case of Corollary 2.4.4. □

Although tensor induction is not an additive functor, the proof of the Mackey formula can easily be adapted to yield a similar formula:

Proposition 2.4.15 *Let H, L be subgroups of a finite group G and let W be a kL-module. We have* $\mathrm{Res}^G_H(\mathrm{Ten}^G_L(W)) \cong \otimes_{x \in [H\backslash G/L]} \mathrm{Ten}^H_{H \cap {}^xL}(\mathrm{Res}^{{}^xL}_{H \cap {}^xL}({}^xW))$.

Proof The result follows from writing H-L-double cosets as union of H-cosets. □

We mention without proof the following generalisation, due to Bouc, of some of the above formulas for tensor products of bimodules of the form $\mathrm{Ind}^{G \times H}_P(M)$ for some subgroup P of $G \times H$ and some kP-module M. With this notation, if $(x, y) \in P$ for some $x \in G$, $y \in H$ and if $m \in M$, we write $x \cdot m \cdot y^{-1} = (x, y) \cdot m$.

Theorem 2.4.16 ([11, Theorem 1.1]) *Let G, H, L be finite groups. Let P be a subgroup of $G \times H$ and Q a subgroup of $H \times L$. Let M be a kP-module and N a kQ-module. We have an isomorphism of kG-kL-bimodules*

$$\begin{aligned} &\mathrm{Ind}^{G \times H}_P(M) \otimes_{kH} \mathrm{Ind}^{H \times L}_Q(N) \\ &\quad \cong \oplus_{t \in [p_2(P)\backslash H/p_1(Q)]} \mathrm{Ind}^{G \times L}_{P * {}^{(t,1)}Q}(M \otimes_{k[i_2(P) \cap {}^t i_1(Q)]} {}^{(t,1)}N) \end{aligned}$$

where

$$p_1(Q) = \{y \in H \mid \exists z \in L, (y, z) \in Q\}, \quad i_1(Q) = Q \cap (H \times 1),$$

$$p_2(P) = \{y \in H \mid \exists x \in G, (x, y) \in P\}, \quad i_2(P) = P \cap (1 \times H),$$

$$P * {}^{(t,1)}Q = \{(x, z) \in G \times L \mid \exists y \in H, (x, h) \in P, ({}^t y, z) \in Q\}.$$

The action of $(x, z) \in P * {}^{(t,1)}Q$ on $M \otimes_{k[i_2(P) \cap {}^t i_1(Q)]} {}^{(t,1)}N$ is given by

$$x \cdot (m \otimes n) \cdot z^{-1} = (x \cdot m \cdot y^{-1}) \otimes (y^t \cdot n \cdot z^{-1}),$$

where $y \in H$ is chosen such that $(x, y) \in P$ and $(y^t, z) \in Q$.

2.5 Relative traces

Relative trace maps generalise the averaging technique encountered earlier, such as in the proofs of Maschke's Theorem 1.11.12, 1.16.13 and 1.11.9.

Definition 2.5.1 (D. G. Higman, J. A. Green [33]) Let G be a group and U a kG-module. Let H be a subgroup of G. We set

$$U^H = \{u \in U \mid yu = u \text{ for all } y \in H\}$$

and call U^H the set of H-fixed points in U. If H has finite index in G, we define a map $\mathrm{Tr}_H^G : U^H \to U^G$ by

$$\mathrm{Tr}_H^G(u) = \sum_{x\in[G/H]} xu$$

for all $u \in U^H$, and we set $U_H^G = \mathrm{Im}(\mathrm{Tr}_H^G)$. The map Tr_H^G is called the *relative trace map from H-fixed points to G-fixed points in U.*

Let G be a group and U a kG-module. Let H be a subgroup of finite index of G. One needs to check that the expression $\sum_{x\in[G/H]} xu$ in the above definition is independent of the choice of $[G/H]$, and that it is a G-fixed point. If $x \in G$, $y \in H$ and $u \in U^H$, then $(xy)u = x(yu) = xu$, so xu is independent of the choice of x in its H-coset. Multiplying $[G/H]$ on the left by an element in G yields another set of representatives of G/H in G, and by the previous independence of the choice of $[G/H]$, this implies that U_H^G is a subset of U^G. Clearly U^H is a k-submodule of U, the map Tr_H^G is k-linear, and hence U_H^G is a k-submodule of U^G. The relative trace map $\mathrm{Tr}_H^G : U^H \to U^G$ is in general neither injective nor surjective. Taking fixed points is functorial: if $\varphi : U \to V$ is a homomorphism of kG-modules, then $\varphi(U^H) \subseteq V^H$. Moreover, we have $\varphi(\mathrm{Tr}_H^G(u)) = \mathrm{Tr}_H^G(\varphi(u))$ for any $u \in U^H$. In particular, we have $\varphi(U_H^G) \subseteq V_H^G$.

Remark 2.5.2 Let G be a finite group, H a subgroup of G, and B a k-algebra.

(a) Let M a kG-module. The action of G on M induces an action of $N_G(H)$ on M^H, and hence we can consider M^H as a $kN_G(H)$-module. Since H acts trivially on M^H, we can also consider M^H as a $kN_G(H)/H$-module. Thus the map sending M to M^H can be regarded as a functor from $\mathrm{Mod}(kG)$ to $\mathrm{Mod}(kN_G(H)/H)$.

(b) Let M a kG-module. We have a canonical k-linear isomorphism $\mathrm{Hom}_{kH}(k, M) \cong M^H$ sending $\varphi \in \mathrm{Hom}_{kH}(k, M)$ to $\varphi(1_k)$. Thus the functor sending M to M^H is isomorphic to the functor $\mathrm{Hom}_{kH}(k, -)$.

(c) Let U, V be kG-B-bimodules. Then G acts on $\mathrm{Hom}_{B^{op}}(U, V)$ by ${}^x\varphi(u) = x\varphi(x^{-1}u)$ for $x \in G$, $u \in U$, and $\varphi \in \mathrm{Hom}_{B^{op}}(U, V)$. With respect to this action, we have $\mathrm{Hom}_{B^{\mathrm{op}}}(U,V)^H = \mathrm{Hom}_{kH\otimes_k B^{\mathrm{op}}}(U,V)$, and for $\psi \in \mathrm{Hom}_{kH\otimes_k B^{\mathrm{op}}}(U,V)$ we have $\mathrm{Tr}_H^G(\psi)(u) = \sum_{x\in[G/H]} x\psi(x^{-1}u)$ for all $u \in U$.

Let G be a finite group. For any kG-module U, we have $(\sum_{x\in G} x)U = U_1^G$, and hence we have the following reformulation of earlier results.

Proposition 2.5.3 *Suppose that k is a field of prime characteristic p. Let P be a finite p-group, and let U be a finitely generated kP-module.*

(i) We have $\mathrm{soc}(U) = U^P$.
(ii) If U has no nonzero projective direct summand, then $U_1^P = \{0\}$.
(iii) The kP-module U is projective if and only if $U^P = U_1^P$.
(iv) If U is projective, then $\dim_k(U^P)$ *is equal to the rank of U as a kP-module.*

Proof By 1.11.1, the socle of U is the sum of all trivial submodules, hence equal to U^P. This shows (i). Statement (ii) is a reformulation of 1.11.3. If U is projective, then $U^P = U_1^P$ by 1.11.2. If U is not projective, then U_1^P is a proper subspace of U^P by (ii), whence (iii). The last statement follows again from 1.11.2. □

Proposition 2.5.4 *Let G be a finite group, let U be a kG-module and let H and L be subgroups of G such that $L \leq H$.*

(i) We have $U^H \leq U^L$.
(ii) We have $\mathrm{Tr}_H^G \circ \mathrm{Tr}_L^H = \mathrm{Tr}_L^G$.
(iii) For any $x \in G$ we have $xU^H = U^{{}^xH}$.
(iv) We have ${}^x(\mathrm{Tr}_L^H(u)) = \mathrm{Tr}_{{}^xL}^{{}^xH}(xu)$ *for any $x \in G$ and any $u \in U^L$.*

Proof All statements are trivial verifications. □

Proposition 2.5.5 *Let G be a finite group, let U be a kG-module and let H and L be subgroups of G. For any $u \in U^L$ we have*

$$\mathrm{Tr}_L^G(u) = \sum_{x\in[H\backslash G/L]} \mathrm{Tr}_{H\cap {}^xL}^H(xu).$$

Proof In the disjoint union $G = \bigcup_{x\in[H\backslash G/L]} HxL$ any double coset HxL is a disjoint union of L-cosets $HxL = \bigcup_{y\in[H/H\cap {}^xL]} yxL$. In other words, we can take for $[G/L]$ the set of all yx, with x running over $[H\backslash G/L]$ and, for any such x, with y running over $[H/H\cap {}^x L]$, which implies the formula. □

The formula in 2.5.5 is also called *Mackey formula*. As in 2.4.1, this formula is a reinterpretation of the decomposition of G as a disjoint union of double cosets.

Proposition 2.5.6 *Let G be a finite group, P a normal subgroup of G and H a subgroup of G containing P. Let U be a kG-module. We have $U_P^G \subseteq U_P^H$.*

Proof Let $u \in U^P$. By the Mackey formula, we have $\mathrm{Tr}_P^G(u) = \sum_{x \in [H\backslash G/P]} \mathrm{Tr}_{P\cap {}^xH}^H(xu)$. Since P is normal in G we have $P \cap {}^xH = P$ and $[H\backslash G/P] = [H\backslash G]$. Thus $\mathrm{Tr}_P^G(u) = \mathrm{Tr}_P^H(\sum_{x\in[H\backslash G]} xu)$. □

Fixed points of induced modules are calculated as follows:

Proposition 2.5.7 *Let G be a finite group and H a subgroup of G. Let V be a kH-module, and set $U = \mathrm{Ind}_H^G(V)$. We have a canonical k-linear isomorphism $V^H \cong U^G$ mapping $v \in V^H$ to $\sum_{x\in[G/H]} x \otimes v$. In particular, we have $U^G = U_H^G$.*

Proof Using 2.5.2 (b) and 2.2.1 we have $V^H \cong \mathrm{Hom}_{kH}(k, V) \cong \mathrm{Hom}_{kG}(k, \mathrm{Ind}_H^G(V)) \cong (\mathrm{Ind}_H^G(V))^G$. From the explicit description of these isomorphisms one gets the k-linear isomorphism as stated. The last statement follows from identifying V^H to the subspace $1 \otimes V^H$ of U^H. □

Let G be a group and let A be a G-algebra over k. Then A is in particular a kG-module via the action of G on A. For any subgroup H in G, then set A^H is a subalgebra of A, since G acts by algebra automorphisms. If $f : A \to B$ is a homomorphism of G-algebras, then f restricts to a homomorphism of algebras $A^H \to B^H$. Relative trace maps on G-algebras have the following additional property.

Proposition 2.5.8 *Let G be a finite group, let A be a G-algebra over k, and let H be a subgroup of G. For any $a \in A^H$ and $b \in A^G$ we have $\mathrm{Tr}_H^G(a)b = \mathrm{Tr}_H^G(ab)$ and $b\mathrm{Tr}_H^G(a) = \mathrm{Tr}_H^G(ba)$. In particular, A_H^G is a two-sided ideal in A^G.*

Proof We have $\mathrm{Tr}_H^G(a)b = \sum_{x\in[G/H]} ({}^xa)b$. Since b is a G-fixed point, this is equal to

$$\sum_{x\in[G/H]} ({}^xa)({}^xb) = \sum_{x\in[G/H]} {}^x(ab) = \mathrm{Tr}_H^G(ab).$$

A similar argument shows the second formula, implying the result. □

Corollary 2.5.9 *Let G be a finite group, A be a G-algebra over k, and H a subgroup of G. If $1_A \in A_H^G$ then $A^G = A_H^G$.*

Proof The set A_H^G is an ideal in A^G by 2.5.8, and thus if $1_A \in A_H^G$, then $a = 1_A \cdot a \in A_H^G$ for all $a \in A^G$. □

If a kG-module U has a k-basis which is permuted by G, then the fixed point spaces are particularly easy to calculate:

Proposition 2.5.10 *Let G be a group and let U be a kG-module such that U is free as a k-module. Suppose U has a k-basis X that is permuted by the action of G on U. Then, for any finite subgroup H of G, the k-module U^H is free, having as a basis the set of H-orbit sums of X in U.*

Proof Let $u \in U$ and write $u = \sum_{x \in X} \alpha_x x$ with coefficients $\alpha_x \in k$. Let $h \in H$. We have $hu = \sum_{x \in X} \alpha_x hx$. Since the action of h on U permutes the elements in X it follows that $u = hu$ if and only if $\alpha_x = \alpha_{hx}$ for all $x \in X$, and hence $u \in U^H$ if and only if $\alpha_x = \alpha_{hx}$ for all $x \in X$ and all $h \in H$. In other words, the map $x \mapsto \alpha_x$ is constant on H-orbits when viewed as a function on X. The result follows. □

When applied to the G-algebra $A = kG$ for some finite group G and $H = G$ we recover 1.5.1.

Example 2.5.11 Let L/K be a finite Galois extension, with Galois group G. The field L can be viewed as G-algebra over K, with the natural action of G on L by field automorphisms. The trace map $\mathrm{tr}_{L/K} : L \to K$ coincides then with the map $\mathrm{Tr}_1^G : L \to L^G = K$ from 2.5.1.

2.6 Higman's criterion

Definition 2.6.1 Let G be a finite group and H a subgroup of G. Let U be a kG-module or a complex of kG-modules. We say that U is *relatively H-projective* if there exists a kH-module or a complex of kH-modules V such that U is isomorphic to a direct summand of $\mathrm{Ind}_H^G(V)$.

Every projective kG-module is relatively 1-projective. The converse need not be true because k may have modules that are not projective. In order to prove that a module or complex U over some k-algebra A is a direct summand of another A-module or complex W, we have to find an A-homomorphism $\pi : W \to U$ and an A-homomorphism $\sigma : U \to W$ such that $\pi \circ \sigma = \mathrm{Id}_U$. Then $W = \sigma(U) \oplus \ker(\pi)$ and $\sigma(U) \cong U$. Note that the equality $\pi \circ \sigma = \mathrm{Id}_U$ implies that π is surjective and σ is injective. The map σ is called a *section* of π, and π is called a *retraction of* σ. As pointed out earlier, one can extend this terminology to any category: an epimorphism is called *split* if it has a section and a monomorphism is called *split* if it has a retraction. See 1.12.20 for a more general variation of this notion. Relative projectivity extends to bimodules and to chain complexes, as follows. Let G be a finite group, H a subgroup of G, B is a k-algebra, and U a kG-B-bimodule or a complex of kG-B-bimodules. Then U is called *relatively kH-projective* if U is isomorphic to a direct summand of $\mathrm{Ind}_H^G(V) = kG \otimes_{kH} V$ for some kH-B-bimodule or complex of kH-B-bimodules V. If U is a kG-B-bimodule, then $\mathrm{End}_{B^{\mathrm{op}}}(U)$ is an interior G-algebra, with structural map sending $x \in G$ to the B^{op}-automorphism of U given by left multiplication with x on U; see 2.5.2 (c) for a detailed description of the induced G-algebra structure on $\mathrm{End}_{B^{\mathrm{op}}}(U)$. Clearly $(\mathrm{End}_{B^{\mathrm{op}}}(U))^H = \mathrm{End}_{kH \otimes_k B^{\mathrm{op}}}(U)$. More generally, if U is a complex of kG-B-bimodules, then $\mathrm{End}_{\mathrm{Ch}(\mathrm{Mod}(B^{\mathrm{op}}))}(U)$ is an interior G-algebra, and $(\mathrm{End}_{\mathrm{Ch}(\mathrm{Mod}(B^{\mathrm{op}}))}(U))^H = \mathrm{End}_{\mathrm{Ch}(\mathrm{Mod}(kH \otimes_k B^{\mathrm{op}}))}(U)$. We state Higman's criterion in slightly greater generality than its traditional version (which is the special case where B is the trivial algebra k and where U is a module).

Theorem 2.6.2 (Higman's criterion) *Let G be a finite group, H a subgroup of G, and B a k-algebra. Let U be a kG-B-bimodule or a complex of kG-B-bimodules. The following statements are equivalent.*

(i) U is relatively H-projective.

(ii) U is isomorphic to a direct summand of the (complex of) kG-B-bimodule(s) $\mathrm{Ind}_H^G(\mathrm{Res}_H^G(U))$.

(iii) The canonical map $\mathrm{Ind}_H^G(\mathrm{Res}_H^G(U)) \to U$ of (complexes of) kG-B-bimodules sending $x \otimes u$ to xu for $x \in G$, $u \in U$ is split.

(iv) For any kG-B-bimodule or complex of bimodules M and any surjective kG-B-bimodule homomorphism $\pi : M \to U$ or chain map, respectively, the following holds: if π splits as a kH-B-bimodule (chain complex) homomorphism, then π splits as a kG-B-bimodule (chain complex) homomorphism.

(v) We have $\mathrm{Id}_U = \mathrm{Tr}_H^G(\varphi)$ for some kH-B-bimodule (chain complex) endomorphism of U.

Proof We prove this for U a bimodule; if U is a complex, one needs to replace bimodule homomorphisms by chain maps of complexes of bimodules. We prove first that (v) implies (iii). Suppose that $\mathrm{Id}_U = \mathrm{Tr}_H^G(\varphi)$ for some $\varphi \in (\mathrm{End}_{B^{\mathrm{op}}}(U))^H = \mathrm{End}_{kH\otimes_k B^{\mathrm{op}}}(U)$. This means that $\sum_{y\in[G/H]} y\varphi(y^{-1}u) = u$ for all $u \in U$. Denote by π the canonical surjective map $\pi : \mathrm{Ind}_H^G(\mathrm{Res}_H^G(U)) = kG \otimes_{kH} U \to U$ mapping $x \otimes u$ to xu. This map is obviously a kG-B-bimodule homomorphism. We show that this map splits as a kG-B-bimodule homomorphism with section $\sigma : U \to kG \otimes_{kH} U$ mapping $u \in U$ to $\sum_{y\in[G/H]} y \otimes \varphi(y^{-1}u)$. Indeed, we have $\pi(\sigma(u)) = \pi(\sum_{y\in[G/H]} y \otimes \varphi(y^{-1}u)) = \sum_{y\in[G/H]} y\varphi(y^{-1}u) = u$ for all $u \in U$. In particular, $U \cong \sigma(U)$ is isomorphic to a direct summand of $kG \otimes_{kH} U = \mathrm{Ind}_H^G\mathrm{Res}_H^G(U) = \sigma(U) \oplus \ker(\pi)$. This shows that (v) implies (iii). Suppose that (iii) holds; that is, the canonical map $kG \otimes_{kH} U \to U$ has a section $\sigma : U \to kG \otimes_{kH} U$ as a

kG-B-bimodule homomorphism. Let $\pi : M \to U$ be a surjective kG-B-bimodule homomorphism. Suppose that π splits as a kH-B-bimodule homomorphism; that is, there is a kH-B-bimodule homomorphism $\tau : U \to M$ satisfying $\pi \circ \tau = \mathrm{Id}_U$. By an appropriate version of Frobenius reciprocity 2.2.1, the map τ corresponds to a kG-B-bimodule homomorphism $\eta : kG \otimes_{kH} U \to M$ satisfying $\eta(x \otimes u) = x\tau(u)$, where $x \in G$ and $u \in U$. We claim that $\eta \circ \sigma$ is a section for π. We have $(\pi \circ \eta)(x \otimes u) = \pi(x\tau(u)) = x\pi(\tau(u)) = xu$, so $\pi \circ \eta$ is equal to the canonical map $kG \otimes_{kH} U \to U$. Since σ is a section for this canonical map, we have $\pi \circ \eta \circ \sigma = \mathrm{Id}_U$, and hence $\eta \circ \sigma$ is a section for π as a kG-B-bimodule homomorphism. This shows that (iii) implies (iv). Suppose that (iv) holds. The canonical map $kG \otimes_{kH} U \to U$ splits as a kH-B-bimodule homomorphism, having as a section the map sending $u \in U$ to $1 \otimes u$. Thus, applying (iv) implies that this map splits as a kG-B-bimodule homomorphism, and hence (iv) implies (iii). Clearly (iii) implies (ii), and (ii) implies (i). In order to prove that (i) implies (v), let V be a kH-B-bimodule. Set $W = \mathrm{Ind}_H^G(V) = kG \otimes_{kH} V$. We first show that statement (iv) holds for W, and then we show that this statement passes down to direct summands. Define $\tau \in \mathrm{End}_{kH \otimes_k B^{\mathrm{op}}}(W)$ by $\tau(x \otimes v) = x \otimes v$ if $x \in H$ and $\tau(x \otimes v) = 0$ if $x \in G \setminus H$, where v runs over V. In other words, τ is the projection of $kG \otimes_{kH} V = (kH \otimes_{kH} V) \oplus (k[G \setminus H] \otimes_{kH} V)$ onto $kH \otimes_{kH} V \cong V$. Then $\mathrm{Tr}_H^G(\tau)(x \otimes v) = \sum_{y \in [G/H]} y\tau(y^{-1}x \otimes v)$. In this sum there is exactly one term for which $y^{-1}x \in H$, and this term is equal to $y(y^{-1}x \otimes v) = x \otimes v$ while all other terms vanish by the definition of τ. This means that $\mathrm{Tr}_H^G(\tau) = \mathrm{Id}_W$. Now if U is isomorphic to a direct summand of W, there are maps $\iota : U \to W$ and $\pi : W \to U$ such that $\pi \circ \iota = \mathrm{Id}_U$. Then $\mathrm{Id}_U = \pi \circ \mathrm{Id}_W \circ \iota = \pi \circ \mathrm{Tr}_H^G(\tau) \circ \iota = \mathrm{Tr}_H^G(\pi \circ \tau \circ \iota)$, where the last equality comes from the fact that π, ι commute with the action of G. □

In most instances where Higman's criterion is used, we have $B = k$. The following consequences of Higman's criterion are formulated using the case $B = k$, but they admit the obvious bimodule versions (left as an exercise below).

Corollary 2.6.3 *Let G be a finite group and H a subgroup of G. Suppose that $|G : H|$ is invertible in k. Then every kG-module and every complex of kG-modules is relatively H-projective. In particular, if M is a kG-module such that $\mathrm{Res}_H^G(M)$ is projective, then M is projective.*

Proof Let M be a kG-module or a complex of kG-modules. Since $|G : H|$ is invertible in k we have $\mathrm{Id}_M = \mathrm{Tr}_H^G(\frac{1}{|G:H|}\mathrm{Id}_M)$, and hence Higman's criterion 2.6.2 implies that M is relatively H-projective. Thus M is isomorphic to a direct summand of $\mathrm{Ind}_H^G(\mathrm{Res}_M^G(M))$. Since induction preserves projective modules, this implies the last statement. □

Corollary 2.6.4 *Let G be a finite group. Suppose that $|G|$ is invertible in k. Every k-free kG-module is projective.*

Proof By 2.6.3, every kG-module M is relatively 1-projective. Thus, if M is a kG-module, then M is isomorphic to a direct summand of $kG \otimes_k M$. If M is k-free, then $kG \otimes_k M$ is free as a kG-module, and hence M is projective as a kG-module. □

Corollary 2.6.5 *Suppose that k is a field. Let G be a finite group and H a subgroup of G. Let U be a kG-module such that every quotient of U is relatively H-projective. If $\mathrm{Res}^G_H(U)$ is semisimple, then U is semisimple.*

Proof Let V be a kG-submodule of U. Since $\mathrm{Res}^G_H(U)$ is semisimple, there is a kH-submodule Y of U such that $U = V \oplus Y$ as kH-modules. Thus the canonical map $U \to U/V$ splits as a kH-homomorphism. Since U/V is assumed to be relatively H-projective, it follows from 2.6.2 that the canonical map $U \to U/V$ is split as a kG-homomorphism, and hence the image of a section is a complement of V in U as a kG-module. Thus U is semisimple. □

Corollary 2.6.6 *Let G be a finite group, H a subgroup of G and P a normal subgroup of G contained in H. Let U be a kG-module or a complex of kG-modules. If U is relatively P-projective, then $\mathrm{Res}^G_H(U)$ is relatively P-projective.*

Proof If U is relatively P-projective, then $\mathrm{Id}_U \in (\mathrm{End}_k(U))^G_P$ by Higman's criterion. By 2.5.6 we have $(\mathrm{End}_k(U))^G_P \subseteq (\mathrm{End}_k(U))^H_P$; in particular we have $\mathrm{Id}_U \in (\mathrm{End}_k(U))^H_P$ which implies, again by Higman's criterion, that $\mathrm{Res}^G_H(U)$ is relatively P-projective. □

Corollary 2.6.7 *Let G be a finite group, H and Q subgroups of G such that $Q \subseteq H$ and let i be an idempotent in $(kG)^H_Q$. For any kG-module U the kH-module iU is relatively Q-projective.*

Proof Let $c \in (kG)^Q$ such that $\mathrm{Tr}^H_Q(c) = i$. After replacing c by ici we may assume that $c \in i(kG)^Q i$. Thus left multiplication by c on iU induces an endomorphism ψ of iU as a kQ-module. An easy verification shows that $\mathrm{Tr}^H_Q(\psi)$ is equal to left multiplication by $\mathrm{Tr}^H_Q(c) = i$, hence equal to the identity on iU. Thus $\mathrm{Id}_{iU} \in (\mathrm{End}_k(iU))^H_Q$, and the result follows from 2.6.2. □

Corollary 2.6.8 *Let G be a finite group and Q a subgroup of G. Let M, N be kG-modules and let $\varphi \in \mathrm{Hom}_{kG}(M, N)$. The following are equivalent.*

(i) $\varphi \in (\mathrm{Hom}_k(M, N))^G_Q$.
(ii) *φ factors through a relatively Q-projective kG-module.*

(iii) *φ factors through the canonical map $\mathrm{Ind}_Q^G(\mathrm{Res}_Q^G(N)) \to N$ sending $x \otimes n$ to xn for all $x \in G$ and $n \in N$.*

Proof Suppose (i) holds. Write $\varphi = \mathrm{Tr}_Q^G(\tau)$ for some $\tau \in (\mathrm{Hom}_k(M, N))^Q = \mathrm{Hom}_{kQ}(M, N)$. Frobenius reciprocity 2.2.1 yields a map $\sigma \in \mathrm{Hom}_{kG}(M, \mathrm{Ind}_Q^G(\mathrm{Res}_Q^G(N)))$ defined by $\sigma(m) = \sum_{x\in[G/Q]} x \otimes \tau(x^{-1}m)$, where $m \in M$. Composing σ with the canonical map $\mathrm{Ind}_Q^G(\mathrm{Res}_Q^G(N)) \to N$ yields φ. Thus (i) implies (iii). Clearly (iii) implies (ii). Suppose finally that (ii) holds. That is, there is a relatively Q-projective kG-module U and kG-homomorphisms $\alpha : M \to U, \beta : U \to N$ such that $\varphi = \beta \circ \alpha$. By Higman's criterion we can write $\mathrm{Id}_U = \mathrm{Tr}_Q^G(\tau)$ for some $\tau \in \mathrm{End}_{kQ}(U)$. Thus $\varphi = \beta \circ \mathrm{Id}_U \circ \alpha = \beta \circ \mathrm{Tr}_Q^G(\tau) \circ \alpha = \mathrm{Tr}_Q^G(\beta \circ \tau \circ \alpha)$, whence (i). □

Corollary 2.6.3 can also be obtained from a stronger result, showing that not only is U isomorphic to a direct summand of $\mathrm{Ind}_H^G(\mathrm{Res}_H^G(U))$ but that the identity functor on $\mathrm{Mod}(kG)$ is isomorphic to a direct summand of the functor $\mathrm{Ind}_H^G \circ \mathrm{Res}_H^G$ on $\mathrm{Mod}(kG)$ whenever the index $|G : H|$ is invertible in k. The argument to prove this is a straightforward extension of the argument used in the proof of Maschke's Theorem in 1.16.13.

Proposition 2.6.9 *Let G be a finite group and H a subgroup of G. Suppose that $|G : H|$ is invertible in k. Then the canonical homomorphism of kG-kG-bimodules $kG \otimes_{kH} kG \to kG$ sending $x \otimes y$ to xy for $x,\ y \in G$, is split. In particular, kG is isomorphic to a direct summand of $kG \otimes_{kH} kG$ as a kG-kG-bimodule.*

Proof One checks that the element $d = \sum_{y\in[G/H]} y \otimes y^{-1}$ in $kG \otimes_{kH} kG$ does not depend on the choice of the set of representatives $[G/H]$ and that $xd = dx$ for all $x \in G$. The image of d under the canonical bimodule homomorphism $kG \otimes_{kH} kG \to kG$ is equal to $|G : H|$, viewed as a scalar in k. Thus the map sending $x \in G$ to $\frac{1}{|G:H|}xd$ is a section of that bimodule homomorphism. □

The fact that a relatively H-projective kG-module U as in 2.6.2 is actually a direct summand of its own restriction to H induced back up to G is a particular case of a more general statement involving adjoint functors. We note here another case of this phenomenon:

Theorem 2.6.10 *Let A, C be k-algebras, let B be a subalgebra of A and let M be an A-C-bimodule. The following are equivalent:*

(i) *There exists a B-C-bimodule N such that M is isomorphic to a direct summand of the A-C-bimodule $A \otimes_B N$.*

(ii) The canonical surjective A-C-bimodule homomorphism $\mu : A \otimes_B M \to M$ sending $a \otimes m$ to am for any $a \in A$ and any $m \in M$ is split.

Proof If (ii) holds then (i) holds with $N = M$, viewed as a B-C-bimodule. Suppose that (i) holds. Let N be a B-C-bimodule such that $A \otimes_B N = M \oplus M'$ for some A-C-bimodule M'. Define a B-C-bimodule homomorphism $\beta : A \otimes_B N \to A \otimes_B M$ by $\beta(m) = 1 \otimes m$ for $m \in M$ and $\beta(m') = 0$ for $m' \in M'$. Denote by $\tau : A \otimes_B N \to M$ the canonical projection of $A \otimes_B N$ onto M with kernel M'. Note that $\mu \circ \beta = \tau$. Indeed, both are the identity on M and zero on M'. Define an A-C-bimodule homomorphism $\alpha : A \otimes_B N \to A \otimes_B M$ by setting $\alpha(a \otimes n) = a\beta(1 \otimes n)$ for all $a \in A$ and $n \in N$. Then, for $a \in A$ and $m \in M$ we have

$$(\mu \circ \alpha)(a \otimes n) = \mu(a\beta(1 \otimes n)) = a\mu(\beta(1 \otimes n)) = a\tau(1 \otimes n) = \tau(a \otimes n).$$

Since τ restricts to the identity on M we get that $\mu \circ \alpha|_M = \mathrm{Id}_M$, whence the result. □

The notion of relative projectivity generalises to arbitrary subalgebras of an algebra.

Definition 2.6.11 Let A be a k-algebra and B a subalgebra of A. An A-module U is called *relatively B-projective* if there is a B-module V such that U is isomorphic to a direct summand of $A \otimes_B V$. Dually, U is called *relatively B-injective* if there is a B-module V such that U is isomorphic to a direct summand of $\mathrm{Hom}_B(A, V)$, viewed as A-module via $(a \cdot \varphi)(a') = \varphi(a'a)$ for a, $a' \in A$ and $\varphi \in \mathrm{Hom}_B(A, V)$.

With this definition, given a finite group G, a subgroup H of G, and a k-algebra B, a kG-B-bimodule is relatively H-projective in the sense of 2.6.1 (extended to kG-B-bimodules) if it is relatively $kH \otimes_k B^{\mathrm{op}}$-projective as a $kG \otimes_k B^{\mathrm{op}}$-module in the sense of 2.6.11.

Theorem 2.6.12 *Let A be a k-algebra, B a subalgebra of A and U an A-module. The following are equivalent:*

(i) U is relatively B-projective.
(ii) The map $A \otimes_B U \to U$ sending $a \otimes u$ to au is split surjective as an A-homomorphism.
(iii) Any short exact sequence of A-modules ending in U whose restriction to B splits is already split as a sequence of A-modules.

Proof The equivalence of (i) and (ii) is the special case of 2.6.10 with $C = k$. The map $A \otimes_B U \to U$ has a section as B-homomorphism, namely the map sending $u \in U$ to $1 \otimes u$. Thus (iii) implies (ii). We need to show that (ii) implies (iii). Let $\psi : W \to U$ be a surjective A-homomorphism such that there is a B-homomorphism $\rho : U \to W$ satisfying $\psi \circ \rho = \mathrm{Id}_U$. Denote by $\pi : A \otimes_B U \to U$ the canonical A-homomorphism sending $a \otimes u$ to au. If (ii) holds, then there is an A-homomorphism $\sigma : U \to A \otimes_B U$ such that $\pi \circ \sigma = \mathrm{Id}_U$. Define an A-homomorphism $\tau : A \otimes_B U \to W$ by $\tau(a \otimes u) = a\rho(u)$. Then $(\psi \circ \tau)(a \otimes u) = \psi(a\rho(u)) = a\psi(\rho(u)) = au = \pi(a \otimes u)$, thus $\psi \circ \tau = \pi$. Precomposing with σ yields $\psi \circ \tau \circ \sigma = \pi \circ \sigma = \mathrm{Id}_U$, and hence $\tau \circ \sigma$ is a section of ψ as an A-homomorphism. This completes the proof. □

The situation considered in 2.6.9 extends analogously to a subalgebra B of a k-algebra A: if A is isomorphic to a direct summand of $A \otimes_B A$ as an A-A-bimodule, then every A-module U is relatively B-projective, since $U \cong A \otimes_A U$ is then isomorphic to a direct summand of $A \otimes_B A \otimes_A U \cong A \otimes_B U$. The following observation is another special case of 2.6.10:

Proposition 2.6.13 *Let A be a k-algebra and B a subalgebra of A. Then A is isomorphic to a direct summand of $A \otimes_B A$ as an A-A-bimodule if and only if the canonical map $A \otimes_B A \to A$ sending $a \otimes a'$ to aa' for all $a, a' \in A$ is a split surjective homomorphism of A-A-bimodules.*

Proof Apply 2.6.10 with $A = C$ and $M = A$ viewed as a B-A-bimodule. □

Definition 2.6.14 Let A be a k-algebra and B a subalgebra of A. We say that A is *relatively B-separable* if A is isomorphic to a direct summand of $A \otimes_B A$ as an A-A-bimodule.

With this terminology, Proposition 2.6.9 is equivalent to asserting that if H is a subgroup of a finite group G such that $|G : H|$ is invertible in k, then kG is relatively kH-separable. The terminology is consistent with its use in the context of field extensions: a finite field extension k'/k is a separable extension if and only if k' is relatively k-separable.

Definition 2.6.15 Two k-algebras A and B are called *separably equivalent* if there is an A-B-bimodule M that is finitely generated projective as a left A-module and as a right B-module and a B-A-bimodule N that is finitely generated projective as a left B-module and as a right A-module, such that A is isomorphic to a direct summand of $M \otimes_B N$ as an A-A-bimodule and such that B is isomorphic to a direct summand of $N \otimes_B M$ as a B-B-bimodule. In that situation we say for short that *A and B are separably equivalent via the bimodules M and N* or that *M and N induce a separable equivalence between A and B.*

This terminology is due to Kadison [44]; we follow the use in [59]. Separable equivalence is an equivalence relation on the class of k-algebras. A k-algebra A is separably equivalent to k if and only if A has a summand k as a k-module and A is isomorphic to a direct summand of the A-A-bimodule $A \otimes_k A$. The second condition is equivalent to requiring that A is separable (in the sense of 1.16.10); that is, A is projective as an $A \otimes_k A^{\mathrm{op}}$-module. One of the main motivations for using this notion in [59] is the fact that a separable equivalence carries just enough information to preserve certain cohomological dimensions.

Lemma 2.6.16 *Let A and B be separably equivalent k-algebras via the bimodules M and N. Then M and N are progenerators as one-sided modules.*

Proof By the definition of separable equivalence, M is finitely generated projective as a left A-module and N is finitely generated projective as a left B-module. Thus N is isomorphic to a direct summand of B^m for some positive integer m. Since A is isomorphic to a direct summand of $M \otimes_B N$ as an A-B-bimodule, it follows that A isomorphic, as a left A-module, to a direct summand of $M \otimes_B B^m \cong M^m$, which shows that M is a progenerator as a left A-module. The rest follows similarly. □

Proposition 2.6.9 implies that if H is a subgroup of a finite group G such that $|G : H|$ is invertible in k, then kG and kH are separably equivalent via the bimodules M and N which are equal to kG viewed as a kG-kH-bimodule and as a kH-kG-bimodule, respectively. The following result generalises Theorem 1.11.9; the underlying argument is a variation of the proof of 2.6.5.

Theorem 2.6.17 *Suppose that k is a field. Let A and B be separably equivalent k-algebras via bimodules M and N. Let U be an A-module. If the B-module $N \otimes_A U$ is semisimple, then U is semisimple.*

Proof Write $M \otimes_B N = A \oplus X$ for some A-A-bimodule X. Let V be a submodule of U. We need to show that V has a complement in U. Since N is finitely generated projective as a right A-module, we can identify $N \otimes_A V$ with its image in $N \otimes_A U$. Since this module is semisimple, we have $N \otimes_A U = N \otimes_A V \oplus W$ for some B-module W. Tensoring by M yields $M \otimes_B N \otimes_A U = M \otimes_B N \otimes_A V \oplus M \otimes_B W$, hence $U \oplus X \otimes_A U = V \oplus X \otimes_A V \oplus M \otimes_B W$. Intersecting U with $X \otimes_A V \oplus M \otimes_B W$ yields a complement of V in U. □

Proposition 2.6.18 *Let A and B be k-algebras and $\alpha : B \to A$ a homomorphism of k-algebras. Consider A as a left or right B-module via α. Suppose that B is isomorphic to a direct summand of A as a B-B-bimodule. Then α is injective and* $\mathrm{Im}(\alpha)$ *is a direct summand of A as a B-B-bimodule.*

Proof The left or right action of an element $b \in B$ on A is given by left or right multiplication with $\alpha(b)$. Let $\iota : B \to A$ and $\pi : A \to B$ be B-B-bimodule homomorphisms satisfying $\pi \circ \iota = \mathrm{Id}_B$. Then $\iota(1_B)$ commutes with $\mathrm{Im}(\alpha)$, the map β sending $a \in A$ to $a\iota(1_B)$ is an A-B-bimodule endomorphism of A, and we have $\beta(\alpha(b)) = \alpha(b)\iota(1_A) = \iota(b)$, hence $\beta \circ \alpha = \iota$. Thus $\pi \circ \beta \circ \alpha = \mathrm{Id}_B$, which shows that as a B-B-bimodule homomorphism, α is split injective with $\pi \circ \beta$ as a retraction. □

One can prove 2.6.18 also as a special case of 2.7.3 below, using that $A \otimes_B -$ is left adjoint to the restriction from A to B via α. Using 2.6.12, Higman's criterion can be generalised essentially verbatim to twisted group algebras. Let G be a finite group and $\alpha \in Z^2(G; k^\times)$. For $x \in G$, denote by x^{-1} the inverse of x in G and by x' the inverse of x in $(k_\alpha G)^\times$; that is, by 1.2.5 we have $x' = \alpha(1,1)^{-1}\alpha(x, x^{-1})^{-1}x^{-1}$. The map sending $a \in k_\alpha G$ to xax' is hence an inner automorphism of $k_\alpha G$. This defines an action of G on $k_\alpha G$ by inner automorphisms, because for any $x, y \in G$, the product xy in G and the product $x \cdot y$ in $k_\alpha G$ differ by a scalar, hence act in the same way on $k_\alpha G$ by conjugation. This G-action does not lift to a group homomorphism $G \to (k_\alpha G)^\times$, unless α represents the trivial class in $H^2(G; k^\times)$. In a similar way, for any two $k_\alpha G$-modules U, V, the group G acts on $\mathrm{Hom}_k(U, V)$ by ${}^x\varphi(u) = x\varphi(x'u)$ for all $u \in U$ and $x \in G$, with x' as above the multiplicative inverse of x in $k_\alpha G$. If H is a subgroup of G, we denote by $k_\alpha H$ the subalgebra of $k_\alpha G$ spanned by the image of H in $k_\alpha G$.

Theorem 2.6.19 *Let G be a finite group, H a subgroup of G and $\alpha \in Z^2(G; k^\times)$. A $k_\alpha G$-module U is relatively $k_\alpha H$-projective if and only if* $\mathrm{Id}_U \in (\mathrm{End}_k(U))_H^G$.

Proof Set $A = k_\alpha G$ and $B = k_\alpha H$. The space C spanned by $G \setminus H$ in A is a complement of B in A as a B-B-bimodule. Thus $A = B \oplus C$ and $A \otimes_B U = B \otimes_B U \oplus C \otimes_B U$. Denote by τ the B-endomorphism of A which is the identity on $B \otimes_B U$ and which is zero on $C \otimes_B U$. The same calculation as in the proof of 2.6.2 shows that $\mathrm{Tr}_H^G(\tau)$ is the identity on $A \otimes_B U$. Suppose that U is relatively B-projective. Then by 2.6.12, the canonical map $\mu : A \otimes_B U \to U$ has a section $\sigma : U \to A \otimes_B U$. Again as in the proof of 2.6.2 one sees that $\mathrm{Tr}_H^G(\mu \circ \tau \circ \sigma) = \mathrm{Id}_U$. Conversely, if $\mathrm{Id}_U = \mathrm{Tr}_H^G(\rho)$ for some B-endomorphism ρ of U, then a section σ for μ is given explicitly by the formula $\sigma(u) = \sum_{x \in [G/H]} x \otimes \rho(x'u)$. □

If k is an algebraically closed field, then 2.6.19 can also be proved by playing it back to the 'untwisted' case of 2.6.2 via 1.2.18. There is an analogous notion of relative injectivity; in the case of finite group algebras (and more generally, symmetric algebras) both notions coincide – this follows from 2.12.7 and 2.7.5 below.

Theorem 2.6.20 *Let A be a k-algebra, B a subalgebra of A and U an A-module. The following are equivalent:*

(i) U is relatively B-injective.
(ii) The map $U \to \mathrm{Hom}_B(A, U)$ sending u to the B-homomorphism $a \mapsto au$ is split injective as an A-homomorphism.
(iii) Any short exact sequence of A-modules starting in U whose restriction to B splits is already split as a sequence of A-modules.

Proof One can show this by dualising the arguments in the previous theorem; alternatively, one can apply the more general 2.7.3 from the next section. □

Exercise 2.6.21 State and prove bimodule versions of the above corollaries to Higman's criterion.

2.7 Relative projectivity and adjoint functors

The terminology of relative projectivity and the arguments in the proofs of previous theorems can be extended further to the situation of the adjunction 2.2.4. If A and B are k-algebras and M is an A-B-bimodule, we say that an A-module U is *relatively M-projective* if the canonical A-homomorphism

$$\begin{cases} M \otimes_B \mathrm{Hom}_A(M, U) & \longrightarrow & U \\ m \otimes \mu & \mapsto & \mu(m) \end{cases}$$

is split surjective as a homomorphism of A-modules. The above map is the adjunction counit of the left adjunction of $M \otimes_B -$ to $\mathrm{Hom}_A(M, -)$. One can show, analogously to the equivalence of the statements (i) and (ii) of 2.6.2, that this is equivalent to saying that U is isomorphic to a direct summand of $M \otimes_B V$ for some B-module V. There is a dual notion of relative injectivity: a B-module V is called *relatively M-injective* if the canonical B-homomorphism

$$\begin{cases} V & \longrightarrow & \mathrm{Hom}_A(M, M \otimes_B V) \\ v & \mapsto & (m \mapsto m \otimes v) \end{cases}$$

is split injective as a homomorphism of B-modules. This map is the adjunction unit of the left adjunction of $M \otimes_B -$ to $\mathrm{Hom}_A(M, -)$. Again, one can show that this is equivalent to saying that V is isomorphic to a direct summand of $\mathrm{Hom}_A(M, U)$ for some A-module U. In fact, one can show this in a general setting of adjoint functors between arbitrary categories, expressed in terms of splitting of adjunction units or counits.

Definition 2.7.1 Let $\mathcal{C}$, $\mathcal{D}$ be categories and $\mathcal{F} : \mathcal{C} \to \mathcal{D}$ a covariant functor. An object M in $\mathcal{C}$ is called *relatively $\mathcal{F}$-projective*, if for any morphism $\pi : V \to W$ in $\mathcal{C}$ and any morphism $\tau : M \to W$ in $\mathcal{C}$ such that there is a morphism $\beta : \mathcal{F}(M) \to \mathcal{F}(V)$ in $\mathcal{D}$ satisfying $\mathcal{F}(\pi) \circ \beta = \mathcal{F}(\tau)$, there is a morphism $\alpha : M \to V$ in $\mathcal{C}$ satisfying $\pi \circ \alpha = \tau$. Dually, M is *relatively $\mathcal{F}$-injective*, if for any morphism $\iota : V \to W$ in $\mathcal{C}$ and any morphism $\kappa : V \to M$ in $\mathcal{C}$ there such that there is a morphism $\delta : \mathcal{F}(W) \to \mathcal{F}(M)$ in $\mathcal{D}$ satisfying $\delta \circ \mathcal{F}(\iota) = \mathcal{F}(\kappa)$, there is a morphism $\gamma : W \to M$ in $\mathcal{C}$ satisfying $\gamma \circ \iota = \kappa$. Equivalently, M is relatively $\mathcal{F}$-injective, if it is relatively $\mathcal{F}$-projective viewed as object in the opposite category $\mathcal{C}^{\mathrm{op}}$ and $\mathcal{F}$ viewed as functor from $\mathcal{C}^{\mathrm{op}}$ to $\mathcal{D}^{\mathrm{op}}$.

The following results show that this notion of relative projectivity with respect to a functor, when applied to functors of the form $\mathrm{Hom}_A(M, -)$ and $M \otimes_B -$ for some A-B-bimodule M as above coincides with the notion of relative projectivity with respect to M described in 2.6.10.

Theorem 2.7.2 *Let $\mathcal{C}$, $\mathcal{D}$ be categories and $\mathcal{F} : \mathcal{C} \to \mathcal{D}$ a functor. Suppose that $\mathcal{F}$ has a left adjoint functor $\mathcal{G} : \mathcal{D} \to \mathcal{C}$. Denote by $f : \mathrm{Id}_{\mathcal{D}} \to \mathcal{F} \circ \mathcal{G}$ and $g : \mathcal{G} \circ \mathcal{F} \to \mathrm{Id}_{\mathcal{C}}$ the unit and counit of some adjunction. For any object M in $\mathcal{C}$ the following statements are equivalent:*

(i) M is relatively $\mathcal{F}$-projective.
(ii) Any morphism $\pi : U \to M$ in $\mathcal{C}$ such that the morphism $\mathcal{F}(\pi) : \mathcal{F}(U) \to \mathcal{F}(M)$ has a right inverse in $\mathcal{D}$ has itself a right inverse in $\mathcal{C}$.
(iii) The morphism $g(M) : (\mathcal{G} \circ \mathcal{F})(M) \to M$ has a right inverse.
(iv) There is an object N in $\mathcal{D}$ and a morphism $\mu : \mathcal{G}(N) \to M$ having a right inverse.

Proof Suppose that (i) holds. Let $\pi : U \to M$ be a morphism in $\mathcal{C}$ such that $\mathcal{F}(\pi)$ has a right inverse $\beta : \mathcal{F}(M) \to \mathcal{F}(U)$. This means that $\mathcal{F}(\pi) \circ \beta = \mathrm{Id}_{\mathcal{F}(M)} = \mathcal{F}(\mathrm{Id}_M)$. Thus (i) implies that there is a morphism $\alpha : M \to U$ satisfying $\pi \circ \alpha = \mathrm{Id}_M$, whence (ii). The composition

$$\mathcal{F} \xrightarrow{f\mathcal{F}} \mathcal{F} \circ \mathcal{G} \circ \mathcal{F} \xrightarrow{\mathcal{F}g} \mathcal{F}$$

is the identity transformation on $\mathcal{F}$. Thus $\mathcal{F}(g(M)) : (\mathcal{F} \circ \mathcal{G} \circ \mathcal{F})(M) \to \mathcal{F}(M)$ has as a right inverse the morphism $f(\mathcal{F}(M))$. Therefore, if (ii) holds, then the morphism $g(M)$ has a right inverse, which proves that (iii) holds. Taking $N = \mathcal{F}(M)$ shows that (iii) implies (iv). Let now N be an object in $\mathcal{D}$ such that there are morphisms $\mu : \mathcal{G}(N) \to M$ and $\iota : M \to \mathcal{G}(N)$ satisfying $\mu \circ \iota = \mathrm{Id}_M$. Let $\pi : V \to W$ be any morphism in $\mathcal{C}$ and $\tau : M \to W$ be a morphism in $\mathcal{C}$ such that there is a morphism $\beta : \mathcal{F}(M) \to \mathcal{F}(V)$ in $\mathcal{D}$ satisfying $\mathcal{F}(\pi) \circ \beta = \mathcal{F}(\tau)$. We define the morphism $\alpha : M \to V$ to be the composition

$$M \xrightarrow{\iota} \mathcal{G}(N) \xrightarrow{\mathcal{G}(f(N))} (\mathcal{G} \circ \mathcal{F} \circ \mathcal{G})(N) \xrightarrow{(\mathcal{G}\circ\mathcal{F})(\mu)} (\mathcal{G} \circ \mathcal{F})(M) \xrightarrow{\mathcal{G}(\beta)} (\mathcal{G} \circ \mathcal{F})(V) \xrightarrow{g(V)} V.$$

It is now a purely formal exercise to show that $\pi \circ \alpha = \tau$. Since g is a natural transformation, we have $\pi \circ g(V) = g(W) \circ (\mathcal{G} \circ \mathcal{F})(\pi)$. Thus

$$\pi \circ g(V) \circ \mathcal{G}(\beta) = g(W) \circ (\mathcal{G} \circ \mathcal{F})(\pi) \circ \mathcal{G}(\beta) = g(W) \circ (\mathcal{G} \circ \mathcal{F})(\tau)$$

since $\mathcal{G}$ is a functor and $\mathcal{F}(\pi) \circ \beta = \mathcal{F}(\tau)$. Therefore we obtain

$$\pi \circ g(V) \circ \mathcal{G}(\beta) \circ (\mathcal{G} \circ \mathcal{F})(\mu) = g(W) \circ (\mathcal{G} \circ \mathcal{F})(\tau \circ \mu).$$

Again, since g is a natural transformation, we have $g(W) \circ (\mathcal{G} \circ \mathcal{F})(\tau \circ \mu) = \tau \circ \mu \circ g(\mathcal{G}(N))$, thus

$$\pi \circ \alpha = \tau \circ \mu \circ g(\mathcal{G}(N)) \circ \mathcal{G}(f(N)) \circ \iota.$$

We have $g(\mathcal{G}(N)) \circ \mathcal{G}(f(N)) = \mathrm{Id}_{\mathcal{G}(N)}$, and hence $\pi \circ \alpha = \tau \circ \mu \circ \iota = \tau$ as claimed, and this shows that (iv) implies (i). □

Dualising the proof of the above theorem yields the analogous result for relative injective objects:

Theorem 2.7.3 *Let $\mathcal{C}$, $\mathcal{D}$, be categories and $\mathcal{F} : \mathcal{C} \to \mathcal{D}$ a functor having a right adjoint functor $\mathcal{G} : \mathcal{D} \to \mathcal{C}$. Denote by $f : \mathrm{Id}_{\mathcal{C}} \to \mathcal{G} \circ \mathcal{F}$ and $g : \mathcal{F} \circ \mathcal{G} \to \mathrm{Id}_{\mathcal{D}}$ the unit and counit of some adjunction. For any object M in $\mathcal{C}$, the following are equivalent:*

(i) M is relatively $\mathcal{F}$-injective.
(ii) Any morphism $\iota : M \to V$ in $\mathcal{C}$ such that $\mathcal{F}(\iota) : \mathcal{F}(M) \to \mathcal{F}(V)$ has a left inverse in $\mathcal{D}$ has itself a left inverse in $\mathcal{C}$.
(iii) The morphism $f(M) : M \to (\mathcal{G} \circ \mathcal{F})(M)$ has a left inverse.
(iv) There is an object N in $\mathcal{D}$ and a morphism $\iota : M \to \mathcal{G}(N)$ having a left inverse.

If two functors are both left and right adjoint to each other, one can use the units and counits of both adjunctions to define relative trace maps, which generalise relative traces in group algebras.

Definition 2.7.4 Let $\mathcal{C}$, $\mathcal{D}$ be categories, and let $\mathcal{F} : \mathcal{C} \to \mathcal{D}$ and $\mathcal{G} : \mathcal{D} \to \mathcal{C}$ be functors such that $\mathcal{G}$ is both right and left adjoint to $\mathcal{F}$. Denote by $f : \mathrm{Id}_{\mathcal{D}} \to$

$\mathcal{F} \circ \mathcal{G}$ and $g : \mathcal{G} \circ \mathcal{F} \to \mathrm{Id}_{\mathcal{C}}$ the unit and counit of a left adjunction of $\mathcal{G}$ to $\mathcal{F}$, respectively, and denote by $f' : \mathrm{Id}_{\mathcal{C}} \to \mathcal{G} \circ \mathcal{F}$ and $g' : \mathcal{F} \circ \mathcal{G} \to \mathrm{Id}_{\mathcal{D}}$ the unit and counit of a right adjunction of $\mathcal{G}$ to $\mathcal{F}$. Let M be an object in $\mathcal{C}$. The *relative trace map*

$$\mathrm{tr}_{\mathcal{F},\mathcal{G}}(M) : \mathrm{Hom}_{\mathcal{D}}(\mathcal{F}(M), \mathcal{F}(M)) \longrightarrow \mathrm{Hom}_{\mathcal{C}}(M, M)$$

is the map sending a morphism $\varphi : \mathcal{F}(M) \to \mathcal{F}(M)$ to the morphism $g(M) \circ \mathcal{G}(\varphi) \circ f'(M) : M \to M$.

The following result generalises Higman's criterion:

Theorem 2.7.5 ([19, 3.2]) *Let $\mathcal{C}$, $\mathcal{D}$ be categories, and let $\mathcal{F} : \mathcal{C} \to \mathcal{D}$ and $\mathcal{G} : \mathcal{D} \to \mathcal{C}$ be functors such that $\mathcal{G}$ is both right and left adjoint to $\mathcal{F}$. Denote by $f : \mathrm{Id}_{\mathcal{D}} \to \mathcal{F} \circ \mathcal{G}$ and $g : \mathcal{G} \circ \mathcal{F} \to \mathrm{Id}_{\mathcal{C}}$ the unit and counit of a left adjunction of $\mathcal{G}$ to $\mathcal{F}$, respectively, and denote by $f' : \mathrm{Id}_{\mathcal{C}} \to \mathcal{G} \circ \mathcal{F}$ and $g' : \mathcal{F} \circ \mathcal{G} \to \mathrm{Id}_{\mathcal{D}}$ the unit and counit of a right adjunction of $\mathcal{G}$ to $\mathcal{F}$. The following statements for an object M in $\mathcal{C}$ are equivalent:*

(i) *M is relatively $\mathcal{F}$-projective.*
(ii) *M is relatively $\mathcal{F}$-injective.*
(iii) *We have $\mathrm{Id}_M \in \mathrm{Im}(\mathrm{tr}_{\mathcal{F},\mathcal{G}}(M))$.*

Proof If (iii) holds, then $g(M)$ has $\mathcal{G}(\varphi) \circ f'(M)$ as right inverse, thus M is relatively $\mathcal{F}$-projective by 2.7.2, and $f'(M)$ has $g(M) \circ \mathcal{G}(\varphi)$ as left inverse, thus M is relatively $\mathcal{F}$-injective by the analogue 2.7.3. Therefore (iii) implies both (i) and (ii). Suppose that (i) holds. By 2.7.2 there is $\iota : M \to (\mathcal{G} \circ \mathcal{F})(M)$ such that $g(M) \circ \iota = \mathrm{Id}_M$. Let $\varphi : \mathcal{F}(M) \to \mathcal{F}(M)$ be the morphism corresponding to ι through the adjunction $\mathrm{Hom}_{\mathcal{C}}(M, (\mathcal{G} \circ \mathcal{F})(M)) \cong \mathrm{Hom}_{\mathcal{D}}(\mathcal{F}(M), \mathcal{F}(M))$. Then by general properties of adjunctions, we have $\iota = \mathcal{G}(\varphi) \circ f'(M)$, and hence $g(M) \circ \mathcal{G}(\varphi) \circ f'(M) = g(M) \circ \iota = \mathrm{Id}_M$. This shows that (i) implies (iii). The dual argument shows that (ii) implies (iii). □

We have encountered earlier a situation where relatively projective and injective objects coincide:

Proposition 2.7.6 *Let $\mathcal{C}$ be an additive category and denote by $\mathcal{F} : \mathrm{Ch}(\mathcal{C}) \to \mathrm{Gr}(\mathcal{C})$ the forgetful functor sending a complex (X, δ) to the underlying graded object X. The following are equivalent for a complex X over $\mathcal{C}$.*

(i) *X is contractible.*
(ii) *X is relatively $\mathcal{F}$-projective.*
(iii) *X is relatively $\mathcal{F}$-injective.*

Proof This is a restatement of 1.18.12. □

2.8 Morita Theory

If two k-algebras are isomorphic, then their module categories are equivalent as k-linear abelian categories, but the converse of this statement is not true in general. The main result in this section, Morita's Theorem 2.8.2 below, characterises equivalences between module categories as being induced by tensoring with suitable bimodules.

Definition 2.8.1 Let A, B be k-algebras. We say that A, B are *Morita equivalent*, if the k-linear abelian categories $\mathrm{Mod}(A)$ and $\mathrm{Mod}(B)$ are equivalent. A *Morita equivalence between A and B* is an equivalence of abelian categories $\mathrm{Mod}(A) \cong \mathrm{Mod}(B)$.

An equivalence between abelian categories preserves kernels and cokernels of morphisms, since these are part of the structural requirements for a category to be abelian. In other words, an equivalence between abelian categories is exact. Morita's Theorem 2.8.2 below implies that in order to verify whether the module categories of two k-algebras are equivalent as abelian categories, it suffices to show that they are equivalent as k-linear categories.

Theorem 2.8.2 (Morita) *Let A, B be k-algebras. The following are equivalent.*

(i) A and B are Morita equivalent.
(ii) The k-linear categories of finitely generated modules $\mathrm{mod}(A)$ *and* $\mathrm{mod}(B)$ *are equivalent.*
(iii) There is a progenerator M of A such that $\mathrm{End}_A(M)^{\mathrm{op}} \cong B$.
(iv) There is an A-B-bimodule M and a B-A-bimodule N such that $M \otimes_B N \cong A$ as A-A-bimodules and $N \otimes_A M \cong B$ as B-B-bimodules, and such that M, N are finitely generated projective as left and right modules. In that case, M and N are left and right progenerators, and we have bimodule isomorphisms:

$$N \cong \mathrm{Hom}_A(M, A) \cong \mathrm{Hom}_{B^{\mathrm{op}}}(M, B),$$
$$M \cong \mathrm{Hom}_B(N, B) \cong \mathrm{Hom}_{A^{\mathrm{op}}}(N, A).$$

(v) There is an A-B-bimodule M and a B-A-bimodule N, both finitely generated projective as left and right modules, and there are bimodule isomorphisms $\Phi : M \otimes_B N \to A$ and $\Psi : N \otimes_A M \to B$ such that

$$\mathrm{Id}_N \otimes \Phi = \Psi \otimes \mathrm{Id}_N, \; \Phi \otimes \mathrm{Id}_M = \mathrm{Id}_M \otimes \Psi,$$

where we identify $N \otimes_A A = N = B \otimes_B N$ and $M \otimes_B B = M = A \otimes_A M$.

An A-B-bimodule M and a B-A-bimodule N are said to *induce a Morita equivalence between A and B* if M, N are finitely generated projective as left and right modules, and if we have bimodule isomorphisms $M \otimes_B N \cong A$ and $N \otimes_B M \cong B$. By 2.6.15, Morita equivalent algebras are separably equivalent. Slightly more general than a Morita equivalence is a *Morita context*; this is a datum (A, B, M, N, Φ, Ψ), where A, B are algebras, M an A-B-bimodule, N a B-A-bimodule, with bimodule homomorphisms $\Phi : M \otimes_B N \to A$, $\Psi : N \otimes_A M \to B$ satisfying $\mathrm{Id}_N \otimes \Phi = \Psi \otimes \mathrm{Id}_N$ and $\Phi \otimes \mathrm{Id}_M = \mathrm{Id}_M \otimes \Psi$. The following two lemmas contain the technicalities of the proof of Morita's Theorem.

Lemma 2.8.3 *Let A be a k-algebra, let M be an A-module, and set $B = \mathrm{End}_A(M)^{\mathrm{op}}$. Consider M as an A-B-bimodule via $a \cdot m \cdot \varphi = a\varphi(m)$ for any $a \in A$, $m \in M$ and $\varphi \in B$. Set $N = \mathrm{Hom}_A(M, A)$. Denote by $\Phi : M \otimes_B N \to A$ the A-A-bimodule homomorphism sending $m \otimes \varphi$ to $\varphi(m)$, and denote by $\Psi : N \otimes_A M \to B$ the B-B-bimodule homomorphism sending $\varphi \otimes m$ to the endomorphism $(m' \mapsto \varphi(m')m)$ of M, where $m, m' \in M$ and $\varphi \in N$. Then the diagram of B-A-bimodules*

$$\begin{array}{ccc} N \otimes_A M \otimes_B N & \xrightarrow{\mathrm{Id}\otimes\Phi} & N \otimes_A A \\ {\scriptstyle \Psi\otimes\mathrm{Id}}\downarrow & & \downarrow{\scriptstyle \cong} \\ B \otimes_B N & \xrightarrow[\cong]{} & N \end{array}$$

and the diagram of A-B-bimodules

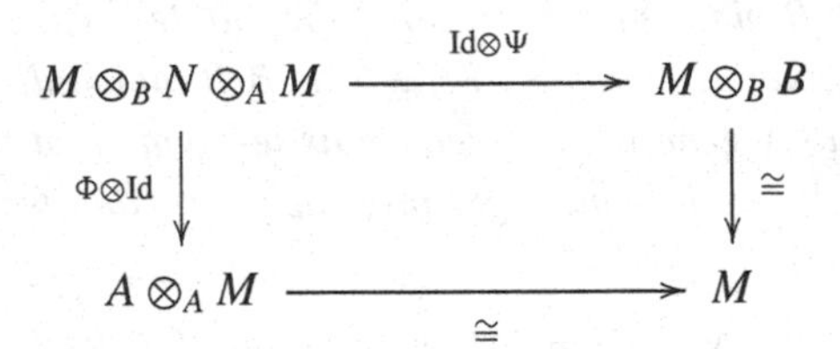

are commutative. Moreover, the following statements are equivalent.

(i) Φ and Ψ are surjective.
(ii) Φ and Ψ are isomorphisms.
(iii) As a left A-module, M is a progenerator of A.

If these equivalent statements hold, then M is a right progenerator of B satisfying $\mathrm{End}_{B^{\mathrm{op}}}(M) \cong A$ and N is a left progenerator of B satisfying $\mathrm{End}_B(N) \cong A^{\mathrm{op}}$.

Proof The commutativity of the diagrams is a straightforward verification. The key fact for proving this lemma is the characterisation of finitely generated projective modules from 1.12.7, implying that M is finitely generated projective as a left A-module if and only if Ψ is surjective, and that this is in turn the case if and only if Ψ is an isomorphism. The implication (ii) $\Rightarrow$ (i) is trivial. Suppose that (i) holds. As just mentioned, 1.12.7 implies that M is finitely generated projective as a left A-module and that Ψ is an isomorphism. The commutativity of the diagram implies that $\mathrm{Id} \otimes \Phi$ is an isomorphism. The surjectivity of Φ implies that there is a finite subset S in M and, for any $s \in S$, an A-homomorphism $\psi_s : M \to A$ such that $1_A = \Phi(\sum_{s\in S} s \otimes \psi_s) = \sum_{s\in S} \psi_s(s)$. Then the A-homomorphism $M^S \to A$ sending $(m_s)_{s\in S} \in M^S$ to $\sum_{s\in S} \psi_s(m_s)$ is surjective, hence split surjective as A is projective (and the map sending $a \in A$ to $(as)_{s\in S}$ is a section), thus M is a progenerator as a left A-module. This shows that (i) implies (iii). Suppose that (iii) holds. Then N is a right progenerator of A, and again by 1.12.7, Ψ is an isomorphism, hence so is $\Psi \otimes \mathrm{Id}$. The commutativity of the diagram in the statement implies that $\mathrm{Id} \otimes \Phi$ is an isomorphism. As a right A-module, A is isomorphic to a direct summand of N^n for some positive integer n. The commutative diagram shows that Φ can be viewed as a direct summand of the direct sum of n copies of $\mathrm{Id} \otimes \Phi$, which implies that Φ is an isomorphism since $\mathrm{Id} \otimes \Phi$ is so. Thus (iii) implies (ii). Suppose finally that these equivalent statements hold. Using again that A is isomorphic to a direct summand of N^n as a right A-module, it follows from the isomorphism Ψ that M is a progenerator as a right B-module. In particular, M is finitely generated projective as a right B-module. Thus the functor $- \otimes_B \mathrm{Hom}_{B^{\mathrm{op}}}(M, B)$ is isomorphic to the functor $\mathrm{Hom}_{B^{\mathrm{op}}}(M, -)$, hence right adjoint to the functor $- \otimes_A M$. Moreover, the functor $- \otimes_A M$ is an equivalence with inverse $- \otimes_B N$ because the maps Φ and Ψ are isomorphisms. In particular, $- \otimes_A M$ induces an isomorphism $\mathrm{End}_{B^{\mathrm{op}}}(M) \cong \mathrm{End}_{A^{\mathrm{op}}}(A) \cong A$. The functor $- \otimes_B \mathrm{Hom}_{B^{\mathrm{op}}}(M, B)$, by virtue of being a right adjoint to $- \otimes_A M$, is an inverse of $- \otimes_A M$, too, and thus $N \cong \mathrm{Hom}_{B^{\mathrm{op}}}(M, B)$, which shows that N is a left progenerator of B. Since $N \otimes_A -$ is an equivalence, we have in particular algebra isomorphisms $\mathrm{End}_B(N) = \mathrm{End}_B(N \otimes_A A) \cong \mathrm{End}_A(A) \cong A^{\mathrm{op}}$. The result follows. □

Lemma 2.8.4 *Let A, B be k-algebras and let $\mathcal{C}$ and $\mathcal{D}$ be full k-linear subcategories of* $\mathrm{Mod}(A)$ *and* $\mathrm{Mod}(B)$ *containing A and B, respectively. Let $\mathcal{F} : \mathcal{D} \to \mathcal{C}$ be a k-linear equivalence with inverse functor $\mathcal{G} : \mathcal{C} \to \mathcal{D}$. Then $\mathcal{F}(B)$ is a progenerator of A, we have* $\mathrm{End}_A(\mathcal{F}(B)) \cong B^{\mathrm{op}}$, *and we have natural isomorphisms of functors $\mathcal{F} \cong \mathcal{F}(B) \otimes_B -$ and $\mathcal{G} \cong \mathrm{Hom}_A(\mathcal{F}(B), -)$.*

Proof Since $\mathcal{F}$ is a k-linear equivalence $\mathcal{D} \cong \mathcal{C}$ and B is an object in $\mathcal{D}$, it follows that $\mathcal{F}$ induces an isomorphism of k-algebras $\mathrm{End}_A(\mathcal{F}(B)) \cong \mathrm{End}_B(B) \cong B^{\mathrm{op}}$. As $\mathrm{Hom}_B(B, -)$ is naturally isomorphic to the identity functor on $\mathrm{Mod}(B)$, we have an isomorphism of functors $\mathcal{G} \cong \mathrm{Hom}_B(B, \mathcal{G}(-))$.

Since $\mathcal{F}$ is an inverse of $\mathcal{G}$, it is in particular a left adjoint of $\mathcal{G}$, which yields together an isomorphism of functors $\mathrm{Hom}_B(B, \mathcal{G}-) \cong \mathrm{Hom}_A(\mathcal{F}(B), -)$. As $\mathcal{F}(B) \otimes_B -$ is in turn left adjoint to $\mathrm{Hom}_A(\mathcal{F}(B), -)$, the uniqueness of left adjoint functors shows that $\mathcal{F} \cong \mathcal{F}(B) \otimes_B -$. Similarly $\mathcal{G} \cong \mathcal{G}(A) \otimes_A - \cong \mathrm{Hom}_A(\mathcal{F}(B), A) \otimes_A -$. Since $\mathcal{G}$ and $\mathcal{F}$ are inverse to each other, we have isomorphisms of B-B-bimodules $B \cong \mathcal{G}(\mathcal{F}(B)) \cong \mathrm{Hom}_A(\mathcal{F}(B), A) \otimes_A \mathcal{F}(B)$, which by 1.12.7 implies that $\mathcal{F}(B)$ is finitely generated projective as an A-module. Similarly, $\mathcal{G}(A)$ is finitely generated projective, hence isomorphic to a direct summand of B^n for some positive integer n. Applying $\mathcal{F}$ shows that A is isomorphic to a direct summand of $\mathcal{F}(B)^n$, which completes the proof. □

Proof of Theorem 2.8.2 The implication (i) $\Rightarrow$ (iii) follows from 2.8.4 applied with $\mathcal{C} = \mathrm{Mod}(A)$ and $\mathcal{D} = \mathrm{Mod}(B)$. The implication (ii) $\Rightarrow$ (iii) follows from 2.8.4 applied with $\mathcal{C} = \mathrm{mod}(A)$ and $\mathcal{D} = \mathrm{mod}(B)$. The implication (iii) $\Rightarrow$ (iv) follows from 2.8.3 with $N = \mathrm{Hom}_A(M, A)$. Suppose that (iv) holds. Then the functors $M \otimes_B -$ and $N \otimes_A -$ are inverse equivalences, implying (i). Since M and N are finitely generated projective as left and right modules, these functors preserve the subcategories of finitely generated modules, and hence (iv) implies also (ii). The bimodule isomorphisms in (iv) have already been hinted at in the proof of 2.8.3: the functors $M \otimes_B -$ and $N \otimes_A -$ are inverse to each other, hence left and right adjoint to each other. A right adjoint of $M \otimes_B -$ is $\mathrm{Hom}_A(M, -) \cong \mathrm{Hom}_A(M, A) \otimes_A -$, where this isomorphism is from 2.9.15. By the uniqueness of adjoint functors from 2.3.7, the functors $N \otimes_A -$ and $\mathrm{Hom}_A(M, A) \otimes_A -$ are isomorphic. But then the bimodules N and $\mathrm{Hom}_A(M, A)$ are isomorphic by standard properties of the tensor product. The equivalence between (v) and the previous statements follows from 2.8.3. □

One consequence of the fact that equivalences between module categories are induced by bimodules is that these equivalences extend to equivalences between categories of bimodules.

Theorem 2.8.5 *Let A, B be k-algebras. Let M be an A-B-bimodule and N a B-A-bimodule such that $M \otimes_B -$ and $N \otimes_A -$ induce inverse equivalences between* $\mathrm{mod}(A)$ *and* $\mathrm{mod}(B)$. *Then the functor $N \otimes_A -$ induces an equivalence* $\mathrm{mod}(A \otimes_k A^{\mathrm{op}}) \cong \mathrm{mod}(B \otimes_k A^{\mathrm{op}})$ *that sends the bimodule A to N, and that has the inverse functor $M \otimes_B -$. Similarly, the functor $- \otimes_A M$ induces an equivalence* $\mathrm{mod}(B \otimes_k A^{\mathrm{op}}) \cong \mathrm{mod}(B \otimes_k B^{\mathrm{op}})$ *that sends the bimodule M to B, with inverse functor $- \otimes_B N$. In particular, the functor $N \otimes_A - \otimes_A M$ induces an equivalence* $\mathrm{mod}(A \otimes_k A^{\mathrm{op}}) \cong \mathrm{mod}(B \otimes_k B^{\mathrm{op}})$ *with inverse $M \otimes_B - \otimes_B N$.*

Proof All statements are immediate consequences of the bimodule isomorphisms $M \otimes_B N \cong A$ and $N \otimes_A M \cong B$. □

Corollary 2.8.6 *Let A, B be k-algebras. Let M be an A-B-bimodule and N a B-A-bimodule such that* $M \otimes_B -$ *and* $N \otimes_A -$ *induce inverse equivalences between* mod(A) *and* mod(B). *Then there is an isomorphism* $Z(A) \cong Z(B)$ *sending* $z \in Z(A)$ *to the unique element* $w \in Z(B)$ *satisfying* $zm = mw$ *for all* $m \in M$. *the inverse of this isomorphism sends* $w \in Z(B)$ *to the unique* $z \in Z(A)$ *satisfying* $wn = nz$ *for all* $n \in N$.

Proof By 2.8.5, we have isomorphisms $\mathrm{End}_{A\otimes_k A^{\mathrm{op}}}(A) \cong \mathrm{End}_{A\otimes_k B^{\mathrm{op}}}(M) \cong \mathrm{End}_{B\otimes B^{\mathrm{op}}}(B)$, given by tensoring with $- \otimes_B M$ and $N \otimes_A -$. We also have $Z(A) \cong \mathrm{End}_{A\otimes_k A^{\mathrm{op}}}(A)$, sending $z \in Z(A)$ to the endomorphism of A given by left multiplication by z on A. Thus $- \otimes_A M$ sends this endomorphism to the endomorphism of M given by left multiplication with z on M. The analogous statements with the remaining functors imply the result. □

An important application of Morita's Theorem arises in the context of the Schur functors, mentioned in §1.10. For A an algebra and e an idempotent in A, the associated *Schur functor* is the covariant exact functor from Mod(A) to Mod(eAe) sending an A-module U to the eAe-module eU and an A-homomorphism $\varphi : U \to V$ to the eAe-homomorphism $eU \to eV$ induced by restricting φ to eU. This functor is isomorphic to the functor $eA \otimes_A -$ as well as to the functor $\mathrm{Hom}_A(Ae, -)$. It follows from 2.2.4 that the Schur functor has a right adjoint $\mathrm{Hom}_{eAe}(eA, -)$ and a left adjoint $Ae \otimes_{eAe} -$. In general the right and left adjoint functors need not be isomorphic. The following result is a criterion for when a Schur functor is an equivalence.

Theorem 2.8.7 *Let A be a k-algebra and e an idempotent in A. The following are equivalent.*

(i) Then the bimodules Ae and eA induce a Morita equivalence between A and eAe.
(ii) We have $AeA = A$.
(iii) For every simple A-module S we have $eS \neq \{0\}$.

Moreover, in that case the left (resp. right) eAe-module eA (resp. Ae) is finitely generated projective.

Proof The map $eA \otimes_A Ae \to eAe$ induced by multiplication in A is always an isomorphism of eAe-eAe-bimodules. The image of the map $Ae \otimes_{eAe} eA \to A$ induced by multiplication in A is equal to AeA, hence surjective if and only if $AeA = A$. The equivalence of (i) and (ii) follows from 2.8.3 applied to Ae

and $eA \cong \mathrm{Hom}_A(Ae, A)$. The implication (ii) $\Rightarrow$ (iii) is trivial. Conversely, if (iii) holds, then AeA annihilates no simple A-module, hence is not contained in any maximal left ideal, and therefore is equal to A. If (i) holds, then eA is isomorphic to the image of A under the equivalence $eA \otimes_A -$, hence finitely generated projective. A similar argument for Ae concludes the proof. □

Theorem 2.8.7 can be proved more directly; we present this here because this is the most frequently needed special case of a Morita equivalence.

Alternative proof of 2.8.7 Suppose that $A = AeA$. We are going to show that the functor $eA \otimes_A - : \mathrm{Mod}(A) \to \mathrm{Mod}(eAe)$ has as inverse the functor $Ae :$ $\mathrm{Mod}(eAe) \to \mathrm{Mod}(A)$ sending an eAe-module N to the A-module $Ae \otimes_{eAe} N$. The functor $Ae \otimes_{eAe} -$ followed by $eA \otimes_A -$ is given by tensoring with $eA \otimes_A$ $Ae \cong eAe$, and clearly $eAe \otimes_{eAe} -$ is the identity functor on $\mathrm{Mod}(eAe)$. Conversely, the functor $eA \otimes_A -$ followed by $Ae \otimes_{eAe} -$ is given by tensoring with $Ae \otimes_{eAe} eA$. In order to show that this is the identity functor on $\mathrm{Mod}(A)$ we have to show that $Ae \otimes_{eAe} eA \cong A$ as A-A-bimodule. What we are going to show, more precisely, is that the map $\mu : Ae \otimes_{eAe} eA \to A$ sending $ce \otimes ed$ to ced is an isomorphism, where $c, d \in A$. Clearly this map is a homomorphism of A-A-bimodules. Its image is $\mathrm{Im}(\mu) = AeA = A$, so this map is surjective. It remains to see that μ is injective. Since $AeA = A$ there is a finite set J and elements $x_j \in Ae, y_j \in eA$, for any $j \in J$, such that $\sum_{j \in J} x_j y_j = 1$. Let $\sum_{s \in S} c_s \otimes d_s$ be in the kernel of this map, where S is a finite indexing set and $c_s \in Ae$, $d_s \in eA$, for $s \in S$. That means that we have $\sum_{s \in S} c_s d_s = 0$. But then also $\sum_{s \in S} y_j c_s d_s = 0$ for any $j \in J$. Note that $y_j c_s \in eAe$. Therefore, tensoring with x_j and taking the sum over all j yields $0 = \sum_{j \in J, s \in S} x_j \otimes y_j c_s d_s = \sum_{j \in J, s \in S} x_j y_j c_s \otimes d_s =$ $\sum_{s \in S} c_s \otimes d_s$, and thus $\ker(\mu) = \{0\}$. This shows the implication (ii) $\Rightarrow$ (i). The implication (i) $\Rightarrow$ (iii) is trivial. For the implication (iii) $\Rightarrow$ (ii), suppose that AeA is a proper ideal. As a left A-module, the quotient A/AeA is generated by the image of 1, hence has a maximal submodule. Thus A/AeA has a simple quotient S. In particular, S is annihilated by e. □

We list some further consequences – in particular, we give a different and slightly more general proof of the Skolem–Noether Theorem.

Corollary 2.8.8 *Let A be a k-algebra and n a positive integer. The algebra A and the matrix algebra $M_n(A)$ over A are Morita equivalent.*

Proof Let e be the idempotent in $M_n(A)$ which is 1_A in the entry $(1, 1)$ and zero in all other entries. We have an obvious isomorphism $A \cong eM_n(A)e$, and it is easy to check that the ideal generated by e in $M_n(A)$ is the whole algebra $M_n(A)$. Thus 2.8.8 follows from 2.8.7 applied to $M_n(A)$ and A instead of A and eAe, respectively. □

Corollary 2.8.9 *Let A be a k-algebra such that every finitely generated projective A-module is free. Then a k-algebra B is Morita equivalent to A if and only if $B \cong M_n(A)$ for some positive integer n.*

Proof Let M be a progenerator for A such that $\mathrm{End}_A(M)^{\mathrm{op}} \cong B$. Then M is finitely generated free as left A-module, hence $M \cong A^n$ for some positive integer n, which implies $\mathrm{End}_A(M)^{\mathrm{op}} \cong M_n(A)$. The converse is clear by 2.8.8. □

Using 2.8.8 and 2.8.9, one can rephrase the definition of the Brauer group in Section 1.16 as follows: the elements in the Brauer group of a field k are the Morita equivalence classes of finite-dimensional division k-algebras D satisfying $Z(D) = k$.

Corollary 2.8.10 *Let A be a k-algebra and e an idempotent in A such that $A = AeA$. Let U be a right A-module and V a left A-module. The inclusion $Ue \times eV \subseteq U \times V$ induces a an isomorphism $Ue \otimes_{eAe} eV \cong U \otimes_A V$.*

Proof By 2.8.7 we have an isomorphism of A-A-bimodules $Ae \otimes_{eAe} eA \cong A$ induced by multiplication in A. Tensoring this isomorphism by $U \otimes_A -$ on the left and by $- \otimes_A V$ on the right yields the result. □

If two algebras are Morita equivalent, then there is a Morita equivalence between these two algebras which can be described in terms of taking matrix algebras and applying Schur functors.

Corollary 2.8.11 *Let A, B be Morita equivalent k-algebras. Then there are positive integers m, n and idempotents $e \in M_m(A)$, $f \in M_n(B)$ such that the ideals generated by e and f are equal to $M_m(A)$ and $M_n(B)$, respectively, and such that we have algebra isomorphisms $B \cong eM_m(A)e$ and $A \cong fM_n(B)f$.*

Proof By Morita's Theorem 2.8.2 there is an A-B-bimodule M and a B-A-bimodule N, both progenerators as one-sided modules, such that $M \otimes_B N \cong A$ as A-A-bimodules and $N \otimes_A M \cong B$ as B-B-bimodules. Then $B \cong \mathrm{End}_{A^{\mathrm{op}}}(N)$. As a right A-module, N is finitely generated projective, hence isomorphic to a direct summand of A^m for some positive integer. Denote by $e : A^m \to A^m$ an idempotent endomorphism that is a projection onto a summand isomorphic to N as a right A-module. Identify $N = e(A^m)$. Then $B \cong \mathrm{End}_{A^{\mathrm{op}}}(N) \cong e\mathrm{End}_{A^{\mathrm{op}}}(A^m)e$, and clearly $\mathrm{End}_{A^{\mathrm{op}}}(A^m) \cong M_m(A)$. Since N is a progenerator as a right A-module, it follows that there is a positive integer s such that A^m is isomorphic to a direct summand of N^s as a right A-module. Thus there are A^{op}-homomorphisms $f : A^m \to N^s$ and $g : N^s \to A^m$ such that $g \circ f = \text{Id}_{A^m}$. Equivalently, there are A^{op}-homomorphisms $f_i : A^m \to N$ and $g_i : N \to A^m$, where $1 \leq i \leq s$, such that $\sum_{i=1}^{s} g_i \circ f_i = \text{Id}_{A^m}$. Since $N = e(A^m)$, we may consider the f_i and g_i as A^{op}-module endomorphisms of A^m satisfying $f_i = e \circ f_i$ and $g_i = g_i \circ e$. This shows that the ideal in the algebra $\text{End}_{A^{\text{op}}}(A^m)$ generated by e contains the unit element Id_{A^m}. Exchanging the roles of A and B concludes the proof. □

Theorem 2.8.12 (Skolem–Noether) *Suppose that every finitely generated projective k-module is free. Let n be a positive integer. Every k-algebra automorphism of the matrix algebra $M_n(k)$ is an inner automorphism.*

Proof We have $M_n(k) \cong \text{End}_k(V)$, where V is a free k-module of rank n, and then V is up to isomorphism the unique finitely generated projective indecomposable $M_n(k)$-module, since $M_n(k)$ is Morita equivalent to k by 2.8.8 and k has k as unique finitely generated projective indecomposable module, up to isomorphism, by the hypothesis on k. Thus if α is an automorphism of $M_n(k)$, there is an isomorphism $f : {}_\alpha V \cong V$ of $\text{End}_k(V)$-modules, where ${}_\alpha V$ is the module equal to V as k-module with $\text{End}_k(V)$-module structure defined by $\varphi \cdot v = \alpha(\varphi)(v)$, where $v \in V$ and $\varphi \in \text{End}_k(V)$; that is, given by restriction through α. Thus $f(\alpha(\varphi)(v)) = \varphi(f(v))$, which means that $\alpha(\varphi) = f^{-1} \circ \varphi \circ f$. □

Let A, B be k-algebras. We briefly mentioned earlier that any k-algebra isomorphism $\alpha : B \cong A$ induces an equivalence from $\text{mod}(A)$ to $\text{mod}(B)$, sending an A-module U to the B-module ${}_\alpha U$, which is equal to U as a k-module and on which $b \in B$ acts as $\alpha(b)$. This functor can be regarded as being given by tensoring with the B-A-bimodule ${}_\alpha A$. Its inverse is given by tensoring with the A-B-bimodule A_α.

Proposition 2.8.13 *Let A, B be k-algebras, and let $\alpha : B \to A$ be an isomorphism of k-algebras. Then α induces an isomorphism of A-B-bimodules ${}_{\alpha^{-1}}B \cong A_\alpha$.*

Proof Let $a \in A$ and $b, c \in B$. We have $\alpha(\alpha^{-1}(a)bc) = a\alpha(b)\alpha(c)$, whence the result. □

Not every Morita equivalence is induced by an algebra isomorphism. We have the following criterion for when this is the case.

Proposition 2.8.14 *Let A and B be k-algebras. Let M be an A-B-bimodule such that the functor $M \otimes_B - : \mathrm{mod}(B) \to \mathrm{mod}(A)$ is an equivalence. The following are equivalent.*

(i) *There is an algebra isomorphism $\alpha : B \to A$ such that $M \cong A_\alpha$ as A-B-bimodules.*
(ii) *There is an algebra isomorphism $\alpha : B \to A$ such that $M \cong {}_{\alpha^{-1}}B$ as A-B-bimodules.*
(iii) *We have $M \cong A$ as left A-modules.*
(iv) *We have $M \cong B$ as right B-modules.*

Proof Clearly (i) implies (iii) and (ii) implies (iv). The statements (i) and (ii) are equivalent by 2.8.13. Suppose that (iii) holds. Let $\varphi : M \cong A$ be an isomorphism of left A-modules. Since $M \otimes_B -$ is an equivalence, we have isomorphisms

$$B^{\mathrm{op}} \cong \mathrm{End}_B(B) \cong \mathrm{End}_A(M \otimes_B B) \cong \mathrm{End}_A(M).$$

The composition of these isomorphisms sends $b \in B^{\mathrm{op}}$ to the endomorphisms ρ_b of M given by right multiplication with b; that is, $\rho_b(m) = mb$ for all $m \in M$. The isomorphism φ induces an algebra isomorphism $\mathrm{End}_A(M) \cong \mathrm{End}_A(A)$ mapping ρ_b to $\varphi \circ \rho_b \circ \varphi^{-1}$. This endomorphism is given by right multiplication with an element in A, denoted $\alpha(b)$. The map sending $b \in A$ to $\alpha(b) \in A$ defined in this way is then an algebra isomorphism, and φ becomes an A-B-bimodule isomorphism $M \cong A_\alpha$. Thus (iii) implies (i). A similar argument shows that (iv) implies (ii), whence the result. □

If A is a k-algebra and M is an A-A-bimodule that is finitely generated projective as a left and right A-module such that the functor $M \otimes_A -$ is an equivalence on $\mathrm{mod}(A)$, then Morita's Theorem implies the existence of an A-A-bimodule N that is finitely generated projective as a left and right A-module and that satisfies bimodule isomorphisms

$$M \otimes_A N \cong A \cong N \otimes_A M.$$

Bimodules M as above are therefore also called *invertible bimodules*. The regular bimodule A is obviously invertible. If M, M' are two invertible A-A-bimodules, then so is $M \otimes_A M'$. The isomorphism classes of invertible bimodules form a group with respect to the group multiplication induced by the tensor product over A, having as unit element the isomorphism class of the regular bimodule A. The inverse of the isomorphism class of an invertible bimodule M is the isomorphism class of the bimodule N as above which satisfies $M \otimes_A N \cong A \cong N \otimes_A M$.

Definition 2.8.15 Let A be a k-algebra. The group of isomorphism classes of invertible A-A-bimodules, with group multiplication induced by $- \otimes_A -$, is called the *Picard group* of A, denoted $\mathrm{Pic}(A)$.

The next result shows that the outer automorphism group of an algebra embeds canonically into its Picard group. Given two k-algebras A, B, an A-B-bimodule M, an algebra automorphism α of A, and an algebra automorphism β of B, we denote as before by ${}_\alpha M$ the A-B-bimodule that is equal to M as a right B-module, such that $a \in A$ acts as $\alpha(a)$ on M. Similarly, we denote by M_β the A-B-bimodule that is equal to M as a left A-module, with $b \in B$ acting as $\beta(b)$ on the right of M.

Proposition 2.8.16 *Let A be a k-algebra, let $\alpha \in \mathrm{Aut}(A)$, and let M be an A-A-bimodule that is finitely generated projective as a left and right A-module, such that the functor $M \otimes_A -$ is an equivalence on* $\mathrm{mod}(A)$.

(i) We have $A \cong A_\alpha$ as A-A-bimodules if and only if α is an inner automorphism of A.
(ii) The map α induces an A-A-bimodule isomorphism ${}_{\alpha^{-1}}A \cong A_\alpha$.
(iii) The A-A-bimodule A_α is invertible, hence induces a Morita equivalence on A.
(iv) The map sending α to A_α induces an injective group homomorphism $\mathrm{Out}(A) \to \mathrm{Pic}(A)$.
(v) There exists $\alpha \in$ $\mathrm{Aut}(A)$ *such that $M \cong A_\alpha$ as A-A-bimodules if and only if $M \cong A$ as left A-modules.*

Proof Suppose that α is an inner automorphism. Let $c \in A^\times$ such that $\alpha(a) = cac^{-1}$ for all $a \in A$. Then the map sending $a \in A$ to ac^{-1} is an A-A-bimodule isomorphism $A \cong A_\alpha$. Conversely, any A-A-bimodule isomorphism $A \cong A_\alpha$ is in particular an automorphism of A as a left A-module, hence given by right multiplication with an invertible element $d \in A^\times$. This map is also a homomorphism of right A-modules, and that means exactly that $abd = ad\alpha(b)$ for all $a, b \in A$; in particular, $d^{-1}bd = \alpha(b)$ for all $b \in A$, hence α is inner. This proves (i). Statement (ii) is the special case $A = B$ of 2.8.13. Using (ii) we have $A_\alpha \otimes_A A_{\alpha^{-1}} \cong A_\alpha \otimes_A {}_\alpha A$, and this is clearly isomorphic to $A \otimes_A A \cong A$, whence (iii). Let β be another automorphism of A. We have bimodule isomorphisms $A_\beta \otimes_A A_\alpha \cong {}_{\beta^{-1}}A \otimes_A A_\alpha \cong ({}_{\beta^{-1}}A)_\alpha \cong (A_\beta)_\alpha = A_{\beta\circ\alpha}$. This proves that the map sending α to A_α induces a group homomorphism $\mathrm{Aut}(A) \to \mathrm{Pic}(A)$. It follows from (i) that this group homomorphism induces an injective group homomorphism $\mathrm{Out}(A) \to \mathrm{Pic}(A)$, whence the result. Statement (v) is the special case $A = B$ of 2.8.14. $\square$

Corollary 2.8.17 *Let A, B, C be k-algebras, let M be an A-B-bimodule, and let N be a B-C-bimodule. Let $\beta \in \mathrm{Aut}(B)$. We have an isomorphism of A-C-bimodules*

$$(M_\beta) \otimes_B N \cong M \otimes_B ({}_{\beta^{-1}}N).$$

Proof Using 2.8.16 yields isomorphisms $(M_\beta) \otimes_B N \cong M \otimes_B (B_\beta) \otimes_B N \cong M \otimes_B ({}_{\beta^{-1}}B) \otimes_B N \cong M \otimes_B ({}_{\beta^{-1}}N)$. □

Corollary 2.8.18 *Let A be a k-algebra, e an idempotent such that $A = AeA$, and let α be an automorphism of A satisfying $\alpha(e) = e$. Denote by β the restriction of α to eAe. Then α is an inner automorphism of A if and only if β is an inner automorphism of eAe.*

Proof By 2.8.7, the algebras A and eAe are Morita equivalent via the Schur functor given by multiplication with e. By 2.8.16, α is inner if and only if $A \cong A_\alpha$ as A-A-bimodules. Since α fixes e and since multiplication by e is a Morita equivalence, this is the case if and only if $eAe_\beta \cong eAe$, whence the result. □

The following result is a slight generalisation of the last statement in 2.4.5.

Proposition 2.8.19 *Let G be a finite group and $\zeta : G \to k^\times$ a group homomorphism. Denote by k_ζ the kG-module that is equal to k as a k-module, with $x \in G$ acting by multiplication with $\zeta(x)$. The unique k-linear map $\eta : kG \to kG$ sending $x \in G$ to $\zeta(x)x$ is a k-algebra automorphism of kG, and we have an isomorphism of kG-kG-bimodules*

$${}_\eta kG \cong \mathrm{Ind}_{\Delta G}^{G\times G}(k_\zeta)$$

sending $x \in G$ to $\zeta(x)^{-1}(x, 1) \otimes 1$.

Proof The fact that η is a k-algebra homomorphism is a trivial verification. It has as an inverse the map that sends $x \in G$ to $\zeta(x)^{-1}x$, so η is an automorphism. We need to check that the k-linear isomorphism ${}_\eta kG \cong \mathrm{Ind}_{\Delta G}^{G\times G}(k_\zeta)$ sending $x \in G$ to $\zeta(x)^{-1}(x, 1) \otimes 1$ is a bimodule homomorphism. Let $x, y \in G$. The left action of y on ${}_\eta kG$ sends x to $\zeta(y)yx$. The given linear isomorphism sends this to $\zeta(y)\zeta(yx)^{-1}(yx, 1) \otimes 1 = \zeta(x)^{-1}(yx, 1) \otimes 1$, which shows that this is a homomorphism of left kG-modules. The right action of y on $\mathrm{Ind}_{\Delta G}^{G\times G}(k_\zeta)$ sends $(x, 1) \otimes 1$ to $(x, y^{-1}) \otimes 1$. The right action of y on ${}_\eta kG$ sends x to xy. The given linear isomorphism sends this to $\zeta(xy)^{-1}(xy, 1) \otimes 1 = \zeta(xy)^{-1}(x, y^{-1}) \otimes \zeta(y) = \zeta(x)^{-1}(x, y^{-1}) \otimes 1$, which shows that this is also a homomorphism of right kG-modules. □

Morita equivalences extend to tensor products.

Proposition 2.8.20 *Let A, B, C, D be k-algebras. Let M be an A-B-bimodule and N a B-A-bimodule inducing a Morita equivalence between A and B. Let U be a C-D-bimodule and V a D-C-bimodule inducing a Morita equivalence between C and D. Then $M \otimes_k U$ and $B \otimes_k V$ induce a Morita equivalence between $A \otimes_k C$ and $B \otimes_k D$. In particular, $M \otimes_k C$ and $N \otimes_k C$ induce a Morita equivalence between $A \otimes_k C$ and $B \otimes_k C$.*

Proof We have a $B \otimes_k D$-balanced map from $M \otimes_k U \times N \otimes_k V$ to $(M \otimes_B N) \otimes_k (U \otimes_C V)$, sending $(m \otimes u, n \otimes v)$ to $(m \otimes n) \otimes (u \otimes v)$. This yields a natural map $M \otimes_k U \otimes_{B \otimes_k D} N \otimes_k V \to (M \otimes_B N) \otimes_k (U \otimes_C V)$. This map is k-linear and natural in M, N, U, V. It is also clearly an isomorphism for $M = B$ as a right B-module, $N = B$ as a left B-module, $U = D$ as a right D-module and $V = D$ as a left D-module. Thus $M \otimes_k U \otimes_{B \otimes_k D} N \otimes_k V \cong (M \otimes_B N) \otimes_k (U \otimes_C V) \cong A \otimes_k B$. A similar argument with exchanged roles shows the first statement, and the second statement is the special case $C = D = U = V$. □

Exercise 2.8.21 Show that Morita equivalent algebras have isomorphic Picard groups.

Exercise 2.8.22 Suppose that k is a field. Consider the algebras $A = k \times k$ and $B = M_2(k) \times k$. Calculate $\mathrm{Pic}(A)$, $\mathrm{Out}(A)$, $\mathrm{Pic}(B)$, and $\mathrm{Out}(B)$.

Exercise 2.8.23 Let A, B be k-algebras. Let M be an A-B-bimodule and N a B-A-bimodule inducing a Morita equivalence between A and B. Use Proposition 2.8.20 to show that the functor $M \otimes_B - \otimes_B N$ induces a bijection between the sets of ideals in A and ideals in B.

2.9 Duality

Duality is a technique which seeks to gain information about a mathematical object by turning it upside down via a contravariant functor. We will consider here two duality functors for module categories over a k-algebra A, namely the duality $\mathrm{Hom}_k(-, k)$ with respect to the base ring k, and the duality $\mathrm{Hom}_A(-, A)$ with respect to the algebra A itself.

Definition 2.9.1 Let A be a k-algebra. For any A-module U denote by $U^* = \mathrm{Hom}_k(U, k)$ the k-dual of U, considered as a right A-module via $(u^* \cdot a)(u) = u^*(au)$ for all $u^* \in U^*$, $a \in A$ and $u \in U$. If W is a right A-module then W^* is a left A-module via $(a \cdot w^*)(w) = w^*(wa)$ for all $w^* \in W^*$, $a \in A$ and $w \in W$. More generally, if A and B are two k-algebras and M is an A-B-bimodule, then the k-dual $M^* = \mathrm{Hom}_k(M, k)$ is a B-A-bimodule via $(b \cdot m^* \cdot a)(m) = m^*(amb)$

for all $m^* \in M^*$, $a \in A$, $b \in B$ and $m \in M$. If $\alpha : U \to V$ is a homomorphism of A-modules, we denote by α^* the unique homomorphism of right A-modules $\alpha^* : V^* \to U^*$ sending $v^* \in V^*$ to $v^* \circ \alpha$.

Duality with respect to the base ring k induces contravariant functors from $\mathrm{Mod}(A)$ to $\mathrm{Mod}(A^{\mathrm{op}})$, and from $\mathrm{Mod}(A^{\mathrm{op}})$ to $\mathrm{Mod}(A)$. Applied to A-B-bimodules, k-duality induces a contravariant functor from $\mathrm{Mod}(A \otimes_k B^{\mathrm{op}})$ to $\mathrm{Mod}(B \otimes_k A^{\mathrm{op}})$. The k-duality functor is *k-linear*; that is, the map $\mathrm{Hom}_A(U, V) \to \mathrm{Hom}_{A^{\mathrm{op}}}(V^*, U^*)$ sending α to α^* is k-linear, and we have a canonical isomorphism of right A-modules

$$(U \oplus V)^* \cong U^* \oplus V^*.$$

If $\alpha : U \to V$ is a surjective A-homomorphism then $\alpha^* : V^* \to U^*$ is injective. However, if α is injective, α^* need not be surjective; this depends on the ring k. If, for instance, k is a field then k-duality is an *exact functor*; that is, if

$$0 \to U \to V \to W \to 0$$

is a short exact sequence of A-modules, then the sequence of right A-modules

$$0 \to W^* \to V^* \to U^* \to 0$$

is exact as well. This is false for $\mathbb{Z}$-modules. There is no nonzero $\mathbb{Z}$-linear map from $\mathbb{Z}/2\mathbb{Z}$ to $\mathbb{Z}$, and hence the $\mathbb{Z}$-dual of the exact sequence

$$0 \longrightarrow \mathbb{Z} \xrightarrow{n \mapsto 2n} \mathbb{Z} \longrightarrow \mathbb{Z}/2\mathbb{Z} \longrightarrow 0$$

is isomorphic to the sequence

$$0 \longrightarrow 0 \longrightarrow \mathbb{Z} \xrightarrow{n \mapsto 2n} \mathbb{Z} \longrightarrow 0$$

which is no longer exact. Taking duality twice yields a covariant functor on $\mathrm{Mod}(A)$ sending an A-module U to its double dual U^{**}. There is a canonical natural transformation from the identity functor on $\mathrm{Mod}(A)$ to the double dual functor on $\mathrm{Mod}(A)$ given by the family of evaluation maps $U \to U^{**}$ mapping $u \in U$ to the map $\mu \mapsto \mu(u)$, where $\mu \in U^*$. This natural transformation is not an equivalence, but if k is a field, it induces an equivalence on the subcategory of A-modules that are finite-dimensional over k. Indeed, by elementary linear algebra, the canonical map induces an isomorphism $U \cong U^{**}$ if and only if $\dim_k(U)$ is finite. Thus k-duality works particularly well for finitely generated modules over algebras that are finite-dimensional over a field, and this is the situation we consider in the next proposition.

Proposition 2.9.2 *Suppose that k is a field, and let A be a finite-dimensional k-algebra. For any finitely generated A-module U, the following hold.*

(i) U is simple if and only if U^* is a simple right A-module.
(ii) U is semisimple if and only if U^* is a semisimple right A-module.
(iii) U is indecomposable if and only if the right A-module U^* is indecomposable.
(iv) U is projective if and only if U^* is injective.
(v) U is injective if and only if U^* is projective.
(vi) We have $(U/\mathrm{rad}(U))^* \cong \mathrm{soc}(U^*)$ and $\mathrm{soc}(U)^* \cong U^*/\mathrm{rad}(U^*)$.
(vii) For any positive integer i we have $(U/\mathrm{rad}^i(U))^* \cong \mathrm{soc}^i(U^*)$ and $\mathrm{soc}^i(U)^* \cong U^*/\mathrm{rad}^i(U^*)$.
(viii) For any positive integer i we have $(U/\mathrm{soc}^i(U))^* \cong \mathrm{rad}^i(U^*)$ and $\mathrm{rad}^i(U)^* \cong U^*/\mathrm{soc}^i(U^*)$.
(ix) $U/\mathrm{rad}(U)$ is simple if and only if $\mathrm{soc}(U^*)$ is simple.
(x) $U/\mathrm{soc}(U)$ is simple if and only if $\mathrm{rad}(U^*)$ is simple.

Proof Consider an exact sequence of A-modules $0 \to V \to U \to U/V \to 0$. Since k is a field, the k-dual of this sequence is an exact sequence of right A-modules $0 \to (U/V)^* \to U^* \to V^* \to 0$. If U is not simple, then choosing for V a proper nonzero submodule of U shows that U^* is not simple either. Applying this argument again to U^* proves (i). Statement (ii) follows from (i) and the fact that k-duality preserves finite direct sums, and this implies (iii) as well. Suppose U^* is injective. Let $\pi : M \to N$ be a surjective A-homomorphism and let $\beta : U \to N$ be an A-homomorphism. In order to show that U is projective we need to find a homomorphism $\alpha : U \to M$ satisfying $\pi \circ \alpha = \beta$. Since π is surjective, its k-dual $\pi^* : N^* \to M^*$ is injective. Since U^* is injective, there is a homomorphism of right A-modules $\gamma : M^* \to U^*$ satisfying $\gamma \circ \pi^* = \beta^*$. Since k-duality is an equivalence on finite-dimensional modules, there is a unique $\alpha : U \to M$ such that $\alpha^* = \gamma$, and then duality yields $\pi \circ \alpha = \beta$ as required. All other parts of (iv) and (v) are proved similarly. For i a positive integer, identify $(U/\mathrm{rad}^i(U))^*$ with the annihilator of $\mathrm{rad}^i(U)$ in U^*. This space consists of all $u^* \in U^*$ such that $u^*(au) = 0$ for all $a \in J(A)^i$ and $u \in U$, or equivalently, such that $u^* \cdot a = 0$ for all $a \in J(A)^i$. Thus this space is equal to $\mathrm{soc}^i(U^*)$ by the analogue of 1.10.22 (ii) for right modules. This shows the first isomorphism in (vii); the second isomorphism follows from applying the dual of this isomorphism to the dual U^* of U. Dualising the inclusion $\mathrm{rad}^i(U) \to U$ yields a surjective map $U^* \to (\mathrm{rad}^i(U)^*$. The kernel of this map consists of all $u^* \in U^*$ such that $u^*(au) = 0$ for all $a \in J(A)^i$ and all $u \in U$, or equivalently, such that $u^* \cdot a = 0$ for all $a \in J(A)^i$, which is equivalent to $u^* \in \mathrm{soc}^i(U^*)$. This proves the second isomorphism in (viii), and the first follows from dualising the second applied to U^* instead of U. The statements (ix) and (x) are the special cases of (vii) and (viii) with $i = 1$. □

Duality in combination with the tensor-Hom adjunction 2.2.4 links tensor products and homomorphism spaces in various ways; we state two cases which we will use later.

Proposition 2.9.3 *Suppose that k is a field. Let A be a finite-dimensional k-algebra, and let U, V be finite-dimensional A-modules. We have a natural isomorphism* $\mathrm{Hom}_A(U, V) \cong (V^* \otimes_A U)^*$. *In particular,* $\dim_k(\mathrm{Hom}_A(U, V)) = \dim_k(V^* \otimes_A U)$.

Proof Since V is finite-dimensional, it follows that $V \cong V^{**} = \mathrm{Hom}_k(V^*, k)$. Thus $\mathrm{Hom}_A(U, V) \cong \mathrm{Hom}_A(U, \mathrm{Hom}_k(V^*, k))$. The adjunction 2.2.4, using the k-A-bimodule V^*, implies that this space is naturally isomorphic to $\mathrm{Hom}_k(V^* \otimes_A U, k)$, whence the result. □

Proposition 2.9.4 *Let A and B be k-algebras, let V be an A-module and let W be a B-module. Suppose that one of V or W is finitely generated projective as a k-module. There is a natural isomorphism of B-A-bimodules*

$$\begin{cases} W \otimes_k V^* & \cong & \mathrm{Hom}_k(V, W) \\ w \otimes v^* & \mapsto & (v \mapsto v^*(v)w) \end{cases},$$

where $w \in W$, $v^* \in V^*$ and $v \in V$. If T is a direct summand of W, then this isomorphism sends $T \otimes_k V^*$ to $\mathrm{Hom}_k(V, T)$.

Proof This follows from 1.12.7, but we will give for convenience a slightly more direct argument. The right A-module structure of $V^* = \mathrm{Hom}_k(V, k)$ is given by $(v^*.a)(v) = v^*(av)$ for all $v^* \in V^*$, $v \in V$, $a \in A$. The B-A-bimodule structure of $W \otimes_K V^*$ is then given by $b(w \otimes_k v^*)a = (bw) \otimes_k (v^*.a)$ for all $a \in A$, $b \in B$, $v^* \in V^*$, $w \in W$. The B-A-bimodule structure of $\mathrm{Hom}_k(V, W)$ is given by $(b.\varphi.a)(v) = b\varphi(av)$ for all $a \in A, b \in B, v \in V$ and $\varphi \in \mathrm{Hom}_k(V, W)$. We checks first that the map $w \otimes v^* \mapsto (v \mapsto v^*(v)w)$ is well-defined; that is, we verifies that the images of $\lambda w \otimes v^*$ and $w \otimes \lambda v^*$ coincide. We checks next that the assignment $w \otimes v^* \mapsto (v \mapsto v^*(v)w)$ is a B-A-bimodule homomorphism. The k-linearity of this map is clear. This map sends $bw \otimes v^*.a$ to the linear map $v \mapsto v^*(av)bw$, which implies that this is a B-A-bimodule homomorphism. In order to see that this is an isomorphism, we only have to show that it is a k-linear isomorphism – so we can forget about the module structures coming from A and B. We first observe that this is true if $V = k$ or if $W = k$; indeed, both sides are canonically isomorphic to W or V^*, respectively. By taking direct sums we get that this map is an isomorphism if one of V or W is free of finite rank as a k-module. The general case follows from taking direct summands. The last statement on T is a trivial consequence of the functoriality of the given map. □

In many applications, V and W are free of finite rank over k. In that case, we can show 2.9.4 as follows: let $\{v_i\}_{i\in I}$ be a k-basis of V, and let $\{w_j\}_{j\in J}$ be a k-basis of W. Let $\{v_i^*\}_{i\in I}$ be the dual basis of V^*; that is, $v_i^*(v_{i'}) = \delta_{ii'}$ for all $i, i' \in I$. Then the set $\{w_j \otimes v_i^*\}_{(i,j)\in I\times J}$ is a k-basis of $W \otimes_k V^*$, and this k-basis is sent to the family of linear maps $v \mapsto v_i^*(v)w_j$, where $(i, j) \in I \times J$. This linear map sends v_i to w_j and all other $v_{i'}$ to zero. Thus this family of linear maps is a k-basis of $\mathrm{Hom}_k(V, W)$, which proves 2.9.4 in this case. In particular, if V is a finite-dimensional vector space over a field K then setting $S = \mathrm{End}_K(V)$ we get an isomorphism

$$S \cong V \otimes_K V^*$$

as S-S-bimodules.

Proposition 2.9.5 *Let U be a finitely generated projective k-module. For any k-modules V, W we have a natural isomorphism* $\mathrm{Hom}_k(U \otimes_k V, W) \cong \mathrm{Hom}_k(V, U^* \otimes_k W)$.

Proof Using 2.2.4, applied to U as a k-k-bimodule, and 2.9.4 yields isomorphisms $\mathrm{Hom}_k(U \otimes_k V, W) \cong \mathrm{Hom}_k(V, \mathrm{Hom}_k(U, W)) \cong \mathrm{Hom}_k(V, U^* \otimes_k W)$ as stated. □

Proposition 2.9.5 can be rephrased as saying that the functor $U \otimes_k -$ is left adjoint to the functor $U^* \otimes_k -$ on $\mathrm{Mod}(k)$. The identity functor on $\mathrm{Mod}(k)$ is isomorphic to $k \otimes_k -$, and the counit of this adjunction is represented by the canonical map $\eta_U : U \otimes_k U^* \to k$ sending $u \otimes u^*$ to $u^*(u)$; this follows from applying the isomorphism in 2.9.5 to $V = U^*$ and $W = k$.

If A is a k-algebra that admits an anti-automorphism, or equivalently, if there is an isomorphism $\iota : A \cong A^{\mathrm{op}}$, then a right A-module can be viewed as a left A-module through restriction via ι. If G is a finite group, then kG has a canonical anti-automorphism sending any $x \in G$ to x^{-1}. Thus, if V is a left kG-module, then its dual V^* can again be viewed as a left kG-module, with the action of an element $x \in G$ on an element $\nu \in V^*$ defined by $(x \cdot \nu)(v) = \nu(x^{-1}v)$ for all $v \in V$. If, for any kG-module V and any $x \in G$, we denote by x_V the endomorphism of V sending $v \in V$ to xv, then the above is equivalent to the equation

$$x_{V^*} = (x_V^{-1})^*$$

for any kG-module V and any $x \in G$. The keyword in the statement of the isomorphism from 2.9.5 is the word 'natural'. In particular, this isomorphism is compatible with the action of a group.

Proposition 2.9.6 *Let G be a finite group and let U be a kG-module that is finitely generated projective as a k-module. For any two kG-modules V, W, the isomorphism $\mathrm{Hom}_k(U \otimes_k V, W) \cong \mathrm{Hom}_k(V, U^* \otimes_k W)$ from 2.9.5 is an isomorphism of kG-modules. In particular, we have a natural isomorphism*

$$\mathrm{Hom}_{kG}(U \otimes_k V, W) \cong \mathrm{Hom}_{kG}(V, U^* \otimes_k W);$$

equivalently, the functors $U \otimes_k -$ and $U^* \otimes_k -$ on $\mathrm{Mod}(kG)$ are biadjoint.

Proof We can verify directly that the isomorphisms in the proof of 2.9.5 are compatible with the actions of G. Alternatively, this follows from taking G-fixed points in the isomorphism from 2.9.5, using its naturality. We spell out the details because this is the prototype for proving similar results 'by naturality'. Let $\varphi \in \mathrm{Hom}_k(U \otimes_k V, W)$, and denote by ψ its image in $\mathrm{Hom}_k(V, U^* \otimes_k W)$ under the isomorphism from 2.9.5. Let $x \in G$. As above, denote by x_U the kG-endomorphism of U induced by left multiplication with x. The dual of the endomorphism x_U of U is the endomorphism $x_{U^*}^{-1}$ of U^*. Since the isomorphism in 2.9.5 is natural, it sends $\varphi \circ (x_U \otimes x_V)$ to $(x_{U^*}^{-1} \circ \mathrm{Id}_W) \circ \psi \circ x_V$, and it sends $x_W \circ \varphi$ to $(\mathrm{Id}_{U^*} \otimes x_W) \circ \psi$. Thus φ is a kG-homomorphism if and only if $\varphi \circ (x_U \otimes x_V) = x_W \circ \varphi$, hence if and only if $(x_{U^*}^{-1} \circ \mathrm{Id}_W) \circ \psi \circ x_V = (\mathrm{Id}_{U^*} \otimes x_W) \circ \psi = (x_{U^*}^{-1} \otimes \mathrm{Id}_W) \circ (x_{U^*} \otimes x_W) \circ \psi$. This is in turn equivalent to $\psi \circ x_V = (x_{U^*} \otimes x_W) \circ \psi$, and that means precisely that ψ is a kG-homomorphism. We will see later that there is an easier way to show this, namely by observing that the functor $U \otimes_k -$ is isomorphic to the functor $\mathrm{Ind}_{\Delta G}^{G \times G}(U) \otimes_{kG} -$ and then applying 2.12.7 below. □

Corollary 2.9.7 *Let G be a finite group and let U be a kG-module that is finitely generated projective as a k-module. Suppose that M is a direct summand of the kG-module $U \otimes_k V$ for some kG-module V. Then M is isomorphic to a direct summand of $U \otimes_k U^* \otimes_k M$. In particular, U is isomorphic to a direct summand of $U \otimes_k U^* \otimes_k U$.*

Proof Consider $\mathcal{F} = U \otimes_k -$ as a functor on $\mathrm{Mod}(kG)$. By 2.9.6, this functor has as an adjoint the functor $\mathcal{G} = U^* \otimes_k -$. Thus, if M is a direct summand of $U \otimes_k V = \mathcal{F}(V)$ for some V, then M is isomorphic to a direct summand of $\mathcal{F}(\mathcal{G}(M)) = U \otimes_k U^* \otimes_k M$, by 2.7.2. Since $U \cong U \otimes_k k$ we get the last statement from the first applied to $M = U$ and $V = k$. □

Corollary 2.9.8 *Let G be a finite group and let U be a kG-module that is finitely generated projective as a k-module. The following are equivalent:*

(i) U is projective.
(ii) $U \otimes_k U^*$ is projective.
(iii) U^* is projective.
(iv) $U \otimes_k U$ is projective.

Proof If U is projective, then so is U^* since kG is symmetric, whence the equivalence of (i) and (iii). If U is projective, so are $U \otimes_k U^*$ and $U \otimes_k U$, by 2.4.10. If $U \otimes_k U^*$ or $U \otimes_k U$ is projective, so is $U \otimes_k U^* \otimes_k U$, hence so is U by 2.9.7. □

Proposition 2.9.9 *Let U, V, W be finitely generated projective k-modules. We have a commutative diagram of k-modules:*

$$\begin{array}{ccc} \mathrm{Hom}_k(U,V) \otimes_k \mathrm{Hom}_k(V,W) & \xrightarrow{\mu} & \mathrm{Hom}_k(U,W) \\ \downarrow & & \downarrow \\ U^* \otimes_k V \otimes_k V^* \otimes_k W & \xrightarrow[\mathrm{Id}\otimes\eta_V\otimes\mathrm{Id}]{} & U^* \otimes_k W \end{array}$$

Here μ is the map given by composition of maps, the vertical arrows are the canonical isomorphisms from 2.9.4, and $\eta_V : V \otimes_k V^* \to k$ sends $v \otimes v^*$ to $v^*(v)$.

Proof The commutativity is easily verified for $U = W = k$. The maps in this diagram are natural in U and W. Thus the commutativity holds for $W = k$ and any finitely generated projective U. Fixing U, the same argument applied again shows the commutativity for any finitely generated projective k-module W. □

Proposition 2.9.10 *Let A and B be k-algebras, let V be an A-module and let W be a right B-module. Suppose that one of V or W is finitely generated projective as a k-module. There is a natural isomorphism of B-A-bimodules*

$$\begin{cases} W^* \otimes_k V^* & \cong & (V \otimes_k W)^* \\ w^* \otimes v^* & \mapsto & (v \otimes w \mapsto v^*(v)w^*(w)) \end{cases}$$

where $w^* \in W^*$, $v^* \in V^*$, $v \in V$ and $w \in W$.

Proof If V or W is k, then both sides are isomorphic to W^* or V^*, respectively. By taking direct sums we get the stated isomorphism if one of V, W is free over k, and by taking direct summands we get the general case. □

Definition 2.9.11 Let A be a k-algebra. For any A-module U denote by $U^\vee = \mathrm{Hom}_A(U, A)$ the A-dual of U; this becomes a right A-module by setting $(\varphi \cdot a)(u) = \varphi(u)a$ for all $\varphi \in \mathrm{Hom}_A(U, A)$, $a \in A$ and $u \in U$. Similarly, if W is a right A-module, then $W^\vee = \mathrm{Hom}_{A^{\mathrm{op}}}(W, A)$ becomes a left A-module by setting $(a \cdot \psi)(w) = a\psi(w)$ for all $\psi \in \mathrm{Hom}_{A^{\mathrm{op}}}(W, A)$, $a \in A$ and $w \in W$. If $\alpha : U \to V$ is a homomorphism of A-modules, then we denote by $\alpha^\vee : V^\vee \to U^\vee$ the unique homomorphism of right A-modules sending $\varphi \in V^\vee$ to $\varphi \circ \alpha$.

Duality with respect to A induces again contravariant functors from $\mathrm{Mod}(A)$ to $\mathrm{Mod}(A^{\mathrm{op}})$ and from $\mathrm{Mod}(A^{\mathrm{op}})$ to $\mathrm{Mod}(A)$. These functors are k-linear, and as before, commute with finite direct sums; that is, for any two A-modules U, V we have a canonical isomorphism

$$(U \oplus V)^\vee \cong U^\vee \oplus V^\vee.$$

If $\alpha : U \to V$ is surjective, then $\alpha^\vee : V^\vee \to U^\vee$ is injective. The A-duality functor is right exact but not exact, in general. By 1.12.16 the functor $\mathrm{Hom}_A(-, A)$ is exact if and only if A is injective as an A-module. We will investigate properties of algebras over a field with this property in the section on selfinjective algebras below. For a finite-dimensional algebra over a field k we saw that k-duality sends finitely generated projective modules to injective modules. Even though A-duality is also contravariant, it still sends finitely generated projective modules to finitely generated projective right modules, and this is regardless of what the base ring is.

Theorem 2.9.12 *Let A be a k-algebra and let P be a finitely generated projective A-module. The following hold.*

(i) $\mathrm{Hom}_A(P, A)$ *is a finitely generated projective* A^{op}*-module.*
(ii) The map $P \to \mathrm{Hom}_{A^{\mathrm{op}}}(\mathrm{Hom}_A(P, A), A)$ *sending* $c \in P$ *to the map which sends* $\varphi \in \mathrm{Hom}_A(P, A)$ *to* $\varphi(c)$ *is an isomorphism of A-modules.*
(iii) For any idempotent $i \in A$ *we have an isomorphism of right A-modules* $iA \cong \mathrm{Hom}_A(Ai, A)$ *sending ia to the unique map sending bi to bia, where* $a, b \in A$.

Proof For $P = A$ as left A-modules the statements (i) and (ii) are clear. Since A-duality commutes with taking finite direct sums and summands both statements hold therefore for any finitely generated projective A-module P. Statement (iii) is a trivial verification: one shows that the given map in (iii) has as inverse the map sending $\varphi \in \mathrm{Hom}_A(Ai, A)$ to $\varphi(i) = \varphi(i^2) = i\varphi(i) \in iA$. □

Statement (ii) in the above theorem is a generalisation of the fact that a finite-dimensional vector space is isomorphic to its the double dual. Statement

(iii) is a particular case of 1.7.4.(v). There are various connections between the two duality functors. One of them relates tensoring with the k-dual of a module with taking the A-dual of another module:

Proposition 2.9.13 *Let A be a k-algebra and let V be a k-module. Suppose that A is finitely generated projective as a k-module. We have a natural isomorphism of right A-modules*

$$\begin{cases} V^* \otimes_k A & \cong & \mathrm{Hom}_A(A \otimes_k V, A) \\ v^* \otimes a & \mapsto & (b \otimes v \mapsto v^*(v)ba) \end{cases}$$

where $v \in V$, $v^* \in V^*$ and $a, b \in A$.

Proof By 2.9.4 we have $V^* \otimes_k A \cong \mathrm{Hom}_k(V, A)$. We show next that we have an isomorphism of right A-modules $\mathrm{Hom}_k(V, A) \cong \mathrm{Hom}_A(A \otimes_k V, A)$ sending $\varphi \in \mathrm{Hom}_k(V, A)$ to the unique map $\psi \in \mathrm{Hom}_A(A \otimes_k V, A)$ defined by $\psi(a \otimes v) = a\varphi(v)$, where $a \in A$ and $v \in V$. Clearly this defines a homomorphism of right A-modules. In order to see that this is an isomorphism one checks that the assignment sending $\psi \in \mathrm{Hom}_A(A \otimes_k V, A)$ to the unique map $\varphi \in \mathrm{Hom}_k(V, A)$ defined by $\varphi(v) = \psi(1_A \otimes v)$, where $v \in V$, is an inverse to the above map. One verifies that this yields the map as stated. □

In terms of relative projectivity, Proposition 2.9.13 implies that A-duality sends a relatively k-projective A-module U to a relatively k-projective right A-module $U^\vee$ provided that A is finitely generated projective as a k-module. Proposition 2.9.4 is a particular case of more general statements.

Proposition 2.9.14 *Let A be a k-algebra and let U and V be A-modules. If one of U or V is finitely generated projective, then the k-linear map*

$$\begin{cases} \mathrm{Hom}_A(U, A) \otimes_A V & \to & \mathrm{Hom}_A(U, V) \\ \varphi \otimes v & \mapsto & (u \mapsto \varphi(u)v) \end{cases}$$

is an isomorphism.

Proof This follows from Theorem 1.12.7, but one can prove this in a slightly more direct way, following the pattern of the proof of 2.9.4. If $U = A$, then both sides are canonically isomorphic to V and the given map is an isomorphism. Since the given map is k-linear it follows that this is an isomorphism for U a free A-module of finite rank. By taking direct summands we get that this is an isomorphism for U a finitely generated projective A-module. Similarly, if $V = A$, then both sides are canonically isomorphic to $\mathrm{Hom}_A(U, A)$, and the same argument as before shows that the given map is an isomorphism whenever V is finitely generated projective. □

Both $\mathrm{Hom}_A(U, A) \otimes_A V$ and $\mathrm{Hom}_A(U, V)$ are contravariant functorial in the variable U and covariant functorial in the variable V. The map in the above proposition is really a k-linear natural transformation of bifunctors. The naturality implies the following result.

Proposition 2.9.15 *Let A and B be k-algebras and M an A-B-bimodule. Suppose that M is finitely generated projective as a left A-module. Then, for any A-module U we have a natural isomorphism of B-modules*

$$\begin{cases} \mathrm{Hom}_A(M, A) \otimes_A U & \cong & \mathrm{Hom}_A(M, U) \\ \varphi \otimes u & \mapsto & (m \mapsto \varphi(m)u) \end{cases}$$

Proof The given map is a k-linear isomorphism by Proposition 2.9.14, and one checks that this is a homomorphism of B-modules. □

Remark 2.9.16 Proposition 2.9.15 describes explicitly an isomorphism of functors $M^\vee \otimes_A - \cong \mathrm{Hom}_A(M, -)$ from $\mathrm{Mod}(A)$ to $\mathrm{Mod}(B)$. From a category theoretic point of view, the importance of this statement is that it shows that not only has $\mathrm{Hom}_A(M, -)$ a left adjoint, namely the functor $M \otimes_B -$ by 2.2.4, but if M is finitely generated projective as left A-module it also has a right adjoint, namely the functor $\mathrm{Hom}_B(M^\vee, -)$, again by 2.2.4.

Remark 2.9.17 Combining the duality functors with respect to k and A is a standard tool in the representation theory of finite-dimensional algebras. If k is a field and A a finite-dimensional k-algebra, then the *Nakayama functor* on $\mathrm{mod}(A)$ is the functor $\mathrm{Hom}_k(-, k) \circ \mathrm{Hom}_A(-, A)$ sending a finitely generated A-module U to $(\mathrm{Hom}_A(U, A))^*$. The results in this Section imply that the Nakayama functor restricts to a k-linear equivalence between the category $\mathrm{proj}(A)$ of finitely generated projective A-modules and the category $\mathrm{inj}(A)$ of finitely generated injective A-modules. If i is an idempotent in A, then the Nakayama functor sends the projective A-module Ai to the injective module $(iA)^*$. We have the obvious analogue for right modules.

Exercise 2.9.18 Let V be a free k-module of finite rank. Set $S = \mathrm{End}_k(V)$. Show that the map sending $\sigma \in S$ to the dual $\sigma^* \in \mathrm{End}_k(V^*)$ is an isomorphism of k-algebras $S^{\mathrm{op}} \cong \mathrm{End}_k(V^*)$. Show that the right S-module structure on V^* induced by this algebra isomorphism is equal to the right S-module structure on V^* induced by the left S-module structure of V.

Exercise 2.9.19 Let A, B be k-algebras, $\alpha \in \mathrm{Aut}(A)$, $\beta \in \mathrm{Aut}(B)$, and let M be an A-B-bimodule. Show that the identity on $M^* = \mathrm{Hom}_k(M, k)$ induces bimodule isomorphisms $({}_\alpha M)^* \cong (M^*)_\alpha$ and $(M_\beta)^* \cong {}_\beta(M^*)$.

2.10 Traces of endomorphisms

Definition 2.10.1 Let n be a positive integer and let V be a free k-module of rank n. The *trace map* $\mathrm{tr}_V\colon \mathrm{End}_k(V) \to k$ is the k-linear map sending a k-linear endomorphism $\varphi : V \to V$ to the trace of a matrix representing φ with respect to a k-basis $\{v_1, v_2, \ldots, v_n\}$ of V. Explicitly, if α_{ij}, $1 \le i, j \le n$ are the unique coefficients in k satisfying $\varphi(v_i) = \sum_{j=1}^{n} \alpha_{ij} v_j$ for $1 \le i \le n$, then we set

$$\mathrm{tr}_V(\varphi) = \sum_{i=1}^{n} \alpha_{ii}.$$

Different choices of a basis of V in Definition 2.10.1 lead to conjugate matrices representing φ. By 1.5.7, conjugate matrices have the same trace. Thus the trace map tr_V does not depend on the choice of a k-basis of V. For the same reason, the determinant of φ, defined by $\det_V(\varphi) = \det(\alpha_{ij})$, does not depend on the choice of a basis of V, where the notation is as in 2.10.1. Trace maps can be interpreted as evaluation maps as follows.

Proposition 2.10.2 *Let V be a k-module that is free of finite rank. Denote by $\sigma : \mathrm{End}_k(V) \cong V \otimes_K V^*$ the isomorphism from 2.9.4 and denote by $\tau : V \otimes_K V^* \to k$ the k-linear map sending $v \otimes v^*$ to $v^*(v)$ for all $v \in V$, $v^* \in V^*$. Then*

$$\mathrm{tr}_V = \tau \circ \sigma.$$

Proof Let $v \in V \setminus \{0\}$ and $v^* \in V^*$. Let $\epsilon \in \mathrm{End}_k(V)$ such that $\sigma(\epsilon) = v \otimes v^*$. By 2.9.4, we have $\epsilon(w) = v^*(w)v$ for all $w \in V$. Let $\{v_1, v_2, \ldots, v_n\}$ be a k-basis of V. Write $v = \sum_{1 \le i \le n} \mu_i v_i$ with coefficients $\mu_i \in k$ for $1 \le i \le n$. Then $\epsilon(v_i) = v^*(v_i)v = \sum_{1 \le j \le n} v^*(v_i)\mu_j v_j$ for all j such that $1 \le j \le n$. Thus ϵ is represented by the matrix $(v^*(v_i)\mu_j)_{1 \le i,j \le n}$ whose trace is $\mathrm{tr}_V(\epsilon) = \sum_{1 \le i \le n} v^*(v_i)\mu_i = v^*(\sum_{1 \le i \le n} \mu_i v_i) = v^*(v) = \tau(v \otimes v^*) = \tau(\sigma(\epsilon))$. It follows that the maps $\tau \circ \sigma$ and tr_V coincide on endomorphisms of V corresponding to elements of the form $v \otimes v^*$ in $V \otimes_k V^*$. Since both maps are k-linear they are equal. □

The maps τ, σ are defined for any V that is finitely generated projective as a k-module, hence we can define in that case the trace map by $\mathrm{tr}_V = \tau \circ \sigma$. Proposition 2.10.2 can be used to show the following commutativity.

Proposition 2.10.3 *Let U and V be free k-modules of finite rank. We have a commutative diagram*

$$\begin{array}{ccc} \mathrm{Hom}_k(U,V) \otimes_k \mathrm{Hom}_k(V,U) & \xrightarrow{\gamma} & \mathrm{End}_k(U) \\ \downarrow & & \downarrow \mathrm{tr}_U \\ \mathrm{End}_k(U \otimes_k V^*) & \xrightarrow[\mathrm{tr}_{U \otimes_k V^*}]{} & k \end{array}$$

where the left vertical map is an isomorphism and γ is induced by composing maps.

Proof The left vertical map is induced by the canonical isomorphisms $\mathrm{Hom}_k(U,V)\otimes_k \mathrm{Hom}_k(V,U)\cong U^*\otimes_k V\otimes_k V^*\otimes_k U\cong \mathrm{End}_k(U\otimes_k V^*)$ from 2.9.4. All maps are given explicitly, so one way to prove 2.10.3 is by a direct calculation. Another proof can be obtained from observing that this is trivial if $V = k$, and then showing that if it holds for V, V', then also for their direct sum. A third proof can be obtained from 2.9.9 applied to $U = W$, combined with 2.10.2. □

We will investigate characters of k-free modules over an algebra in Chapter 3, with an emphasis on finite group algebras. We note here one special case.

Definition 2.10.4 Let A be a k-algebra that is free of finite rank as a k-module. The *regular character of* A is the k-linear map $\rho : A \to k$ which sends $a \in A$ to the trace $\mathrm{tr}_A(l_a)$ of the endomorphism l_a of A given by left multiplication with a on A; that is, $l_a(b) = ab$ for all $a, b \in A$.

There is an obvious right analogue of this definition. The regular character of A satisfies $\rho(ab) = \rho(ba)$ for all $a, b \in A$, since the k-linear endomorphisms $l_a \circ l_b$ and $l_b \circ l_a$ have the same trace, as a consequence of 1.5.7. Trace maps can be extended to finitely generated projective modules over any k-algebra, by making use of Proposition 2.9.15.

Definition 2.10.5 Let A be a k-algebra and let M be a finitely generated projective A-module. We define the *trace of* M *over* A to be the k-linear map

$$\mathrm{tr}_{A/M} : \mathrm{End}_A(M) \to A/[A, A]$$

which is the composition of the isomorphism $\mathrm{End}_A(M) \cong \mathrm{Hom}_A(M, A) \otimes_A M$ from Proposition 2.9.15, followed by the evaluation map $\mathrm{Hom}_A(M, A) \otimes_A M \to A$ sending $\mu \otimes m$ to the image of $\mu(m)$ in $A/[A, A]$, where $\mu \in \mathrm{Hom}_A(M, A)$ and $m \in M$.

The evaluation map in this definition sends $(\mu \cdot a) \otimes m$ to $(\mu \cdot a)(m) = \mu(m)a$, while it sends $\mu \otimes am$ to $\mu(am) = a\mu(m)$, where $a \in A$. The difference of these two elements is in the commutator subspace $[A, A]$, and hence $\mathrm{tr}_{A/M}$ is well-defined. If $A = k$ and $M = V$ is a free k-module of finite rank, then this definition coincides by Proposition 2.10.2 with the trace tr_V defined above.

2.11 Symmetric algebras

Let A and B be k-algebras, and let M be an A-B-bimodule. Then $M^* = \mathrm{Hom}_k(M, k)$ is a B-A-bimodule, with bimodule structure defined by $(b \cdot \alpha \cdot a)(m) = \alpha(amb)$ for all $m \in M$, $a \in A$, $b \in B$ and $\alpha \in \mathrm{Hom}_k(M, k)$. In particular, A^* is again an A-A-bimodule. In general A^* need not be isomorphic to A as an A-A-bimodule. An element $t \in A^*$ is called *symmetric* or a *central function* if $t(ab) = t(ba)$ for all $a, b \in A$.

Definition 2.11.1 Let A be a k-algebra. We say that A is *symmetric* if A is finitely generated projective as a k-module and if $A \cong A^*$ as A-A-bimodules. A linear map $s : A \to k$ is called a *symmetrising form of* A if there is an A-A-bimodule isomorphism $A \cong A^*$ that maps 1_A to s.

Since A is generated as a left and as a right A-module by 1_A, any symmetrising form in A^* generates A^* both as a left and as a right A-module. A symmetrising form s of a symmetric k-algebra A is automatically symmetric: if $\Phi : A \cong A^*$ is an A-A-bimodule isomorphism such that $s = \Phi(1_A)$, then for any $a \in A$, we have $a \cdot 1_A = a = 1_A \cdot a$, hence applying Φ yields $a \cdot s = s \cdot a$, which is equivalent to $s(ab) = s(ba)$ for all $a, b \in A$. The definition of symmetric algebras requires the existence of a bimodule isomorphism $A \cong A^*$ but does not specify a particular choice for such an isomorphism. Choosing a bimodule isomorphism $A \cong A^*$ is equivalent to choosing a symmetrising form. Two different isomorphisms $A \cong A^*$ will 'differ' by an automorphism of A as an A-A-bimodule. Every automorphism of A as an A-A-bimodule is in particular an automorphism of A as a left A-module, hence of the form $a \mapsto az$ for some $z \in A^\times$. This map is a right A-homomorphism as well if and only if $z \in Z(A)$. This argument yields an algebra isomorphism $Z(A) \cong \mathrm{End}_{A \otimes_k A^{\mathrm{op}}}(A)$ mapping $z \in Z(A)$ to $a \mapsto az = za$, which induces a group isomorphism $Z(A)^\times \cong \mathrm{Aut}_{A \otimes_k A^{\mathrm{op}}}(A)$; see 2.11.5 below. One can define symmetric algebras in terms of symmetric nondegenerate bilinear forms. More precisely, elementary linear algebra shows that fixing a bimodule isomorphism $\Phi : A \cong A^*$ is equivalent to fixing a nondegenerate symmetric bilinear form $\langle -, - \rangle$ on A satisfying $\langle ac, b \rangle = \langle a, cb \rangle$ for all $a, b, c \in A$. The form $\langle -, - \rangle$ is related to the symmetrising form $s = \Phi(1)$ by the formula $\langle a, b \rangle = s(ab)$ for all $a, b \in A$. If k is a field, then the existence of an isomorphism $A \cong A^*$ implies automatically that A is finite-dimensional. Examples of symmetric algebras include finite group algebras and matrix algebras. We will see that the fact that finite group algebras and matrix algebras have distinguished bases implies that there are canonical choices of symmetrising forms.

Theorem 2.11.2 *Let G be a finite group. Then the group algebra kG is symmetric. More precisely, there is an isomorphism of kG-kG-bimodules* $(kG)^* \cong kG$ *sending a k-linear map* $\mu : kG \to k$ *to the element* $\mu^0 = \sum_{x \in G} \mu(x^{-1})x$ *in kG. This map restricts to a Z(kG)-module isomorphism* $\mathrm{Cl}_k(G) \cong Z(kG)$. *The symmetrising form on kG corresponding to this isomorphism is the linear map* $s : kG \to k$ *defined by* $s(\sum_{x \in G} \lambda_x x) = \lambda_1$, *where* $\lambda_x \in k$ *for* $x \in G$.

Proof The map sending μ to μ^0 is easily checked to be an isomorphism of kG-kG-bimodules. We have $s^0 = 1$, hence s is the symmetrising form corresponding to this isomorphism. Moreover, μ^0 belongs to the centre $Z(kG)$ if and only if the values of μ at group elements depend only on their conjugacy class; that is, if and only if μ induces a class function from G to k. This yields the isomorphism $\mathrm{Cl}_k(G) \cong Z(kG)$; alternatively, this isomorphism is a restatement of the isomorphism in 1.5.5 (ii). □

Theorem 2.11.3 *Let let n be a positive integer. The matrix algebra* $M_n(k)$ *is symmetric with symmetrising form the trace map* $\mathrm{tr} : M_n(k) \to k$.

Proof For $1 \leq i, j \leq n$ denote by $E_{i,j}$ the matrix whose coefficient at (i, j) is equal to 1 and all of whose other coefficients are zero. The set $\{E_{i,j}\}_{1 \leq i,j \leq n}$ is obviously a k-basis of $M_n(k)$; in particular, $M_n(k)$ is finitely generated projective as k-module. We denote by $E_{i,j}^*$ the dual basis element in $M_n(k)^*$, sending $E_{i,j}$ to 1 and $E_{i',j'}$ to 0 for $1 \leq i', j' \leq n$ such that $(i', j') \neq (i, j)$. One checks that the map sending $E_{i,j}$ to $E_{j,i}^*$ is a bimodule isomorphism $M_n(k) \cong M_n(k)^*$. This isomorphism maps the identity matrix $\mathrm{Id}_n = \sum_{i=1}^n E_{i,i}$ to the trace map $\mathrm{tr} = \sum_{i=1}^n E_{i,i}^*$. □

Lemma 2.11.4 *Let A be a k-algebra and* $t \in A^*$. *The map* $\Phi : A \to A^*$ *sending* $a \in A$ *to* $a \cdot t \in A^*$ *(resp.* $t \cdot a$*) is a homomorphism of left (resp. right) A-modules. Moreover,* Φ *is injective if and only if* $\ker(t)$ *contains no nonzero left (resp. right) ideal, and* Φ *is a homomorphism of A-A-bimodules if and only if t is symmetric.*

Proof Consider the case $\Phi(a) = a \cdot t$, for $a \in A$. Clearly Φ is a homomorphism of left A-modules. We have $a \in \ker(\Phi)$ if and only if $a \cdot t = 0$, hence if and only if $t(ba) = 0$ for all $b \in A$, which is equivalent to $Aa \subseteq \ker(t)$. Thus Φ is injective if and only if $\ker(t)$ contains no nonzero left ideal. For $a \in A$, we have $a \cdot 1_a = 1_A \cdot a$ and we have $\Phi(1_A) = t$. Therefore Φ is a bimodule homomorphism if and only if $a \cdot t = t \cdot a$, which is equivalent to $t(ba) = t(ab)$ for all $a, b \in A$. □

Theorem 2.11.5 *Let A be a symmetric k-algebra, let* $\Phi : A \cong A^*$ *be an isomorphism of A-A-bimodules and set* $s = \Phi(1_A)$.

(i) *We have* $\Phi(a) = a \cdot s = s \cdot a$ *for all* $a \in A$. *The form s is symmetric, and* $\ker(s)$ *contains no nonzero left or right ideal.*

(ii) *If* s' *is another symmetrising forms of A then there is a unique element* $z \in Z(A)^\times$ *such that* $s' = z \cdot s$.

Proof Statement (i) follows immediately from 2.11.4. Let s' be another symmetrising form for A. Since 1_A generates A as left A-module and since s is the image of 1_A of some A-A-bimodule isomorphism, s generates A^* as left A-module. Thus there is a unique element $z \in A$ satisfying $s' = z \cdot s$. Exchanging the roles of s and s' yields a unique element $y \in Z(A)$ such that $s = y \cdot s'$. Then $s = y \cdot s' = (yz) \cdot s$, hence $yz = 1$, which shows that z is invertible. □

Proposition 2.11.6 *Let A be a symmetric k-algebra and let* $s : A \to k$ *be a symmetrising form. Let* $z \in A$ *and set* $t = z \cdot s$. *The following are equivalent:*

(i) *The form t is symmetric.*

(ii) *We have* $[A, A] \subseteq \ker(t)$.

(iii) *We have* $z \in Z(A)$.

Proof The equivalence between (i) and (ii) is trivial. The form $t = z \cdot s$ is symmetric if and only if $s(zbc) = s(zcb)$ for all $b, c \in A$. Using the symmetry of s, this is equivalent to $0 = s(zbc - zcb) = s(czb - zcb) = s((cz - zc)b)$, hence to $(cz - zc) \cdot s = 0$. Since the map sending $a \in S$ to $a \cdot s \in A^*$ is an isomorphism, this is in turn equivalent to $cz - zc = 0$ for all $c \in A$, hence to $z \in Z(A)$. This shows the equivalence between (i) and (iii). □

The k-submodule $[A, A]$ is a $Z(A)$-submodule of A, and hence the quotient $A/[A, A]$ and its dual $(A/[A, A])^*$ are $Z(A)$-modules.

Corollary 2.11.7 *Let A be a symmetric k-algebra with a symmetrising form s. The A-A-bimodule isomorphism* $A \cong A^*$ *sending* $a \in A$ *to* $a \cdot s$ *induces an isomorphism of* $Z(A)$*-modules* $Z(A) \cong (A/[A, A])^*$.

Proof By 2.11.6, the isomorphism $A \cong A^*$ sending a to $a \cdot s$ maps $Z(A)$ isomorphically to the subspace of A^* consisting of all symmetric forms. Again by 2.11.6, this space can be identified with the space $(A/[A, A])^*$. One easily checks that this isomorphism is compatible with the $Z(A)$-module structures. □

Definition 2.11.8 Let A be a symmetric k-algebra with a symmetrising form $s : A \to k$. For any k-submodule U of A we define $U^\perp = \{a \in A \mid s(au) = 0 \text{ for all } u \in U\}$.

In other words, $U^\perp$ corresponds to the annihilator of U in A^* through the isomorphism $A \cong A^*$ sending $a \in A$ to $a \cdot s$. In particular, $U^\perp$ is again a k-submodule of A, and $U^\perp$ depends in general on the choice of the symmetrising form s. We have $U \subseteq (U^\perp)^\perp$, but this inclusion need not be an equality in general (it is an equality if k is a field, for instance, because both spaces have the same dimension). For two k-submodules U, V of A one verifies easily that $(U+V)^\perp = U^\perp \cap V^\perp$ and that $U^\perp + V^\perp \subseteq (U \cap V)^\perp$. Again, this inclusion need not be an equality, but it is an equality if k is a field. The following two propositions describe special cases of subspaces U for which $U^\perp$ does not depend on the symmetrising form.

Proposition 2.11.9 *Let A be a symmetric k-algebra with a symmetrising form s and I an ideal in A. The left and right annihilator of I in A are both equal to $I^\perp$; in particular, $I^\perp$ does not depend on the choice of s.*

Proof Clearly the left and right annihilator of I are both contained in $I^\perp$. If $a \in I^\perp$ then $s(aI) = \{0\} = s(Ia)$. But aI is a right ideal and Ia is a left ideal, so this forces $aI = \{0\} = Ia$, hence a belongs to the left and right annihilator of I. □

Proposition 2.11.10 *Let A be a symmetric k-algebra with a symmetrising form s. We have $[A, A]^\perp = Z(A)$.*

Proof For $z \in A$ we have $z \in [A, A]^\perp$ if and only if $[A, A] \subseteq \ker(z \cdot s)$. By 2.11.6 this is the case if and only if $z \in Z(A)$. □

The following results collect basic construction principles for symmetric algebras.

Theorem 2.11.11 *Let A be a symmetric k-algebra with symmetrising forms $s \in A^*$ and $t \in B^*$.*

(i) For any idempotent $e \in A$, the algebra eAe is symmetric with symmetrising form $s|_{eAe}$.
(ii) The opposite algebra A^{op} is symmetric with symmetrising form s.
(iii) For any homomorphism of commutative rings $k \to k'$, the k'-algebra $k' \otimes_k A$ is symmetric.

Proof Clearly eAe is finitely generated projective as k-module because it is a direct summand of A as k-module. Any bimodule isomorphism $A \cong A^*$ restricts to a bimodule isomorphism $eAe \cong e \cdot A^* \cdot e$, and any element in $e \cdot A^* \cdot e$ can be identified with an element in $(eAe)^*$, whence (i). Statement (ii) is trivial.

For statement (iii), note that tensoring a free k-module by $k' \otimes_k -$ yields a free k'-module, hence tensoring a finitely generated projective k-module yields a finitely generated projective k'-module. Moreover, using the standard adjunction, we get that

$$\begin{aligned}(k' \otimes_k A)^* &= \mathrm{Hom}_{k'}(k' \otimes_k A, k') \cong \mathrm{Hom}_k(A, \mathrm{Hom}_{k'}(k', k'))\\ &\cong \mathrm{Hom}_k(A, k') \cong A^* \otimes_k k' \cong A \otimes_k k'\end{aligned}$$

as required. □

Theorem 2.11.12 *Let A and B be symmetric k-algebras with symmetrising forms $s \in A^*$ and $t \in B^*$.*

(i) The k-algebra $A \otimes_k B$ is symmetric with symmetrising form $s \otimes t$, mapping $a \otimes b$ to $s(a)t(b)$ for all $a \in A$ and $b \in B$.
(ii) The k-algebra $A \times B$ is symmetric, with symmetrising form $s \times t$ mapping $(a, b) \in A \times B$ to $s(a) + t(b)$.

Proof Statement (i) follows from the isomorphism $(A \otimes_k B)^* \cong A^* \otimes_k B^*$ in 2.9.10 and statement (ii) follows from the obvious isomorphism $(A \times B)^* \cong A^* \oplus B^*$ as $(A \times B)$-$(A \times B)$-bimodules. □

Corollary 2.11.13 *Let A be a k-algebra and n a positive integer. Then A is symmetric if and only if $M_n(A)$ is symmetric.*

Proof If A is symmetric then $M_n(A) \cong M_n(k) \otimes_k A$ is symmetric by 2.11.3 and 2.11.12 (ii). Conversely, the matrix E which is 1 in the 1-1-entry and zero everywhere else is an idempotent satisfying $EM_n(A)E \cong A$, hence A is symmetric by 2.11.11 (i). □

Theorem 2.11.14 *Let A be a finite-dimensional semisimple algebra over a field k. Then A is symmetric. In particular, every finite-dimensional division k-algebra is symmetric.*

Proof By Wedderburn's Theorem, A is a direct product of matrix algebras over finite-dimensional division rings. Thus, by 2.11.12, we may assume that A is a matrix algebra over a finite-dimensional division k-algebra D. It follows from 2.11.13 that we may assume that $A = D$. By 1.15.5, the space $[D, D]$ is a proper subspace of D. Thus there exists a nonzero k-linear map $t : D \to k$ with $[D, D] \subseteq \ker(t)$, or equivalently, such that t is symmetric. Since D is simple as a left or right module, it follows that $\ker(t)$ contains no nonzero left or right ideal, and hence t is a symmetrising form on D. □

Remark 2.11.15 Other examples of symmetric algebras include Iwahori–Hecke algebras and cohomology algebras of Poincaré duality spaces.

Exercise 2.11.16 Adapt the proof of 2.11.2 to show that twisted finite group algebras are symmetric.

Exercise 2.11.17 Let A be a symmetric k-algebra, and let i, j be idempotents in A. Show that a bimodule isomorphism $A \cong A^*$ induces an isomorphism of iAi-jAj-bimodules $iAj \cong (jAi)^*$.

2.12 Symmetry and adjunction

Let A be a k-algebra. Any k-linear map $s : A \to k$ induces for any A-module U a homomorphism of right A-modules from $\mathrm{Hom}_A(U, A)$ to $\mathrm{Hom}_k(U, k)$, sending $\varphi \in \mathrm{Hom}_A(U, A)$ to $s \circ \varphi$. This family of maps is a natural transformation of functors $\mathrm{Hom}_A(-, A) \to \mathrm{Hom}_k(-, k)$. If A is symmetric and s is a symmetrising form of A, then this natural transformation is an isomorphism of functors.

Theorem 2.12.1 *Let A be a symmetric k-algebra and $s : A \to k$ a symmetrising form. For any A-module U we have a natural isomorphism of right A-modules*

$$\mathrm{Hom}_A(U, A) \cong \mathrm{Hom}_k(U, k)$$

sending $\varphi \in \mathrm{Hom}_A(U, A)$ to $s \circ \varphi$. Equivalently, s induces an isomorphism $\mathrm{Hom}_A(-, A) \cong \mathrm{Hom}_k(-, k)$ of functors from $\mathrm{Mod}(A)$ to $\mathrm{Mod}(A^{\mathrm{op}})$.

Proof Let U be an A-module. Let $\Phi : A \cong A^*$ be the isomorphism of A-A-bimodules mapping 1_A to the symmetrising form s. Thus Φ induces a natural isomorphism

$$\mathrm{Hom}_A(U, A) \cong \mathrm{Hom}_A(U, A^*) = \mathrm{Hom}_A(U, \mathrm{Hom}_k(A, k)).$$

This isomorphism sends $\varphi \in \mathrm{Hom}_A(U, A)$ to the map $\psi \in \mathrm{Hom}_A(U, \mathrm{Hom}_k(A, k))$ defined by $\psi(u)(a) = s(\varphi(u)a)$. We apply now to the right side the adjunction 2.2.4 to the algebras A, k and to A, viewed as a k-A-bimodule. This yields a natural isomorphism

$$\mathrm{Hom}_A(U, \mathrm{Hom}_k(A, k)) \cong \mathrm{Hom}_k(A \otimes_A U, k) \cong \mathrm{Hom}_k(U, k).$$

This isomorphism is given explicitly: it sends ψ as above to the map sending $u \in U$ to $\psi(u)(1_A) = s(\varphi(u))$, hence it maps φ to $s \circ \varphi$ as claimed. □

The isomorphism in the above theorem is determined by the choice of a symmetrising form of A. This is the reason why the various adjunction isomorphisms in this section depend on the choice of symmetrising forms. In practice, this tends to be a minor technical issue, because any two different symmetrising forms of a symmetric algebra 'differ' by an invertible central element, which gives us sufficient control over the effect of different choices of symmetrising forms on adjunction isomorphisms.

Corollary 2.12.2 *Let A, B be symmetric k-algebras with symmetrising forms $s : A \to k$ and $t : B \to k$. For any A-B-bimodule M we have natural isomorphisms of B-A-bimodules*

$$\mathrm{Hom}_A(M, A) \cong M^* \cong \mathrm{Hom}_{B^{\mathrm{op}}}(M, B)$$

given by the maps sending $\varphi \in \mathrm{Hom}_A(M, A)$ to $s \circ \varphi$ and $\psi \in \mathrm{Hom}_{B^{\mathrm{op}}}(M, B)$ to $t \circ \psi$.

Proof This follows from applying 2.12.1 twice, with A and B^{op}, respectively. □

Corollary 2.12.3 *Let A be a symmetric k-algebra with symmetrising form $s : A \to k$, let B be a k-algebra and M an A-B-bimodule that is finitely generated projective as a left A-module. For any A-module U we have natural isomorphisms of B-modules*

$$\mathrm{Hom}_A(M, U) \cong \mathrm{Hom}_A(M, A) \otimes_A U \cong M^* \otimes_A U$$

where the inverse of the first isomorphism maps $\mu \otimes u$ to the map $m \mapsto \mu(m)u$, and the second isomorphism maps $\mu \otimes u$ to $(s \circ \mu) \otimes u$, where $\mu \in \mathrm{Hom}_A(M, A)$, $u \in U$ and $m \in M$.

Proof The first isomorphism follows from 1.12.7 and the second from the previous corollary. □

Corollary 2.12.4 *Let A be a symmetric k-algebra with symmetrising form $s : A \to k$, let B be a k-algebra and M an A-B-bimodule that is finitely generated projective as a left A-module. We have isomorphisms of B-B-bimodules*

$$\mathrm{End}_A(M) \cong \mathrm{Hom}_A(M, A) \otimes_A M \cong M^* \otimes_A M$$

mapping $\mu \otimes u$ to the map $m \mapsto \mu(m)u$ and the element $(s \circ \mu) \otimes u$ respectively, where $\mu \in \mathrm{Hom}_A(M, A)$, and $u, m \in M$. In particular, the image of Id_M in $M^* \otimes_A M$ under this isomorphism is an expression of the form

$$\sum_{i \in I} (s \circ \mu_i) \otimes m_i,$$

where I is a finite indexing set, $\mu_i \in \mathrm{Hom}_A(M, A)$ and $m_i \in M$ such that for all $m \in M$ we have

$$m = \sum_{i \in I} \mu_i(m)m_i.$$

Proof The isomorphisms $\mathrm{End}_A(M) \cong \mathrm{Hom}_A(M, A) \otimes_A M \cong M^* \otimes_A M$ are obtained as a special case of 2.12.3 with $U = M$. By the formulas given in 2.12.3, the inverse of these isomorphisms send an expression of the form $\sum_{i \in I}(s \circ \mu_i) \otimes m_i$ to the endomorphism $m \mapsto \sum_{i \in I} \mu_i(m)m_i$. This endomorphism is the identity precisely if $m = \sum_{i \in I} \mu_i(m)m_i$ for all $m \in M$, whence the last statement. □

Corollary 2.12.5 *Let A, B, C be k-algebras such that B is symmetric. Let M be an A-B-bimodule and N a B-C-bimodule such that N is finitely generated projective as a left B-module. Any choice of a symmetrising form of B determines a natural isomorphism of C-A-bimodules*

$$(M \otimes_B N)^* \cong N^* \otimes_B M^*.$$

Proof The standard adjunction yields $(M \otimes_B N)^* = \mathrm{Hom}_k(M \otimes_B N, k)) \cong \mathrm{Hom}_B(N, \mathrm{Hom}_k(M, k)) = \mathrm{Hom}_B(N, M^*)$, and since N is finitely generated projective as left B-module, we get from 2.12.3 that the last expression is isomorphic to $N^* \otimes_B M^*$. □

Corollary 2.12.6 *Let A be a symmetric k-algebra and B a k-algebra that is Morita equivalent to A. Then B is symmetric.*

Proof Let M be an A-B-bimodule inducing a Morita equivalence between A and B. Then $B^{\mathrm{op}} \cong \mathrm{End}_A(M)$ as algebras, hence as B-B-bimodules. By 2.12.4, we have $B^{\mathrm{op}} \cong M^* \otimes_A M$ as B-B-bimodules. This bimodule is selfdual by 2.12.5 (applied with A instead of B). Thus B^{op} is symmetric, and hence so is B, by 2.11.11 (ii). □

Theorem 2.12.7 *Let A, B be symmetric k-algebras and M an A-B-bimodule such that M is finitely generated projective as a left A-module and as a right B-module. Then M^* is finitely generated as a left B-module and as a right A-module. Any choice of symmetrising forms of A and B determines for any A-module U and any B-module V natural isomorphisms*

$$\mathrm{Hom}_A(M \otimes_B V, U) \cong \mathrm{Hom}_B(V, M^* \otimes_A U),$$

$$\mathrm{Hom}_A(U, M \otimes_B V) \cong \mathrm{Hom}_B(M^* \otimes_A U, V).$$

Equivalently, the functors $M \otimes_B -$ and $M^* \otimes_A -$ are biadjoint, and any choice of symmetrising forms of A and B determines isomorphisms of bifunctors

$$\mathrm{Hom}_A(M \otimes_B -, -) \cong \mathrm{Hom}_B(-, M^* \otimes_A -),$$
$$\mathrm{Hom}_A(-, M \otimes_B -) \cong \mathrm{Hom}_B(M^* \otimes_A -, -).$$

Proof The isomorphisms $A \cong A^*$ and $B \cong B^*$ imply that taking duals preserves finitely generated projective modules. The functor $M \otimes_B -$ has as a right adjoint the functor $\mathrm{Hom}_A(M, -)$. By 2.12.3 this functor is isomorphic to $M^* \otimes_A -$, and any choice of a symmetrising form s on A determines an isomorphism between these two functors. The same argument applied to M^* shows that $M^* \otimes_A -$ is also left adjoint to $M \otimes_B -$. □

Proposition 2.12.8 *Suppose that k is a field. Let A be a split semisimple finite-dimensional k-algebra. Let X, Y be simple A-modules. We have $X^* \otimes_A Y \cong k$ if $X \cong Y$ and $X^* \otimes_A Y \cong \{0\}$ otherwise.*

Proof Since A is split semisimple, it is a direct product of matrix algebras over k; in particular, A is symmetric and all A-modules are projective. By 2.12.3 we have $X^* \otimes_A Y \cong \mathrm{Hom}_A(X, Y)$, hence Schur's Lemma implies the result. □

Proposition 2.12.9 *Let A, B be symmetric separably equivalent k algebras via bimodules M and N. Then A, B are separably equivalent via M and M^*.*

Proof View $M \otimes_B -$ as a functor from $\mathrm{Mod}(B \otimes_k A^{\mathrm{op}})$ to $\mathrm{Mod}(A \otimes_k A^{\mathrm{op}})$. This functor has $M^* \otimes_A -$ as a left and right adjoint. Since A is isomorphic to a direct summand of $M \otimes_B N$, it follows from the implication (iv) $\Rightarrow$ (iii) in 2.7.2 that A is isomorphic to a direct summand of $M \otimes_B M^* \otimes_A A \cong M \otimes_B M^*$. A similar argument applied to the functor $- \otimes_A M$ and its left and right adjoint $- \otimes_B M^*$ concludes the proof. □

Let A, B be symmetric k-algebras and M an A-B-bimodule such that M is finitely generated projective as a left A-module and as a right B-module. It is sometimes useful to describe the adjunction isomorphisms and their units and counits in Theorem 2.12.7 more explicitly. The unit of the adjoint pair $(M \otimes_B -, M^* \otimes_A -)$ is a natural transformation from the identity functor $B \otimes_B -$ on $\mathrm{Mod}(B)$ to the functor $M^* \otimes_A M \otimes_B -$, and the corresponding counit is a natural transformation from $M \otimes_B M^* \otimes_A -$ to $A \otimes_A -$. Evaluating these natural transformations at B and A, respectively, yields bimodule homomorphisms $B \to M^* \otimes_A M$ and $M \otimes_B M^* \to A$ representing these transformations; see 2.3.3. The bimodule homomorphism $B \to M^* \otimes_A M$ representing the unit of the adjunction is the image of Id_M under the adjunction isomorphism

$$\mathrm{Hom}_A(M, M) \cong \mathrm{Hom}_B(B, M^* \otimes_A M),$$

where on the left side, we have identified $M = M \otimes_B B$. The bimodule homomorphism representing the counit is the image of Id_{M^*} under the adjunction isomorphism

$$\mathrm{Hom}_B(M^*, M^*) \cong \mathrm{Hom}_A(M \otimes_B M^*, A),$$

where on the left side we have identified $M^* = M^* \otimes_A A$. The fact that the images of Id_M and Id_{M^*} are indeed bimodule homomorphisms follows from the naturality of these maps. Combining Example 2.3.3 and the appropriate versions of the isomorphisms from 2.12.4 yields the following description of the adjunction isomorphisms and the related units and counits in Theorem 2.12.7. For $\gamma \in \mathrm{Hom}_A(M \otimes_B V, A)$, $v \in V$, $u \in U$, and $m \in M$, define $\gamma_v \in \mathrm{Hom}_A(M, A)$ by $\gamma_v(m) = \gamma(m \otimes v)$ and $\lambda_{\gamma,u} \in \mathrm{Hom}_A(M \otimes_B V, U)$ by $\lambda_{\gamma,u}(m \otimes v) = \gamma_v(m)u$. Let I be a finite indexing set, $\mu_i \in \mathrm{Hom}_A(M, A)$, $m_i \in M$, such that $\sum_{i \in I} \mu_i(m)m_i = m$ for all $m \in M$. Combining the isomorphism $\mathrm{Hom}_A(M, A) \cong M^*$ induces by composing with the symmetrising form s and the standard tensor-Hom adjunction isomorphism shows that the adjunction isomorphism

$$\mathrm{Hom}_A(M \otimes_B V, U) \cong \mathrm{Hom}_B(V, M^* \otimes_A U) \tag{2.12.1}$$

from Theorem 2.12.7 sends $\lambda_{\gamma,u}$ to the map $v \mapsto s \circ \gamma_v \otimes u$. The unit and counit of this adjunction are represented by the bimodule homomorphisms

$$\epsilon_M : B \to M^* \otimes_A M,\ 1_B \mapsto \sum_{i \in I} (s \circ \mu_i) \otimes m_i,$$

$$\eta_M : M \otimes_B M^* \to A,\ m \otimes (s \circ \alpha) \mapsto \alpha(m), \tag{2.12.2}$$

where I, μ_i, m_i are as before, and where $m \in M$ and $\alpha \in \mathrm{Hom}_A(M, A)$. Similarly, we have an adjunction isomorphism

$$\mathrm{Hom}_B(M^* \otimes_A U, V) \cong \mathrm{Hom}_A(U, M \otimes_B V) \tag{2.12.3}$$

obtained from Equation (2.12.1) by exchanging the roles of A and B and using M^* instead of M together with the canonical double duality $M^{**} \cong M$. The adjunction unit and counit of this adjunction are represented by bimodule homomorphisms

$$\epsilon_{M^*} : A \to M \otimes_B M^*,\ 1_A \mapsto \sum_{j \in J} m_j \otimes (t \circ \beta_j),$$

$$\eta_{M^*} : M^* \otimes_A M \to B,\ (t \circ \beta) \otimes m \mapsto \beta(m), \tag{2.12.4}$$

where J is a finite indexing set, $\beta_j \in \mathrm{Hom}_{B^{\mathrm{op}}}(M, B)$, $m_j \in M$, such that $\sum_{j \in J} m_j \beta_j(m) = m$ for all $m \in M$, where $\beta \in \mathrm{Hom}_{B^{\mathrm{op}}}(M, B)$. Note the slight abuse of notation: for the maps ϵ_{M^*} and η_{M^*} in Equation (2.12.4) to coincide with those obtained from Equation (2.12.2) applied to M^* instead of M we need to identify M and M^{**}. One could avoid this here and in the exercises below by replacing the pair of bimodules (M, M^*) by a pair of bimodules (M, N) that are dual to each other through a fixed choice of a nondegenerate bilinear map $M \times N \to k$. Using these explicit descriptions of the adjunction maps, their units and counits, one can prove a series of useful formal properties of these maps – see the exercises at the end of this section.

The adjunction units and counits above give rise to relative trace maps, as described in 2.7.4. We will describe later one special case in detail, namely the case where $B = k$ and $M = A$, viewed as an A-k-bimodule, since this will be useful in the context of stable categories of symmetric algebras.

Exercise 2.12.10 Let A, B be symmetric k-algebras with symmetrising forms s and t, respectively. Let M be an A-B-bimodule that is finitely generated projective as a left A-module and as a right B-module. Show that the adjunction units and counits of the adjunctions (2.12.1) and (2.12.3) are also the units and counits of the corresponding adjunctions for right modules. More precisely, show that the maps ϵ_M and η_M represent the unit and counit of the adjoint pair $(- \otimes_B M^*, - \otimes_A M)$, and the maps ϵ_{M^*} and η_{M^*} represent the unit and counit of the adjoint pair $(- \otimes_A M, - \otimes_B M^*)$.

Exercise 2.12.11 Let A, B, C be symmetric k-algebras. Let t be a symmetrising form of B. Let M be an A-B-bimodule that is finitely generated projective as a left A-module and as a right B-module. Let N be a B-C-bimodule that is finitely generated projective as a left B-module and as a right C-module. Show that the natural isomorphism of C-A-bimodules

$$N^* \otimes_B M^* \cong (M \otimes_B N)^*$$

from 2.12.5 sends $(t \circ \beta) \otimes \mu$ to the map $m \otimes n \mapsto \mu(m\beta(n))$, where $\mu \in M^*$, $\beta \in \mathrm{Hom}_B(N, B)$, $m \in M$, and $n \in N$.

Exercise 2.12.12 Let A, B be symmetric k-algebras with symmetrising forms s and t, respectively. Let M be an A-B-bimodule that is finitely generated projective as a left A-module and as a right B-module. Using the previous exercise, applied to $C = A$ and $N = M^*$, show that the adjunction units and counits from the left and right adjunction of the functors $M^* \otimes_A -$ and $M \otimes_B -$ are dual to each other. More precisely, show that we have a commutative diagram of A-A-bimodules

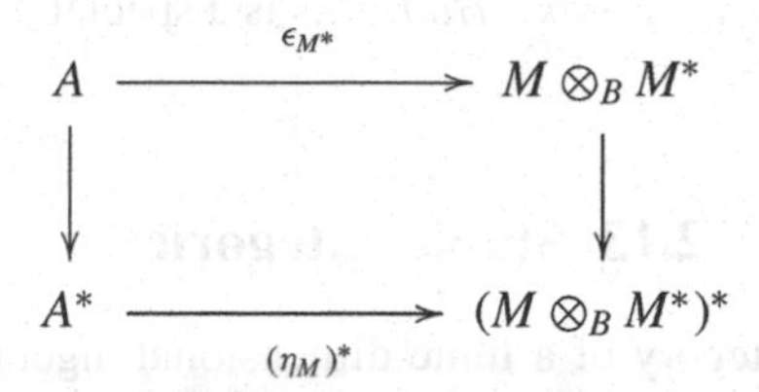

where the left vertical isomorphism is induced by s (sending $a \in A$ to the linear map $a \cdot s$ defined by $(a \cdot s)(a') = s(aa')$ for all $a \in A$) and where the right vertical isomorphism combines the isomorphism $(M \otimes_B M^*)^* \cong M^{**} \otimes_B M^*$ from the previous exercise and the canonical isomorphism $M^{**} \cong M$. *Hint:* The commutativity is verified by chasing 1_A through this diagram. Similarly, show that we have a commutative diagram of B-B-bimodules

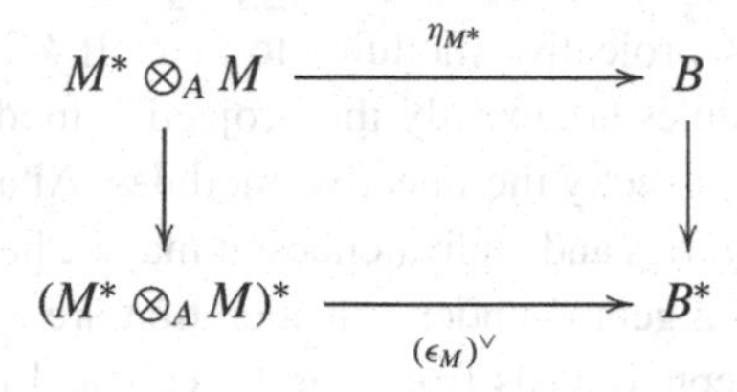

where the right vertical isomorphism is induced by t and the left vertical isomorphism is as before from the previous exercise combined with $M^{**} \cong M$.

Exercise 2.12.13 Let A be a symmetric k-algebra with symmetrising form s. We denote by Ind_k^A the induction functor $A \otimes_k -$ and by Res_k^A the restriction functor, which we identify with the functor $A \otimes_A -$, where here A is viewed as a k-A-bimodule. Thus $\mathrm{Ind}_k^A \circ \mathrm{Res}_k^A$ is the functor obtained from tensoring over A by the A-A-bimodule $A \otimes_k A$ and $\mathrm{Res}_k^A \circ \mathrm{Ind}_k^A$ is the functor obtained from tensoring over k by A, where here A is viewed as a k-k-bimodule. Show that for any A-module U and any k-module V there is an adjunction isomorphism

$$\mathrm{Hom}_A(\mathrm{Ind}_k^A(V), U) \cong \mathrm{Hom}_k(V, \mathrm{Res}_k^A(U))$$

whose unit is represented by the canonical map $k \to A$ and whose counit is represented by the map $A \otimes_k A \to A$ induced by multiplication in A. Show that there is an adjunction isomorphism

$$\mathrm{Hom}_k(\mathrm{Res}^A_k(U), V) \cong \mathrm{Hom}_A(U, \mathrm{Ind}^A_k(V))$$

whose unit is represented by the map $\tau_A : A \to A \otimes_k A$ obtained from dualising the multiplication map $\mu_A : A \otimes_k A \to A$, and whose counit is represented by the symmetrising form $s : A \to k$. (*Hint:* this is a special case of 2.12.12.)

2.13 Stable categories

The stable module category of a finite-dimensional algebra A over a field is the category obtained from the module category of A by identifying projective modules to zero. There is an obvious variation identifying injective modules to zero instead. If A is selfinjective, then the classes of projective and injective modules coincide, and hence so do the two versions of stable categories. Since finite group representation theory requires more general coefficient rings in order to relate character theory (in characteristic zero) to modular representations (in prime characteristic), we will need to develop these concepts in slightly greater generality: instead of just identifying projective modules to zero, we identify all relatively k-projective modules to zero. If k is a field, then relatively k-projective modules are exactly the projective modules, and relatively k-injective modules are exactly the injective modules. At a first reading of the sections on stable categories and equivalences, it may be helpful to assume that k is a field and that all algebras under consideration are finite-dimensional in order to separate conceptual ideas from purely technical issues related to the degree of generality we ultimately need to strive for. Further results on stable categories and equivalences that hold specifically for algebras over fields and complete discrete valuation rings are described in the sections 4.13, 4.14, and 4.15 below.

Following the terminology introduced in 2.6.11, an A-module U is *relatively k-projective* if U is isomorphic to a direct summand of $A \otimes_k V$ for some k-module V, and by 2.6.12, this is the case if and only if the map $A \otimes_k U \to U$ sending $a \otimes u$ to au is a split surjective A-homomorphism. Similarly, U is *relatively k-injective* if U is isomorphic to a direct summand of $\mathrm{Hom}_k(A, V)$ for some k-module V, and this is the case if and only if the map $U \to \mathrm{Hom}_k(A, U)$ sending $u \in U$ to the map $a \to au$ is a split injective A-homomorphism. We say that a homomorphism of A-modules $\varphi : U \to V$ *factors through a relatively k-projective (resp. k-injective) module* if there is a relatively k-projective (resp. k-injective) A-module P and A-homomorphisms $\alpha : U \to P$ and $\beta : P \to V$ such that $\varphi = \beta \circ \alpha$. We denote by $\mathrm{Hom}_A^{\mathrm{pr}}(U, V)$ the subset of $\mathrm{Hom}_A(U, V)$ consisting of all A-homomorphisms from U to V that factor through a relatively k-projective module and by $\mathrm{Hom}_A^{\mathrm{inj}}(U, V)$ the subset of $\mathrm{Hom}_A(U, V)$ consisting of all A-homomorphisms from U to V that factor through a relatively k-injective module. If $\varphi, \varphi' \in \mathrm{Hom}_A(U, V)$ factor through relatively k-projective modules P, P', respectively, then for any scalar $\lambda \in k$ the homomorphism $\lambda\varphi$ factors through P and the sum $\varphi + \varphi'$ factors through the direct sum $P \oplus P'$. Thus $\mathrm{Hom}_A^{\mathrm{pr}}(U, V)$ is a k-submodule of $\mathrm{Hom}_A(U, V)$; similarly, $\mathrm{Hom}_A^{\mathrm{inj}}(U, V)$ is a k-submodule of $\mathrm{Hom}_A(U, V)$. Given two composable A-homomorphisms φ, ψ, if one of them factors through a relatively k-projective or k-injective A-module, then so does their composition $\psi \circ \varphi$. Thus the following definition makes sense:

Definition 2.13.1 Let A be a k-algebra. The *k-stable category of* $\mathrm{Mod}(A)$, denoted $\underline{\mathrm{Mod}}(A)$, is the category whose objects are the A-modules and, for any two A-modules U, V, the space of morphisms from U to V in $\underline{\mathrm{Mod}}(A)$ is the quotient space

$$\underline{\mathrm{Hom}}_A(U, V) = \mathrm{Hom}_A(U, V)/\mathrm{Hom}_A^{\mathrm{pr}}(U, V)$$

with composition of morphisms in $\underline{\mathrm{Mod}}(A)$ induced by the composition of A-homorphisms in $\mathrm{Mod}(A)$. Similarly, the *injective k-stable category of* $\mathrm{Mod}(A)$, denoted $\overline{\mathrm{Mod}}(A)$, is the category whose objects are the A-modules and, for any two A-modules U, V, the space of morphisms from U to V in $\overline{\mathrm{Mod}}(A)$ is the quotient space

$$\overline{\mathrm{Hom}}_A(U, V) = \mathrm{Hom}_A(U, V)/\mathrm{Hom}_A^{\mathrm{inj}}(U, V)$$

with composition of morphisms in $\overline{\mathrm{Mod}}(A)$ induced by the composition of A-homomorphisms in $\mathrm{Mod}(A)$.

Thus a morphism from U to V in $\underline{\mathrm{Mod}}(A)$ is a class of A-homomorphisms, usually denoted $\underline{\varphi}$, of an A-homomorphism $\varphi : U \to V$ modulo A-homomorphisms that factors through a relatively k-projective A-module. The kernel and cokernel of a representative φ of its class $\underline{\varphi}$ will depend, in general, on the choice of φ. Thus there is no well-defined notion of a kernel and cokernel for an arbitrary morphism in $\underline{\mathrm{Mod}}(A)$, and hence the category $\underline{\mathrm{Mod}}(A)$ is no longer abelian. It is still a k-linear category: the morphism spaces in $\underline{\mathrm{Mod}}(A)$ are k-modules, composition in $\underline{\mathrm{Mod}}(A)$ is clearly k-bilinear, and an easy verification shows that the direct sum in $\mathrm{Mod}(A)$ induces a direct sum in $\underline{\mathrm{Mod}}(A)$.

Remark 2.13.2 It is not known to what extent the k-stable category $\underline{\mathrm{mod}}(A)$ of a k-algebra determines structural or numerical invariants of A. A conjecture of Auslander and Reiten predicts that if A and B are two finite-dimensional algebras over a field k such that $\underline{\mathrm{mod}}(A) \cong \underline{\mathrm{mod}}(B)$ as k-linear categories, then A and B have the same number of isomorphism classes of nonprojective simple modules. There are many stable equivalences in block theory for which this is not known – and this is amongst the biggest obstacles in the structure theory of blocks of finite groups. We will show in 4.13.23 below that the Auslander–Reiten conjecture holds if one of the two algebras is a finite p-group algebra over a field of prime characteristic p.

If A has the property that the classes of relatively k-projective and relatively k-injective modules coincide, then the k-stable and injective k-stable categories coincide. In that case, the images in $\underline{\mathrm{Mod}}(A)$ of a short exact sequences in $\mathrm{Mod}(A)$ induce a structure of *triangulated category* on $\mathrm{Mod}(A)$. This notion, which goes back to work of Verdier and Puppe in the 1960s, has become fundamental in algebraic topology and homological algebra in general, and for cohomological aspects of finite group representation theory in particular. We refer to the appendix for more details.

Since the identity of any relatively k-projective A-module factors trivially through a relatively k-projective module it follows that every relatively k-projective A-module is isomorphic to zero in $\underline{\mathrm{Mod}}(A)$. Thus, if P, Q are relatively k-projective A-modules then $\underline{\mathrm{Hom}}_A(U \oplus P, V \oplus Q) \cong \underline{\mathrm{Hom}}_A(U, V)$. In other words, the k-module $\underline{\mathrm{Hom}}_A(U, V)$ is invariant under adding or deleting relatively k-projective direct summands to or from U and V. The next result shows that the relatively k-projective modules are exactly the modules that get identified to zero in the k-stable category.

Proposition 2.13.3 *Let A be a k-algebra and U an A-module. The following are equivalent.*

(i) U is isomorphic to the zero object in $\underline{\mathrm{Mod}}(A)$.
(ii) $\underline{\mathrm{Id}_U}$ is the zero endomorphism of U in $\underline{\mathrm{End}}_A(U)$.
(iii) U is relatively k-projective.

Proof The zero object has the zero endomorphism as its only endomorphism, and hence (i) implies trivially (ii). If (ii) holds then Id_U factors through a relatively k-projective A-module P; that is, $\mathrm{Id}_U = \beta \circ \alpha$ for some A-homomorphisms $\alpha : U \to P$ and $\beta : P \to U$. Equivalently, β is split surjective, with α as a section. This implies that U is isomorphic to a direct summand of P, hence U is relatively k-projective. Finally, (iii) implies (i) by the remarks preceding this proposition. □

We have an analogous result in the injective k-stable category. An A-homomorphism $\varphi : U \to V$ is called *k-split surjective* if there is a k-linear map $\sigma : V \to U$ such that $\varphi \circ \sigma = \mathrm{Id}_V$; similarly, φ is called *k-split injective* if there is a k-linear map $\rho : V \to U$ such that $\rho \circ \varphi = \mathrm{Id}_U$. A short exact sequence of A-modules is called *k-split* if its restriction to k is a split exact sequence. Note that any A-module U is a quotient of a relatively k-projective A-module via the map $A \otimes_k U \to U$ sending $a \otimes u$ to au; this map is k-split surjective with section sending $u \in U$ to $1 \otimes u$. Similarly, U can be embedded in a relatively k-injective A-module via the map $U \to \mathrm{Hom}_k(A, U)$ sending $u \in U$ to the map $a \mapsto au$, and this is a k-split injective map with retraction sending $\mu \in \mathrm{Hom}_k(A, U)$ to $\mu(1)$.

Proposition 2.13.4 *Let A be a k-algebra and $\varphi : U \to V$ a homomorphism of A-modules. The following are equivalent.*

(i) φ factors through a relatively k-projective A-module.
(ii) φ factors through the canonical A-homomorphism $A \otimes_k V \to V$ sending $a \otimes v$ to av.
(iii) φ factors through any k-split surjective A-homomorphism $\delta : W \to V$.

Proof The implications (iii)$\Rightarrow$ (ii) $\Rightarrow$ (i) are trivial. Suppose that (i) holds. That is, there is a relatively k-projective A-module P and A-homomorphisms $\alpha : U \to P$ and $\beta : P \to V$ such that $\varphi = \beta \circ \alpha$. Since δ is k-split, it follows that β factors through δ as a k-linear map. But then since P is relatively k-projective, it follows further that β factors through δ as an A-homomorphism. That is, there is an A-homomorphism $\gamma : P \to W$ such that $\beta = \delta \circ \gamma$. Thus φ factors through δ, which shows that (i) implies (iii). □

A consequence of this is that if A is finitely generated over a Noetherian ring k, then the k-stable category $\underline{\mathrm{mod}}(A)$ of finitely generated A-modules can be identified with the full subcategory of $\underline{\mathrm{Mod}}(A)$ of all finitely generated A-modules.

Corollary 2.13.5 *Suppose that the ring k is Noetherian. Let A be a k-algebra such that A is finitely generated as a k-module. Let U, V be A-modules. Suppose that V is finitely generated. A homomorphism $\varphi : U \to V$ factors through a relatively k-projective A-module if and only if it factors through a finitely generated relatively k-projective A-module.*

Proof If φ factors through a relatively k-projective A-module, then it factors through $A \otimes_k V$, by 2.13.4. The hypotheses imply that V is finitely generated as a k-module and hence that $A \otimes_k V$ is finitely generated as an A-module. The result follows. □

Let A be a k-algebra and let U, V be A-modules. The following result characterises $\mathrm{Hom}_A^{\mathrm{pr}}(U, V)$ in certain cases; this extends 2.9.14.

Proposition 2.13.6 *Let A be a k-algebra and let U, V be A-modules. Suppose that V has a relatively k-projective cover which is finitely generated projective as an A-module. Then $\mathrm{Hom}_A^{\mathrm{pr}}(U, V)$ is equal to the image of the canonical map $\mathrm{Hom}_A(U, A) \otimes_A V \to \mathrm{Hom}_A(U, V)$ which sends $\varphi \otimes v$ to the map $u \mapsto \varphi(u)v$, where $\varphi \in \mathrm{Hom}_A(U, A)$, $u \in U$ and $v \in V$.*

Proof By 2.13.4, an A-homomorphism $\psi : U \to V$ is in $\mathrm{Hom}_A^{\mathrm{pr}}(U, V)$ if and only if ψ factors through a relatively k-projective cover. By the assumptions on V it follows that $\psi \in \mathrm{Hom}_A^{\mathrm{pr}}(U, V)$ if and only if ψ factors through a free A-module of finite rank, thus if and only if ψ is a finite sum of homomorphisms that factor through A. Any homomorphism $A \to V$ is determined by its image of 1, hence is of the form $a \mapsto av$ for all $a \in A$ and some fixed element $v \in V$. Precomposing this map with a homomorphism $\varphi : U \to A$ yields the map $u \mapsto \varphi(u)v$, where $u \in U$. This is the image of $\varphi \otimes v$ under the canonical map $\mathrm{Hom}_A(U, A) \otimes_A V \to \mathrm{Hom}_A(U, V)$, considered in 1.12.6. The result follows. □

If A is a finite-dimensional algebra over a field, then any finite-dimensional A-module V satisfies the assumption on V in 2.13.6.

Proposition 2.13.7 *Let A be a k-algebra and $\varphi : U \to V$ a homomorphism of A-modules. The following are equivalent.*

(i) φ factors through a relatively k-injective A-module.
(ii) φ factors through the canonical A-homomorphism $U \to \mathrm{Hom}_k(A, U)$ sending $u \in U$ to the map $a \mapsto au$.
(iii) φ factors through any k-split injective A-homomorphism $\delta : U \to W$.

Proof Dualise the arguments in the proof of 2.13.4. □

Proposition 2.13.8 *Let A be a k-algebra and $\varphi : U \to V$ a homomorphism of A-modules. If φ is k-split surjective, then the image of φ in $\underline{\mathrm{Hom}}_A(U, V)$ is an isomorphism if and only if φ is split surjective and $\ker(\varphi)$ is relatively k-projective. In particular, U and V are isomorphic in $\underline{\mathrm{Mod}}(A)$ if and only if there are relatively k-projective A-modules P, Q such that $U \oplus P \cong V \oplus Q$ as A-modules.*

Proof Suppose that φ is k-split surjective and that the class $\underline{\varphi}$ is an isomorphism in $\underline{\mathrm{Mod}}(A)$. Thus there is an A-homomorphism $\psi : V \to U$ such that $\underline{\varphi} \circ \underline{\psi} = \underline{\mathrm{Id}}_V$, or equivalently, such that $\mathrm{Id}_V - \varphi \circ \psi$ factors through a relatively k-projective A-module. But then, by 2.13.4, $\mathrm{Id}_V - \varphi \circ \psi$ factors through φ; say,

$\mathrm{Id}_V - \varphi \circ \psi = \varphi \circ \alpha$ for some $\alpha : V \to U$. Thus $\mathrm{Id}_V = \varphi \circ (\psi + \alpha)$, which shows that φ is split surjective with section $\psi + \alpha$. Then $\ker(\varphi)$ must be zero in $\underline{\mathrm{Mod}}(A)$ for $\underline{\varphi}$ to be an isomorphism (because otherwise it would not even be a monomorphism in $\underline{\mathrm{Mod}}(A)$), hence $\ker(\varphi)$ must be relatively k-projective by 2.13.3. This proves the first statement. Suppose now that $\varphi : U \to V$ induces an isomorphism in $\underline{\mathrm{Mod}}(A)$. Set $P = A \otimes_k V$ and denote by $\pi : P \to V$ the canonical map sending $a \otimes v$ to av. Then the homomorphism $(\varphi, \pi) : U \oplus P \to V$ is k-split surjective because π is so. This homomorphism still induces an isomorphism in $\underline{\mathrm{Mod}}(A)$ since P is zero in $\underline{\mathrm{Mod}}(A)$. By the first statement, φ is split surjective with $Q = \ker(\varphi)$ relatively k-projective, hence $U \oplus P \cong V \oplus Q$ as required. The converse is trivial. □

If the Krull–Schmidt Theorem holds in the module category under consideration, then in any isomorphism $U \oplus P \cong V \oplus Q$ we can 'cancel' common projective summands on both sides. In that case, if U, V have no nonzero projective direct summands, an isomorphism $U \cong V$ in $\underline{\mathrm{Mod}}(A)$ is therefore equivalent to an isomorphism $U \cong V$ in $\mathrm{Mod}(A)$. As we will see later, this is the case if U, V are finitely generated modules over an $\mathcal{O}$-algebra A, where $\mathcal{O}$ is a complete local Noetherian ring and A is finitely generated as an $\mathcal{O}$-module. We have as usual a dual version of the above proposition:

Proposition 2.13.9 *Let A be a k-algebra and $\varphi : U \to V$ a homomorphism of A-modules. If φ is k-split injective, then the image of φ in $\overline{\mathrm{Hom}}_A(U, V)$ is an isomorphism if and only if φ is split injective and* $\mathrm{coker}(\varphi)$ *is relatively k-injective. In particular, U and V are isomorphic in $\overline{\mathrm{Mod}}(A)$ if and only if there are relatively k-injective A-modules I, J such that $U \oplus I \cong V \oplus J$ as A-modules.*

Proof Dualise the proof of 2.13.8. □

Tensoring with bimodules that are finitely generated projective on both sides induces functors between k-stable categories:

Proposition 2.13.10 *Let A, B be k-algebras, let M be an A-B-bimodule, and let $\varphi : V \to V'$ be a B-homomorphism.*

(i) If M is finitely generated projective as a left A-module and if V is a relatively k-projective B-module, then the A-module $M \otimes_B V$ is relatively k-projective.
(ii) If M is finitely generated projective as a right B-module and if $\varphi : V \to V'$ is a k-split B-homomorphism, then $\mathrm{Id}_M \otimes \varphi : M \otimes_B V \to M \otimes_B V'$ is a k-split A-homomorphism.

(iii) If M is finitely generated projective as a left A-module and as a right B-module, then the functor $M \otimes_B -$ induces a functor $\underline{\mathrm{Mod}}(B) \to \underline{\mathrm{Mod}}(A)$ as k-linear triangulated categories.

Proof Suppose that M is finitely generated projective as a left A-module and that V is relatively k-projective. Then V is isomorphic to a direct summand of the B-module $B \otimes_k W$ for some k-module W. Thus $M \otimes_B V$ is isomorphic to a direct summand of $M \otimes_k W$. Since M is finitely generated projective as a left A-module, it follows that the A-module $M \otimes_k W$ is isomorphic to a direct summand of a finite number of copies of $A \otimes_k W$, hence relatively k-projective. This shows (i). Suppose now that φ is k-split and that M is finitely generated projective as a right B-module; that is, as a right B-module, M is a direct summand of a free right B-module Y of finite rank. By the assumptions on φ the map $\mathrm{Id}_B \otimes \varphi : B \otimes_B V \to B \otimes_B V'$ is k-split. Taking finitely many copies of this map implies that $\mathrm{Id}_Y \otimes \varphi$ is k-split. A direct summand of a k-split map is k-split (cf. 1.12.24), and hence $\mathrm{Id}_M \otimes \varphi$ is k-split, whence (ii). Statement (iii) follows from combining (i) and (ii), observing that the exact triangles in $\underline{\mathrm{Mod}}(A)$ are induced by k-split exact sequences (see A.3.1). □

Let G be a finite group. A kG-module is relatively k-projective if and only if it is relatively $\{1\}$-projective in the sense of Definition 2.6.1. In conjunction with Higman's criterion and earlier results this yields the following characterisation of kG-homomorphisms that factor through relatively k-projective modules.

Proposition 2.13.11 *Let G be a finite group and H a subgroup of G. Let U and V be kG-modules.*

(i) A kG-homomorphism $\varphi : U \to V$ factors through a relatively kH-projective module if and only if $\varphi \in (\mathrm{Hom}_k(U, V))^G_H$.
(ii) We have $\mathrm{Hom}^{\mathrm{pr}}_{kG}(U, V) = (\mathrm{Hom}_k(U, V))^G_1$.
(iii) We have $\underline{\mathrm{Hom}}_{kG}(U, V) \cong (\mathrm{Hom}_k(U, V)^G)/(\mathrm{Hom}_k(U, V))^G_1$.
(iv) If U, V are finitely generated projective as k-modules, then

$$\underline{\mathrm{Hom}}_{kG}(U, V) \cong (U^* \otimes_k V)^G/(U^* \otimes_k V)^G_1.$$

Proof Statement (i) is a restatement of 2.6.8. Statement (ii) is the special case of (i) with $H = \{1\}$, and (iii) is a trivial consequence of (ii). Statement (iv) follows from (iii) and 2.9.4. □

Corollary 2.13.12 *Suppose that k is a field of prime characteristic p. Let P be a finite p-group and let U, V be finitely generated kP-modules. We have $\underline{\mathrm{Hom}}_{kP}(U, V) = \{0\}$ if and only if the kP-module $U^* \otimes_k V$ is projective.*

Proof By 2.13.11, we have $\underline{\mathrm{Hom}}_{kP}(U, V) = \{0\}$ if and only if $(U^* \otimes_k V)^P = (U^* \otimes_k V)_1^P$. By 2.5.3, this is the case if and only if $U^* \otimes_k V$ is projective. □

2.14 The Heller operator on stable categories

Definition 2.14.1 Let A be a k-algebra. For any A-module U choose pairs (P_U, π_U) and (I_U, ι_U) consisting of a relatively k-projective A-module P_U, a k-split surjective A-homomorphism $\pi_U : P_U \to U$, a relatively k-injective A-module I_U and a k-split injective A-homomorphism $\iota_U : U \to I_U$. We set $\Omega_A(U) = \ker(\pi_U)$ and $\Sigma_A(U) = \mathrm{coker}(\iota_U)$.

The operator Ω_A on $\mathrm{Mod}(A)$ defined in this way is frequently called *Heller operator*. The modules $\Omega_A(U)$ and $\Sigma_A(U)$ make the sequences of A-modules

$$0 \longrightarrow \Omega_A(U) \longrightarrow P_U \xrightarrow{\pi_U} U \longrightarrow 0$$

$$0 \longrightarrow U \xrightarrow{\iota_U} I_U \longrightarrow \Sigma_A(U) \longrightarrow 0$$

exact, and in fact, both sequences are k-split exact. The operators Ω_A and Σ_A defined on $\mathrm{Mod}(A)$ depend on the choices of (P_U, π_U) and (I_U, ι_U), and therefore will not, in general be functorial on $\mathrm{Mod}(A)$. They become however functorial on the k-stable category and injective k-stable category. We will see that if A has the property that the classes of relatively k-projective modules and relatively k-injective modules coincide, then Ω_A and Σ_A induce inverse equivalences on $\underline{\mathrm{Mod}}(A)$.

Theorem 2.14.2 *Let A be a k-algebra. For any A-module U choose a pair (P_U, π_U) consisting of a relatively k-projective A-module P_U and a k-split surjective A-homomorphism $\pi_U : P_U \to U$. Then, for any homomorphism of A-modules $\alpha : U \to V$ there are homomorphisms of A-modules $\beta : P_U \to P_V$ and $\gamma : \Omega_A(U) \to \Omega_A(V)$ making the diagram*

$$\begin{array}{ccccccccc} 0 & \longrightarrow & \Omega_A(U) & \longrightarrow & P_U & \xrightarrow{\pi_U} & U & \longrightarrow & 0 \\ & & \downarrow{\scriptstyle\gamma} & & \downarrow{\scriptstyle\beta} & & \downarrow{\scriptstyle\alpha} & & \\ 0 & \longrightarrow & \Omega_A(V) & \longrightarrow & P_V & \xrightarrow[\pi_V]{} & V & \longrightarrow & 0 \end{array}$$

commutative. Moreover, if α factors through a relatively k-projective A-module then so does γ, and the map sending α to γ induces a map $\underline{\mathrm{Hom}}_A(U, V) \to$

$\underline{\mathrm{Hom}}_A(\Omega_A(U), \Omega_A(V))$ *through which* Ω_A *becomes a functor, which is independent, up to unique isomorphism of functors, of the choice of the pairs* (P_U, π_U).

Proof We first observe that the operator Ω_A preserves relatively k-projective modules. Indeed, if U is relatively k-projective then π_U is split surjective because it is k-split surjective. Thus $\Omega_A(U)$ is a direct summand of P_U, showing that $\Omega_A(U)$ is relatively k-projective. If $\alpha : U \to V$ is an A-homomorphism, then $\alpha \circ \pi_U : P_U \to V$ lifts through π_V because π_V is k-split surjective and P_U is relatively k-projective. Thus there is an A-homomorphism $\beta : P_U \to P_V$ making the right square in the diagram commutative. But then β must send $\ker(\pi_U)$ to $\ker(\pi_V)$, and hence restricts to an A-homomorphism $\gamma : \Omega_A(U) \to \Omega_A(V)$ making the left square commutative. Suppose that α factors through a relatively k-projective A-module. Then, by 2.13.4, α factors through π_V; say $\alpha = \pi_V \circ \rho$ for some A-homomorphism $\rho : U \to P_V$. Then $\pi_V \circ (\beta - \rho \circ \pi_U) = \pi_V \circ \beta - \pi_V \circ \rho \circ \pi_U = \pi_V \circ \beta - \alpha \circ \pi_U = 0$, or equivalently, $\mathrm{Im}(\beta - \rho \circ \pi_U) \subseteq \ker(\pi_V) = \Omega_A(V)$. Since γ is the restriction of β and $\ker(\pi_U) = \Omega_A(U)$ this shows that γ is equal to the restriction of $\beta - \rho \circ \pi_U$, hence factors through the inclusion $\Omega_A(U) \to P_U$. This implies that Ω_A induces a functor on $\underline{\mathrm{Mod}}(A)$. To see that this functor is unique up to unique isomorphism we apply the diagram in the statement to $U = V$ but with possibly different pairs (P_U, π_U), (P_V, π_V); then in particular Id_U induces a unique isomorphism, in the stable category, between $\Omega_A(U)$ and $\Omega_A(V)$, showing that Ω_A is unique up to unique isomorphism of functors. □

The Heller operator 'commutes' with automorphisms; this is a special case of more general results in 2.14.6 and 2.17.7 below.

Corollary 2.14.3 *Let A be a k-algebra, U an A-module, and* $\alpha \in \mathrm{Aut}(A)$. *We have an isomorphism* ${}_\alpha\Omega(U) \cong \Omega({}_\alpha U)$ *in* $\underline{\mathrm{Mod}}(A)$.

Proof Write $\Omega(U) = \ker(\pi)$, where $\pi : P \to U$ is a relatively k-projective cover of U. The kernel of the map ${}_\alpha P \to {}_\alpha U$ induced by π is equal to ${}_\alpha\Omega(U)$. This map is also a relatively k-projective cover of ${}_\alpha U$, and hence its kernel is isomorphic to $\Omega({}_\alpha U)$ in $\underline{\mathrm{Mod}}(A)$. □

Since Ω_A is unique up to unique isomorphism of functors on $\underline{\mathrm{Mod}}(A)$, we will very often not specify the choice of the pairs (P_U, π_U), but rather implicitly assume such a choice whenever we use the notation Ω_A. We have, of course, the dual version:

Theorem 2.14.4 *Let A be a k-algebra. For any A-module U choose a pair* (I_U, ι_U) *consisting of a relatively k-injective A-module I_U and a k-split* *injective A-homomorphism $\iota_U : U \to I_U$. Then, for any homomorphism of A-modules $\alpha : U \to V$ there are homomorphisms of A-modules $\beta : I_U \to I_V$ and $\gamma : \Sigma_A(U) \to \Sigma_A(V)$ making the diagram*

$$\begin{array}{ccccccccc} 0 & \longrightarrow & U & \xrightarrow{\iota_U} & I_U & \longrightarrow & \Sigma_A(U) & \longrightarrow & 0 \\ & & \downarrow{\scriptstyle\alpha} & & \downarrow{\scriptstyle\beta} & & \downarrow{\scriptstyle\gamma} & & \\ 0 & \longrightarrow & V & \xrightarrow[\iota_V]{} & I_V & \longrightarrow & \Sigma_A(V) & \longrightarrow & 0 \end{array}$$

commutative. Moreover, if α factors through a relatively k-injective A-module then so does γ, and the map sending α to γ induces a map $\overline{\mathrm{Hom}}_A(U, V) \to \overline{\mathrm{Hom}}_A(\Sigma_A(U), \Sigma_A(V))$ through which Σ_A becomes a functor, which is independent, up to unique isomorphism of functors, of the choice of the pairs (I_U, ι_U).

Proof Dualise the arguments of the proof of 2.14.2. □

Theorem 2.14.5 *Let A be a k-algebra such that the classes of relatively k-projective and relatively k-injective A-modules coincide. Then the functors Ω_A and Σ_A on the k-stable category $\underline{\mathrm{Mod}}(A)$ are inverse equivalences.*

Proof Consider a k-split short exact sequence of A-modules

$$0 \longrightarrow \Omega_A(U) \longrightarrow P_U \longrightarrow U \longrightarrow 0$$

with P_U relatively k-projective. By the assumptions, P_U is also relatively k-injective, hence we get a uniquely determined isomorphism $U \cong \Sigma_A(\Omega_A(U))$ in $\underline{\mathrm{Mod}}(A)$. Similarly $U \cong \Omega_A(\Sigma_A(U))$ in $\underline{\mathrm{Mod}}(A)$, which implies the result. □

As mentioned before, in the situation of Theorem 2.14.5, the k-stable category $\underline{\mathrm{Mod}}(A)$ is *triangulated*, with the shift functor Σ_A, and exact triangles induced by short exact sequences in $\mathrm{Mod}(A)$; see Theorem A.3.2 below.

Proposition 2.14.6 *Let A, B be k-algebras, and let N be a B-A-bimodule that is finitely generated projective as a left B-module and as a right A-module. Let U be an A-module. For any $n \geq 0$ we have an isomorphism $M \otimes_A \Omega_A^n(U) \cong \Omega_B^n(M \otimes_A U)$ in the k-stable category $\underline{\mathrm{Mod}}(B)$. If A, B are finitely generated projective as k-modules and have the property that the classes of relatively k-projective modules and relatively k-injective modules coincide, then there is such an isomorphism for all integers n.*

Proof The result is trivial for $n = 0$. By the hypotheses and 2.13.10, tensoring with $M \otimes_A -$ is an exact functor preserving relatively k-projective modules and the property of a map being k-split. Thus this functor maps a relatively k-projective cover of U to a relatively k-projective cover of $M \otimes_A U$. This shows the result for $n = 1$; iterating yields the result for $n \geq 1$. Under the additional hypotheses in the second part of the statement, the Heller operator becomes an equivalence (cf. 2.14.5). The result follows. □

Corollary 2.14.7 *Let A be a k-algebra and B a subalgebra of A such that A is finitely generated projective as a left B-module. Let U be an A-module. For any $n \geq 0$ there is an isomorphism $\mathrm{Res}^A_B(\Omega^n_A(U)) \cong \Omega^n_B(\mathrm{Res}^A_B(U))$ in the k-stable category $\underline{\mathrm{Mod}}(B)$. If A, B are finitely generated projective as k-modules and have the property that the classes of relatively k-projective modules and relatively k-injective modules coincide, then there is such an isomorphism for all integers n.*

Proof This is 2.14.6 applied to $N = A$ as a B-A-bimodule. □

See 2.17.7 below for a bimodule version of 2.14.6.

Exercise 2.14.8 Suppose that k is a field. Let A be a k-algebra, and let $x \in A$. Set $U = A/Ax$ and $V = \{a \in A | ax = 0\}$. Show that $V \cong \Omega^2_A(U)$ in $\underline{\mathrm{Mod}}(A)$.

2.15 Stable categories of symmetric algebras

One of the crucial properties of symmetric k-algebras is that the classes of relatively k-projective and relatively k-injective modules coincide. If k is a field this just says that projective and injective modules coincide, leading to the definition of the slightly more general class of selfinjective algebras over fields in 4.11.1. Recall from 2.6.11, that an A-module U is *relatively k-projective* if there is a k-module V such that U is isomorphic to a direct summand of $A \otimes_k V$, and U is *relatively k-injective* if there is a k-module V such that U is isomorphic to a direct summand of $\mathrm{Hom}_k(A, V)$; the left A-module structure of $\mathrm{Hom}_k(A, V)$ is induced by the right A-module structure of A. By 2.6.12 and 2.6.20, in both cases one can choose V to be the restriction to k of U.

Theorem 2.15.1 *Let A be a symmetric k-algebra. For any k-module V there is a natural isomorphism of A-modules $A \otimes_k V \cong \mathrm{Hom}_k(A, V)$. In particular, an A-module U is relatively k-projective if and only if U is relatively k-injective, and the functors Σ_A and Ω_A on $\underline{\mathrm{Mod}}(A)$ induce inverse equivalences.*

Proof Since $A \cong A^*$ as A-A-bimodules we have $A \otimes_k V \cong A^* \otimes_k V$. Since also A is finitely generated projective as k-module we have $A^* \otimes_k V \cong \mathrm{Hom}_k(A, k)$ by 2.9.4 applied to A and V instead of V and W. This shows that the classes of relatively k-projective and relatively k-injective A-modules coincide. Thus the last statement is a special case of Theorem 2.14.5. □

For the sake of generality, we characterise algebras whose classes of relatively k-projective and relatively k-injective modules coincide.

Proposition 2.15.2 *Let A be a k-algebra such that A is finitely generated projective as a k-module. The following are equivalent:*

(i) The classes of relatively k-projective A-modules and relatively k-injective A-modules coincide.
(ii) The classes of finitely generated relatively k-projective A-modules and finitely generated relatively k-injective A-modules coincide.
(iii) As a left A-module, the k-dual $A^* = \mathrm{Hom}_A(A, k)$ is a progenerator.
(iv) The classes of relatively k-projective right A-modules and relatively k-injective right A-modules coincide.
(v) The classes of finitely generated relatively k-projective right A-modules and finitely generated relatively k-injective right A-modules coincide.
(vi) As a right A-module, the k-dual $A^* = \mathrm{Hom}_A(A, k)$ is a progenerator.

Proof The implication (i) ⇒ (ii) is trivial. Suppose that (ii) holds. Then in particular every finitely generated projective A-module is relatively k-injective. Thus A is relatively k-injective as a left A-module, and hence A is isomorphic to a direct summand of $\mathrm{Hom}_k(A, A)$, where the left A-module structure of $\mathrm{Hom}_k(A, A)$ is induced by the right A-module structure on the first argument. Since A is finitely generated projective as a k-module, it is isomorphic to a direct summand of a free k-module k^m, for some positive integer m. Thus $\mathrm{Hom}_A(A, A)$ is isomorphic, as a left A-module, to a direct summand of $\mathrm{Hom}_A(A, k^m) \cong (A^*)^m$. This shows that A itself is isomorphic to a direct summand of $(A^*)^m$ as a left A-module. As a left A-module, $A^* = \mathrm{Hom}_k(A, k)$ is trivially relatively k-injective, hence relatively k-projective. Since A, hence A^*, is finitely generated projective as a k-module it follows that A^* is finitely generated projective as a left A-module. Since A is a summand of $(A^*)^m$ as a left A-module this shows that A^* is a progenerator as a left A-module, hence (ii) implies (iii). Suppose that (iii) holds. Thus A^* is finitely generated projective as an A-module and A is isomorphic to a direct summand of $(A^*)^m$ for some positive integer m. Let U be an A-module. Since A is finitely generated projective as a k-module, by

2.9.4 we have an isomorphism of left A-modules $\mathrm{Hom}_k(A, U) \cong A^* \otimes_k U$; the left A-module structure on $\mathrm{Hom}_k(A, U)$ is induced by the right A-module structure on the first argument A. Thus $\mathrm{Hom}_k(A, U)$ is isomorphic to a direct summand of $\mathrm{Hom}_A((A^*)^n, U) \cong A \otimes_k U^n$ for some positive integer n, hence every relatively k-injective A-module is relatively k-projective. Similarly, $A \otimes_k U \cong$ is a direct summand of $(A^*)^m \otimes_k U \cong \mathrm{Hom}_k(A, U^m)$, hence any relatively k-projective A-module is relatively k-injective. This shows (iii) $\Rightarrow$ (i). A similar argument shows that (iii), (iv) and (v) are equivalent. We show next that (i) implies (iv). Let W be a finitely generated k-module. The relatively k-projective right A-module $W \otimes_k A$ is, by 2.9.4, isomorphic to $\mathrm{Hom}_k(A^*, W)$, where the right A-module structure is induced by the left A-module structure of A^*. But then $\mathrm{Hom}_k(A^*, W)$ is isomorphic to a direct summand of $\mathrm{Hom}_k(A^m, W) \cong \mathrm{Hom}_k(A, W)^m \cong \mathrm{Hom}_k(A, W^m)$, hence relatively k-injective. This shows that every relatively k-projective right A-module is relatively k-injective. Similarly, again by 2.9.4, the relatively k-injective right A-module $\mathrm{Hom}_k(A, W)$ is isomorphic to $W \otimes_k A^*$, which by the above is a direct summand of $W \otimes_k A^m \cong W^m \otimes_k A$, hence relatively k-projective. This shows indeed that (i) implies (iv). But then the right analogue of this implication yields the implication (iv) $\Rightarrow$ (i). □

Corollary 2.15.3 *Let A, B be k-algebras such that A, B are finitely generated projective as k-modules. Suppose that both algebras A and B have the property that their classes of relatively k-projective and relatively k-injective modules coincide. Then the tensor product $A \otimes_k B$ has this property, too.*

Proof This follows from the previous proposition and the fact that if A^*, B^* are progenerators as left modules then $A^* \otimes_k B^*$ is a progenerator as a left $A \otimes_k B$-module. □

Frobenius reciprocity for finite group algebras, and more generally, adjunction isomorphisms for functors induced by bimodules over symmetric algebras induce adjunction isomorphisms at the level of stable categories. For convenience, we state first the special case of finite group algebras, and then the general result for symmetric algebras.

Proposition 2.15.4 *Let G be a finite group and H a subgroup of G. Let U be a kG-module and V a kH-module. Frobenius reciprocity induces natural isomorphisms*

$$\underline{\mathrm{Hom}}_{kG}(\mathrm{Ind}_H^G(V), U) \cong \underline{\mathrm{Hom}}_{kH}(V, \mathrm{Res}_H^G(U)),$$

$$\underline{\mathrm{Hom}}_{kG}(U, \mathrm{Ind}_H^G(V)) \cong \underline{\mathrm{Hom}}_{kH}(\mathrm{Res}_H^G(U), V).$$

Proof One verifies directly that the natural isomorphisms given in 2.2.1 preserve homomorphisms that factor through relatively k-projective modules, using the fact that the functors Ind_H^G and Res_H^G preserve relatively k-projective modules. Alternatively, in the same way that Frobenius reciprocity arises as special cases of adjunction isomorphisms from 2.12.7, the present proposition is a special case of the next result. □

Proposition 2.15.5 *Let A, B be symmetric k-algebras and M an A-B-bimodule that is finitely generated projective as a left A-module and as a right B-module. Let U be an A-module and V a B-module. The adjunction isomorphism from the adjunction 2.12.7 induces a natural k-linear isomorphisms*

$$\underline{\mathrm{Hom}}_A(M \otimes_B V, U) \cong \underline{\mathrm{Hom}}_B(V, M^* \otimes_A U),$$

$$\underline{\mathrm{Hom}}_A(U, M \otimes_B V) \cong \underline{\mathrm{Hom}}_B(M^* \otimes_A U, V).$$

Proof One way to prove this is by abstract nonsense. By 2.13.10, the functors $M \otimes_B -$ and $M^* \otimes_A -$ induce functors between $\underline{\mathrm{Mod}}(A)$ and $\underline{\mathrm{Mod}}(B)$. By 2.12.7 the functors $M \otimes_B -$ and $M^* \otimes_A -$ between $\mathrm{Mod}(A)$ and $\mathrm{Mod}(B)$ are biadjoint, giving rise to adjunction units and counits, which are determined by bimodule homomorphisms between A, $M \otimes_B M^*$ and B, $M^* \otimes_A M$. It follows from the characterisation of adjoint functors in terms of units and counits in 2.3.5 that the functors induced by $M \otimes_B -$ and $M^* \otimes_A -$ on k-stable categories remain biadjoint. One can prove this also by a more direct verification. Suppose that $\alpha \in \mathrm{Hom}_A(M \otimes_B V, U)$ factors through $\psi : Y \to U$, where Y is a relatively k-projective A-module; say $\alpha = \psi \circ \gamma$ for some $\gamma \in \mathrm{Hom}_A(M \otimes_B V, Y)$. Denote $\beta \in \mathrm{Hom}_B(V, M^* \otimes_A U)$ and $\delta \in \mathrm{Hom}_B(V, M^* \otimes_A Y)$ the maps corresponding to α and γ through the adjunction isomorphisms. The naturality of the adjunction isomorphisms implies that $\alpha = \psi \circ \gamma$ corresponds to $\beta = (\mathrm{Id}_{M^*} \otimes \psi) \circ \delta$. By 2.13.10, the B-module $M^* \otimes_A Y$ is relatively k-projective, and hence β factors through a relatively k-projective B-module. A similar argument shows that if β factors through a relatively k-projective module, then so does α. This implies the first of the two isomorphisms, and the second follows from exchanging the roles of A and B. □

Exercise 2.15.6 By adapting the proof of Proposition 2.15.5, show the following statement. Let A, B be k-algebras that are finitely generated projective as k-modules and that have the property that their classes of relatively k-projective and relatively k-injective modules coincide. Let M be an A-B-bimodule that is finitely generated projective as a left A-module and as a right B-module. Show that for any A-module U and any B-module V we have a natural isomorphism

$$\underline{\mathrm{Hom}}_A(M \otimes_B V, U) \cong \underline{\mathrm{Hom}}_B(V, \mathrm{Hom}_A(M, U)).$$

2.16 Relative traces for symmetric algebras

Gaschütz [30] and Ikeda [40] developed an analogue for Frobenius algebras of the relative trace map Tr_1^G on a finite group algebra. We describe this for symmetric algebras. The starting point is the map obtained from dualising the multiplication map $A \otimes_k A \to A$ and using the symmetry of A.

Definition 2.16.1 Let A be a symmetric k-algebra with symmetrising form s. We denote by $\tau_A : A \to A \otimes_k A$ the A-A-bimodule homomorphism making the following diagram commutative:

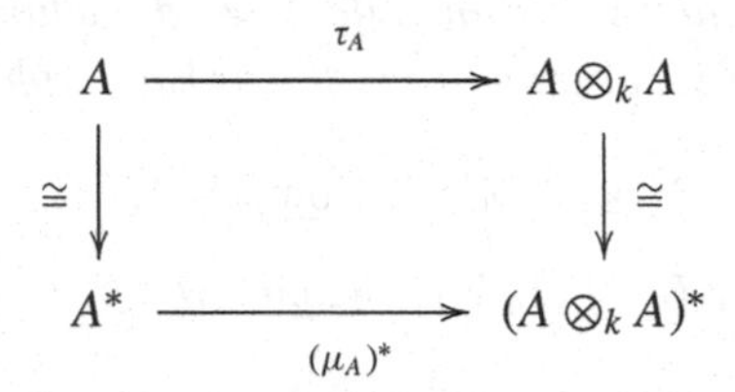

where $\mu_A : A \otimes_k A \to A$ is the A-A-bimodule homomorphism induced by multiplication in A, and where the vertical isomorphisms are induced by s; that is, the isomorphism $A \cong A^*$ sends $a \in A$ to s_a defined by $s_a(b) = s(ab)$ for all $b \in A$, and the isomorphism $A \otimes_k A \cong (A \otimes_k A)^*$ sends $a \otimes b$ to the linear map $s_b \otimes s_a$, where we identify $k \otimes_k k = k$.

The fact that the order of a and b gets reversed in the isomorphism $A \otimes_k A \cong (A \otimes_k A)^*$ comes from 2.12.5.

Remark 2.16.2 The map τ_A depends on the choice of the symmetrising form s of A. If s' is another symmetrising form of A, then by 2.11.5 there is a unique element $z \in Z(A)^\times$ such that $s' = s_z$. The corresponding bimodule homomorphism $\tau'_A : A \to A \otimes_k A$ satisfies $\tau'(a) = z^{-1}\tau_A(a)$ for all $a \in A$; this can be easily verified directly, or using Proposition 2.16.4 below.

The map τ_A yields the following description of relatively injective envelopes of modules.

Proposition 2.16.3 *Let A be a symmetric k-algebra with symmetrising form s.*

(i) The map $\tau_A : A \to A \otimes_k A$ is a relatively k-injective envelope of A as an $A \otimes_k A^{\mathrm{op}}$-module.

(ii) The map τ_A is split injective as a homomorphism of left A-modules and right A-modules.

(iii) For any A-module U, the map $\tau_A \otimes \mathrm{Id}_U : U \to A \otimes_k U$ is a relatively k-injective envelope of U.

Proof The multiplication map $\mu_A : A \otimes_k A \to A$ is surjective, thus split surjective as a homomorphism of left A-modules and as a homomorphism of right A-modules. Its dual is therefore split injective as a homomorphism of left A-modules and as a homomorphism of right A-modules, proving (ii). But then τ_A is in particular split injective as a homomorphism of k-modules. Since $A \otimes_k A$ is the regular $A \otimes_k A^{\text{op}}$-module, this module is also relatively k-injective, by 2.15.1. This shows (i). The fact that τ_A is split injective as a homomorphism of right A-modules implies that $\tau_A \otimes \text{Id}_U$ is split injective as a k-homomorphism. Moreover, $A \otimes_k U$ is relatively k-projective, hence relatively k-injective by 2.15.1 again. This completes the proof. □

In order to describe τ_A it suffices to describe its value at 1, since τ_A is a bimodule homomorphism and 1 generates A as a bimodule (even as a left or right A-module).

Proposition 2.16.4 *Let A be a symmetric k-algebra with symmetrising form s. Let X be a finite subset of A and for any $x \in X$ let $x' \in A$. The following are equivalent.*

(i) $\tau_A(1) = \sum_{x\in X} x' \otimes x$.
(ii) $\tau_A(1) = \sum_{x\in X} x \otimes x'$.
(iii) $\sum_{x\in X} s(x'a)x = a$ *for all* $a \in A$.
(iv) $\sum_{x\in X} s(xa)x' = a$ *for all* $a \in A$.

Moreover, if these conditions hold, then the sets X and $X' = \{x'\}_{x\in X}$ both generate A as a k-module, the bimodule isomorphism $A \cong A^*$ sending 1 to s has as its inverse the map sending $\mu : A \to k$ to the element $\sum_{x\in X} \mu(x')x$, and for all $a \in A$ we have

$$\sum_{x\in X} ax' \otimes x = \sum_{x\in X} x' \otimes xa.$$

Proof The isomorphism $A \cong A^*$ which is used in the definition of τ_A sends 1 to s. The isomorphism $A \otimes_k A \cong (A \otimes_k A)^*$ used in the definition of τ_A sends $\sum_{x\in X} x \otimes x'$ to the linear map $\sum_{x\in X} s_{x'} \otimes s_x$. This linear map sends $a \otimes b$ to $\sum_{x\in X} s(x'a)s(xb) = \sum_{x\in X} s(s(x'a)xb)$, where $a, b \in A$. Therefore (i) holds if and only if $s(ab) = \sum_{x\in X} s(s(x'a)xb)$ for all $a, b \in A$. This is equivalent to $s(((\sum_{x\in X} s(x'a)x) - a)b) = 0$ for all $a, b \in A$, which in turn is equivalent to $(\sum_{x\in X} s(x'a)x) - a = 0$ for all $a \in A$. This shows the equivalence of (i) and (iii). A similar argument shows the equivalence of (ii) and (iv). A variation of the first argument shows the equivalence of (i) and (iv) as follows. Using the equality $s(x'a)s(xb) = s(s(xb)x'a)$, we get that (i) holds if and only if $s(ab) = s(ba) = \sum_{x\in X} s(s(xb)x'a)$ for all $a, b \in A$, hence if and only if $s(((\sum_{x\in X} s(xb)x') - b)a) = 0$ for all a, $b \in A$. This is equivalent to $(\sum_{x\in X} s(xb)x') - b = 0$ for all $b \in A$, hence equivalent to (iv). If (iii) holds, then X generates A as a k-module. If (iv) holds, then X' generates A as a k-module. Statement is (iii) applied to s_a, where $a \in A$, yields the inverse of the isomorphism $A \cong A^*$ as stated. Finally, τ_A is a bimodule homomorphism, and hence $a\tau_A(1) = \tau_A(a) = \tau_A(1)a$ for all $a \in A$, which is the last statement. □

Remark 2.16.5 One can use Proposition 2.16.4 to solve Exercise 2.12.13. In order to show that τ_A and s represent the unit and counit of a left adjunction of Res_k^A to $\mathrm{Ind}_k^A = A \otimes_k -$, it suffices to show that the two compositions

$$A \xrightarrow[\tau_A]{} A \otimes_k A \xrightarrow[\mathrm{Id}\otimes s]{} A, \qquad A \xrightarrow[\tau_A]{} A \otimes_k A \xrightarrow[s\otimes \mathrm{Id}]{} A$$

are the identity on A. This is equivalent to Proposition 2.16.4 (iv) and (iii), respectively.

If A is free over k, then we may choose a subset X of A satisfying the conclusions of Proposition 2.16.4 to be a basis of A. The set X' is then automatically the dual basis of A with respect to s.

Proposition 2.16.6 *Let A be a symmetric k-algebra with symmetrising form s. Suppose that A is free as a k-module. Let X be a k-basis of A. Let $X' = \{x' | x \in X\}$ be the dual k-basis of A with respect to s; that is, $s(xx') = 1$ for all $x \in A$ and $s(xy') = 0$ for all x, $y \in X$ such that $x \neq y$. Then $\tau_A(1) = \sum_{x\in X} x' \otimes x = \sum_{x\in X} x \otimes x'$.*

Proof Let $y \in X$. We have $\sum_{x\in X} s(x'y)x = y$. By linearity we get that $\sum_{x\in X} s(x'a)x = a$ for all $a \in A$. The result follows from 2.16.4. □

Note that the formula for $\tau_A(1)$ in 2.16.6 does not depend on the choice of a k-basis X of A, but it does depend on the choice of the symmetrising form s. The explicit description of τ_A leads to the following explicit description of relatively k-injective envelopes.

Corollary 2.16.7 *Let A be a symmetric k-algebra with symmetrising form s. Let X be a finite subset of A and for $x \in X$ let $x' \in A$ such that $\tau_A(1) = \sum_{x\in X} x' \otimes x$. Let U be an A-module. The map $U \to A \otimes_k U$ sending $u \in U$ to $\sum_{x\in X} x' \otimes xu$ is a relatively k-injective envelope of U.*

Proof This follows from 2.16.3 (iii). □

Proposition 2.16.8 *Let G be a finite group and A a G-algebra that is free of finite rank over k. Suppose that A is symmetric with a G-stable symmetrising form s; that is, $s({}^g a) = s(a)$ for all $a \in A$ and all $g \in G$. Then the map τ_A :*

$A \to A \otimes_k A$ is a G-homomorphism, where G acts diagonally on $A \otimes_k A$. In particular, we have $\tau_A(1) \in (A \otimes_k A)^G$.

Proof Let X be a k-basis of A, and let $X' = \{x' | x \in X\}$ be the dual k-basis of A with respect to s; that is, $s(xx') = 1$ for all $x \in A$ and $s(xy') = 0$ for all $x, y \in X$ such that $x \neq y$. Since s is G-stable, it follows that for any $g \in G$, the set ${}^gX$ is a k-basis of A with dual basis ${}^gX'$. The independence of τ_A of the choice of a basis of A implies that ${}^g\tau_A(a) = \sum_{x \in X} {}^gx' \otimes {}^gx{}^ga = \sum_{x \in X} x' \otimes x{}^ga = \tau_A({}^ga)$ for all $g \in G$ and all $a \in A$. □

The G-stability of a symmetrising form s of a symmetric G-algebra holds in particular if G acts by inner automorphisms on A, since $s(a) = s(uau^{-1})$ for any $a \in A$ and any $u \in A^\times$. The map in the following Definition is known as the Gaschütz–Ikeda map.

Definition 2.16.9 Let A be a symmetric k-algebra with symmetrising form s. Let X be a finite subset of A and for $x \in X$ let $x' \in A$ such that $\tau_A(1) = \sum_{x \in X} x' \otimes x$. Let M be an A-A-bimodule. We define the *projective trace map* $\mathrm{Tr}_1^A : M \to M$ by setting $\mathrm{Tr}_1^A(m) = \sum_{x \in X} x'mx$ for all $m \in M$.

The map Tr_1^A does not depend on the choice of X, because the expression $\sum_{x \in X} x'mx$ is the image of m under the action of the element $\tau(1)$, viewed as an element of $A \otimes_k A^{\mathrm{op}}$. The map Tr_1^A does, however, depend on the choice of s. Whenever using the map Tr_1^A we implicitly assume a choice of a symmetric form. The image of Tr_1^A has an interpretation similar to that in the case of a finite group action.

Definition 2.16.10 (cf. [14, §5.B]) Let A be a k-algebra, and let M be an A-A-bimodule. We denote by M_1^A the image in M^A of $\mathrm{Hom}^{\mathrm{pr}}_{A \otimes_k A^{\mathrm{op}}}(A, M)$ under the canonical isomorphism $\mathrm{Hom}_{A \otimes_k A^{\mathrm{op}}}(A, M) \cong M^A$ from 1.5.11. We set $\bar{M}_1^A = M^A/M_1^A$. We set $Z^{\mathrm{pr}}(A) = A_1^A$ and call $Z^{\mathrm{pr}}(A)$ the *projective ideal* in $Z(A)$. The *stable centre of* A is the quotient $\underline{Z}(A) = Z(A)/Z^{\mathrm{pr}}(A)$.

Note that $\bar{M}_1^A \cong \underline{\mathrm{Hom}}_{A \otimes_k A^{\mathrm{op}}}(A, M)$; in particular, $\underline{Z}(A) \cong \underline{\mathrm{End}}_{A \otimes_k A^{\mathrm{op}}}(A)$. The projective ideal $Z^{\mathrm{pr}}(A)$ in $Z(A)$ consists of all elements $z \in Z(A)$ such that the A-A-bimodule endomorphism of A sending $a \in A$ to za belongs to $\mathrm{End}^{\mathrm{pr}}_{A \otimes_k A^{\mathrm{op}}}(A)$.

Proposition 2.16.11 *Let A be a symmetric k-algebra and M an A-A-bimodule. We have $M_1^A = \mathrm{Im}(\mathrm{Tr}_1^A)$. In particular, $\mathrm{Im}(\mathrm{Tr}_1^A)$ is a $Z(A)$-submodule of M^A, and this submodule is independent of the choice of a symmetrising form of A.*

Proof Note that $A \otimes_k A^{\mathrm{op}}$ is again symmetric, so that the classes of relative k-projective and relative k-injective $A \otimes_k A^{\mathrm{op}}$-modules coincide. Thus, using 2.13.7 and 2.16.3, an $A \otimes_k A^{\mathrm{op}}$-homomorphism $A \to M$ factors through a relatively k-projective module if and only if it factors through τ_A. Any $A \otimes_k A^{\mathrm{op}}$-homomorphism $A \otimes_k A \to M$ is determined by its value at $1 \otimes 1$, hence of the form ψ_m for some $m \in M$, where $\psi_m(a \otimes b) = amb$ for all a, $b \in A$. Thus $M_1^A = \{(\psi_m \circ \tau_A)(1) | m \in M\}$. Let X be a subset of A as in 2.16.9. Then $\psi_m(\tau_A(1)) = \psi_m(\sum_{x \in X} x' \otimes x) = \sum_{x \in X} x'mx = \mathrm{Tr}_1^A(m)$, whence the result. □

Corollary 2.16.12 *Let A be a symmetric k-algebra. We have $Z^{\mathrm{pr}}(A) = \mathrm{Im}(\mathrm{Tr}_1^A : A \to Z(A))$.*

Proof This follows from 2.16.10 applied to $M = A$. □

Remark 2.16.13 Let A be a symmetric k-algebra. Let M and N be A-A-bimodules. As mentioned in 1.5.9, a surjective bimodule homomorphism $M \to N$ need not induce a surjective map $M^A \to N^A$, but it still induces a surjective map $M_1^A \to N_1^A$.

We state now the generalisation of 2.13.11 to symmetric algebras mentioned earlier.

Proposition 2.16.14 *Let A be a symmetric k-algebra. Let U, V be A-modules. We have $\mathrm{Hom}_A^{\mathrm{pr}}(U, V) = (\mathrm{Hom}_k(U, V))_1^A$. In particular, U is relatively k-projective if and only if $\mathrm{Id}_U \in (\mathrm{End}_k(U))_1^A$.*

Proof An A-homomorphism $U \to V$ factors through a relatively k-projective module if and only if it factors through the relatively k-injective envelope $\tau_U : U \to A \otimes_k U$ from 2.16.3. Any A-homomorphism $\psi : A \otimes_k U \to V$ is, by a standard adjunction, equal to $\psi(a \otimes u) = a\mu(u)$ for some k-linear map $\pi : U \to V$. With the notation from 2.16.4, we have $\psi \circ \tau_U(u) = \psi(\sum_{x \in X} x' \otimes xu) = \sum_{x \in X} x'\mu(xu) = \sum_{x \in X}(x' \cdot \mu \cdot x)(u) = \mathrm{Tr}_1^A(\mu)(u)$. The result follows. □

Proposition 2.16.15 *Let A be a symmetric k-algebra and M an A-A-bimodule. We have $[A, M] \subseteq \ker(\mathrm{Tr}_1^A : M \to M^A)$.*

Proof Write $\tau_A(1) = \sum_{x \in X} x' \otimes x$, where the notation is as in 2.16.4. Let $a \in A$ and $m \in M$. We need to show that $\mathrm{Tr}_1^A(am) = \mathrm{Tr}_1^A(ma)$. We have $\mathrm{Tr}_1^A(am) = \sum_{x \in X} x'amx$. This is the image of m under the action of the element $\sum_{x \in X} x'a \otimes x$, where now M is viewed as an $A \otimes_k A^{\mathrm{op}}$-module. Similarly, $\mathrm{Tr}_1^A(ma)$ is the image of m under the action of $\sum_{x \in X} x' \otimes ax$. Thus it suffices to show that we have an equality $\sum_{x \in X} x'a \otimes x = \sum_{x \in X} x' \otimes ax$ in $A \otimes_k A$. We send both elements to $(A \otimes_k A)^*$ under the isomorphism $A \otimes_k A \cong (A \otimes_k A)^*$ induces by s, as used in Definition 2.16.1. The image of the element $\sum_{x\in X} x'a \otimes x$ in $(A \otimes_k A)^*$ is the linear form sending $c \otimes d$ to $\sum_{x\in X} s(x'ad)s(xc) = \sum_{x\in X} s(x'ads(xc)) = \sum_{x\in X} s(ads(xc)x') = s(cad)$, where we use the k-linearity of s, the symmetry of s, and the formula 2.16.4(iv). Similarly, the image of $\sum_{x\in X} x' \otimes ax$ in $(A \otimes_k A)^*$ is the form sending $c \otimes d$ to $\sum_{x\in X} s(x'd)s(axc) = \sum_{x\in X} s(x'ds(cax)) = s(cad)$, by 2.16.4 (iii). The result follows. □

Proposition 2.16.16 *Let A be a symmetric algebra with symmetrising form s.*

(i) For any two elements $i, j \in A$ satisfying $ji = 0$ we have $iAj \subseteq \ker(\mathrm{Tr}_1^A)$.
(ii) For any $a, b \in A$ we have $s(\mathrm{Tr}_1^A(a)b) = s(a\mathrm{Tr}_1^A(b))$.

Proof Let $a, i, j \in A$ such that $ji = 0$. Then $iaj = jia + [ia, j] = [ia, j]$, and hence $\mathrm{Tr}_1^A(iaj) = 0$ by 2.16.15. Write $\tau_A(1) = \sum_{x\in X} x' \otimes x$ for some finite subset X of A and elements $x' \in A$. By 2.16.4, we have $\tau_A(1) = \sum_{x\in X} x \otimes x'$. Since s is symmetric, it follows that $s(\mathrm{Tr}_1^A(a)b) = \sum_{x\in X} s(x'axb) = \sum_{x\in X} s(axbx') = s(a\mathrm{Tr}_1^A(b))$. This proves (ii). □

Proposition 2.16.17 *Let G be a finite group and A a symmetric G-algebra that is free of finite rank over k. Suppose that A has a G-stable symmetrising form s and that there is $c \in A^G$ such that $\mathrm{Tr}_1^A(c) = 1$. Then the bimodule homomorphism $\mu : A \otimes_k A \to A$ given by multiplication in A has a G-stable section $\sigma : A \to A \otimes_k A$, where G acts diagonally on $A \otimes_k A$.*

Proof Write $\tau_A(1) = \sum_{x\in X} x \otimes x'$ for some k-basis X of A with dual basis X' with respect to s. The map σ sending $a \in A$ to $a \sum_{x\in X} x \otimes cx'$ is the composition of τ_A followed by the A-A-bimodule endomorphism of $A \otimes_k A$ sending $b \otimes b'$ to $b \otimes cb$ for all $b, b' \in A$. We have $\mu(\sigma(1)) = \sum_{x\in X} xcx' = \mathrm{Tr}_1^A(c) = 1$, and hence σ is a section of μ as an A-A-bimodule homomorphism. By 2.16.8, the element $\tau_A(1)$ is G-stable, and c is G-stable by the assumptions. Thus σ commutes with the action of G. □

Proposition 2.16.18 *Let n be a positive integer. Set $S = M_n(k)$, considered as a symmetric algebra with the trace map $\mathrm{tr} : S \to k$. We have $\mathrm{Tr}_1^S(1_S) = n \cdot 1_k$.*

Proof For $1 \leq s, t \leq n$, denote by e_{st} the matrix in S with (s, t)-entry equal to 1_k and all other entries 0. Then $\{e_{st}\}_{1\leq s,t\leq n}$ is a k-basis of S whose dual basis with respect to tr is $\{e_{ts}\}_{1\leq s,t\leq n}$. Since $e_{st}e_{ts} = e_{ss}$ for all t, it follows that $\mathrm{Tr}_1^S(1_S) = \sum_{1\leq s,t\leq n} e_{st}e_{ts} = n \cdot \sum_{1\leq s\leq n} e_{ss} = n \cdot 1_k$. □

Exercise 2.16.19 Let A be a symmetric k-algebra with symmetrising form. Let $k \to k'$ be a homomorphism of commutative rings. Set $A' = k' \otimes_k A$ and $s' = \mathrm{Id} \otimes s$. Let M be an A-A-bimodule, and set $M' = k' \otimes_k A$. Show that $\mathrm{Tr}_1^{A'} =$

$\mathrm{Id} \otimes \mathrm{Tr}_1^A(M)$ as maps from M' to $(M')^{A'}$. Deduce that $(M')_1^{A'}$ is equal to the image of $k' \otimes_k M_1^A$ in M'.

2.17 Stable equivalences of Morita type

An equivalence between the stable categories of two algebras need not be induced by tensoring with a bimodule. This is in contrast to Morita's Theorem, stating that an equivalence between the module categories of two algebras is always induced by tensoring with a bimodule. In the context of block theory, however, all known stable equivalences are indeed induced by suitable bimodules. This motivates the following definition, due to Broué.

Definition 2.17.1 ([14, §5.A]) Let A, B be k-algebras, let M be an A-B-bimodule and N a B-A-bimodule. We say that *M and N induce a stable equivalence of Morita type between A and B* if M, N are finitely generated projective as left and right modules with the property that $M \otimes_B N \cong A$ in $\underline{\mathrm{Mod}}(A \otimes_k A^{\mathrm{op}})$ and $N \otimes_A M \cong B$ in $\underline{\mathrm{Mod}}(B \otimes_k B^{\mathrm{op}})$.

Remark 2.17.2 An isomorphism $M \otimes_B N \cong A$ in $\underline{\mathrm{Mod}}(A \otimes_k A^{\mathrm{op}})$ is by 2.13.8 equivalent to an isomorphism $M \otimes_B N \oplus X' \cong A \oplus X$ in $\mathrm{Mod}(A \otimes_k A^{\mathrm{op}})$ for some finitely generated relatively k-projective $A \otimes_k A^{\mathrm{op}}$-modules X', X. If A has no nonzero relatively k-projective direct summand as an $A \otimes_k A^{\mathrm{op}}$-module and if $\mathrm{mod}(A \otimes_k A^{\mathrm{op}})$ satisfies the Krull–Schmidt property, then the summand X' on the left side must be isomorphic to a direct summand of X on the right side. Thus we may cancel X', and so in this situation, the isomorphism $M \otimes_B N \cong A$ in $\underline{\mathrm{Mod}}(A \otimes_k A^{\mathrm{op}})$ is equivalent to an isomorphism

$$M \otimes_B N \cong A \oplus X$$

in $\mathrm{mod}(A \otimes_k A^{\mathrm{op}})$ for some finitely generated relatively k-projective $A \otimes_k A^{\mathrm{op}}$-module X. If in addition A is finitely generated projective as a k-module, then X is a projective $A \otimes_k A^{\mathrm{op}}$-module. The analogous hypotheses for B imply that there is an isomorphism

$$N \otimes_A M \cong B \oplus Y$$

for some relatively k-projective $B \otimes_k B^{\mathrm{op}}$-module Y. This is the situation we will typically encounter in the context of the block theory of finite groups. In that situation, the bimodules M and N induce a separable equivalence between A and B (cf. 2.6.15), and in particular, M and N are progenerators as one-sided modules (cf. 2.6.16).

Remark 2.17.3 Suppose that B is a subalgebra of A, and that both algebras are finitely generated projective as k-modules. We say that *induction and restriction induce a stable equivalence of Morita type between A and B* if $A = B \oplus Y$ for some projective $B \otimes_k B^{op}$-submodule Y of A and if $A \otimes_B A \cong A \oplus X$ for some projective $A \otimes_k A^{op}$-module X. This is indeed a special case of a stable equivalence of Morita type, given by the A-B-bimodule $M = A_B$ (obtained from restricting A on the right to B) and the B-A-bimodule $N = {}_BA$ (obtained from restricting A on the left to B).

Example 2.17.4 The 'smallest' example of a stable equivalence of Morita type between two different finite group algebras is as follows. Suppose that k is a field of characteristic 2. Let C_2 be the cyclic subgroup of order 2 of the symmetric group S_3 generated by the transposition $(1, 2)$. Then induction and restriction between the algebra $A = kS_3$ and its subalgebra $B = kC_2$ induce a stable equivalence of Morita type. One way to see this consists of making use of Exercise 1.7.11, showing that A is isomorphic to the direct product of B and a matrix algebra, and then using the fact that all modules over a matrix algebra are projective (hence zero in the stable category). An alternative approach is as follows. Observe first that S_3 has exactly two C_2-C_2-double cosets, namely C_2 itself and C_2sC_2, where $s = (1, 2, 3)$. The double coset C_2sC_2 has four elements. Thus, as a B-B-bimodule, we have $A = B \oplus Y$, where $Y = k[C_2sC_2]$. Using that $C_2 \cap {}^sC_2$ is trivial, it follows from 2.4.6 (or by an easy direct computation) that Y is isomorphic to the projective B-B-bimodule $B \otimes_k B$. In order to show that $A \otimes_B A \cong A \oplus X$ for some projective $A \otimes_k A$-module X, we first show that A is isomorphic to a direct summand of $A \otimes_B A$. Since $|S_3 : C_2| = 3$ is invertible in k, this is a special case of 2.6.9. Thus $A \otimes_B A \cong A \oplus X$ for some $A \otimes_k A^{op}$-bimodule. Restricting this isomorphism to B on both sides and using $A = B \oplus Y$ yields $(B \oplus Y) \otimes_B (B \oplus Y) \cong B \oplus Y \oplus \mathrm{Res}^{A \otimes_k A^{op}}_{B \otimes_k B^{op}}(X)$. The left side in this isomorphism is of the form B plus a projective $B \otimes_k B^{op}$-module. But then the right side must be of this form, too, and hence the restriction of X to $B \otimes_k B^{op}$ is projective. But the index of $C_2 \times C_2$ in $S_3 \times S_3$ is odd, hence invertible in k. It follows from 2.6.3 that X is projective as a $k(S_3 \times S_3)$-module, or equivalently, as an $A \otimes_k A^{op}$-module. The stable equivalence of Morita type in this example is in fact induced by an isomorphism between kC_2 and the principal block of kS_3, but the arguments used in this example extend to blocks with cyclic defect groups, and more generally, to stable equivalences of Morita type in 'trivial intersection' cases; see 5.2.5 and 9.8.6.

Following [14, 5.1], tensoring by $M \otimes_B -$ and $N \otimes_A -$ induces k-linear inverse equivalences between the k-stable categories $\underline{\mathrm{Mod}}(A)$ and $\underline{\mathrm{Mod}}(B)$; moreover, if X and Y are zero this is a Morita equivalence.

Proposition 2.17.5 *Let A, B be k-algebras. Let M be an A-B-bimodule and N a B-A-bimodule such that M and N induce a stable equivalence of Morita type between A and B. Then the functors $M \otimes_B -$ and $N \otimes_A -$ induce inverse equivalences $\underline{\mathrm{Mod}}(A) \cong \underline{\mathrm{Mod}}(B)$.*

Proof By 2.13.10 both functors preserve relatively k-projective modules and the classes of k-split maps. Thus $M \otimes_B -$ induces a functor from $\underline{\mathrm{Mod}}(B)$ to $\underline{\mathrm{Mod}}(A)$. Similarly for N. If X is a relatively k-projective $A \otimes_k A^{\mathrm{op}}$-module, then X is a summand of $A \otimes_k W \otimes_k A$ for some k-module W. It follows that the functor $X \otimes_A -$ sends any A-module to a relatively k-projective A-module, and therefore induces the zero functor on $\underline{\mathrm{Mod}}(A)$. Composing $M \otimes_B -$ and $N \otimes_A -$ yields a functor $M \otimes_B N \otimes_A -$. Since $M \otimes_B N \oplus X \cong A \oplus X'$ for some relatively k-projective $A \otimes_k A^{\mathrm{op}}$-modules X, X', it follows that the functor on $\underline{\mathrm{Mod}}(A)$ induced by $M \otimes_B N \otimes_A -$ is isomorphic to the identity functor on $\underline{\mathrm{Mod}}(A)$. A similar argument shows that tensoring by $N \otimes_A M$ induces the identity functor on $\underline{\mathrm{Mod}}(B)$, whence the result. □

If A and B have the property that their classes of relatively k-projective modules and relatively k-injective modules coincide, then the equivalence $\underline{\mathrm{Mod}}(A) \cong \underline{\mathrm{Mod}}(B)$ induced by a stable equivalence of Morita type is in fact an equivalence of triangulated categories. As mentioned before, we discuss this in more detail in the appendix. For a k-algebra A whose relatively k-projective and relatively k-injective modules coincide, the functors Ω_A and its inverse Σ_A are induced by a stable equivalence of Morita type. To see this, we define bimodules as follows:

Definition 2.17.6 Let A be a k-algebra. Denote by $\mu_A : A \otimes_k A \to A$ the $A \otimes_k A^{\mathrm{op}}$-homomorphism sending $a \otimes a'$ to aa', for any $a, a' \in A$. Set $\boldsymbol{\Omega} A = \ker(\mu_A)$ and $\boldsymbol{\Sigma} A = \mathrm{Hom}_A(\boldsymbol{\Omega} A, A)$.

The following well-known result implies that Heller translates commute, as functors up to natural equivalence, with stable equivalences of Morita type (over complete local rings this is, for instance, proved in [55, 2.9]). By more general results of Auslander and Reiten in [3, 4.2, 4.4], Heller translates commute on modules (but not necessarily on homomorphism spaces) with arbitrary stable equivalences between symmetric algebras or selfinjective algebras with no direct factor of Loewy length 2.

Proposition 2.17.7 *Let A, B be k-algebras and let M be an A-B-bimodule.*

(i) If M is finitely generated relatively k-projective as a right B-module, then there is a canonical isomorphism $\boldsymbol{\Omega} A \otimes_A M \cong \Omega_{A \otimes_k B^{\mathrm{op}}}(M)$ in $\underline{\mathrm{Mod}}(A \otimes_k B^{\mathrm{op}})$.

(ii) *If M is finitely generated relatively k-projective as a left A-module, then there is a canonical isomorphism $\Omega_{A\otimes_k B^{\mathrm{op}}}(M) \cong M \otimes_B \mathbf{\Omega}B$ in $\underline{\mathrm{Mod}}(A \otimes_k B^{\mathrm{op}})$.*

Proof The short exact sequence of $A \otimes_k A^{\mathrm{op}}$-modules

$$0 \longrightarrow \mathbf{\Omega}A \longrightarrow A \otimes_k A \xrightarrow{\mu_A} A \longrightarrow 0$$

splits as a sequence of left A-modules and as a sequence of right A-modules because the right term is projective as a left and right A-module. Tensoring this sequence by $- \otimes_A M$ yields therefore a k-split short exact sequence of $A \otimes_k B^{\mathrm{op}}$-modules of the form

$$0 \longrightarrow \mathbf{\Omega}A \otimes_A M \longrightarrow A \otimes_k M \longrightarrow M \longrightarrow 0.$$

The middle term of this sequence is relatively k-projective as an $A \otimes_k B^{\mathrm{op}}$-module because M is finitely generated relatively k-projective as a right B-module, and hence the left term is canonically isomorphic, in $\underline{\mathrm{Mod}}(A \otimes_k B^{\mathrm{op}})$, to the A-B-bimodule $\Omega_{A\otimes_k B^{\mathrm{op}}}(M)$. A similar argument shows the remaining isomorphism. □

Corollary 2.17.8 *Let A, B, C be k-algebras, let M be an A-B-bimodule and N a B-C-bimodule. Suppose that M and N are finitely generated relatively k-projective as left and right modules. We have canonical isomorphisms in $\underline{\mathrm{Mod}}(A \otimes_k C^{\mathrm{op}})$*

$$\Omega_{A\otimes_k B^{\mathrm{op}}}(M) \otimes_B N \cong \Omega_{A\otimes_k C^{\mathrm{op}}}(M \otimes_B N) \cong M \otimes_B \Omega_{B\otimes_k C^{\mathrm{op}}}(N).$$

Proof Applying four times the previous proposition yields a sequence of bimodule isomorphisms $\Omega_{A\otimes_k B^{\mathrm{op}}}(M) \otimes_B N \cong \mathbf{\Omega}A \otimes_A M \otimes_B N \cong \Omega_{A\otimes_k C^{\mathrm{op}}}(M \otimes_B N) \cong M \otimes_B N \otimes_C \mathbf{\Omega}C \cong M \otimes_B \Omega_{B\otimes_k C^{\mathrm{op}}}(N)$, whence the result. □

Theorem 2.17.9 *Let A be a k-algebra such that A is finitely generated projective as a k-module. Suppose that the classes of relatively k-projective and relatively k-injective A-modules coincide.*

(i) *The $A \otimes_k A^{\mathrm{op}}$-modules $\mathbf{\Omega}A$ and $\mathbf{\Sigma}A$ induce a stable equivalence of Morita type on A.*
(ii) *We have natural isomorphism $\Omega_A \cong \mathbf{\Omega}A \otimes_A -$ and $\Sigma_A \cong \mathbf{\Sigma}A \otimes_A -$ as functors on $\underline{\mathrm{Mod}}(A)$.*
(iii) *We have isomorphisms $\mathbf{\Omega}A \cong \Omega_{A\otimes_k A^{\mathrm{op}}}(A)$ and $\mathbf{\Sigma}A \cong \Sigma_{A\otimes_k A^{\mathrm{op}}}(A)$ in $\underline{\mathrm{Mod}}(A \otimes_k A^{\mathrm{op}})$.*

Proof Since A is finitely generated projective as a k-module, the $A \otimes_k A^{\mathrm{op}}$-module $A \otimes_k A$, upon restriction to its left and right A-module structure, becomes finitely generated projective as an A-module. The short exact sequence

$$0 \longrightarrow \mathbf{\Omega} A \longrightarrow A \otimes_k A \xrightarrow{\mu_A} A \longrightarrow 0$$

is k-split and the middle term is projective as an $A \otimes_k A^{\mathrm{op}}$-module, hence $\mathbf{\Omega} A \cong \Omega_{A\otimes_k A^{\mathrm{op}}}(A)$ in $\underline{\mathrm{Mod}}(A \otimes_k A^{\mathrm{op}})$. As in the proof of 2.17.7, the right term in this short exact sequence is projective as a left and right A-module, and hence this sequence splits upon restriction to the left or right A-module structure. Since A is finitely generated projective as a k-module, this implies that $\mathbf{\Omega} A$ is finitely generated projective as a left and right A-module. Any functor sends a split exact sequence to a split exact sequence. Applying the contravariant functor $\mathrm{Hom}_A(-A)$ to the above sequence yields therefore again an exact sequence of $A \otimes_k A^{\mathrm{op}}$-modules of the form

$$0 \longrightarrow \mathrm{Hom}_A(A, A) \longrightarrow \mathrm{Hom}_A(A \otimes_k A, A) \longrightarrow \mathrm{Hom}_A(\mathbf{\Omega} A, A) \longrightarrow 0\,.$$

The right term in this sequence is $\mathbf{\Sigma} A$, and the left term is canonically isomorphic to A. Using the standard adjunction (applied to $B = k$ and $M = A$) one sees that the term in the middle is isomorphic to $\mathrm{Hom}_k(A, A)$ as an A-A-bimodule. Since A is finitely generated projective as a k-module, we have, by 2.9.13 applied to A instead of V, an isomorphism of A-A-bimodules $\mathrm{Hom}_k(A, A) \cong \mathrm{Hom}_k(A, k) \otimes_k A$. By 2.15.2, $\mathrm{Hom}_k(A, k)$ is finitely generated projective as a left A-module, and hence $\mathrm{Hom}_k(A, k) \otimes_k A$ is finitely generated projective as a left and right A-module. Thus the previous short exact sequence splits as a short exact sequence of both left and right A-modules. This shows that $\mathbf{\Sigma} A$ is finitely generated projective as a left and right A-module and that $\Omega_{A\otimes_k A^{\mathrm{op}}}(\mathbf{\Sigma} A) \cong A$ in $\underline{\mathrm{Mod}}(A \otimes_k A^{\mathrm{op}})$. Proposition 2.17.7, applied to $\mathbf{\Sigma} A$ instead of M, shows that $\mathbf{\Omega} A \otimes_A \mathbf{\Sigma} A \cong \Omega_{A\otimes_k A^{\mathrm{op}}}(\mathbf{\Sigma} A) \cong A \cong \mathbf{\Sigma} A \otimes_A \mathbf{\Omega} A$ in $\underline{\mathrm{Mod}}(A \otimes_k A^0)$. This shows (i). The isomorphism $\Omega_A \cong \mathbf{\Omega} A \otimes_A -$ follows from 2.17.7, applied to $B = k$. The isomorphism $\Sigma_A \cong \mathbf{\Sigma} A \otimes_A -$ follows from the fact that the functors Σ_A and $\mathbf{\Sigma} A \otimes_A -$ are both inverses of Ω_A, proving (ii). The algebra $A \otimes_k A^{\mathrm{op}}$ has again the property that the classes of relatively k-injective and relatively k-projective modules coincide, and hence the functors $\Omega_{A\otimes_k A^{\mathrm{op}}}$ and $\Sigma_{A\otimes_k A^{\mathrm{op}}}$ are inverse on $\underline{\mathrm{Mod}}(A \otimes_k A^{\mathrm{op}})$. Statement (iii) follows from the preceding remarks. □

Given two algebras A, B over a commutative ring k and an A-B-bimodule, the functor $M \otimes_B -$ from $\mathrm{Mod}(B)$ to $\mathrm{Mod}(A)$ determines M up to isomorphism; indeed, $M \otimes_B B$ is a left A-module whose right B-module structure is induced by $\mathrm{End}_B(B) \cong B^{\mathrm{op}}$. If M is finitely generated projective as a left A-module then $M \otimes_B -$ induces a functor from $\underline{\mathrm{Mod}}(B)$ to $\underline{\mathrm{Mod}}(A)$, but this functor does not in general determine M. Nonetheless, if M, N induce a stable equivalence of Morita type between symmetric algebras A and B then N is determined by M, up to isomorphism in the k-stable category $\underline{\mathrm{mod}}(B \otimes_k A^{\mathrm{op}})$. This will follow from extending stable equivalences of Morita type to certain categories of bimodules, for which we introduce the following notation.

Definition 2.17.10 Let A, B be k-algebras. We denote by $\mathrm{Perf}(A, B)$ the full subcategory of $\mathrm{Mod}(A \otimes_k B^{\mathrm{op}})$ consisting of all A-B-bimodules that are projective as a left A-module and as a right B-module, and we denote by $\underline{\mathrm{Perf}}(A, B)$ the image of $\mathrm{Perf}(A, B)$ in $\underline{\mathrm{Mod}}(A \otimes_k B^0)$. Similarly, we denote by $\mathrm{perf}(A, B)$ the full subcategory of $\mathrm{mod}(A \otimes_k B^{\mathrm{op}})$ consisting of all A-B-bimodules that are finitely generated projective as left A-module and as right B-module, and we denote by $\underline{\mathrm{perf}}(A, B)$ the image of $\mathrm{perf}(A, B)$ in $\underline{\mathrm{mod}}(A \otimes_k B^{\mathrm{op}})$.

The following result, which is mentioned in the proof of [14, 5.4], is a replacement for the bimodule version of Morita equivalences in 2.8.5.

Proposition 2.17.11 *Let A, B be k-algebras, let M be an A-B-bimodule and N a B-A-bimodule. Suppose that M and N induce a stable equivalence of Morita type between A and B.*

(i) The functor $M \otimes_B -$ induces equivalences $\underline{\mathrm{perf}}(B, A) \cong \underline{\mathrm{perf}}(A, A)$ and $\underline{\mathrm{perf}}(B, B) \cong \underline{\mathrm{perf}}(A, B)$ with inverses induced by the functor $N \otimes_A -$.
(ii) The functor $- \otimes_A M$ induces equivalences $\underline{\mathrm{perf}}(A, A) \cong \underline{\mathrm{perf}}(A, B)$ and $\underline{\mathrm{perf}}(B, A) \cong \underline{\mathrm{perf}}(B, B)$ with inverses induced by the functor $- \otimes_B N$.
(iii) The functor $N \otimes_A - \otimes_A M$ induces an equivalence $\underline{\mathrm{perf}}(A, A) \cong \underline{\mathrm{perf}}(B, B)$, with inverse induced by the functor $M \otimes_B - \otimes_B N$.

Proof By 2.17.1, M belongs to $\mathrm{perf}(A, B)$ and N belongs to $\mathrm{perf}(B, A)$. Let U be in $\mathrm{perf}(B, A)$. As a left B-module, U is a direct summand of B^n for some positive integer n, hence $M \otimes_B U$ is a direct summand of M^n as a left A-module. This shows that $M \otimes_B U$ is finitely generated as a left A-module. A similar argument shows that $M \otimes_B U$ is finitely generated projective as a right A-module. Thus $M \otimes_B -$ induces an exact functor from $\mathrm{perf}(B, A)$ to $\mathrm{perf}(A, A)$. Similarly for the other functors occurring in the statements. Since $M \otimes_B N$ is isomorphic, in $\underline{\mathrm{mod}}(A \otimes_k A^{\mathrm{op}})$ to A, we have $M \otimes_B N \oplus X \cong A \oplus X'$ for some relatively k-projective $A \otimes_k A^{\mathrm{op}}$-modules X, X'. It suffices to show that $X \otimes_A -$ induces the zero functor on $\underline{\mathrm{perf}}(A, A)$. Now X is isomorphic to a direct summand of $(A \otimes_k A^{\mathrm{op}}) \otimes_k V \cong A \otimes_k V \otimes_k A$ as an $A \otimes_k A^{\mathrm{op}}$-module for some k-module V. Thus, for any A-module U, the A-module $X \otimes_A U$ is isomorphic to a direct summand of $A \otimes_k V \otimes_k U$, hence relatively k-projective. The result follows. □

Corollary 2.17.12 *Let A, B be k-algebras, let M be an A-B-bimodule and N a B-A-bimodule. Suppose that M and N induce a stable equivalence of Morita type between A and B. Suppose that there is a positive integer n such that $\Omega^n_{A\otimes_k A^{\mathrm{op}}}(A) \cong A$. Then $\Omega^n_{B\otimes_k B^{\mathrm{op}}}(B) \cong B$.*

Proof By 2.17.8 we have an isomorphism $N \otimes_A \Omega^n_{A\otimes_k A^{\mathrm{op}}}(A) \otimes_A M \cong \Omega_{B\otimes_k B^{\mathrm{op}}}(N \otimes_A M)$ in $\underline{\mathrm{perf}}(B, B)$. Since $N \otimes_A M \cong B$ in $\underline{\mathrm{perf}}(B, B)$, the result follows. □

An easy variation of these arguments shows that stable equivalences of Morita type can be composed.

Proposition 2.17.13 *Let A, B, C be k-algebras, M an A-B-bimodule, N a B-A-bimodule, U a B-C-bimodule, and V a C-B-bimodule. Suppose that M, N induce a stable equivalence of Morita type between A and B, and that U, V induce a stable equivalence of Morita type between B and C. Then $M \otimes_B U$ and $V \otimes_B N$ induce a stable equivalence of Morita type between A and C.*

Proof Write $M \otimes_B N \cong A \oplus X$ for some projective $A \otimes_k A^{\mathrm{op}}$-module X, and $U \otimes_C V \cong B \oplus W$ for some projective $B \otimes_k B^{\mathrm{op}}$-module W. Then $(M \otimes_B U) \otimes_C (V \otimes_B N) \cong M \otimes_B (B \oplus W) \otimes_B N \cong A \oplus X \oplus M \otimes_B W \otimes_B N$. Since W is a projective $B \otimes_k B^{\mathrm{op}}$, it follows that $M \otimes_B W \otimes_B N$ is a projective $A \otimes_k A^{\mathrm{op}}$-module. The same argument with reversed roles of A and C concludes the proof. □

There is no analogue of Proposition 2.8.20 for stable equivalences of Morita type; that is, stable equivalences of Morita type do not extend in general to tensor products.

Let A be a k-algebra. As mentioned after the Definition 2.16.10, the projective ideal $Z^{\mathrm{pr}}(A)$ in $Z(A)$ is mapped onto $\mathrm{End}^{\mathrm{pr}}_{A\otimes_k A^{\mathrm{op}}}(A)$ under the canonical k-algebra isomomorphism $Z(A) \cong \mathrm{End}_{A\otimes_k A^{\mathrm{op}}}(A)$, and hence taking quotients induces an isomorphism

$$\underline{Z}(A) \cong \underline{\mathrm{End}}_{A\otimes_k A^{\mathrm{op}}}(A).$$

This identification of the stable centre $\underline{Z}(A)$ of A as the stable endomorphism algebra of the A-A-bimodule of A, together with statement (iii) in 2.17.11 has the following immediate consequence:

Corollary 2.17.14 ([14, 5.4]) *Let A, B be k-algebras, let M be an A-B-bimodule and N a B-A-bimodule. Suppose that M and N induce a stable equivalence of Morita type between A and B. Then the functor $N \otimes_A - \otimes_A M$ induces a k-algebra isomorphism $\underline{Z}(A) \cong \underline{Z}(B)$.*

Proof The functor $N \otimes_A - \otimes_A M$ sends A to $N \otimes_A M \cong B \oplus Y$ for some projective B-B-bimodule Y, and hence the equivalence $\underline{\mathrm{perf}}(A, A) \cong \underline{\mathrm{perf}}(B, B)$ induced by this functor sends A to B, up to isomorphism, in $\underline{\mathrm{perf}}(B, B)$. In particular, this functor induces an isomorphism between the stable endomorphism algebras of the bimodules A and B, whence the result. □

Proposition 2.17.15 *Let A, B be symmetric k-algebras, let M be an A-B-bimodule and N a B-A-bimodule. Suppose that M and N induce a stable equivalence of Morita type between A and B. Then N is isomorphic, in $\underline{\mathrm{perf}}(B, A)$, to the k-dual $M^* = \mathrm{Hom}_k(M, k)$ of M.*

Proof By 2.12.7, the functor $M^* \otimes_A -$ from $\mathrm{Perf}(A, A)$ to $\mathrm{Perf}(B, A)$ is right and left adjoint to the functor $M \otimes_B -$. Since the induced functor by $M \otimes_B -$ is an equivalence from $\underline{\mathrm{Perf}}(B, A)$ to $\underline{\mathrm{Perf}}(A, A)$, its right adjoint is also an inverse, and hence the functors induced by $M^* \otimes_A -$ and $N \otimes_A -$ are both inverses. In particular, they send the A-A-bimodule A to isomorphic bimodules in $\underline{\mathrm{Perf}}(B, A)$. Since M and N are finitely generated as left and right modules, the result follows. □

Stable equivalences of Morita type are given by bimodules that are 'stably invertible', and we can define the analogue of the Picard group for stable equivalences.

Definition 2.17.16 Let A be a k-algebra. The *stable Picard group*, denoted $\mathrm{StPic}(A)$, is the group of isomorphism classes of A-A-bimodules M for which there exists an A-A-bimodule N such that M and N induce a stable equivalence of Morita type on A, with product induced by the tensor product over A. The unit element of this group is the isomorphism class of the regular bimodule A, and if M and N induce a stable equivalence of Morita type on A, then the isomorphism class of N is the inverse in $\mathrm{StPic}(A)$ of the isomorphism class of M.

Since a Morita equivalence is a stable equivalence of Morita type, it follows that the stable Picard group $\mathrm{StPic}(A)$ contains the Picard group $\mathrm{Pic}(A)$ as a subgroup. While the isomorphism class of an A-A-bimodule M inducing a Morita equivalence is determined by the isomorphism class of the functor $M \otimes_A -$ on $\mathrm{Mod}(A)$, it is not the case, in general, that the isomorphism class of an A-A-bimodule M inducing a stable equivalence of Morita type is determined by the isomorphism class of the functor on $\underline{\mathrm{Mod}}(A)$ induced by $M \otimes_A -$.

In other words, the group homomorphism from StPic(A) to the group of self equivalences of $\underline{\text{Mod}}(A)$ need not be injective.

Proposition 2.17.17 (cf. [56, 11.4.1, 11.4.5]) *Let A, B be k-algebras that are finitely generated projective as a k-module and such that the classes of relatively k-projective modules and relatively k-injective modules coincide. Let M be an A-B-bimodule and N a B-A-bimodule inducing a stable equivalence of Morita type.*

(i) The functor $M \otimes_B - \otimes_B N$ induces an isomorphism of stable Picard groups $\mathrm{StPic}(B) \cong \mathrm{StPic}(A)$, *with inverse induced by the functor $N \otimes_A - \otimes_A M$.*
(ii) The groups $\langle \mathbf{\Omega} A\rangle$ and $\langle \mathbf{\Omega} B\rangle$ are central subgroups of $\mathrm{StPic}(A)$ *and* $\mathrm{StPic}(B)$, *respectively, and the functor $M \otimes_B - \otimes_B N$ induces and isomorphism $\langle \mathbf{\Omega} A\rangle \cong \langle \mathbf{\Omega} A\rangle$.*

Proof If V and V' are B-B-bimodules that induce a self stable equivalence of Morita type on B, then it follows from 2.17.13 that $M \otimes_B V \otimes_B N$ and $M \otimes_B V' \otimes_B N$ induce a self stable equivalence of Morita type on A. An easy and strictly formal verification shows that this yields a group homomorphism $\mathrm{StPic}(B) \to \mathrm{StPic}(A)$ with inverse as stated. This proves (i). Statement (ii) is an immediate consequence of 2.17.9 and 2.17.7. □

Remark 2.17.18 Let A be a symmetric k-algebra. In view of 2.17.15 one should think of bimodules inducing equivalences of Morita type on A as bimodule analogues of endotrivial kG-modules, where G is a finite group. These will be considered in more detail in §7.5.

2.18 Projective and injective resolutions

Let A be a k-algebra and U an A-module. Informally, a bounded below chain complex of A-modules of the form

$$\cdots \longrightarrow P_2 \xrightarrow{\delta_2} P_1 \xrightarrow{\delta_1} P_0 \xrightarrow{\pi} U \longrightarrow 0$$

is called a *projective resolution of U* if it is exact and all P_i are projective. It is called a *relatively k-projective resolution of U* if it is contractible as a complex of k-modules (or equivalently, k-split acyclic, by 1.18.15) and if all P_i are relatively k-projective. If U is projective as a k-module, then both notions coincide; in particular, if k is a field, both notions coincide for all A-modules. An exact complex as above can be viewed as a chain map obtained from 'bending down' the map π and viewing U as a chain complex concentrated in degree zero:

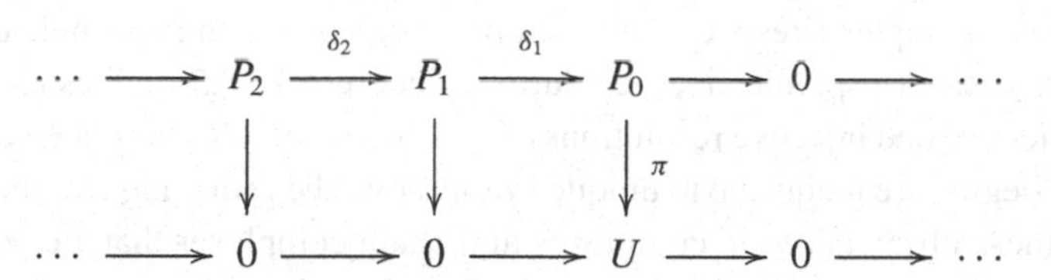

This chain map is then a quasi-isomorphism because the homology of both rows is concentrated in degree 0, where it is isomorphic to $P_0/\mathrm{Im}(\delta_0) = P_0/\ker(\pi) \cong U$. Similarly, the informal version of an *injective resolution of U* is an exact bounded below cochain complex of the form

$$0 \longrightarrow U \xrightarrow{\iota} I^0 \xrightarrow{\delta^0} I^1 \xrightarrow{\delta^1} I^2 \xrightarrow{\delta^2} \cdots$$

where the modules I^i are injective, and such a complex is called a *relatively k-injective resolution of U* if it is contractible as a complex of k-modules and all I^i are relatively k-injective. As before, we view ι as a quasi-isomorphism of cochain complexes. from U to $I^0 \xrightarrow{\delta^0} I^1 \xrightarrow{\delta^1} \cdots$. This definition of projective and injective resolutions extends to resolutions to complexes.

Definition 2.18.1 If X is a bounded below complex over an abelian category $\mathcal{A}$, then a *projective resolution of X* is a pair (P, π) consisting of a bounded below complex P whose components are projective objects in $\mathcal{A}$, together with a chain map $\pi : P \to X$ such that the mapping cone $\mathrm{cone}(\pi)$ of π is acyclic, or equivalently, such that π is a quasi-isomorphism. If X is a bounded below cochain complex over $\mathcal{A}$, then an *injective resolution* of X is a pair (I, ι) consisting of a bounded below cochain complex of injective objects in $\mathcal{A}$ and a chain map $\iota : X \to I$ such that $\mathrm{cone}(\iota)$ is acyclic, or equivalently, such that ι is a quasi-isomorphism.

Let A be a k-algebra. A complex X of A-modules is called *k-split* if X is split as a complex of k-modules, and *k-split acyclic* if it is k-split and acyclic, or equivalently, if it is contractible as a complex of k-modules.

Definition 2.18.2 Let A be a k-algebra. If X is a bounded below complex of A-modules, then a *relatively k-projective resolution of X* is a pair (P, π) consisting of a bounded below complex P whose components are relatively k-projective A-modules, together with a chain map $\pi : P \to X$ such that $\mathrm{cone}(\pi)$ is k-split acyclic. If X is a bounded below cochain complex of A-modules, then a *relatively k-injective resolution* of X is a pair (I, ι) consisting of a bounded below cochain complex of relatively k-injective A-modules and a cochain map $\iota : X \to I$ such that $\mathrm{cone}(\iota)$ is k-split acyclic.

If $\mathcal{A}$ has enough projective (resp. injective) objects, then every object in $\mathcal{A}$, viewed as a complex (iresp. cochain complex) concentrated in a single degree, has a projective (resp. injective) resolution. Theorem 1.18.5 implies moreover that projective and injective resolutions of (cochain) complexes concentrated in a single degree are unique up to unique homotopy. The following two theorems extend these observations to complexes and chain complexes that are bounded below, as well as to their relatively k-projective and injective analogues for k-split complexes of modules over a k-algebra.

Theorem 2.18.3 *Let A be an algebra over a commutative ring.*

(i) Every bounded below chain complex X of A-modules has a projective resolution (P_X, π_X). Moreover, if A is Noetherian and all terms of X are finitely generated, then one can choose all terms of P_X to be finitely generated.
(i) Every bounded below chain complex X of A-modules has a bounded below relatively k-projective resolution (P_X, π_X). Moreover, if A is Noetherian and all terms of X are finitely generated, then one can choose all terms of P_X to be finitely generated.
(iii) Every bounded below cochain complex of A-modules has an injective resolution (I_X, ι_X).
(iv) Every bounded below k-split cochain complex of A-modules has a relatively k-injective resolution (I_X, ι_X). Moreover, if A is Noetherian and all terms of X are finitely generated, then one can choose all terms of I_X to be finitely generated.

Proof Let (X, δ) be a bounded below chain complex of A-modules. By an appropriate shift of X, we may assume that $X_n = 0$ for any negative integer n. We set $P_n = 0$ and $\pi_n = 0$ for any negative integer n. For n non negative, we construct P_n and π_n inductively. Assume that for $i < n$ we have already constructed a projective module P_i, a map $\pi_i : P_i \to X_i$ and a map $\epsilon_i : P_i \to P_{i-1}$ with the following properties:

(1) $\pi_{i-1} \circ \epsilon_i = \delta_i \circ \pi_i$, for $i < n$;
(2) the map π_i is surjective and induces an isomorphism $\ker(\epsilon_i)/\mathrm{Im}(\epsilon_{i+1}) \cong H_i(X)$ for $i < n - 1$;
(3) π_{n-1} maps $\ker(\epsilon_{n-1})$ onto $\ker(\delta_{n-1})$; and
(4) $\ker(\epsilon_{n-1})$ is the inverse image of $\ker(\delta_{n-1})$ in P_{n-1} under π_{n-1}.

We construct P_n and π_n as follows. Let U be the inverse image of $\mathrm{Im}(\delta_n)$ in P_{n-1}. Thus π_{n-1} induces an isomorphism $\ker(\epsilon_{n-1})/U \cong H_{n-1}(X)$. Let V be the submodule of $U \oplus X_n$ consisting of all $(u, x) \in U \oplus X_n$ satisfying $\pi_{n-1}(u) = \delta_n(x)$.

That is, V is the kernel of the map $(-\pi_{n-1}, \delta_n) : U \oplus X_n \to X_{n-1}$, or equivalently, V is the pullback of the maps $\pi_{n-1}|_U$ and δ_n. Take a projective cover $\pi_V : P_n \to V$ of V. We define $\epsilon_n : P_n \to P_{n-1}$ to be the composition of π_V followed by the projection $V \to P_{n-1}$ mapping $(u, x) \in V$ to u. We define $\pi_n : P_n \to X_n$ to be the composition of π_V followed by the projection $V \to X_n$ mapping $(u, x) \in V$ to x. Then π_n is surjective, by construction. Moreover, we have $\mathrm{Im}(\epsilon_n) = U$, thus π_{n-1} induces an isomorphism $\ker(\epsilon_{n-1})/\mathrm{Im}(\epsilon_n) \cong H_{n-1}(X)$. Also, by the construction of π_n and ϵ_n we have $\pi_{n-1} \circ \epsilon_n = \delta_n \circ \pi_n$. Finally, since $\ker(\epsilon_n)$ is the inverse image in P_n of the submodule $\{(0, x) | x \in \ker(\delta_n)\}$ of V, the map π_n sends $\ker(\epsilon_n)$ onto $\ker(\delta_n)$ and $\ker(\epsilon_n)$ is the inverse image of $\ker(\delta_n)$ in P_n. Suppose now that A is Noetherian. If all components of X and the P_i for $i < n$ are finitely generated, so is V, and hence P_n can be chosen to be finitely generated. This concludes the proof of (i). For the proof of (ii) we proceed as before, with the following changes. We require the P_i to be relatively k-projective, in (3) we require the restriction of π_{n-1} to $\ker(\epsilon_{n-1})$ to be k-split surjective. We take for P_n a relatively k-projective cover of the pullback V constructed above, and define π_n as before. One verifies that then π_n induces a k-split surjective map from $\ker(\epsilon_n)$ to $\ker(\delta_n)$. This implies that for any relatively k-projective A-module Y, the induced chain map $\mathrm{Hom}_A(Y, P) \to \mathrm{Hom}_A(Y, X)$ is a quasi-isomorphism. Applied to $Y = A \otimes_k T$ for some k-module T it follows from a standard adjunction that the chain map $\mathrm{Hom}_k(T, P) \to \mathrm{Hom}_k(T, X)$ is a quasi-isomorphism for any k-module T. Thus the cone of π has the property that $\mathrm{Hom}_k(T, \mathrm{cone}(\pi))$ is acyclic for any k-module T. This forces $\mathrm{cone}(\pi)$ to be contractible as a complex of k-modules, whence (ii). Statements (iii) and (iv) are proved analogously. □

Note that in statement (iii) of Theorem 2.18.3, even if A and k are Noetherian, we do not make a statement on injective resolutions consisting of finitely generated injective modules, because the algebra A may not have any finitely generated injective modules – whereas it always has enough finitely generated relatively k-injective modules. If (P, π) is a projective resolution of a bounded complex X, then P is bounded below and has nonzero homology in at most finitely many degrees. A similar statement holds for injective resolutions of bounded complexes.

The following result shows that quasi-isomorphic bounded below chain complexes have homotopy equivalent projective resolutions, and that taking projective resolutions is a functorial construction. Similar statements hold for injective resolutions, and there are analogous versions for relatively k-projective and injective resolutions. We give the statements and proofs for projective resolutions in module categories.

Theorem 2.18.4 *Let A be an algebra over a commutative ring, let X, Y be bounded below chain complexes of A-modules and let (P_X, π_X), (P_Y, π_Y) be projective resolutions of X, Y, respectively. Let $f : X \to Y$ be a chain map. Then the following hold.*

(i) There is, up to homotopy, a unique chain map $P_f : P_X \to P_Y$ such that $\pi_Y \circ P_f \simeq f \circ \pi_X$.
(ii) f is a quasi-isomorphism if and only if P_f is a homotopy equivalence.
(iii) If $f \sim 0$ then $P_f \sim 0$.

Proof The existence and uniqueness, up to homotopy, of P_f making the diagram in (i) commutative follows from 1.18.5. Since π_X, π_Y are quasi-isomorphisms, it follows that f is a quasi-isomorphism if and only if P_f is a quasi-isomorphism. By 1.18.20, P_f is a homotopy equivalence if and only if P_f is a quasi-isomorphism, whence (ii). Statement (iii) follows from (ii) and the fact that a homotopy equivalence is a quasi-isomorphism. □

There are canonical resolutions associated with modules and bimodules. Let A be a k-algebra. For $n \geq 0$ denote by $A^{\otimes n} = A \otimes_k A \otimes_k \cdots \otimes_k A$ the tensor product over k of n copies of A, with the convention $A^{\otimes 0} = k$. For $n \geq 1$, we consider $A^{\otimes n}$ as an A-A-bimodule, or equivalently, as an $A \otimes_k A^{\mathrm{op}}$-module, with $a \in A$ acting by left and right multiplication on the first and last copy of A, respectively. If A is projective as a k-module, then $A^{\otimes n}$ is projective as a left and as a right A-module for $n \geq 1$. If A is finitely generated projective as a k-module, then $A^{\otimes n}$ is finitely generated projective as a left and as a right A-module. The A-A-bimodule $A^{\otimes 2} = A \otimes_k A$ becomes the free $A \otimes_k A^{\mathrm{op}}$-module of rank one through the canonical identification of A-A-bimodules and $A \otimes_k A^{\mathrm{op}}$-modules.

Proposition 2.18.5 *Let A be a k-algebra. For $n \geq -1$ set $X_n = A^{\otimes n+2}$ and for $n \geq 0$ denote by $d_n : X_n \to X_{n-1}$ the $A \otimes_k A^{\mathrm{op}}$-homomorphism given by*

$$d_n(a_0 \otimes a_1 \otimes \cdots \otimes a_{n+1}) = \sum_{i=0}^{n} (-1)^i a_0 \otimes a_1 \otimes \cdots \otimes a_i a_{i+1} \otimes \cdots \otimes a_{n+1}.$$

Set $X_n = 0$ for $n \leq -2$ and $d_n = 0$ for $n \leq -1$. Then $(X_n, d_n)_{n \in \mathbb{Z}}$ is contractible as a complex of right A-modules and as a complex of left A-modules. In particular, this complex is an acyclic complex of $A \otimes_k A^{\mathrm{op}}$-modules. The bounded below complex $X = (X_n, d_n)_{n \geq 0}$ together with the map $d_0 : X_0 = A \otimes_k A \to X_{-1} = A$ given by multiplication in A is a relatively k-projective resolution of the $A \otimes_k A^{\mathrm{op}}$-module A.

Proof For $n \geq -1$ and i satisfying $0 \leq i \leq n$ define the $A \otimes_k A^{\mathrm{op}}$-homomorphism $d_{n,i} : A^{\otimes(n+2)} \to A^{\otimes(n+1)}$ by setting

$$d_{n,i}(a_0 \otimes a_1 \otimes \cdots \otimes a_{n+1}) = a_0 \otimes a_1 \otimes \cdots \otimes a_i a_{i+1} \otimes \cdots \otimes a_{n+1}.$$

Then $d_n = \sum_{i=0}^{n}(-1)^i d_{n,i}$, and for $n \geq 0$ we have

$$d_{n-1} \circ d_n = \sum_{j=0}^{n-1}\sum_{i=0}^{n}(-1)^{i+j} d_{n-1,j} \circ d_{n,i}.$$

We show that the terms in this sum can be paired with opposite signs. If $j \geq i$, then $d_{n-1,j} \circ d_{n,i} = d_{n-1,i} \circ d_{n,j+1}$. If $j < i$, then $d_{n-1,j} \circ d_{n,i} = d_{n-1,i-1} \circ d_{n,j}$. Thus pairing the summand indexed (i, j) with that indexed by $(j+1, i)$ if $j \geq i$ and with $(j, i-1)$ if $j < i$ shows that all summands cancel. This shows that (X_n, d_n) is a chain complex. Define homomorphisms of right A-modules $h_n : X_n \to X_{n+1}$ by

$$h_n(a_0 \otimes a_1 \otimes \cdots \otimes a_{n+1}) = 1 \otimes a_0 \otimes a_1 \otimes \cdots \otimes a_{n+1}$$

for $n \geq -1$ and $h_n = 0$ for $n \leq -2$. One verifies that (X_n, d_n) is contractible as a complex of right A-modules with the homotopy h satisfying $\mathrm{Id}_{X_n} = d_{n+1} \circ h_n + h_{n-1} \circ d_n$ for all $n \in \mathbb{Z}$. In a similar way one shows that this complex is contractible as a complex of left A-modules. Since a contractible complex is acyclic, all statements follow. □

If A is projective as a k-module, then the complex in Proposition 2.18.5 yields a projective resolution of A as an $A \otimes_k A^{\mathrm{op}}$-module, and this is a projective resolution of finitely generated modules if A is finitely generated projective as a k-module. Since this complex is contractible as a complex of right A-modules, it remains k-split exact upon tensoring with $- \otimes_A U$ for any A-module U. This implies the following observation.

Corollary 2.18.6 *With the notation of Proposition 2.18.5, let U be an A-module. Then $X \otimes_A U$, together with the canonical map $X_0 \otimes_A U = A \otimes_k U \to U$ given by multiplication on U is a relatively k-projective resolution of U.*

We show next that relatively k-projective or relatively k-injective resolutions of a non relatively k-projective or injective module over a symmetric k-algebra is always infinite.

Proposition 2.18.7 *Let A be a symmetric k-algebra, and let X a bounded k-split complex of A-modules. Suppose that X_i is relatively k-projective for all nonzero integers i, and that $H_i(X) = \{0\}$ for all nonzero integers i. Then the complex X is split, or equivalently, X is homotopy equivalent to the module*

$H_0(X)$ viewed as a complex concentrated in degree 0. In particular, $H_0(X)$ is relatively k-projective.

Proof If X is concentrated in degree zero, then $X \cong H_0(X)$, so there is nothing to prove. Assume that X has at least two nonzero terms. Let m be the smallest integer such that X_m is nonzero. Suppose first that $m < 0$. Then X_m is relatively k-projective. Since $H_m(X) = \{0\}$, it follows that the differential $\delta_{m+1} : X_{m+1} \to X_m$ is surjective. By the assumptions, δ_{m+1} is k-split surjective, hence split surjective as a homomorphism of A-modules. Thus, as a complex of A-modules, X is a direct sum of a complex $X_m \cong X_m$, with X_m in the degrees $m + 1$ and m, and a complex X' with $X'_m = \{0\}$. All terms of X' are still relatively k-projective, X' is k-split, the homology of X' is zero in all nonzero degrees, and $H_0(X') \cong H_0(X)$. Arguing by induction over the length of the involved complexes, it follows that X' is split as a complex of A-modules, hence so is X. Suppose now that $m \geq 0$; that is, X_i is zero for all $i < m$. We dualise the previous argument. The largest integer n satisfying $X_n \neq \{0\}$ is positive. Then X_n is relatively k-injective. By the assumptions, the map $\delta_n : X_n \to X_{n-1}$ is k-split injective, hence split injective as an A-homomorphism. Thus X is a direct summand of the contractible complex $X_n \cong X_n$, with X_n in degrees n and $n - 1$, and a complex X'' which, arguing as before, has the property that X'' is split. This shows that X is split. The last statement follows from 1.18.15 (i). □

Corollary 2.18.8 *Let A be a symmetric k-algebra, and let X be a k-split acyclic bounded complex of A-modules. Suppose that X_i is relatively k-projective for all nonzero i. Then X_0 is relatively k-projective, and X is contractible as a complex of A-modules.*

Proof By 2.18.7, the complex X is split. It is also acyclic, hence contractible by 1.18.15. □

Corollary 2.18.9 *Let A be a symmetric k-algebra. Let U be an A-module. Suppose that U is not relatively k-projective. Let*

$$\cdots \longrightarrow P_2 \xrightarrow{\delta_2} P_1 \xrightarrow{\delta_1} P_0 \xrightarrow{\pi} U \longrightarrow 0$$

be a relatively k-projective resolution of U, and let

$$0 \longrightarrow U \xrightarrow{\iota} I^0 \xrightarrow{\delta^0} I^1 \xrightarrow{\delta^1} I^2 \xrightarrow{\delta^2} \cdots$$

be a relatively k-injective resolution of U. Then P_n and I^n are nonzero for any integer $n \geq 0$.

Proof Clearly $P_0 \neq \{0\}$ because $U \neq \{0\}$. Suppose that $P_n = \{0\}$ for some positive integer n. Then the complex

$$X = \; 0 \longrightarrow P_{n-1} \longrightarrow \cdots \longrightarrow P_1 \xrightarrow{\delta_1} P_0 \xrightarrow{\pi} U \longrightarrow 0$$

is bounded, k-split exact, and all but one of its terms are relatively k-projective. It follows from 2.18.8 that U has to be relatively k-projective, too, contradicting the assumptions. The second statement is proved analogously, using that the classes of relatively k-injective and relatively k-projective A-modules coincide. □

2.19 Derived categories

Derived categories provide the structural framework for the functoriality properties of projective and injective resolutions.

Definition 2.19.1 Let $\mathcal{A}$ be an abelian category. If $\mathcal{A}$ has enough projective objects, then the *bounded below derived category* $D^-(\mathcal{A})$ is the full additive subcategory of $K(\mathcal{A})$ consisting of all bounded below chain complexes of projective objects in $\mathcal{A}$, and the *bounded derived category* $D^b(\mathcal{A})$ is the full additive subcategory of $D^-(\mathcal{A})$ consisting of all bounded below chain complexes of projective objects with nonzero homology in at most finitely many degrees. Similarly, if $\mathcal{A}$ has enough injective objects, then the *injective bounded below derived category* $D^+(\mathcal{A})$ is the homotopy category of bounded below cochain complexes of injective objects in $\mathcal{A}$, and the *injective bounded derived category* is the homotopy category of bounded below cochain complexes of injective objects in $\mathcal{A}$ with nonzero cohomology in at most finitely many degrees.

If $\mathcal{A}$ has both enough projectives *and* injectives, we will see that the two versions of bounded derived categories are canonically equivalent, modulo the standard convention of switching between chain and cochain complexes. It will depend on the circumstances which version of the derived category is more convenient. For instance, the category of sheaves over a topological space has enough injectives but not enough projectives in general. The category of finitely generated modules over a Noetherian ring has enough projectives, but not always enough injectives (for instance, there are no finitely generated injective $\mathbb{Z}$-modules).

Theorem 2.19.2 *Let A be an algebra over a commutative ring.*

(i) There is, up to unique isomorphism of functors, a unique functor $\mathcal{D}$: $K^-(\mathrm{Mod}(A)) \to D^-(\mathrm{Mod}(A))$ sending a bounded below complex X to a

projective resolution P_X and sending the homotopy class of a chain map $f : X \to Y$ to the homotopy class of $P_f : P_X \to P_Y$.

(ii) The functor $\mathcal{D} : K^-(\mathrm{Mod}(A)) \to D^-(\mathrm{Mod}(A))$ is right adjoint to the inclusion functor $D^-(\mathrm{Mod}(A)) \to K^-(\mathrm{Mod}(A))$, and $\mathcal{D}$ restricts to the identity functor on $D^-(\mathrm{Mod}(A))$.

(iii) The functor $\mathcal{D}$ is universal; that is, given an additive functor $\mathcal{D}'$: $K^-(\mathrm{Mod}(A)) \to \mathcal{C}$ which sends quasi-isomorphisms to isomorphisms, there is a unique functor $\mathcal{E} : K^-(\mathrm{Mod}(A)) \to \mathcal{C}$, up to isomorphism of functors, satisfying $\mathcal{D}' \cong \mathcal{E} \circ \mathcal{D}$.

Proof Statement (i) is a reformulation of 2.18.4. Statement (ii) is a consequence of 1.18.5: since π_Y is a quasi-isomorphism, it induces an isomorphism

$$\mathrm{Hom}_{D^-(\mathrm{Mod}(A))}(P_X, P_Y) \cong \mathrm{Hom}_{K^-(\mathrm{Mod}(A))}(P_X, Y).$$

The functor $\mathcal{E}$ in statement (iii) is constructed by setting $\mathcal{E}(X) = \mathcal{D}'(P_X)$. □

Thus taking projective resolutions of bounded complexes of modules is a functor which sends quasi-isomorphisms to homotopy equivalences. A quasi-isomorphism need not be invertible in the homotopy category, so another way of looking at this construction is that this is a localisation that is universal subject to transforming homotopy equivalence classes of quasi-isomorphisms into invertible morphisms. The analogous results for injective resolutions show that the injective bounded derived category has the same universal property, hence is canonically equivalent to $D^b(\mathcal{A})$. The characterisation of $D^b(\mathcal{A})$ and $D^-(\mathcal{A})$ as a localisation has under suitable hypotheses a solution even for unbounded homotopy categories $K(\mathcal{A})$ over more general abelian categories $\mathcal{A}$. Restricting our attention to bounded categories gains us the above easy description of derived categories, but we forfeit in the process some more powerful tools involving unbounded derived categories. See Keller [45] for an approach using unbounded derived categories. The analogous versions of derived categories relative to k are obtained by identifying to zero not all acyclic complexes but only those that are k-split acyclic.

Remark 2.19.3 A projective resolution P_X of a bounded complex X is bounded below but need not be bounded above. Since the bounded complex X, and hence also P_X, has nonzero homology only in finitely many degrees, there is an integer n such that $H_i(P_X)$ vanishes for all $i \geq n$. Write P_X in the form

$$\cdots \longrightarrow P_{n+1} \xrightarrow{\delta_{n+1}} P_n \xrightarrow{\delta_n} P_{n-1} \longrightarrow \cdots \longrightarrow P_t \longrightarrow 0 \longrightarrow \cdots.$$

Since $H_n(P_X)$ is zero, we have $\mathrm{Im}(\delta_{n+1}) = \ker(\delta_n)$. Denoting by $P_X^{(n)}$ the 'truncated' complex

$$\cdots \longrightarrow 0 \longrightarrow \ker(\delta_n) \longrightarrow P_n \xrightarrow{\delta_n} P_{n-1} \longrightarrow \cdots \longrightarrow P_t \longrightarrow 0 \longrightarrow \cdots$$

we obtain a quasi-isomorphism $P_X \to P_X^{(n)}$ given by the obvious commutative diagram

$$\begin{array}{ccccccccccccccc}
\cdots & \longrightarrow & P_{n+2} & \xrightarrow{\delta_{n+2}} & P_{n+1} & \xrightarrow{\delta_{n+1}} & P_n & \xrightarrow{\delta_n} & P_{n-1} & \longrightarrow & \cdots & \longrightarrow & P_t & \longrightarrow 0 & \longrightarrow \cdots \\
 & & \downarrow & & \downarrow{\scriptstyle \delta_{n+1}} & & \downarrow & & \downarrow & & & & \downarrow & \downarrow & \\
\cdots & \longrightarrow & 0 & \longrightarrow & \ker(\delta_n) & \longrightarrow & P_n & \xrightarrow{\delta_n} & P_{n-1} & \longrightarrow & \cdots & \longrightarrow & P_t & \longrightarrow 0 & \longrightarrow \cdots .
\end{array}$$

In this way we have associated with a bounded complex X of A-modules a quasi-isomorphic complex with at most one nonprojective term, hence a single A-module, namely $\ker(\delta_n)$. This construction involves a choice of the integer n. The exactness of the the complex P_X in degree $i \geq n$ implies in particular that we have an exact sequence

$$0 \longrightarrow \ker(\delta_{n+1}) \longrightarrow P_{n+1} \xrightarrow{\delta_{n+1}} \ker(\delta_n) \longrightarrow 0$$

Thus $\ker(\delta_{n+1}) \cong \Omega(\ker(\delta_n)$ in the stable category. If Ω is an equivalence, with inverse Σ, then $\Sigma^n(\ker(\delta_n)) \cong \Sigma^{n+1}(\ker(\delta_{n+1}))$ depends no longer on n as an object in the stable category – and this leads to a functor from $D^b(\mathrm{Mod}(A))$ to $\underline{\mathrm{Mod}}(A)$. The problem here is that Ω is not an equivalence unless $\underline{\mathrm{Mod}}(A)$ is the relatively k-stable category and A relatively k-injective. Therefore, in order to construct a functor $D^b(A) \to \underline{\mathrm{Mod}}(A)$ it is necessary to replace $D^b(A)$ by the relatively k-projective version. See Buchweitz [16] for a more general treatment of this theme, as well as Grime [35] for a presentation closer to the needs of the present book.

2.20 Derived functors and cohomology

One of the fundamental construction principles in homological algebra is that replacing an object by a projective or injective resolution leads to new invariants of functors. Let k be a commutative ring.

Definition 2.20.1 Let A, B be k-algebras A, B and let $\mathcal{F} : \text{Mod}(A) \to \text{Mod}(B)$ be a functor. Let n be an integer. The *n-th left derived functor* $L_n(\mathcal{F}) : \text{Mod}(A) \to \text{Mod}(B)$ of $\mathcal{F}$ is the composition of the functors sending an A-module U to the n-th homology $H_n(\mathcal{F}(P_U))$ of the chain complex obtained from applying $\mathcal{F}$ to a projective resolution P_U of U. The *n-th right derived functor* $R^n(\mathcal{F}) : \text{Mod}(A) \to \text{Mod}(B)$ is the composition of functors sending an A-module U to the n-th cohomology $H^n(\mathcal{F}(I_U))$ of the cochain complex obtained from applying $\mathcal{F}$ to a injective resolution I_U of U.

The above constructions are well-defined functors by Proposition 1.18.3 and Theorem 2.18.4. We normally consider $L_n(\mathcal{F})$ only if $\mathcal{F}$ is right exact, because then $L_0(\mathcal{F}) \cong \mathcal{F}$. Similarly, we consider $R^n(\mathcal{F})$ only when $\mathcal{F}$ is left exact, because then $R^0(\mathcal{F}) \cong \mathcal{F}$. Note that $L_n(\mathcal{F})$ and $R^n(\mathcal{F})$ are zero for n negative, since a projective resolution of V and an injective resolution of U are zero in negative degree. We use analogous notation if $\mathcal{G} : \text{Mod}(A) \to \text{Mod}(B)$ is contravariant; in that case, $\mathcal{G}$ sends a projective resolution P_U of U to a cochain complex $\mathcal{G}(P_U)$. This yields thus a functor $L^n(\mathcal{G}) : \text{Mod}(A) \to \text{Mod}(B)$ sending U to $H^n(\mathcal{G}(P_U))$, and similarly we get a functor $R_n(\mathcal{G})$ sending U to $H_n(\mathcal{G}(I_U))$. These construction principles of left and right derived functors extend verbatim to functors between abelian categories having enough projective and injective objects, respectively. If $\mathcal{F} : \mathcal{A} \to \mathcal{B}$ is a contravariant functor, then $\mathcal{F}$ can be viewed as a covariant functor $\mathcal{F} : \mathcal{A}^{\text{op}} \to \mathcal{B}$. Under the passage from $\mathcal{A}$ to $\mathcal{A}^{\text{op}}$ a projective object becomes injective and vice versa; similarly, a projective resolution becomes an injective resolution and vice versa.

The left and right derived functors of tensor products and homomorphism functors are denoted by Tor_n^A and Ext_A^n, respectively.

Definition 2.20.2 Let A be a k-algebra, U a right A-module, and n an integer. We define a functor $\text{Tor}_n^A(U, -) : \text{Mod}(A) \to \text{Mod}(k)$ by $\text{Tor}_n^A(U, -) = L_n(U \otimes_A -)$.

This functor sends a left A-module V to the k-module $H_n(U \otimes_A P_V)$, where P_V is a projective resolution of V. Since $U \otimes_A -$ is right exact, it follows that $\text{Tor}_0^A(U, -) = U \otimes_A -$. For $n < 0$, the functor $\text{Tor}_n^A(U, -)$ is zero.

Definition 2.20.3 Let A be a k-algebra, U an A-module and n an integer. We define a functor $\text{Ext}_A^n(U, -) : \text{Mod}(A) \to \text{Mod}(k)$ by $\text{Ext}_A^n(U, -) = R^n(\text{Hom}_A(U, -))$.

This functor sends an A-module V to the k-module $H^n(\text{Hom}_A(U, I_V))$, where I_V is an injective resolution of V. Since $\text{Hom}_A(U, -)$ is left exact, it follows that $\text{Ext}_A^0(U, V) = \text{Hom}_A(U, V)$. For $n < 0$, the functor $\text{Ext}_A^n(U, -)$ is zero. One of the fundamental observations about Ext and Tor is that they are *balanced*: we have defined $\mathrm{Tor}_n^A(U, V)$ as the left derived functor $L_n(U \otimes_A -)$ evaluated at V, but we could have defined this equally well as the left derived functor $L_n(- \otimes_A V)$ evaluated at U; both constructions are naturally isomorphic. Thus $\mathrm{Tor}_n^A(-, -)$ is a bifunctor that is covariant in both arguments. Similarly, we have defined $\mathrm{Ext}_A^n(U, V)$ as the right derived functor $R^n(\mathrm{Hom}_A(U, -))$ evaluated at V, but this is also naturally isomorphic to the right derived functor $R^n(\mathrm{Hom}_A(-, V))$ evaluated at U (see Proposition 2.20.4 below). Note that because $\mathrm{Hom}_A(-, V)$ is contravariant, the right derived functor $R^n(\mathrm{Hom}_A(-, V))$ evaluated at U is calculated as the cohomology of $H^n(\mathrm{Hom}(P_U, V))$ for a *projective* resolution P_U of U in $\mathrm{Mod}(A)$ (which is an injective resolution of U in $\mathrm{Mod}(A)^{\mathrm{op}}$). Combining earlier results, we have the following identifications for $\mathrm{Ext}_A^n(U, V)$.

Proposition 2.20.4 *Let A be a k-algebra, let U, V be A-modules, and let n be an integer. Let P_U, P_V be projective resolutions of U, V, respectively, and let I_V be an injective resolution of V. We have canonical isomorphisms*

$$\begin{aligned}
\mathrm{Ext}_A^n(U, V) = H^n(\mathrm{Hom}_A(U, I_V)) &\cong \mathrm{Hom}_{K(\mathrm{Mod}(A))}(U, I_V[n]) \\
&\cong \mathrm{Hom}_{K(\mathrm{Mod}(A))}(P_U, I_V[n]) \cong \mathrm{Hom}_{K(\mathrm{Mod}(A))}(P_U, V[n]) \\
&\cong H^n(\mathrm{Hom}_A(P_U, V)) \cong \mathrm{Hom}_{K(\mathrm{Mod}(A))}(P_U, P_V[n]).
\end{aligned}$$

Proof There is some minor abuse of notation in that we regard I_V as a cochain complex in the second and third term, but as a chain complex in the fourth term. The first equality is the definition of Ext. The isomorphisms follow from applying 1.18.11 to the quasi-isomorphisms $P_U \to U$ and $P_V \to V \to I_V$, combined with 1.18.4. □

As before, $\mathrm{Ext}_A^n(-, -)$ is a bifunctor which is contravariant in the first argument and covariant in the second argument. More often than not, we will use the description of $\mathrm{Ext}_A^n(U, V)$ as $H^n(\mathrm{Hom}_A(P_U, V))$. The last description of this bifunctor in Proposition 2.20.4 in terms of chain homotopy classes leads to further structural properties of Ext. This description implies that for three A-modules U, V, W, the composition of chain maps between projective resolutions of these modules induces bilinear maps

$$\mathrm{Ext}_A^n(U, V) \times \mathrm{Ext}_A^m(V, W) \to \mathrm{Ext}_A^{m+n}(U, W).$$

For $U = V = W$ this yields a graded and associative multiplication on $\mathrm{Ext}_A^*(U, U)$.

Proposition 2.20.5 *Let A be a k-algebra and let U, V be A-modules. If U is projective or if V is injective, then $\mathrm{Ext}_A^n(U, V) = \{0\}$ for any positive integer n.*

Proof If U is projective, then it is its own projective resolution, concentrated in degree 0. If $n > 0$, then there is no nonzero chain map $U \to V[n]$. The conclusion follows from 2.20.4. Similarly, if V injective, then it is its own injective resolution, concentrated in degree 0, whence the result. □

Proposition 2.20.6 *Let A, B be k-algebras, let U be an A-module, V a B-module, and M an A-B-bimodule. Suppose that M is finitely generated projective as a left A-module and as a right B-module. For any $n \geq 0$ we have a natural isomorphism*

$$\mathrm{Ext}_A^n(M \otimes_B V, U) \cong \mathrm{Ext}_B^n(V, \mathrm{Hom}_A(M, U)).$$

Proof Let P be a projective resolution of the B-module V. By the assumptions on M, it follows that $M \otimes_B P$ is a projective resolution of $M \otimes_B V$. The tensor-Hom adjunction for complexes in 2.2.7 implies that we have a natural isomorphism of cochain complexes $\mathrm{Hom}_A(M \otimes_B P, U) \cong \mathrm{Hom}_B(P, \mathrm{Hom}_A(M, U))$. Taking cohomology in degree n yields the result. □

The special case of the restriction to a subgroup of a finite group is known as *Eckmann–Shapiro Lemma*. We spell this out for future reference.

Proposition 2.20.7 *Let G be a finite group and H a subgroup of G. Let U be a kG-module and V a kH-module. For any $n \geq 0$ we have natural isomorphisms*

$$\mathrm{Ext}_{kG}^n(\mathrm{Ind}_H^G(V), U) \cong \mathrm{Ext}_{kH}^n(V, \mathrm{Res}_H^G(U)),$$
$$\mathrm{Ext}_{kG}^n(U, \mathrm{Ind}_H^G(V)) \cong \mathrm{Ext}_{kH}^n(\mathrm{Res}_H^G(U), V).$$

Proof The first isomorphism follows from the first isomorphism in 2.2.1 applied to a projective resolution of V instead of V, and then taking cohomology in degree n, as in the proof of 2.20.6. The second isomorphism follows from the second isomorphism in 2.2.1 applied to a projective resolution of U instead of U, and taking cohomology. □

For more detailed discussions on Ext and Tor, see any standard reference on homological algebra, such as [93], notably [93, §2.7]. We mention two special cases.

Definition 2.20.8 Let A be a k-algebra that is projective as a k-module. The *Hochschild cohomology of A with coefficients in an A-A-bimodule M in degree*

$n \geq 0$ is the k-module

$$HH^n(A;M) = \mathrm{Ext}^n_{A\otimes_k A^{\mathrm{op}}}(A;M).$$

We set $HH^n(A) = HH^n(A;A)$.

This definition makes sense without the hypothesis that A is projective as a k-module, but Hochschild's original definition uses a relatively k-projective resolution, and so the definition above coincides with Hochschild's only if A is projective as a k-module. The functor $HH^n(A;-)$ is the n-th right derived functor of the functor $\mathrm{Hom}_{A\otimes_k A^{\mathrm{op}}}(A;-)$, and the latter is naturally isomorphic to the A-fixed point functor, sending an A-A-bimodule M to M^A. We can describe $HH^*(A;M)$ more explicitly as follows. Applying the functor $\mathrm{Hom}_{A\otimes_k A^{\mathrm{op}}}(-,M)$ to the projective $A\otimes_k A^{\mathrm{op}}$-module resolution X of A constructed in Proposition 2.18.5 yields a cochain complex of k-modules of the form

$$0 \longrightarrow \mathrm{Hom}_{A^e}(A^{\otimes 2},M) \longrightarrow \mathrm{Hom}_{A^e}(A^{\otimes 3},M) \longrightarrow \mathrm{Hom}_{A^e}(A^{\otimes 4},M) \longrightarrow \cdots$$

with differential induced by that of the complex X. The cohomology of this cochain complex is $HH^*(A;M)$. Writing

$$A^{\otimes n+2} = A\otimes_k A^{\otimes n}\otimes_k A$$

for $n \geq 0$ one observes that an $A\otimes_k A^{\mathrm{op}}$-homomorphism $A^{\otimes n+2}\to M$ is determined by its restriction to the subspace $1\otimes A^{\otimes n}\otimes 1$, and any k-linear map $1\otimes A^{\otimes n}\otimes 1\to M$ extends uniquely to an $A\otimes_k A^{\mathrm{op}}$-homomorphism $A^{\otimes n+2}\to M$ (this is a special case of the tensor-Hom adjunction). Thus we have a canonical isomorphism

$$\mathrm{Hom}_{A\otimes_k A^{\mathrm{op}}}(A^{\otimes n+2},M) \cong \mathrm{Hom}_k(A^{\otimes n},M).$$

Chasing the differential of the cochain complex $\mathrm{Hom}_{A\otimes_k A^{\mathrm{op}}}(X,M)$ through these isomorphisms yields the following cochain complex of k-modules.

Proposition 2.20.9 *Let A be a k-algebra. Suppose that A is projective as a k-module. For $n \geq 0$, define the k-linear map $\delta^n : \mathrm{Hom}_k(A^{\otimes n},M)\to \mathrm{Hom}_k(A^{\otimes n+1},M)$ by setting*

$$\begin{aligned}
&(\delta^n f)(a_0\otimes a_1\otimes\cdots\otimes a_n)\\
&\quad = a_0 f(a_1\otimes\cdots\otimes a_n) + \sum_{i=1}^{n}(-1)^i f(a_0\otimes\cdots\otimes a_{i-1}a_i\otimes\cdots\otimes a_n)\\
&\qquad + (-1)^{n+1} f(a_0\otimes\cdots\otimes a_{n-1})a_n.
\end{aligned}$$

Then $HH^n(A; M) \cong \ker(\delta^n)/\mathrm{Im}(\delta^{n-1})$ *for any* $n \geq 0$, *with the convention* $\delta^{-1} = 0$.

Proof This is a straightforward verification. □

By results of Gerstenhaber, the algebra $HH^*(A) = \mathrm{Ext}^*_{A\otimes_k A^{\mathrm{op}}}(A, A)$ is also graded-commutative, and it has a Lie algebra structure of degree -1. We have $HH^0(A) \cong Z(A)$, and one can show, using Proposition 2.20.9, that $HH^1(A)$ is the quotient of the space of derivations on A by the subspace of inner derivations on A. See [93] for more details and proofs. If U is an A-module, then by Corollary 2.18.6, the functor $- \otimes_k U$ sends a projective resolution A as an $A \otimes_k A^{\mathrm{op}}$-module to a projective resolution of U as an A-module, and hence induces a homomorphism of graded algebras

$$HH^*(A) \to \mathrm{Ext}^*_A(U, U).$$

Note that $\mathrm{Ext}^*_A(U, U)$ need not be graded-commutative in general.

Proposition 2.20.10 *Let A and B be k-algebras. Suppose that the A-B-bimodule M and the B-A-bimodule N induce a Morita equivalence between A and B. The functor* $N \otimes_A - \otimes_A M$ *induces an isomorphism of graded k-algebras* $HH^*(A) \cong HH^*(B)$.

Proof Clearly A belongs to $\mathrm{perf}(A, A)$, and the functor $N \otimes_A - \otimes_A M$ induces, by 2.17.11, an equivalence $\underline{\mathrm{perf}}(A, A) \cong \underline{\mathrm{perf}}(B, B)$ sending A to B. This equivalence sends therefore a projective bimodule resolution P of A to a projective bimodule resolution Q of B. Therefore, this equivalence induces isomorphisms $HH^n(A) = \mathrm{Hom}_{K(\mathrm{Mod}(A\otimes_k A^{\mathrm{op}}))}(P, P[n]) \cong HH^n(B) = \mathrm{Hom}_{K(\mathrm{Mod}(B\otimes_k B^{\mathrm{op}}))}(Q, Q[n])$. The functoriality implies that this becomes a graded algebra isomorphism as stated. □

By a result of Happel, Hochschild cohomology is invariant under derived equivalences. We will show this in 2.21.9 for the special case of Rickard equivalences between symmetric algebras.

Definition 2.20.11 Let G be a group. The *cohomology of G with coefficients in a kG-module U in degree* $n \geq 0$ is the k-module defined by

$$H^n(G; U) = \mathrm{Ext}^n_{kG}(k, U).$$

The Hochschild resolution of kG tensored by $- \otimes_{kG} k$ yields a projective resolution of k as a kG-module. Using this particular resolution, one verifies that Definition 2.20.11 coincides with the definitions for $n = 1$ and $n = 2$ in Section 1.2; see Remark 1.2.11. By construction, the functor $H^n(G; -)$ is the n-th right derived functor of the functor $\mathrm{Hom}_{kG}(k, -)$. The latter is naturally isomorphic to the fixed point functor, sending a kG-module U to U^G. For $U = k$, group cohomology yields a graded algebra $H^*(G; k) = \mathrm{Ext}^*_{kG}(k, k)$, which is *graded-commutative*; that is, for $\zeta \in H^n(G; k)$ and $\eta \in H^m(G; k)$ we have a product on $H^*(G; k) = \mathrm{Ext}^*_{kG}(k, k)$ induced by composition of chain maps between a projective resolution of k and its shifts, which is associative, such that $\zeta\eta \in H^{n+m}(G; k)$ and $\zeta\eta = (-1)^{mn}\eta\zeta$, for any $m, n \geq 0$. Since the functor $- \otimes_{kG} k$ sends a projective bimodule resolution of kG to a projective resolution of k, it follows that this functor induces a homomorphism of graded k-algebras $HH^*(kG) \to H^*(G; k)$. This algebra homomorphism is split surjective, with section constructed as follows. The $k(G \times G)$-bimodule isomorphism $kG \cong \mathrm{Ind}_{\Delta G}^{G\times G}(k)$ from Corollary 2.4.5 implies that the functor $\mathrm{Ind}_{\Delta G}^{G\times G}$, precomposed with the obvious functor given by the canonical isomorphism $G \cong \Delta G$, induces a graded algebra homomorphism $H^*(G; k) \to HH^*(kG)$, and one checks that this is a section as claimed.

Proposition 2.20.12 *Let G be a finite group and b an idempotent in $Z(kG)$. Set $B = kGb$. We have a canonical graded isomorphism $HH^*(B) \cong H^*(G; B)$, where on the right side B is regarded as a kG-module with G acting by conjugation on B.*

Proof We note first that $HH^*(B) \cong HH^*(kG; B)$, Indeed, setting $B' = kG(1 - b)$, regarded as a kG-kG-bimodule, we have $HH^*(kG; B) = \mathrm{Ext}^*_{k(G\times G)}(B' \oplus B; B)$. Since b and $1 - b$ are orthogonal central idempotents, it follows that a minimal projective resolution of B' as a kG-kG-bimodule is annihilated by b, and hence $\mathrm{Ext}^*_{k(G\times G)}(B'; B) = \{0\}$. The isomorphism $kG \cong \mathrm{Ind}_{\Delta G}^{G\times G}(k)$ from 2.4.5 together with 2.20.7 applied with $G \times G$ and ΔG instead of G and H, respectively, yield a graded isomorphism $HH^*(kG; B) \cong H^*(\Delta G; B)$. Clearly $(x, x) \in \Delta G$ acts on B as conjugation by x, whence the result. □

2.21 Derived equivalences and Rickard complexes

Rickard proved in [77] an analogue of Morita's Theorem for derived equivalences.

Definition 2.21.1 For A a k-algebra, a *tilting complex of A-modules* is a bounded complex T of finitely generated projective A-modules satisfying $\mathrm{Hom}_{K^b(\mathrm{Mod}(A))}(T, T[n]) = \{0\}$ for any nonzero integer n, such that the direct summands of T as a complex generate the homotopy category $K^b(\mathrm{proj}(A))$ of bounded complexes of finitely generated projective A-modules as a triangulated category.

If U is a B-module, viewed as a complex concentrated in a single degree, then the condition $\mathrm{Hom}_{K^b(\mathrm{Mod}(B))}(U, U[n]) = \{0\}$ for $n \neq 0$ holds trivially. Thus the image T of a B-module U under an equivalence $K^b(\mathrm{Mod}(B)) \cong K^b(\mathrm{Mod}(A))$ satisfies $\mathrm{Hom}_{K^b(\mathrm{Mod}(A))}(T, T[n]) = \{0\}$ for $n \neq 0$.

Theorem 2.21.2 ([77]) *Let A, B be k-algebras. The following are equivalent.*

(i) There is an equivalence of triangulated categories $D^b(\mathrm{Mod}(A)) \cong D^b(\mathrm{Mod}(B))$.
(ii) There is an equivalence of triangulated categories $K^b(\mathrm{Mod}(A)) \cong K^b(\mathrm{Mod}(B))$.
(iii) There is a tilting complex of A-modules T such that $\mathrm{End}_{K^b(\mathrm{Mod}(A))}(T) \cong B^{\mathrm{op}}$.
(iv) If A is noetherian, any of the above statements is also equivalent to the existence of a derived equivalence $D^b(\mathrm{mod}(A)) \cong D^b(\mathrm{mod}(B))$.

See [48] for proofs and for further references. The tilting complex T in (iii) corresponds to the progenerator in Morita's Theorem 2.8.2 (iii). If (ii) holds, then taking projective resolutions implies (i), and then the preimage of the regular B-module B under an equivalence as in (ii) yields a tilting complex T as in (iii). The most difficult implication is (iii) $\Rightarrow$ (i). This is because a complex T as in (iii) does not automatically become a complex of A-B-bimodules, as the right 'action' of B on T through the isomorphism $\mathrm{End}_{K^b(\mathrm{Mod}(A))}(T) \cong B^{\mathrm{op}}$ is defined only 'up to homotopy'. For symmetric algebras, Rickard showed in [78] that the bimodule version of Morita's Theorem for symmetric algebras holds verbatim, with bimodules replaced by bounded complexes. We denote as usual by X^* the k-dual of a complex X.

Theorem 2.21.3 ([78]) *Let A and B be symmetric k-algebras. The following are equivalent.*

(i) There is an equivalence of triangulated categories $D^b(\mathrm{Mod}(A)) \cong D^b(\mathrm{Mod}(B))$.
(ii) There is a bounded complex of A-B-bimodules X that are finitely generated projective as left A-modules and as right B-modules, such that we have homotopy equivalences of complexes of bimodules $X \otimes_B X^* \simeq A$ *and* $X^* \otimes_A X \simeq B$. *Moreover, one can choose X such that all but possibly one component are finitely projective as A-B-bimodules.*

As for the analogous statements of Morita's Theorem, the implication (ii) $\Rightarrow$ (i) in Theorem 2.21.3 is immediate, and this is the only implication we will need in this book. Indeed, if X is as in (ii) above, then the functors $X \otimes_B -$ and $X^* \otimes_A -$ induce inverse equivalences $D^b(\mathrm{Mod}(A)) \cong D^b(\mathrm{Mod}(B))$ and $K^b(\mathrm{Mod}(A)) \cong K^b(\mathrm{Mod}(B))$. In the main application of this result in the context of blocks with cyclic and Klein four groups, we will directly construct complexes of bimodules as in 2.21.3 (ii). In particular, we will not need the converse, which is the difficult part of the proof, and for which we refer to Rickard's paper. The statement that X can be chosen such that all but possibly one of the bimodules in X are projective follows from Proposition 2.21.5. This is based on the technique, described in 2.19.3, replacing a bounded complex by a quasi-isomorphic bounded complex with at most one nonprojective term.

Definition 2.21.4 Let A and B be symmetric k-algebras. Let X be a bounded complex of A-B-bimodules that are finitely generated projective as left A-modules and as right B-modules. Suppose that $X \otimes_B X^* \simeq A$ as complexes of A-A-bimodules and that $X^* \otimes_A X \simeq B$ as complexes of B-B-bimodules. Then X is called a *two-sided tilting complex* or a *Rickard complex* of A-B-bimodules.

A relatively k-projective resolution of a Rickard complex X for symmetric algebras A and B is also a projective resolution. Indeed, the terms of a Rickard complex are projective as left and right modules, hence as k-modules, where we use the fact that A and B are finitely generated projective as k-modules. Thus the terms of relatively k-projective resolution of X are projective as k-modules and relatively k-projective as $A \otimes_k B^{\mathrm{op}}$-modules, hence projective as $A \otimes_k B^{\mathrm{op}}$-modules.

Proposition 2.21.5 *Let A, B be symmetric k-algebras, X a Rickard complex of A-B-bimodules and $\pi : P \to X$ a bounded below projective resolution of X. Let Y be a bounded complex of A-B-bimodules such that Y_i is projective as an $A \otimes_k B^{\mathrm{op}}$-module for all nonzero integers i and such that Y_0 is finitely generated projective as a left and right module. Suppose that there is a quasi-isomorphism $\tau : P \to Y$. Then Y is a Rickard complex.*

Proof The chain map $P \to X$ is degreewise split as a chain map of complexes of k-modules, because it can be chosen degreewise surjective (by adding a contractible summand, if necessary) and the terms of X are projective as k-modules. Hence $\ker(P \to X)$ is a contractible complex of projective k-modules. Therefore the k-dual of this map is also a quasi-isomorphism. Similarly, the k-dual of τ yields a quasi-isomorphism $Y^* \to P^*$. Let $Q \to X^*$ be a projective resolution of X^*. Composed with $X^* \to P^*$, this yields a quasi-isomorphism $Q \to P^*$. By 1.18.11, this lifts up to homotopy through the quasi-isomorphism

$Y^* \to P^*$, and hence there is a quasi-isomorphism $Q \to Y^*$. Using repeatedly 1.17.11 yields quasi-isomorphisms $Y^* \otimes_A Y \leftarrow P \otimes_A Y^* \leftarrow P \otimes_A Q \to P \otimes_A X^* \to X \otimes_A X^* \simeq A$. Thus $Y \otimes_B Y^*$ has homology concentrated in degree zero, where it is isomorphic to A. All terms and the homology of $Y \otimes_B Y^*$ are projective as k-modules, and hence, by 1.18.17, this complex is k-split. By the assumptions on Y, all terms in nonzero degrees of $Y \otimes_B Y^*$ are projective as $A \otimes_k A^{\text{op}}$-modules. It follows from 2.18.7 that $Y \otimes_B Y^* \simeq A$. The same argument shows that $Y^* \otimes_A Y \simeq B$, hence that Y is indeed a Rickard complex. □

A derived equivalence given by a Rickard complex with at most one nonprojective term yields a stable equivalence of Morita type.

Proposition 2.21.6 ([14], [78]) *Let A and B be symmetric k-algebras without nonzero projective summands as modules over $A \otimes_k A^{\text{op}}$ and $B \otimes_k B^{\text{op}}$, respectively. Let X be a Rickard complex of A-B-bimodules such that all terms of X except X_0 are projective. Then X_0 and its dual $(X_0)^*$ induce a stable equivalence of Morita type between A and B.*

Proof We have $(X^*)_i = (X_{-i})^*$, for any $i \in \mathbb{Z}$. Thus the terms of the complex $X \otimes_B X^*$ in nonzero degrees are direct sums of modules of the form $X_i \otimes_B (X_j)^*$ for some indices $i \neq j$, and hence they are projective as $A \otimes_k A^{\text{op}}$-modules. The degree zero term of $X \otimes_B X^*$ is equal to $\oplus_{i \in \mathbb{Z}} X_i \otimes_B (X_i)^*$. Since $X \otimes_B X^* \simeq A$, it follows that A is a direct summand of $\oplus_{i \in \mathbb{Z}} X_i \otimes_B (X_i)^*$, hence of $X_0 \otimes_B (X_0)^*$, since A has no nonzero projective summand as an $A \otimes_k A^{\text{op}}$-module. A complement W of A in $X \otimes_B X^*$ is contractible, and all terms in nonzero degrees of W are projective. Since a contractible complex is a direct sum of complexes of the form $U \xrightarrow{\cong} U$, it follows that all terms of W are projective. This shows in particular that all summands of $X_0 \otimes_B (X_0)^*$ other than A must be projective, and hence that $X_0 \otimes_B (X_0)^* \cong A \oplus P$ for some projective $A \otimes_k A^{\text{op}}$-module P. The same argument shows $(X_0)^* \otimes_A X_0 \cong B \oplus Q$ for some projective $B \otimes_k B^{\text{op}}$-module Q. The result follows. □

Proposition 2.21.6 holds more generally if X has a unique nonprojective term since X may always be shifted so that this term is in degree 0. It is far more difficult – and not always possible – to reconstruct a derived equivalence from a stable equivalence of Morita type. The first major obstacle is that it is not known in general whether a stable equivalence of Morita type between two finite-dimensional algebras over a field preserves the number of isomorphism classes of nonprojective simple modules, while this is known to hold for derived equivalences. To complicate matters, there are examples of stably equivalent block algebras that are not derived equivalent. If a Rickard complex has nonzero homology in a single degree, then it induces a Morita equivalence. To see this, we will require the following special case of what is known as Künneth's formula, itself a special case of a spectral sequence. See [93, 3.6.3, 5.6.4] for more details.

Proposition 2.21.7 *Let A, B, C be k-algebras, X a bounded below complex of A-B-bimodules and Y a bounded below complex of B-C-bimodules. Suppose that Y is split as a complex of left B-modules and that $H_i(Y)$ is finitely generated projective as a left B-module, for any integer i, or that X is split as a complex of right B-modules and that $H_i(X)$ is finitely generated projective as a right B-module, for any integer i. Then for any integer n, we have a natural isomorphism of A-C-bimodules*

$$H_n(X \otimes_B Y) \cong \oplus_{i+j=n} H_i(X) \otimes_B H_j(Y),$$

where in the direct sum (i, j) runs over all pairs of integers satisfying $i + j = n$.

Proof Let i, j, n be integers such that $i + j = n$. Denote by δ, ϵ, ζ the differentials of X, Y, $X \otimes_B Y$, respectively. Let $x \in \ker(\delta_i)$ and $y \in \ker(\epsilon_j)$. Then $x \otimes y \in \ker(\zeta)$. Moreover, if $x \in \text{Im}(\delta_{i+1})$ or $y \in \text{Im}(\epsilon_{j+1})$, then $x \otimes y \in \text{Im}(\zeta_{n+1})$. Thus the assignment sending a pair $(x + \text{Im}(\delta_{i+1}), y + \text{Im}(\epsilon_{j+1}))$ to $x \otimes y$ induces an A-C-bimodule homomorphism $H_i(X) \otimes_B H_j(Y) \to H_n(X \otimes_A Y)$. We will show that the direct sum of these maps, taken over all pairs (i, j) satisfying $i + j = n$ yields an isomorphism as stated. The map constructed in this way is an A-C-bimodule homomorphism which is natural in X and Y. In order to show that it is an isomorphism, it suffices to show that this is an isomorphism of k-modules. In other words, we may ignore the left A-module structure and the right C-module structure. Suppose that Y is split as a complex of left B-modules and that $H_j(Y)$ is a finitely generated projective B-module for all integers j. By 1.18.15 we have $Y \simeq H_*(Y)$, where $H_*(Y)$ is considered as a complex with zero differential. This homotopy equivalence yields a homotopy equivalence $X \otimes_B Y \simeq X \otimes_B H_*(Y)$. This is isomorphic, as a complex, to $\oplus_{j \in \mathbb{Z}} X \otimes_B H_j(Y)$, where $H_j(Y)$ is viewed as a complex concentrated in degree j. Since $H_j(Y)$ is a finitely generated projective left B-module, it follows that $H_n(X \otimes_B H_j(Y)) \cong H_{n-j}(X) \otimes_B H_j(Y)$. Taking the direct sum over all j implies the result in this case. The case where X is split as a complex of right B-modules and $H_i(X)$ finitely generated projective as a right B-module, for all i, is proved similarly. □

Proposition 2.21.8 *Let A, B be symmetric k-algebras, and let X be a Rickard complex of A-B-bimodules. Suppose that X is k-split and that there is a unique* *integer n such that $H_n(X) \neq \{0\}$. Then $H_n(X)$ and its k-dual $H_n(X)^*$ induce a Morita equivalence between A and B. In particular, $H_0(X)$ is finitely generated projective as a left A-module and as a right B-module.*

Proof The complex X is bounded, k-split, and its terms are projective as left A-modules. By 2.18.7, X is split as a complex of left A-modules, hence homotopy equivalent to its homology $H_n(X)$. This shows that $H_n(X)$ is finitely generated projective as a left A-module. The same argument shows that $H_n(X)$ is finitely generated projective as a right B-module. Since A and B are symmetric, it follows that $H_n(X)^*$ is finitely generated projective as a left B-module and as a right A-module. Künneth's formula 2.21.7 implies that $H_n(X) \otimes_B H_n(X)^* \cong H_0(X \otimes_B X^*) \cong A$. Similarly, $H_n(X)^* \otimes_A H_n(X) \cong B$, whence the result. □

Proposition 2.21.9 *Let A and B be symmetric algebras, and let X be a Rickard complex of A-B-bimodules. The functor $X^* \otimes_A - \otimes_A X$ from $K(A \otimes_k A^{op})$ to $K(B \otimes_k B^{op})$ induces an isomorphism of graded k-algebras $HH^*(A) \cong HH^*(B)$.*

Proof This follows as in the proof of 2.20.10, with N and M replaced by X^* and X, respectively. □

3
Character Theory

Let A be an algebra over a commutative ring k, and let V be an A-module such that V is free of finite rank over k. The *character of* V is the k-linear map $\chi_V : A \to k$ sending $a \in A$ to the trace of the endomorphism of V induced by left multiplication with a on V. The character theory of modules over a finite group algebra kG in the case where k is a field of characteristic zero, also called *ordinary representation theory*, was initiated by Frobenius and Schur during the last decade of the 19th century. The choice of topics in this chapter is guided by applications to block theory in later chapters, except perhaps Burnside's p^aq^b-Theorem, included here simply for its classic beauty. See [41] for a broader treatment.

3.1 Characters of modules and finite groups

Let k be a commutative ring.

Definition 3.1.1

(a) Let A be a k-algebra. A *central function on* A is a k-linear map $\tau : A \to k$ satisfying $\tau(ab) = \tau(ba)$ for all $a, b \in A$. We denote by $\mathrm{CF}(A)$ the set of all central functions on A.

(b) Let G be a group. A *class function on* G *with values in* k is a map $\tau : G \to k$ satisfying $\tau(x) = \tau(yxy^{-1})$ for all $x, y \in G$. We denote by $\mathrm{Cl}_k(G)$ the set of k-valued class functions on G.

The set $\mathrm{CF}(A)$ is a k-submodule of the k-module $A^* = \mathrm{Hom}_k(A, k)$ of all k-linear maps from A to k. It follows from 1.5.7 that the trace $\mathrm{tr} : M_n(k) \to k$ is a central function whose kernel is equal to the subspace of additive commutators. In particular, conjugate matrices have the same trace. Thus if V is a free

k-module of rank n, then the trace map $\mathrm{tr}_V : \mathrm{End}_k(V) \to k$, defined in 2.10.1, is a central function. If $A = kG$ for some group G, then any map $G \to k$ extends uniquely to a k-linear map $kG \to k$. This correspondence sends class functions on G to central functions on kG, and we identify $\mathrm{Cl}_k(G) = \mathrm{CF}(kG)$ through this correspondence whenever this is useful. If G is a finite group, then $\mathrm{Cl}_k(G)$ is free of k-rank equal to the number of conjugacy classes of G. The k-module $\mathrm{Cl}_k(G)$ is in fact a k-algebra. Explicitly, for ψ, ψ' in $\mathrm{Cl}_k(G)$ and $\lambda \in k$ we define $\lambda\psi$ by $(\lambda\psi)(x) = \lambda\psi(x)$, the sum $\psi + \psi'$ by $(\psi + \psi')(x) = \psi(x) + \psi'(x)$, and the product by $(\psi\psi')(x) = \psi(x)\psi'(x)$, for all $x \in G$.

Definition 3.1.2 Let A be a k-algebra and V an A-module. Suppose that V is free of finite rank over k. The *character of* V is the k-linear map $\chi_V : A \to k$ sending $a \in A$ to the trace $\mathrm{tr}_V(\rho(a))$ of the linear endomorphism $\rho(a)$ of V defined by $\rho(a)(v) = av$ for all $v \in V$. A map $\chi : A \to k$ is called a *character of A over k* if χ is the character of some A-module which is free of finite rank over k.

Equivalently, χ_V is the composition of the structural algebra homomorphism $\rho : A \to \mathrm{End}_k(V)$ and the trace map $\mathrm{tr}_V : \mathrm{End}_k(V) \to k$. Characters of right A-modules are defined analogously. It is easy to verify that isomorphic modules will yield the same characters. A character is a k-linear map but need not be multiplicative. As above, if $A = kG$ for some finite group G, then χ_V is determined by its values on G, and we may view χ_V as function from G to k. With the notation as in 3.1.2, we define the *determinant of the A-module V* to be the map $\det_V : A \to k$ sending $a \in A$ to the determinant, denoted $\det_V(a)$, of the endomorphism of V sending $v \in V$ to av. Since the determinant is multiplicative, we have $\det_V(ab) = \det_V(a)\det_V(b)$ for all $a, b \in A$.

Proposition 3.1.3 *Let A be a k-algebra and let V be an A-module that is free of finite rank over K. The following hold.*

(i) We have $\chi_V(1_A) = \mathrm{rank}_k(V) \cdot 1_k$.
(ii) We have $\chi_V(ab) = \chi_V(ba)$ for all $a, b \in A$.

Proof For (i) we observe that 1_A acts as identity on V, hence is represented by the identity matrix in $M_n(k)$, where $n = \mathrm{rank}_k(V)$, and the trace of this matrix is equal to $n \cdot 1_k$. Statement (ii) follows from the symmetry properties of traces mentioned at the beginning of this section. □

The first statement in 3.1.3 implies that if k is a field of positive characteristic p, then $\chi_V(1_A) = 0$ whenever the dimension of V is divisible by p, and so the dimension of V is not an invariant of its character. The second statement in 3.1.3 says that a character χ_V is a central function. If $A = kG$ for a finite group G, then via the identification $\mathrm{Cl}_k(G) = \mathrm{CF}(kG)$, every character of a kG-module that is free of finite rank over k can be regarded as a class function. Not every class function on G is a character, however, and we will later prove some deep results, due to Brauer, which characterise those class functions over a field of characteristic zero that are characters.

Definition 3.1.4 Suppose that k is a field. Let A be a finite-dimensional k-algebra. A central function $\chi : A \to k$ is called an *irreducible character of A* if χ is the character of a simple A-module. We denote by $\mathrm{Irr}(A)$ the set of irreducible characters of A. We denote by $\mathbb{Z}\mathrm{Irr}(A)$ the subgroup of $\mathrm{CF}(A)$ generated by $\mathrm{Irr}(A)$. The elements of $\mathbb{Z}\mathrm{Irr}(A)$ are called *generalised characters of A*. If $A = kG$ for some finite group G we write $\mathrm{Irr}_k(G)$ instead of $\mathrm{Irr}(kG)$.

Proposition 3.1.5 *Suppose that k is a field. Let A be a split finite-dimensional k-algebra and let $a \in A$. The following are equivalent.*

(i) *We have $a \in [A, A] + J(A)$.*
(ii) *We have $\chi(a) = 0$ for any character of A.*
(iii) *We have $\chi(a) = 0$ for any $\chi \in \mathrm{Irr}(A)$.*

Proof Suppose that $a \in J(A)$. Then a is nilpotent. Thus the endomorphism of a finite-dimensional A-module V given by $v \mapsto av$ can be represented by a strict upper diagonal matrix, hence has trace zero. If $a, b \in A$, then the endomorphism of V given by the action of the additive commutator $ab - ba$ is a commutator in the matrix algebra $\mathrm{End}_k(V)$, hence has trace zero by 1.5.7. This shows that (i) implies (ii). Clearly (ii) implies (iii). Suppose that (iii) holds. For any simple A-module V, denote by φ_a the endomorphism of V sending $v \in V$ to av. By the assumptions, this endomorphism has trace zero, hence belongs by 1.5.7 to $[\mathrm{End}_k(V), \mathrm{End}_k(V)]$. Since $A/J(A) \cong \prod_V \mathrm{End}_k(V)$, with V running over a set of representatives of the isomorphism classes of simple A-modules, it follows that the image of a in $A/J(A)$ is contained in $[A/J(A), A/J(A)]$, and hence that $a \in [A, A] + J(A)$. This completes the proof. □

Proposition 3.1.6 *Suppose that k is a field. Let A be a split finite-dimensional k-algebra.*

(i) *The map sending a simple A-module V to its character χ_V induces a bijection between the isomorphism classes of simple A-modules and* $\mathrm{Irr}(A)$*; in particular,* $|\mathrm{Irr}(A)|$ *is equal to the number of isomorphism classes of simple A-modules.*
(ii) *The set* $|\mathrm{Irr}(A)|$ *is linearly independent in* $\mathrm{CF}(A)$.
(iii) *Suppose in addition A is semisimple. The set* $\mathrm{Irr}(A)$ *is a k-basis of* $\mathrm{CF}(A)$. *In particular, we have* $|\mathrm{Irr}(A)| = \dim_k(Z(A))$.

Proof Since $J(A)$ annihilates all simple A-modules and since $J(A)$ is in the kernel of any character of A, we may assume that A is split semisimple. Then we have an algebra isomorphism $\alpha : A \cong \prod_V \mathrm{End}_k(V)$ mapping $a \in A$ to the tuple of endomorphisms $v \mapsto av$, where V runs over a set $\mathcal{R}$ of representatives of the isomorphism classes of simple A-modules. Since the trace map tr_V on $\mathrm{End}_k(V)$ is nonzero, there is $\varphi_V \in \mathrm{End}_k(V)$ such that $\mathrm{tr}(\varphi_V) = 1$. Denote by $a_V \in A$ the unique element in A such that $\alpha(a_V)$ corresponds to the element $(0, \ldots, 0, \varphi_V, 0, \ldots, 0)$ in the product $\prod_V \mathrm{End}_k(V)$. That is, a_V acts as φ_V on V and annihilates all simple A-modules not isomorphic to V. Thus $\chi_V(a_V) = 1$ and $\chi_W(a_V) = 0$ for any simple A-module W that is not isomorphic to V. Since $Z(\mathrm{End}_k(V)) \cong k$, the dimension of $Z(A)$ is equal to the number $|\mathcal{R}|$ of isomorphism classes of simple A-modules. The result follows. □

If k is a field, then the trivial kG-module k is simple. Its character belongs hence to $\mathrm{Irr}_k(G)$.

Proposition 3.1.7 *Let G be a finite group. The character of the trivial kG-module k is the constant function sending every $x \in G$ to 1_k.*

Proof The trace of the identity map on k is 1_k, whence the result. □

If A is a k-algebra and if V, W are two A-modules, then the direct sum $V \oplus W$ is an A-module, with $a \in A$ acting on $(v, w) \in V \oplus W$ componentwise by $a(v, w) = (av, aw)$. If both V, W are free of finite rank over k then so is $V \oplus W$, and the character of $V \oplus W$ is the sum of the characters of V and W:

Proposition 3.1.8 *Let A be a k-algebra and let V, W be A-modules that are free of finite rank as k-modules. Then $V \oplus W$ is free of finite rank as a k-module, and for any $a \in A$ we have $\chi_{V\oplus W}(a) = \chi_V(a) + \chi_W(a)$ and $\mathrm{det}_{V\oplus W}(a) = \mathrm{det}_V(a)\,\mathrm{det}_W(a)$.*

Proof Let $\{v_1, v_2, \ldots, v_n\}$ be a k-basis of V and let $\{w_1, w_2, \ldots, w_m\}$ be a k-basis of W. Then the set $\{(v_i, 0) | 1 \leq i \leq n\} \cup \{(0, w_j) | 1 \leq j \leq m\}$ is a k-basis of $V \oplus W$. Let $a \in A$ and let λ_i, $\mu_j \in K$ such that $av_i = \sum_{1\leq k\leq n} \lambda_{i,k} v_k$ and $aw_j = \sum_{1\leq k\leq n} \mu_{j,k} w_k$. Then the matrix that represents the action of a on $V \oplus W$ is the block diagonal matrix with two blocks, consisting of the matrices $(\lambda_{i,k})$ and $(\mu_{j,k})$. The trace (resp. determinant) of this block diagonal matrix is the sum of the traces (resp. product of the determinants) of the two matrices just described. □

If k is a field, then the character of a module depends only on its composition factors. More generally, the character of an A-module V depends only on the image of V in the Grothendieck group of finitely generated A-modules (this generalises the statement on characters in the previous proposition):

Proposition 3.1.9 *Let A be a k-algebra and let*

$$0 \longrightarrow U \longrightarrow V \longrightarrow W \longrightarrow 0$$

be a short exact sequence of A-modules. Suppose that U, W are free of finite rank as k-modules. Then V is free of finite rank as k-module, and for any $a \in A$ we have $\chi_V(a) = \chi_U(a) + \chi_W(a)$.

Proof Since W is free this short exact sequence splits as sequence of k-modules, and hence V is free of finite rank as k-module. Moreover, this implies that if $\{u_1, u_2, \ldots, u_n\}$ is a k-basis of U and $\{w_1, w_2, \ldots, w_m\}$ a subset of V whose image in W is a k-basis of W then the set $\{u_i | 1 \leq i \leq n\} \cup \{w_j | 1 \leq j \leq m\}$ is a k-basis of V. As before, the matrix representing the action of a on V with respect to this basis has diagonal elements which are the diagonal elements of the matrices representing the action of a on U, V in the above bases, whence the result. □

Corollary 3.1.10 *Suppose that k is a field and that A is a finite-dimensional k-algebra. Every character of A is a $\mathbb{Z}$-linear combination with nonnegative coefficients of* $\mathrm{Irr}(A)$. *In particular, every character of A is a generalised character of A.*

Proof Let U be a finite-dimensional A-module and V a maximal submodule. Then S is a simple A-module, hence $\chi_S \in \mathrm{Irr}(A)$. By 3.1.9 we have $\chi_U = \chi_V + \chi_S$. Arguing by induction over $\dim_k(U)$, we may assume that χ_V is a $\mathbb{Z}$-linear combination of $\mathrm{Irr}(A)$, whence the result. □

Let A, B be k-algebras, V an A-module, and W a B-module. Then $V \otimes_k W$ is an $A \otimes_k B$-module with $a \otimes b$ acting on $v \otimes w$ by $(a \otimes b) \cdot (v \otimes w) = av \otimes bw$. Applied to $A = B = kG$ for some finite group G we get that $V \otimes_k W$ is a $kG \otimes_K kG$-module, hence a $k(G \times G)$-module with (x, y) acting on $v \otimes w$ by $(x, y) \cdot (v \otimes w) = xv \otimes yw$. Restricting this back to kG via the diagonal algebra homomorphism $\Delta : kG \to k(G \times G)$ yields a kG-module structure on $V \otimes_k W$ with $x \in G$ acting on $v \otimes w$ diagonally by $x \cdot (v \otimes w) = xv \otimes xw$. In this way the tensor product of two kG-modules is again an kG-module. This reasoning extends in an obvious way to modules over Hopf algebras.

Proposition 3.1.11 *Let G be a finite group and let V, W be kG-modules that are free of finite rank as k-modules. Then $V \otimes_k W$ is free of finite rank as a k-module and for any $x \in G$ we have $\chi_{V \otimes_k W}(x) = \chi_V(x)\chi_W(x)$.*

Proof Let $\{v_i | 1 \leq i \leq n\}$ be a k-basis of V and let $\{w_j | 1 \leq j \leq m\}$ be a k-basis of W. Then the set $\{v_i \otimes w_j | 1 \leq i \leq n, 1 \leq j \leq m\}$ is a k-basis of $V \otimes_k W$. We have $x(v_i \otimes w_j) = xv_i \otimes xw_j = \sum_{1 \leq r \leq n} \sum_{1 \leq s \leq m} \lambda_{i,r}\mu_{j,s} v_r \otimes w_s$. The corresponding $nm \times nm$-matrix of the coefficients $\lambda_{i,r}\mu_{j,s}$ has trace $\chi_{V \otimes_k W}(x) = \sum_{1 \leq i \leq n} \sum_{1 \leq j \leq m} \lambda_{i,i}\mu_{j,j} = (\sum_{1 \leq i \leq n} \lambda_{i,i})(\sum_{1 \leq j \leq m} \mu_{j,j}) = \chi_V(x)\chi_W(x)$. □

This shows that the subgroup $\mathbb{Z}\mathrm{Irr}_k(G)$ of $\mathrm{Cl}_k(G)$ is in fact a subring of the k-algebra $\mathrm{Cl}_k(G)$. We calculate next the regular character of a finite group algebra. Recall from 2.10.4 that if A is a k-algebra that is free of finite rank as a k-module, then the character of the regular left A-module A is called the *regular character of* A.

Proposition 3.1.12 *Let G be a finite group and let $\rho : kG \to k$ be the regular character of kG. We have $\rho(1_G) = |G|1_k$ and for every $x \in G \setminus \{1_G\}$ we have $\rho(x) = 0_k$.*

Proof As a k-module, kG is free of rank $|G|$ and hence its character value at 1_G is equal to $|G|1_k$ by 3.1.3 (i). Let $x \in G \setminus \{1_G\}$. Then, for any $y \in G$, we have $xy \neq y$. Thus the matrix representing the action of x on kG with respect to the k-basis G is a permutation matrix having no nonzero entry in the diagonal, and so the character of kG vanishes at all nontrivial group elements. □

Proposition 3.1.13 *Suppose that k is a field. Let A be a finite-dimensional semisimple k-algebra. Write $A \cong \oplus_{i=1}^r (V_i)^{n_i}$ where $\{V_i | 1 \leq i \leq r\}$ is a set of representatives of the isomorphism classes of simple A-modules, and where the n_i are positive integers. Denote by ρ the regular character of A, and for $1 \leq i \leq r$ denote by χ_i the character of V_i.*

(i) We have $\rho = \sum_{i=1}^r n_i\chi_i$.
(ii) If A is split, then $\rho = \sum_{i=1}^r \chi_i(1)\chi_i$; in particular, we have $\dim_k(A) \cdot 1_k = \sum_{i=1}^r \chi_i(1)^2$.

Proof Statement (i) follows from 3.1.8. Wedderburn's Theorem 1.13.3 implies that if A is split, then $n_i = \dim_k(V_i)$. Since $\chi_i(1) = \dim_k(V_i) \cdot 1_k$ and $\rho(1) = \dim_k(A) \cdot 1_k$, statement (ii) follows. □

For finite group algebras, this reads as follows.

Proposition 3.1.14 *Let G be a finite group and ρ the regular character of kG. Suppose that k is a splitting field for kG and that the image of $|G|$ in k is invertible in k. Then $\rho = \sum_{\chi \in \mathrm{Irr}_k(G)} \chi(1)\chi$. In particular, we have $|G| \cdot 1_k = \sum_{\chi \in \mathrm{Irr}_K(G)} \chi(1)^2$.*

Proof This follows from 3.1.13. □

Corollary 3.1.15 *Let G be a finite group. Suppose that k is a splitting field for kG and that the image of $|G|$ in k is invertible in k. Let $s : kG \to k$ be the canonical symmetrising form of kG mapping 1_G to 1_k and $x \in G \setminus \{1_G\}$ to 0. Then*

$$s = \sum_{\chi \in \mathrm{Irr}_k(G)} \frac{\chi(1)}{|G|}\chi.$$

Proof Denote by ρ the regular character of kG. By 3.1.12 we have $s = \frac{1}{|G|}\rho$. The result follows from 3.1.14. □

The regular character of a finite group G is a special case of the character of a permutation kG-module. The next result shows that characters of permutation modules can be read off the action of G on a permutation basis.

Proposition 3.1.16 *Let G be a finite group and let M be a finite G-set. For any $x \in G$, the character χ of the permutation module kM is given by the formula*

$$\chi(x) = |\{m \in M | xm = m\}| \cdot 1_k;$$

that is, $\chi(x)$ is the image in k of the number of fixed points of x in M. In particular, the values of χ are the images in k of nonnegative integers.

Proof Label the elements $M = \{m_1, m_2, \ldots, m_t\}$, where $t = |M|$. Let $x \in G$. Then, for any i such that $1 \leq i \leq t$ there is a unique integer j such that $1 \leq j \leq t$ and such that $xm_i = m_j$. Thus the matrix representing the action of x on kM is a permutation matrix, having an entry 1 in (i, j) if $xm_i = m_j$ and an entry 0 everywhere else. Since only diagonal entries contribute to the trace, it follows that the value of the trace of this matrix is the number of fixed points of x on M. □

Corollary 3.1.17 *Let G be a finite group and H a subgroup of G. The character of the transitive permutation kG-module kG/H sends $x \in G$ to the number of cosets yH in G/H satisfying $y^{-1}xy \in H$.*

Proof For $x, y \in G$, we have $xyH = yH$ if and only if $y^{-1}xy \in H$. The result follows from 3.1.16. □

If k is a domain or field of characteristic zero, then the character of a permutation kG-module kM contains as information the number of fixed points in M of all elements in G. This information is enough to determine M as a G-set, up to isomorphism. If a finite p-group P acts on a finite set M, where p is a prime, then all nontrivial P-orbits in M have lengths divisible by M, and hence $|M| \equiv |M^P| \pmod{p}$. This has the following immediate consequence:

Corollary 3.1.18 *Let p be a prime, P a finite p-group and M a finite P-set. Suppose that k is an integral domain of characteristic zero. Denote by χ the character of the permutation kP-module kM. For any $y \in P$ we have $\chi(y) \equiv |M| \pmod p$.*

A transitive permutation kG-module is isomorphic to $kG/H \cong \mathrm{Ind}_H^G(k)$ for some subgroup H of G. The calculation of the character of $\mathrm{Ind}_H^G(k)$ in 3.1.17 is a special case of calculating characters of induced modules.

Definition 3.1.19 Let G be a finite group and H a subgroup of G. For a class function $\varphi : G \to k$ we denote by $\mathrm{Res}_H^G(\varphi) : H \to k$ the class function obtained from restricting φ to H. For a class function $\psi : H \to k$ we define a function $\mathrm{Ind}_H^G(\psi) : G \to k$ by setting

$$\mathrm{Ind}_H^G(\psi)(x) = \sum_{y \in [H\backslash G]} \psi^0(yxy^{-1}),$$

where $\psi^0(z) = \psi(z)$ if $z \in H$ and $\psi^0(z) = 0$ if $z \in G \setminus H$.

Note that ψ^0 is invariant under conjugation with elements in H and hence the value $\psi^0(yxy^{-1})$ depends only on the image of y in $[H\backslash G]$, which shows in particular that this definition does not depend on the choice of $[H\backslash G]$.

Theorem 3.1.20 *Let G be a finite group and H a subgroup of G. Let $\psi : H \to k$ be a class function.*

(i) The function $\mathrm{Ind}_H^G(\psi)$ is a class function.
(ii) If $|H|$ is invertible in k then $\mathrm{Ind}_H^G(\psi)(x) = \frac{1}{|H|}\sum_{y\in G} \psi^0(yxy^{-1})$ for any $x \in G$.
(iii) We have $\mathrm{Ind}_H^G(\psi)(1) = [G : H]\psi(1)$.

Proof Let $z \in G$. If y runs over a set of representatives in G of $H\backslash G$ then so does yz. Thus $\mathrm{Ind}_H^G(\psi)(zxz^{-1}) = \sum_{y\in[H\backslash G]} \psi^0(yzxz^{-1}y^{-1}) = \sum_{y\in[H\backslash G]} \psi^0(yxy^{-1}) = \mathrm{Ind}_H^G(\psi)(x)$. This shows (i), and (ii) follows from the fact that ψ^0 is invariant under conjugation with elements in H. Statement (iii) is obvious. □

Theorem 3.1.21 *Let G be a finite group and H a subgroup of G. Let M be a k-free kH-module of finite rank over k. Let $\psi : H \to k$ be the character of M. Then $\mathrm{Ind}_H^G(\psi)$ is the character of $\mathrm{Ind}_H^G(M)$.*

Proof Write $\mathrm{Ind}_H^G(M) = \oplus_{y\in[G/H]} y \otimes M$. Let $x \in G$. The action of x permutes the summands of this direct sum as follows. If $y \in [G/H]$ then $xy = y'h$ for a unique $y' \in [G/H]$ and $h \in H$. Thus $xy \otimes M = y'h \otimes M = y' \otimes hM = y' \otimes M$. If $y' \neq y$ then the summand $y \otimes M$ will not contribute to the character of

$\mathrm{Ind}_H^G(M)$. Thus we only have to consider those y for which $xy = yh$ for some $h \in H$ or equivalently, for which $y^{-1}xy = h \in H$. On such a summand, the contribution to the character of $\mathrm{Ind}_H^G(M)$ will be the character of M evaluated at $y^{-1}xy = h$. Thus $\mathrm{Ind}_H^G(M)$ has as character the function sending $x \in G$ to $\sum_{y \in [G/H]} \psi^0(y^{-1}xy)$. Note that if y runs over $[G/H]$ then y^{-1} runs over $[H\backslash G]$, so this formula coincides with the formula given in 3.1.20 (ii). □

The transitivity of induction extends from characters of modules to arbitrary class functions:

Proposition 3.1.22 *Let G be a finite group, and let H, L be subgroups of G such that $L \subseteq H \subseteq G$. For any class function $\psi : L \to k$ we have $\mathrm{Ind}_L^G(\psi) = \mathrm{Ind}_H^G(\mathrm{Ind}_L^H(\psi))$.*

Proof Straightforward verification. If k is splitting a field, this follows also from observing that this holds for characters of modules (by the transitivity of induction of modules) and then using the fact that $\mathrm{Irr}_k(G)$ is a k-basis of $\mathrm{Cl}_k(G)$. □

One can prove a general version of Mackey's formula for class functions; we will only need the following special case:

Proposition 3.1.23 *Let G be a finite group and let H, L be subgroups of G such that $G = HL$. For any class function $\psi : H \to k$ we have $\mathrm{Res}_K^G(\mathrm{Ind}_H^G(\psi)) = \mathrm{Ind}_{H \cap L}^L(\mathrm{Res}_{H \cap L}^H(\psi))$.*

Proof As in the previous result, this is a straightforward verification, which in that case where k is a splitting field, can be deduced from the Mackey formula for the induction and restriction of modules. □

Proposition 3.1.24 *Let H be a normal subgroup of a finite group G. For any class function $\psi : H \to k$ and any $x \in G \setminus H$ we have $\mathrm{Ind}_H^G(\psi)(x) = 0$.*

Proof If $x \in G \setminus H$, then H contains no G-conjugate of x as H is assumed to be normal in G. Thus the sum on the right side in the second statement of 3.1.20 is empty, whence the result. □

The tensor product reciprocity formula from 2.4.8 carries over to class functions as well. The easy proof is again left to the reader:

Proposition 3.1.25 *Let G be a finite group and H a subgroup of G. Let $\psi : H \to k$ and $\varphi : G \to k$ be class functions. We have $\chi \cdot \mathrm{Ind}_G^H(\psi) = \mathrm{Ind}_H^G(\mathrm{Res}_H^G(\chi) \cdot \psi)$.*

Remark 3.1.26 Let k be a subring of a commutative ring k', and let γ be a k-algebra automorphism of k'. Let A be a k-algebra, and set $A' = k' \otimes_k A$. As described briefly before 1.14.14, the automorphism γ induces a ring automorphism of A' sending $\lambda \otimes a$ to $\gamma(\lambda) \otimes a$, where $\lambda \in k'$ and $a \in A$. Restriction along this automorphism induces an equivalence on $\mathrm{Mod}(A')$ as an abelian category. This equivalence sends an A'-module M to the A'-module ${}_\gamma M$, which is equal to M as a k-module, with $\lambda \otimes a$ acting as $\gamma(\lambda) \otimes a$ on M. In particular, the scalar λ acts on ${}_\gamma M$ as the scalar multiplication by $\gamma(\lambda)$ on M; we denote this by $\lambda \cdot m = \gamma(\lambda)m$. If M is free of finite rank n as a k'-module, then we have

$$\chi_{{}_\gamma M}(1 \otimes a) = \gamma^{-1}(\chi_M(1 \otimes a)),$$

$$\det{}_{{}_\gamma M}(1 \otimes a) = \gamma^{-1}(\det{}_M(1 \otimes a)).$$

To see this, choose a k'-basis $\{m_i | 1 \leq i \leq n\}$ of M, and write $(1 \otimes a)m_i = \sum_{j=1}^{n} \alpha_{ij} m_j$, to be understood as the action of $1 \otimes a$ on M. The element $1 \otimes a$ is fixed under the automorphism induced by γ, and hence this is also the action on ${}_\gamma M$. In ${}_\gamma M$, the sum $\sum_{j=1}^{n} \alpha_{ij} m_j$ is equal to $\sum_{j=1}^{n} \gamma^{-1}(\alpha_{ij}) \cdot m_j$, whence the above formulae.

3.2 Characters and duality

Let A be a k-algebra and V an A-module. The k-dual $V^* = \mathrm{Hom}_k(V, k)$ is a right A-module via $(\mu \cdot a)(v) = \mu(av)$ for all $\mu \in \mathrm{Hom}_k(V, k)$, $a \in A$ and $v \in V$. If V is free of finite rank over k then V^* is again free over k, with rank equal to that of V.

Proposition 3.2.1 *Let A be a k-algebra and V an A-module that is free of finite rank n as a k-module. Then the character of the right A-module V^* is equal to the character of V.*

Proof Let $\{v_i | 1 \leq i \leq n\}$ be a k-basis of V. Denote by $v_i^* : V \to k$ the unique k-linear map satisfying $v_i^*(v_i) = 1$ and $v_i^*(v_j) = 0$ for $j \neq i$, where $1 \leq i \leq n$. That is, the set $\{v_i^* | 1 \leq i \leq n\}$ is the dual basis of $\{v_i | 1 \leq i \leq n\}$. Let $x \in A$. Let $\lambda_{i,j} \in k$ such that $xv_i = \sum_{1 \leq j \leq n} \lambda_{i,j} v_j$ for all i, j such that $1 \leq i, j \leq n$. Then $\chi_V(x) = \sum_{1 \leq i \leq n} \lambda_{i,i}$. For any i, j such that $1 \leq i, j \leq n$ we have $(v_i^* \cdot x)(v_j) = v_i^*(xv_j) = \sum_{1 \leq t \leq n} \lambda_{j,t} v_i^*(v_t) = \lambda_{j,i} = \sum_{1 \leq t \leq n} \lambda_{t,i} v_t^*(v_j)$, thus $v_i^* \cdot x = \sum_{1 \leq t \leq n} \lambda_{t,i} v_t^*$, and so the character of V^* as right A-module evaluated at x is equal to $\sum_{1 \leq i \leq n} \lambda_{i,i}$, which is the character of V as left A-module evaluated at x. □

If $A = kG$ for some finite group G and if V is a finitely generated kG-module, then we can view V^* again as a left kG-module via the isomorphism $kG \cong (kG)^{\mathrm{op}}$ sending $x \in G$ to x^{-1}. Explicitly, the left kG-module structure of V^* is given by $(x \cdot \mu)(v) = \mu(x^{-1}v)$ for all $\mu \in V^*$, $x \in G$ and $v \in V$. The characters of V and its dual are related as follows:

Proposition 3.2.2 *Let G be a finite group and let V be a kG-module that is free of finite rank as a k-module. Denote be χ_{V^*} the character of V^* as left kG-module. For any $x \in G$ we have $\chi_{V^*}(x) = \chi_V(x^{-1})$.*

Proof By 3.2.1, the character of V^* as right kG-module is equal to the character χ_V of V. Since the left kG-module structure on V^* is obtained via the map sending $x \in G$ to x^{-1}, the result follows. □

This shows that taking k-duals of modules induces to an involution on the group $\mathbb{Z}\mathrm{Irr}_k(G)$. This involution extends to $\mathrm{Cl}_k(G)$ by defining for every $\psi \in \mathrm{Cl}_k(G)$ a class function ψ^* satisfying $\psi^*(x) = \psi(x^{-1})$ for all $x \in G$. If the base ring is a subfield K of the complex number field then the character of the dual of a KG-module is obtained by complex conjugation:

Proposition 3.2.3 *Let G be a finite group and let K be a subfield of $\mathbb{C}$. Let χ be a character of G over K. Then $\chi(x^{-1}) = \overline{\chi(x)}$, the complex conjugate of $\chi(x)$, for any $x \in G$.*

Proof Suppose χ is the character of the KG-module V. By replacing KG by $\mathbb{C}G$ and V by $\mathbb{C} \otimes_K V$ we may assume that $K = \mathbb{C}$. Let $x \in G$ and denote by $\rho(x)$ the K-linear endomorphism of V given by the action of x on V; that is, $\rho(x)(v) = xv$ for all $x \in G$ and $v \in V$. Choose a K-basis of V in such a way that $\rho(x)$ is represented by an upper triangular matrix

$$M = \begin{pmatrix} \lambda_1 & * & \cdots & * \\ 0 & \lambda_2 & \cdots & * \\ \cdots & \cdots & \cdots & \cdots \\ 0 & 0 & \cdots & \lambda_n \end{pmatrix}.$$

Then $\chi(x) = \sum_{1 \leq i \leq n} \lambda_i$. The group element x has finite order, say $x^m = 1$, where m is a positive integer, and hence M^m must be the identity matrix. This force $(\lambda_i)^m = 1$ for $1 \leq i \leq m$. In other words, the λ_i are m-th roots of unity. But for roots of unity, the complex conjugate and the inverse coincide in $\mathbb{C}$; that is, $(\lambda_i)^{-1} = \overline{\lambda_i}$. Now $\rho(x^{-1}) = \rho(x)^{-1}$ is represented by M^{-1}, which is an upper triangular matrix whose diagonal elements are the $(\lambda_i)^{-1}$, hence the trace of M^{-1} is the complex conjugate of the trace of M. The result follows. □

Proposition 3.2.4 *Let G, H be finite groups. Suppose that k is a field. Let V be a finite-dimensional kG-module, and let W be a finite-dimensional kH-module. Then $\mathrm{Hom}_k(V, W)$ becomes a $k(H \times G)$-module with $(y, x) \in H \times G$ acting on $\varphi \in \mathrm{Hom}_k(V, W)$ by $((y, x).\varphi)(v) = y\varphi(x^{-1}v)$ for all $v \in V$. The character ψ of of the $k(H \times G)$-module $\mathrm{Hom}_k(V, W)$ is given by the formula*

$$\psi(y, x) = \chi_W(y)\chi_V(x^{-1})$$

for all $(y, x) \in H \times G$.

Proof By 2.9.4 we have $\mathrm{Hom}_k(V, W) \cong W \otimes_k V^*$, and the character of this module is as claimed by 3.2.2 and 3.1.11. □

We have defined characters only for modules that are free of finite rank over the base ring because this hypothesis makes it possible to define trace maps. One can extend this to modules that are finitely generated projective over k. This is a consequence of the canonical isomorphism $\mathrm{End}_k(V) \cong V \otimes_k V^*$ from 2.9.4, which holds whenever V is finitely generated projective as a k-module, together with 2.10.2.

3.3 The orthogonality relations

Let G be a finite group. By the theorems of Maschke and Wedderburn, if k is a splitting field for G such that $|G|$ is invertible in k, then $kG \cong \prod_V \mathrm{End}_k(V)$, with V running over a set of representatives of the isomorphism classes of simple kG-modules. Through this isomorphism, for any V in this product, the identity Id_V corresponds to the unique primitive idempotent in $Z(kG)$ which acts as identity on every simple kG-module isomorphic to V. Moreover, this idempotent annihilates every simple kG-module not isomorphic to V. We determine the primitive idempotents in $Z(kG)$ explicitly.

Theorem 3.3.1 *Let G be a finite group. Suppose that k is a splitting field for kG and that $|G|$ is invertible in k. Let V be a simple kG-module, and denote by χ the character of V. Then the unique primitive idempotent in $Z(kG)$ which acts as identity on V depends only on the character χ of V and is equal to*

$$e(\chi) = \frac{\chi(1)}{|G|} \sum_{x \in G} \chi(x^{-1})x.$$

In particular, nonisomorphic simple kG-modules have different characters, and the set $\{e(\chi)\}$ with χ running over the characters of simple kG-modules is a primitive decomposition of 1 in $Z(kG)$. We have $\chi(e(\chi)) = \chi(1)$ and

$\chi(e(\chi')) = 0$, where χ, χ' are the characters of two nonisomorphic simple kG-modules.

Proof By Maschke's Theorem 1.11.12, the algebra kG is semisimple. Thus, as left kG-modules, we have $kG \cong \oplus_{i=1}^{r}(V_i)^{n_i}$, with $\{V_i\}_{1\leq i\leq r}$ a set of representatives of the isomorphism classes of simple kG-modules and positive integers n_i. Denote by e_V the primitive idempotent in kG acting as identity on V. Write $e_V = \sum_{y\in G} \lambda_y y$ for some coefficients $\lambda_y \in k$. Then for any $x \in G$ we have

$$e_V x^{-1} = \sum_{y\in G} \lambda_y y x^{-1}.$$

Using 3.1.14 it follows that $\rho(e_V x^{-1}) = \sum_{y\in G} \lambda_y \rho(yx^{-1}) = |G| \cdot \lambda_x$. If V_i is not isomorphic to V, then e_V annihilates V_i, and hence $\chi_i(e_V x^{-1}) = 0$. If $V_i \cong V$, then $\chi = \chi_i$, and e_V acts as identity on V_i, hence $\chi_i(e_V x^{-1}) = \chi_i(x^{-1})$. Using the equation $\rho = \sum_{i=1}^{r} \chi_i(1)\chi_i$ from 3.1.14 we get that

$$\rho(e_V) = \sum_{i=1}^{r} \chi_i(1)\chi_i(e_V) = \chi(1)\chi(x^{-1}).$$

Together we get that $|G|\lambda_x = \chi(1)\chi(x^{-1})$, hence $\lambda_x = \frac{\chi(1)}{|G|}\chi(x^{-1})$. Thus $e_V = e(\chi)$ as in the statement, and all parts of the theorem are proved. □

Using 1.14.6 yields the following reformulation.

Corollary 3.3.2 *Let G be a finite group. Suppose that k is a splitting field for kG and that $|G|$ is invertible in k. We have $kG = \prod_{\chi\in \mathrm{Irr}_k(G)} kGe(\chi)$, and for each $e(\chi)$, the algebra $kGe(\chi)$ is a matrix algebra isomorphic to* $\mathrm{End}_K(V)$, *where V is a simple kG-module with χ as character.*

In our terminology from §1.7 the set $\{e(\chi) | \chi \in \mathrm{Irr}_k(G)\}$ is a primitive decomposition of 1 in $Z(kG)$, hence determines the block decomposition of kG. The corresponding block algebras $kGe(\chi)$ are isomorphic to the matrix algebras $M_{\chi(1)}(k)$.

Corollary 3.3.3 *Let G be a finite group. Suppose that k is a splitting field for kG and that $|G|$ is invertible in k. Let V be a finite-dimensional kG-module, and let $\chi \in \mathrm{Irr}_k(G)$. Then $e(\chi)V$ is the sum of all simple kG-submodules of V with character χ. The character of the submodule $e(\chi)V$ is equal to $n(V, \chi)\chi$ for some integer $n(V, \chi) \geq 0$, and we have $V = \oplus_{\chi\in \mathrm{Irr}_K(G)} e(\chi)V$.*

Proof By 3.3.1, multiplication by e_χ on V acts as identity on all simple submodules of V with χ as character and annihilates all other simple submodules. The result follows. □

The submodules $e(\chi)V$ of V in the preceding corollary are the *isotypic components of V* considered in 1.9.8. In particular, any kG-module can be decomposed *uniquely* as a direct sum of its isotypic components. Given a family of matrices, one in each matrix algebra $kGe(\chi)$, one can get an explicit description of this family as an element of kG; this is also called the *Fourier transformation.*

Proposition 3.3.4 *Let G be a finite group. Suppose that k is a splitting field for kG and that $|G|$ is invertible in k. Let $\chi \in \mathrm{Irr}_k(G)$. For any $s \in kGe_\chi$ we have*

$$s = \frac{\chi(1)}{|G|} \sum_{x \in G} \chi(x^{-1}s)x.$$

Proof Since s can be written as a linear combination of elements in G it suffices to prove this equality for $s = ye(\chi)$ for some $y \in G$. Since $e(\chi)$ acts as identity on every simple kG-module with χ as character we have $\chi(x^{-1}ye(\chi)) = \chi(x^{-1}y)$. Thus $\sum_{x \in G} \chi(x^{-1}ye(\chi))x = \sum_{x \in G} \chi(x^{-1}y)xy^{-1}y = \frac{|G|}{\chi(1)} e(\chi)y$. Multiplying by $\frac{\chi(1)}{|G|}$ yields the result. □

If k is a splitting field for a finite group G such that $|G|$ is invertible in k, then there is a scalar product on the space of k-valued central functions $\mathrm{CF}(kG)$ on kG that can be explicitly described in terms of the values of class functions on the group elements.

Definition 3.3.5 Let G be a finite group and k a commutative ring such that $|G|$ is invertible in k. We define a scalar product

$$\langle -, - \rangle_G : \mathrm{Cl}_k(G) \times \mathrm{Cl}_k(G) \to k$$

by setting

$$\langle \alpha, \beta \rangle_G = \frac{1}{|G|} \sum_{x \in G} \alpha(x)\beta(x^{-1})$$

for any $\alpha, \beta \in \mathrm{Cl}_k(G)$. If no confusion arises, we write $\langle \alpha, \beta \rangle$ instead of $\langle \alpha, \beta \rangle_G$.

The first orthogonality relations show that the characters of the simple kG-modules form an orthonormal basis of $\mathrm{CF}(kG)$ with respect to this scalar product.

Theorem 3.3.6 (First Orthogonality Relations) *Let G be a finite group. Suppose that k is a splitting field for kG and that $|G|$ is invertible in k. The map sending a simple kG-module to its character induces a bijection between the set of isomorphism classes of simple kG-modules and the set $\mathrm{Irr}_k(G)$ of irreducible* *characters of G over k. Moreover, the set $\mathrm{Irr}_k(G)$ is an orthonormal k-basis of $\mathrm{Cl}_k(G)$ and an orthonormal $\mathbb{Z}$-basis of $\mathbb{Z}\mathrm{Irr}_k(G)$. Explicitly,*

(i) the set $\mathrm{Irr}_k(G)$ is a k-basis of $\mathrm{Cl}_k(G)$ and a $\mathbb{Z}$-basis of $\mathbb{Z}\mathrm{Irr}_k(G)$;
(ii) we have $\langle \chi, \chi \rangle = 1$ for any $\chi \in \mathrm{Irr}_k(G)$, and
(iii) we have $\langle \chi, \chi' \rangle = 0$ for any two different $\chi, \chi' \in \mathrm{Irr}_k(G)$.

Proof The first statement is a special case of 3.1.6. Let V, W be simple kG-modules with characters χ, ψ, respectively. By 3.3.1, the element $e_V = \frac{\chi(1)}{|G|} \sum_{x \in G} \chi(x^{-1})x$ is the primitive idempotent in $Z(kG)$ which acts as identity on V. Thus

$$\chi(1) = \chi(e_V) = \frac{\chi(1)}{|G|} \sum_{x \in G} \chi(x^{-1})\chi(x).$$

Dividing both sides by $\chi(1)$ yields $\langle \chi, \chi \rangle = 1$. Suppose that W is not isomorphic to V. Then e_V annihilates W, and hence

$$0 = \psi(e_V) = \frac{\chi(1)}{|G|} \sum_{x \in G} \chi(x^{-1})\psi(x),$$

whence $\langle \chi, \psi \rangle = 0$. Orthonormal sets are automatically linearly independent. It follows again from 3.1.6 that $\mathrm{Irr}_k(G)$ is a basis of $\mathrm{Cl}_k(G)$. □

The above proof of 3.3.6 is based on 3.3.1, and hence uses implicitly Wedderburn's Theorem. For a more elementary proof, using only some linear algebra but not Wedderburn's Theorem, see Serre [85, Ch. 2]. In conjunction with Schur's Lemma, we obtain the following:

Corollary 3.3.7 *Let G be a finite group. Suppose that k is a splitting field for kG and that $|G|$ is invertible in k. Let χ, χ' be the characters of two finite-dimensional kG-modules X, X', respectively. We have*

$$\langle \chi, \chi' \rangle_G = \dim_k(\mathrm{Hom}_{kG}(X, X')) \cdot 1_k.$$

This formula is the scalar product, introduced in 1.14.13 of the images of X, X' in the Grothendieck group $R(kG)$ of finite-dimensional kG-modules. In other words, we have:

Corollary 3.3.8 *Let G be a finite group. Suppose that k is a splitting field for kG and that $|G|$ is invertible in k. The map sending a finite-dimensional kG-module to its character induces an isomorphism of abelian groups*

$$R(kG) \cong \mathbb{Z}\mathrm{Irr}_k(G)$$

which preserves the scalar products on these groups.

This means that the Grothendieck group $R(kG)$, together with its scalar product, can be canonically identified with the group of generalised characters $\mathbb{Z}\mathrm{Irr}_k(G)$. In particular, $R(kG)$ is a ring, with product induced by the tensor product of kG-modules, and endowed with an involution induced by taking k-duals of modules. Another consequence is that the 'orthogonality part' of the first orthogonality relations remains true over fields of characteristic zero that are not splitting fields:

Corollary 3.3.9 *Let G be a finite group. Suppose that k is a field such that $|G|$ is invertible in k. Let $\chi, \psi \in \mathrm{Irr}_k(G)$. If $\chi \neq \psi$ then $\langle \chi, \psi \rangle = 0$.*

Proof Let k' be a splitting field for G containing k. Let V, W be simple kG-modules with characters χ, ψ, respectively. By 1.12.10 we have $\mathrm{Hom}_{k'G}(k' \otimes_k V, k' \otimes_k W) \cong k' \otimes_k \mathrm{Hom}_{kG}(V, W) = \{0\}$, since V, W are non isomorphic simple modules. Thus $\langle \chi, \psi \rangle = 0$ by 3.3.7. □

One of the very useful applications of the orthogonality relations and Frobenius' reciprocity is a criterion for when an irreducible character is induced from a proper subgroup:

Proposition 3.3.10 *Let G be a finite group. Suppose that k is a splitting field for all subgroups of G and that $|G|$ is invertible in k. Let N be a normal subgroup of G. Let $\chi \in \mathrm{Irr}_k(G)$ and $\psi \in \mathrm{Irr}_k(N)$ such that $\langle \mathrm{Res}^G_N(\chi), \psi \rangle_N \neq 0$. Denote by H the subgroup of all $x \in G$ satisfying $\psi(xyx^{-1}) = \psi(y)$ for all $y \in N$. There is an irreducible character $\tau \in \mathrm{Irr}_k(H)$ such that $\chi = \mathrm{Ind}^G_H(\tau)$ and $\mathrm{Res}^H_N(\tau) = a \cdot \psi$ for some positive integer a.*

Proof Let V be a simple kG-module with character χ, and let $W = e_\psi \mathrm{Res}^G_N(V)$ be the isotypic component of $\mathrm{Res}^G_N(V)$ consisting of the sum of simple kN-submodules with character ψ; in particular, the character of W is equal to $a \cdot \psi$ for some integer $a \geq 0$. Since $\langle \mathrm{Res}^G_N(\chi), \psi \rangle_N \neq 0$, it follows that the submodule W is nonzero, hence that a is positive. Note that the action of G on V permutes the isotypic components of $\mathrm{Res}^G_N(V)$, and it permutes them transitively because V is simple. Thus $\dim_k(V) = |G : H| \dim_k(W)$. Since the elements in H stabilise ψ they commute with e_ψ and hence W is in fact a kH-submodule of $\mathrm{Res}^G_H(V)$. Let τ be the character of W as kH-module. Since H extends the action of N on $e_\psi \mathrm{Res}^G_N(V)$ the character of W as a kH-module restricts to the character $a \cdot \psi$ of N on W. Then $\langle \tau, \mathrm{Res}^G_H(\chi) \rangle_H \neq 0$. It follows from 3.3.7 and Frobenius' reciprocity 2.2.1 (or from the reciprocity for class functions 3.3.15 below) that $\langle \mathrm{Ind}^G_H(\tau), \chi \rangle_G \neq 0$. Since V and $\mathrm{Ind}^G_H(W)$ have the same dimension this implies that they are isomorphic, whence the result. □

Theorem 3.3.11 (Second orthogonality relations) *Let G be a finite group. Suppose that k is a splitting field for G and that $|G|$ is invertible in k. For any two elements $x, y \in G$ we have*

$$\sum_{\chi \in \mathrm{Irr}_K(G)} \chi(x)\chi(y^{-1}) = |C_G(x)|$$

if x and y are conjugate in G, and we have

$$\sum_{\chi \in \mathrm{Irr}_K(G)} \chi(x)\chi(y^{-1}) = 0$$

if x and y are not conjugate in G.

Proof Let $y \in G$. Let $\gamma : G \to k$ be the function that maps every conjugate of y to 1 and every other element in G to zero. Clearly γ is a class function. By 3.3.6, there are unique coefficients $\alpha_\chi \in k$ for every $\chi \in \mathrm{Irr}_k(G)$ such that $\gamma = \sum_{\chi \in \mathrm{Irr}_K(G)} \alpha_\chi \chi$. By 3.3.6 again, we get that $\langle \gamma, \chi \rangle = \alpha_\chi$. But we also have $\langle \gamma, \chi \rangle = \frac{1}{|G|} \sum_{x \in G} \gamma(x)\chi(x^{-1})$. In this sum all summands are zero except those for which x is conjugate to y. If x is conjugate to y then $\chi(x^{-1}) = \chi(y^{-1})$. The number of different conjugates of y is $|G : C_G(y)|$. Thus this sum is equal to $|G : C_G(y)| \frac{1}{|G|} \chi(y^{-1}) = \frac{1}{|C_G(y)|} \chi(y^{-1})$ and hence $\alpha_\chi = \frac{1}{|C_G(y)|} \chi(y^{-1})$ for all $\chi \in \mathrm{Irr}_k(G)$. For every $x \in G$ we therefore have $\gamma(x) = \sum_{\chi \in \mathrm{Irr}_k(G)} \frac{1}{|C_G(y)|} \chi(x)\chi(y^{-1})$. If x is not conjugate to y, then this sum is zero. If x and y are conjugate, then this sum is 1. The equalities in the statement follow. □

The following result restates 3.3.11 in a slightly more precise way:

Theorem 3.3.12 *Let G be a finite group. Suppose that k is a splitting field for G and that $|G|$ is invertible in k. Let $\{V_i | 1 \leq i \leq h\}$ be a complete set of representatives of the isomorphism classes of simple kG-modules, and let χ_i be the character of V_i, where $1 \leq i \leq h$. We have an isomorphism of $k(G \times G)$-modules*

$$kG \cong \oplus_{1 \leq i \leq h} V_i \otimes_k V_i^*;$$

in particular, the character β of kG as a $k(G \times G)$-module is given by the formula

$$\beta(x, y) = \sum_{1 \leq i \leq h} \chi_i(x)\chi_i(y^{-1})$$

for all $x, y \in G$. Moreover, we have $\beta(x, y) = |C_G(x)|$ if x and y are conjugate in G, and $\beta(x, y) = 0$ if x and y are not conjugate in G.

Proof By 1.16.13 the algebra kG is semisimple. Wedderburn's Theorem for split algebras 1.14.6 implies that the product of the structural homomorphisms $kG \to \mathrm{End}_k(V_i)$ induces an algebra isomorphism

$$kG \cong \prod_{1 \le i \le h} \mathrm{End}_k(V_i).$$

It follows from 2.9.4 that

$$\mathrm{End}_K(V_i) \cong V_i \otimes_k V_i^*$$

as $k(G \times G)$-modules, whence the isomorphism in 3.3.12. The formula for the character of β is then just a particular case of the formula in 3.1.11 for characters of tensor products. But kG is also a permutation module for $k(G \times G)$, obtained from the action of $G \times G$ on G by $(x, y) \cdot z = xzy^{-1}$. Thus, by 3.1.16, $\beta(x, y)$ is equal to the number of different elements z such that $xzy^{-1} = z$, or equivalently, such that $x = zyz^{-1}$. In particular, if x and y are not conjugate, then there is no such z, hence $\beta(x, y) = 0$ in that case. If x and y are conjugate and $x = zyz^{-1}$ for some $z \in G$, then any other $z' \in G$ with the property $x = z'y(z')^{-1}$ satisfies $z' = cz$ for some $c \in C_G(x)$. Thus the number of different elements z satisfying $x = zyz^{-1}$ is equal to $|C_G(x)|$, whence the equality $\beta(x, y) = |C_G(x)|$ in that case. □

Remark 3.3.13 The regular character ρ in 3.1.14 is obtained by restricting β to the subgroup $G \times 1$ of $G \times G$, which yields the formula in 3.1.14. In fact, the first and second orthogonality relations are equivalent to each other. To see this, let $\mathcal{K}$ be a set of representatives of the conjugacy classes of G. Let $\chi, \chi' \in \mathrm{Irr}_K(G)$. Since χ and χ' are class functions and since the number of elements in the conjugacy class of an element $x \in G$ is equal to $|G : C_G(x)|$ we have

$$\frac{1}{|G|} \sum_{x \in G} \chi(x)\chi'(x^{-1}) = \sum_{x \in \mathcal{K}} \frac{\chi(x)}{|C_G(x)|} \chi'(x^{-1}).$$

Let X be the square matrix $X = (\frac{\chi(x)}{|C_G(x)|})$ and let Y be the square matrix $Y = (\chi(x^{-1}))^t$, with χ running over $\mathrm{Irr}_K(G)$ and x running over $\mathcal{K}$. The above equation shows that the First Orthogonality Relations are equivalent to $XY = \mathrm{Id}$. But then X and Y are inverse to each other, so this is equivalent to $YX = \mathrm{Id}$. This, in turn, is equivalent to the second orthogonality relations.

Theorem 3.3.14 *Let G be a finite group. Suppose that k is a splitting field for G and that $|G|$ is invertible in k. The group G is abelian if and only if $\chi(1) = 1$ for every $\chi \in \mathrm{Irr}_k(G)$, or equivalently, if and only if every simple kG-module is 1-dimensional.*

Proof Let h be the number of conjugacy classes of G. By 3.3.6 we have $h = |\mathrm{Irr}_K(G)|$. The group G is abelian if and only if $h = |G|$. Since $|G| = \sum_{\chi \in \mathrm{Irr}_K(G)} \chi(1)^2$ this is equivalent to $\chi(1)^2 = 1$, hence $\chi(1) = 1$ for all $\chi \in \mathrm{Irr}_K(G)$. □

The following result relates the scalar product and the induction/restriction functors in a way that mirrors the adjunction of induction and restriction in 2.2.1.

Theorem 3.3.15 (Frobenius' reciprocity for class functions) *Let G be a finite group, let H be a subgroup of G. Suppose that $|G|$ is invertible in k. Let $\psi : H \to k$ and $\varphi : G \to k$ be class functions. We have*

$$\langle \psi, \mathrm{Res}^G_H(\varphi)\rangle_H = \langle \mathrm{Ind}^G_H(\psi), \varphi\rangle_G,$$

$$\langle \mathrm{Res}^G_H(\varphi), \psi\rangle_H = \langle \varphi, \mathrm{Ind}^G_H(\psi)\rangle_G.$$

Proof Straightforward computation. If k is a splitting field for G and H, then this follows also from 2.2.1 and 3.3.7. □

A finite group G is called *nilpotent* if G is the direct product of its Sylow subgroups. By elementary group theory, this is equivalent to the condition that the normaliser $N_G(H)$ of any proper subgroup H of G is strictly larger than H, and also to the condition that the quotient G/N by any proper normal subgroup N of G has a nontrivial centre. Subgroups and quotients of finite nilpotent groups are again nilpotent, and finite p-groups are nilpotent, for any prime p. We combine 3.3.10 and 3.3.14 to show that every irreducible character of a nilpotent finite group is induced from a character of degree 1 of some subgroup; finite groups with this property are also called *M-groups*, and a character that is induced from a degree 1 character of some subgroup is called *monomial*. We will need the elementary group theoretic fact that a nonabelian nilpotent finite group G has an abelian normal subgroup A that is not contained in $Z(G)$; the proof of the following result works therefore more generally for *supersolvable* finite groups, but we will not need this.

Theorem 3.3.16 *Let G be a finite nilpotent group and k a splitting field for all subgroups of G. Suppose that $|G|$ is invertible in k. For any $\chi \in \mathrm{Irr}_k(G)$ there is a subgroup H and $\psi \in \mathrm{Irr}_k(H)$ such that $\chi = \mathrm{Ind}^G_H(\psi)$ and such that $\psi(1) = 1$.*

Proof We argue by induction over the order of G; we may assume that G is not abelian by 3.3.14. Let V be a simple kG-module with character χ. Again by induction, we may assume that the structural map $G \to \mathrm{GL}_k(V)$ is injective. Since G has a normal abelian subgroup A that is not contained in $Z(G)$

there is an element $y \in A$ such that the action of y on V is not of the form $\lambda \mathrm{Id}_V$ for some $\lambda \in k^\times$; in other words, the restriction $\mathrm{Res}^G_A(V)$ has at least two different isotypic components. Thus, by 3.3.10, χ is induced from some proper subgroup H of G, and hence the result follows by induction applied to that subgroup H. □

Let G be a group. A G-set M is called 2-*transitive* if G acts transitively on M and if the stabiliser G_m in G of some (hence any) element $m \in M$ acts transitively on $M \setminus \{m\}$. We denote abusively the character of the trivial kG-module k (which is the constant function on G with value 1_k) by 1 again.

Theorem 3.3.17 *Let G be a finite group, and let M be a finite G-set. Suppose that k is a field of characteristic zero. Let t be the number of G-orbits in M. Denote by $\pi : G \to k$ the character of the permutation kG-module kM. Then the following hold:*

(i) We have $\sum_{x\in G} \pi(x) = t|G|$.
(ii) We have $\langle \pi, 1\rangle = 1$ if and only if G is transitive on M.
(iii) Suppose that G is transitive on M, let $m \in M$, let H be the stabiliser of m in G and let s be the number of H-orbits in M. Then $\sum_{x\in G} \pi(x)^2 = s|G|$.
(iv) Suppose that G is transitive on M. Then G is 2-transitive on M if and only if $\pi = 1 + \chi$ for some $\chi \in \mathrm{Irr}_k(G) \setminus \{1\}$.

Proof Suppose first that G is transitive on M. Then, for any two different m_1, m_2 in M there is $x \in G$ such that $m_2 = xm_1$. Thus a group element y lies in the stabiliser G_{m_2} of m_2 if and only if $yxm_1 = xm_1$, or if and only if $x^{-1}yx$ lies in the stabiliser of G_{m_1}, which shows that the stabilisers of different elements in M are G-conjugate, and in particular, all have the same order. We count the number of elements in the subset $\{(x, m) \in G \times M \mid xm = m\}$ in two different ways: on one hand, this is equal to $\sum_{x\in G} |\{m \in M \mid xm = m\}| = \sum_{x\in G} \pi(x)$, and on the other hand, this is $\sum_{m\in M} |\{x \in G \mid xm = m\}| = \sum_{m\in M} |G_m| = |M||G_m| = |G|$, because the index of the stabiliser $|G : G_m|$ is equal to the length of the G-orbit of m in M, and the latter is all of M as G is transitive on M. This proves (i) if G is transitive on M. If $M = M_1 \cup M_2 \cup \cdots \cup M_t$ is the union of G-orbits M_i, then the permutation character of kM is the sum of the permutation characters of the kM_i, whence (i). By observing that $\sum_{x\in G} \pi(x) = \langle \pi, 1\rangle|G|$, statement (ii) is a trivial consequence of (i). Suppose that G is transitive on M. Then the stabilisers G_m of elements $m \in M$ in G are all conjugate. In particular, they all have the same number, say s, of orbits on M. Thus $\sum_{x\in G} \pi(x)^2 = \sum_{x\in G} |\{(m, n) \in M \times M \mid xm = m, xn = n\}| = \sum_{x\in G} \sum_{m\in M;\ xm=m} |\{n \in M \mid xn = n\}| = \sum_{m\in M} \sum_{x\in G_m} |\{n \in M \mid xn = n\}| = \sum_{m\in M} \sum_{x\in G_m} \pi(x) = \sum_{m\in M} s|G_m| = s|G|$, where we used the formula in (i) applied to G_m. This shows (iii). Write $\pi = \sum_{\chi \in \mathrm{Irr}_k(G)} a_\chi \chi$ with $a_\chi \in K$. By the first orthogonality relations we get that $a_\chi = \langle \pi, \chi \rangle$. Since G is transitive on M, it follows from (ii) that $a_1 = 1$. By (iii), G is 2-transitive on M if and only if $2 = \langle \pi, \pi \rangle = 1 + \sum_{\chi \in \mathrm{Irr}_k(G)-\{1\}} (a_\chi)^2$. This, in turn, happens if and only exactly one of the a_χ's is 1 and all others are zero. This proves (iv). □

3.4 Character tables of finite groups

Let G be a finite group and let K be a splitting field of characteristic zero for G. The *character table of G over K* is the square matrix $(\chi_i(x_j))_{1 \leq i,j \leq h}$, where $\{x_j\}_{1 \leq j \leq h}$ is a set of representatives of the conjugacy classes of G and where $\{\chi_i\}_{1 \leq i \leq h} = \mathrm{Irr}_K(G)$. This matrix is defined up to permutations of the rows and columns.

Theorem 3.4.1 *Let G be a finite group and let K be a splitting field of characteristic zero for G. The character table of G over K is a nondegenerate square matrix.*

Proof By the Second Orthogonality Relations 3.3.11, the product of the character table with its transpose is a diagonal matrix whose diagonal entries are the orders of centralisers, hence nonzero. □

The character table of finite cyclic groups. Let n be a positive integer, let G be a cyclic group of order n and let y be a generator of G. Since G is abelian, every simple KG-module has dimension 1 and every irreducible character χ of G over K is a group homomorphism $\chi : G \to K^\times$. Since $y^n = 1$ we get $\chi(y)^n = 1$, and χ is determined by its value at y, because G is generated by y. Let ζ be a primitive n-th root of unity in K. Then $\{\zeta^k | 0 \leq k \leq n-1\}$ is the set of all n-th roots of unity. For any i there is a unique irreducible character $\chi_i : G \to K^\times$ mapping y to ζ^i. Since χ_i is then a group homomorphism, we get $\chi_k(y^j) = \zeta^{ij}$. Thus the character table of G is the matrix

$$(\zeta^{ij})_{1 \leq i,j \leq n-1}.$$

Note that the first row, for $i = 0$, is constant equal to 1. It is customary to write the trivial character as the first row of the character table.

The Klein four group. Let $G = \langle x, y | x^2 = y^2 = 1, xy = yx \rangle \cong \langle x \rangle \times \langle y \rangle$ with $x^2 = y^2 = 1$ be a Klein four group. Any irreducible character χ of G is a group homomorphism from G to $K^\times$, hence determined by its values on x and y. Moreover, these values have to be 1 or -1 because x, y have order 2. Thus the character table of G in this case is the matrix:

	1	x	y	xy
χ_1	1	1	1	1
χ_2	1	-1	1	-1
χ_3	1	1	-1	-1
χ_4	1	-1	-1	1

Finite abelian groups. If G is abelian, then G is a direct product of cyclic groups, say

$$G = \langle y_1 \rangle \times \langle y_2 \rangle \times \cdots \times \langle y_m \rangle.$$

Then

$$\mathrm{Irr}_K(G) = \mathrm{Hom}(G, K^\times) = \prod_{1 \leq i \leq m} \mathrm{Hom}(\langle y_i \rangle, K^\times)$$

and thus the character table of G can be obtained from that of cyclic groups.

If G is not abelian, then the first step towards the character table of G consists of determining the characters of degree 1, or equivalently, group homomorphisms $\chi : G \to K^\times$. If χ is such a group homomorphism then $\chi(xyx^{-1}y^{-1}) = \chi(x)\chi(y)\chi(x)^{-1}\chi(y)^{-1} = 1$, hence the commutator subgroup $[G, G]$ of G is in the kernel of χ. Thus χ can be viewed as a group homomorphism from $G/[G, G]$ to $K^\times$, sending a coset $x[G, G]$ to $\chi(x)$. Conversely, every group homomorphisms $G/[G, G] \to K^\times$ can be composed with the canonical surjection $G \to G/[G, G]$, yielding a group homomorphism $G \to K^\times$. Thus we have a canonical bijection $\mathrm{Hom}(G, K^\times) \cong \mathrm{Hom}(G/[G, G], K^\times)$. For instance, if S_n is the symmetric group on n letters for some positive integer n, we have $[S_n, S_n] = A_n$, and S_n/A_n is cyclic of order 2. Thus S_n has exactly two isomorphism classes of KG-modules of dimension 1, corresponding to the trivial group homomorphism $S_n \to K^\times$ mapping every element in S_n to 1, and the sign map $\mathrm{sgn} : S_n \to K^\times$ mapping every element π in S_n to either 1 or -1, according to whether π is an even or odd permutation. This is also called the *sign representation of* S_n.

The symmetric group S_3. The symmetric group S_3 is the smallest nonabelian finite group. In particular, S_3 has at least one irreducible character of degree greater than one. The group S_3 has two characters ψ_1, ψ_2 of degree 1, namely the trivial character and the character of the sign representation. Since $6 = 1 + 1 + 2^2$, it follows that the only way to have an irreducible character χ with $\chi(1) > 1$ is $\chi(1) = 2$. If ρ is the regular character of S_3, then $\rho = \psi_1 + \psi_2 +$

2χ, hence $\chi = \frac{1}{2}(\rho - \psi_1 - \psi_2)$. Thus the character table of S_3 is:

	(1)	(12)	(123)
ψ_1	1	1	1
ψ_2	1	−1	1
χ	2	0	−1

Notice that we calculated the character table of S_3 without explicitly describing the simple KS_3-modules.

Dihedral groups. Let n be a positive integer and let

$$D_{2n} = \langle x, t | x^n = t^2 = 1, \; txt^{-1} = x^{-1} \rangle.$$

Every element of D_{2n} is either of the form x^k for some unique integer k, $0 \leq k \leq n-1$, or of the form $x^k t$ for some unique integer k, $0 \leq k \leq n-1$. The group $\langle x \rangle$ is cyclic of order n, hence of index 2 in D_{2n}, and thus $G/\langle x \rangle$ is abelian, which implies $[D_{2n}, D_{2n}] \subseteq \langle x \rangle$. We have $xtx^{-1}t^{-1} = x^2$. If n is odd, then $\langle x^2 \rangle = \langle x \rangle$, and if n is even, then $\langle x^2 \rangle$ has index 2 in $\langle x \rangle$. It follows that

$$[D_{2n}, D_{2n}] = \begin{cases} \langle x \rangle, & n \text{odd}; \\ \langle x^2 \rangle, & n \text{even}. \end{cases}$$

Thus $D_{2n}/[D_{2n}, D_{2n}]$ is cyclic of order 2 if n is odd, and a Klein four group if n is even. It follows that for n odd, the set $\mathrm{Hom}(D_{2n}, K^\times)$ has two elements ψ_1, ψ_2 defined by $\psi_1(x) = \psi_1(t) = 1$ and $\psi_2(x) = 1$, $\psi_2(t) = -1$. If n is even, then $\mathrm{Hom}(D_{2n}, K^\times)$ has four elements $\psi_1, \psi_2, \psi_3, \psi_4$ defined by the pattern of the character table of the Klein four group; that is:

	x^k	$x^k t$
ψ_1	1	1
ψ_2	$(-1)^k$	$(-1)^k$
ψ_3	1	−1
ψ_4	$(-1)^k$	$(-1)^{k+1}$

There are simple 2-dimensional KD_{2n}-modules V_i given, in matrix form, by the group homomorphisms $\rho_i : D_{2n} \to M_2(K)^\times$ defined as follows. Let ζ be a primitive n-th root of unity in K. For any positive integer i set

$$\rho_i(x) = \begin{pmatrix} \zeta^i & 0 \\ 0 & \zeta^{-i} \end{pmatrix}, \; \rho_i(t) = \begin{pmatrix} 0 & \zeta^{-i} \\ \zeta^i & 0 \end{pmatrix}.$$

Geometrically, $\rho_i(x)$ corresponds to the rotation of the regular n-gon by the angle $e^{\frac{2\pi i}{n}}$ and $\rho_i(t)$ corresponds to the reflection of the regular n-gon at the x-axis. One checks that this defines indeed a group homomorphism from D_{2n}

to $M_2(K)^\times$. In particular, for any integer $k \geq 0$ we have $\rho_i(x^k) = \begin{pmatrix} \zeta^{ik} & 0 \\ 0 & \zeta^{ik} \end{pmatrix}$ and $\rho_i(x^k t) = \begin{pmatrix} 0 & \zeta^{-ik} \\ \zeta^{ik} & 0 \end{pmatrix}$. Thus the character χ_i of this 2-dimensional KD_{2n}-module V_i is given by

$$\chi_i(x^k) = \zeta^{ik} + \zeta^{-ik},\ \chi_i(x^k t) = 0$$

for all integers $k \geq 0$. There are several possibilities to proceed from here: one can either verify directly that $\langle \chi_i, \chi_i \rangle = 1$ for $0 < i < \frac{n}{2}$ and that $\langle \chi_i, \chi_j \rangle = 0$ for $0 < i < j < \frac{n}{2}$, which shows that the characters χ_i, with $0 < i < \frac{n}{2}$ are irreducible and pairwise different. One can also use the explicit description of ρ_i to see that $V_i = K^2$ has no invariant 1-dimensional subspace: the only space fixed by $\rho_i(t)$ is the 'diagonal' subspace $\{(\lambda, \lambda) | \lambda \in K\}$ of K^2, but $\rho_i(x)$ sends (λ, λ) to $(\zeta^i \lambda, \zeta^{-i} \lambda)$, and this is not an element of the diagonal subspace unless $\zeta^i = \zeta^{-i}$, or equivalently, unless $\zeta^{2i} = 1$, which in turn happens if and only if n divides $2i$. In particular, the ρ_i defined simple KD_{2n}-modules for $0 < i < \frac{n}{2}$. A similar argument shows that they are pairwise nonisomorphic. To see that D_{2n} has no other simple modules, up to isomorphism, we add up the squares of the character degrees of the characters we have constructed so far. If $n = 2m$ is even, then we have four characters of degree 1 and $m - 1$ characters of degree 2; hence the sum of the squares of their degrees is $4 + 2^2(m-1) = 2n = |D_{2n}|$. Similarly, if $n = 2m + 1$ is odd, then we have 2 irreducible characters of degree 1 and m irreducible characters of degree 2; thus in this case the sum of the squares of the character degrees is $2 + 2^2 m = 2n = |D_{2n}|$. Thus, in both cases, we have constructed all irreducible characters of D_{2n}.

Alternatively, we could have described the two-dimensional simple modules as induced modules as follows. For any integer $i \geq 0$ let W_i be the 1-dimensional $K\langle x \rangle$-module with character ψ_i sending x to ζ^i. Set $V_i = \mathrm{Ind}_{\langle x \rangle}^{D_{2n}}(W_i)$. We will show that V_i is the 2-dimensional KD_{2n}-module with character χ_i satisfying $\chi(x^k) = \zeta^{ik} + \zeta^{-ik}$ and $\chi(x^k t) = 0$ for any integer $k \geq 0$. As K-vector spaces we have $W_i = K$ and $V_i = KD_{2n} \otimes_{K\langle x \rangle} W_i = 1 \otimes W_i \oplus t \otimes W_i$. The action of t exchanges the two summands, so this action is represented by the matrix $\begin{pmatrix} 0 & 1 \\ 1 & 0 \end{pmatrix}$ with respect to the K-basis $\{1 \otimes 1_k, t \otimes 1_K\}$. Let $w \in W_i$. The action of x on $1 \otimes w$ is given by $x(1 \otimes w) = x \otimes w = 1 \otimes xw = 1 \otimes \zeta^i w = \zeta^i(1 \otimes w)$, and the action of x on $t \otimes w$ is given by $x(t \otimes w) = xt \otimes w = ttxt \otimes w = tx^{-1} \otimes w = t \otimes x^{-1} w = t \otimes \zeta^{-i} w = \zeta^{-i}(t \otimes w)$. Thus the action of x on W_i is represented by the matrix $\begin{pmatrix} \zeta^i & 0 \\ 0 & \zeta^{-i} \end{pmatrix}$. This is exactly the KD_{2n}-module structure as defined above. The simplicity of the V_i follows from 2.4.7.

The dihedral group D_8 and the quaternion group Q_8. The groups $D_8 = \langle x, t | x^4 = t^2 = 1, txt = x^{-1} \rangle$ and $Q_8 = \langle x, y | x^4 = 1, y^2 = x^2, yxy^{-1} = x^{-1} \rangle$ are non isomorphic but they have the same character table:

ψ_1	1	1	1	1	1
ψ_2	1	1	-1	-1	1
ψ_3	1	-1	1	-1	1
ψ_4	1	-1	-1	1	1
χ	2	0	0	0	-2

The character table of D_8 is calculated above, but one can also treat D_8 and Q_8 simultaneously as follows. If G is one of D_8 or Q_8 then $Z = Z(G)$ has order 2 and G/Z is a Klein four group. In both cases, the last column corresponds to the character values at the unique involution in Z. Thus the four irreducible characters of G/Z inflated to G via the canonical map $G \to G/Z$ yield the irreducible characters $\psi_1, \psi_2, \psi_3, \psi_4$. Since G is non abelian there is an irreducible character χ of degree at least 2. Since $8 = 1 + 1 + 1 + 1 + 2^2$ there is exactly one irreducible character of degree 2, and if we denote by ρ the regular character of G then $2\chi = \rho - \psi_1 - \psi_2 - \psi_3 - \psi_4$, which implies the last row. This is another example of the character table being determined without calculating explicitly all underlying simple KG-modules.

Frobenius groups. A *Frobenius group* is a finite group of the form $G = H \rtimes E$, where E, H are finite groups such that E acts freely on $H \setminus \{1\}$. The smallest Frobenius group with nontrivial E is $S_3 \cong C_3 \rtimes C_2$. If n is an odd positive integer, then the dihedral group D_{2n} of order $2n$ is a Frobenius group, isomorphic to the semidirect product $C_n \rtimes C_2$ with the nontrivial element of C_2 acting by inverting every element in C_n. By a theorem of Thompson, if $G = H \rtimes E$ is a Frobenius group, then H is nilpotent and the Sylow subgroups of E are either cyclic or (at the prime 2) generalised quaternion (cf. [32, 10.3.1]). The free action of E on $H \setminus \{1\}$ implies that $C_G(y) \subseteq H$, for any nontrivial y in H, which in turn implies that E acts freely on the set of nontrivial conjugacy classes of H, and hence E acts freely on $\mathrm{Irr}_K(H) \setminus \{1_H\}$ (see [32, 4.5.3]). This implies that for any nontrivial λ, $\lambda' \in \mathrm{Irr}_K(H)$, we have $\langle \mathrm{Ind}_H^G(\lambda), \mathrm{Ind}_H^G(\lambda') \rangle = \langle \lambda, \sum_{x \in E} {}^x\lambda' \rangle_H$, which is equal to 1 if λ, λ' belong to the same E-orbit in $\mathrm{Irr}_K(H)$, and zero otherwise. This shows that if Λ is a set of representatives of the E-orbits in $\mathrm{Irr}_K(H) \setminus \{1_H\}$, then the characters $\chi_\lambda = \mathrm{Ind}_H^G(\lambda)$, with $\lambda \in \Lambda$, are pairwise different irreducible characters of G. For any $\mu \in \mathrm{Irr}_K(E)$ we have an irreducible character $\chi_\mu \in \mathrm{Irr}_K(G)$ defined by $\chi_\mu(yx) = \mu(x)$, for all $x \in E$, $y \in H$; that is, χ_μ is obtained from inflating μ

along the canonical surjection $G \to E$ with kernel H. By taking the sum of the squares of the character degrees, one sees that there are no other irreducible characters; that is, we have

$$\mathrm{Irr}_K(G) = \{\chi_\mu | \mu \in \mathrm{Irr}_K(E)\} \cup \{\chi_\lambda | \lambda \in \Lambda\}.$$

Mackey's formula yields $\mathrm{Res}^G_E(\chi_\lambda) = \mathrm{Ind}^E_1 \mathrm{Res}^H_1(\lambda)$, hence χ_λ vanishes on E. The character table of G is determined by those of E and H.

The alternating group A_4. The alternating group A_4 is isomorphic to a semidirect product $V_4 \rtimes C_3$, which is a Frobenius group with C_3 permuting transitively the three nontrivial elements in V_4. It has four conjugacy classes, represented by (1), $(1,2)(3,4)$, $(1,2,3)$, $(1,3,2)$. Its character table can be calculated using the character table of V_4 and the considerations on Frobenius groups above. Let ζ be a primitive cube root of unity.

η_1	1	1	1	1
η_2	1	1	ζ	ζ^2
η_3	1	1	ζ^2	ζ
η_4	3	-1	0	0

The characters η_1, η_2, η_3 are the characters of C_3 inflated to A_4, and η_4 is induced from a nontrivial character of V_4.

The alternating group A_5. The group A_5 has five conjugacy classes, represented by (1), $(1,2,3)$, $(1,2,3,4,5)$, $(1,3,4,5,2)$, $(1,2)(3,4)$. Setting $\gamma = \frac{1+\sqrt{5}}{2}$ and $\bar{\gamma} = \frac{1-\sqrt{5}}{2}$, the character table of A_5 is as follows:

χ_1	1	1	1	1	1
χ_2	3	0	γ	$\bar{\gamma}$	-1
χ_3	3	0	$\bar{\gamma}$	γ	-1
χ_4	4	1	-1	-1	0
χ_5	5	-1	0	0	1

The character χ_4 is obtained from taking the quotient of the natural 5-dimensional permutation module of A_5 by the trivial submodule (generated by the sum of the elements of a permutation basis). The character χ_5 is induced from a nontrivial character of degree one of A_4. The character χ_2 is obtained by inducing a degree 2 character of the dihedral subgroup of order 10 generated by the 5-cycle $(1,2,3,4,5)$ and the involution $(1,2)(3,5)$ in A_5; this yields a character of degree 12, equal to $\chi_2 + \chi_4 + \chi_5$. Conjugating this with an element in $S_5 \setminus A_5$ yields $\chi_3 + \chi_4 + \chi_5$.

Remark 3.4.2 We have assumed in this section that K is a splitting field of characteristic zero for any of the finite groups, without elaborating exactly what that means for K. If G is cyclic of order n, then the calculations of the character table of G imply that K is a splitting field for G if and only if K contains a primitive n-th root of unity. If G is abelian, then K is a splitting field for G if and only if K is a splitting field for the cyclic direct factors of G, hence if and only if K contains a primitive m-th root of unity, where m is the exponent of G (the least common multiple of the orders of the elements in G). If K contains a primitive n-th root of unity, then K is a splitting field for the dihedral group D_{2n}. We will prove later a theorem of Brauer stating that given any finite group G, if K contains a primitive $|G|$-th root of unity, then K is a splitting field for KG.

3.5 Integrality of character values

Based on integrality considerations we will show that for any finite group G and an irreducible character χ of G, the character degree $\chi(1)$ divides the group order $|G|$. Any field K of characteristic zero has a subfield isomorphic to $\mathbb{Q}$, and we will usually identify $\mathbb{Q}$ to its image in K.

Definition 3.5.1 Let K be a field of characteristic zero. A number $\alpha \in K$ is called an *algebraic integer* or *integral over* $\mathbb{Z}$ if α is the root of a nonconstant polynomial of the form $f(X) = X^n + a_{n-1}X^{n-1} + \cdots + a_1X + a_0$ with coefficients $a_i \in \mathbb{Z}$; that is, α is the root of a polynomial in $\mathbb{Z}[X]$ with leading coefficient equal to 1.

Every n-th root of unity $\zeta \in K$ is an algebraic integer, since ζ is a root of the polynomial $X^n - 1$. We collect a few standard facts about algebraic integers:

Proposition 3.5.2 *Let K be a field of characteristic zero and let S be the set of all algebraic integers in K.*

(i) Every subring R of K that is finitely generated as a $\mathbb{Z}$-module is contained in S.
(ii) The set S is a subring of K.
(iii) We have $S \cap \mathbb{Q} = \mathbb{Z}$.

Proof Let R be a subring of K that is finitely generated as a $\mathbb{Z}$-module. Let $\gamma \in R$. Since R is a subring of K, it contains all powers of γ. Let N be the $\mathbb{Z}$-submodule of R generated by the set $\{\gamma^k | k \geq 0\}$ of all powers of γ. Since R is finitely generated as a $\mathbb{Z}$-module and $\mathbb{Z}$ is Noetherian, it follows that N is finitely generated as a $\mathbb{Z}$-module as well. Thus there is a positive integer t such that N is generated by the finite set $\{\gamma^k | 0 \leq k \leq t-1\}$. But then γ^t, which belongs to N, is a $\mathbb{Z}$-linear combination of $1, \gamma, \ldots, \gamma^{t-1}$, which means that γ is the root of a polynomial in $\mathbb{Z}[X]$ with leading coefficient 1, hence γ is an algebraic integer. This proves (i). Let α, β be algebraic integers. Let $f(X) = X^m + a_{m-1}X^{m-1} + \cdots + a_1X + a_0$ and $g(X) = X^n + b_{n-1}X^{n-1} + \cdots + b_1X + b_0$ be polynomials in $\mathbb{Z}[X]$ such that $f(\alpha) = 0 = g(\beta)$. Consider the $\mathbb{Z}$-module R in K generated by the finite set $\{\alpha^i\beta^j | 0 \le i \le m-1, 0 \le j \le n-1\}$. Since $\alpha^m = -a_{m-1}\alpha^{m-1} - \cdots - a_1\alpha - a_0$ it follows that R contains all powers of α, β and their products; thus R is in fact a subring of K. In particular, R contains $\alpha + \beta$ and $\alpha\beta$. Since R is finitely generated as a $\mathbb{Z}$-module, every element in R is an algebraic integer. Thus (i) implies (ii). Suppose now that $\alpha \in \mathbb{Q}$. Write $\alpha = \frac{p}{q}$ with integers p, q such that $q \neq 0$; we may choose p and q to be coprime, after cancelling common factors. The equation $f(\frac{p}{q}) = 0$ translates to $p^m = -q(a_{m-1}p^{m-1} + a_{m-2}qp^{m-2} + \cdots + a_0q^{m-1})$. Thus q divides p^m. Since p, q were chosen coprime, this forces $q = 1$ or $q = -1$, hence $\alpha \in \mathbb{Z}$, completing the proof. □

Note that the ring S in 3.5.2 need not be finitely generated as a $\mathbb{Z}$-module.

Proposition 3.5.3 *Let G be a finite group and K be a splitting field of characteristic zero for G. Let $\chi \in \mathrm{Irr}_K(G)$ and V a simple KG-module with character χ. Let $x \in G$, denote by m the order of x, and by ζ_m a primitive m-th root of unity in an extension field of K. Then $\chi(x) \in \mathbb{Z}[\zeta_m]$; in particular, $\chi(x)$ is an algebraic integer. If moreover K is a subfield of $\mathbb{C}$, then the following hold:*

(i) $|\chi(x)| \le \chi(1)$.
(ii) $|\chi(x)| = \chi(1)$ *if and only if x acts as $\lambda\mathrm{Id}_V$ on V for some root of unity $\lambda \in \mathbb{C}$.*
(iii) The set $Z(\chi) = \{y \in G | |\chi(y)| = \chi(1)\}$ *is a normal subgroup of G.*
(iv) We have $Z(\chi) = G$ *if and only if* $\chi(1) = 1$.

Proof The proof of the first statement employs an argument already applied in the proof of 3.2.3. Up to extending scalars, we may assume that K is algebraically closed. If V is a KG-module with character χ we can choose a K-basis $\{v_1, v_2, \ldots, v_n\}$ of V in such a way that the matrix M representing the action of x on V is an upper diagonal matrix; that is, M is of the form

$$M = \begin{pmatrix} \lambda_1 & * & \cdots & * \\ 0 & \lambda_2 & \cdots & * \\ 0 & 0 & \cdots & \lambda_n \end{pmatrix}.$$

Since $x^m = 1$, it follows that $M^m = I$, the identity matrix, and hence $\lambda_i^m = 1$ for $1 \le i \le n$. Thus all λ_i are in $\mathbb{Z}[\zeta_m]$, hence so is their sum $\chi(x)$. By Proposition 3.5.2, all elements in $\mathbb{Z}[\zeta_m]$ are algebraic integers. This proves the first statement. Assume now that $K \subseteq \mathbb{C}$. Since the absolute value of a root of unity is 1, we get, using the triangle inequality, that

$$|\chi(x)| = |\sum_{1\leq i\leq n} \lambda_i| \leq \sum_{1\leq i\leq n} |\lambda_i| = \chi(1),$$

which proves (i). The equality $|\chi(x)| = \chi(1)$ holds if and only if all λ_i are equal to a root of unity λ, or equivalently, if x acts as $\lambda\mathrm{Id}_V$ on V, whence (ii). In that case, x^{-1} acts as $\lambda^{-1}\mathrm{Id}_V$ on V, and if furthermore $y \in G$ acts as $\mu\mathrm{Id}_V$ for some $\mu \in K$ then their product xy acts as $\lambda\mu\mathrm{Id}_V$. Thus $Z(\chi)$ is a subgroup of G. The fact that χ is a class function implies that $Z(\chi)$ is in fact a normal subgroup of G, which proves (iii). If $\chi(1) = 1$ then $\chi(y)$ is a root of unity for all $y \in G$, hence $Z(\chi) = G$. Conversely, if $Z(\chi) = G$ then every $y \in G$ acts as $\lambda_y\mathrm{Id}_V$ for some $\lambda_y \in K$. But then every one-dimensional subspace of V is a KG-submodule of V, and since V was simple to begin with, this forces $\dim_K(V) = 1$, which completes the proof. □

If a character of a finite group G takes values in $\mathbb{Z}$, then the next result implies that its value at an element $x \in G$ depends only on the cyclic subgroup generated by x. This applies in particular to characters of permutation modules.

Proposition 3.5.4 *Let G be a finite group and let K be a field of characteristic zero. Let χ be the character of a finitely generated KG-module and let $x \in G$.*

(i) We have $\prod_y \chi(y) \in \mathbb{Z}$, where y runs over the set of all elements in G such that $\langle y\rangle = \langle x\rangle$.
(ii) We have $\sum_y \chi(y) \in \mathbb{Z}$, where y runs over the set of all elements in G such that $\langle y\rangle = \langle x\rangle$.
(iii) If $\chi(x) \in \mathbb{Z}$ then for all $y \in G$ such that $\langle y\rangle = \langle x\rangle$ we have $\chi(x) = \chi(y)$.

Proof Since the statements involve only the cyclic group $\langle x\rangle$ we may assume that $G = \langle x\rangle$. Set $n = |G|$. Since G is cyclic, we may assume that $K = \mathbb{Q}(\zeta)$, where ζ is a primitive n-th root of unity. Let $y \in G$ such that $\langle y\rangle = G$. Then $y = x^k$ for some integer k such that $(n,k) = 1$. Let $\sigma \in \mathrm{Gal}(\mathbb{Q}(\zeta) : \mathbb{Q})$ be defined by $\sigma(\zeta) = \zeta^k$. The eigenvalues of x are powers of ζ, and hence $\sigma(\chi(x)) = \chi(x^k) = \chi(y)$. In particular, if $\chi(x) \in \mathbb{Z}$ then $\chi(x) = \chi(y)$, which shows (iii). If k runs over the integers between 1 and n that are coprime to n then $y = x^k$ runs over the set of generators of G and σ as above runs over the Galois group $\mathrm{Gal}(K : \mathbb{Q})$. Thus $\prod_y \chi(y) = \prod_\sigma \sigma(\chi(x))$ and $\sum_y \chi(y) = \sum_\sigma \sigma(\chi(x))$. Both of these numbers are fixed by the Galois group of $K : \mathbb{Q}$, hence are in $\mathbb{Q}$. They are also algebraic integers, hence in $\mathbb{Z}$. This implies (i) and (ii). □

Theorem 3.5.5 *Let G be a finite group and let K be a splitting field of characteristic zero for G. Let $\chi \in \mathrm{Irr}_K(G)$ and V a simple KG-module with character χ.*

(i) An element z in $Z(KG)$ acts on V as multiplication by the scalar $\frac{\chi(z)}{\chi(1)}$.
(ii) The map $\omega : Z(KG) \to K$ defined by $\omega(z) = \frac{\chi(z)}{\chi(1)}$ for all $z \in Z(KG)$ is a K-algebra homomorphism.
(iii) We have $\omega(e(\chi)) = 1$ and $\omega(e(\chi')) = 0$ for any $\chi' \in \mathrm{Irr}_K(G)$ such that $\chi' \neq \chi$.
(iv) For any $z \in Z(KG)$ and $a \in KG$ we have $\chi(za) = \omega(z)\chi(a)$.
(v) For any $z \in Z(KG)$ we have $ze(\chi) = \frac{\chi(z)}{\chi(1)}e(\chi)$.

Proof Let V be a simple KG-module having χ as its character. Let $z \in Z(KG)$. The map sending $v \in V$ to zv is a KG-endomorphism of V because z commutes with all elements in KG. By Schur's Lemma and the assumptions on K being a splitting field, this endomorphism is of the form $\lambda_z \mathrm{Id}_V$ for some $\lambda_z \in K$. The map sending $z \in Z(KG)$ to λ_z is obviously a K-algebra homomorphism. Moreover, $\chi(z) = \lambda_z \mathrm{dim}_K(V) = \lambda_z \chi(1)$, and therefore $\lambda_z = \frac{\chi(z)}{\chi(1)} = \omega(z)$. This proves (i) and (ii). Statement (iii) follows from the last statement in 3.3.1. If $a \in KG$, then za acts as $\lambda_z a$ on V, and hence $\chi(za) = \lambda_z \chi(a) = \omega(z)\chi(a)$. This shows (iv). By Corollary 3.3.3, as a left KG-module, $KGe(\chi)$ is isomorphic to a direct sum of copies of V. Statement (i) implies that $z \in Z(KG)$ acts as multiplication by $\frac{\chi(z)}{\chi(1)}$ on $KGe(\chi)$, whence (v). □

Theorem 3.5.6 *Let G be a finite group, let K be a splitting field of characteristic zero for G, and let $\chi \in \mathrm{Irr}_K(G)$. Let $c \in Z(KG)$ be the sum of all conjugates of an element $x \in G$. Then $\omega(c) = \frac{\chi(c)}{\chi(1)} = \frac{|G|\chi(x)}{|C_G(x)|\chi(1)}$ is an algebraic integer.*

Proof Let $\{x_i | 1 \leq i \leq h\}$ be a set of representatives of the conjugacy classes in G, and let c_i be the sum in KG of all conjugates of x_i, where $1 \leq i \leq h$. There are integers a_{ijk} for $1 \leq i, j, k \leq h$ such that $c_i c_j = \sum_{1 \leq k \leq h} a_{ijk} c_k$. Since ω is an algebra homomorphism, we get

$$\omega(c_i)\omega(c_j) = \sum_{1 \leq k \leq h} a_{ijk}\omega(c_k)$$

for all i, j such that $1 \leq i, j \leq h$. Thus the $\mathbb{Z}$-submodule R of K generated by the finite set $\{\omega(c_i) | 1 \leq i \leq h\}$ is a subring of K, hence contained in the ring of algebraic integers in K by 3.5.2. Since χ is a class function and the number of different G-conjugates of x is equal to $|G : C_G(x)|$ we get the equality $\chi(c) = |G : C_G(x)|\chi(x)$, whence the second equality in the statement. □

Theorem 3.5.7 *Let G be a finite group, let K be a splitting field of characteristic zero of G and let $\chi \in \mathrm{Irr}_K(G)$. The degree $\chi(1)$ divides the group order $|G|$.*

Proof Let $\{x_i | 1 \leq i \leq h\}$ be a set of representatives of the conjugacy classes in G, and let c_i be the sum in KG of all conjugates of x_i, where $1 \leq i \leq h$. Since χ is a class function, we get that

$$|G| = \sum_{x \in G} \chi(x)\chi(x^{-1}) = \sum_{1 \leq i \leq h} |G : C_G(x_i)|\chi(x_i)\chi(x_i^{-1})$$
$$= \sum_{1 \leq i \leq h} \chi(c_i)\chi(x_i^{-1}).$$

Dividing by $\chi(1)$ yields that

$$\frac{|G|}{\chi(1)} = \sum_{1 \leq i \leq h} \omega(c_i)\chi(x_i^{-1}),$$

where $\omega(c_i) = \frac{\chi(c_i)}{\chi(1)}$ for $1 \leq i \leq h$, and by 3.5.6, these numbers are algebraic integers. Thus, by 3.5.2 (ii), the rational number $|G|/\chi(1)$ is an algebraic integer, hence an integer by 3.5.2 (iii). □

This result can be strengthened in various ways. If A is an abelian normal subgroup of G, one can show that $\chi(1) \mid |G : A|$, and if $Z(\chi)$ is the subgroup of all $x \in G$ satisfying $|\chi(x)| = \chi(1)$, then $\chi(1) \mid |G : Z(\chi)|$. Note that $Z(\chi)$ is a normal subgroup of G containing $Z(G)$.

If K is the quotient field of a principal ideal domain $\mathcal{O}$, then the next result implies that $\mathcal{O}$ contains all algebraic integers in K, and hence any K-valued character of a finite group G has values in $\mathcal{O}$. We will see in 4.16.5, that if $\mathcal{O}$ is a discrete valuation ring, then there is a stronger structural fact behind this observation: for any finite-dimensional KG-module M there is an $\mathcal{O}G$-module U such that U is $\mathcal{O}$-free and $M \cong K \otimes_{\mathcal{O}} U$.

Theorem 3.5.8 *Let $\mathcal{O}$ be a principal ideal domain and let K be the quotient field of $\mathcal{O}$. Then $\mathcal{O}$ is integrally closed; that is, if $\alpha \in K$ is a root of a non zero polynomial $f \in \mathcal{O}[X]$ with leading coefficient 1 then $\alpha \in \mathcal{O}$. In particular, $\mathcal{O}$ contains the ring of algebraic integers in K.*

Proof Let $f = X^n + a_{n-1}X^{n-1} + \cdots + a_1X + a_0$ be a polynomial in $\mathcal{O}[X]$ and let $\alpha \in K$ such that $f(\alpha) = 0$. Write $\alpha = \frac{\beta}{\gamma}$ with $\beta, \gamma \in \mathcal{O}$. Since $\mathcal{O}$ is a principal ideal domain there is $\delta \in \mathcal{O}$ such that $\mathcal{O}\beta + \mathcal{O}\gamma = \mathcal{O}\delta$. In particular, $\beta, \gamma \in \mathcal{O}\delta$. Thus, after cancelling δ we may assume that $\mathcal{O}\beta + \mathcal{O}\gamma = \mathcal{O}$. Multiplying the equation $f(\alpha) = 0$ by γ^n yields the equation

$$\beta^n + a_{n-1}\gamma\beta^{n-1} + \cdots + a_1\gamma^{n-1}\beta + a_0\gamma^n = 0.$$

Then β^n is in the ideal $\mathcal{O}\gamma$. Taking the n-th power of the equation of ideals $\mathcal{O} = \mathcal{O}\beta + \mathcal{O}\gamma$ yields $\mathcal{O} = \mathcal{O}^n = (\mathcal{O}\beta + \mathcal{O}\gamma)^n \subseteq \mathcal{O}\beta^n + \mathcal{O}\gamma = \mathcal{O}\gamma$, and thus γ is invertible in $\mathcal{O}$, which implies that $\alpha \in \mathcal{O}$. □

Corollary 3.5.9 *Let G be a finite group and $\mathcal{O}$ a principal ideal domain such that the quotient field K of $\mathcal{O}$ is a splitting field of characteristic zero for G. For any $\chi \in \mathrm{Irr}_K(G)$, any $a \in \mathcal{O}G$, and any $z \in Z(\mathcal{O}G)$ we have $\chi(a) \in \mathcal{O}$ and $\frac{\chi(z)}{\chi(1)} \in \mathcal{O}$.*

Proof It follows from 3.5.3 that if $x \in G$, then $\chi(x)$ is an algebraic integer, and hence we have $\chi(x) \in \mathcal{O}$ by 3.5.8. Since χ extends $\mathcal{O}$-linearly to $\mathcal{O}G$, the first statement follows. Similarly, by 3.5.6, if z is a conjugacy class sum of an element in G, then $\frac{\chi(z)}{\chi(1)}$ is an algebraic integer, hence in $\mathcal{O}$ by 3.5.8. Since any element in $Z(\mathcal{O}G)$ is an $\mathcal{O}$-linear combination of conjugacy class sums, the result follows. □

3.6 Burnside's $p^a q^b$-Theorem

A finite group G is called *solvable* if there is a finite sequence of subgroups

$$\{1\} = G_0 \subseteq G_1 \subseteq \cdots \subseteq G_n = G$$

with the property that G_{i-1} is normal in G_i and G_i/G_{i-1} is abelian, for $1 \leq i \leq n$. Equivalently, G is solvable if all composition factors of G in a composition series of G are cyclic of (possibly different) prime orders. If N is a normal subgroup of G such that N and G/N are solvable then G itself is solvable. Any finite p-group for some prime number p is solvable because one of the standard properties of a finite p-group P is that if $P \neq \{1\}$ then $Z(P) \neq \{1\}$. Burnside's $p^a q^b$-Theorem states that any finite group whose order has at most two prime divisors is solvable. The proof is again based on some of the integrality considerations from the previous section.

Theorem 3.6.1 (Burnside) *Let G be a finite group. Suppose that $|G| = p^a q^b$ for prime numbers p, q and integers $a, b \geq 0$. Then G is solvable.*

Finite groups with three prime divisors need not be solvable: the alternating group A_5 has order $60 = 2^2 \cdot 3 \cdot 5$ and is simple nonabelian, hence not solvable. In order to prove 3.6.1 we need the following two results:

Theorem 3.6.2 (Burnside) *Let G be a finite group, let $x \in G$ and let C be the conjugacy class of x in G. Let $\chi \in \mathrm{Irr}_{\mathbb{C}}(G)$ such that $(\chi(1), |C|) = 1$. Then either $\chi(x) = 0$ or $|\chi(x)| = \chi(1)$.*

Proof The number $\frac{\chi(x)|C|}{\chi(1)}$ is an algebraic integer, by 3.5.6. Since $(\chi(1), |C|) = 1$ there are rational integers a, b such that $a\chi(1) + b|C| = 1$. Thus $\frac{\chi(x)(1-a\chi(1))}{\chi(1)} = b\frac{\chi(x)|C|}{\chi(1)}$ is an algebraic integer. Since $a\chi(1)$ is an algebraic integer as well, it follows that $\alpha = \frac{\chi(x)}{\chi(1)}$ is an algebraic integer. Suppose $|\chi(x)| < \chi(1)$. Then $|\alpha(x)| < 1$. Let m be the order of x in G, and let K be the splitting field in $\mathbb{C}$ of the polynomial $X^m - 1$; that is, $K = \mathbb{Q}[\zeta]$ for some primitive m-th root of unity ζ in $\mathbb{C}$. Let $\mathrm{Gal}(K/\mathbb{Q})$ be the Galois group of K over $\mathbb{Q}$, and let $\sigma \in \mathrm{Gal}(K/\mathbb{Q})$. Since $\chi(x)$ is a sum of $\chi(1)$ roots of unity not all of which are equal, the same is true for $\sigma(\chi(x))$. It follows that $|\sigma(\chi(x))| < \chi(1)$. Thus $|\sigma(\alpha)| < 1$. Also, if α is a root of a polynomial $f \in \mathbb{Z}[X]$, then $\sigma(\alpha)$ is a root of the polynomial f as well, because the coefficients of f being in $\mathbb{Z}$ are invariant under σ. In particular, $\sigma(\alpha)$ is again an algebraic integer. Setting $\beta = \prod_{\sigma\in\mathrm{Gal}(K/\mathbb{Q})} \sigma(\alpha)$, we get that β is an algebraic integer such that $|\beta| < 1$. But since β is fixed by all elements of the Galois group of K over $\mathbb{Q}$, it follows that $\beta \in \mathbb{Q}$. Therefore, by 3.5.2 (iii), β is an integer. However, the only integer whose absolute value is smaller than 1 is zero. This shows that $\alpha = 0$, hence $\chi(x) = 0$. □

Theorem 3.6.3 *Let G be a finite nonabelian simple group. Then 1 is the only conjugacy class in G whose size is the power of some prime number.*

Proof Let $x \in G$ and let C be the conjugacy class of x in G. Suppose that $x \neq 1$ and that $|C| = p^a$ for some prime number p. Let $\chi \in \mathrm{Irr}_{\mathbb{C}}(G)$, $\chi \neq 1$. We first rule out the possibility $|\chi(x)| = \chi(1)$. Since G is simple nonabelian, we have $G = [G, G]$, hence the trivial character 1 is the only irreducible character having degree 1. Thus $\chi(1) > 1$. Therefore, by 3.5.3 (iii), the set $Z(\chi)$ of $y \in G$ satisfying $|\chi(y)| = \chi(1)$ is a proper normal subgroup of G, hence trivial by the simplicity of G. Thus 3.6.2 implies that if p does not divide $\chi(1)$, then $\chi(x) = 0$. Consider the character ρ of the regular $\mathbb{C}G$-module $\mathbb{C}G$. We have

$$0 = \rho(x) = \sum_{\chi\in\mathrm{Irr}_{\mathbb{C}}(G)} \chi(1)\chi(x) = 1 + \sum_{\chi\in\mathrm{Irr}_{\mathbb{C}}(G);\, p|\chi(1)} \chi(1)\chi(x).$$

But then $-\frac{1}{p} = \sum_{\chi\in\mathrm{Irr}_{\mathbb{C}}(G);\, p|\chi(1)} \frac{\chi(1)}{p}\chi(x)$ is an algebraic integer, contradicting 3.5.2 (iii). □

Proof of 3.6.1 Arguing by contradiction, let G be a counterexample of minimal order. Then G must be simple and nonabelian; indeed, if G had a nontrivial proper normal subgroup N, then by the minimality of G, both N and G/N would be solvable, hence so would be G. Since nontrivial finite groups of prime power order have a nontrivial centre, it follows that $p \neq q$, and that a and b are both positive. Let P be a Sylow p-subgroup of G and let $y \in Z(P)$, $y \neq 1$. Let C be the conjugacy class of y in G. Then $|C| = |G : C_G(y)|$ is prime to p because $P \subseteq C_G(y)$. By the assumptions on the order of G we get that $|C|$ is a power of q. This contradicts, however, 3.6.3. □

3.7 Brauer's characterisation of characters

The purpose of this section is to prove a fundamental result of Brauer which characterises characters of a finite group in terms of restrictions to a certain class of subgroups.

Definition 3.7.1 A finite group H is called *p-elementary*, where p is a prime, if $H = C \times P$ for some cyclic p'-subgroup C of H and a Sylow p-subgroup P of H. Moreover, H is called *elementary* if H is p-elementary for some prime p.

The class of elementary finite groups is closed under taking subgroups and quotient groups. Any elementary finite group is in particular nilpotent; that is, a direct product of its Sylow subgroups. Brauer's characterisation of characters states that the property of a class function to be a virtual character is detected upon restriction to elementary subgroups. Given a finite group G and a field K we denote as before by $\mathrm{Cl}_K(G)$ the K-vector space of K-valued class functions on G. If K has characteristic zero, we denote by $\mathbb{Z}\mathrm{Irr}_K(G)$ the abelian subgroup of $\mathrm{Cl}_K(G)$ generated by the set $\mathrm{Irr}_K(G)$ of K-valued irreducible characters of G. The K-vector space $\mathrm{Cl}_K(G)$ becomes a ring via multiplication in K; the unit element with respect to this multiplication is the trivial character, abusively denoted 1_G, of G. Since the character of the tensor product of two KG-modules is the product of the characters of the two modules, the abelian group $\mathbb{Z}\mathrm{Irr}_K(G)$ is in fact a subring of $\mathrm{Cl}_K(G)$.

Theorem 3.7.2 *Let G be a finite group, K a splitting field of characteristic zero for all subgroups of G and let χ be a K-valued class function on G. The following are equivalent:*

(i) $\chi \in \mathbb{Z}\mathrm{Irr}_K(G)$.
(ii) $\mathrm{Res}^G_H(\chi) \in \mathbb{Z}\mathrm{Irr}_K(H)$ *for any elementary subgroup H of G.*

The proof we present here requires the consideration of a slightly larger class of subgroups of G.

Definition 3.7.3 A finite group H is called *p-quasi-elementary*, where p is a prime, if $H = E \rtimes P$ for some cyclic normal p'-subgroup E of H and a Sylow

p-subgroup P of H. Moreover, H is called *quasi-elementary* if H is p-quasi-elementary for some prime p.

One verifies easily that the class of quasi-elementary finite groups is closed under taking subgroups and quotients. Any p-quasi-elementary finite group is in particular p-nilpotent; that is, a semidirect product of a Sylow p-subgroup acting on a normal p'-subgroup. Given a finite group G and a field K of characteristic zero, we denote by $\mathrm{perm}_K(G)$ the subgroup of $\mathbb{Z}\mathrm{Irr}_K(G)$ generated by the characters of the transitive permutation modules $\mathrm{Ind}_H^G(K)$, where H runs over the subgroups of G. That is, the elements of $\mathrm{perm}_K(G)$ are of the form $\sum_H a_H \mathrm{Ind}_H^G(1_H)$, where H runs over the subgroups of G and $a_H \in \mathbb{Z}$. The tensor product of two permutation module is again a permutation module by 1.1.9, and hence $\mathrm{perm}_K(G)$ is a subring of $\mathbb{Z}\mathrm{Irr}_K(G)$. We denote by $\mathrm{qelem}_K(G)$ the subgroup of $\mathrm{perm}_K(G)$ generated by the characters of the permutation modules $\mathrm{Ind}_H^G(K)$, where H runs over all quasi-elementary subgroups of G. That is, the elements of $\mathrm{qelem}_K(G)$ are of the form $\sum_H a_H \mathrm{Ind}_H^G(1_H)$, where H runs over the quasi-elementary subgroups of G and $a_H \in \mathbb{Z}$. Thanks to the formula for the tensor product of induced modules in 2.4.11, the subgroup $\mathrm{qelem}_K(G)$ is actually an ideal in the ring $\mathrm{perm}_K(G)$. It turns out that this ideal is equal to $\mathrm{perm}_K(G)$. To prove this, we need the following technical observation.

Lemma 3.7.4 *Let G be a finite group, p a prime, and K a field of characteristic zero. For any $x \in G$ there is a p-quasi-elementary subgroup H of G such that the character $\mathrm{Ind}_H^G(1_H)$ of the permutation module $\mathrm{Ind}_H^G(K)$ evaluated at x is not divisible by p.*

Proof The cyclic group generated by x decomposes uniquely as a direct product $\langle x\rangle = Q \times E$, where Q is the Sylow p-subgroup of $\langle x\rangle$ and E its complement. Set $N = N_G(E)$. We have $\langle x\rangle/E \cong Q$, so this is a p-subgroup of N/E. A Sylow p-subgroup of N/E is of the form H/E, where H is a p-quasi-elementary subgroup of N. By 3.1.16, the character value $\mathrm{Ind}_H^G(1_H)(x)$ is equal to the number of cosets gH, with $g \in G$, satisfying $xgH = gH$, or equivalently, satisfying $g^{-1}xg \in H$. In particular, any such g satisfies $g^{-1}Eg \subseteq H$. Since H/E is a p-group but E has order prime to p, this forces $g^{-1}Eg = E$, hence $g \in N$. Thus $\mathrm{Ind}_H^G(1_H)(x)$ is in fact equal to the number of cosets gH with $g \in N$ such that $xgH = gH$. For any $g \in N$ we have $EgH = gEH = gH$, and hence all elements in E act trivially on N/H via left multiplication. Since $\langle x\rangle/E$ is a p-group, it follows from 3.1.18 that

$$|\{gH \in N/H \mid xgH = gH\}| \equiv |N/H| \not\equiv 0 \pmod{p}$$

whence the result. □

Theorem 3.7.5 (L. Solomon) *Let G be a finite group and K a field of characteristic zero. We have* $\mathrm{perm}_K(G) = \mathrm{qelem}_K(G)$*; equivalently, there are integers a_H such that*

$$1_G = \sum_H a_H \mathrm{Ind}_H^G(1_H)$$

where H runs over the quasi-elementary subgroups of G.

Proof Since $\mathrm{qelem}_K(G)$ is an ideal in $\mathrm{perm}_K(G)$ by the remarks above, it is indeed sufficient to show that 1_G belongs to $\mathrm{qelem}_K(G)$. Let $x \in G$. Set $Z_x = \{\psi(x) | \psi \in \mathrm{qelem}_K(G)\}$. This is a subgroup of $\mathbb{Z}$ which by 3.7.4 is not contained in $p\mathbb{Z}$ for any prime p, and hence $Z_x = \mathbb{Z}$. Thus there is $\psi_x \in \mathrm{qelem}_K(G)$ such that $\psi_x(x) = 1$, or equivalently, such that the virtual character $1_G - \psi_x$ vanishes at x. Taking the product over all $x \in G$ yields the equation $\prod_{x\in G}(1_G - \psi_x) = 0$. Developing this expression and bringing all terms but 1_G to the right side yields an equality as in the statement. □

If one allows all characters rather than permutation characters, one can replace quasi-elementary subgroups by the smaller class of elementary subgroups:

Theorem 3.7.6 (Brauer) *Let G be a finite group and K a field of characteristic zero. Suppose that K is a splitting field for all subgroups of G. Let $\chi \in \mathrm{Irr}_K(G)$. There are integers $a(H, \psi)$ such that*

$$\chi = \sum_{(H,\psi)} a(H, \psi)\mathrm{Ind}_H^G(\psi)$$

where (H, ψ) runs over all pairs consisting of an elementary subgroup H of G and an irreducible K-valued character ψ of degree 1 of H.

Proof An elementary finite group is nilpotent, and hence any irreducible character of an elementary subgroup H of G is induced from a character of degree 1 of some subgroup of H by 3.3.16. It suffices therefore to show the slightly weaker statement of this theorem where we allow the pairs (H, ψ) to run over elementary subgroups H of G and all irreducible characters ψ of H. Furthermore, it suffices to show this for $\chi = 1_G$, since an equation as in the theorem for 1_G yields such an equation for any $\chi \in \mathrm{Irr}_K(G)$ by multiplying both sides by χ. Suppose that G is a counterexample of minimal order. Then, by 3.7.5, G is quasi-elementary for some prime p, but G is not elementary. Write $G = E \rtimes P$, where E is a cyclic normal p'-subgroup of G and P a Sylow p-subgroup of G. Since G is not elementary, it follows that

P acts nontrivially on E, and so $Z = C_E(P)$ is a proper subgroup of E. Then $H = Z \times P$ is a proper subgroup of p'-index in G. Since, by Frobenius' Reciprocity, we have $\langle \mathrm{Ind}_H^G(1_H), 1_G\rangle_G = \langle 1_H, \mathrm{Res}_H^G(1_G)\rangle_H = 1$, it follows that the class function $\psi = \mathrm{Ind}_H^G(1_H) - 1_G$ is an actual character of G. By the assumptions on G, the character ψ has at least one irreducible component χ that is not induced from a proper subgroup of G. Since $G = EH$ and $E \cap H = Z$, Mackey's formula yields $\mathrm{Res}_E^G\mathrm{Ind}_H^G(1_H) = \mathrm{Ind}_Z^E(1_Z)$. Thus $\langle \mathrm{Res}_E^G(\psi), 1_E\rangle = \langle \mathrm{Ind}_Z^E(1_Z) - 1_E, 1_E\rangle = 0$, where we use again Frobenius' Reciprocity. Let now $\lambda \in \mathrm{Irr}_K(E)$ be an irreducible component of $\mathrm{Res}_E^G(\chi)$; since E is abelian, λ is a group homomorphism from E to $K^\times$. By the above, 1_E is not a component of $\mathrm{Res}_E^G(\psi)$, hence not a component of $\mathrm{Res}_E^G(\chi)$, and hence $\lambda \neq 1_E$. Moreover, by 3.3.10, the character λ is G-stable because otherwise χ would be induced from a proper subgroup of G. Since Z is normal in G, the elements in Z act as identity on G/H by left multiplication, hence as identity on a module affording the character χ, and hence on a module affording the character λ. This means that Z is contained in the kernel $N = \ker(\lambda)$ of λ. Since λ is G-stable, for $y \in E$ and $u \in P$ we have $\lambda(u^{-1}yu) = \lambda(y)$, hence $u^{-1}yuy^{-1} \in N$, or equivalently, $u^{-1}yu \in yN$. This shows that P acts by conjugation on any coset yN. By 3.1.18 we have

$$|yN \cap Z| \equiv |yN| \equiv |N| \not\equiv 0 \,(\mathrm{mod}\; p).$$

In particular, the set $yN \cap Z$ is non empty, so $yN \cap N$ is non empty, and hence $y \in N$. This implies $N = E$. But then λ is a trivial character of E, a contradiction. □

The integers $a(H, \psi)$ in Brauer's Theorem are not unique, in general, but they can be made canonical by requiring suitable naturality properties – this is due to independent work of Boltje [9] and Snaith [86].

Proof of Theorem 3.7.2 We consider the set $\mathcal{E}$ in $\mathrm{Cl}_K(G)$ consisting of all class functions ψ with the property that $\mathrm{Res}_H^G(\psi) \in \mathbb{Z}\mathrm{Irr}_K(H)$ for all elementary subgroups H of G. Since restriction to a subgroup induces a ring homomorphism on class functions, the set $\mathcal{E}$ is a subring of $\mathrm{Cl}_K(G)$. This ring contains the set $\mathcal{I}$ of all elements of the form $\sum_{(H,\psi)} a(H, \psi)\mathrm{Ind}_H^G(\psi)$ where the $a(H, \psi)$ are integers and where (H, ψ) runs over all pairs consisting of an elementary subgroup H of G and an irreducible K-valued character ψ of H. By using the reciprocity property for products of induced class functions 3.1.25 one sees that the set $\mathcal{I}$ is in fact an ideal in $\mathcal{E}$. This ideal contains 1_G by 3.7.6. Thus $\mathcal{E} = \mathcal{I}$, which shows that (ii) implies (i). The converse is trivial. □

3.8 Splitting fields for group algebras

An important consequence of 3.7.6 is another theorem of Brauer which yields a sufficient condition for when a field of characteristic zero is a splitting field for a finite group G. The *exponent* of a finite group G is the least common multiple of the orders of the elements in G, in particular, the exponent of G divides the order of G.

Theorem 3.8.1 *Let G be a finite group and n the exponent of G. Let K be a field of characteristic zero containing a primitive n-th root of unity. Then K is a splitting field for all subgroups of G.*

Proof It suffices to show that K is a splitting field for G since the exponent of any subgroup of G divides the exponent n of G. Let K'/K be a field extension such that K' is a splitting field for G. By 1.14.9 we need to show that if V' is a simple $K'G$-module V' then there is a simple KG-module V such that $K' \otimes_K V$. Let χ be the character of V'. By 3.7.6 there are integers $a(H, \psi)$ such that $\chi = \sum_{(H,\psi)} a(H, \psi)\mathrm{Ind}_H^G(\psi)$ where (H, ψ) runs over all pairs consisting of an elementary subgroup H of G and an irreducible K-valued character ψ of degree 1 of H. Since the orders of all elements in H divide n, the image of any group homomorphism $\psi : H \to K'$ is contained in K, as K contains a primitive n-th root of unity. Thus, for any such ψ, the character $\mathrm{Ind}_H^G(\psi)$ is the character of a KG-module. This shows that χ is in fact in $\mathbb{Z}\mathrm{Irr}_K(G)$, so $\chi = \sum_\eta a_\eta \eta$ for some integers a_η, where η runs over $\mathrm{Irr}_K(G)$. The integers a_η are nonnegative, since χ is the character of a $K'G$-module and since the characters η, upon extension to K', have pairwise different constituents by 3.3.9. Thus χ is the character of a KG-module V. Since χ is the character of a simple $K'G$-module, it follows that V is necessarily simple, whence the result. □

The converse of this theorem holds as well: if K is a splitting field of characteristic zero for all subgroups of a finite group G, then K is in particular a splitting field for all cyclic subgroups of G, and hence K contains a primitive n-th root of unity, where n is the exponent of G. It is possible though for K to be a splitting field for G and not contain a primitive n-th root of unity; for instance, the rational number field is a splitting field for all symmetric groups (but not for all of their subgroups). The above theorem holds also for fields of positive characteristic:

Theorem 3.8.2 *Let G be a finite group, p a prime and n the p'-part of the exponent of G. Let k be a field of characteristic p containing a primitive n-th root of unity. Then k is a splitting field for all subgroups of G.*

Proof Let ζ be a primitive n-th root of unity in k. It suffices to show that the finite field $\mathbb{F}_p(\zeta)$ is a splitting field for G. By 1.14.10 there is a finite extension k' of $\mathbb{F}_p(\zeta)$ such that k' is a splitting field for G. Proceeding inductively it suffices to show that then a maximal subfield k of k' containing $\mathbb{F}_p(\zeta)$ is still a splitting field for G. Let S' be a simple $k'G$-module. Set $E' = \mathrm{End}_{k'}(S')$. Note that for any $x \in G$ we have $\mathrm{tr}_{S'}(x) \in \mathbb{F}_p(\zeta) \subseteq k$. Since k' is a splitting field for G, the structural map $k'G \to E'$ is surjective. Denote by E the image of kG in E' under this map. Thus E is a k-subalgebra of E' generated by the image of G and thus satisfying $k' \cdot E = E'$, and for any $e \in E$ we have $\mathrm{tr}_{S'}(e) \in k$. In particular, $k \cdot \mathrm{Id}_{S'} \subseteq Z(E) \subseteq Z(E') = k' \cdot \mathrm{Id}_{S'}$. Since k is a maximal subfield of k', either $Z(E) \cong k'$ or $Z(E) \cong k$. If $Z(E) \cong k'$ then $E = E'$. Since E' is a full matrix algebra over k', for any $\alpha \in k' \setminus k$ there is $e \in E'$ such that $\mathrm{tr}_{S'}(e) = \alpha$, contradicting the fact that $\mathrm{tr}_{S'}(e) \in k$ for all $e \in E = E'$. Thus $Z(E) \cong k$. Since $k' \cdot J(E)$ is a nilpotent ideal in the matrix algebra E' we have $J(E) = \{0\}$; that is, E is semisimple, hence a direct product of matrix rings over division algebras over k. Using again $Z(E) \cong k$ we get that $E \cong M_n(D)$ for some positive integer n and some division ring D which is a finite extension of the finite field k, hence commutative by a theorem of Wedderburn. But then $D \cong Z(E) \cong k$, so $D = k$. As $E' = k \cdot E$, the algebra E' is a quotient of $k \otimes_k E \cong k' \otimes_k M_n(k) \cong M_n(k')$, which is a simple algebra, and hence $E' \cong M_n(k')$. The unique simple $M_n(k)$-module S, viewed as kG-module through the above maps $kG \to E \cong M_n(k)$ satisfies therefore $k' \otimes_k S \cong S'$. It follows from 1.14.9 that k is a splitting field for G, whence the result. □

This proof yields a more precise statement: it shows that if a subfield k of a field k' of prime characteristic p contains the character values of an absolutely simple $k'G$-module S' then there is a simple kG-module S satisfying $k' \otimes_k S \cong S'$. Using some Galois Theory one can refine this further, dropping the assumption to S' being simple (but not necessarily absolutely simple). This stronger statement is not true over fields of characteristic zero; for instance, if G is the quaternion group of order 8, then the character values of all simple $\mathbb{C}G$-modules are in $\mathbb{Q}$, but $\mathbb{Q}$ is not a splitting field for G because the 2-dimensional simple $\mathbb{C}G$-modules cannot be realised over $\mathbb{Q}$.

3.9 Integral group rings

To what extent does the group algebra $\mathbb{Z}G$ determine the structure of a finite group G? Zassenhaus conjectured that a ring isomorphism $\mathbb{Z}G \cong \mathbb{Z}H$ should imply a group isomorphism $G \cong H$, where G, H are finite groups. This is not the case: M. Hertweck found a first counterexample to this conjecture in 1997. What is still true is that an isomorphism of integral group rings $\mathbb{Z}G \cong \mathbb{Z}H$ implies that G and H have the same character tables. We use the following terminology and notation. Given a finite group G and a commutative ring R, a *conjugacy class sum* in RG is the sum in RG of all G-conjugates of an element $x \in G$. For c the conjugacy class sum of an element $x \in G$, we denote by $|c|$ the number of different conjugates of x in G; that is, $|c| = |G : C_G(x)|$. By 1.5.1, the set of conjugacy class sums in RG is a R-basis of $Z(RG)$. The following theorem, known as the *class sum correspondence*, is due to Glauberman in the case $R = \mathbb{Z}$, and has been generalised in various sources such as Passman [69], Saksonov [83], Roggenkamp and Scott [81].

Theorem 3.9.1 *Let G, H be finite groups and R an integral domain of characteristic zero such that no prime divisor of $|G|$ is invertible in R. Suppose that there is an R-algebra isomorphism $\alpha : RG \cong RH$. Then, for any conjugacy class sum c in RG there is a conjugacy class sum d in RH and an element $\delta_c \in R^\times$ such that $\alpha(c) = \delta_c d$ and $|c| = |d|$. The map sending c to d defined in this way induces a bijection between the sets of conjugacy classes of G and of H. Moreover, α can be chosen in such a way that $\delta_c = 1$ for all conjugacy class sums c in RG. In particular, G and H have the same character tables over a splitting field containing R.*

Proof Denote by $\mathcal{C}$ the set of conjugacy class sums of G in RG and by $\mathcal{D}$ the set of conjugacy class sums of H in RH. The sets $\mathcal{C}$ and $\mathcal{D}$ are R-bases of $Z(RG)$ and $Z(RH)$, respectively, by 1.5.1. Since $RG \cong RH$ we have $|G| = |H|$ and $Z(RG) \cong Z(RH)$; thus $\mathcal{C}$ and $\mathcal{D}$ have the same number of elements. Let K be a field containing R and a primitive $|G|$-th root of unity ζ; by 3.8.1, K is a splitting field for G and H. The isomorphism $\alpha : RG \cong RH$ extends to an isomorphism of K-algebras $KG \cong KH$, abusively still denoted by the same letter α. Let V be a simple KH-module and let $\eta \in \mathrm{Irr}_K(H)$ be its character, viewed as linear map from KH to K. Through the isomorphism α we can consider V as simple KG-module, denoted by $\mathrm{Res}_\alpha(V)$, and then the character of $\mathrm{Res}_\alpha(V)$ is $\eta \circ \alpha$. Since α is an isomorphism, the map sending η to $\eta \circ \alpha$ is a bijection $\mathrm{Irr}_K(H) \cong \mathrm{Irr}_K(G)$. If c is the conjugacy class sum in RG of all conjugates of an element $x \in G$, we denote by c^{-1} the conjugacy class sum of x^{-1}. Moreover, we have $\chi(c) = |c|\chi(x)$, for any character χ of G. The sets $\{\alpha(c)|c \in \mathcal{C}\}$ and $\mathcal{D}$ are both R-bases of $Z(RH)$. Write

$$\alpha(c) = \sum_{d \in \mathcal{D}} \mu(c, d)d$$

for some $\mu(c,d) \in R$. We show that $\mu(c,d) \in \mathbb{Z}[\zeta]$. Applying $\eta \in \mathrm{Irr}_K(H)$ yields $\eta(\alpha(c)) = \sum_{d\in\mathcal{D}} \mu(c,d)\eta(d)$. Since $\eta(\alpha(c))$ and $\eta(d)$ belong to $\mathbb{Q}(\zeta)$ so does $\mu(c,d)$. Multiplying the previous equation by $\eta(d')$ for some fixed $d' \in \mathcal{D}$ yields $\eta(\alpha(c))\eta(d') = \sum_{d\in\mathcal{D}} \mu(c,d)\eta(d)\eta(d')$. We take now the sum over all $\eta \in \mathrm{Irr}_K(H)$ of this equation. The Second Orthogonality Relations 3.3.11 imply that the sum for $d' \neq d^{-1}$ vanishes, and for $d' = d^{-1}$ yields

$$\sum_{\eta\in\mathrm{Irr}_K(H)} \eta(\alpha(c))\eta(d^{-1}) = \mu(c,d)\cdot|d|\cdot|H|.$$

Since $\eta\circ\alpha$ and η are irreducible characters of G and H, respectively, the left side is an algebraic integer in $\mathbb{Q}(\zeta)$, hence in $\mathbb{Z}[\zeta]$. Applying the norm N with respect to the Galois extension $\mathbb{Q}(\zeta):\mathbb{Q}$ yields thus an integer, equal to $N(\mu(c,d))\cdot|d|^m\cdot|H|^m$, where m is the order of $\mathrm{Gal}(\mathbb{Q}(\zeta):\mathbb{Q})$. Now $N(\mu(c,d))$ is an element in $R\cap\mathbb{Q}$, and no prime divisor of $|G|$, thus no prime divisor of $|d|\cdot|H|$, is invertible in R. Thus $R\cap\mathbb{Q}$ is a subring of $\{\frac{a}{b}|a,b\in\mathbb{Z},(b,|G|)=1\}$. This forces $N(\mu(c,d))\in\mathbb{Z}$, hence $\mu(c,d)\in\mathbb{Z}[\zeta]$. We may assume without loss of generality that $R\subseteq K=\mathbb{Q}(\zeta)$. Then $\eta(d^{-1})$ and $\eta(\alpha(c^{-1}))$ are the complex conjugates of $\eta(d)$ and $\eta(\alpha(c))$, respectively, and hence $\mu(c^{-1},d^{-1})$ is the complex conjugate $\bar{\mu}(c,d)$ of $\mu(c,d)$. Again by the Second Orthogonality Relations 3.3.11 we have

$$\begin{aligned}|G||c| &= \sum_{\eta}\eta(\alpha(c))\bar{\eta}(\alpha(c)) = \sum_{\eta}\sum_{d\in\mathcal{D}}\sum_{d'\in\mathcal{D}}\mu(c,d)\bar{\mu}(c,d')\eta(d)\bar{\eta}(d')\\ &= \sum_{d\in\mathcal{D}}|\mu(c,d)|^2|H||d|\end{aligned}$$

where η runs over $\mathrm{Irr}_K(H)$. Since $|G|=|H|$, comparing these two expressions yields

$$|c| = \sum_{d\in\mathcal{D}}|\mu(c,d)|^2|d|$$

for all $c\in\mathcal{C}$. Taking the sum over all conjugacy classes of G of this equation yields

$$|G| = \sum_{c\in\mathcal{C}}\sum_{d\in\mathcal{D}}|\mu(c,d)|^2|d| = \sum_{d\in\mathcal{D}}\Big(\sum_{c\in\mathcal{C}}|\mu(c,d)|^2\Big)|d|.$$

Applying $\sigma\in\mathrm{Gal}(\mathbb{Q}(\zeta):\mathbb{Q})$ to this equation yields a similar equation with $\mu(c,d)$ replaced by $\sigma(\mu(c,d))$. Taking the sum over all $\sigma\in\mathrm{Gal}(\mathbb{Q}(\zeta):\mathbb{Q})$ of these equations yields

$$|G|m = \sum_{d\in\mathcal{D}}\Big(\sum_{c\in\mathcal{C}}\sum_{\sigma} |\sigma(\mu(c,d))|^2\Big)|d|$$

and this is also equal to $|H|m$ since $|H| = |G|$. Since the $\mu(c, d)$ are algebraic integers, their norms are rational integers. The well-known inequality between the arithmetic and geometric means (see e.g. [41, (4.10)] for a proof) applied to the m numbers $|\sigma(\mu(c, d))|^2$, for fixed c, d, and with σ running over $\mathrm{Gal}(\mathbb{Q}(\zeta) : \mathbb{Q})$ yields

$$\sum_{\sigma} |\sigma(\mu(c,d))|^2 \geq m \cdot |N(\mu(c,d))|^{\frac{2}{m}}.$$

Equality holds if and only if all $|\sigma(\mu(c, d))|$ are equal. If $\mu(c, d) \neq 0$ then $|N(\mu(c, d))| \geq 1$, hence $\sum_\sigma |\sigma(\mu(c, d))|^2 \geq m$. Since for any $d \in \mathcal{D}$ there is at least one $c \in \mathcal{C}$ satisfying $\mu(c, d) \neq 0$, this forces that for any $d \in \mathcal{D}$ there is exactly one $c = c(d)$ in $\mathcal{C}$ satisfying $\sum_\sigma |\sigma(\mu(c, d))|^2 = m$, hence $|\mu(c, d)| = 1$, and $\mu(c', d) = 0$ for $c' \neq c$ in $\mathcal{C}$. But G and H have the same number of conjugacy classes, and hence, for any $c \in \mathcal{C}$ there is a unique $d = d(c) \in \mathcal{D}$ such that $\alpha(c) = \delta_c d(c)$, where $\delta_c = \mu(c, d(c)) \in R^\times$. Since $|\delta_c| = 1$, a previous equality implies, $|c| = |d|$. In order to show that we can always find an isomorphism α for which all δ_c are 1, we define an automorphism τ of RH as follows. Let $\epsilon : RH \to R$ be the augmentation homomorphism; that is, $\epsilon(\sum_{y\in H} m_y y) = \sum_{y\in H} m_y$ for all $\sum_{y\in H} m_y y \in RH$. In particular, $\epsilon(d) = |d|$ for all $d \in \mathcal{D}$. Since ϵ is a ring homomorphism it maps every invertible element in RH to an element in $R^\times$. Thus the map $\epsilon \circ \alpha : RG \to R$ is a ring homomorphism with the property that $\epsilon(\alpha(x)) \in R^\times$ for any $x \in G$. Hence the unique R-linear map $\beta : RG \to RH$ defined by

$$\beta(x) = \epsilon(\alpha(x))^{-1}\alpha(x)$$

for all $x \in G$ is a ring isomorphism. Note also that the value $\epsilon(\alpha(x))$ of an element $x \in G$ depends only on the conjugacy class c of x in G. We have $\alpha(c) = \delta_c d$ for a unique $d \in \mathcal{D}$ and some $\delta_c \in R^\times$. Thus $\epsilon(\alpha(c)) = \delta_c|d|$ and hence $\delta_c = \epsilon(\alpha(x))$. Taking the sum over all conjugates of x yields

$$\beta(c) = \epsilon(\alpha(x))^{-1}\alpha(c) = \epsilon(\alpha(x))^{-1}\delta_c d = d$$

as required. Since $\mathrm{Irr}_K(G) = \{\eta \circ \alpha | \eta \in \mathrm{Irr}_K(H)\}$ the character tables of G and H over K are equal. □

In the situation of the above theorem, if $z \in Z(G)$ then $\alpha(z) = \delta_z y$ for some $y \in Z(H)$ and some $\delta_z \in R^\times$. One can show more generally that any central unit of finite order in RH has this form (it does not need to be part of a group basis); see e.g. [37], [8], [42].

Corollary 3.9.2 *Let G be an abelian finite group and R an integral domain of characteristic zero in which no prime divisor of $|G|$ is invertible. For any R-algebra automorphism α of RG there is a unique group automorphism φ of G and a unique group homomorphism $\mu : G \to R^\times$ such that $\alpha(x) = \mu(x)\varphi(x)$ for any $x \in G$. This correspondence induces an isomorphism of groups* $\mathrm{Aut}(RG) \cong \mathrm{Hom}(G, R^\times) \rtimes \mathrm{Aut}(G)$.

Proof Since G is abelian, every element of G is its own conjugacy class. The previous theorem implies that for any $x \in G$ there is a unique $\varphi(x) \in G$ and a unique element $\mu(x) \in R^\times$ satisfying $\alpha(x) = \mu(x)\varphi(x)$. A trivial verification shows that μ is a group homomorphism and that φ is an automorphism of G. Conversely, for any group homomorphism $\mu : G \to R^\times$ and any group automorphism φ of G, the unique R-linear map β defined, for $x \in G$, by $\beta(x) = \mu(x)\varphi(x)$ is an algebra automorphism, with inverse $\beta^{-1}(x) = \mu(\varphi^{-1}(x))^{-1}\varphi^{-1}(x)$. The result follows. □

Here is a sufficient criterion for the existence of an isomorphism $\mathbb{Z}G \cong \mathbb{Z}H$.

Theorem 3.9.3 *Let G be a finite group and let H be a subgroup of $(\mathbb{Z}G)^\times$ such that H is a $\mathbb{C}$-basis of $\mathbb{C}G$. Then $\mathbb{Z}H = \mathbb{Z}G$.*

Proof Since both G, H are $\mathbb{C}$-bases of $\mathbb{C}G = \mathbb{C}H$ we have $|G| = |H|$. The inclusion $\mathbb{Z}H \subseteq \mathbb{Z}G$ implies that any element $t \in H$ can be written uniquely in the form

$$t = \sum_{x \in G} \beta(t, x)x$$

for some integers $\beta(t, x)$. In order to prove the reverse inclusion $\mathbb{Z}G \subseteq \mathbb{Z}H$ we need to show that the inverse of the matrix $(\beta(t, x))_{t \in H, x \in G}$ is again a matrix with integer coefficients. For $s \in H$ we have $s^{-1} = \sum_{x \in G} \beta(s^{-1}, x)x$, and hence, for any two elements $t, s \in H$ we have

$$ts^{-1} = \sum_{x,y \in G} \beta(t, x)\beta(s^{-1}, y^{-1})xy^{-1}.$$

The coefficient at 1_G on the right side of this expression is $\sum_{x \in G} \beta(t, x)\beta(s^{-1}, x^{-1})$. Thus, if we denote by ρ the regular character of $\mathbb{C}G = \mathbb{C}H$ then

$$\sum_{x \in G} \beta(t, x)\beta(s^{-1}, x^{-1}) = \rho(ts^{-1}) = \begin{cases} 1 & \text{if} t = s \\ 0 & \text{if} t \neq s \end{cases}.$$

This shows that the inverse of the matrix $(\beta(t,x))_{t\in H, x\in G}$ is the transpose of the matrix with integer coefficients $(\beta(t^{-1},x^{-1}))_{t\in H,x\in G}$, whence the equality $\mathbb{Z}G = \mathbb{Z}H$ as stated. □

We mention, without proof, some more results around this theme. By a result of Roggenkamp and Scott [81], the isomorphism problem holds for finite p-groups, even in a slightly stronger version over p-local rings. Given a prime number p, let $\mathbb{Z}_{(p)} = \{\frac{a}{b} | a, b \in \mathbb{Z}, p \nmid b\}$; this is the localisation of $\mathbb{Z}$ at the prime p.

Theorem 3.9.4 *Let p be a prime and let P, Q be finite p-groups. If $\mathbb{Z}_{(p)}P \cong \mathbb{Z}_{(p)}Q$ then $P \cong Q$.*

The *modular isomorphism problem* is still open: given a prime p and finite p-groups P, Q, does an algebra isomorphism $\mathbb{F}_pP \cong \mathbb{F}_pQ$ imply a group isomorphism $P \cong Q$? Yes, if one of P, Q is abelian by a result of Deskins [26], but unknown in general. There are examples, due to Dade [23], of finite non-isomorphic groups G, H such that $kG \cong kH$ for every field k; the groups G, H in those examples are not p-groups. See [36] for an overview and further results. The following result on normalisers in group rings is essentially due to Coleman [21]; it is stated in this form in [51].

Proposition 3.9.5 *Let $\mathcal{O}$ be an integral domain and p a rational prime number such that $p \in J(\mathcal{O})$. Let G be a finite group, R a p-subgroup of G, and Q a subgroup of R. Consider $\mathcal{O}^\times$ and G as subgroups of $X = (\mathcal{O}G)^\times$, and set $Y = Z(\mathcal{O}Q)^\times$.*

(i) We have $N_X(\mathcal{O}^\times \cdot R) = N_X(R) = N_G(R) \cdot ((\mathcal{O}G)^R)^\times$.
(ii) We have $N_Y(\mathcal{O}^\times \cdot R) = N_Y(R) = Z(Q) \cdot ((\mathcal{O}G)^R)^\times$.

Proof Let $x \in N_X(\mathcal{O}^\times \cdot R)$. Thus if $u \in R$, then $xux^{-1} \in \mathcal{O}^\times \cdot R$. Since u, and hence also xux^{-1}, are mapped to 1 under the augmentation homomorphism $\mathcal{O}G \to \mathcal{O}$, it follows that $xux^{-1} \in R$. This shows the first equalities in (i) and (ii). Write $x = \sum_{g\in G} \lambda_g g$ for for some coefficients $\lambda_g \in \mathcal{O}$. The group $G \times G$ acts on G via $(g, h) \cdot w = gwh^{-1}$, where g, h, $w \in G$. Restricting this action along the group homomorphism $R \to G \times G$ sending $u \in R$ to (xux^{-1}, u) yields an action of R on G given by ${}^u g = xux^{-1}gu^{-1}$, for $u \in R$ and $g \in G$. We have $\sum_{g\in G} \lambda_g g = x = (xux^{-1})xu^{-1} = \sum_{g\in G} \lambda_g({}^u g)$. This shows that the function $g \mapsto \lambda_g$ is constant on R-orbits. The image of x in $\mathcal{O}$ under the augmentation map $\mathcal{O}G \to \mathcal{O}$ is equal to $\sum_{g\in G} \lambda_g$, and this must be invertible in $\mathcal{O}$ as x is invertible in $\mathcal{O}G$. Since p is in the radical of $\mathcal{O}$, this implies that R has an orbit of length 1, and that such an orbit can be chosen of the form $\{y\}$ for some $y \in G$ with $\lambda_y \in \mathcal{O}^\times$. The property ${}^u y = y$ is equivalent to $y = xux^{-1}yu^{-1}$, hence to $yuy^{-1} = xux^{-1}$, for all $u \in R$. Thus $y^{-1}x$ centralises R, hence belongs to $((\mathcal{O}G)^R)^\times$. Writing $x = yy^{-1}x$ proves (i). For (ii) we observe that if x belongs to $Y = Z(\mathcal{O}Q)^\times$, then the above argument shows that $y \in Q$. Since $y^{-1}x$ and x centralise Q in that case, it follows that $y \in Z(Q)$, proving (ii). □

Coleman's result can be interpreted as detecting p-fusion in group algebras. This theme is carried further in Section 8.7 in the context of the local structure of a p-block of a finite group.

4
Algebras over p-Local Rings

About three decades after Frobenius and Schur had laid the foundations of 'classical' finite group representation theory over the complex numbers, R. Brauer started in the late 1920s a systematic investigation of group representations over fields of positive characteristic. In order to relate group representations over fields of positive characteristic to character theory in characteristic zero, Brauer chose a *p-modular system* $(K, \mathcal{O}, k)$ consisting of a complete discrete valuation ring $\mathcal{O}$ with a residue field $k = \mathcal{O}/J(\mathcal{O})$ of prime characteristic p and a quotient field K of characteristic zero. The present chapter contains a summary of general properties of algebras over complete discrete valuation rings and their module categories as needed later on, including the Krull–Schmidt Theorem, lifting theorems for idempotents, a generalisation of a theorem of Wedderburn–Malcev due to Külshammer, Okuyama, Watanabe, and specialisations of earlier results to algebras over complete discrete valuation rings. For A an $\mathcal{O}$-algebra extending scalars yields a functor $K \otimes_{\mathcal{O}} -$ from $\mathrm{Mod}(A)$ to $\mathrm{Mod}(K \otimes_{\mathcal{O}} A)$, and reducing scalars modulo $J(\mathcal{O})$ yields a functor $k \otimes_{\mathcal{O}} -$ from $\mathrm{Mod}(A)$ to $\mathrm{Mod}(k \otimes_{\mathcal{O}} A)$. The fundamental techniques relating modules over $K \otimes_{\mathcal{O}} A$ and $k \otimes_{\mathcal{O}} A$ are discussed in the sections on decomposition matrices and decomposition maps.

4.1 Local rings and algebras

A commutative ring $\mathcal{O}$ is called *local* if $\mathcal{O}$ has a unique maximal ideal. If $\mathcal{O}$ is a commutative local ring, then the unique maximal ideal of $\mathcal{O}$ is equal to its Jacobson radical $J(\mathcal{O})$ because $J(\mathcal{O})$ is the intersection of all maximal ideals in $\mathcal{O}$ by 1.10.6. If $\mathcal{O}$ is a local commutative ring, then $k = \mathcal{O}/J(\mathcal{O})$ is a field, called the *residue field of the local ring* $\mathcal{O}$. Thus, for F a finitely generated free $\mathcal{O}$-module we have $F \cong \mathcal{O}^n$, where n is the dimension of the k-vector space

$k \otimes_{\mathcal{O}} F$; in particular, n is uniquely determined by F. We call n the *rank of* F, denoted by $\mathrm{rk}_{\mathcal{O}}(F)$.

Examples 4.1.1 Let p be a prime.

(a) The ring $\mathbb{Z}_{(p)} = \{\frac{a}{b} \mid a, b \in \mathbb{Z}, p \nmid b\}$ is commutative local with unique maximal ideal $p\mathbb{Z}_{(p)} = \{\frac{a}{b} \mid a, b \in \mathbb{Z}, p \nmid b, p|a\}$.
(b) The group algebra kP of a finite abelian p-group P over a field k of characteristic p is commutative local; here the augmentation ideal $I(kP)$ is the unique maximal ideal by 1.11.1. Note that $\mathbb{Z}_{(p)}$ is an integral domain, but kP is not unless P is trivial, because the ideal $I(kP)$ is nilpotent.
(c) Any field k is a local commutative ring with unique maximal ideal $\{0\}$.

Proposition 4.1.2 *Let $\mathcal{O}$ be a commutative ring. Then $\mathcal{O}$ is local if and only if $\mathcal{O}^\times = \mathcal{O} \setminus J(\mathcal{O})$.*

Proof The inclusion $\mathcal{O}^\times \subseteq \mathcal{O} \setminus J(\mathcal{O})$ holds always because $J(\mathcal{O})$ is a proper ideal, hence contains no invertible element. If $\mathcal{O}^\times = \mathcal{O} \setminus J(\mathcal{O})$ then $J(\mathcal{O})$ is the unique maximal ideal of $\mathcal{O}$ for the same reason. Conversely, if $J(\mathcal{O})$ is the unique maximal ideal in $\mathcal{O}$ then $k = \mathcal{O}/J(\mathcal{O})$ is a field. Thus if $\lambda \in \mathcal{O} \setminus J(\mathcal{O})$, then the image $\bar{\lambda}$ of λ in k is non zero, hence invertible, hence $k = k\bar{\lambda}$. This implies $\mathcal{O} = \mathcal{O}\lambda + J(\mathcal{O})$. Nakayama's Lemma applied to $\mathcal{O}$ yields $\mathcal{O} = \mathcal{O}\lambda$. Thus $\lambda \in \mathcal{O}^\times$. □

Proposition 4.1.3 *Let $\mathcal{O}$ be a commutative local Noetherian ring and M a finitely generated $\mathcal{O}$-module. We have $\cap_{k \geq 1} J(\mathcal{O})^k M = \{0\}$. In particular, we have $\cap_{k \geq 1} J(\mathcal{O})^k = \{0\}$.*

Proof Set $N = \cap_{k \geq 1} J(\mathcal{O})^k M$. Since $\mathcal{O}$ is Noetherian and M finitely generated, it follows that M is Noetherian, and hence N is finitely generated as an $\mathcal{O}$-module. We are going to prove that $J(\mathcal{O})N = N$. Let U be a submodule of M that is maximal with respect to the property $U \cap N = J(\mathcal{O})N$; such a U exists because M is Noetherian, hence every ascending chain of submodules of M becomes constant. Let $\lambda \in J(\mathcal{O})$. Our next goal is to show that there exists a positive integer t such that $\lambda^t M \subseteq U$. For any integer $k \geq 1$ define $U_k = \{m \in M | \lambda^k m \in U\}$. Clearly U_k is a submodule of M, and we have $U_k \subseteq U_{k+1}$. Again since M is Noetherian there is a positive integer t such that $U_t = U_{t+1}$. We have $J(\mathcal{O})N = U \cap N \subseteq (U + \lambda^t M) \cap N$. We are going to show that the second inclusion is an equality; the maximality assumption on U implies then $U = U + \lambda^t M$, or equivalently, $\lambda^t M \subseteq U$. Let $x \in (U + \lambda^t M) \cap N$. That is, $x = u + \lambda^t y \in N$ for some $u \in U$ and $y \in M$. Then $\lambda x = \lambda u + \lambda^{t+1} y \in J(\mathcal{O})N \subseteq U$, hence $\lambda^{t+1} y \in U$, which means that $y \in U_{t+1} = U_t$. Therefore $\lambda^t y \in U$. As mentioned before, this shows that $\lambda^t M \subseteq U$. Using yet again that $\mathcal{O}$ is Noetherian, $J(\mathcal{O})$ is finitely generated as an $\mathcal{O}$-module. Write $J(\mathcal{O}) = \sum_{1\leq i\leq n} \lambda_i \mathcal{O}$ for some positive integer n and some elements $\lambda_i \in J(\mathcal{O})$, where $1 \leq i \leq n$. By the above, there is a positive integer t such that $\lambda_i^t M \subseteq U$ for $1 \leq i \leq n$. Then $N \subseteq J(\mathcal{O})^{nt} M \subseteq \sum_{1\leq i\leq n} \lambda_i^t M \subseteq U$. Together we get that $J(\mathcal{O})N = U \cap N = N$. Thus $N = \{0\}$ by Nakayama's Lemma 1.10.4. This shows the first equality, and the second equality is the special case $M = \mathcal{O}$ of the first. □

Theorem 4.1.4 *Let $\mathcal{O}$ be a commutative local ring. Every finitely generated projective $\mathcal{O}$-module is free.*

Proof Let M be a finitely generated projective $\mathcal{O}$-module. Set $k = \mathcal{O}/J(\mathcal{O})$. Since $\mathcal{O}$ is local, k is a field. Set $\bar{M} = M/J(\mathcal{O})M$. Then $\bar{M}$ is an $\mathcal{O}$-module that is annihilated by $J(\mathcal{O})$, hence $\bar{M}$ can be viewed as a k-vector space. Since M is finitely generated as an $\mathcal{O}$-module, M is finite-dimensional as a k-vector space ; say, $\bar{M} \cong k^n$ for some integer $n \geq 0$. Consider the canonical map $\mathcal{O}^n \twoheadrightarrow k^n$. Since $\mathcal{O}^n$ is projective, there is an $\mathcal{O}$-homomorphism $\alpha : \mathcal{O}^n \to M$ which induces an isomorphism $k^n \cong \bar{M}$. Thus $\mathrm{Im}(\alpha) + J(\mathcal{O})M = M$, and hence α is surjective by Nakayama's Lemma. Since M is projective this implies that α splits. Let $\beta : M \to \mathcal{O}^n$ be a homomorphism satisfying $\alpha \circ \beta = \mathrm{Id}_M$. Since α induces an isomorphism $k^n \cong \bar{M}$, it follows that the map β induces the inverse of this isomorphism. But then again $\mathrm{Im}(\beta) + J(\mathcal{O})^n = \mathcal{O}^n$, which by Nakayama's Lemma 1.10.4 forces β to be surjective. Thus α and β are inverse to each other, and hence $M \cong \mathcal{O}^n$ is free. □

Corollary 4.1.5 *Let $\mathcal{O}$ be a commutative local ring and n a positive integer. Every $\mathcal{O}$-algebra automorphism of the matrix algebra $M_n(\mathcal{O})$ is an inner automorphism.*

Proof By 4.1.4 the ring $\mathcal{O}$ satisfies the hypotheses, hence the conclusion, of 2.8.12, which yields the result. □

Corollary 4.1.6 *Let $\mathcal{O}$ be a commutative local ring and n a positive integer. Let A be an $\mathcal{O}$-algebra that has a unitary subalgebra S isomorphic to $M_n(\mathcal{O})$. If $a \in A^\times$ satisfies $aSa^{-1} = S$, then $a = sc$ for some $s \in S^\times$ and some $c \in C_A(S)^\times$.*

Proof The assumptions imply that conjugation by a induces an automorphism of S. Thus, by Corollary 4.1.5 there is $s \in S^\times$ such that $ata^{-1} = sts^{-1}$ for all $t \in S$. Then $c = s^{-1}a \in C_A(S)^\times$ satisfies $a = sc$. □

Proposition 4.1.7 *Let $\mathcal{O}$ be a commutative local ring with residue field k, and let W be an $\mathcal{O}$-module. The map sending $w \in W$ to $1_k \otimes w$ in $k \otimes_{\mathcal{O}} W$ induces a k-vector space isomorphism $W/J(\mathcal{O})W \cong k \otimes_{\mathcal{O}} W$ which is natural in W.*

Proof This is a special case of 2.3.8. □

Proposition 4.1.8 *Let $\mathcal{O}$ be a commutative local ring with residue field k. Let A be an $\mathcal{O}$-algebra that is finitely generated as an $\mathcal{O}$-module. Let J be an ideal in A such that $J \subseteq J(A)$.*

(i) We have $J(\mathcal{O})A \subset J(A)$.
(ii) There is a positive integer n such that $J(A)^n \subset J(\mathcal{O})A$. In particular, if $\mathcal{O}$ is Noetherian, then $\cap_{n>0} J(A)^n = \{0\}$.
(iii) If k has at least three elements, then the $\mathcal{O}$-submodule of A generated by $A^\times$ is equal to A.
(iv) The set $1+J$ is a subgroup of $A^\times$, for any positive integer n the set $1+J^n$ is a normal subgroup of $1+J$ and the group $1+J$ acts trivially on the quotient $(1+J^n)/(1+J^{n+1})$. In particular, the group $(1+J^n)/(1+J^{n+1})$ is abelian.
(v) If p is a prime number such that $p \in J$, then for any positive integer n the group $(1+J^n)/(1+J^{n+1})$ is abelian of exponent dividing p. In particular, any element of finite order in $1+J$ has a power of p as order.
(vi) If $\mathcal{O} = k$ has prime characteristic p, then $1+J$ has finite exponent dividing p^{m-1}, where m is the smallest positive integer satisfying $J^m = \{0\}$.

Proof Statement (i) follows from 1.10.10. Set $\bar{A} = A/J(\mathcal{O})A$. Then $\bar{A}$ is a finite-dimensional k-algebra. The image of $J(A)$ in $\bar{A}$ is $J(\bar{A})$ by 1.10.12, and this is a nilpotent ideal by 1.10.8. Thus there is a positive integer n such that $J(\bar{A})^n = \{0\}$, or equivalently, such that $J(A)^n$ is in the kernel $J(\mathcal{O})A$ of the canonical map $A \to A/J(\mathcal{O})A$. This proves the inclusion $J(A)^n \subseteq J(\mathcal{O})A$, and the second statement in (ii) follows from 4.1.3. In order to prove (iii) we may assume that $\mathcal{O} = k$, by Nakayama's Lemma. The statement follows then from 1.13.7. By the assumptions on J we have $1+J \subseteq 1+J(A) \subseteq A^\times$. Clearly $1+J^n$ is closed under multiplication in $1+J$. If $a \in J^n$ and $a' \in J(A)$ such that $(1+a)(1+a') = 1$, then $a' = -a - aa' \in J^n$, showing that $1+J^n$ is also closed under taking inverses, hence a subgroup of $1+J(A)$. If a, $a' \in J$ such that $(1+a)(1+a') = 1$ and if $b \in J^n$, then $(1+a)(1+b)(1+a') = 1+(1+a)b(1+a') = 1+b+r$ for some $r \in J^{n+1}$. Since $1+b+r = (1+b)(1+(1+b)^{-1}r) \in (1+b)(1+J^{n+1})$, this completes the proof of (iv). If $a \in J^n$, then $(1+a)^p = 1+pa+r$ for some $r \in J^{n+1}$. Therefore, if $p \in J$, then $pa \in J^{n+1}$, which implies (v), and (vi) is an immediate consequence of (v). □

The hypothesis on k containing at least three elements in 4.1.8 (iii) is necessary for the reduction to direct factors in the above proof: the 2-dimensional $\mathbb{F}_2$-algebra $A = \mathbb{F}_2 \times \mathbb{F}_2$ has 1 as its unique invertible element, and hence $A^\times$ does not contain a basis of A.

Corollary 4.1.9 *Let $\mathcal{O}$ be a commutative local ring with residue field k of prime characteristic p. Let G be a finite group having a normal subgroup N of index prime to p. We have $J(\mathcal{O}G) = J(\mathcal{O}N)\mathcal{O}G$.*

Proof By 4.1.8 (i) we have $J(\mathcal{O})\mathcal{O}G \subseteq J(\mathcal{O}G)$. Thus we may assume that $\mathcal{O} = k$. In that case, the result is a special case of 1.11.10. □

Statement 4.1.8 (i) implies that the simple A-modules are in fact exactly the simple $\bar{A}$-modules, where $\bar{A}$ is the finite-dimensional k-algebra $A/J(\mathcal{O})A$. We use this to extend earlier terminology from 1.14.1 as follows:

Definition 4.1.10 Let $\mathcal{O}$ be a commutative local ring with residue field k. Let A be an $\mathcal{O}$-algebra that is finitely generated as an $\mathcal{O}$-module. We say that the $\mathcal{O}$-algebra A is *split* if $A/J(A)$ is a split semisimple k-algebra.

Equivalently, the $\mathcal{O}$-algebra A is split if and only if the k-algebra $\bar{A} = k \otimes_{\mathcal{O}} A$ is split.

Proposition 4.1.11 *Let $\mathcal{O}$ be a commutative local ring. Let A be an $\mathcal{O}$-algebra that is finitely generated as an $\mathcal{O}$-module, and let U be a finitely generated A-module.*

(i) We have $J(A)U = \mathrm{rad}(U)$.
(ii) We have $J(A)U = \{0\}$ *if and only if U is a finite direct sum of simple A-module.*

Proof Set $k = \mathcal{O}/J(\mathcal{O})$, $\bar{U} = U/J(\mathcal{O})U$ and $\bar{A} = A/J(\mathcal{O})A$. Since $\bar{U}$ is annihilated by $J(\mathcal{O})A$, we may consider $\bar{U}$ as $\bar{A}$-module. By 4.1.8 (i) we have $J(\mathcal{O})U \subseteq J(A)U$, and hence $J(\mathcal{O})U$ is contained in every maximal submodule of U. Thus the canonical map $U \to \bar{U}$ induces a bijection between the sets of maximal submodules of U and of $\bar{U}$. Since U is finitely generated as A-module, $\bar{U}$ has finite dimension over k. By 1.10.18 (ii), we have $J(\bar{A})\bar{U} = \mathrm{rad}(\bar{U})$. The left side in this equality is equal to $J(A)U/J(\mathcal{O})U$, and the right side is equal to $\mathrm{rad}(U)/J(\mathcal{O})U$. Thus $J(A)U = \mathrm{rad}(U)$, which proves (i). Statement (ii) follows from (i) and 1.10.18 (i). □

Proposition 4.1.12 *Let $\mathcal{O}$ be a Noetherian commutative ring. Let A be an $\mathcal{O}$-algebra that is finitely generated as an $\mathcal{O}$-module, and let U be a finitely generated A-module. The algebra* $\mathrm{End}_A(U)$ *is finitely generated as an $\mathcal{O}$-module.*

Proof The statement is true for $U = A$, because in that case $\mathrm{End}_A(U) \cong A^{\mathrm{op}}$, and A is finitely generated as an $\mathcal{O}$-module by the assumptions. If U is free of finite rank n then $\mathrm{End}_A(U) \cong M_n(A^{\mathrm{op}})$, which is again finitely generated as an $\mathcal{O}$-module. In general, U is a quotient F/V of a free A-module F of finite rank, and a submodule V of F. Since F is in particular projective, every endomorphism of U lifts to an endomorphism of F. Thus, if we denote by E the subalgebra of $\mathrm{End}_A(F)$ consisting of all A-endomorphisms ψ of F satisfying $\psi(V) \subseteq V$, then the canonical map $E \to \mathrm{End}_A(U)$ sending ψ to the induced endomorphism of U is a surjective algebra homomorphism. Now E is finitely generated as an $\mathcal{O}$-module because it is a submodule of the finitely generated $\mathcal{O}$-module $\mathrm{End}_A(F)$, and $\mathcal{O}$ is assumed to be Noetherian. But then any quotient of E is finitely generated as well. □

4.2 Discrete valuation rings

Definition 4.2.1 A commutative ring $\mathcal{O}$ is called a *discrete valuation ring* if $\mathcal{O}$ is a local principal ideal domain such that $J(\mathcal{O}) \neq \{0\}$.

Example 4.2.2 Let p be a prime. The subring $\mathbb{Z}_{(p)} = \{\frac{a}{b} \mid a, b \in \mathbb{Z}, (b, p) = 1\}$ of the rational number field is a discrete valuation ring, with unique maximal ideal $J(\mathbb{Z}_{(p)}) = p\mathbb{Z}_{(p)}$.

This example is a special case of a more general construction principle for discrete valuation rings, which consists of taking $\mathcal{O} = R_{\mathfrak{p}}$, where R is the ring of algebraic integers of an algebraic number field K and where $R_{\mathfrak{p}}$ is the localisation of R at a nonzero prime ideal $\mathfrak{p}$ in R.

Theorem 4.2.3 *Let $\mathcal{O}$ be a discrete valuation ring and let $\pi \in \mathcal{O}$ such that $J(\mathcal{O}) = \pi\mathcal{O}$.*

(i) For every $\lambda \in \mathcal{O} \setminus \{0\}$ there is a unique maximal integer $\nu(\lambda)$ such that $\lambda \in \pi^{\nu(\lambda)}\mathcal{O}$.
(ii) For any $\lambda, \mu \in \mathcal{O} \setminus \{0\}$ we have $\nu(\lambda\mu) = \nu(\lambda) + \nu(\mu)$ and $\nu(\lambda + \mu) \geq \min\{\nu(\lambda), \nu(\mu)\}$.
(iii) Every non zero ideal in $\mathcal{O}$ is of the form $\pi^n\mathcal{O}$ for some unique integer $n \geq 0$.

Proof Let $\lambda \in \mathcal{O} \setminus \{0\}$. Since $\bigcap_{k\geq 1} \pi^k\mathcal{O} = \{0\}$ by 4.1.3 there is a unique maximal integer $\nu(\lambda) \geq 0$ such that $\lambda = \pi^{\nu(\lambda)}\lambda'$ for some $\lambda' \in \mathcal{O}$. The maximality of $\nu(\lambda)$ forces $\lambda' \in \mathcal{O} \setminus \pi\mathcal{O} = \mathcal{O}^\times$, where the last equality follows from 4.1.2. Thus $\lambda\mathcal{O} = \pi^{\nu(\lambda)}\mathcal{O}$, which proves both (i) and (iii). Statement (ii) is a trivial consequence. □

In the situtation of 4.2.3, the map $\nu : \mathcal{O} \setminus \{0\} \to \mathbb{Z}$ is called the *valuation* of the ring $\mathcal{O}$. One can use valuations to give an alternative definition of valuation rings. Suppose that K is a field and that $\nu : K^\times \to \mathbb{Z}$ is a surjective map satisfying $\nu(\lambda\mu) = \nu(\lambda) + \nu(\mu)$ and $\nu(\lambda + \mu) \geq \min\{\nu(\lambda), \nu(\mu)\}$ for all $\lambda, \mu \in K^\times$. For notational convenience, set $\nu(0) = \infty$. Then the set $\mathcal{O} = \{\lambda \in K | \nu(\lambda) \geq 0\}$ is a discrete valuation ring, and K is the quotient field of $\mathcal{O}$. The unique maximal ideal in $\mathcal{O}$ is $J(\mathcal{O}) = \{\lambda \in K | \nu(\lambda) \geq 1\}$. Take for π any element in $\mathcal{O}$ such that $\nu(\pi) = 1$. One easily checks that $\mathcal{O}$ has the properties stated in the theorem above. A module M over some commutative ring $\mathcal{O}$ is called *torsion-free* if, for any $\lambda \in \mathcal{O}$ and any $m \in M$ the equation $\lambda m = 0$ implies $\lambda = 0$ or $m = 0$. In particular, the regular $\mathcal{O}$-module $\mathcal{O}$ is torsion-free if and only if $\mathcal{O}$ is an integral domain.

Theorem 4.2.4 *Let $\mathcal{O}$ be a discrete valuation ring and let M be a finitely generated $\mathcal{O}$-module. The following are equivalent.*

(i) M is free.
(ii) M is projective.
(iii) M is torsion-free.

Proof By 4.1.4 the module M is free if and only if M is projective. Since $\mathcal{O}$ is an integral domain, every finitely generated free module is torsion free. Conversely, let M be a finitely generated torsion free $\mathcal{O}$-module, and let W be a subset in M whose image $\overline{W}$ in $\overline{M} = M/J(\mathcal{O})M$ is a k-basis. Then $\mathcal{O}W + J(\mathcal{O})M = M$. By Nakayama's Lemma, W generates M as $\mathcal{O}$-module. For any $w \in W$ let $\lambda_w \in \mathcal{O}$ such that $\sum_{w \in W} \lambda_w w = 0$. We are going to show inductively that $\lambda_w \in J(\mathcal{O})^n$ for any $w \in W$ and any positive integer n. For $n = 1$ this follows from the fact that $\overline{W}$ is a basis of $\overline{M}$. Since $\mathcal{O}$ is a principal ideal domain, we have in particular $J(\mathcal{O}) = \pi\mathcal{O}$ for some $\pi \in J(\mathcal{O})$. Thus if $\lambda_w \in J(\mathcal{O})^n$ we have $\lambda_w = \pi^n \lambda'_w$ for some suitable $\lambda'_w \in \mathcal{O}$, where $w \in W$. But then π^n annihilates the element $\sum_{w \in W} \lambda'_w w$. As M is torsion free, this implies that the latter sum is actually zero. But then, by the first argument, we have $\lambda'_w \in J(\mathcal{O})$, and thus $\lambda_w \in J(\mathcal{O})^{n+1}$, where $w \in W$. Using 4.1.3 this implies that $\lambda_w = 0$ for any $w \in W$, and therefore W is a basis of M. □

Corollary 4.2.5 *Let $\mathcal{O}$ be a discrete valuation ring and let M be a finitely generated free $\mathcal{O}$-module. Then every submodule of M is free.*

Proof Every submodule of a free module is torsion free. The result follows from 4.2.4. □

Corollary 4.2.5 holds more generally for arbitrary principal ideal domains (this follows from the classification of finitely generated modules of principal ideal domains). It does not imply that every submodule of a finitely generated module over a discrete valuation ring is a direct summand of that module. A submodule N of an $\mathcal{O}$-module M is called *pure* if for every $\mathcal{O}$-module W the induced map $W \otimes_{\mathcal{O}} N \to W \otimes_{\mathcal{O}} M$ is injective. If N is a direct summand of M then N is clearly pure in M. For discrete valuation rings, this characterises pure submodules of finitely generated free modules:

Proposition 4.2.6 *Let $\mathcal{O}$ be a discrete valuation ring, let M be a free $\mathcal{O}$-module of finite rank and let N be a submodule of M. The following are equivalent.*

(i) *N is a pure submodule of M.*
(ii) *N is a direct summand of M.*
(iii) *M/N is torsion free.*
(iv) *M/N is free.*
(v) *$J(\mathcal{O})N = J(\mathcal{O})M \cap N$.*
(vi) *The canonical map $k \otimes_{\mathcal{O}} N \to k \otimes_{\mathcal{O}} M$ is injective.*

Proof We clearly have $J(\mathcal{O})N \subseteq J(\mathcal{O})M \cap N$. If N is a direct summand of M there is a submodule N' of M such that $N \oplus N' = M$. Then $J(\mathcal{O})M = J(\mathcal{O})N \oplus J(\mathcal{O})N'$. Since $J(\mathcal{O})N' \subseteq N'$ and $J(\mathcal{O})N \subseteq N$, intersecting with N yields $J(\mathcal{O})M \cap N = J(\mathcal{O})N$. Thus (ii) implies (v). Suppose that $J(\mathcal{O})M \cap N = J(\mathcal{O})N$. We show that M/N is torsion free. Let $\pi \in \mathcal{O}$ such that $J(\mathcal{O}) = \pi\mathcal{O}$. Suppose $\lambda \in \mathcal{O}$ and $m \in M$ are such that $\lambda(m+N) = 0+N$, or equivalently, such that $\lambda m \in N$. We have to show that then $\lambda = 0$ or $m \in N$. Suppose that $\lambda \neq 0$ and $m \notin N$. Then λ is not invertible because that would imply $m \in N$, so $\lambda = \pi^n\lambda'$ for some positive integer n and some invertible element λ' in $\mathcal{O}$. Thus $\pi^n m \in N$. Let $t \geq 0$ be the smallest integer such that $\pi^t m \in N$. Since $m \notin N$ we have $t \geq 1$. Thus $\pi^t m \in J(\mathcal{O})M \cap N = J(\mathcal{O})N$. Write $\pi^t m = \pi n$ for some $n \in N$. But since M is free, hence torsion free, this equation implies that $\pi^{t-1}m = n \in N$, contradicting the minimality of t with this property. Therefore, M/N is torsion free, and hence (v) implies (iii). If (iii) holds, then M/N is free by 4.2.4, so (iii) implies (iv). If M/N is free, then the canonical map $M \to M/N$ splits, and hence N is a direct summand of M, so (iv) implies (ii). Clearly (ii) implies (i) because tensor products commute with direct sums, and (i) implies (vi). It remains to show that (vi) implies (v). Using the natural isomorphism $k \otimes_{\mathcal{O}} M \cong M/J(\mathcal{O})M$, statement (vi) is equivalent to the injectivity of the map $N/J(\mathcal{O})N \to M/J(\mathcal{O})M$. This is clearly equivalent to (v). □

Corollary 4.2.7 *Let $\mathcal{O}$ be a discrete valuation ring, and let M be a free $\mathcal{O}$-module of finite rank. The intersection of any family of pure submodules of M is a pure submodule of M.*

Proof Let I be an indexing set and $\{M_i\}_{i\in I}$ a family of pure submodules of M_i. Set $N = \cap_{i\in I} M_i$. By 4.2.6, it suffices to show that M/N is torsion free. Let $m \in M$ and $\lambda \in \mathcal{O}$ such that $\lambda \neq 0$ and such that $\lambda m \in N$. Then $\lambda m \in M_i$ for all $i \in I$, hence $m \in M_i$ for all $i \in I$, implying the result. □

Proposition 4.2.8 *Let $\mathcal{O}$ be a discrete valuation ring with residue field $k = \mathcal{O}/J(\mathcal{O})$, let M be a free $\mathcal{O}$-module of finite rank and N a submodule of M. Then N is free, and the following hold.*

(i) We have $\mathrm{rk}_{\mathcal{O}}(M) - \mathrm{rk}_{\mathcal{O}}(N) \leq \dim_k(k \otimes_{\mathcal{O}} (M/N))$. This is an equality if and only if N is pure in M.
(ii) Denote by $\bar{N}$ the image of N in $k \otimes_{\mathcal{O}} M$. We have $\dim_k(\bar{N}) \leq \mathrm{rk}_{\mathcal{O}}(N)$. This is an equality if and only if N is pure in M.
(iii) The set N' of all $m \in M$ for which there exists a nonzero element $\lambda \in \mathcal{O}$ satisfying $\lambda m \in N$ is the unique minimal $\mathcal{O}$-pure submodule of M containing N.

Proof By 4.2.5, N is free. Let S be a subset of M such that the image of S in $k \otimes_{\mathcal{O}} M/N$ is a k-basis. Denote by W the $\mathcal{O}$-submodule of M spanned by the set S; this is a free module of rank at most $|S|$. We have $M = W + N + J(\mathcal{O})M$, hence $M = W + N$ by Nakayama's Lemma. Thus $\mathrm{rk}_{\mathcal{O}}(M) \leq \mathrm{rk}_{\mathcal{O}}(W) + \mathrm{rk}_{\mathcal{O}}(N) \leq \dim_k(k \otimes_{\mathcal{O}} (M/N)) + \mathrm{rk}_{\mathcal{O}}(N)$, whence the inequality in (i). This is an equality if and only if $rk_{\mathcal{O}}(W) = \dim_k(k \otimes_{\mathcal{O}} M/N) = \mathrm{rk}_{\mathcal{O}}(M) - \mathrm{rk}_{\mathcal{O}}(N)$, so if and only if W is a complement to N in M. This shows (i). We have $\bar{N} = N/(N \cap J(\mathcal{O})M)$. The inequality in (ii) follows from the inclusion $J(\mathcal{O})N \subseteq N \cap J(\mathcal{O})M$. The equality in (ii) is equivalent to $J(\mathcal{O})N = N \cap J(\mathcal{O})M$, hence to N being pure in M by 4.2.6. By construction, N' is the $\mathcal{O}$-submodule of M containing N such that N'/N is the torsion submodule of M/N, and hence M/N' is the largest torsion free quotient of M/N. Statement (iii) follows from 4.2.6. □

Exercise 4.2.9 Let $\mathcal{O}$ be a discrete valuation ring, and let A be an $\mathcal{O}$-algebra that is free of finite rank as an $\mathcal{O}$-module. Show that $\mathcal{O} \cdot 1_A$ is $\mathcal{O}$-pure in A.

The ideals I in an $\mathcal{O}$-free $\mathcal{O}$-algebra A of finite $\mathcal{O}$-rank that is pure in A are exactly the ideals I such that A/I is $\mathcal{O}$-free. If G is a finite group and N a normal subgroup of G, then the kernel of the canonical map $\mathcal{O}G \to \mathcal{O}G/N$ is $\mathcal{O}$-pure. In general, not every $\mathcal{O}$-pure ideal in $\mathcal{O}G$ is of this form.

Exercise 4.2.10 Let $\mathcal{O}$ be a discrete valuation ring, and let A be an $\mathcal{O}$-algebra that is free of finite rank as an $\mathcal{O}$-module. Show that A has a unique minimal $\mathcal{O}$-pure ideal I such that A/I is commutative.

Exercise 4.2.11 Let $\mathcal{O}$ be a discrete valuation ring, and let G be a finite group. Let I be the minimal $\mathcal{O}$-pure ideal in $\mathcal{O}G$ such that $\mathcal{O}G/I$ is commutative. Show that $\mathcal{O}G/I \cong \mathcal{O}G/G'$, where G' is the derived subgroup of G.

Exercise 4.2.12 Let $\mathcal{O}$ be a discrete valuation ring, and let A be an $\mathcal{O}$-algebra that is free of finite rank as an $\mathcal{O}$-module. Let $\mathcal{O}'$ be a discrete valuation ring containing $\mathcal{O}$ as a subring. Let I be the unique minimal $\mathcal{O}$-pure ideal I such that A/I is commutative. Show that $I' = \mathcal{O}' \otimes_{\mathcal{O}} I$ is the unique minimal $\mathcal{O}'$-pure ideal in $A' = \mathcal{O}' \otimes_{\mathcal{O}} A$ such that A'/I' is commutative.

Exercise 4.2.13 The purpose of this exercise is to illustrate that the sum of two pure submodules need not be pure. Let $\mathcal{O}$ be a discrete valuation ring. Let $\pi \in \mathcal{O}$ such that $\pi\mathcal{O} = J(\mathcal{O})$. Consider $U = \{(\lambda, 0) \mid \lambda \in \mathcal{O}\}$ and $V = \{(\lambda, \pi\lambda) \mid \lambda \in \mathcal{O}\}$. Show that U and V are $\mathcal{O}$-pure submodules of $\mathcal{O} \oplus \mathcal{O}$, both isomorphic to $\mathcal{O}$. Show that their sum $U + V$ in $\mathcal{O} \oplus \mathcal{O}$ is equal to $\mathcal{O} \oplus \pi\mathcal{O}$ and show that this is not a pure submodule of $\mathcal{O} \oplus \mathcal{O}$.

4.3 Complete discrete valuation rings

A sequence $(\lambda_m)_{m\geq 1}$ of elements in a commutative local ring $\mathcal{O}$ is called a *Cauchy sequence* if for every $a \geq 1$ there exists $N \geq 1$ such that $\lambda_m - \lambda_n \in J(\mathcal{O})^a$ for all $m, n \geq N$. Two sequences $(\lambda_m)_{m\geq 1}$, $(\mu_m)_{m\geq 1}$ are called *equivalent* if for any $a \geq 1$ there exists $N \geq 1$ such that $\lambda_m - \mu_m \in J(\mathcal{O})^a$ for all $m \geq N$. In that case one verifies that if one of the two sequences is a Cauchy sequence, then so is the other.

Definition 4.3.1 A commutative local Noetherian ring $\mathcal{O}$ is called *complete* if for every Cauchy sequence $(\lambda_m)_{m\geq 1}$ in $\mathcal{O}$ there is $\lambda \in \mathcal{O}$ such that for any $a \geq 1$ there exists $N \geq 1$ such that $\lambda - \lambda_m \in J(\mathcal{O})^a$ for $m \geq N$. In that case, we call λ a *limit* of the Cauchy sequence $(\lambda_m)_{m\geq 1}$.

With the notation and hypotheses of 4.3.1, if a limit λ exists, it is unique. Indeed, if $\lambda, \mu \in \mathcal{O}$ both satisfy $\lambda - \lambda_m, \mu - \lambda_m \in J(\mathcal{O})^a$ for all $m \geq N$, then $\lambda - \mu = (\lambda - \lambda_m) - (\mu - \lambda_m) \in J(\mathcal{O})^a$ for all $m \geq N$. It follows from 4.1.3 that $\lambda - \mu \in \bigcap_{a\geq 1} J(\mathcal{O})^a = \{0\}$. A field k is trivially a complete local ring with maximal ideal $\{0\}$. Every Cauchy sequence in k is just a constant sequence $\lambda_m = \lambda$ for all $m \geq 1$, and has therefore λ as limit.

Theorem 4.3.2 *Let $\mathcal{O}$ be a commutative local Noetherian ring. There is a complete commutative local Noetherian ring $\hat{\mathcal{O}}$ and a ring homomorphism $\sigma : \mathcal{O} \to \hat{\mathcal{O}}$ with the following universal property: we have $\sigma(J(\mathcal{O})) \subseteq J(\hat{\mathcal{O}})$, and for every complete commutative local Noetherian ring $\mathcal{O}'$ and every ring homomorphism $\rho : \mathcal{O} \to \mathcal{O}'$ satisfying $\rho(J(\mathcal{O})) \subseteq J(\mathcal{O}')$ there is a unique ring homomorphism $\tau : \hat{\mathcal{O}} \to \mathcal{O}'$ such that $\rho = \tau \circ \sigma$. In particular, the pair $(\hat{\mathcal{O}}, \sigma)$ is unique up to unique isomorphism.*

Proof Let $\hat{\mathcal{O}}$ be the set of equivalence classes of Cauchy sequences in $\mathcal{O}$. Denote by $[\lambda_m]_{m\geq 1}$ the equivalence class of a Cauchy sequence $(\lambda_m)_{m\geq 1}$. We define a ring structure on $\hat{\mathcal{O}}$ by taking the componentwise sum and componentwise product of Cauchy sequences. More precisely, if $(\lambda_m)_{m\geq 1}$, $(\mu_m)_{m\geq 1}$ are two Cauchy sequences, one checks that $(\lambda_m + \mu_m)_{m\geq 1}$ and $(\lambda_m\mu_m)_{m\geq 1}$ are Cauchy sequences whose equivalence classes depend only on the equivalence classes of the sequences $(\lambda_m)_{m\geq 1}$, $(\mu_m)_{m\geq 1}$. Thus, setting $[\lambda_m]_{m\geq 1} + [\mu_m]_{m\geq 1} = [\lambda_m + \mu_m]_{m\geq 1}$ and $[\lambda_m]_{m\geq 1}[\mu_m]_{m\geq 1} = [\lambda_m\mu_m]_{m\geq 1}$ defines a ring structure on $\hat{\mathcal{O}}$. One verifies that the set $J(\hat{\mathcal{O}})$, consisting of equivalence classes of Cauchy sequences $(\lambda_m)_{m\geq 1}$ with all but finitely many elements in $J(\mathcal{O})$, is the unique maximal ideal in $\hat{\mathcal{O}}$. The map $\sigma : \mathcal{O} \to \hat{\mathcal{O}}$ sending $\lambda \in \mathcal{O}$ to the equivalence class of the constant Cauchy sequence $\lambda_m = \lambda$ for $m \geq 1$ is a ring homomorphism sending $J(\mathcal{O})$ to $J(\hat{\mathcal{O}})$. One checks next that $\hat{\mathcal{O}}$ is complete and Noetherian. Finally, if $\mathcal{O}'$ is a complete commutative local Noetherian ring and $\rho : \mathcal{O} \to \mathcal{O}'$ a ring homomorphism sending $J(\mathcal{O})$ to $J(\mathcal{O}')$, we define $\tau : \hat{\mathcal{O}} \to \mathcal{O}'$ by mapping the equivalence class $[\lambda_m]_{m\geq 1}$ of a Cauchy sequence $(\lambda_m)_{m\geq 1}$ to the limit in $\mathcal{O}'$ of the Cauchy sequence $(\rho(\lambda_m))_{m\geq 1}$ in $\mathcal{O}'$. The result follows. □

A subsequence of a Cauchy sequence is again a Cauchy sequence, with the same limit if any. Thus in the construction of $\hat{\mathcal{O}}$, it would be sufficient to consider sequences $(\lambda_m)_{m\geq 1}$ with the property $\lambda_{m+1} - \lambda_m \in J(\mathcal{O})^m$; any sequence with this property is automatically a Cauchy sequence. This amounts to describing $\hat{\mathcal{O}}$ as the inverse limit $\varprojlim \mathcal{O}/J(\mathcal{O})^m$, with respect to the system of canonical maps $\mathcal{O}/J(\mathcal{O})^{m+1} \to \mathcal{O}/J(\mathcal{O})^m$.

Example 4.3.3 Let p be a prime. The commutative local ring $\mathbb{Z}_{(p)} = \{\frac{a}{b} \mid a, b \in \mathbb{Z}, p \nmid b\}$ is not complete; its completion, denoted $\hat{\mathbb{Z}}_p$, is the ring of p-adic integers.

The notions of Cauchy sequences and completeness extend to modules. Let $\mathcal{O}$ be a commutative local Noetherian ring and let U be an $\mathcal{O}$-module. A *Cauchy sequence in* U is a sequence $(u_m)_{m\geq 1}$ of elements $u_m \in U$ such that for any $a \geq 1$ there is $N \geq 1$ with the property $u_m - u_n \in J(\mathcal{O})^a U$ for all $m, n \geq N$. An element $u \in U$ is a *limit of a Cauchy sequence* $(u_m)_{m\geq 1}$ if for any $a \geq 1$ there is $N \geq 1$ such that $u - u_m \in J(\mathcal{O})^a M$ for all $m \geq N$. The module U is called *complete* if every Cauchy sequence in U has a limit. If U is finitely generated and a Cauchy sequence in U has a limit, then this limit is unique because $\cap_{m\geq 1} J(\mathcal{O})^m U = \{0\}$. One extends the notion of equivalent Cauchy sequences in $\mathcal{O}$ to a notion of equivalent Cauchy sequences in U, and proceeds as in Theorem 4.3.2 to define the completion $\hat{U}$ of U as the inverse limit $\varprojlim U/J(\mathcal{O})^m U$, using those Cauchy sequences $(u_m)_{m\geq 1}$ that satisfy $u_{m+1} - u_m \in J(\mathcal{O})^k U$ for $m \geq 1$. One verifies that $\hat{U} \cong \hat{\mathcal{O}} \otimes_{\mathcal{O}} U$; that is, completion is a functor.

The above considerations on completion extend to finitely generated modules over any $\mathcal{O}$-algebra A and any ideal I in A satisfying $\cap_{a\geq 1} I^a = \{0\}$. If a finitely generated A-module U is complete with respect to I, then U is also complete with respect to any ideal J that has the property that $J^n \subseteq I$ for some positive integer n. The completions of U with respect to two ideals I, J in A coincide if there exist positive integers n, m such that $I^m \subseteq J$ and $J^n \subseteq I$. We will make use of this observation in the context of Proposition 4.1.8: if A is finitely generated as a module over a Noetherian commutative local ring $\mathcal{O}$, and if U is a finitely generated A-module that is complete as an $\mathcal{O}$-module, or equivalently, complete with respect to the ideal $J(\mathcal{O})A$, then U is complete with respect to any ideal J of A contained in $J(A)$. If J is an ideal in A satisfying $J(\mathcal{O})A \subseteq J \subseteq J(A)$, then the completions of a finitely generated A-module U with respect to $J(\mathcal{O})A$ and J coincide. In particular, if the A-module U is complete as an $\mathcal{O}$-module, then $U = \varprojlim U/J^m U$. For further details we refer standard sources such as [65, §8]. If $\mathcal{O}$ is complete, and then all finitely generated modules are complete as well by the next result.

Theorem 4.3.4 *Let $\mathcal{O}$ be a complete commutative local Noetherian ring. Then every finitely generated $\mathcal{O}$-module is complete.*

Proof Since $\mathcal{O}$ is complete as regular $\mathcal{O}$-module, every finitely generated free $\mathcal{O}$-module is complete (the direct sum of two complete modules is complete). Let U be a finitely generated $\mathcal{O}$-module and let $(u_m)_{m\geq 1}$ be a Cauchy sequence in U such that $u_{m+1} - u_m \in J(\mathcal{O})^m U$. Since U is finitely generated, there is a surjective $\mathcal{O}$-homomorphism $\mu : F \to U$ for some free $\mathcal{O}$-module F of finite rank. Let $\alpha_1 \in F$ such that $\mu(\alpha_1) = u_1$. Since μ is surjective, μ maps $J(\mathcal{O})^m F$ onto $J(\mathcal{O})^m U$. Thus, for $m \geq 1$, there are $\gamma_m \in J(\mathcal{O})^m F$ such that

$\mu(\gamma_m) = u_{m+1} - u_m$. Define α_m inductively by $\alpha_{m+1} = \alpha_m + \gamma_m$. Then $(\alpha_m)_{m\geq 1}$ is a Cauchy sequence in F that is mapped onto the Cauchy sequence $(u_m)_{m\geq 1}$ in U. Since F is free of finite rank, the sequence $(\alpha_m)_{m\geq 1}$ has a limit α in F. Then $\mu(\alpha)$ is the required limit of $(u_m)_{m\geq 1}$ in U. □

Proposition 4.3.5 (Hensel's Lemma; cf. [84, §4, Proposition 7]) *Let $\mathcal{O}$ be a complete discrete valuation ring with residue field $k = \mathcal{O}/J(\mathcal{O})$. Let $f \in \mathcal{O}[x]$, and denote by $\bar{f}$ the image of f in $k[x]$. Suppose that $\bar{f}$ has a simple root λ in k. Then f has a unique root μ in $\mathcal{O}$ such that $\bar{\mu} = \lambda$.*

Proof Let μ, μ' be roots of f whose images in k are both equal to λ. Write $f(x) = (x - \mu)g(x)$ for some $g \in \mathcal{O}[x]$. Then $0 = f(\mu') = (\mu' - \mu)g(\mu')$. Since λ is a simple root of $\bar{f}$, we have $\bar{g}(\bar{\mu}') = \bar{g}(\lambda) \neq 0$, hence $g(\mu') \neq 0$. Since $f(\mu') = 0$ this forces $\mu' = \mu$. This shows the uniqueness statement. For the existence part, we construct a Cauchy sequence whose limit will be a root of f lifting λ. Choose $\mu_1 \in \mathcal{O}$ with $\bar{\mu}_1 = \lambda$. Then $f(\mu_1) \in J(\mathcal{O})$ because the image λ of μ_1 in k is a root of $\bar{f}$. Let $n \geq 1$. Suppose we have constructed an element $\mu_n \in \mathcal{O}$ that lifts λ such that $f(\mu_n) \in J(\mathcal{O})^n$. We will construct an element $\mu_{n+1} \in \mathcal{O}$ that lifts λ, such that $f(\mu_{n+1}) \in J(\mathcal{O})^{n+1}$ and such that $\mu_{n+1} - \mu_n \in J(\mathcal{O})^n$. Let $\eta \in J(\mathcal{O})^n$. The Taylor approximation yields

$$f(\mu_n + \eta) = f(\mu_n) + \eta f'(\mu_n) + \eta^2\tau$$

for some $\tau \in \mathcal{O}$. Then $\eta^2\tau \in J(\mathcal{O})^{n+1}$. Since μ_n lifts the simple root λ of $\bar{f}$, we have $f'(\mu_n) \in \mathcal{O}^\times$. Thus, for a suitable choice of η, we get that $\mu_{n+1} = \mu_n + \eta$ satisfies $f(\mu_{n+1}) \in J(\mathcal{O})^{n+1}$. The limit μ of the Cauchy sequence $(\mu_n)_{n\geq 1}$ still lifts λ and satisfies $f(\mu) \in J(\mathcal{O})^n$ for all $n \geq 1$, hence $f(\mu) = 0$ by 4.1.3. □

Corollary 4.3.6 *Let $\mathcal{O}$ be a complete discrete valuation ring with a residue field k of prime characteristic p, and let m be a positive integer prime to p. The map sending $\tau \in 1 + J(\mathcal{O})$ to τ^m is an automorphism of the abelian group $1 + J(\mathcal{O})$, inducing automorphisms of the subgroups $1 + J(\mathcal{O})^n$ for any $n \geq 1$.*

Proof Let $\tau \in 1 + J(\mathcal{O})$, and set $f(x) = x^m - \tau$. Then $\bar{f}(x) = x^m - \bar{1}$. Since m is prime to p, the unit element of k is a simple root of $\bar{f}$. By 4.3.5 there is a unique $\mu \in \mathcal{O}$ satisfying $\mu^m = \tau$ and such that $\bar{\mu} = \bar{1}$, or equivalently, such that $\mu \in 1 + J(\mathcal{O})$. Thus the map sending τ to τ^m is an automorphism of $1 + J(\mathcal{O})$, which clearly preserves the subgroups $1 + J(\mathcal{O})^n$, for $n \geq 1$. For $\rho \in J(\mathcal{O})$ we have $(1 + \rho)^m = 1 + m\rho + \rho^2\sigma$ for some $\sigma \in \mathcal{O}$. Therefore, if $m\rho + \rho^2\sigma \in J(\mathcal{O})^n$, then also $\rho \in J(\mathcal{O})^n$, which shows that the map sending τ to τ^m maps $1 + J(\mathcal{O})^n$ onto $1 + J(\mathcal{O})^n$. □

Corollary 4.3.7 *Let $\mathcal{O}$ be a complete discrete valuation ring with a residue field k of prime characteristic p, and let m be a positive integer prime to p. Then for any m-th root of unity $\lambda \in k$ there is a unique m-th root of unity $\mu \in \mathcal{O}$ whose canonical image in k is λ.*

Proof This follows from 4.3.5 applied to the polynomial $f(x) = x^m - 1$. □

Corollary 4.3.8 *Let $\mathcal{O}$ be a complete discrete valuation ring with a residue field of prime characteristic p. Let G be a finite group, and let P a Sylow p-subgroup of G. The restriction map $\mathrm{res}^G_P : H^2(G; 1+J(\mathcal{O})) \to H^2(P; 1+J(\mathcal{O}))$ is injective.*

Proof Set $m = |G : P|$; this is a positive integer that is prime to p. Let $\alpha \in H^2(G; 1+J(\mathcal{O}))$ such that $\mathrm{res}^G_P(\alpha)$ is trivial. By 1.2.15, the class α^m is trivial. But then α is trivial because taking m-th powers is an automorphism of $1+J(\mathcal{O})$, hence induces an automorphism of $H^2(G; 1+J(\mathcal{O}))$. □

A field k of prime characteristic p is called *perfect* if the map sending $\alpha \in k$ to α^p is surjective (that is, an automorphism of the field). If k is finite or algebraically closed then k is perfect.

Theorem 4.3.9 *Let $\mathcal{O}$ be a complete discrete valuation ring with residue field $k = \mathcal{O}/J(\mathcal{O})$ of prime characteristic p. Suppose that k is perfect. The group homomorphism $f : \mathcal{O}^\times \to k^\times$ induced by the canonical surjective map $\mathcal{O} \to k$ has a unique section $g : k^\times \to \mathcal{O}^\times$. In particular, there is a canonical group isomorphism $\mathcal{O}^\times \cong k^\times \times (1+J(\mathcal{O}))$.*

Proof Since $\mathcal{O}^\times = \mathcal{O} \setminus J(\mathcal{O})$, the group homomorphism $f : \mathcal{O}^\times \to k^\times$ is surjective and $\ker(f) = 1+J(\mathcal{O})$. Let $\pi \in \mathcal{O}$ such that $J(\mathcal{O}) = \pi\mathcal{O}$. We start with a preliminary observation. If $\delta, \gamma \in \mathcal{O}$ such that $\delta - \gamma \in \pi^n\mathcal{O}$ for some positive integer n, then $\delta^p - \gamma^p \in \pi^{n+1}\mathcal{O}$. This follows from the fact that $p \in \pi\mathcal{O}$ combined with the binomial formula: for $p = 2$ we get

$$(\delta - \gamma)^2 = \delta^2 - \gamma^2 + 2\delta(\delta - \gamma)$$

and since $\delta - \gamma \in \pi^n\mathcal{O}$ and $2 \in \pi\mathcal{O}$ it follows that $\delta^2 - \gamma^2 \in \pi^{n+1}\mathcal{O}$. For p odd we get

$$(\delta - \gamma)^p = \delta^p - \gamma^p + \sum_{1 \le n \le p-1} \binom{p}{n} \delta^n \gamma^{p-n} (-1)^{p-n}$$

and since $\binom{p}{n} = \binom{p}{p-n}$ and $(-1)^{p-n} = -(-1)^n$, the summands in the latter sum come in pairs, so that we get

$$(\delta-\gamma)^p = \delta^p - \gamma^p + \sum_{1\le n\le \frac{p}{2}} \binom{p}{n}(\delta^n\gamma^{p-n} - \delta^{p-n}\gamma^n)(-1)^n$$

$$= \delta^p - \gamma^p + \sum_{1\le n\le \frac{p}{2}} \binom{p}{n}\delta^n\gamma^n(\gamma^{p-2n} - \delta^{p-2n})(-1)^n.$$

Since $\gamma^{p-2n} - \delta^{p-2n}$ is divisible by $\delta-\gamma$ for $1 \le n \le \frac{p}{2}$ we get $\delta^p - \gamma^p \in \pi^n\mathcal{O}$ as claimed. We define g as follows. Let $\alpha \in k^\times$. Since k is perfect, for any integer $n \ge 0$ there is $\alpha_n \in k$ such that $(\alpha_n)^{p^n} = \alpha$. Set $U_n = U_n(\alpha) = \{\beta^{p^n} \mid \beta \in \mathcal{O}, f(\beta) = \alpha_n\}$. Then $U_{n+1} \subseteq U_n$ and every element in U_n is mapped onto α by f. If $\delta, \gamma \in \mathcal{O}$ such that $f(\delta) = \alpha_n = f(\gamma)$, then in particular $\delta - \gamma \in \pi\mathcal{O}$. Thus, by our preliminary observation, $\delta^{p^n} - \gamma^{p^n} \in \pi^n\mathcal{O}$. In other words, the difference of any two elements in U_n lies in $\pi^n\mathcal{O}$. Let now $\beta_n \in U_n$, for $n \ge 0$. Since $U_{n+1} \subseteq U_n$ we get that $\beta_{n+1} - \beta_n \in \pi^n\mathcal{O}$. Thus $(\beta_n)_{n\ge 1}$ is a Cauchy sequence in $\mathcal{O}$. We define $g(\alpha)$ to be the limit in $\mathcal{O}$ of this sequence. Since $g(\alpha) - \beta_n \in \pi^n J(\mathcal{O})$ we get that $f(g(\alpha)) = f(\beta_n) = \alpha$. This value does not depend on the choice of the β_n since for any other $\beta'_n \in U_k$ we have $\beta_n - \beta'_n \in \pi^n\mathcal{O}$. Now the set $U_n(\alpha)$ depends multiplicatively on α; that is, $U_n(\alpha\alpha') = U_n(\alpha)U_n(\alpha')$ for $\alpha, \alpha' \in k^\times$. This implies that $g(\alpha\alpha') = g(\alpha)g(\alpha')$. Thus the group homomorphism f is split with g as section. In particular we have $\mathcal{O}^\times \cong k^\times \times \ker(f) = k^\times \times (1 + J(\mathcal{O}))$. Let $h : k^\times \to \mathcal{O}^\times$ be another section of f. Then the map sending $\alpha \in k^\times$ to $g(\alpha)h(\alpha^{-1})$ is a group homomorphism from $k^\times$ to $1 + \pi\mathcal{O}$. Since $(\alpha_n)^{p^n} = \alpha$, the image of this group homomorphism is contained in $(1+\pi\mathcal{O})^{p^n}$, which by the binomial formula, is contained in $1 + \pi^n\mathcal{O}$. However, $\bigcap_{n\ge 1}(1 + \pi^n\mathcal{O}) = \{1\}$ by 4.1.3, hence $g(\alpha) = h(\alpha)$, showing the uniqueness of g. □

If $\mathcal{O}$ also has characteristic p, then the sets $U_n(\alpha)$ depend also additively on α. Thus, in that case, g extends to a ring homomorphism $k \to \mathcal{O}$ which is a section of the canonical map $\mathcal{O} \to k$. In particular, $\mathcal{O}$ contains a copy of k.

Corollary 4.3.10 *Let $\mathcal{O}$ be a complete discrete valuation ring with a perfect residue field of prime characteristic p. Identify $k^\times$ with its canonical preimage in $\mathcal{O}^\times$. Let G be a finite group, P a Sylow p-subgroup of G, and $\alpha \in H^2(G; \mathcal{O}^\times)$. If $\mathrm{res}^G_P(\alpha)$ is trivial, then $\alpha \in H^2(G; k^\times)$.*

Proof By 4.3.9 we have $\mathcal{O}^\times = k^\times \times (1 + J(\mathcal{O}))$, hence $H^2(G; \mathcal{O}^\times) = H^2(G; k^\times) \times H^2(G; 1 + J(\mathcal{O}))$. The result follows from 4.3.8. □

We mention without proof the following existence and uniqueness result for complete discrete valuation rings:

Theorem 4.3.11 (cf. [84, II.§5 Théorème 3]) *Let k be a perfect field of positive characteristic p. Then there is, up to isomorphism, a unique complete discrete valuation ring $\mathcal{O}$ having characteristic zero and residue field k such that $J(\mathcal{O}) = p\mathcal{O}$.*

A complete discrete valuation ring $\mathcal{O}$ of characteristic zero with residue field k of positive characteristic p is called *unramified* if $J(\mathcal{O}) = p\mathcal{O}$; this amounts to saying that p remains a prime element in $\mathcal{O}$. It is also true that a field k admits up to isomorphism a unique complete discrete valuation ring R having the same characteristic as its residue field k. In fact, in this case R can be explicitly described as the ring of formal power series $R = k[[X]]$ over the indeterminate X with coefficients in k. The unique maximal ideal is in this case $Xk[[X]]$; see [84, II.§4]. We mention further – and again without proof – the following result, due to Thévenaz.

Proposition 4.3.12 ([90, (51.10)]) *Let $\mathcal{O}$ be an unramified complete discrete valuation ring with an algebraically closed residue field k of prime characteristic p and quotient field K of characteristic zero. Then every finite-dimensional division K-algebra is commutative.*

4.4 Local algebras over complete local rings

Let $\mathcal{O}$ be a commutative local ring with residue field k, and let A be an $\mathcal{O}$-algebra that is finitely generated as an $\mathcal{O}$-module. Extending earlier terminology, the algebra A is called *local* if $A/J(A)$ is a division ring. Thus A is *split local* if and only if $A/J(A) \cong k$. If A is local, then its unit element 1 is the unique idempotent in A, because an idempotent $i \neq 1$ is neither invertible nor in the radical. One of the main results of this section is Theorem 4.4.4 which shows that if $\mathcal{O}$ is complete, then the converse holds as well; that is, if A has no idempotent besides 1, then A is local. If the residue field k is algebraically closed and A is local, then A is split local. Group algebras of finite p-groups over an arbitrary field of characteristic p are split local by 1.11.1. The following theorem generalises this:

Theorem 4.4.1 *Let $\mathcal{O}$ be a commutative local ring with residue field $k = \mathcal{O}/J(\mathcal{O})$ of prime characteristic p, and let P be a finite group. The group algebra $\mathcal{O}P$ is local if and only if P is a p-group. Moreover, if P is a p-group, then $\mathcal{O}P$ is split local.*

Proof Suppose first that P is a p-group. Since $J(\mathcal{O})(\mathcal{O}P) \subseteq J(\mathcal{O}P)$ we have $\mathcal{O}P/J(\mathcal{O}P) \cong kP/J(kP) \cong k$, where the last isomorphism is from 1.11.1. Therefore $\mathcal{O}P$ is split local. Conversely, if P is not a p-group, then P has a nontrivial subgroup H of order prime to p. Then $|H|$ is invertible in $\mathcal{O}$, and hence $e_H = \frac{1}{|H|}\sum_{y\in H} y$ is an idempotent in $\mathcal{O}P$ by 1.1.2. Thus $\mathcal{O}P$ has an idempotent different from 1, and so cannot be local by the remarks above. □

We show now that if $\mathcal{O}$ is complete local Noetherian, then we can lift idempotents from the finite-dimensional k-algebra $A/J(\mathcal{O})A$ to the $\mathcal{O}$-algebra A. We derive from this a characterization of local $\mathcal{O}$-algebras. According to our conventions, an idempotent in A is a nonzero element $i \in A$ satisfying $i^2 = i$. Recall that two idempotents i, j are called *orthogonal* if $ij = 0 = ji$. In that case their sum $i + j$ is again an idempotent. An idempotent $i \in A$ is called *primitive* if it cannot be written as the sum of two orthogonal idempotents. We start with an elementary polynomial identity that we need in the proof of 4.4.3 below.

Lemma 4.4.2 *Let $f \in \mathbb{Z}[t]$ be the polynomial defined by $f(t) = 3t^2 - 2t^3$. We have $f(t) - t = t(1-t)(2t-1)$ and $f(t)^2 - f(t) = t^2(1-t)^2(4t^2 - 4t - 3)$.*

Proof We have $t(1-t)(2t-1) = (t-t^2)(2t-1) = 2t^2 - t - 2t^3 + t^2 = 3t^2 - 2t^3 - t = f(t) - t$, which shows the first equality. We have $f(t)^2 - f(t) = 9t^4 - 12t^5 + 4t^6 - 3t^2 + 2t^3$, and we have $t^2(1-t)^2(4t^2 - 4t - 3) = (t^2 - 2t^3 + t^4)(4t^2 - 4t - 3) = 4t^4 - 4t^3 - 3t^2 - 8t^5 + 8t^4 + 6t^3 + 4t^6 - 4t^5 - 3t^4 = 9t^4 - 12t^5 + 4t^6 - 3t^2 + 2t^3$, whence the Lemma. □

Theorem 4.4.3 *Let $\mathcal{O}$ be a complete commutative local Noetherian ring. Let A be an $\mathcal{O}$-algebra that is finitely generated as an $\mathcal{O}$-module and j an idempotent in $A/J(\mathcal{O})A$. Then there is an idempotent i in A whose image is $j = i + J(\mathcal{O})A$. Moreover, any such idempotent i is primitive in A if and only if j is primitive in $A/J(\mathcal{O})A$.*

Proof Let e_1 be any element in A whose image in $A/J(\mathcal{O})A$ is equal to j. For $k \geq 1$ define inductively e_k by $e_{k+1} = f(e_k)$, where f is the polynomial from 4.4.2 whose coefficients are identified to their images in $\mathcal{O}$. We show inductively that $e_k^2 - e_k \in J(\mathcal{O})^k A$ and $e_{k+1} - e_k \in J(\mathcal{O})^k$ for all $k \geq 1$. For $k = 1$ we have $e_1^2 - e_1 \in J(\mathcal{O})A$, since e_1 lifts j and $j^2 - j = 0$. For $k \geq 1$ we get inductively $e_{k+1}^2 - e_{k+1} = f(e_k)^2 - f(e_k) = e_k^2(1-e_k)^2(4e_k^2 - 4e_k - 3) \in J(\mathcal{O})^{k+1}A$ by 4.4.2, and then also $e_{k+1} - e_k = f(e_k) - e_k = e_k(1-e_k)(2e_k - 1) \in J(\mathcal{O})^k A$. In particular, the sequence $(e_k)_{k\geq 1}$ is a Cauchy sequence in A. Since $\mathcal{O}$ is complete, there is a unique element $i \in A$ such that $i - e_k \in J(\mathcal{O})^k A$ for all $k \geq 1$. Since all e_k lift j, so does i. Moreover, for all $k \geq 1$ we have $i^2 - i = (i - e_k + e_k)^2 - i = (i-e_k)^2 + (i-e_k)e_k + e_k(i-e_k) +$

$e_k^2 - e_k - (i - e_k)$. This expression is contained in $J(\mathcal{O})^k A$ because $i - e_k$ and $e_k^2 - e_k$ are contained in $J(\mathcal{O})^k$. Thus $i^2 - i = 0$. If j is primitive, clearly i is so, since $J(\mathcal{O})A$ contains no idempotent. If j is not primitive, then the algebra $j(A/J(\mathcal{O})A)j \cong iAi/J(\mathcal{O})iAi$ contains an idempotent different from j; thus, by the first statement, iAi contains an idempotent i_1 different from i. But then $i = (i - i_1) + i_1$, and so i is not primitive. □

Theorem 4.4.4 *Let $\mathcal{O}$ be a complete commutative local Noetherian ring. Let A be an $\mathcal{O}$-algebra that is finitely generated as an $\mathcal{O}$-module. The following statements are equivalent.*

(i) The algebra A is local.
(ii) The unit element 1_A is a primitive idempotent in A.
(iii) We have $A^\times = A \setminus J(A)$.

Proof Since $J(A)$ contains no idempotent, it follows that if 1_A is not primitive in A, then its image in $A/J(A)$ is not primitive, and therefore $A/J(A)$ cannot be a division algebra. Thus (i) implies (ii). If $A^\times = A \setminus J(A)$ then every non zero element of $A/J(A)$ is invertible, so (iii) implies (i). In order to show that (ii) implies (iii), we first observe that we can reduce the situation to $\mathcal{O} = k$. Indeed, this follows from the fact that on one hand, 1_A is primitive if and only if its image in $A/J(\mathcal{O})A$ is primitive (by 4.4.3), and on the other hand that $A^\times$ is the inverse image of $(A/J(\mathcal{O})A)^\times$. Thus we may assume that A is a finite-dimensional k-algebra whose unit element 1_A is primitive.

Let $x \in A$. For any $k \geq 1$ we have $Ax^{k+1} \subset Ax^k$. Since A has finite dimension, it is Artinian, and so the descending chain of left submodules Ax^k becomes eventually constant. Similarly, the ascending chain of left annihilators of x^k, $k \geq 1$, becomes eventually constant. This means that there is a positive integer n such that $Ax^n = Ax^{2n}$ and such that the left annihilators of x^n and of x^{2n} coincide. In other words, $x^n = ux^{2n}$ for some $u \in A$, and for any $a \in A$ we have $ax^n = 0$ if and only if $ax^{2n} = 0$. Then $(1_A - ux^n)ux^{2n} = (1_A - ux^n)x^n = x^n - ux^{2n} = 0$, so $(1_A - ux^n)u$ annihilates x^{2n} and therefore annihilates x^n. But then $(1_A - ux^n)ux^n = 0$, or equivalently, $(ux^n)^2 = ux^n$. Since 1_A is the unique idempotent in A either $ux^n = 1_A$, in which case x is invertible, or $ux^n = 0$, in which case $x^n = ux^{2n} = 0$, thus x is nilpotent. In other words, all noninvertible elements in A are nilpotent. Therefore, if a, $b \in A$ are noninvertible, then ay and by are non invertible, hence nilpotent for all $y \in A$. Thus the sum $a + b$ is again noninvertible; indeed, if y is an inverse of $a + b$, then $ay = 1_A - by$, but ay, by are nilpotent and hence $1_A - by$ is invertible. This contradiction shows that the set $A \setminus A^\times$ of noninvertible elements in A is an ideal. Since a proper left or right ideal in A contains no invertible element, it follows that $J(A)$ is the unique maximal left or right ideal in A, hence equal to $J(A)$. Therefore (ii) implies (iii). □

Corollary 4.4.5 *Let $\mathcal{O}$ be a complete commutative local Noetherian ring. Let A be an $\mathcal{O}$-algebra that is finitely generated as an $\mathcal{O}$-module. Suppose that A is local. Then $J(A)$ is the unique maximal ideal of A, every non zero quotient algebra and every unital subalgebra of A are local.*

Proof Since a proper ideal in A contains no invertible element, it follows from 4.4.4 that $J(A)$ is the unique maximal ideal in A. Hence if I is a proper ideal in A, then $I \subset J(A)$ and therefore $(A/I)/J(A/I) \cong A/J(A)$ is a division algebra. Thus A/I is local. If B is a unital subalgebra of A, then 1_A remains primitive in B, hence B is local. □

Corollary 4.4.6 *Let $\mathcal{O}$ be a complete commutative local Noetherian ring. Let A be an $\mathcal{O}$-algebra that is finitely generated as an $\mathcal{O}$-module. An idempotent i in A is primitive if and only if iAi is local.*

Proof Clearly i is primitive if and only if there is no idempotent in iAi other than i itself, thus if and only if iAi is local by 4.4.4. □

Corollary 4.4.7 *Let $\mathcal{O}$ be a complete commutative local Noetherian ring. Let A be an $\mathcal{O}$-algebra that is finitely generated as an $\mathcal{O}$-module and let U be a finitely generated A-module. Then U is indecomposable if and only if* $\mathrm{End}_A(U)$ *is local.*

Proof Suppose that $U = V \oplus V'$ for some submodules V, V' of U such that V is nonzero. Let $f : U \to U$ be the projection onto V with kernel V'. Then f is an idempotent in $\mathrm{End}_A(U)$ and we have $f = \mathrm{Id}_U$ if and only if $V' = 0$. Thus U is indecomposable if and only if Id_U is the unique idempotent in $\mathrm{End}_A(U)$. By 4.1.12, the algebra $\mathrm{End}_A(U)$ is finitely generated as an $\mathcal{O}$-module. The statement follows from 4.4.4. □

Corollary 4.4.8 (Rosenberg's Lemma) *Let $\mathcal{O}$ be a complete commutative local Noetherian ring. Let A be an $\mathcal{O}$-algebra that is finitely generated as an $\mathcal{O}$-module and let i be a primitive idempotent in A. Let $\mathcal{N}$ be a family of ideals in A. If $i \in \sum_{N \in \mathcal{N}} N$ then $i \in M$ for some $M \in \mathcal{N}$.*

Proof If $i \in \sum_{N \in \mathcal{N}} N$, then $i \in \sum_{N \in \mathcal{N}} iNi$, since i is an idempotent. By 4.4.4, the algebra iAi is local, and therefore each of the ideals iNi, where $N \in \mathcal{N}$, is either contained in $J(iAi)$ or equal to iAi. Thus there must be at least one $M \in \mathcal{N}$ such that $iMi = iAi$, which is equivalent to $i \in M$. □

Proposition 4.4.9 *Let $\mathcal{O}$ be a local commutative ring and A a split local $\mathcal{O}$-algebra. Then $[A, A] = [J(A), J(A)] \subseteq J(A)^2$.*

Proof Let $a, b \in A$. Since A is split local there are $\lambda, \mu \in \mathcal{O}$ and $c, d \in J(A)$ such that $a = \lambda \cdot 1_A + c$ and $b = \mu \cdot 1_A + d$. A short calculation shows that $[a, b] = ab - ba = cd - dc = [c, d]$, hence $[a, b] \in [J(A), J(A)] \subseteq J(A)^2$. □

4.5 Projective covers and injective envelopes

Throughout this section, $\mathcal{O}$ is a complete local commutative Noetherian ring with residue field $\mathcal{O}/J(\mathcal{O}) = k$. We use the results of the previous section to investigate the structure of finitely generated projective and injective indecomposable modules.

Proposition 4.5.1 *Let A be an $\mathcal{O}$-algebra that is finitely generated as $\mathcal{O}$-module and let U, V be finitely generated A-modules. Let $\varphi : U \to V$ be an A-homomorphism. Then $\varphi(\mathrm{rad}(U)) \subseteq \mathrm{rad}(V)$ and the A-homomorphism $\bar{\varphi} : U/\mathrm{rad}(U) \to V/\mathrm{rad}(V)$ induced by φ is surjective if and only if φ is surjective.*

Proof We have $\mathrm{rad}(U) = J(A)U$ by 4.1.11, hence $\varphi(\mathrm{rad}(U)) \subseteq J(A)V = \mathrm{rad}(V)$. If φ is surjective, so is clearly $\bar{\varphi}$. If $\bar{\varphi}$ is surjective then $\mathrm{Im}(\varphi) + \mathrm{rad}(V) = V$. Thus φ is surjective by Nakayama's Lemma. □

This Proposition has a dual version for socles of modules.

Proposition 4.5.2 *Let A be a finite-dimensional k-algebra and let U, V be finite-dimensional A-modules. Let $\varphi : U \to V$ be an A-homomorphism. Then $\varphi(\mathrm{soc}(U)) \subseteq \mathrm{soc}(V)$ and the A-homomorphism $\underline{\varphi} : \mathrm{soc}(U) \to \mathrm{soc}(V)$ induced by φ is injective if and only if φ is injective.*

Proof If S is a simple submodule of U then either $\varphi(S) = \{0\}$ or $\varphi(S)$ is a simple submodule of V isomorphic to S; in particular, $\varphi(\mathrm{soc}(U)) \subseteq \mathrm{soc}(V)$. If φ is injective, then its restriction $\underline{\varphi}$ to $\mathrm{soc}(U)$ is injective. Conversely, if $\underline{\varphi}$ is injective, then $\ker(\varphi) \cap \mathrm{soc}(U) = \{0\}$. If $\ker(\varphi)$ were non zero, then $\ker(\varphi)$ would have a simple submodule – but any such simple submodule would also be contained in $\mathrm{soc}(U)$, so this is not possible. Thus φ must be injective as well. □

Theorem 4.5.3 *Let A be an $\mathcal{O}$-algebra that is finitely generated as an $\mathcal{O}$-module.*

(i) *If U is a finitely generated projective indecomposable A-module, then $U/\mathrm{rad}(U)$ is simple; in particular, $\mathrm{rad}(U)$ is the unique maximal submodule of U.*
(ii) *If U, V are finitely generated projective A-modules, then we have $U/\mathrm{rad}(U) \cong V/\mathrm{rad}(V)$ if and only if $U \cong V$.*
(iii) *For every simple A-module S there is a projective indecomposable A-module U such that $S \cong U/\mathrm{rad}(U)$.*

In other words, the map sending U to $U/\mathrm{rad}(U)$ induces a bijection between the set of isomorphism classes of projective indecomposable A-modules and the set of isomorphism classes of simple A-modules.

Proof Let U be a finitely generated A-module. By 4.1.11 we have $\mathrm{rad}(U) = J(A)U$. Thus every endomorphism φ of U maps $\mathrm{rad}(U)$ to itself, hence induces an endomorphism $\overline{\varphi}$ of $U/\mathrm{rad}(U)$ sending $u + \mathrm{rad}(U)$ to $\varphi(u) + \mathrm{rad}(U)$. The map sending φ to $\bar{\varphi}$ defined in this way is an algebra homomorphism $\mathrm{End}_A(U) \to \mathrm{End}_A(U/\mathrm{rad}(U))$. If U is projective, then the map sending φ to $\overline{\varphi}$ is surjective because the canonical map $U \to U/\mathrm{rad}(U)$ is surjective. Thus, if U is projective indecomposable, then $\mathrm{End}_A(U)$ is local by 4.4.7 and its quotient algebra $\mathrm{End}_A(U/\mathrm{rad}(U))$ is still local by 4.4.5. It follows from 4.4.7 that $U/\mathrm{rad}(U)$ is indecomposable. But $U/\mathrm{rad}(U)$ is also a semisimple. This is only possible if $U/\mathrm{rad}(U)$ is in fact a simple module, whence (i). With the notation as in (ii), any A-homomorphism $\varphi : U \to V$ sends $\mathrm{rad}(U)$ to $\mathrm{rad}(V)$, hence induces a homomorphism $\bar{\varphi} : U/\mathrm{rad}(U) \to V/\mathrm{rad}(V)$. If φ is an isomorphism then clearly $\bar{\varphi}$ is an isomorphism. Conversely, any isomorphism $\bar{\varphi} : U/\mathrm{rad}(U) \cong V/\mathrm{rad}(V)$ lifts to an A-homomorphism $\varphi : U \to V$ since U is projective and since the canonical map $V \to V/\mathrm{rad}(V)$ is surjective. It follows from 4.5.1 that φ is surjective. Then φ is split surjective as V is projective. But then φ is an isomorphism as $U/\mathrm{rad}(U) \cong V/\mathrm{rad}(U) \oplus \ker(\varphi)/\mathrm{rad}(\ker(\varphi)) \cong V/\mathrm{rad}(V)$ forces $\ker(\varphi) = \{0\}$. This proves (ii). Let $s \in S \setminus \{0\}$. Then the map $A \to S$ sending $a \in A$ to as is a surjective A-homomorphism. Thus there is an indecomposable direct summand U of A whose image in S is non zero. Since S is simple this implies that U maps onto S. As $U/\mathrm{rad}(U)$ is simple, it follows that $U/\mathrm{rad}(U) \cong S$, which completes the proof of (iii). □

By 1.7.4, if i is a primitive idempotent in an $\mathcal{O}$-algebra A then Ai is projective indecomposable. For the particular ring $\mathcal{O}$ chosen in this section the converse of this statement is true, as well.

Proposition 4.5.4 *Let A be an $\mathcal{O}$-algebra that is finitely generated as an $\mathcal{O}$-module, and let U be a finitely generated projective indecomposable*

A-module. There is a primitive idempotent i in A such that $U \cong Ai$ as left A-modules.

Proof Set $S = U/\text{rad}(U)$ and denote by $\rho : U \to S$ the canonical surjective A-homomorphism. By 4.5.3 (i) the module S is simple. Let $s \in S \setminus \{0\}$ and let $\pi : A \to S$ be the A-homomorphism defined by $\pi(a) = as$. The simplicity of S implies that π is surjective. Since A is free, hence projective, there is an A-homomorphism $\sigma : A \to U$ satisfying $\rho \circ \sigma = \pi$. Since π is surjective we have $\text{Im}(\sigma) + \text{rad}(U) = U$. Since $\text{rad}(U) = J(A)U$, Nakayama's Lemma implies that σ is surjective. Then σ is split surjective because U is projective. Thus U is a direct summand of A as an A-module. By 1.7.4 (ii) we have $U \cong Ai$ for some primitive idempotent i in A. □

Theorem 4.5.3 has a dual version for socles of injective modules provided that socles and finitely generated injective modules exist.

Theorem 4.5.5 *Let A be a finite-dimensional k-algebra.*

(i) If U is a finitely generated injective indecomposable A-module, then $\text{soc}(U)$ is simple; in particular, $\text{soc}(U)$ is the unique simple submodule of U.

(ii) If U, V are finitely generated injective A-modules, then we have $\text{soc}(U) \cong \text{soc}(V)$ if and only if $U \cong V$.

(iii) For every simple A-module S there is an injective indecomposable A-module U such that $S \cong \text{soc}(U)$, and then $U \cong V^*$, where V is a projective cover of the simple right A-module S^*.

In other words, the map sending U to $\text{soc}(U)$ induces a bijection between the set of isomorphism classes of injective indecomposable A-modules and the set of isomorphism classes of simple A-modules.

Proof We dualise the arguments in the proof of 4.5.3. If U is finitely generated injective indecomposable, then any endomorphism φ of $\text{soc}(U)$ extends to an endomorphism ψ of U. Thus the restriction to $\text{soc}(U)$ induces a surjective k-algebra homomorphism $\text{End}_A(U) \to \text{End}_A(\text{soc}(U))$; in particular, $\text{End}_A(\text{soc}(U))$ is again local. But $\text{soc}(U)$ is also semisimple, and hence $\text{soc}(U)$ is simple. This proves (i). If $U \cong V$ then $\text{soc}(U) \cong \text{soc}(V)$ is clear. Conversely, with the notation of (ii), if there is an isomorphism $\varphi : \text{soc}(U) \cong \text{soc}(V)$, then since V is injective, φ extends to an A-homomorphism $\psi : U \to V$. It follows from 4.5.2 that ψ is injective. But then ψ is split injective since U is an injective A-module. The fact that ψ restricted to $\text{soc}(U)$ yields the isomorphism $\varphi : \text{soc}(U) \cong \text{soc}(V)$ implies that ψ is an isomorphism, whence (ii). For (iii), let S be a simple A-module. Then its k-dual $S^* = \text{Hom}_A(S, k)$ is a simple right A-module. By 4.5.3 for right modules there is a projective indecomposable right A-module V such that we have a short exact sequence of right A-modules $0 \to \mathrm{rad}(V) \to V \to S^* \to 0$. Since the functor $\mathrm{Hom}_A(-, k)$ is contravariant exact it follows that there is an exact sequence of left A-modules $0 \to S \to V^* \to \mathrm{rad}(V)^* \to 0$. Moreover, V^* is then injective as V was projective. This proves (iii). □

Proposition 4.5.6 *Let A be a finite-dimensional k-algebra and U a finitely generated A-module. Suppose that U has no simple direct summand. We have* $\mathrm{soc}(U) \subseteq \mathrm{rad}(U)$.

Proof Arguing by contradiction, suppose that U has a simple submodule S that is not contained in $\mathrm{rad}(U)$. By 4.5.5 there is an injective A-module I such that $\mathrm{soc}(I) = S$. Extend the inclusion $S \to I$ to an A-homomorphism $\varphi : \mathrm{rad}(U) \oplus S \to I$ such that $\mathrm{rad}(U) \subseteq \ker(\varphi)$. Since S is not contained in $\mathrm{rad}(U)$, the direct sum $\mathrm{rad}(U) \oplus S$ is a submodule of U. Since I is injective, we can extend φ to an A-homomorphism, still called φ, from U to I. Then $\ker(\varphi)$ contains $\mathrm{rad}(U)$, hence $\mathrm{Im}(\varphi)$ is a semisimple submodule of I. Since $\mathrm{soc}(I)$ is simple, we have $\mathrm{Im}(\varphi) \cong S$, and hence $\ker(\varphi)$ is a complement of S in U. This contradicts the assumption on U. □

Corollary 4.5.7 *Let A be a finite-dimensional k-algebra such that $J(A)^2 = \{0\}$. For any finitely generated indecomposable nonsimple A-module U we have* $\mathrm{rad}(U) = \mathrm{soc}(U)$*; that is, the socle and radical series of U coincide.*

Proof We have $\mathrm{soc}(U) \subseteq \mathrm{rad}(U)$ from 4.5.6. We have further $J(A)\mathrm{rad}(U) = J(A)^2U = \{0\}$, hence $\mathrm{rad}(U) \subseteq \mathrm{soc}(U)$ by 1.10.18 (i). □

Definition 4.5.8 Let A be an $\mathcal{O}$-algebra that is finitely generated as an $\mathcal{O}$-module. A *projective cover of a* finitely generated A-*module* U is a pair (P_U, π_U) consisting of a projective A-module P_U and a surjective A-homomorphism $\pi_U : P_U \to U$ which induces an isomorphism $P_U/\mathrm{rad}(P_U) \cong U/\mathrm{rad}(U)$.

An alternative definition would be to define a projective cover of a (not necessarily finitely generated) A-module U as a pair (P_U, π_U) consisting of a projective A-module P_U and a surjective A-homomorphism $\pi_U : P_U \to U$ with the property that the restriction of π_U to any proper submodule of P_U is no longer surjective. For finitely generated modules the two definitions are easily seen to coincide.

Theorem 4.5.9 (Existence and uniqueness of projective covers) *Let A be an $\mathcal{O}$-algebra that is finitely generated as an $\mathcal{O}$-module and let U be a finitely generated A-module.*

(i) *The module* U *has a projective cover* (P_U, π_U).
(ii) *We have* $\ker(\pi_U) \subset \mathrm{rad}(P_U)$ *and* P_U *is finitely generated.*
(iii) *For any other pair* (P, π) *consisting of a projective* A*-module* P *and a surjective* A*-homomorphism* $\pi : P \to U$ *there is a split surjection* $\tau : P \to P_U$ *such that* $\pi_U \circ \tau = \pi$.
(iv) *If* (P'_U, π'_U) *is another projective cover of* U*, there is an isomorphism* $\tau : P'_U \to P_U$ *such that* $\pi_U \tau = \pi'_U$.

Proof For every simple A-module S there is, by 4.5.3, a projective indecomposable A-module P_S such that $P_S/\mathrm{rad}(P_S) \cong S$. Thus every simple A-module has a finitely generated projective cover. It follows that any finite direct sum of simple A-modules has a finitely generated projective cover; in particular, $U/\mathrm{rad}(U)$ has a finitely generated projective cover $\overline{\pi}_U : P_U \to U/\mathrm{rad}(U)$. As P_U is projective, $\overline{\pi}_U$ lifts through the canonical surjection $U \to U/\mathrm{rad}(U)$ to an A-homomorphism $\pi_U : P_U \to U$. As $\overline{\pi}_U$ is surjective, we have $\mathrm{Im}(\pi_U) + \mathrm{rad}(U) = U$, hence π_U is surjective by Nakayama's Lemma. Thus (P_U, π_U) is a finitely generated projective cover of U, proving (i). As π_U induces an isomorphism $P_U/\mathrm{rad}(P_U) \cong U/\mathrm{rad}(U)$ we have $\ker(\pi_U) \subset \mathrm{rad}(P_U)$, which completes the proof of (ii). Since P is projective and π_U is surjective, there is an A-homomorphism $\tau : P \to P_U$ such that $\pi_U \tau = \pi$. Since π is surjective, we have $\mathrm{Im}(\tau) + \ker(\pi_U) = P_U$. Nakayama's Lemma and statement (ii) imply that τ is surjective. Then τ is split as P_U is projective, which proves (iii). Statement (iv) follows from applying (iii) twice. □

One useful consequence of the preceding theorem is that we can lift finitely generated projective $k \otimes_{\mathcal{O}} A$-modules uniquely to projective A-modules:

Proposition 4.5.10 *Let* A *be an* $\mathcal{O}$*-algebra that is finitely generated as an* $\mathcal{O}$*-module and set* $\bar{A} = A/J(\mathcal{O})A$. *For any* A*-module* U *denote by* $\bar{U}$ *the* $\bar{A}$*-module* $k \otimes_{\mathcal{O}} U$.

(i) *For any finitely generated projective* $\bar{A}$*-module* Q *there is, up to isomorphism, a unique finitely generated projective* A*-module* P *such that* $\bar{P} \cong Q$ *as* $\bar{A}$*-modules.*
(ii) *Let* U *be a finitely generated* A*-module and* P *a finitely generated projective* A*-module. Then* P *is a projective cover of* U *if and only if* $\bar{P}$ *is a projective cover of* $\bar{U}$.
(iii) *Suppose that* A *is free of finite rank as an* $\mathcal{O}$*-module. Let* U *be a finitely generated* $\mathcal{O}$*-free* A*-module. Then* U *is a projective* A*-module if and only if* $\bar{U}$ *is a projective* $\bar{A}$*-module.*

Proof Consider Q as an A-module via the canonical surjection $A \to \bar{A}$. Let P be a projective cover of Q as an A-module. By 4.5.9 this yields a projective A-module P satisfying $P/\text{rad}(P) \cong Q/\text{rad}(Q)$. But then $\bar{P} \cong P/J(\mathcal{O})P$ is a projective cover of Q as an $\bar{A}$-module, hence isomorphic to Q since Q is projective. This shows (i). Statement (ii) follows from (i) and the observation that $U/\text{rad}(U) \cong \bar{U}/\text{rad}(\bar{U})$; since $J(\mathcal{O})A \subseteq J(A)$, this isomorphism can be regarded as an isomorphism of modules over either A or $\bar{A}$. We show (iii). If U is projective as an A-module, then clearly $\bar{U}$ is projective as an $\bar{A}$-module. For the converse, suppose that $\bar{U}$ is a projective $\bar{A}$-module, and let P be a projective cover of U. By (ii), $\bar{P}$ is a projective cover of $\bar{U}$, hence isomorphic to $\bar{U}$. Since A is assumed in (iii) to be $\mathcal{O}$-free, so is P. Thus P and U have the same $\mathcal{O}$-rank, equal to $\dim_k(\bar{U})$, whence $P \cong U$. This shows (iii). □

The last statement in Proposition 4.5.10 holds for orders over not necessarily complete discrete valuation rings; see [22, (30.11)]. We reformulate some of the previous results.

Proposition 4.5.11 *Let A be an $\mathcal{O}$-algebra that is finitely generated as an $\mathcal{O}$-module. Let i be a primitive idempotent in A, and let S be a simple A-module. Denote by P_S a projective cover S. Set $S^* = \text{Hom}_k(S, k)$. The following are equivalent.*

(i) $P_S \cong Ai$.
(ii) $S \cong Ai/J(A)i$.
(iii) $iS \neq \{0\}$.
(iv) $P_{S^*} \cong iA$.
(v) $S^* \cong iA/iJ(A)$.
(vi) $S^*i \neq \{0\}$.

Proof Since S is simple, its projective cover P_S is indecomposable by 4.5.3, and isomorphic to Aj for some primitive idempotent j in A by 4.5.4. It follows that $P_S \cong Ai$ if and only if the unique simple quotient $Ai/J(A)i$ of Ai is isomorphic to S, whence the equivalence of (i) and (ii). The analogous argument with right modules yields the equivalence of (iv) and (v). Since $Ai/J(A)i$ is the unique simple quotient of Ai, it follows that $S \cong Ai/J(A)i$ if and only if $\text{Hom}_A(Ai, S) \neq \{0\}$. By 1.7.4 (v), this space is isomorphic to iS, whence the equivalence of (ii) and (iii). The analogous argument with right modules yields the equivalence of (v) and (vi). Duality (with respect to k) yields an isomorphism $(iS)^* \cong S^*i$, whence the equivalence of (iii) and (vi), which concludes the proof. □

The following result, which describes projective covers of bimodules, is due to Rouquier.

Proposition 4.5.12 ([82, Lemma 10.2.12]) *Let A, B be split $\mathcal{O}$-algebras that are free of finite rank as $\mathcal{O}$-modules. Let M be a finitely generated A-B-bimodule. A projective cover of M is isomorphic to*

$$\oplus_T P_{M\otimes_B T} \otimes_{\mathcal{O}} P_{T^*}$$

where T runs over a set of representatives of the isomorphism classes of simple B-modules, where $T^* = \mathrm{Hom}_k(T, k)$, and where $P_{M\otimes_B T}$, P_{T^*} denote projective covers of the A-module $M \otimes_B T$ and right B-module T^*, respectively.

Proof By 4.5.10 we may assume that $\mathcal{O} = k$. Let S be a simple A-module and T a simple B-module. The standard tensor-Hom adjunction applied to tensoring on the right by the B-k-bimodule T yields an isomorphism

$$\mathrm{Hom}_k(M \otimes_B T, S) \cong \mathrm{Hom}_{B^{\mathrm{op}}}(M, \mathrm{Hom}_k(T, S)) \cong \mathrm{Hom}_{B^{\mathrm{op}}}(M, S \otimes_k T^*),$$

where the second isomorphism uses 2.9.4. These are isomorphisms of A-A-bimodules; thus, taking A-fixed points yields an isomorphism

$$\mathrm{Hom}_A(M \otimes_B T, S) \cong \mathrm{Hom}_{A\otimes_k B^{\mathrm{op}}}(M, S \otimes_k T^*).$$

The hypothesis that A and B are split implies that if S and T run of sets of representatives of the isomorphism classes of simple A-modules and simple B-modules, respectively, then $S \otimes_k T^*$ runs over a set of representatives of the isomorphism classes of simple $A \otimes_k B^{\mathrm{op}}$-modules. The dimension of the term on the left side is the number of copies of S in a decomposition of $M \otimes_B T/\mathrm{rad}(M \otimes_B T)$, hence the number of copies of P_S in a projective cover of $M \otimes_B T$. Similarly, the dimension of the term on the right side is the number of copies of $P_{S\otimes_k T^*}$ in a projective cover of M. The result follows. □

For a finite-dimensional algebra A over a field k, the notion of a projective cover admits the following dual.

Definition 4.5.13 Let A be a finite-dimensional k-algebra. An *injective envelope of a* finitely generated *A-module* U is a pair (I_U, ι_U) consisting of an injective A-module I_U and an injective A-homomorphism $\iota_U : U \to I_U$ which induces an isomorphism $\mathrm{soc}(U) \cong \mathrm{soc}(I_U)$.

As in the case of projective covers, one can define an injective envelope for a (not necessarily finitely generated) A-module U as a pair (I_U, ι_U) consisting of an injective A-module I_U and an injective A-homomorphism $\iota : U \to I_U$ such that every nontrivial submodule of I_U has a nontrivial intersection with $\mathrm{Im}(\iota_U)$.

Theorem 4.5.14 *Let A be a finite-dimensional k-algebra and U a finitely generated A-module.*

(i) *The module U has an injective envelope (I_U, ι_U), and we have $I_U \cong (P_{U^*})^*$. In particular, I_U is finitely generated.*
(ii) *The map ι_U induces an isomorphism $\mathrm{soc}(U) \cong \mathrm{soc}(I_U)$.*
(iii) *For any pair (I, ι) consisting of an injective A-module I and an injective A-homomorphism $\iota : U \to I$ there is a split injective A-homomorphism $\tau : I_U \to I$ such that $\iota = \tau\iota_U$.*
(iv) *For any other injective envelope (I'_U, ι'_U) of U there is an isomorphism $\tau : I_U \cong I'_U$ such that $\iota'_U = \tau\iota_U$.*

Proof By 4.5.9 applied to right modules, the dual U^* has a projective cover (P_{U^*}, π_{U^*}). Dualising this map and composing it with the natural isomorphism $U \cong U^{**}$ yields an A-homomorphism $U \cong U^{**} \to (P_{U^*})^*$. This is a finitely generated injective envelope of U. The rest of the proof of 4.5.14 consists of dualising the proof of 4.5.9. □

Exercise 4.5.15 Let A be a finite-dimensional k-algebra. Consider A^* as an A-A-bimodule. Show that the functors $A^* \otimes_A -$ and $\mathrm{Hom}_A(A^*, -)$ induce inverse k-linear equivalences between the categories $\mathrm{proj}(A)$ and $\mathrm{inj}(A)$ consisting of finitely generated projective and injective A-modules, respectively.

4.6 The Krull–Schmidt Theorem

Throughout this Section $\mathcal{O}$ is a complete commutative local Noetherian ring with residue field $k = \mathcal{O}/J(\mathcal{O})$. Recall that a *primitive decomposition of an idempotent e in an $\mathcal{O}$-algebra A* is a finite set I of pairwise orthogonal primitive idempotents in A such that $\sum_{i \in I} i = e$.

Examples 4.6.1

(1) Let n be a positive integer. For $1 \le i \le n$ denote by E_i the matrix in $M_n(\mathcal{O})$ with entry 1 in (i, i) and entry 0 everywhere else. Then $\{E_i\}_{1 \le i \le n}$ is a primitive decomposition of 1 in the matrix algebra $M_n(\mathcal{O})$. A trivial verification shows that the idempotents E_i are all conjugate by invertible matrices in $M_n(\mathcal{O})$.
(2) Let A be an $\mathcal{O}$-algebra, let U be an A-module and let I be a primitive decomposition of Id_U in $\mathrm{End}_A(U)$. Then $U = \oplus_{\pi \in I} \pi(U)$ is a decomposition of U as a direct sum of indecomposable A-modules. Not every A-module has such a decomposition; a sufficient condition for the existence of such a decomposition is that U is finitely generated as an $\mathcal{O}$-module.

(3) Let G be a finite group and let K be a splitting field of characteristic zero for G. For any $\chi \in \mathrm{Irr}_K(G)$ set $e(\chi) = \frac{\chi(1)}{|G|}\sum_{x\in G}\chi(x^{-1})x$. Then the set $\{e(\chi) \mid \chi \in \mathrm{Irr}_K(G)\}$ is a primitive decomposition of 1 in $Z(KG)$.

The Krull–Schmidt Theorem for algebras states that under suitable hypotheses, primitive decompositions of idempotents exist and are unique up to conjugacy.

Theorem 4.6.2 (Krull–Schmidt) *Let A be an $\mathcal{O}$-algebra that is finitely generated as an $\mathcal{O}$-module and let e be an idempotent in A. Then e has a primitive decomposition, and any two primitive decompositions I, J of e are conjugate in A; that is, there is $u \in A^\times$ such that $J = uIu^{-1}$. Moreover, we may choose u such that u centralises $I \cap J$.*

Proof The existence of a primitive decomposition of e follows from 1.7.6. Since $e = \sum_{i\in I} i = \sum_{j\in J} j$ we have two decompositions of the left A-module Ae as a direct sum of indecomposable A-modules $Ae = \oplus_{i\in I}Ai = \oplus_{j\in J}Aj$. Dividing by the radical yields $Ae/J(A)e = \oplus_{i\in I}Ai/J(A)i = \oplus_{j\in J}Aj/J(A)j$. The modules Ai, Aj are projective (as they are direct summands of the free A-module A of rank 1) and indecomposable, as i, j are primitive idempotents. Thus, by 4.5.3, the modules $Ai/J(A)i$, $Aj/J(A)j$ are simple. It follows from 1.9.7 that there is a bijective map $\pi : I \to J$ such that $Ai/J(A)i \cong A\pi(i)/J(A)\pi(i)$ for all $i \in I$. We may choose π to be the identity on $I \cap J$. But then 4.5.3 implies that $Ai \cong A\pi(i)$ for all $i \in A$. Applying 1.7.4 yields elements $c_i \in iA\pi(i)$, $d_i \in \pi(i)Ai$, such that $c_id_i = i$ and $d_ic_i = \pi(i)$ for all $i \in A$. Set now $u = 1 - e + \sum_{i\in I} d_i$ and $v = 1 - e + \sum_{i\in I} c_i$. Since the elements of I (resp. J) are pairwise orthogonal, we have $uv = vu = 1$ and $uiv = \pi(i)$ for all $i \in I$. Since π is the identity on $I \cap J$, the result follows. □

The product of two idempotents in an algebra need not be an idempotent. The product of two *commuting* idempotents is either zero or an idempotent. The following two easy consequences of the above theorem are useful technical tools for achieving commutation of idempotents by replacing them with suitable conjugates.

Corollary 4.6.3 *Let A be an $\mathcal{O}$-algebra that is finitely generated as an $\mathcal{O}$-module and let e, f be idempotents in A. There is $u \in A^\times$ such that e and ufu^{-1} commute.*

Proof We may assume that e, f are different from 1_A. Let I, I', J, J' be primitive decompositions of e, $1 - e$, f, $1 - f$, respectively. Then $I \cup I'$ and $J \cup J'$ are primitive decompositions of 1_A in A. By 4.6.2, there is $u \in A^\times$ such that $I \cup I' =$

$u(J \cup J')u^{-1}$. The elements in $I \cup I'$ are pairwise orthogonal, hence commute pairwise, and therefore e and ufu^{-1} commute as well. □

Corollary 4.6.4 *Let A be an $\mathcal{O}$-algebra that is finitely generated as an $\mathcal{O}$-module, let e be an idempotent in A and I a not necessarily primitive decomposition of 1_A in A. There is $u \in A^\times$ such that ueu^{-1} commutes with all elements in I.*

Proof We may assume that e is different from 1_A. By refining I, if necessary, we may assume that I is a primitive decomposition of 1_A in A. Let J, J' be primitive decompositions of e, $1_A - e$, respectively. Then $J \cup J'$ is also a primitive decomposition of 1_A, and hence, by 4.6.2, there is $u \in A^\times$ such that $I = u(J \cup J')u^{-1}$. But then ueu^{-1} is a sum of elements in I, hence commutes with all elements in I. □

Corollary 4.6.5 *Let n be a positive integer. Any two primitive idempotents in $M_n(\mathcal{O})$ are conjugate.*

Proof Let e be a primitive idempotent in $M_n(\mathcal{O})$. A primitive decomposition of $1 - e$ together with e is a primitive decomposition of 1, hence conjugate to the canonical primitive decomposition $\{E_i\}_{1 \leq i \leq n}$ from 4.6.1. Since the E_i are all conjugate, it follows that e is conjugate to the matrix E_1, whence the result. □

Theorem 4.6.6 *Let A be an $\mathcal{O}$-algebra that is finitely generated free as an $\mathcal{O}$-module, and let S be a unitary subalgebra of A that is isomorphic to a matrix algebra $M_n(\mathcal{O})$ for some positive integer n. Denote by $C_A(S)$ the centraliser in A of S. For any primitive idempotent $i \in S$, multiplication by i induces an an $\mathcal{O}$-algebra isomorphism*

$$C_A(S) \cong iAi,$$

and multiplication in A induces an $\mathcal{O}$-algebra isomorphism

$$S \otimes_{\mathcal{O}} C_A(S) \cong A.$$

In particular, A and $C_A(S)$ are Morita equivalent.

Proof Let I be a primitive decomposition of 1 in S. Since S is a matrix algebra, we have $iSj \cong \mathcal{O}$ for all i, $j \in I$. The map sending $a \in A$ to the A^{op}-endomorphism of A given by left multiplication with a on A is an algebra isomorphism $A \cong \mathrm{End}_{A^{\mathrm{op}}}(A)$. An A^{op}-endomorphism given by left multiplication with a is a homomorphism of left S-modules if and only if $a \in C_A(S)$. Thus the previous isomorphism restricts to an isomorphism $C_A(S) \cong \mathrm{End}_{S \otimes_{\mathcal{O}} A^{\mathrm{op}}}(A)$. By 4.6.5, the elements of I are all conjugate in S. Thus $SiS = S$ and $AiA = A$ for any

$i \in I$. By 2.8.7, the algebra $S \otimes_{\mathcal{O}} A^{\mathrm{op}}$ is Morita equivalent to the algebra $iSi \otimes_{\mathcal{O}} (iAi)^{\mathrm{op}} \cong (iAi)^{\mathrm{op}}$. Under this Morita equivalence, the $S \otimes_{\mathcal{O}} A^{\mathrm{op}}$-module A corresponds to the $(iAi)^{\mathrm{op}}$-module iAi. Thus we have $C_A(S) \cong \mathrm{End}_{(iAi)^{\mathrm{op}}}(iAi) \cong iAi$, and one sees that this is the first isomorphism in the statement given by multiplication with i. We have a direct sum decomposition $A = \oplus_{i,j\in I} iAj$. Since all elements in I are conjugate, the spaces iAj have all the same $\mathcal{O}$-rank. For $i \in I$, we have $\mathrm{rk}_{\mathcal{O}}(A) = \mathrm{rk}_{\mathcal{O}}(S) \cdot \mathrm{rk}_{\mathcal{O}}(iAi) = \mathrm{rk}_{\mathcal{O}}(S) \cdot \mathrm{rk}_{\mathcal{O}}(C_A(S))$. This shows that the algebras $S \otimes_{\mathcal{O}} C_A(S)$ and A have the same $\mathcal{O}$-rank. It suffices therefore to show that the map $S \otimes_{\mathcal{O}} C_A(S) \to A$ given by multiplication in A is surjective. By Nakayama's Lemma, we may assume $\mathcal{O} = k$. Since both sides have the same dimension, it suffices to show that the map $S \otimes_k C_A(S) \to A$ is injective. Choose a basis $\{e_{i,j}\}_{i,j\in I}$ of S such that $e_{i,j} \in iSj$; after possibly adjusting by scalars, we may assume $e_{i,i} = i = e_{i,j}e_{j,i}$ for $i, j \in I$. Any element in $S \otimes_k C_A(S)$ is of the form $\sum_{i,j\in I} e_{i,j} \otimes c_{i,j}$ for some $c_{i,j} \in C_A(S)$. Suppose that the image of this element in A is zero; that is, $\sum_{i,j\in I} e_{i,j}c_{i,j} = 0$. Since $c_{i,j}$ centralises S, we have $e_{i,j}c_{i,j} = c_{i,j}e_{i,j} \in iAj$. Since A is the direct sum of the spaces iAj, it follows that $e_{i,j}c_{i,j} = 0$ for all i, j in I. Then $jc_{i,j} = e_{j,i}e_{i,j}c_{i,j} = 0$. Since j is conjugate in S to all elements in I and since conjugation by elements in S fixes $c_{i,j}$, it follows that $j'c_{i,j} = 0$ for all $j' \in I$. Taking the sum over all $j' \in I$ yields $c_{i,j} = 0$, whence the second isomorphism in the statement. We already noted that A and $C_A(S) \cong iAi$ are Morita equivalent. □

Applied to endomorphism algebras of modules, the algebra version of the Krull–Schmidt Theorem above takes the following form for modules. In this form, this theorem is also called the Krull–Remak–Schmidt–Azumaya Theorem; see [28, §2.13] for a brief history of this theorem.

Theorem 4.6.7 (Krull–Schmidt for modules) *Let A be an $\mathcal{O}$-algebra that is finitely generated as an $\mathcal{O}$-module and let U be a finitely generated A-module. Then U is a direct sum of finitely many indecomposable submodules of U. Suppose that $U = \oplus_{1\le i\le n} U_i = \oplus_{1\le j\le m} V_j$, where n, m are positive integers and U_i, V_j are non zero indecomposable submodules of U for any i, j. Then $n = m$, and there is a permutation π on the set $\{1, 2, \ldots, n\}$ such that $U_i \cong V_{\pi(i)}$ for all i, $1 \le i \le n$.*

Proof Observe that a decomposition of U as direct sum of indecomposable modules amounts to choosing a primitive decomposition of Id_A in the algebra $\mathrm{End}_A(U)$ (cf. Example 4.6.1 (2)). Since Id_A has a primitive decomposition by 4.6.2, the module U has a decomposition as a direct sum of finitely many indecomposable A-modules. For $1 \le i \le n$ let $\beta_i : U \to U$ be the canonical projection of U onto U_i, and for $1 \le j \le m$ let $\gamma_j : U \to U$ be the canonical projection of U onto V_j. Then the sets $\{\beta_i\}_{1\le i\le n}$ and $\{\gamma_j\}_{1\le j\le m}$ are two primitive decompositions of Id_U in $\mathrm{End}_A(U)$. Thus, by 4.6.2, there is a bijective map $\pi : I \to J$ and an automorphism α of U such that $\gamma_{\pi(i)} = \alpha \circ \beta_i \circ \alpha^{-1}$. It follows that $V_{\pi(i)} = \gamma_{\pi(i)}(U) = \alpha(\beta_i(U)) = \alpha(U_i)$, hence α induces an isomorphism $U_i \cong V_{\pi(i)}$ as required. □

Corollary 4.6.8 *Let A be an $\mathcal{O}$-algebra that is finitely generated as an $\mathcal{O}$-module and let i, j be idempotents in A. We have $Ai \cong Aj$ as A-modules if and only if i and j are conjugate in A.*

Proof Suppose that $Ai \cong Aj$. It follows from 4.6.7 applied to the left A-module A, that then also $A(1-i) \cong A(1-j)$. Thus, by 1.7.4, there are elements $c \in iAj$, $d \in jAi$ satisfying $cd = i$, $dc = j$ and elements $u \in (1-i)A(1-j)$, $v \in (1-j)A(1-i)$ satisfying $uv = 1-i$ and $vu = 1-j$. Set $a = d+v$. The element a is invertible with inverse $a^{-1} = c+u$; indeed, $(d+v)(c+u) = dc + vu = j + (1-j) = 1$. Moreover, we have $ai = (d+v)i = di = jd = j(d+v) = ja$, hence $aia^{-1} = j$. Conversely, if there is $a \in A^\times$ such that $aia^{-1} = j$, then right multiplication with a^{-1} sends $bi \in Ai$ to $bia^{-1} = ba^{-1}aia^{-1} = ba^{-1}j \in Aj$, and hence this map is an isomorphism from Ai to Aj with inverse given by right multiplication with a. □

Corollary 4.6.9 *Let A be an $\mathcal{O}$-algebra that is finitely generated as an $\mathcal{O}$-module.*

(i) The map sending an idempotent $i \in A$ to the A-module Ai induces a bijection between the set of conjugacy classes of primitive idempotents in A and the set of isomorphism classes of projective indecomposable A-modules.
(ii) The map sending an idempotent $i \in A$ to the A-module $Ai/J(A)i$ induces a bijection between the set of conjugacy classes of primitive idempotents in A and the set of isomorphism classes of simple A-modules.
(iii) If i, j are primitive idempotents in A, then $iAj \subseteq J(A)$ if and only if i, j are not conjugate in A.

Proof Statement (i) follows from 4.6.8 applied to primitive idempotents, and statement (ii) follows then from 4.5.3. Let i, j be primitive idempotents in A. If i, j are conjugate, then $j = uiu^{-1}$ for some $u \in A^\times$. Thus $iu^{-1}ju = i$ is contained in $iAju$ but not in $J(A)$, since $J(A)$ contains no idempotent. Since $J(A)$ is an ideal, it follows that iAj is not contained in $J(A)$. Conversely, if i and j are not conjugate, then by 4.6.8 the projective indecomposable modules Ai and Aj are not isomorphic. Again using 4.5.3, the image of any A-homomorphism $Ai \to Aj$ is contained in the unique maximal submodule $J(A)j$ of Aj. Since any

A-homomorphism $Ai \to Aj$ is induced by right multiplication with an element in iAj, it follows that $iAj \subseteq J(A)j \subseteq J(A)$. This proves (iii). □

Corollary 4.6.10 *Let A be an $\mathcal{O}$-algebra that is finitely generated as an $\mathcal{O}$-module and let M be a finitely generated A-module. Let τ, τ' be idempotents in* $\mathrm{End}_A(M)$. *Then $\tau(M)$, $\tau'(M)$ are direct summands of M, and we have $\tau(M) \cong \tau'(M)$ if and only if τ, τ' are conjugate in* $\mathrm{End}_A(M)$.

Proof Clearly $\tau(M)$ is a direct summand of M with complement $\ker(\tau) = (\mathrm{Id}_M - \tau)(M)$. Similarly for $\tau'(M)$. Suppose that τ, τ' are conjugate; that is, there is an automorphism α of M such that $\tau' = \alpha \circ \tau \circ \alpha$. Then $\tau'(M) = \alpha(\tau(M))$, and hence α induces an isomorphism $\tau(M) \cong \tau'(M)$. Conversely, suppose that there is an isomorphism $\tau(M) \cong \tau(M')$. The Krull–Schmidt Theorem applied to decompositions of M implies that there is also an isomorphism $(\mathrm{Id}_M - \tau)(M) \cong (\mathrm{Id}_M - \tau')(M)$. The sum of these two isomorphisms yields an automorphism α of M satisfying $\alpha(\tau(M)) = \tau'(M)$ and $\alpha((\mathrm{Id}_M - \tau)(M)) = (\mathrm{Id}_M - \tau')(M)$. Thus $\alpha \circ \tau \circ \alpha^{-1}$ is an idempotent endomorphism of M with the same image and same kernel as the idempotent τ', hence equal to τ'. □

The algebra theoretic version 4.6.2 of the Krull–Schmidt Theorem implies that the Krull–Schmidt property holds for any object X in an idempotent split $\mathcal{O}$-linear category such that the endomorphism algebra of X is finitely generated as an $\mathcal{O}$-module. In particular, it holds for categories of bounded complexes of finitely generated module. We state this for future reference.

Theorem 4.6.11 (Krull–Schmidt for bounded complexes) *Let A be an $\mathcal{O}$-algebra that is finitely generated as an $\mathcal{O}$-module and let U be a bounded complex of finitely generated A-module. Then U is a direct sum of finitely many indecomposable subcomplexes of U. Suppose that $U = \oplus_{1 \leq i \leq n} U_i = \oplus_{1 \leq j \leq m} V_j$, where n, m are positive integers and U_i, V_j non zero indecomposable subcomplexes of U for any i, j. Then $n = m$, and there is a permutation π on the set $\{1, 2, \ldots, n\}$ such that $U_i \cong V_{\pi(i)}$ for all i, $1 \leq i \leq n$.*

Proof The assumptions imply that the algebra $\mathrm{End}_{\mathrm{Ch}(\mathrm{Mod}(A))}(U)$ is finitely generated as an $\mathcal{O}$-module. The result follows exactly as in the proof of 4.6.7, with $\mathrm{End}_A(U)$ replaced by the algebra $\mathrm{End}_{\mathrm{Ch}(\mathrm{Mod}(A))}(U)$. □

Corollary 4.6.12 *Let A be an $\mathcal{O}$-algebra that is finitely generated as an $\mathcal{O}$-module and let X be a bounded complex of finitely generated A-module. Let τ, τ' be idempotents in* $\mathrm{End}_{\mathrm{Ch}(\mathrm{Mod}(A))}(X)$. *Then $\tau(X)$, $\tau'(X)$ are direct summands of X, and we have an isomorphism of complexes $\tau(X) \cong \tau'(X)$ if and only if τ, τ' are conjugate in* $\mathrm{End}_{\mathrm{Ch}(\mathrm{Mod}(A))}(X)$.

Proof This is the same proof as that of 4.6.10, with the reference 4.6.11 instead of 4.6.7. □

Corollary 4.6.13 *Let A be an $\mathcal{O}$-algebra that is finitely generated as an $\mathcal{O}$-module, and let U, V be finitely generated A-modules. Suppose that the sets of isomorphism classes of indecomposable direct summands of U and V coincide. Then* $\mathrm{End}_A(U)$ *and* $\mathrm{End}_A(V)$ *are Morita equivalent.*

Proof Let τ be the projection of $U \oplus V$ onto U, viewed as an endomorphism of $U \oplus V$. Let ι be a primitive idempotent in $\mathrm{End}_A(U \oplus V)$. Then $\iota(U \oplus V)$ is an indecomposable direct summand of $U \oplus V$. The Krull–Schmidt Theorem implies that $\iota(U \oplus V)$ is isomorphic to a direct summand of U or V, and the hypotheses on U and V imply that $\iota(U \oplus V)$ is isomorphic to a direct summand of U. It follows from 4.6.10 that ι is conjugate to an idempotent in $\mathrm{End}_A(U) \cong \tau \circ \mathrm{End}_A(U \oplus V) \circ \tau$. Thus the 2-sided ideal generated by τ contains all primitive idempotents in $\mathrm{End}_A(U \oplus V)$, hence the identity element. It follows from 2.8.7 that $\mathrm{End}_A(U \oplus V)$ and $\mathrm{End}_A(U)$ are Morita equivalent. The same argument shows that $\mathrm{End}_A(U \oplus V)$ and $\mathrm{End}_A(V)$ are Morita equivalent, whence the result. □

Exercise 4.6.14 Let A be an $\mathcal{O}$-algebra that is finitely generated as an $\mathcal{O}$-module. Let $\alpha \in \mathrm{Aut}(A)$. Show that if I is a primitive decomposition of 1 in A, then so is $\alpha(I)$.

The notation of automorphisms as subscripts to modules in the next exercise is as in the comments preceding 2.8.16.

Exercise 4.6.15 Let A be an $\mathcal{O}$-algebra that is finitely generated as an $\mathcal{O}$-module. Let J be a set of representatives of the conjugacy classes of primitive idempotents in A. Let $\alpha \in \mathrm{Aut}(A)$. Show that there is a permutation τ of I such that ${}_{\alpha}Ai \cong A\tau(i)$ for all $i \in J$. Show that $A\tau(i) \cong A\alpha^{-1}(i)$, and that if α is inner, then τ is the identity on I. Deduce that the map sending α to τ^{-1} induces a group homomorphism from $\mathrm{Out}(A)$ to the symmetric group S_I on I.

4.7 Lifting idempotents and points

Let $\mathcal{O}$ be a complete commutative local Noetherian ring with residue field $k = \mathcal{O}/J(\mathcal{O})$. If A is an $\mathcal{O}$-algebra that is finitely generated as an $\mathcal{O}$-module, then any idempotent in the k-algebra $\bar{A} = k \otimes_{\mathcal{O}} A$ 'lifts' uniquely, up to conjugation, to an idempotent in A, and primitive idempotents lift to primitive idempotents.

By invoking the Theorems of Wedderburn and Krull–Schmidt, we generalise this to arbitrary surjective algebra homomorphisms.

Theorem 4.7.1 (Lifting Theorem of idempotents) *Let A, B be $\mathcal{O}$-algebras that are finitely generated as $\mathcal{O}$-modules and let $f : A \to B$ be a surjective algebra homomorphism.*

(i) The homomorphism f maps $J(A)$ onto $J(B)$ and $A^\times$ onto $B^\times$.
(ii) For any primitive idempotent i in A either $i \in \ker(f)$ or $f(i)$ is a primitive idempotent in B.
(iii) For any primitive idempotent j in B there is a primitive idempotent i in A such that $f(i) = j$.
(iv) Any two primitive idempotents i, i' in A not contained in $\ker(f)$ are conjugate in A if and only if $f(i)$, $f(i')$ are conjugate in B.

Proof Since f is surjective, it follows that $f(J(A))$ is an ideal in B and that $B/f(J(A))$ is a quotient of $A/J(A)$, hence semisimple. Thus $J(B) \subseteq f(J(A))$. Then $f(J(A))/J(B)$ is a nilpotent ideal in $B/J(B)$, hence contained in the radical of $B/J(B)$, which is zero. This shows that $f(J(A)) = J(B)$. In order to show that f maps $A^\times$ onto $B^\times$, we first note that if A and B are semisimple, this follows from Wedderburn's Theorem because in that case, f is a projection of A onto a subset of its simple direct factors. The general case follows from this and the fact that $A^\times$ is the inverse image of $(A/J(A))^\times$. Indeed, denote by $\bar{f} : A/J(A) \to B/J(B)$ the algebra homomorphism induced by f. As just mentioned, the map $\bar{f}$ sends $(A/J(A))^\times$ onto $(B/J(A))^\times$. Thus if $v \in B^\times$, then there is $u \in A^\times$ such that $f(u) - v \in J(B)$. But then there is $w \in J(A)$ such that $f(w) = f(u) - v$. Thus $u - w \in A^\times$ satisfies $f(u - w) = v$. This proves (i). If i is a primitive idempotent in A such that $f(i) \neq 0$, then iAi is a local algebra by 4.4.4, hence $f(iAi) = f(i)Bf(i)$ is a local algebra by 4.4.5, and thus $f(i)$ is primitive in B. This proves (ii). Let I be a primitive decomposition of 1_A in A. It follows from (ii) that $f(I) \setminus \{0\}$ is a primitive decomposition of 1_B in B. Thus $B = \oplus_{i \in I, f(i) \neq 0} Bf(i)$ as a left B-module. If j is a primitive idempotent in B, then Bj is an indecomposable direct summand of B, and hence $Bj \cong Bf(i)$ for some $i \in I$ by the Krull–Schmidt Theorem 4.6.7. Therefore, by 4.6.8, there is $v \in B^\times$ such that $vjv^{-1} = f(i)$. By (i) there is $u \in A^\times$ such that $f(u) = v$, and then $u^{-1}iu$ is a primitive idempotent in A such whose image under f in B is j. This shows (iii). Let i, i' be primitive idempotents in A not contained in $\ker(f)$ such that $f(i)$, $f(i')$ are conjugate in B. Since f is surjective, f induces a surjective map $Ai \to Bf(i)$. Since f maps $J(A)$ to $J(B)$ this induces a surjective map $Ai/J(A)i \to Bf(i)/J(B)f(i)$. Since $Ai/J(A)i$ is simple, this map is in fact an isomorphism (of A-modules, where we view any B-module as A-module with

$a \in A$ acting as $f(a)$). Thus $Ai/J(A)i \cong Bf(i)/J(B)f(i) \cong Bf(i')/J(B)f(i') \cong Ai'/J(A)i'$, hence $Ai \cong Ai'$ by 4.5.3(ii), and so i, i' are conjugate by 4.6.8, which completes the proof. □

The above theorem implies in particular that two primitive idempotents in A lifting the same idempotent in B are conjugate. We can in that case be more precise regarding the elements that conjugate the two idempotents in A.

Corollary 4.7.2 *Let A, B be $\mathcal{O}$-algebras that are finitely generated as $\mathcal{O}$-modules and let $f : A \to B$ be a surjective algebra homomorphism. Let i, i' be primitive idempotents in A not contained in* $\ker(f)$ *such that $f(i) = f(i')$. Then there is an element $u \in A^\times \cap (1 + \ker(f))$ such that $i' = uiu^{-1}$.*

Proof By 4.7.1 (iv), we have $i' = wiw^{-1}$ for some $w \in A^\times$. Set $j = f(i) = f(i')$ and $y = f(w)$. Then $j = f(i') = yjy^{-1}$. Thus $y \in B^\times$ centralises j and $1 - j$, whence $y = jyj + (1 - j)y(1 - j) \in (jBj)^\times \times ((1 - j)B(1 - j))^\times$. By 4.7.1 (i) applied to the algebras iAi and $(1 - i)A(1 - i)A$ there is an element $x \in (iAi)^\times \times ((1 - i)A(1 - i))^\times \subseteq A^\times$ such that $f(x) = y$. Moreover, x centralises i. Thus $wx^{-1}ixw^{-1} = i'$ and $f(wx^{-1}) = yy^{-1} = 1$, which shows that $wx^{-1} \in 1 + \ker(f)$. □

Primitive idempotents remain primitive under suitable coefficient ring extensions.

Corollary 4.7.3 *Let $\mathcal{O}'/\mathcal{O}$ be an extension of complete commutative local Noetherian rings such that $J(\mathcal{O}) \subseteq J(\mathcal{O}')$ and $\mathcal{O}' = \mathcal{O} + J(\mathcal{O}')$. Let A be an $\mathcal{O}$-algebra that is finitely generated as an $\mathcal{O}$-module. Let j be a primitive idempotent in A. Then $1 \otimes j$ is primitive in the $\mathcal{O}'$-algebra $A' = \mathcal{O}' \otimes_{\mathcal{O}} A$.*

Proof The hypotheses imply that $\mathcal{O}'$ and $\mathcal{O}$ have the same residue field $k = \mathcal{O}/J(\mathcal{O}) \cong \mathcal{O}'/J(\mathcal{O}')$. It follows that the canonical map $A \to A'$ induces a surjective k-algebra homomorphism $A/J(\mathcal{O})A \to A'/J(\mathcal{O}')A'$. Applying 4.7.1 to this map as well as the canonical surjections $A \to A/J(\mathcal{O})A$ and $A' \to A'/J(\mathcal{O}')A'$ yields the result. □

Remark 4.7.4 The fact that p'-roots of unity in k lift uniquely to p'-roots of unity in $\mathcal{O}$, proved in 4.3.7, is equivalent to the fact that for a cyclic p'-group $H = \langle x \rangle$, primitive idempotents in kH lift uniquely to primitive idempotents in $\mathcal{O}H$ (note that $\mathcal{O}H$ is commutative, so conjugation is trivial). To see this, denote by n the order of x and let λ be a primitive n-th root of unity in $k^\times$. Let $\eta : H \to k^\times$ be the group homomorphism defined by $\eta(x) = \lambda$. By 3.3.1, the element $e(\eta) = \frac{1}{n}\sum_{y \in H} \eta(y^{-1})y$ is a primitive idempotent in kH. By 4.7.1 there is a unique primitive idempotent $i \in \mathcal{O}H$ that lifts $e(\eta)$. Since $kHe(\eta)$ is one-dimensional, $\mathcal{O}Hi$ has $\mathcal{O}$-rank one, hence determines a group homomorphism $\theta : H \to \mathcal{O}^\times$ that lifts η. In particular, $\mu = \theta(x)$ is the unique primitive n-th root of unity in $\mathcal{O}^\times$ that lifts λ.

The following theorem is a refinement of the lifting theorem for idempotents.

Theorem 4.7.5 *Let A, B be $\mathcal{O}$-algebras that are finitely generated as $\mathcal{O}$-modules, let I be an ideal in A, let J be an ideal in B, and let $f : A \to B$ be an algebra homomorphism such that $f(I) = J$.*

(i) For any primitive idempotent i in A contained in I either $i \in \ker(f)$ or $f(i)$ is a primitive idempotent in B contained in J.
(ii) For any primitive idempotent j in B contained in J there is a primitive idempotent i in A such that $f(i) = j$, and then any such i is contained in I.
(iii) Any two primitive idempotents i, i' in A contained in I but not contained in $\ker(f)$ are conjugate in A if and only if $f(i)$, $f(i')$ are conjugate in B.

Proof Let i be a primitive idempotent in A such that $f(i) \neq 0$. By 4.7.1, $f(i)$ is primitive in the subalgebra $f(A)$ of B. In order to prove (i), it suffices to observe that an idempotent j in J that is primitive in $f(A)$ remains primitive in B. Indeed, if $j = j_1 + j_2$ with orthogonal idempotents j_1, j_2 in B which commute with j, then $j_1 = jj_1 \in J$ and similarly $j_2 \in J$, contradicting the fact that j is primitive in $f(A)$. This proves (i). Again by 4.7.1 there is a primitive idempotent $i \in A$ such that $f(i) = j$. Then i belongs to the inverse image $I + \ker(f)$ of J in A. Since i is not contained in $\ker(f)$, it follows from Rosenberg's Lemma 4.4.8 that $i \in I$. This shows (ii). In order to show (iii), observe first that $Ai = Ai^2 = Ii$ because i is an idempotent contained in I. Suppose that $f(i)$, $f(i')$ are conjugate in B. Then $Bf(i) \cong Bf(i')$ as B-modules. It follows that $Bf(i) = Jf(i) = f(I)f(i) = f(A)f(i)$. Similarly for $f(i')$. Thus $f(A)f(i) \cong f(A)f(i')$ as $f(A)$-modules, and hence $f(i)$, $f(i')$ are conjugate in $f(A)$. It follows from 4.7.1 that i, i' are conjugate in A. The converse is clear. □

Corollary 4.7.6 *Let A be an $\mathcal{O}$-algebra that is finitely generated as an $\mathcal{O}$-module and let I be an ideal of A. Then I contains an idempotent if and only if I is not contained in $J(A)$.*

Proof If I is contained in $J(A)$, then I contains no idempotent by 1.10.5. Suppose that I is not contained in $J(A)$. The image $J = (I + J(A))/J(A)$ of I in $A/J(A)$ is a nonzero ideal. By 1.13.3, the quotient $A/J(A)$ is a direct product of simple algebras. Thus J is a direct product of some of these simple algebras, hence contains an idempotent. By 4.7.5, any primitive idempotent in J lifts to an idempotent in I, whence the result. □

Corollary 4.7.7 *Let A be an $\mathcal{O}$-algebra that is finitely generated as an $\mathcal{O}$-module and let I be a nonzero ideal of A. We have $I^2 = I$ if and only if $I = AeA$ for some idempotent e in A.*

Proof Suppose that $I^2 = I \neq \{0\}$. If $I \subseteq J(A)$, then $I^2 \subseteq J(A)I$, which is strictly smaller than I by Nakayama's Lemma, a contradiction. This shows that I is not contained in $J(A)$. (This follows also from 4.1.8.) By 4.7.6 the ideal I contains an idempotent e. Since I is an ideal, we have $AeA \subseteq I$. Choose e such that AeA is maximal with respect to the inclusion and subject to being contained in I. Suppose that AeA is strictly smaller than I. Then $J = I/AeA$ is a nonzero ideal in $B = A/AeA$ satisfying $J^2 = J$. By 4.7.6 there is a primitive idempotent $j \in J$. By 4.7.5 there is a primitive idempotent i in I lifting j; in particular, i is not contained in AeA. By 4.6.3, after possibly replacing i by a conjugate, we may assume that e and i commute. Since i is primitive we have either $i = ei = ie$ or $ei = ie = 0$. The first case is impossible as i is not in Ae. Thus $e + i$ is an idempotent in I, and the ideal $A(e + i)A$ is strictly bigger than AeA, a contradiction. This shows that $I = AeA$. The converse is trivial. □

Corollary 4.7.8 *Let A be an $\mathcal{O}$-algebra that is finitely generated as an $\mathcal{O}$-module and let I be an ideal of A contained in $J(A)$. Let $a \in A$ and let i be an idempotent in A.*

(i) We have $a \in A^\times$ if and only if $a + I \in (A/I)^\times$.
(ii) The idempotent i is primitive in A if and only if $i + I$ is primitive in A/I.

Proof If $a + I$ is invertible in A/I then $A = aA + I$. Since $I \subseteq J(A)$ it follows from Nakayama's Lemma 1.10.4 that $A = aA$. Similarly, $Aa = A$, and hence a is invertible in A. The converse is trivial, whence (i). Since $J(A)$ contains no idempotent, statement (ii) follows from 4.7.1 (ii). □

Corollary 4.7.9 *Let A be an $\mathcal{O}$-algebra that is finitely generated as an $\mathcal{O}$-module and let B be a subalgebra of A. We have $J(A) \cap B \subseteq J(B)$.*

Proof The ideal $J(A) \cap B$ of B contains no idempotent, hence is contained in $J(B)$ by 4.7.6. □

Proposition 4.7.10 *Let A be an $\mathcal{O}$-algebra that is finitely generated as an $\mathcal{O}$-module and let B be a subalgebra of A. We have $A = B + J(A)$ if and only if $J(B) \subseteq J(A)$ and the inclusion $B \subseteq A$ induces an isomorphism of k-algebras $B/J(B) \cong A/J(A)$. In that case, the following hold.*

(i) We have $J(B) = J(A) \cap B$.
(ii) Every primitive idempotent i in B is primitive in A, and two primitive idempotents i, j in B are conjugate in B if and only if they are conjugate in A.

(iii) *For every simple A-module S, the B-module* $\mathrm{Res}^A_B(S)$ *is simple, and restriction induces a bijection between the isomorphism classes of simple A-modules and of simple B-modules.*

Moreover, if k is a splitting field for A and B, and if (ii) or (iii) holds, then $A = B + J(A)$.

Proof If the inclusion $B \subseteq A$ induces an isomorphism $B/J(B) \cong A/J(A)$, then clearly $A = B + J(A)$. Conversely, if $A = B + J(A)$, then the inclusion $B \subseteq A$ induces a surjective algebra homomorphism $B \to A/J(A)$ with kernel $B \cap J(A)$. By 4.7.9 we have $B \cap J(A) \subseteq J(B)$. Since $A/J(A) \cong B/B \cap J(A)$ is semisimple, it follows that $J(B) = B \cap J(A)$, proving the equivalence as stated, and proving also that (i) holds in that case. By 4.7.8, the canonical maps $A \to A/J(A)$ and $B \to B/J(B)$ induce bijections on conjugacy classes of primitive idempotents. Thus if $B/J(B) \cong A/J(A)$, then (ii) holds. But then also (iii) holds, because simple A-modules correspond bijectively to the simple $A/J(A)$-modules, and similarly for B. If k is a splitting field, then the simple factors of $A/J(A)$ are matrix algebras whose dimensions are determined by the dimensions of the simple A-modules. Thus either (ii) or (iii) imply that A and B have isomorphic semisimple quotients, whence the result. □

Corollary 4.7.11 *Let A be a commutative $\mathcal{O}$-algebra that is finitely generated as an $\mathcal{O}$-module and let B be a subalgebra of A such that* $A = B + J(A)$. *Then B contains every idempotent in A.*

Proof By 4.7.10, a primitive decomposition of 1 in B remains a primitive decomposition of 1 in A, and hence, by 4.6.2, every primitive idempotent in A is conjugate to a primitive idempotent in B. Since A is commutative it follows that every primitive idempotent of A is contained in B. Since any idempotent in A has a primitive decomposition, the result follows. □

Corollary 4.7.12 *Let A be an $\mathcal{O}$-algebra that is free of finite rank as an $\mathcal{O}$-module such that $k \otimes_{\mathcal{O}} A$ is isomorphic to a direct product of matrix algebras over k. Then A is isomorphic to a direct product of matrix algebras over $\mathcal{O}$.*

Proof By the assumptions, any simple $k \otimes_{\mathcal{O}} A$-module S is projective, hence lifts by 4.7.1, up to isomorphism, uniquely to a projective A-module U_S satisfying $k \otimes_{\mathcal{O}} U_S \cong S$. By Wedderburn's Theorem 1.14.6, the canonical map $k \otimes_{\mathcal{O}} A \to \prod_S \mathrm{End}_k(S)$ is an isomorphism, where S runs over a set of representatives of the isomorphism classes of simple $k \otimes_{\mathcal{O}} A$-modules. It follows from Nakayama's Lemma 1.10.4 that the canonical map $A \to \prod_S \mathrm{End}_{\mathcal{O}}(U_S)$

is surjective, hence an isomorphism, since both sides are $\mathcal{O}$-free of the same rank. □

Corollary 4.7.13 *Suppose that k has prime characteristic p. Let G be a finite p′-group. If k is a splitting field for G, then $\mathcal{O}G$ is isomorphic to a finite direct product of matrix algebras over $\mathcal{O}$.*

Proof If k is a splitting field for G, then kG is isomorphic to a direct product of matrix algebras as a consequence of Maschke's Theorem 1.16.13. Thus 4.7.13 follows from 4.7.12. □

Applying the lifting theorem to the canonical surjective algebra homomorphism

$$\mathrm{End}_{\mathrm{Ch}^b(\mathrm{mod}(A))}(X) \to \mathrm{End}_{K^b(\mathrm{mod}(A))}(X),$$

for X a bounded complex of finitely generated modules over an $\mathcal{O}$-algebra A, yields the following criterion for when X is indecomposable in $K^b(\mathrm{mod}(A))$.

Proposition 4.7.14 *Let A be an $\mathcal{O}$-algebra that is finitely generated as an $\mathcal{O}$-module, and let X be a noncontractible bounded complex of finitely generated A-modules. Then X is indecomposable in $K^b(\mathrm{mod}(A))$ if and only if in $\mathrm{Ch}^b(\mathrm{mod}(A))$ we have $X = Y \oplus Z$ for some indecomposable bounded noncontractible complex Y and some contractible bounded complex Z. In particular, if X is indecomposable noncontractible in $\mathrm{Ch}^b(\mathrm{mod}(A))$, then X remains indecomposable in $K^b(\mathrm{mod}(A))$.*

Proof The algebra $\mathrm{End}_{K(\mathrm{mod}(A))}(X)$ is finitely generated as an $\mathcal{O}$-module because X is a bounded complex of finitely generated A-modules. Thus any primitive idempotent $\bar{\pi}$ in $\mathrm{End}_{K(\mathrm{mod}(A))}(X)$ lifts to a primitive idempotent π in $\mathrm{End}_{\mathrm{Ch}(\mathrm{mod}(A))}(X)$ that is not homotopic to zero, hence which has the property that $\pi(X)$ is a noncontractible indecomposable direct summand of X in $\mathrm{Ch}(\mathrm{mod}(A))$. Thus X is indecomposable in $K^b(\mathrm{mod}(A))$ if and only if $\mathrm{End}_{K(\mathrm{mod}(A))}(X)$ is local, and this holds if and only if $\mathrm{Id}_X = \pi + \pi'$ for some orthogonal idempotents π, π' in $\mathrm{End}_{\mathrm{Ch}(\mathrm{mod}(A))}(X)$ such that π is primitive and the image of π in $\mathrm{End}_{K(\mathrm{mod}(A))}(X)$ is equal to the image of Id_X, in which case the image of π' in $\mathrm{End}_{K(\mathrm{mod}(A))}(X)$ is necessarily zero. In that case the summands $Y = \pi(X)$ and $Z = \pi'(X)$ of X satisfy the conclusion. □

The argument shows slightly more generally that a decomposition of a bounded complex X in $K^b(\mathrm{mod}(A))$ lifts to a decomposition in $\mathrm{Ch}^b(\mathrm{mod}(A))$ plus a contractible complex, and hence the Krull–Schmidt property holds in $K^b(\mathrm{mod}(A))$. This can also be deduced from 1.18.19 and the Krull–Schmidt property in $\mathrm{Ch}^b(\mathrm{mod}(A))$.

Remark 4.7.15 It is possible for an algebra homomorphism to preserve primitive idempotents without being surjective modulo radicals. The diagonal embedding $k \times k$ to $M_2(k)$ sending (λ, μ) to the diagonal matrix with entries λ and μ sends the two primitive idempotents $(1, 0)$ and $(0, 1)$ in $k \times k$ to primitive idempotents in $M_2(k)$. These are the only primitive idempotents of $M_2(k)$ that lift to idempotents in $k \times k$.

The algebra version of the Krull–Schmidt Theorem 4.6.2, its corollaries 4.6.8, 4.6.9 and the Lifting Theorem for idempotents 4.7.1 all suggest that from a conceptual point of view, we need to consider primitive idempotents within their conjugacy classes. This motivates the following terminology.

Definition 4.7.16 ([71]) Let A be an $\mathcal{O}$-algebra that is finitely generated as an $\mathcal{O}$-module, and let I be a primitive decomposition of 1_A in A. A *point of* A is an $A^\times$-conjugacy class α of primitive idempotents in A. The *multiplicity of a point* α *on the algebra* A is the cardinal $m_\alpha = |I \cap \alpha|$ of the set $I \cap \alpha$.

Any two primitive decompositions of 1_A in A are conjugate by 4.6.2, and hence the multiplicity $|I \cap \alpha|$ of a point α on A does not depend on the choice of a primitive decomposition I of 1_A. Since $A = \oplus_{i \in I} Ai$ as left A-modules, the multiplicity $m_\alpha = |I \cap \alpha|$ is also the number of indecomposable direct summands of A isomorphic to Aj, where $j \in \alpha$, in a decomposition of A as direct sum of indecomposable left modules. By 4.6.5 a matrix algebra over $\mathcal{O}$ has a unique point. See 4.7.21 below for a more general statement.

Proposition 4.7.17 *Let A be an $\mathcal{O}$-algebra that is finitely generated as an $\mathcal{O}$-module.*

(i) The map sending a primitive idempotent $i \in A$ to the projective A-module Ai induces a bijection between the set of points of A and the set of isomorphism classes of projective indecomposable A-modules.
(ii) The map sending a primitive idempotent $i \in A$ to the simple A-module $Ai/J(A)i$ induces a bijection between the set of points of A and the set of isomorphism classes of simple A-modules.
(iii) The number of points of A is equal to the number of isomorphism classes of simple A-modules.
(iv) If A is split, then for any point α on A with corresponding simple A-module S we have $m_\alpha = \dim_k(S)$.

Proof The statements (i) and (ii) are reformulations of 4.6.8 and 4.6.9, and (iii) is an immediate consequence of (ii). By Wedderburn's Theorem 1.13.3, if A is split, then the multiplicity of S as a direct summand of the left A-module $A/J(A)$ is equal to $\dim_k(S)$. This is equal to the multiplicity of a projective cover P_S of S as a direct summand of the left A-module A, and since any direct summand of A isomorphic to P_S is equal to Ai for some $i \in \alpha$, the last statement follows. □

One can use the last statement to give another characterisation of the image of the canonical map $\mathrm{Out}(A) \to \mathrm{Pic}(A)$, previously considered in Proposition 2.8.16.

Proposition 4.7.18 *Let A be an $\mathcal{O}$-algebra that is finitely generated as an $\mathcal{O}$-module. Suppose that A is split. Let M be an A-A-bimodule inducing a Morita equivalence. There is an automorphism $\alpha \in \mathrm{Aut}(A)$ such that $M \cong A_\alpha$ as A-A-bimodules if and only if $\dim_k(M \otimes_A S) = \dim_k(S)$ for any simple A-module S.*

Proof By 2.8.16, we have $M \cong A_\alpha$ for some $\alpha \in \mathrm{Aut}(A)$ if and only if $M \cong A$ as left A-modules. This is equivalent to requiring that for any simple A-module T, the multiplicity of a projective cover P_T of T in A is equal to that of P_T as a summand of M. Since M induces a Morita equivalence, we have $P_T \cong M \otimes_A P_S$ for some simple module S satisfying $T \cong M \otimes_A S$. Thus the multiplicity of P_T as a summand of M is equal to that of P_S as a summand of A. Since A is split, it follows from the last statement of 4.7.17 that this is equivalent to $\dim_k(T) = \dim_k(S)$, whence the result. □

The refined version of the Lifting Theorem for idempotents 4.7.5 reads as follows:

Theorem 4.7.19 *Let A, B be $\mathcal{O}$-algebras that are finitely generated as $\mathcal{O}$-modules, let I be an ideal in A and J an ideal in B. Let $f : A \to B$ be an algebra homomorphism such that $f(I) = J$.*

(i) For every point α of A contained in I, either $\alpha \subset \ker(f)$ or $f(\alpha)$ is a point of B contained in J.
(ii) For every point β of B contained in J there is a unique point α of A contained in I such that $f(\alpha) = \beta$.
(iii) If α is a point of A contained in I and β a point of B contained in J such that $f(\alpha) = \beta$, then $m_\alpha = m_\beta$.

Proof The statements (i) and (ii) are reformulations of the corresponding statements in 4.7.5. If E is a primitive decomposition of 1_A in A, then the set $F = f(E) \setminus \{0\}$ is a decomposition of 1_B in B. It need not be a primitive decomposition of 1_B because idempotents in A outside of I need not remain primitive when mapped to B under f. By decomposing the idempotents in F we get a primitive decomposition F' of 1_B. It follows from (i) and (ii) that the elements in $\beta \cap F'$ are all in the image of $\alpha \cap E$, and hence both sets have the same cardinality. □

Theorem 4.7.19 applies in particular if $B = A/J(A)$ and f is the canonical surjection, yielding the following reformulation of 4.7.8:

Corollary 4.7.20 *Let A be an $\mathcal{O}$-algebra that is finitely generated as an $\mathcal{O}$-module. The canonical algebra homomorphism $A \to A/J(A)$ induces a bijection between the sets of points of A and of $A/J(A)$.*

Corollary 4.7.21 *Let A be an $\mathcal{O}$-algebra that is finitely generated as an $\mathcal{O}$-module. The following are equivalent:*

(i) The algebra A has a unique point.
(ii) The algebra A has a unique isomorphism class of simple modules.
(iii) The algebra $A/J(A)$ is simple.

Proof The equivalence of (i) and (ii) is a particular case of 4.7.17 (ii). Now $A/J(A)$ is a finite direct product of simple Artinian algebras, and by Wedderburn's Theorem 1.13.3, every simple factor corresponds to a unique isomorphism class of simple modules, whence the equivalence of (ii) and (iii). □

Corollary 4.7.22 *Let A be an $\mathcal{O}$-algebra that is finitely generated as an $\mathcal{O}$-module. For any point α of A there is a unique maximal ideal M_α of A such that α is not contained in M_α, and the map sending α to M_α is a bijection between the sets of points and of maximal ideals of A.*

Proof Write $A/J(A) = B_1 \times B_2 \times \cdots \times B_r$ as direct product of simple algebras B_s, $1 \le s \le r$. Every maximal ideal M of A contains $J(A)$, hence its image in $A/J(A)$ is a maximal ideal of $A/J(A)$, and hence equal to

$$B_1 \times B_2 \times \cdots \times B_{t-1} \times B_{t+1} \times \cdots \times B_r$$

for a unique integer t such that $1 \le t \le r$. Since B_t has a unique point by the previous Corollary, this lifts to a unique point in A not contained in M. Every point of A arises in this way by 4.7.20, whence the result. □

The bijection between points and maximal ideals is the reason for this terminology: the points of an affine variety correspond to the maximal ideals of the associated affine algebra. As an application of the above material, we conclude this section with a theorem of Dade, which generalises Corollary 4.3.6, and which will be used in the section on extensions of nilpotent blocks.

Theorem 4.7.23 ([24, Proposition 1.16]) *Let A be an $\mathcal{O}$-algebra that is free of finite rank as an $\mathcal{O}$-module. Suppose that k is algebraically closed of* *prime characteristic p. Let m be a positive integer that is prime to p. For any $y \in 1 + J(A)$ there is a unique $z \in 1 + J(A)$ such that $y = z^m$. In particular, if A is commutative, then the map sending $y \in 1 + J(A)$ to y^m is a group automorphism of $1 + J(A)$.*

Proof Note that $1 + J(A)$ is a subgroup of $A^\times$, and so the last statement follows immediately from the first. In order to prove the first statement, we start by considering the case where A is commutative and local (hence split local, as k is algebraically closed). Let $y \in 1 + J(A)$. Let X be an independent variable. Since A is commutative, so is $A[X]$, and hence $(X^m - y)A[X]$ is an ideal in $A[X]$. Consider the algebra $B = A[X]/(X^m - y)A[X]$. Denote by x the image of X in B. Then B is free as an A-module, of rank m, with basis $\{1, x, x^2, \ldots, x^{m-1}\}$. In particular, B is $\mathcal{O}$-free. Identifying A to its image in B, we have $J(A)B = BJ(A)$, so this is a 2-sided ideal in B that becomes nilpotent upon reduction modulo $J(\mathcal{O})$, and hence this ideal is contained in $J(B)$. Note that $A/J(A) \cong k$. Thus reducing B modulo $J(A)$ yields

$$B/J(A)B \cong k[X]/(X^m - 1)k[X]$$

where we use the fact that $y \in 1 + J(A)$. This is the group algebra of the cyclic group of order m generated by the image $\bar{x}$ of X, hence isomorphic to a direct product of copies of k which correspond to the m-th roots of unity in k (and there are m different m-th roots of unity as k is algebraically closed of characteristic p prime to m). Thus

$$B/J(A)B \cong \prod_{i=1}^{m} k\bar{e}_i$$

for some idempotents $\bar{e}_i$ in $B/J(A)B$. Since $J(A)B$ is contained in $J(B)$, and since B is commutative, each idempotent $\bar{e}_i$ lifts uniquely to a primitive idempotent e_i in B. Nakayama's Lemma (applied to B as an A-module) implies that

$$B = \prod_{i=1}^{m} Ae_i$$

and since B is free of rank m as an A-module, it follows further (using the Krull–Schmidt Theorem and the fact that A is indecomposable as an A-module since A is assumed local) that multiplication by e_i is an algebra isomorphism $A \to Ae_i$. Thus we have

$$x = \sum_{i=1}^{m} x_i e_i$$

for uniquely determined elements $x_i \in A$. Thus also $x^m = \sum_{i=1}^m x_i^m e_i$. Using $x^m = y = \sum_{i=1}^m y e_i$, it follows that $x_i^m = y$, and that the x_i are exactly the m-th roots of y in $A^\times$. If ζ is a primitive m-th root of 1 in $\mathcal{O}$, then $\zeta^i x_1$ is also an m-th root of y in $A^\times$. Thus we may choose notation such that $x_1 \in 1 + J(A)$ and $x_i = \zeta^{i-1} x_1$ for $1 \leq i \leq m$. In particular, $z = x_1$ is the unique m-th root of y in $1 + J(A)$. This proves the theorem for A local commutative. Consider now the general case; that is, A is an arbitrary $\mathcal{O}$-algebra that is free of finite rank as an $\mathcal{O}$-module. As before, let $y \in 1 + J(A)$. Denote by $\mathcal{O}[y]$ the unitary subalgebra of A generated by y. By 4.7.9 we have $J(A) \cap \mathcal{O}[y] \subseteq J(\mathcal{O}[y])$. Since $y \in 1 + J(A)$, we have $y - 1 \in J(A) \cap \mathcal{O}[y]$, and hence $\mathcal{O}[y]/(J(A) \cap \mathcal{O}[y]) \cong k$, which shows that $J(\mathcal{O}[y]) = J(A) \cap \mathcal{O}[y]$; in particular, $\mathcal{O}[y]$ is a commutative local subalgebra of A. Thus there is a unique element $z \in 1 + J(\mathcal{O}[y]) = (1 + J(A)) \cap \mathcal{O}[y]$ satisfying $z^m = y$. Let $w \in 1 + J(A)$ such that $w^m = y$. The argument above shows that $\mathcal{O}[w]$ is a local commutative subalgebra of A, and since $w^m = y$, we have $\mathcal{O}[y] \subseteq \mathcal{O}[w]$. Thus w and z are both m-th roots of y in $(1 + J(A)) \cap J(\mathcal{O}[w]) = 1 + J(\mathcal{O}[w])$, which shows that $w = z$ by the uniqueness statement for the commutative local algebra $\mathcal{O}[w]$. □

We use this theorem to show the following generalisation of Corollary 4.3.8.

Corollary 4.7.24 *Suppose that k is algebraically closed of prime characteristic p. Let G be a finite group and let P be a Sylow p-subgroup of G. Let A be commutative G-algebra over $\mathcal{O}$ that is free of finite rank as an $\mathcal{O}$-module. The restriction map $H^2(G; 1 + J(A)) \to H^2(P; 1 + J(A))$ is injective.*

Proof Set $m = |G : P|$; this is a positive integer that is prime to p. Let $\alpha \in H^2(G; 1 + J(A))$ such that the restriction to P of α is the trivial class in $H^2(P; 1 + J(A))$. By 1.2.15 the class α^m is trivial in $H^2(G; 1 + J(A))$. But then α is trivial, since taking m-th powers is an automorphism of the group $1 + J(A)$ by 4.7.23. □

4.8 The Wedderburn–Malcev Theorem

Let $\mathcal{O}$ be a complete local commutative Noetherian ring with residue field $k = \mathcal{O}/J(\mathcal{O})$. The Wedderburn–Malcev Theorem says that the canonical surjection from a split finite-dimensional k-algebra onto its semisimple quotient has a section that is unique up to conjugacy.

Theorem 4.8.1 *Let A be a split finite-dimensional k-algebra. There is a unitary subalgebra S of A such that $A = S \oplus J(A)$ as k-vector spaces. If T is another* *unitary subalgebra satisfying $A = T \oplus J(A)$ then there is an element $u \in 1 + J(A)$ such that $T = uSu^{-1}$.*

Instead of proving this directly, we prove a generalisation of this theorem due to Külshammer, Okuyama and Watanabe [50]. We slightly extend the definition of relative separability in 2.6.14 as follows: if A, C are algebras over some commutative ring and $\gamma : C \to A$ is an algebra homomorphism, we say that A is *relatively C-separable* if A is isomorphic to a direct summand of $A \otimes_C A$ as an A-A-bimodule, where the left and right C-module structure on A is induced by γ. By 2.6.10 this is equivalent to requiring that the canonical surjection $A \otimes_C A \to A$ given by multiplication in A splits as a homomorphism of A-A-bimodules. If N is a C-C-bimodule, we write as before $N^C = \{n \in N \mid cn = nc \text{ for all } c \in C\}$. The following theorem states roughly speaking that if a certain algebra homomorphism lifts to a bimodule homomorphism then under suitable hypotheses it lifts to an algebra homomorphism.

Theorem 4.8.2 ([50]) *Let A, B, C be $\mathcal{O}$-algebras that are finitely generated as $\mathcal{O}$-modules. Let $\alpha : C \to A$ and $\beta : C \to B$ be algebra homomorphisms. Let J be an ideal of B contained in $J(B)$ and denote by $\pi : B \to B/J$ the canonical surjection. Suppose that A is relatively C-separable. Let $\rho : A \to B/J$ be an algebra homomorphism such that $\rho \circ \alpha = \pi \circ \beta$. Suppose there is a homomorphism of C-C-bimodules $\tau : A \to B$ satisfying $\pi \circ \tau = \rho$. Then there is an algebra homomorphism $\sigma : A \to B$ satisfying $\sigma \circ \alpha = \beta$ and $\pi \circ \sigma = \rho$. Moreover, σ is unique up to an inner automorphism of B induced by conjugation with an element in $1_B + J^C$.*

Proof The picture to have in mind is this:

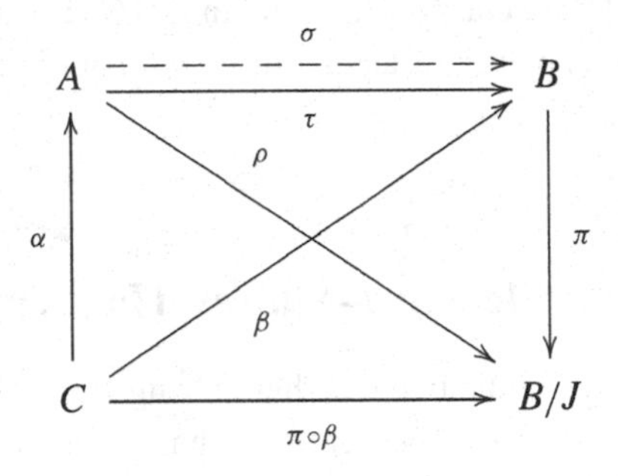

Since A is relatively C-separable, the map $\mu : A \otimes_C A \to A$ induced by multiplication in A splits as a homomorphism of A-A-bimodules. Any section of μ sends 1_A to an element $w \in (A \otimes_C A)^A$ satisfying $\mu(w) = 1_A$. More explicitly, there is an element $w \in A \otimes_C A$ satisfying $aw = wa$ for all

$a \in A$ such that, if we write $w = \sum_{i=1}^{r} x_i \otimes y_i$ with suitable $x_i, y_i \in A$, we have $\sum_{i=1}^{r} x_i y_i = 1_A$. We are going to construct σ as the limit of a sequence of C-C-bimodule homomorphisms $\tau_n : A \to B$ satisfying $\tau_{n+1}(a) - \tau_n(a) \in J^{2^n}$ and $\tau_n(ab) - \tau_n(a)\tau_n(b) \in J^{2^n}$ for all $a, b \in A$ and all integers $n \geq 0$. Set $\tau_0 = \tau$. Suppose we have already constructed $\tau_0, \tau_1, \ldots, \tau_n$ for some $n \geq 0$ such that $\tau_i(ab) - \tau_i(a)\tau_i(b) \in J^{2^i}$ for $0 \leq i \leq n$ and such that $\tau_{i+1}(a) - \tau_i(a) \in J^{2^i}$ for $0 \leq i < n$. Then in particular

$$(\tau_n(1_A) - 1_B)\tau_n(1_A) = \tau_n(1_A)^2 - \tau_n(1_A) \in J^{2^n}.$$

Since $\tau_n(1_A) \in 1_B + J \subseteq B^\times$ this implies that $\tau_n(1_A) - 1_B \in J^{2^n}$. Define a C-C-bimodule homomorphism $\theta : A \otimes_C A \to J^{2^n}$ by setting

$$\theta(a \otimes b) = \tau_n(ab) - \tau_n(a)\tau_n(b)$$

for all $a, b \in A$. The following equation measures how far θ is from being associative. We have

$$\theta(ab \otimes c) - \tau_n(a)\theta(b \otimes c) = \theta(a \otimes bc) - \theta(a \otimes b)\tau_n(c)$$

for all $a, b, c \in A$; indeed, both sides are equal to $\tau_n(abc) - \tau_n(ab)\tau_n(c) - \tau_n(a)\tau_n(bc) + \tau_n(a)\tau_n(b)\tau_n(c)$. Define $\lambda : A \otimes_C A \otimes_C A \to J^{2^n}$ by setting

$$\lambda(a \otimes b \otimes c) = \theta(a \otimes b)\tau_n(c)$$

for all $a, b, c \in A$ and define $\eta : A \to J^{2^n}$ by

$$\eta(a) = \lambda(a \otimes w) = \sum_{i=1}^{r} \theta(a \otimes x_i)\tau_n(y_i)$$

for all $a \in A$. Finally, set $\tau_{n+1} = \tau_n + \eta$. Then, in particular, $\tau_{n+1}(a) - \tau_n(a) = \eta(a) \in J^{2^n}$ for all $a \in A$. A straightforward computation yields

$$\eta(a)\tau_n(b) - \lambda(a \otimes wb) = -\sum_{i=1}^{r} \theta(a \otimes x_i)\theta(y_i \otimes b) \in J^{2^{n+1}}$$

for all $a, b \in A$, or equivalently,

$$\lambda(a \otimes wb) \equiv \eta(a)\tau_n(b) \pmod{J^{2^{n+1}}}.$$

Consider next the expression

$$\begin{aligned}\eta(ab) - \tau_n(a)\eta(b) &= \lambda(ab \otimes w) - \tau_n(a)\lambda(b \otimes w)\\ &= \sum_{i=1}^{r} (\theta(ab \otimes x_i) - \tau_n(a)\theta(b \otimes x_i))\tau_n(y_i).\end{aligned}$$

Using the above equation regarding the associativity of θ it follows that this sum is equal to

$$\sum_{i=1}^{r}(\theta(a \otimes bx_i) - \theta(a \otimes b)\tau_n(x_i))\tau_n(y_i)$$
$$= \lambda(a \otimes bw) - \sum_{i=1}^{r}\theta(a \otimes b)\tau_n(x_i)\tau_n(y_i).$$

Using $\theta(x_i \otimes y_i) = \tau_n(x_iy_i) - \tau_n(x_i)\tau_n(y_i)$ this expression is equal to

$$\lambda(a \otimes bw) + \sum_{i=1}^{r}\theta(a \otimes b)\theta(x_i \otimes y_i) - \sum_{i=1}^{r}\theta(a \otimes b)\tau_n(x_iy_i).$$

The sum in the middle of this expression is in $J^{2^{n+1}}$, and the sum on the right is $\theta(a \otimes b)\tau_n(1_A)$ because $\sum_{i=1}^{r} x_iy_i = 1_A$. Thus we get

$$\eta(ab) - \tau_n(a)\eta(b) \equiv \lambda(a \otimes bw) - \theta(a \otimes b)\tau_n(1_A) \,(\mathrm{mod}\, J^{2^{n+1}}).$$

Consider now

$$\tau_{n+1}(ab) - \tau_{n+1}(a)\tau_{n+1}(b) = (\tau_n(ab) + \eta(ab)) - (\tau_n(a) + \eta(a))(\tau_n(b) + \eta(b))$$
$$\equiv \tau_n(ab) + \eta(ab) - \eta(a)\tau_n(b) - \tau_n(a)\tau_n(b) - \tau_n(a)\eta(b) \,(\mathrm{mod}\, J^{2^{n+1}})$$

where the last congruence holds because both $\eta(a)$ and $\eta(b)$ are in J^{2^n}, hence their product is in $J^{2^{n+1}}$. Using $\theta(a \otimes b) = \tau_n(ab) - \tau_n(a)\tau_n(b)$ and the congruence $\eta(ab) - \tau_n(a)\eta(b) \equiv \lambda(a \otimes bw) - \theta(a \otimes b)\tau_n(1_A) \,(\mathrm{mod}\, J^{2^{n+1}})$ from above we get that

$$\tau_{n+1}(ab) - \tau_{n+1}(a)\tau_{n+1}(b) \equiv \theta(a \otimes b) + \lambda(a \otimes bw) - \theta(a \otimes b)\tau_n(1_A)$$
$$- \eta(a)\tau_n(b) \,(\mathrm{mod}\, J^{2^{n+1}}).$$

Using the congruence $\lambda(a \otimes wb) \equiv \eta(a)\tau_n(b) \,(\mathrm{mod}\, J^{2^{n+1}})$ and the fact that $wb = bw$, this simplifies to

$$\tau_{n+1}(ab) - \tau_{n+1}(a)\tau_{n+1}(b)) \equiv \theta(a \otimes b)(1_B - \tau_n(1_A)) \,(\mathrm{mod}\, J^{2^{n+1}}).$$

Now both $\theta(a \otimes b)$ and $1_B - \tau_n(1_A)$ are in J^{2^n}, and thus we finally get that

$$\tau_{n+1}(ab) - \tau_{n+1}(a)\tau_{n+1}(b) \in J^{2^{n+1}}$$

for all $a, b \in A$. Define now $\sigma : A \otimes_C A \to B$ by setting

$$\sigma = \lim_n \tau_n$$

that is, for any $a \in A$ we denote by $\sigma(a)$ the unique element in B such that $\sigma(a) - \tau_n(a) \in J^{2^n}$ for all $n \geq 0$. (We use here the completeness with respect to J – which is possible since J is contained in $J(B)$, hence some power of J is contained in $J(\mathcal{O})B$.) The map σ has all the required properties: all τ_n are C-C-bimodule homomorphisms, hence so is σ. All τ_n composed with π yield ρ, hence the same is true for σ. Finally, σ is also an algebra homomorphism because $\sigma(ab) - \sigma(a)\sigma(b) \in J^{2^n}$ for all $a, b \in A$ and all $n \geq 0$ by construction, hence $\sigma(ab) = \sigma(a)\sigma(b)$ for all $a, b \in A$. Since σ is both an algebra homomorphism and a C-C-bimodule homomorphism we get $\sigma \circ \alpha = \beta$. This proves the existence of σ as claimed. For the uniqueness of σ up to conjugacy by an element in $1_B + J^C$, let $\sigma' : A \to B$ be another algebra homomorphism satisfying $\sigma \circ \alpha = \beta$ and $\pi \circ \sigma' = \rho$. Set $\delta = \sigma - \sigma'$. Then $\delta : A \to J$ is a homomorphism of C-C-bimodules satisfying

$$\delta(ab) = \sigma(a)\delta(b) + \delta(a)\sigma'(b)$$

for all $a, b \in A$. Define the C-C-bimodule homomorphism $\phi : A \otimes_C A \to J$ by setting

$$\phi(a \otimes b) = \sigma(a)\delta(b)$$

for $a, b \in A$. Set $v = \phi(w) = \sum_{i=1}^r \sigma(x_i)\delta(y_i)$. Since $aw = wa$ for all $a \in A$ and since ϕ is a homomorphism of C-C-bimodules we have $cv = vc$ for all $c \in C$, where we identify c to its image $\beta(c) \in B$. Equivalently, we have $v \in J^C$. Then, for any $a \in A$, we have

$$\begin{aligned}
\sigma(a)v &= \sum_{i=1}^r \sigma(ax_i)\delta(y_i) = \phi(aw) = \phi(wa) = \sum_{i=1}^r \sigma(x_i)\delta(y_i a) \\
&= \sum_{i=1}^r (\sigma(x_i)\sigma(y_i)\delta(a) + \sigma(x_i)\delta(y_i)\sigma'(a)) = \delta(a) + v\sigma'(a) \\
&= \sigma(a) - \sigma'(a) + v\sigma'(a)
\end{aligned}$$

where we used that $\sum_{i=1}^r \sigma(x_i)\sigma(y_i) = \sigma(\sum_{i=1}^r x_i y_i) = \sigma(1_A) = 1_B$. In other words, $\sigma(a)(1 - v) = (1 - v)\sigma'(a)$ for all $a \in A$. The proof is complete. □

Remark 4.8.3 If $\mathcal{O} = k$ in the above theorem, then the ideal J in B is nilpotent, and hence the inductive procedure to construct σ requires only a finite number of steps; in other words, $\sigma = \tau_n$ for n large enough. Thus this proof provides, at least in principle, an explicit recipe how to construct σ.

Proof of 4.8.1 Since A is split, its semisimple quotient $A/J(A)$ is a direct product of finitely many matrix algebras. Thus $A/J(A)$ is relatively k-separable. The map $A \to A/J(A)$ splits as a k-linear map. Applying the previous

Theorem 4.8.2 to $A/J(A)$, A, k and the ideal $J(A)$ instead of A, B, C and the ideal J, respectively, yields the result. □

Theorem 4.8.2 has the following equivariant version. For G a finite group, A, C two G-algebras, with G acting trivially on C, and a G-algebra homomorphism $C \to A$, we say that A is *G-stably C-separable*, if the canonical A-A-bimodule homomorphism $\mu : A \otimes_C A \to A$ has G-stable section $\nu : A \to A \otimes_C A$, where $A \otimes_C A$ is considered with the diagonal G-action on the two factors A. That is, the element $w = \nu(1_A)$ belongs to $(A \otimes_C A)^A$ and to $(A \otimes_C A)^G$, and satisfies $\mu(w) = 1_A$. Note that $J(A)$ is G-stable, by 1.10.5 and that $J(A)^G = A^G \cap J(A) \subseteq J(A^G)$, by 4.7.9.

Theorem 4.8.4 *Let G be a finite group. Let A, B, C be G-algebras over $\mathcal{O}$ that are finitely generated as $\mathcal{O}$-modules, such that G acts trivially on C. Let $\alpha : C \to A$ and $\beta : C \to B$ be G-algebra homomorphisms. Let J be a G-stable ideal of B contained in $J(B)$ and denote by $\pi : B \to B/J$ the canonical surjection. Suppose that A is G-stably C-separable. Let $\rho : A \to B/J$ be a G-algebra homomorphism such that $\rho \circ \alpha = \pi \circ \beta$. Suppose there is a G-stable homomorphism of C-C-bimodules $\tau : A \to B$ satisfying $\pi \circ \tau = \rho$. Then there is G-algebra homomorphism $\sigma : A \to B$ satisfying $\sigma \circ \alpha = \beta$ and $\pi \circ \sigma = \rho$. Moreover, σ is unique up to an inner automorphism of B induced by conjugation with an element in $1_B + (J^C)^G$.*

Proof The proof is identical to that of 4.8.2, checking at each step that G-invariance is preserved. By the assumptions, the element $w \in (A \otimes_C A)^A$ is in $(A \otimes_C A)^G$. The maps τ_n constructed in the proof of 4.8.2 are all G-stable; indeed, $\tau_0 = \tau$ is G-stable by the assumptions, and in the inductive step constructing τ_{n+1}, the maps θ, λ, and η are G-stable. Thus the map σ, which arises as the limit of the τ_n, is a G-algebra homomorphism. Similarly, for the uniqueness statement one notes that the maps δ and Φ in the proof of 4.8.2 are G-stable. Thus $v = \Phi(w)$ is G-stable, hence in $(J^C)^G$, and the result follows. □

The following lemma makes 2.16.18 slightly more precise.

Lemma 4.8.5 *Let V be a free $\mathcal{O}$-module of finite rank n. Set $S = \mathrm{End}_{\mathcal{O}}(V)$, considered as a symmetric algebra with the trace form $\mathrm{tr}_V : S \to \mathcal{O}$. For any idempotent $e \in S$ we have $\mathrm{Tr}_1^S(e) = \mathrm{rk}_{\mathcal{O}}(eV) \cdot 1_{\mathcal{O}}$. In particular, if e is primitive, then $\mathrm{Tr}_1^S(e) = 1_{\mathcal{O}}$, and $\mathrm{Tr}_1^S(1_S) = n \cdot 1_{\mathcal{O}}$. Moreover, we have $\mathrm{Tr}_1^S(e) \in \mathcal{O}^\times$ if and only if the integer $\mathrm{rk}_{\mathcal{O}}(eSe)$ is invertible in k.*

Proof By 2.16.18, we have $\mathrm{Tr}_1^S(1_S) = n \cdot 1_{\mathcal{O}}$. Since all primitive idempotents in S are conjugate and since a primitive decomposition of 1_S has exactly n elements, it follows that $\mathrm{Tr}_1^S(i) = 1_{\mathcal{O}}$ for any primitive idempotent i in S. Thus, for e an arbitrary idempotent in S, we have $\mathrm{Tr}_1^S(e) = m \cdot 1_{\mathcal{O}}$, where m is the number of primitive idempotents in a decomposition of e. Since the primitive idempotents in S are exactly the projections onto direct summands of rank 1 of V, it follows that $m = \mathrm{rk}_{\mathcal{O}}(eV)$. Thus $\mathrm{Tr}_1^S(e)$ is invertible in $\mathcal{O}$ if and only if $\mathrm{rk}_{\mathcal{O}}(eV)$ is invertible in k. Since $eSe \cong \mathrm{End}_{\mathcal{O}}(eV)$, we have $\mathrm{rk}_{\mathcal{O}}(eSe) = (\mathrm{rk}_{\mathcal{O}}(eV))^2$, whence the last statement. □

Proposition 4.8.6 *Let G be a finite group and A a G-algebra over $\mathcal{O}$ that is finitely generated free as an $\mathcal{O}$-module. Suppose that $A = S \oplus J$ for some G-stable matrix subalgebra S of A and some G-stable ideal J contained in $J(A)$. Suppose that S has a G-stable idempotent e such that $\mathrm{rk}_{\mathcal{O}}(eSe)$ is invertible in k. Let T be a G-stable matrix subalgebra of A that is isomorphic to S as an $\mathcal{O}$-algebra. Then $T = uSu^{-1}$ for some $u \in 1 + J^G$.*

Proof Let $e \in S^G$ be an idempotent such that $\mathrm{rk}_{\mathcal{O}}(eSe)$ is prime to p. By Lemma 4.8.5, we have $\mathrm{Tr}_1^S(e) \in \mathcal{O}^\times$. It follows from 2.16.17 that S is G-stably $\mathcal{O}$-separable; that is, the multiplication map $S \otimes_{\mathcal{O}} S \to S$ has a G-equivariant section. Since S, T are G-stable, we may consider them as a G-subalgebras of A. Since J is G-stable, the composition of the canonical maps $T \to A \to A/J \cong S$ is a G-algebra isomorphism. Indeed, this map is nonzero over k, hence injective, and thus an isomorphism over k as both sides have the same dimension. Nakayama's Lemma implies that this map is an isomorphism over $\mathcal{O}$, and we observed already that this is a G-algebra homomorphism. Thus the inclusion $S \to A$ and the isomorphism $S \cong T$ followed by the inclusion $T \to A$ both lift the isomorphism $S \cong A/J$ through the canonical map $A \to A/J$. It follows from 4.8.4 that T and S are conjugate by an element in $1 + J^G$. □

Exercise 4.8.7 Let A be an $\mathcal{O}$-algebra that is free of finite rank as an $\mathcal{O}$-module. Show that the existence of a matrix subalgebra S and an ideal J contained in $J(A)$ satisfying $A = S \oplus J$ is equivalent to asserting that A has a unique isomorphism class of simple modules, and that any simple A-module lifts to an $\mathcal{O}$-free A-module.

4.9 Basic algebras and quivers

Let $\mathcal{O}$ be a complete local Noetherian commutative ring with residue field $k = \mathcal{O}/J(\mathcal{O})$.

Definition 4.9.1 Let A be an $\mathcal{O}$-algebra that is finitely generated as an $\mathcal{O}$-module. We say that A is a *basic $\mathcal{O}$-algebra* if $A/J(A)$ is a direct product of division rings.

Proposition 4.9.2 *Let A be an $\mathcal{O}$-algebra that is finitely generated as an $\mathcal{O}$-module. The following are equivalent.*

(i) The $\mathcal{O}$-algebra A is basic.
(ii) For any point α on A we have $m_\alpha = 1$.
(iii) The elements of a primitive decomposition of 1_A are pairwise nonconjugate.
(iv) As a left A-module, A is a direct sum of pairwise nonisomorphic projective indecomposable A-modules.

Proof The equivalence of (ii) and (iii) is immediate. The equivalence of (iii) and (iv) follows from 4.6.8 (i). Since the canonical map $A \to A/J(A)$ induces a multiplicity preserving bijection between the sets of points of A and of $A/J(A)$, we may assume that $J(A) = \{0\}$. Then, by 1.13.3, the algebra A is a direct product of matrix algebras $M_{n_i}(D_i)$ over division rings D_i, hence basic if and only of all n_i are equal to 1. In that case the points of A are the sets $\{1_{D_i}\}$, so they all have multiplicity 1. Conversely, if one of the n_i is greater than 1, then $M_{n_i}(D_i)$ contains two orthogonal conjugate idempotents, and hence there is at least one point with multiplicity greater than 1. Thus (i) and (ii) are equivalent. □

Proposition 4.9.3 *Let A be an $\mathcal{O}$-algebra that is finitely generated as an $\mathcal{O}$-module. Then A is split basic if and only if all simple $k \otimes_{\mathcal{O}} A$-modules have dimension* 1.

Proof By definition, A is basic if and only if $A/J(A)$ is a direct product of division k-algebras. Thus A is basic and split if and only if $A/J(A)$ is a direct product of copies of k. The result follows. □

Proposition 4.9.4 *Let A be an $\mathcal{O}$-algebra that is finitely generated as an $\mathcal{O}$-module. Suppose that A is split and basic. Then $[A, A] \subseteq J(A)$.*

Proof Since A is split basic, it follows that $A/J(A)$ is a direct product of copies of k; in particular, $A/J(A)$ is commutative. Thus $[A/J(A), A/J(A)] = \{0\}$, or equivalently, $[A, A] \subseteq J(A)$. □

Not every Morita equivalence between $\mathcal{O}$-algebras is induced by an algebra isomorphism, but for basic algebras, this is the case.

Proposition 4.9.5 *Let A and B be $\mathcal{O}$-algebras that are finitely generated as $\mathcal{O}$-modules. Suppose that A and B are basic. Let M be an A-B-bimodule and N a B-A-bimodule inducing a Morita equivalence between A and B. Then there is an algebra isomorphism $\alpha : B \cong A$ such that $M \cong A_\alpha \cong {}_{\alpha^{-1}}B$ as A-B-bimodules.*

Proof The algebra B is basic, hence isomorphic, as a left B-module, to a direct sum of pairwise nonisomorphic projective indecomposable B-modules. Since

$M \otimes_B -$ is an equivalence, it follows that $M \cong M \otimes_B B$ is isomorphic, as a left A-module, to a direct sum of pairwise nonisomorphic projective indecomposable A-modules. The fact that A and B are Morita equivalent implies that they have the same number of isomorphism classes of projective indecomposable modules. Thus $M \cong A$ as a left A-module. It follows from 2.8.16 (v) that there is an algebra isomorphism $\alpha : B \cong A$ such that $M \cong A_\alpha$ as A-B-bimodules. The second isomorphism follows from 2.8.13. □

Corollary 4.9.6 *Let A and B be $\mathcal{O}$-algebras that are finitely generated as $\mathcal{O}$-modules. Suppose that A and B are basic. Then A and B are Morita equivalent if and only if A and B are isomorphic.*

Proof This is a trivial consequence of 4.9.5. □

Corollary 4.9.7 *Let A be an $\mathcal{O}$-algebra that is finitely generated as an $\mathcal{O}$-module. If A is basic, then the map sending $\alpha \in \mathrm{Aut}(A)$ to the bimodule A_α induces a group isomorphism $\mathrm{Out}(A) \cong \mathrm{Pic}(A)$.*

Proof By 2.8.16 the given map $\alpha \mapsto A_\alpha$ induces an injective group homomorphism $\mathrm{Out}(A) \to \mathrm{Pic}(A)$. It follows from 4.9.5 that this map is surjective. □

Corollary 4.9.8 *Let A be an $\mathcal{O}$-algebra that is finitely generated as an $\mathcal{O}$-module. Let I be a set of pairwise orthogonal representatives of the conjugacy classes of primitive idempotents in A. Set $e = \sum_{i \in I} i$. Then eAe is basic and Morita equivalent to A. Any basic algebra that is Morita equivalent to A is isomorphic to eAe.*

Proof The algebra eAe is basic by 4.9.2: Since every primitive idempotent in A is conjugate to an idempotent in I, it follows that $1_A \in AeA$, hence $A = AeA$. By 2.8.7, the algebras A and eAe are Morita equivalent. By 4.9.6 any basic algebra that is Morita equivalent to A is isomorphic to eAe, whence the result. □

The basic algebras that are Morita equivalent to an $\mathcal{O}$-algebra A are called the *basic algebras of A*; they are unique up to isomorphism by 4.9.8. In order to determine module categories of $\mathcal{O}$-algebras, it suffices to consider basic $\mathcal{O}$-algebras, although one should be aware that other structures, such as group actions, or Hopf algebra structures, may be lost in the passage to basic algebras. One of the key observations to describe a basic algebra in terms of generators and relations is the following.

Lemma 4.9.9 *Let A be an $\mathcal{O}$-algebra that is finitely generated as $\mathcal{O}$-module and let B be a unitary subalgebra of A such that $A = B + J(A)^2$. Then $A = B$.*

Proof Suppose we know that $A = B + J(A)^n$ for some $n \geq 2$. Then $J(A) = (B \cap J(A)) + J(A)^n$. Thus $J(A)^n = ((B \cap J(A)) + J(A)^n)^n \subseteq B + J(A)^{n+1}$. Thus $A = B + J(A)^{n+1}$. Since some power of $J(A)$ is contained in $J(\mathcal{O})A$, we have $A = B + J(\mathcal{O})A$. Nakayama's Lemma forces $A = B$. □

Proposition 4.9.10 *Let A be a split basic $\mathcal{O}$-algebra that is finitely generated as an $\mathcal{O}$-module, and set $\bar{A} = k \otimes_{\mathcal{O}} A$. Let I be a primitive decomposition of 1 and let E be a subset of $J(A)$ such that the image of E in $J(\bar{A})/J(\bar{A})^2$ is a k-basis. The subalgebra of A generated by $I \cup E$ is equal to A.*

Proof Let B be the subalgebra of A generated by $I \cup E$. Since A is split basic, it follows that $A/J(A) \cong \bar{A}/J(\bar{A})$ is a direct product of $|I|$ copies of k, and hence the image of I in $A/J(A)$ is a k-basis. The assumptions on E imply that the image of $I \cup E$ in $\bar{A}/J(\bar{A})^2$ is a k-basis. It follows from 4.9.9, applied to $\bar{A}$ and the subalgebra generated by the image of $I \cup E$ in $\bar{A}$ that the image of B in $\bar{A}$ is equal to $\bar{A}$. Thus $A = B + J(\mathcal{O})A$, and Nakayama's Lemma implies that $A = B$. □

The ingredients to describe a split basic $\mathcal{O}$-algebra in terms of generators and relations as in the previous proposition can be organised in terms of the path algebra of a finite directed graph; this goes back to work of P. Gabriel, and is one of the fundamental tools in the representation theory of finite-dimensional algebras. We sketch this briefly, and refer to [4, Ch. I, §1] for more details. Let A be an $\mathcal{O}$-algebra that is finitely generated as an $\mathcal{O}$-module. Suppose that A is split basic. Let I be a primitive decomposition of 1 in A, and let E be a subset of $J(A)$ whose image in $J(\bar{A})/J(\bar{A})^2$ is a k-basis. For $i \in I$ denote by $\bar{i}$ the image of i in $\bar{A}$, and set $S_i = Ai/J(A)i \cong \bar{A}\bar{i}/J(\bar{A})\bar{i}$. Since we have a vector space decomposition

$$J(\bar{A})/J(\bar{A})^2 = \oplus_{i,j \in I} \bar{i}J(\bar{A})\bar{j}/\bar{i}J(\bar{A})^2\bar{j}$$

we may choose E to be a disjoint union of subsets $E_{i,j}$ of $iJ(A)j$ such that the image of $E_{i,j}$ in $\bar{i}J(\bar{A})\bar{j}/\bar{i}J(\bar{A})^2\bar{j}$ is a k-basis. We adopt the convention that $E_{i,j}$ is the empty set if the space $\bar{i}J(\bar{A})\bar{j}/\bar{i}J(\bar{A})^2\bar{j}$ is zero.

Definition 4.9.11 With the notation and hypotheses above, the *graph of A* is the finite directed graph $Q = Q(A)$ defined as follows. The vertex set of $Q(A)$ is I. For any $i, j \in I$, the set of arrows from i to j is labelled by the set $E_{i,j}$.

The graph of a split basic $\mathcal{O}$-algebra A is unique up to isomorphism of graphs, because any two primitive decompositions of I in A are conjugate. Note that A and $k \otimes_{\mathcal{O}} A$ have the same graph. The graph of A has a single vertex if and only if 1_A is primitive in A; that is, if and only if A is split local. In that case, any edge is a loop at that vertex, and the number of loops is equal to $\dim_k(J(\bar{A})/J(\bar{A})^2)$.

Example 4.9.12 Let P be a finite p-group. Then $\mathcal{O}P$ is split local. Thus the graph $Q(\mathcal{O}P)$ has a single vertex. Any edge in $Q(\mathcal{O}P)$ is thus a loop at that vertex. The number of loops is equal to $\dim_k(J(kP)/J(kP)^2)$. It is well-known and easy to verify that this is also the rank of the elementary abelian p-group $P/\Phi(P)$, where $\Phi(P)$ is the Frattini subgroup of P.

The graph of $\mathcal{O}P$ in the previous example is connected, since it has a single vertex. A connected finite directed graph is called a *quiver*. Algebras whose graph is a quiver are exactly the algebras that are indecomposable.

Proposition 4.9.13 *Let A be an $\mathcal{O}$-algebra that is finitely generated as an $\mathcal{O}$-module. Suppose that A is split basic. The following are equivalent.*

(i) *The graph of A is a quiver.*
(ii) *The $\mathcal{O}$-algebra A is indecomposable.*
(iii) *The k-algebra $k \otimes_{\mathcal{O}} A$ is indecomposable.*

Proof If $\bar{A} = k \otimes_{\mathcal{O}} A$ decomposes as a direct product of two k-algebras A_0 and A_1, then $Q(A)$ is the disjoint union of the graphs of A_0 and A_1. Thus (i) implies (iii). If A is decomposable as an $\mathcal{O}$-algebra then so is the k-algebra $\bar{A}$, and hence (iii) implies (ii). We need to show that (ii) implies (i). Let I be a primitive decomposition of 1 in A. Let E be a disjoint union of subsets $E_{i,j}$ of $iJ(A)j$ such that the image of $E_{i,j}$ in $\bar{i}J(\bar{A})\bar{j}/\bar{i}J(\bar{A})^2\bar{j}$ is a k-basis. Arguing by contradiction, suppose that the graph $Q(A)$ can be written as the disjoint union of two subgraphs Q_0 and Q_1. That is, I is the disjoint union of the vertex sets I_0 of Q_0 and I_1 of Q_1, and for $i \in I_0$, $j \in I_1$ the sets $E_{i,j}$ and $E_{j,i}$ are empty. Set $e = \sum_{i \in I_0} i$ and $f = \sum_{j \in I_1} j$, the sums in A. Thus e and f are orthogonal idempotents in A such that $e + f = 1$. It follows that we have an $\mathcal{O}$-module decomposition $A = eAe \oplus eAf \oplus fAe \oplus fAf$. It suffices to show that eAf and fAe are zero; indeed, this would imply that $A \cong eAe \times fAf$ is decomposable as an algebra. In order to show that eAf and fAe are zero we may assume that $\mathcal{O} = k$. Since A is basic, the elements in I are pairwise nonconjugate. By 4.6.9, both eAf and fAe are contained in $J(A)$. In fact, both of these spaces are contained in $J(A)^2$, because otherwise there would be edges between a vertex in I_0 and a vertex in I_1. Thus the subalgebra $B = eAe + fAf$ satisfies $A = B + J(A)^2$. It follows from 4.9.9 that $A = B = eAe + fAf$, hence eAf and fAe are zero. Thus (ii) implies (i). □

Any finite directed graph $Q = (I, E)$ with vertex set I and set of edges $E = \cup_{i \in I} E_{i,j}$ gives rise to its *path category*. This is the category whose object set is the set of vertices I, and whose morphism set from a vertex i to another vertex j consists of all directed paths starting at i and ending at j. The path of length zero at a vertex i is the identity morphism of i viewed as an object in the path category. The composition in the path category is given by concatenating paths. If the graph Q has a circular path or a loop (that is, an edge from $i \in I$ to itself) then the morphism set in the path category is infinite because a circular path or loop can be composed with itself. The category algebra over $\mathcal{O}$ of the path category is called the *path algebra* and is denoted by $\mathcal{O}Q$. This is an algebra having as an $\mathcal{O}$-basis the paths in Q. The vertices, regarded as paths of length zero, are idempotents in $\mathcal{O}Q$, and their sum, taken over the vertex set, is the unit element of $\mathcal{O}Q$; see §1.4 for more details on category algebras. The path algebra of the quiver $Q(A)$ of a split basic algebra A is obtained by taking the generating set $I \cup E$ of A but keeping only the relations involving the idempotents. It follows that A is a quotient of the path algebra of its graph.

Theorem 4.9.14 *Let A be an $\mathcal{O}$-algebra that is free of finite rank as an $\mathcal{O}$-module. Suppose that A is split basic. Let I be a primitive decomposition of 1 in A. Set $\bar{A} = k \otimes_{\mathcal{O}} A$. For $i, j \in I$, let $E_{i,j}$ be a subset of $iJ(A)j$ such that the image of $E_{i,j}$ in $\bar{i}J(\bar{A})\bar{j}/\bar{i}J(\bar{A})^2\bar{j}$ is a k-basis, where $\bar{i}$, $\bar{j}$ are the images of i, j in $\bar{A}$. The map $Q(A) \to A$ sending a vertex labelled by $i \in I$ to i as an element of A and sending an edge labelled by an element $e \in E_{i,j}$ to e as an element in A induces surjective $\mathcal{O}$-algebra homomorphisms $\gamma : \mathcal{O}Q(A) \to A$ and $\bar{\gamma} : kQ(A) \to \bar{A}$. Set $N = \ker(\gamma)$ and $\bar{N} = \ker(\bar{\gamma})$. The following hold.*

(i) The canonical map $\mathcal{O}Q(A) \to kQ(A)$ sends N onto $\bar{N}$.
(ii) The ideal $\bar{N}$ in $kQ(A)$ is finitely generated and contained in the ideal generated by all paths of length 2. Moreover, if $n \geq 2$ is an integer satisfying $J(\bar{A})^n = \{0\}$, then $\bar{N}$ contains the ideal generated by all paths of length n.

Proof Define γ as the unique algebra homomorphism sending the vertex set I to I viewed as a subset of A and sending a path of n consecutive edges, labelled by elements $e_1, e_2, \ldots, e_n$ in the set of edges E to the product of these elements in A. The map γ is surjective by 4.9.10. Thus γ splits as an $\mathcal{O}$-linear map as A is assumed to be $\mathcal{O}$-free. Statement (i) follows. In order to prove (ii) we may assume that $\mathcal{O} = k$. The map induced by γ from $kQ(A)$ to $A/J(A)^2$ sends E to a k-basis of $J(A)/J(A)^2$. This implies that N is contained in the ideal generated by the paths of length 2. Let $n \geq 2$ such that $J(A)^n = \{0\}$. Since E is finite, there are finitely many paths of any given length. Thus the ideal P generated by all paths of length n is finitely generated. It follows that $P \subseteq N$, and that $kQ(A)/P$ is finite-dimensional, having as a basis the paths of lengths at most $n - 1$. But then N/P is finite-dimensional, and hence N is generated by the paths of length

n and a finite set whose image in N/P is a basis; in particular, N is finitely generated. This shows (ii). $\square$

The graph $Q(A)$ of a split basic algebra A is uniquely determined by $\bar{A}$. The homomorphism γ in the previous theorem depends, however, on the choices of the subsets $E_{i,j}$. In other words, the ideal $N = \ker(\gamma)$ of the path algebra $\mathcal{O}Q(A)$ is not uniquely determined by A. Statement (ii) in the above theorem need not hold for the ideal N. In general, N is not contained in the ideal generated by the paths of length 2, and N need not contain any path of positive length; see for instance [38, Theorem C]. Even the finite generation of N seems to be unclear: by lifting a finite generating set of $\bar{N}$ to N we obtain a finitely generated ideal N' contained in N, satisfying $N = N' + J(\mathcal{O})N$. However, N is not finitely generated as an $\mathcal{O}$-module, and so Nakayama's Lemma does not immediately apply; that is, we cannot conclude that $N = N'$.

Exercise 4.9.15 Let A be an $\mathcal{O}$-algebra that is finitely generated as an $\mathcal{O}$-module. Let $i, j \in A$ be idempotents. Denote by $\bar{i}, \bar{j}$ their images in $\bar{A} = k \otimes_{\mathcal{O}} A$. Consider $\bar{A}$ as an A-A-bimodule via the canonical surjection $A \to \bar{A}$. Then $J(\bar{A})/J(\bar{A})^2$ is an A-A-bimodule. Show that $i \cdot J(\bar{A})/J(\bar{A})^2 \cdot j$ can be canonically identified with $\bar{i}J(\bar{A})\bar{j}/\bar{i}J(\bar{A})^2\bar{j}$.

Exercise 4.9.16 Let A be an $\mathcal{O}$-algebra that is finitely generated as an $\mathcal{O}$-module. Show that if A is indecomposable as an algebra and has at least two isomorphism classes of simple modules, then $[A, A] \not\subseteq J(A)^2$. (This is in contrast to 4.4.9.)

Exercise 4.9.17 Suppose that Q is a graph with a single vertex v and a single loop e as its unique edge. Show that there is an isomorphism $\mathcal{O}Q \cong \mathcal{O}[x]$ that sends the path e to the indeterminate x.

Exercise 4.9.18 Let n be a positive integer. Denote by $T_n(k)$ the subalgebra of $M_n(k)$ consisting of all upper triangular matrices. Show that the quiver of $T_n(k)$ is of the form

$$1 \longrightarrow 2 \longrightarrow \cdots \longrightarrow n.$$

Show that the path algebra of this quiver is isomorphic to $T_n(k)$.

Exercise 4.9.19 Suppose that $\mathrm{char}(k) = p > 0$. Let P be a finite p-group, E a p'-subgroup of $\mathrm{Aut}(P)$, and $\alpha \in H^2(E; k^\times)$. Suppose that k is a splitting field for $k_\alpha(P \rtimes E)$. Show that $k_\alpha(P \rtimes E)$ is basic if and only if α is trivial and E is abelian.

4.10 The Cartan matrix

Let k be a field.

Definition 4.10.1 Let A be a finite-dimensional k-algebra and let I be a set of representatives of the conjugacy classes of primitive idempotents in A. The *Cartan matrix of* A is the square matrix of non negative integers $C = (c_{ij})_{i,j \in I}$ where c_{ij} is the number of composition factors isomorphic to the simple A-module $S_i = Ai/J(A)i$ in a composition series of the projective indecomposable A-module Aj.

As a consequence of the Jordan–Hölder Theorem, any two different composition series of Aj are equivalent, and hence the integer c_{ij} is independent of the choice of such a composition series. It is useful to extend this terminology to algebras over a complete local commutative Noetherian ring $\mathcal{O}$ with k as residue field: if A is an $\mathcal{O}$-algebra that is finitely generated as $\mathcal{O}$-module, then the Cartan matrix of A is defined to be the Cartan matrix of the k-algebra $k \otimes_{\mathcal{O}} A$. Morita equivalent algebras have the same Cartan matrices, because a Morita equivalence is an exact functor that sends simple modules to simple modules and projective indecomposable modules to projective indecomposable modules. Derived equivalences do not preserve Cartan matrices in general. For split algebras, the Cartan matrix admits the following description:

Theorem 4.10.2 *Let A be a finite-dimensional split k-algebra, and let I be a set of representatives of the conjugacy classes of primitive idempotents in A. Let $C = (c_{ij})_{i,j \in I}$ be the Cartan matrix of A. For any $i, j \in I$ we have $c_{ij} = \dim_k(iAj)$.*

In order to prove this theorem we prove a more general statement which includes an alternative proof of the Jordan–Hölder Theorem for finite-dimensional split algebras. As before, we make use of 4.6.9, saying that for any simple A-module S there is a primitive idempotent i such that $Ai/J(A)i \cong S$, and that i is unique up to conjugacy with this property.

Theorem 4.10.3 *Let A be a finite-dimensional split k-algebra, let U be a finitely generated A-module, let S be a simple A-module and let i be a primitive idempotent in A such that $Ai/J(A)i \cong S$. The number of composition factors isomorphic to S in any composition series of U is equal to $\dim_k(iU)$. In particular, this number does not depend on the chosen composition series.*

Proof By 1.7.4 we have a k-linear isomorphism $iU \cong \mathrm{Hom}_A(Ai, U)$ sending iu to $ai \mapsto aiu$, where $u \in U$ and $a \in A$; the inverse of this map sends $\varphi \in \mathrm{Hom}_A(Ai, U)$ to $\varphi(i)$. Since $i^2 = i$ we have $\varphi(i) = i\varphi(i) \in iU$, so this makes sense. If U is simple and not isomorphic to S we have $\mathrm{Hom}_A(Ai, U) = \{0\}$ because the only simple quotient of Ai is isomorphic to S. Since every homomorphism from Ai to S is either zero or has kernel $J(A)i$ we get that $\mathrm{Hom}_A(Ai, U) \cong \mathrm{Hom}_A(S, U) \cong k$ if $U \cong S$. This proves 4.10.3 in case that U is simple. We proceed now by induction over $\dim_k(U)$. Let

$$U = U_0 \supset U_1 \supset U_2 \supset \cdots \supset U_n = \{0\}$$

be a composition series of U. The exact sequence

$$0 \to U_1 \to U \to U/U_1 \to 0$$

gives rise to an exact sequence

$$0 \to iU_1 \to iU \to i(U/U_1) \to 0.$$

Thus $\dim_k(iU) = \dim_k(iU_1) + \dim_k(i(U/U_1))$. By induction, $\dim_k(iU_1)$ is the number of factors in the composition series of U_1 isomorphic to S. By the first argument, $\dim_k(i(U/U_1))$ is either one or zero, depending on whether the simple A-module U/U_1 is isomorphic to S or not. The result follows. □

Proof of 4.10.2 Applying 4.10.3 to the A-module Aj yields the equality $c_{ij} = \dim_k(iAj)$ as stated. □

Proposition 4.10.4 *Let A be a finite-dimensional k-algebra. The Cartan matrix C of A is a block diagonal matrix whose blocks are the Cartan matrices C_b of the block algebras Ab, with b running over the set of blocks of A.*

Proof Let I be a set of representatives of the conjugacy classes of primitive idempotents in A. Let $i, j \in I$. Suppose that i, j belong to two different blocks of A. Then, by 1.7.9 (iii) we have $\mathrm{Hom}_A(Ai, Aj) = \{0\}$. It follows that Aj has no composition factor isomorphic to the simple modules $S_i = Ai/J(A)i$, and hence $c_{ij} = 0$. The result follows. □

The reasoning in the proof of the previous result can be extended to obtain a criterion for when the projective indecomposable modules Ai, and Aj belong to the same block.

Theorem 4.10.5 (Brauer) *Let A be a finite-dimensional k-algebra and let i, j be primitive idempotents in A. The following are equivalent.*

(i) *The modules Ai and Aj belong to the same block of A.*
(ii) *There is a finite sequence $i = i_0, i_1, \ldots, i_n = j$ of primitive idempotents i_k in A such that, for every integer k, $0 \leq k \leq n-1$, at least one of the homomorphism spaces $\mathrm{Hom}_A(Ai_k, Ai_{k+1})$ or $\mathrm{Hom}_A(Ai_{k+1}, Ai_k)$ is non zero.*

Proof If (ii) holds then Ai_k and Ai_{k+1} belong to the same block of A for any k, $0 \leq k \leq n-1$, by 1.7.9. Thus Ai and Aj belong to the same block. This shows that (ii) implies (i). In order to show the converse, define a relation on the set of primitive idempotents in A by $i \sim j$, if there is a finite sequence of primitive idempotents in A satisfying (ii). This is obviously an equivalence relation. Let I be a primitive decomposition of 1_A in A. Let $i \in I$, and let e be the sum of all $j \in J$ such that $i \sim j$. Thus $\mathrm{Hom}_A(Ae, A(1-e)) = \{0\} = \mathrm{Hom}_A(A(1-e), Ae)$. By 1.7.4, this translates to $eA(1-e) = \{0\} = (1-e)Ae$. Therefore $A = eAe \oplus (1-e)A(1-e)$. Since e and $1-e$ are orthogonal, this shows that e and $1-e$ lie in the centre of A. Moreover, e has to be primitive in $Z(A)$ because it is the sum of primitive idempotents belonging to a single orbit of the equivalence relation $\sim$. In other words, e is a block of A. Thus (i) implies (ii). □

For A a finite-dimensional k-algebra, we denote $R(A)$ the Grothendieck group of finite-dimensional A-modules; this is a free abelian group having the isomorphism classes of simple A-modules as a basis (cf. 1.8.7). We denote by $\mathrm{Pr}(A)$ the subgroup of $R(A)$ generated by the isomorphism classes of finite-dimensional projective A-modules.

Proposition 4.10.6 *Let A be a finite-dimensional k-algebra. Suppose that the Cartan matrix C of A is nonsingular. Then the group $R(A)/\mathrm{Pr}(A)$ is finite, and we have $|R(A)/\mathrm{Pr}(A)| = |\det(C)|$.*

Proof The subgroup $\mathrm{Pr}(A)$ is generated by the isomorphism classes $[Ai]$ of the projective indecomposable A-modules Ai, with i running over a set of representatives I of the conjugacy classes of primitive idempotents in A. Setting $S_i = Ai/J(A)i$ for $i \in I$, we have

$$[Ai] = \sum_{j \in I} c_{ij}[S_j]$$

in $R(A)$. Since C is nondegenerate, it follows that the $[Ai]$ form a basis of $\mathrm{Pr}(A)$ and that $\mathrm{Pr}(A)$ has the same rank as $R(A)$. Thus the quotient $R(A)/\mathrm{Pr}(A)$ is finite. The statement on $\det(C)$ follows from elementary facts on subgroups of finitely generated free abelian groups. □

4.11 Selfinjective algebras

Let k be a field.

Definition 4.11.1 A finite-dimensional k-algebra A is *selfinjective* if A is injective as a left A-module.

Any algebra that is Morita equivalent to a selfinjective algebra is again selfinjective, because a Morita equivalence preserves projective and injective modules.

Theorem 4.11.2 *Let G be a finite group. The group algebra kG is selfinjective.*

Proof One way to prove this is to use the fact that kG is symmetric (cf. 2.11.2) and then apply 4.12.1 below. But one can prove 4.11.2 also easily directly. Let U be a kG-module and let $\alpha : kG \to U$ be an injective homomorphism of kG-modules. We have to show that α splits. Let $\gamma : U \to kG$ be any k-linear map with 1-dimensional image $k \cdot 1$ in kG. such that $\gamma(\alpha(1)) = 1$ and $\gamma(\alpha(x)) = 0$ for $x \neq 1$. Define $\beta : U \to kG$ by setting $\beta(u) = \sum_{x \in G} x\gamma(x^{-1}u)$. By our standard argument from the proof of Maschke's Theorem, β is a kG-homomorphism. For any $y \in G$ we have $\beta(\alpha(y)) = \sum_{x \in G} x\gamma(x^{-1}\alpha(y)) = \sum_{x \in G} x\gamma(\alpha(x^{-1}y)) = y\gamma(\alpha(1)) = y$. Thus $\beta \circ \alpha = \mathrm{Id}_{kG}$ as required. □

Theorem 4.11.3 *Let A be a finite-dimensional k-algebra. The following statements are equivalent.*

(i) A is selfinjective.
(ii) A is injective as a right A-module.
(iii) Every finitely generated projective A-module is injective.
(iv) Every finitely generated injective A-module is projective.
(v) The k-dual $A^* = \mathrm{Hom}_k(A, k)$ is a progenerator of A as a left A-module.
(vi) The k-dual $A^* = \mathrm{Hom}_k(A, k)$ is a progenerator of A as a right A-module.
(vii) The classes of projective and injective left (resp. right) A-modules coincide.

Proof The numbers of isomorphism classes of finitely generated projective indecomposable and injective indecomposable modules coincide since they are by 4.5.3 and 4.5.5 both equal to the number of isomorphism classes of simple modules. Thus (i) and (ii) are equivalent to (vi) and (v), respectively. For the same reason, (iii) and (iv) are equivalent. The equivalence of (i) and (iii) follows trivially from the fact, that any finitely generated projective A-module is a finite direct sum of modules isomorphic to direct summands of A as left A-module. By applying k-duality to the statements (iii), (iv) we conclude that

(iii) and (iv) are equivalent to the corresponding statements for right modules, and hence equivalent to (ii). Since k is a field, the relatively k-projective (resp. relatively k-injective) A-modules are exactly the projective (resp. injective) A-modules. Thus the equivalence of (v), (vi), (vii) follows from 2.15.2. □

Corollary 4.11.4 *Let A and B be finite-dimensional selfinjective k-algebras. Then the k-algebras A^{op} and $A \otimes_k B$ are selfinjective.*

Proof The first two equivalent statements in 4.11.3 imply that that A^{op} is selfinjective. Since A^* and B^* are progenerators of A and B, respectively, it follows that $A^* \otimes_k B^* \cong (A \otimes_k B)^*$ is a progenerator of $A \otimes_k B$, whence the result. □

Theorem 4.11.5 *Let A be a finite-dimensional selfinjective k-algebra. For any finitely generated projective indecomposable A-module P, both $P/\mathrm{rad}(P)$ and $\mathrm{soc}(P)$ are simple, and the map sending $P/\mathrm{rad}(P)$ to $\mathrm{soc}(P)$ induces a permutation on the set of isomorphism classes of simple A-modules.*

Proof By 4.11.3, the classes of finitely generated projective and injective A-modules coincide. The result follows from combining 4.5.3 and 4.5.5. □

The permutation on the set of isomorphism classes of simple modules defined in 4.11.5 need not be the identity. It is the identity if A is symmetric, as we will see in 4.12.2. In general the socles of a finite-dimensional algebra A as left and right module need not coincide. For selfinjective algebras they do:

Theorem 4.11.6 *Let A be a finite-dimensional selfinjective k-algebra. The socle of A as a left A-module is equal to the socle of A as a right A-module.*

Proof Let $i, j \in A$ be primitive idempotents. The projective indecomposable A-module Ai is also injective, hence has a simple socle S. Let $a \in iAj$. Right multiplication by a induces an A-homomorphism $\varphi : Ai \to Aj$. If $Sa \neq \{0\}$ then φ is injective by 4.5.2. But then φ is split injective as Ai is injective. Since Aj is indecomposable this implies that φ is an isomorphism. Thus there is $c \in jAi$ such that right multiplication by c yields the inverse of φ. But then right multiplication by ac is the identity on Ai, hence $ac = i \notin J(A)$, and so $a \notin J(A)$. This shows that if $a \in J(A)$ then $\mathrm{soc}(A)a = \{0\}$. Thus $\mathrm{soc}(A)$ is contained in the left annihilator of $J(A)$, which by the right module version of 1.10.18, is equal to the socle of A as right A-module. The same argument shows that the socle of A as right A-module is contained in the socle of A as left A-module. □

Note that these two theorems are a generalisation of 1.11.2, describing the socle of finite p-group algebras.

Proposition 4.11.7 *Let A be a finite-dimensional selfinjective k-algebra. Let U be a finitely generated A-module. The following are equivalent.*

(i) *For any finitely generated projective A-module P and any A-homomorphism* $\pi : P \to U$ *we have* $\mathrm{soc}(P) \subseteq \ker(\pi)$.

(ii) *For any finitely generated injective A-module I and any A-homomorphism* $\iota : U \to I$ *we have* $\mathrm{Im}(\iota) \subseteq \mathrm{rad}(I)$.

(iii) *We have* $\mathrm{soc}(A)U = \{0\}$.

(iv) *The A-module U has no nonzero projective direct summand.*

(v) *The right A-module* U^* *has no nonzero projective direct summand.*

Proof The dual of a finitely generated projective module is injective, and vice versa. Since A is selfinjective, it follows that duality preserves finitely generated projective and injective modules. Thus (iv) and (v) are equivalent. Suppose that (iv) holds. Let P be a finitely generated projective A-module and let $\pi : P \to U$ be an A-homomorphism. Suppose that S is a simple submodule of P not contained in $\ker(\pi)$. Let $S \to I_S$ be an injective envelope of S. Since P is also injective, the inclusion $S \subseteq P$ extends to a map $\tau : I_S \to P$. The socle of I_S is simple, and hence τ is injective. Thus I_S is isomorphic to a direct summand of P which is mapped injectively into U. As I_S is injective, this implies that U has a direct summand isomorphic to I_S, which is a contradiction since I_S is also projective. This shows that (iv) implies (i). The converse is trivial. Dually, let I be a finitely generated injective A-module and $\iota : U \to I$ an A-homomorphism such that $\mathrm{Im}(\iota)$ is not contained in $\mathrm{rad}(I)$. Then $\mathrm{Im}(\iota)$ is not contained in every maximal submodule of I, or equivalently, there is a surjective A-homomorphism $\alpha : I \to S$ for some simple A-module S such that $\alpha \circ \iota : U \to S$ is nonzero, hence surjective as S is simple. Denote by $\pi_S : P_S \twoheadrightarrow S$ a projective cover of S. Since I is also projective, there is $\beta : I \to P_S$ such that $\pi_S \circ \beta = \alpha$. Precomposing with ι shows that $\pi_S \circ \beta \circ \iota : U \to S$ is surjective. Thus the image of $\beta \circ \iota$ is not contained in the unique maximal submodule $\mathrm{rad}(P_S) = \ker(\pi_S)$ of P_S, and hence $\beta \circ \iota$ is surjective. But then this map is split surjective, contradicting the fact that U has no nonzero projective direct summand. This shows that (iv) implies (ii). Again, the converse is trivial. Suppose that (iv) holds. Let I be an injective envelope of U. Then, using (ii), U is isomorphic to a submodule of $\mathrm{rad}(I) = J(A)I$. Since $\mathrm{soc}(A)$ annihilates $J(A)$ it follows that $\mathrm{soc}(A)$ annihilates $J(A)I$, hence also U. This shows that (iv) implies (iii). For the converse implication, suppose that U is a projective indecomposable A-module. Then $U \cong Ai$ for some primitive idempotent $i \in A$, and hence $\mathrm{soc}(A)U = \mathrm{soc}(A)i = \mathrm{soc}(U)$, which is clearly non zero. □

The earlier result 1.11.3 on finite p-group algebras is a special case of 4.11.7.

Proposition 4.11.8 *Let A be a finite-dimensional selfinjective k-algebra such that* $J(A)^3 = \{0\}$. *For any indecomposable nonprojective nonsimple A-module U we have* $\mathrm{rad}(U) = \mathrm{soc}(U)$; *that is, the socle and radical series of U coincide.*

Proof By 4.11.6, $\mathrm{soc}(A)$ is a 2-sided ideal in A that, by 4.11.7, annihilates U. Thus U can be viewed as a module over the algebra $B = A/\mathrm{soc}(A)$. Clearly U remains indecomposable and nonsimple as a B-module. Since $J(A)^3$ is zero we have $J(A)^2 \subseteq \mathrm{soc}(A)$. By 1.10.18 (iii), $J(B)$ is the image of $J(A)$ in B and hence $J(B)^2 = \{0\}$. The result follows from 4.5.7. □

Proposition 4.11.9 *Let A be a finite-dimensional selfinjective k-algebra such that $J(A)^3 = \{0\}$. Let U be an indecomposable nonprojective A-module and let V be an indecomposable nonsimple submodule of U. Then $\Omega(V)$ is isomorphic to a submodule of $\Omega(U)$.*

Proof Since V is not simple, we have $\mathrm{rad}(U) \cap V = \mathrm{soc}(U) \cap V = \mathrm{soc}(V) = \mathrm{rad}(V)$, by 4.11.8. Thus there is a projective cover $\pi : P \to U$ of U of the form $P = Q \oplus R$, where $\pi|_Q : Q \to V$ is a projective cover of V. It follows that $\Omega(V) \cong \ker(\pi|_Q)$ is a submodule of $\Omega(U) = \ker(\pi)$. □

Remark 4.11.10 Let A be a finite-dimensional selfinjective k-algebra. Since A is injective as a left and right module, it follows that the A-duality functor $\mathrm{Hom}_A(-, A)$ is exact, hence an equivalence from $\mathrm{mod}(A)$ to $\mathrm{mod}(A^{\mathrm{op}})$, with inverse the A-duality functor $\mathrm{Hom}_{A^{\mathrm{op}}}(-, A)$ for right A-modules. Thus the Nakayama functor $\mathrm{Hom}_k(-, k) \circ \mathrm{Hom}_A(-, A)$ is an equivalence on $\mathrm{mod}(A)$. If A is basic, then by Corollary 4.9.7, the Nakayama functor is induced by an algebra automorphism of A, uniquely determined up to inner automorphisms, called the *Nakayama automorphism.* If A is symmetric, then by Theorem 2.12.1 the k-duality and A-duality functors are isomorphic, and hence the Nakayama automorphism is inner, or equivalently, can be chosen to be the identity on A. Thus in that case we have an isomorphism of A-modules $Ai \cong (iA)^*$ for any idempotent i in A; see Remark 2.9.17. This isomorphism is induced by the bimodule isomorphism $A \cong A^*$ sending 1 to a symmetrising form on A; see Exercise 2.11.17.

4.12 Symmetric algebras over fields and local rings

Let $\mathcal{O}$ be a complete discrete valuation ring with quotient field K and residue field k. We allow the case $K = \mathcal{O} = k$ unless stated otherwise. We collect in this section some properties of finite group algebras over $\mathcal{O}$ that hold more generally for symmetric algebras. We use notation and terminology introduced in §2.11. Given two $\mathcal{O}$-algebras A, B and an A-B-bimodule M, the $\mathcal{O}$-dual $M^* = \mathrm{Hom}_{\mathcal{O}}(M, \mathcal{O})$ becomes a B-A-bimodule by setting $(b \cdot \alpha \cdot a)(m) = \alpha(amb)$ for all $m \in M$, $a \in A$, $b \in B$ and $\alpha \in \mathrm{Hom}_{\mathcal{O}}(M, \mathcal{O})$. In particular, A^* is again an A-A-bimodule. The algebra A is symmetric if $A \cong A^*$ as A-A-bimodules and if A is free of finite rank as an $\mathcal{O}$-module. If $\mathcal{O} = k$, then the condition $A \cong A^*$ automatically implies that A is finite-dimensional over k. Examples of symmetric algebras include group algebras of finite groups and matrix algebras; see §2.11.

Theorem 4.12.1 *Let A be a symmetric k-algebra. Then A is selfinjective.*

Proof A bimodule isomorphism $A^* \cong A$ implies in particular that A^* is a progenerator as a left A-module. Thus A is selfinjective by 4.11.3. □

Theorem 4.12.2 *Let A be a symmetric k-algebra. For any projective indecomposable A-module U we have* $\mathrm{soc}(U) \cong U/\mathrm{rad}(U)$.

Proof Let U be a projective indecomposable A-module. Then U is also injective by 4.11.3, hence $\mathrm{soc}(U)$ is simple by 4.11.5. Let I be a primitive decomposition of 1 in A. Let e be the sum of all $i \in I$ such that $U \cong Ai$. Then Ae is a direct sum of copies of U while $A(1-e)$ has no direct summand isomorphic to U. Let T be the sum of all simple submodules of A isomorphic to $\mathrm{soc}(U)$. By 4.11.5 the left A-module $A(1-e)$ has no submodule isomorphic to $\mathrm{soc}(U)$, and thus $T \subseteq Ae$. Note that T is a 2-sided ideal in A; indeed, if S is a submodule of A isomorphic to $\mathrm{soc}(U)$ then either $Sa = \{0\}$ or $Sa \cong S$ for every $a \in A$. Let $s : A \to k$ be a symmetrising form for A. For any $a \in A$ we have $s((1-e)ae) = s(e(1-e)a) = s(0) = 0$, hence $(1-e)Ae \subseteq \ker(s)$. Now $T \subseteq Ae$, hence $(1-e)T \subseteq \ker(s)$. But $(1-e)T$ is a right ideal in A, hence $(1-e)T = \{0\}$, or equivalently, $\mathrm{Hom}_A(A(1-e), T) = \{0\}$. Thus $\mathrm{Hom}_A(A, T) = \mathrm{Hom}_A(Ae, T)$. This means that T is a quotient of a finite direct sum of copies of U. As T is semisimple this means that T is isomorphic to a direct sum of copies of $U/\mathrm{rad}(U)$. But T was also defined as sum of simple modules isomorphic to $\mathrm{soc}(U)$. Thus $\mathrm{soc}(U)$ is isomorphic to $U/\mathrm{rad}(U)$. □

Theorem 4.12.3 *Let A be a symmetric k-algebra. For any two idempotents i, j in A be have* $\dim_k(iAj) = \dim_k(jAi)$. *In particular, if A is split, then the Cartan matrix of A is a symmetric matrix.*

Proof Let i, j be idempotents in A. Since A is symmetric we have $A \cong A^*$ as A-A-bimodules. Thus $iAj \cong iA^*j$ as k-vector spaces. Given $\tau \in A^*$, the element $i\tau j \in A^*$ is defined by $(i\tau j)(a) = \tau(jai)$. Thus $i\tau j$ is completely determined by the restriction of τ to jAi. This implies that $iA^*j \cong (jAi)^*$. Thus $\dim_k(iAj) = \dim_k((jAi)^*) = \dim_k(jAi)$, where the second equality uses the fact that all involved vector spaces are finite-dimensional and hence their dimensions are preserved under taking duals. For the second statement, let I

be a system of representatives of the conjugacy classes of primitive idempotents in A. By 4.10.2, if A is split, then the Cartan matrix of A is equal to $C = (\dim_k(iAj))_{i,j \in I}$. Thus C is symmetric by the first statement. □

Let A be a symmetric k-algebra with a symmetrising form $s : A \to k$. For any k-subspace U of A we consider as in 2.11.8 the space $U^\perp = \{a \in A \mid s(aU) = 0\}$ corresponding to the annihilator of U in A^* through the isomorphism $A \cong A^*$ induced by s (sending $a \in A$ to the linear map $b \mapsto s(ab)$). The proof of the following lemma is elementary linear algebra.

Lemma 4.12.4 *Let A be a symmetric k-algebra with a symmetrising form s. For any two k-subspaces U, V of A we have:*

(i) $\dim_k(U) + \dim_k(U^\perp) = \dim_k(A)$.
(ii) $(U^\perp)^\perp = U$.
(iii) $(U + V)^\perp = U^\perp \cap V^\perp$.
(iv) $(U \cap V)^\perp = U^\perp + V^\perp$.
(v) If $U \subseteq V$, then $V^\perp \subseteq U^\perp$.

In a selfinjective k-algebra A the left and right annihilators of $J(A)$ coincide by 4.11.6. If A is symmetric this holds for any ideal I since both annihilators are equal to $I^\perp$, by 2.11.9. This can be used to give a simpler proof of 4.11.6 in this case.

Theorem 4.12.5 ([49, §2]) *Let A be a symmetric algebra over a field k with symmetrising form s. We have*

(i) $[A, A]^\perp = Z(A)$.
(ii) $J(A)^\perp = \mathrm{soc}(A)$.
(iii) $(\mathrm{soc}(A) \cap Z(A))^\perp = [A, A] + J(A)$.

Proof Statement (i) follows from 2.11.10. Since A is symmetric, every simple A-module is isomorphic to a submodule of A by 4.12.2, and hence the left annihilator of $\mathrm{soc}(A)$ is $J(A)$. But $\mathrm{soc}(A)$ is an ideal, and hence $J(A) = \mathrm{soc}(A)^\perp$ by 2.11.9, whence (ii). Statement (iii) follows from (i), (ii), and 4.12.4 (iv). □

Corollary 4.12.6 *Let A be a symmetric k-algebra. The socles of A as a left A-module and as a right A-module are equal.*

Proof It follows from 4.12.5 (ii) that the socles of A as a left or right module are both equal to $J(A)^\perp$, for any choice of a symmetrising form of A. □

Corollary 4.12.7 *Let A be a split basic symmetric k-algebra. We have* $\mathrm{soc}(A) \subseteq Z(A)$.

Proof By 4.9.4 we have $[A,A] \subseteq J(A)$. Passing to perpendicular spaces reverses the inclusion, so the result follows from 4.12.5. □

We denote as before by $\ell(A)$ the number of isomorphism classes of simple modules of a finite-dimensional k-algebra A.

Corollary 4.12.8 *Let A be a split symmetric algebra over a field k. We have* $\ell(A) = \dim_k(Z(A) \cap \mathrm{soc}(A))$.

Proof By 1.15.4, the number of isomorphism classes of simple A-modules is equal to the codimension in A of $[A,A] + J(A)$. By 4.12.5 (iii) this is equal to the dimension of $Z(A) \cap \mathrm{soc}(A)$ as claimed. □

Corollary 4.12.9 *Let A be an indecomposable nonsimple split symmetric k-algebra. We have* $\dim_k(Z(A)) > \ell(A)$.

Proof The hypotheses imply that 1 belongs to $Z(A)$ but not to $\mathrm{soc}(A)$. The result follows from 4.12.8. □

Matrix algebras over k are symmetric and have one-dimensional centres. The above observation can be used to show the converse.

Corollary 4.12.10 *Let A be a split symmetric k-algebra such that* $\dim_k(Z(A)) = 1$. *Then $A \cong M_n(k)$ for some positive integer n.*

Proof It follows from 4.12.9 that A has no nonsimple direct factor. Thus A is semisimple, hence a direct product of matrix algebras. The hypothesis $\dim_k(Z(A)) = 1$ implies that A is a matrix algebra. □

Lemma 4.12.11 *Let A be a symmetric k-algebra such that $A/J(A)$ is separable. Then the socle* $\mathrm{soc}(A)$ *of A as a left and right A-module is equal to the socle of A as an $A \otimes_k A^{\mathrm{op}}$-module. Moreover, we have an isomorphism of $A \otimes_k A^{\mathrm{op}}$-modules* $\mathrm{soc}(A) \cong A/J(A)$. *In particular,* $\mathrm{soc}(A)$ *is semisimple as an $A \otimes_k A^{\mathrm{op}}$-module.*

Proof Since $A/J(A)$ is separable, it follows from 1.16.16 that $J(A)$ is the radical of A as an $A \otimes_k A^{\mathrm{op}}$-module. Denote by $\mathrm{soc}(A)$ the socle of A as an $A \otimes_k A^{\mathrm{op}}$-module. The algebra A is symmetric by the assumptions, and $A/J(A)$ is symmetric by 2.11.14. We get from 2.9.2 that $\mathrm{soc}(A) \cong \mathrm{soc}(A^*) \cong (A/J(A))^* \cong A/J(A)$. Since $J(A)$ is also the radical of A as a left or right module, it follows from restricting these isomorphisms to one side that $\mathrm{soc}(A)$ is the socle of A as a left and as a right A-module, whence the result. □

Lemma 4.12.12 *Let A be a split symmetric k-algebra, and let I be a nonzero ideal in A. Then* $I \cap Z(A) \neq \{0\}$.

Proof Let e be an idempotent in A such that eAe is a basic algebra of A. Since A and eAe are Morita equivalent, it follows that multiplication by e induces an isomorphism $Z(A) \cong Z(eAe)$, and that eIe is a nonzero ideal in eAe. Thus we may assume that A is basic. Since A is split, it follows that $A/J(A)$ remains semisimple as an A-A-bimodule, and hence $\mathrm{soc}(A)$ is semisimple as A-A-bimodule, thus equal to the sum of all minimal two-sided ideals in A. Thus $I \cap \mathrm{soc}(A) \neq \{0\}$. Since A is basic, we have $\mathrm{soc}(A) \subseteq Z(A)$, whence the result. □

Lemma 4.12.13 *Let A be a finite-dimensional commutative k-algebra. If* $\dim_k(\mathrm{soc}(A)) = 1$ *then A is split local symmetric.*

Proof Any linear map $s : A \to k$ is symmetric as A is commutative. If s does not vanish on $\mathrm{soc}(A)$, then s is nondegenerate because $\mathrm{soc}(A)$ is in that case the unique minimal ideal. Thus A is symmetric. But then $J(A) = \mathrm{soc}(A)^{\perp}$ has codimension 1 in A, and hence A is split local. □

Theorem 4.12.14 (cf. [2], [68, Proposition 1]) *Let A be a split indecomposable symmetric k-algebra. The following are equivalent.*

(i) *$Z(A)$ is symmetric.*
(ii) $\dim_k(\mathrm{soc}(Z(A))) = 1$.
(iii) *A is Morita equivalent to a commutative k-algebra.*
(iv) *$A \cong M_n(Z(A))$ for some positive integer n.*
(v) $J(A) = J(Z(A))A$.

Proof The algebra $Z(A)$ is split by 1.14.11, and local because A is indecomposable as an algebra. Thus if $Z(A)$ is symmetric, then its socle is one-dimensional. This shows that (i) implies (ii). It follows from 4.12.13, applied to $Z(A)$, that (ii) implies (i). Suppose that (ii) holds. Since $\mathrm{soc}(Z(A))$ contains $Z(A) \cap \mathrm{soc}(A)$, it must be equal to $Z(A) \cap \mathrm{soc}(A)$ and $\ell(A) = 1$ by 4.12.8. Thus A has a unique conjugacy class of primitive idempotents, hence is isomorphic to a matrix algebra $M_n(C)$ over its basic algebra $C = iAi$, where i is a primitive idempotent. Note that C is symmetric split local. Since A and C are Morita equivalent, multiplication by i induces an algebra isomorphism $Z(A) \cong Z(C)$. By 4.4.9 we have $[C, C] \subseteq J(C)^2$. Since C is symmetric split local, $\mathrm{soc}(C)$ is also 1-dimensional, hence equal to $\mathrm{soc}(Z(C))$. The kernel of any symmetrising form on C contains $[C, C]$ but not the ideal $\mathrm{soc}(C)$. Thus $[C, C] \cap \mathrm{soc}(Z(C)) = \{0\}$. Since $\mathrm{soc}(Z(C))$ is the unique minimal ideal in $Z(C)$, it follows that $[C, C] \cap Z(C) = \{0\}$. By 2.11.10 we have $Z(C) = [C, C]^{\perp}$. This implies that $C = Z(C) + [C, C] = Z(C) + J(C)^2$. Thus $C = Z(C)$ by 4.9.9. This shows that (ii) implies (iii) and (iv). Since a Morita equivalence preserves centres (by 2.8.6) and symmetry (by 2.12.6), statement (iv) implies both (ii) and (iii). Clearly

(iv) implies (v). Suppose that (v) holds. We first show that A has a unique conjugacy class of primitive idempotents. Arguing by contradiction, suppose that i, j are two idempotents that are not conjugate in A. Using 4.6.8, we have $iAj \subseteq J(A) = J(Z(A))A$. Thus $iAj \subseteq J(Z(A))iAj$. Nakayama's Lemma applied to the $Z(A)$-module iAj implies that $iAj = \{0\}$. Since A is indecomposable, Brauer's Theorem 4.10.5 yields a contradiction. This shows that A has a unique conjugacy class of primitive idempotents. Arguing as earlier, it follows that A is isomorphic to a matrix algebra over its basic algebra iAi, where i is a primitive idempotent, and we have $Z(iAi) = Z(A)i \cong Z(A)$. Moreover, we have $J(iAi) = iJ(A)i = J(Z(A))iAi$. Thus $iAi = ki + J(iAi) = Z(A)i + J(Z(A))iAi$. Nakayama's Lemma applied to the $Z(A)$-module iAi implies that $iAi = Z(A)i$. Thus (v) implies (iv). □

For A a symmetric k-algebra, we denote by $\mathrm{soc}^2(A)$ the inverse image in A of $\mathrm{soc}(A/\mathrm{soc}(A))$.

Lemma 4.12.15 *Let A be a split local symmetric k-algebra. Then* $\mathrm{soc}^2(A) \subseteq Z(A)$.

Proof By 4.4.9 we have $[A, A] \subseteq J(A)^2$. Since $[A, A]^\perp = Z(A)$ and $(J(A)^2)^\perp = \mathrm{soc}^2(A)$, the result follows. □

Let A be a symmetric $\mathcal{O}$-algebra with symmetrising form $s : A \to \mathcal{O}$. For any $\mathcal{O}$-submodule U of A we define as before the $\mathcal{O}$-submodule $U^\perp = \{a \in A \mid s(aU) = 0\}$. Some, but not all statements of 4.12.4 carry over to symmetric $\mathcal{O}$-algebras; this is related to purity considerations – see 4.2.6 for basic properties of pure submodules.

Lemma 4.12.16 *Let A be a symmetric $\mathcal{O}$-algebra with a symmetrising form s. Let U, V be $\mathcal{O}$-submodules of A and let U' be the smallest pure $\mathcal{O}$-submodule containing U; that is, U' consists of all $a \in A$ such that $\lambda a \in U$ for some nonzero $\lambda \in \mathcal{O}$.*

(i) The $\mathcal{O}$-submodule $U^\perp$ is pure in A and $U^\perp = (U')^\perp$.
(ii) We have $\mathrm{rk}_\mathcal{O}(U) + \mathrm{rk}_\mathcal{O}(U^\perp) = \mathrm{rk}_\mathcal{O}(A)$.
(iii) We have $U \subseteq (U^\perp)^\perp$, and the equality holds if and only if U is a pure $\mathcal{O}$-submodule of A.
(iv) We have $(U + V)^\perp = U^\perp \cap V^\perp$.

Proof Let $b \in A$ and $\mu \in \mathcal{O}$ such that $\mu \neq 0$ and $\mu b \in U^\perp$. Thus $s(\mu bU) = \{0\}$. Since s is $\mathcal{O}$-linear and $\mathcal{O}$ is a domain, it follows that $s(bU) = \{0\}$, hence $b \in U^\perp$. This shows that $U^\perp$ is $\mathcal{O}$-pure in A. Since $U \subseteq U'$ we have $(U')^\perp \subseteq U^\perp$. Conversely, let $c \in U^\perp$ and let $a \in U'$. Then $\lambda a \in U$ for some nonzero

$\lambda \in \mathcal{O}$, hence $0 = s(\lambda ac) = \lambda s(ac)$. Since $\lambda \neq 0$ it follows that $s(ac) = 0$, hence $c \in (U')^{\perp}$. This shows (i). Note that U and U' have the same $\mathcal{O}$-rank, and therefore, in order to prove (ii) we may thus assume that U is pure. Then every linear map $U \to \mathcal{O}$ extends to a linear map $A \to \mathcal{O}$, and hence the map $A \to U^*$ sending $a \in A$ to $(a \cdot s)|_U$ is surjective. The kernel of this map is clearly $U^{\perp}$. Statement (ii) follows. Statement (iii) follows from the previous statements and (iv) is a trivial verification. □

We observed in 2.11.14 that if K has characteristic zero, then any finite-dimensional division algebra over K is symmetric, hence any finite-dimensional semisimple K-algebra is symmetric by Wedderburn's Theorem and 2.12.6. We use this to identify symmetric $\mathcal{O}$-subalgebras in semisimple algebras over K, motivated by the fact that if G is a finite group, then $\mathcal{O}G$ is a symmetric $\mathcal{O}$-algebra whose coefficient extension to the K-algebra KG is semisimple.

Proposition 4.12.17 *Let A be an $\mathcal{O}$-algebra such that A is free of finite rank as an $\mathcal{O}$-module and such that the K-algebra $K \otimes_{\mathcal{O}} A$ is symmetric. Let $t : K \otimes_{\mathcal{O}} A \to K$ be a symmetrising form of $K \otimes_{\mathcal{O}} A$. Identify A to its canonical image $1 \otimes A$ in $K \otimes_{\mathcal{O}} A$.*

(i) The $\mathcal{O}$-algebra A is symmetric if and only if there exists an element $z \in Z(K \otimes_{\mathcal{O}} A)^{\times}$ with the property that for any $c \in K \otimes_{\mathcal{O}} A$ we have $t(zcA) \subseteq \mathcal{O}$ if and only if $c \in A$.

(ii) If A is symmetric and z satisfies (i), then the map $s : A \to \mathcal{O}$ defined by $s(a) = t(za)$ for all $a \in A$ is a symmetrising form of A.

Proof Suppose that A is symmetric. Let s be a symmetrising form of A. Extend s to a K-linear map $K \otimes_{\mathcal{O}} A \to K$, abusively still denoted by s. Then s is a symmetrising form of $K \otimes_{\mathcal{O}} A$ by 2.11.11 (iii), hence $s = z \cdot t$ for some $z \in Z(K \otimes_{\mathcal{O}} A)^{\times}$, by 2.11.5. In particular, $z \cdot t$ sends A to $\mathcal{O}$, whence $t(zcA) \subseteq \mathcal{O}$ for any $c \in A$. Conversely, if $c \in K \otimes_{\mathcal{O}} A$ such that $t(zcA) \subseteq \mathcal{O}$, then $(zc) \cdot t$ induces a linear map from A to $\mathcal{O}$, hence is equal to $a \cdot s = (az) \cdot t$ for some $a \in A$. Thus the K-linear map $(az - cz) \cdot t$ on $K \otimes_{\mathcal{O}} A$ has A in its kernel, hence is zero as A contains a K-basis of $K \otimes_{\mathcal{O}} A$. Since t is a symmetrising form for $K \otimes_{\mathcal{O}} A$ this forces $az - cz = 0$, hence $a = c \in A$ as z is invertible. This shows one implication in (i), and it also shows that if A is symmetric, then (ii) holds. For the converse in (i), let $z \in Z(K \otimes_{\mathcal{O}} A)$ such that for any $c \in K \otimes_{\mathcal{O}} A$ we have $t(zcA) \subseteq \mathcal{O}$ if and only if $c \in A$. Then the map $s = z \cdot t : K \otimes_{\mathcal{O}} A \to K$ is symmetric and sends A to $\mathcal{O}$. In order to show that s induces a symmetrising form on A, it suffices to show that for any $\mathcal{O}$-linear map $w : A \to \mathcal{O}$ there is an element $a \in A$ such that $w = a \cdot s$ on A. Extending w to $K \otimes_{\mathcal{O}} A$ in conjunction with the fact that $z \cdot t$ is a symmetrising form for $K \otimes_{\mathcal{O}} A$ yields an element

$c \in K \otimes_{\mathcal{O}} A$ such that $w = (cz) \cdot t$. But then $t(czA) = w(A) \subseteq \mathcal{O}$, and therefore $c \in A$. This shows the converse in (i). □

Corollary 4.12.18 *Let A be a symmetric $\mathcal{O}$-algebra with symmetrising form $s : A \to \mathcal{O}$. Denote by the same letter s the K-linear extension of s to a map $K \otimes_{\mathcal{O}} A \to K$. For any $c \in K \otimes_{\mathcal{O}} A$ we have $s(cA) \subseteq \mathcal{O}$ if and only if $c \in A$.*

Proof This is the special case of 4.12.17 with t equal to the K-linear extension of s to $K \otimes_{\mathcal{O}} A$ and $z = 1$. □

Lemma 4.12.19 *Let A be an $\mathcal{O}$-algebra that is free of finite rank as an $\mathcal{O}$-module and let U be an $\mathcal{O}$-submodule of A. Identify A to its canonical image in $K \otimes_{\mathcal{O}} A$ and denote by W the K-subspace of $K \otimes_{\mathcal{O}} A$ generated by U. Then U is pure in A if and only if $U = A \cap W$.*

Proof Suppose that $U = A \cap W$, let $a \in A$ and $\lambda \in \mathcal{O}$ such that $\lambda \neq 0$ and such that $\lambda a \in U$. Then $a = \lambda^{-1}\lambda a \in A \cap W = U$, hence U is pure. Conversely, suppose that U is pure in A. We clearly have $U \subseteq A \cap W$. Let $w \in A \cap W$; that is, w is a finite K-linear combination of elements in U. Since K is the quotient field of $\mathcal{O}$, the product of the denominators in the coefficients of such a linear combination yields a nonzero $\lambda \in \mathcal{O}$ such that $\lambda w \in U$. But then $w \in U$ since U is pure. □

Proposition 4.12.20 *Let A be a symmetric $\mathcal{O}$-algebra such that $K \otimes_{\mathcal{O}} A$ is a direct product of matrix algebras over K. Identify A to its canonical image $1 \otimes A$ in $K \otimes_{\mathcal{O}} A$. The following are equivalent.*

(i) We have $[A, A] = A \cap [K \otimes_{\mathcal{O}} A, K \otimes_{\mathcal{O}} A]$.

(ii) We have $[A, A] = A \cap (\cap_\chi \ker(\chi))$, where χ runs over the characters of the simple $K \otimes_{\mathcal{O}} A$-modules.

(iii) The $\mathcal{O}$-module $[A, A]$ is a pure submodule of A.

(iv) We have $Z(A)^\perp = [A, A]$.

(v) The canonical map $Z(A) \to Z(k \otimes_{\mathcal{O}} A)$ is surjective.

Proof Let $s : A \to \mathcal{O}$ be a symmetrising form of A. By the assumptions, $K \otimes_{\mathcal{O}} A$ is isomorphic to $\prod_V \mathrm{End}_K(V)$, where V runs over a set of representatives of the isomorphism classes of simple $K \otimes_{\mathcal{O}} A$-modules. Through this isomorphism, $[K \otimes_{\mathcal{O}} A, K \otimes_{\mathcal{O}} A]$ is mapped to the tuples of endomorphisms of the simple modules V having trace zero (cf. 1.5.7). This shows the equivalence of (i) and (ii). Since $[K \otimes_{\mathcal{O}} A, K \otimes_{\mathcal{O}} A]$ is the K-subspace of $K \otimes_{\mathcal{O}} A$ generated by $[A, A]$, the equivalence of (i) and (iii) follows from 4.12.19. By 2.11.10 we have $[A, A]^\perp = Z(A)$, hence the equivalence of (iii) and (iv) follows from 4.12.16 (iii). The canonical map $A \to k \otimes_{\mathcal{O}} A$ maps

$[A, A]$ onto $[k \otimes_{\mathcal{O}} A, k \otimes_{\mathcal{O}} A]$. The induced map $Z(A) \to Z(k \otimes_{\mathcal{O}} A)$ is surjective if and only if $\mathrm{rk}_{\mathcal{O}}(Z(A)) = \dim_k(Z(k \otimes_{\mathcal{O}} A))$, thus by 4.12.16 if and only if $\mathrm{rk}_{\mathcal{O}}([A, A]) = \dim_k([k \otimes_{\mathcal{O}} A, k \otimes_{\mathcal{O}} A])$. By 4.2.8 (ii) this is equivalent to $[A, A]$ being pure in A, whence the equivalence between (iii) and (v). □

The equivalences between (i), (ii), (iii) in the previous proposition do not require A to be symmetric. This proposition applies in particular to finite group algebras, because if G is a finite group, then the canonical map $Z(\mathcal{O}G) \to Z(kG)$ is surjective. Thus we have the following:

Corollary 4.12.21 *Let G be a finite group and suppose that K is splitting field of characteristic zero for KG. Let $a \in \mathcal{O}G$. We have $a \in [\mathcal{O}G, \mathcal{O}G]$ if and only if $\chi(a) = 0$ for all $\chi \in \mathrm{Irr}_K(G)$.*

4.13 Stable categories for $\mathcal{O}$-injective algebras

Let $\mathcal{O}$ be a complete local commutative Noetherian ring with residue field k and field of fractions K. For A an $\mathcal{O}$-algebra that is finitely generated as an $\mathcal{O}$-module, we denote by $\underline{\mathrm{mod}}(A)$ the $\mathcal{O}$-stable category of finitely generated A-modules, as defined in 2.13.1. That is, the objects of $\underline{\mathrm{mod}}(A)$ are the finitely generated A-modules, and for any two finitely generated A-modules U, V, the morphism space in $\underline{\mathrm{mod}}(A)$ from U to V is the $\mathcal{O}$-module $\underline{\mathrm{Hom}}_A(U, V) = \mathrm{Hom}_A(U, V)/\mathrm{Hom}_A^{\mathrm{pr}}(U, V)$, where $\mathrm{Hom}_A^{\mathrm{pr}}(U, V)$ is the $\mathcal{O}$-submodule of $\mathrm{Hom}_A(U, V)$ consisting of all A-homomorphisms from U to V that factor through a finitely generated relatively $\mathcal{O}$-projective A-module. The composition of morphisms in $\underline{\mathrm{mod}}(A)$ is induced by the usual composition of A-homomorphisms. If the classes of relatively $\mathcal{O}$-projective and relatively $\mathcal{O}$-injective modules coincide, then the category $\underline{\mathrm{mod}}(A)$ is triangulated, with the Heller operator as shift functor and exact triangles induced by $\mathcal{O}$-split short exact sequences of A-modules; see A.3.2 for details. Similar to the characterisation of selfinjective algebras, we have the following characterisation of $\mathcal{O}$-algebras whose classes of relatively $\mathcal{O}$-projective and relatively $\mathcal{O}$-injective modules coincide.

Proposition 4.13.1 *Let A be an $\mathcal{O}$-algebra that is finitely generated free as an $\mathcal{O}$-module. The following are equivalent.*

(i) *The k-algebra $k \otimes_{\mathcal{O}} A$ is selfinjective.*
(ii) *As a left A-module, A is relatively $\mathcal{O}$-injective.*
(iii) *As a right A-module, A is relatively $\mathcal{O}$-injective.*

(iv) *The classes of finitely generated relatively $\mathcal{O}$-projective and relatively $\mathcal{O}$-injective left A-modules coincide.*
(v) *The classes of finitely generated relatively $\mathcal{O}$-projective and relatively $\mathcal{O}$-injective right A-modules coincide.*
(vi) *The classes of relatively $\mathcal{O}$-projective and relatively $\mathcal{O}$-injective left (resp. right) A-modules coincide.*

Proof The equivalence of (iv) and (v) is from 2.15.2. If U is a relatively $\mathcal{O}$-injective A-module, then $k \otimes_{\mathcal{O}} U$ is an injective $k \otimes_{\mathcal{O}} A$-module. Thus the implications (ii) $\Rightarrow$ (i) and (iii) $\Rightarrow$ (i) are trivial consequences of 4.11.3. Clearly (iv) and (v) imply (ii) and (iii), respectively. If (i) holds, then 4.11.3 implies that the k-dual of $k \otimes_{\mathcal{O}} A$ is a progenerator for $k \otimes_{\mathcal{O}} A$. By the standard lifting theorems, the $\mathcal{O}$-dual of A is a progenerator for A. It follows from 2.15.2 that (i) implies (iv), (v), (vi), completing the proof. □

It follows from 4.11.4 that if A and B are two $\mathcal{O}$-algebras satisfying the hypotheses and the equivalent conditions in 4.13.1, then so do A^{op} and $A \otimes_{\mathcal{O}} B$.

Proposition 4.13.2 *Let A be an $\mathcal{O}$-algebra that is finitely generated as an $\mathcal{O}$-module and let U be a finitely generated A-module that is not relatively $\mathcal{O}$-projective. Then U is indecomposable as an object in the $\mathcal{O}$-stable category* $\underline{\mathrm{mod}}(A)$ *if and only if $U = V \oplus W$ for some indecomposable A-module V that is not relatively $\mathcal{O}$-projective, and some relatively $\mathcal{O}$-projective A-module W. In particular, if U is indecomposable in* $\mathrm{mod}(A)$ *and not relatively $\mathcal{O}$-projective, then U is indecomposable in* $\underline{\mathrm{mod}}(A)$.

Proof The proof is similar to that of 4.7.14. The algebra $\underline{\mathrm{End}}_A(U)$ is a quotient of $\mathrm{End}_A(U)$ by the ideal $\mathrm{End}_A^{\mathrm{pr}}(A)$. Thus any primitive idempotent in $\underline{\mathrm{End}}_A(U)$ lifts to a primitive idempotent in $\mathrm{End}_A(U)$. It follows that U is indecomposable in $\underline{\mathrm{mod}}(A)$ if and only if $\underline{\mathrm{End}}_A(U)$ is local, hence if and only if $\mathrm{Id}_U = \eta + \eta'$ for some primitive idempotent $\pi \in \mathrm{End}_A(U)$ whose image in $\underline{\mathrm{End}}_A(U)$ is equal to that of Id_U, and some idempotent $\eta' \in \mathrm{End}_A^{\mathrm{pr}}(U)$. In that case, $V = \eta(U)$ and $W = \eta'(U)$ satisfy the conclusion. □

More generally, a direct sum decomposition of a finitely generated A-module U in $\underline{\mathrm{mod}}(A)$ lifts to a decomposition in $\mathrm{mod}(A)$ plus a relatively $\mathcal{O}$-projective direct summand. In this way, the Krull–Schmidt property passes down to $\underline{\mathrm{mod}}(A)$, a fact that can also be deduced from the Krull–Schmidt property in $\mathrm{mod}(A)$ together with 2.13.8. The arguments in the above proof of 4.13.2 are similar to those in the proof of 4.7.14 – both are special cases of a more general situation in which the Krull–Schmidt property in an $\mathcal{O}$-linear category passes down to a quotient category. The following result is one of the crucial tools for detecting homomorphisms that factor through a projective A-module.

Proposition 4.13.3 *Let A be a finite-dimensional selfinjective k-algebra and let U, V be finitely generated indecomposable nonprojective A-modules. If $\alpha : U \to V$ is surjective or injective then α does not factor through a projective module.*

Proof Suppose that α is surjective and that α factors through a projective module. Then α factors through a projective cover $\tau : P \to V$ of V, by 2.13.4. That is, there is a homomorphism $\gamma : U \to P$ satisfying $\tau \circ \gamma = \alpha$. Since α is surjective we have $\mathrm{Im}(\gamma) + \ker(\tau) = P$. But $\ker(\tau) \subseteq \mathrm{rad}(P)$, and thus $\mathrm{Im}(\gamma) = P$ by Nakayama's Lemma. As P is a projective module, this implies that γ is split surjective, and hence U has a direct summand isomorphic to P, contradicting the assumptions on U. (Up to this point we have not used the assumption that A is selfinjective.) Dualising the previous argument, using an injective envelope I of U and the assumption that I is also projective as A is selfinjective, we get that α cannot be injective either. □

Corollary 4.13.4 *Let A be a finite-dimensional selfinjective k-algebra and let U, V be finitely generated indecomposable nonprojective A-modules. If one of U or V is simple then $\underline{\mathrm{Hom}}_A(U, V) \cong \mathrm{Hom}_A(U, V)$.*

Proof Since U or V is simple, any nonzero A-homomorphism from U to V is surjective or injective, thus does not factor through a projective module by 4.13.3. □

Corollary 4.13.5 *Let A be a finite-dimensional selfinjective k-algebra and let U, V be finitely generated indecomposable nonprojective A-modules. Let $\alpha : U \to V$ be an A-homomorphism. If α factors through a projective A-module, then $\mathrm{Im}(\alpha) \subseteq \mathrm{rad}(V)$ and $\mathrm{soc}(U) \subseteq \ker(\alpha)$.*

Proof If $\mathrm{Im}(\alpha)$ is not contained in $\mathrm{rad}(V)$, then there is a maximal submodule M of V that does not contain $\mathrm{Im}(\alpha)$, and hence the composition of α with the canonical map $V \to V/M$ is a nonzero map from U to the simple A-module V/M. By 4.13.4 this composition does not factor through a projective A-module, and hence α does not factor through a projective A-module. Similarly, if $\ker(\alpha)$ does not contain $\mathrm{soc}(U)$, then there is a simple submodule S of U such that the inclusion $S \subseteq U$ composed with α is a nonzero map from S to V. By 4.13.4, this map does not factor through a projective A-module, and hence neither does α. □

Corollary 4.13.6 *Let A be a finite-dimensional selfinjective k-algebra. Let U, V, W be finitely generated indecomposable nonprojective A-modules such that $\mathrm{soc}(U)$ is simple. Let $\alpha : U \to V$, $\beta : V \to W$ and $\gamma : U \to W$ be A-homomorphisms such that γ is injective and $\gamma - \beta \circ \alpha$ factors through a projective A-module. Then α is injective. If also $\mathrm{soc}(V)$ is simple then β is injective.*

Proof It follows from 4.13.3 that $\gamma - \beta \circ \alpha$ is not injective. Since γ is injective and $\mathrm{soc}(U)$ is simple this implies that $\beta \circ \alpha$ must be injective, too. In particular, α is injective. If also $\mathrm{soc}(V)$ is simple then either β is injective or $\ker(\beta)$ contains $\mathrm{soc}(V)$. Since $\mathrm{Im}(\alpha)$ contains $\mathrm{soc}(V)$ and since $\alpha(\mathrm{soc}(U)) \subseteq \mathrm{soc}(V)$, the latter case is impossible because it would imply that $\beta \circ \alpha$ is not injective. □

In order to illustrate some of the terminology on the general theme of derived functors, we include the following cohomological fact.

Proposition 4.13.7 *Let A be a finite-dimensional selfinjective k-algebra, and let U, V be A-modules. Let n be a positive integer. We have a natural isomorphism*

$$\mathrm{Ext}_A^n(U, V) \cong \underline{\mathrm{Hom}}_A(\Omega^n(U), V).$$

In particular, if one of U, V is projective, then $\mathrm{Ext}_A^n(U, V) = \{0\}$.

Proof The space $\mathrm{Ext}_A^n(U, V)$ is the n-th left derived functor of $\mathrm{Hom}_A(-, V)$ evaluated at U, hence is the cohomology in degree n of the cochain complex obtained from applying $\mathrm{Hom}_A(-, V)$ to a projective resolution

$$\cdots \longrightarrow P_{n+1} \xrightarrow{\delta_{n+1}} P_n \xrightarrow{\delta_n} P_{n-21} \longrightarrow \cdots \longrightarrow P_0 \longrightarrow U \longrightarrow 0$$

of U. This yields a cochain complex of the form

$$\cdots \longrightarrow \mathrm{Hom}_A(P_{n-1}, V) \xrightarrow{\epsilon^{n-1}} \mathrm{Hom}_A(P_n, V) \xrightarrow{\epsilon^n} \mathrm{Hom}_A(P_{n+1}, V) \longrightarrow \cdots$$

where ϵ^{n-1} and ϵ^n are induced by precomposition with δ_n and δ_{n+1}, respectively. The cohomology in degree n of this cochain complex is $\ker(\epsilon^n)/\mathrm{Im}(\epsilon^{n-1})$. The space $\ker(\epsilon^n)$ consists of all A-homomorphisms $P_n \to V$ that vanish on $\mathrm{Im}(\delta_{n+1})$, or equivalently, that factor through $P_n/\mathrm{Im}(\delta_{n+1})$. Note that $\mathrm{Im}(\delta_{n+1}) = \ker(\delta_n)$, so δ_n induces an isomorphism $P_n/\mathrm{Im}(\delta_{n+1}) \cong \mathrm{Im}(\delta_n) = \Omega^n(U)$. The space $\mathrm{Im}(\epsilon^{n-1})$ consists of all A-homomorphisms $P_n \to V$ that factor through $\delta_n : P_n \to P_{n-1}$, and hence can be identified with all A-homomorphisms $P_n/\mathrm{Im}(\delta_{n+1}) \to V$ that factor through the inclusion $P_n/\mathrm{Im}(\delta_{n+1}) \to P_{n-1}$. Since A is selfinjective, this yields the space of all A-homomorphisms $P_n/\mathrm{Im}(\delta_{n+1}) \to V$ that factor through a projective A-module. It follows that the above isomorphism $P_n/\mathrm{Im}(\delta_{n+1}) \cong \mathrm{Im}(\delta_n) = \Omega^n(U)$ identifies $\ker(\epsilon^n)$ with $\mathrm{Hom}_A(\Omega^n(U), V)$ and $\mathrm{Im}(\epsilon^{n-1})$ with the subspace $\mathrm{Hom}_A^{\mathrm{pr}}(\Omega^n(U), V)$. Thus the corresponding quotient is $\underline{\mathrm{Hom}}_A(\Omega^n(U), V)$ as stated. The last statement follows immediately. Alternatively, the last statement follows from 2.20.5 and the fact that projective A-modules are injective. □

Corollary 4.13.8 *Let A be a finite-dimensional selfinjective k-algebra, U a finitely generated A-module, and let S be a nonprojective simple A-module. For any positive integer n we have*

$$\mathrm{Ext}_A^n(U, S) \cong \mathrm{Hom}_A(\Omega^n(U)/\mathrm{rad}(\Omega^n(U)), S),$$

where the notation is chosen such that $\Omega^n(U)$ has no nonzero projective direct summand. In particular, if A is split, then $\dim_k(\mathrm{Ext}_A^n(U, S))$ is equal to the number of summands isomorphic to S in a decomposition of $\Omega^n(U)/\mathrm{rad}(\Omega^n(U))$ as a direct sum of simple modules.

Proof The kernel of a nonzero A-homomorphism from $\Omega^n(U)$ to the simple A-module S is a maximal submodule of $\Omega^n(U)$, hence contains $\mathrm{rad}(\Omega^n(U))$. Thus $\mathrm{Hom}_A(\Omega^n(U), S)$ can be identified with $\mathrm{Hom}_A(\Omega^n(U)/\mathrm{rad}(\Omega^n(U)), S)$. The first statement follows from 4.13.7 together with 4.13.4. The second statement follows from the first and Schur's Lemma. □

If A is selfinjective, then the Heller operator is an equivalence on the stable category of A-modules, and hence the expression $\underline{\mathrm{Hom}}_A(\Omega^n(U), V)$ in Proposition 4.13.7 makes sense for all integers n. We introduce the following notation. See 2.20.8 for the definition of Hochschild cohomology $HH^n(A)$ of an algebra A.

Definition 4.13.9 Let A be a finite-dimensional selfinjective k-algebra, and let U, V be A-modules. For any integer n we set

$$\widehat{\mathrm{Ext}}_A^n(U, V) = \underline{\mathrm{Hom}}_A(\Omega^n(U), V),$$
$$\widehat{HH}^n(A) = \underline{\mathrm{Hom}}_{A\otimes_k A^{\mathrm{op}}}(\Omega^n_{A\otimes_k A^{\mathrm{op}}}(A), A).$$

Remark 4.13.10 With the notation of Definition 4.13.9, the graded space $\widehat{\mathrm{Ext}}_A^*(U, V)$ is sometimes called *Tate-Ext of U and V*, and $\widehat{HH}^*(A)$ is called the *Tate–Hochschild cohomology of A*. Since $\widehat{\mathrm{Ext}}$ is defined in terms of the stable module category of A, it follows trivially that if one of U, V is projective, then $\widehat{\mathrm{Ext}_A}^n(U, V) = \{0\}$ for all integers n. Proposition 4.13.7 implies that for positive n we have $\widehat{\mathrm{Ext}}_A^n(U, V) \cong \mathrm{Ext}_A^n(U, V)$. In degree 0 we have

$\mathrm{Ext}^0_A(U,V) = \mathrm{Hom}_A(U,V)$, while $\widehat{\mathrm{Ext}}^0_A(U,V) = \underline{\mathrm{Hom}}_A(U,V)$, so this is the quotient of $\mathrm{Ext}^0_A(U,V)$ by $\mathrm{Hom}^{\mathrm{pr}}_A(U,V)$. For n negative, we have $\mathrm{Ext}^n_A(U,V) = \{0\}$, but $\widehat{\mathrm{Ext}}^n_A(U,V)$ need not be zero. If A is symmetric and if U, V are finitely generated, then $\widehat{\mathrm{Ext}}^{n-1}_A(U,V)$ is dual to $\widehat{\mathrm{Ext}}^{-n}_A(V,U)$ for any integer n; this is known as *Tate duality*. Similarly, for n positive, we have $\widehat{HH}^n(A) \cong HH^n(A)$. In degree 0 we have $HH^0(A) \cong Z(A)$ and $\widehat{HH}^0(A) \cong \underline{Z}(A)$, which is the quotient of $Z(A)$ by $Z^{\mathrm{pr}}(A)$. For n negative, we have $HH^n(A) = \{0\}$, but $\widehat{HH}^n(A)$ need not be zero. If A is symmetric, then Tate duality yields a duality between $\widehat{HH}^{n-1}(A)$ and $\widehat{HH}^{-n}(A)$. See for instance [60] for proofs of Tate duality for symmetric algebras over fields. Tate duality can be regarded as a special case of *Auslander–Reiten duality*, which is at the heart of the notion of an *almost split sequence*. See [4] for details and references.

Amongst the many variations of the tensor-Hom adjunction is the following version of Eckmann–Shapiro 2.20.7 for Tate-Ext.

Proposition 4.13.11 *Let G be a finite group and H a subgroup of G. Let U be a kG-module, V a kH-module, and n an integer. We have natural isomorphisms*

$$\widehat{\mathrm{Ext}}^n_{kG}(\mathrm{Ind}^G_H(V), U) \cong \widehat{\mathrm{Ext}}^n_{kH}(V, \mathrm{Res}^G_H(U)),$$

$$\widehat{\mathrm{Ext}}^n_{kG}(U, \mathrm{Ind}^G_H(V)) \cong \widehat{\mathrm{Ext}}^n_{kH}(\mathrm{Res}^G_H(U), V),$$

Proof This follows from 2.15.4 applied to the appropriate Heller translates of U and V, together with the fact, from 2.14.6, that taking Heller translates commutes with restriction and induction as functors on stable categories. □

Corollary 4.13.12 *Let G be a finite group and b an idempotent in $Z(kG)$. Set $B = kGb$. For any integer n we have an isomorphism $\widehat{HH}^n(B) \cong \hat{H}^n(G; B)$, where G acts by conjugation on B.*

Proof As in the proof of 2.20.12, we have $\widehat{HH}^n(B) \cong \widehat{HH}^n(kG; B)$. By 2.4.5, we have $kG \cong \mathrm{Ind}^{G\times G}_{\Delta G}(k)$. Thus 4.13.11, applied with $G \times G$, and ΔG instead of G and H, respectively, yields the result as in the proof of 2.20.12. □

For the remainder of this section, we assume that $\mathcal{O}$ is a complete local principal ideal domain with residue field $k = \mathcal{O}/J(\mathcal{O})$ and field of fractions K. Let $\pi \in \mathcal{O}$ such that $J(\mathcal{O}) = \pi\mathcal{O}$. The next proposition generalises 4.5.10 (iii) in the case of relatively $\mathcal{O}$-injective algebras.

Proposition 4.13.13 *Let A be an $\mathcal{O}$-algebra that is finitely generated free as an $\mathcal{O}$-module such that $k \otimes_{\mathcal{O}} A$ is a selfinjective k-algebra. Let U be a finitely generated $\mathcal{O}$-free A-module. Suppose that $k \otimes_{\mathcal{O}} U$ has a direct summand Y that is projective as a $k \otimes_{\mathcal{O}} A$-module. Then U has a direct summand X that* *is projective as an A-module such that* $k \otimes_{\mathcal{O}} X = Y$*. In particular, if* $k \otimes_{\mathcal{O}} U$ *is projective as a* $k \otimes_{\mathcal{O}} A$*-module, then U is projective as an A-module.*

Proof Let X be a projective A-module satisfying $k \otimes_{\mathcal{O}} X \cong Y$. The canonical map $X \to k \otimes_{\mathcal{O}} X \cong Y \subseteq k \otimes_{\mathcal{O}} U$, viewed as an A-homomorphism, lifts through the canonical map $U \to k \otimes_{\mathcal{O}} U$, yielding an A-homomorphism $\varphi : X \to U$ such that the image of $\varphi(Y)$ in $k \otimes_{\mathcal{O}} U$ is Y. The $\mathcal{O}$-rank of X is equal to the k-dimension of Y, and $\varphi(X)$ is $\mathcal{O}$-free because it is a submodule of the $\mathcal{O}$-free module U. Thus φ is injective, and $\varphi(X)$ is an $\mathcal{O}$-pure submodule. But $\varphi(X) \cong X$ is projective, hence relatively $\mathcal{O}$-injective by 4.13.1, and therefore a direct summand of U as an A-module. Applying this to $Y = k \otimes_{\mathcal{O}} U$ yields the last statement (which follows also from 4.5.10). □

Corollary 4.13.14 *Let A be an* $\mathcal{O}$*-algebra that is finitely generated free as an* $\mathcal{O}$*-module such that* $k \otimes_{\mathcal{O}} A$ *is a selfinjective k-algebra. Let W be a finitely generated* $k \otimes_{\mathcal{O}} A$*-module and let Y be a finitely generated projective* $k \otimes_{\mathcal{O}} A$*-module. Let V be an* $\mathcal{O}$*-free A-module such that* $k \otimes_{\mathcal{O}} V \cong W \oplus Y$*. Then* $V \cong U \oplus X$ *for some* $\mathcal{O}$*-free A-module U satisfying* $k \otimes_{\mathcal{O}} U \cong W$ *and some projective A-module X satisfying* $k \otimes_{\mathcal{O}} X \cong Y$*. Moreover, the correspondence* $V \mapsto U$ *induces a bijection between the isomorphism classes of finitely generated* $\mathcal{O}$*-free A-modules that lift* $W \oplus Y$ *and W, respectively.*

Proof By 4.13.13, V has a direct summand isomorphic to a projective A-module X that satisfies $k \otimes_{\mathcal{O}} X \cong Y$. The Krull–Schmidt Theorem implies that a complement U of X in V satisfies $k \otimes_{\mathcal{O}} U \cong W$. The isomorphism class of X is uniquely determined by that of Y, and hence the isomorphism class of U is uniquely determined by that of V, which shows the last statement. □

Finite group algebras over $\mathcal{O}$ have the property that their extensions to K become semisimple. A consequence of this feature is that homomorphism spaces in the $\mathcal{O}$-stable category are torsion $\mathcal{O}$-modules. The following observation shows that in order to bound the $\mathcal{O}$-torsion of stable homomorphism spaces, it suffices to consider identity homomorphisms on modules.

Lemma 4.13.15 *Let A be an* $\mathcal{O}$*-algebra that is finitely generated as an* $\mathcal{O}$*-module. Suppose that* $\mathrm{char}(K) = 0$*. Let U, V be finitely generated A-modules and let d be a positive integer. If* π^d *annihilates the image of* Id_U *in* $\underline{\mathrm{End}}_A(U)$ *or if* π^d *annihilates the image of* Id_V *in* $\underline{\mathrm{End}}_A(V)$*, then* π^d *annihilates* $\underline{\mathrm{Hom}}_A(U, V)$*.*

Proof If π^d annihilates the image of Id_U in $\underline{\mathrm{End}}_A(U)$, then $\pi^d\mathrm{Id}_U$ factors through a relatively $\mathcal{O}$-projective module. Composing this with an arbitrary homomorphism $\varphi : U \to V$ shows that $\pi^d\varphi$ factors through a relatively

$\mathcal{O}$-projective A-module. A similar argument shows that this conclusion also holds if $\pi^d \mathrm{Id}_V$ factors through a relatively $\mathcal{O}$-projective module. □

Proposition 4.13.16 *Let A be an $\mathcal{O}$-algebra that is finitely generated free as an $\mathcal{O}$-module. Suppose that* $\mathrm{char}(K) = 0$ *and that $K \otimes_{\mathcal{O}} A$ is semisimple. Set $A^e = A \otimes_{\mathcal{O}} A^{\mathrm{op}}$. There is a positive integer d such that π^d annihilates $\underline{\mathrm{End}}_{A^e}(A)$, and then π^d annihilates $\underline{\mathrm{Hom}}_A(U, V)$ for any two finitely generated A-modules U, V.*

Proof We first show that $\underline{\mathrm{Hom}}_A(U, V)$ is a torsion $\mathcal{O}$-module, where U, V are finitely generated A-modules. Let $\varphi \in \mathrm{Hom}_A(U, V)$. Let $\tau : P \to V$ be a finitely generated relatively $\mathcal{O}$-projective cover of V. Then $\mathrm{Id}_K \otimes \tau : K \otimes_{\mathcal{O}} P \to K \otimes_{\mathcal{O}} V$ is split surjective, since $K \otimes_{\mathcal{O}} A$ is semisimple. Thus there is $\psi : K \otimes_{\mathcal{O}} U \to K \otimes_{\mathcal{O}} P$ such that $\mathrm{Id}_K \otimes \varphi = (\mathrm{Id}_K \otimes \tau) \circ \psi$. Since P, identified to its image $1 \otimes P$ in $K \otimes_{\mathcal{O}} P$, contains a K-basis of $K \otimes_{\mathcal{O}} P$, it follows that there is an integer d such that $\pi^d \mathrm{Im}(\psi) \subseteq P$. Denote by $\beta : U \to P$ the homomorphism obtained from restricting ψ to U. By the construction of β we have $\pi^d \varphi = \tau \circ \beta$, so $\pi^d \varphi$ factors through P and hence the image in $\underline{\mathrm{Hom}}_A(U, V)$ of $\pi^d \varphi$ is zero. This shows that $\underline{\mathrm{Hom}}_A(U, V)$ is annihilated by some power of π. Applied to A^e and $U = V = A$, this shows that there is a positive integer d such that π^d annihilates $\underline{\mathrm{End}}_{A^e}(A)$. In particular, $\pi^d \mathrm{Id}_A$ factors through the projective cover $A \otimes_{\mathcal{O}} A \to A$ of A as an A^e-module. Applying the identity functor $A \otimes_A -$ to an arbitrary A-homomorphism $\varphi : U \to V$ shows that $\pi^d \varphi$ factors through the relatively $\mathcal{O}$-projective cover $A \otimes_{\mathcal{O}} V \to V$, and hence π^d annihilates $\underline{\mathrm{Hom}}_A(U, V)$ for all finitely generated A-modules U and V. □

The smallest positive power of π that annihilates all homomorphism spaces in an $\mathcal{O}$-stable category is invariant under separable equivalences (cf. 2.6.15).

Proposition 4.13.17 *Let A and B be $\mathcal{O}$-algebras that are finitely generated free as $\mathcal{O}$-modules. Suppose that A and B are separably equivalent. Let d be a positive integer. If π^d annihilates $\underline{\mathrm{Hom}}_A(U, U')$ for all finitely generated A-modules U, U', then π^d annihilates $\underline{\mathrm{Hom}}_B(V, V')$ for any two finitely generated B-modules V, V'.*

Proof Let M be an A-B-bimodule and N a B-A-bimodule, both finitely generated projective as left and right modules, such that A is isomorphic to a direct summand of $M \otimes_B N$ and such that B is isomorphic to a direct summand of $N \otimes_A M$. Suppose that π^d annihilates $\underline{\mathrm{Hom}}_A(U, U')$ for all finitely generated A-modules U, U'. Let V be a finitely generated B-module. By 4.13.15 it suffices to show that π^d annihilates the image of Id_V in $\underline{\mathrm{End}}_B(V)$. By the assumptions, π^d annihilates the image of $\mathrm{Id}_{M\otimes_B V}$ in $\underline{\mathrm{End}}_A(N \otimes_B V)$. Applying the functor $M \otimes_A -$, which preserves projective modules, implies that π^d annihilates the image of the identity map in $\underline{\mathrm{End}}_A(M \otimes_B N \otimes_A V)$. Since A is isomorphic to a direct summand of $M \otimes_B N$, it follows that V is isomorphic to a direct summand of $M \otimes_B N \otimes_A V$, and hence π^d annihilates the image of Id_V in $\underline{\mathrm{End}}_B(V)$. □

In the case of finite group algebras, one can be more precise.

Lemma 4.13.18 ([55, Lemma 3.7]) *Let G be a finite group and V a finitely generated $\mathcal{O}$-free $\mathcal{O}G$-module. Suppose that V has no nonzero projective direct summand. We have $V_1^G \subseteq J(\mathcal{O})V^G$.*

Proof It follows from 4.13.13 that the kG-module $\bar{V} = k \otimes_{\mathcal{O}} V$ has no nonzero projective direct summand. Since $\sum_{x\in G} x$ is contained in $\mathrm{soc}(kG)$, it follows from 4.11.7 that $\sum_{x\in G} x$ annihilates $\bar{V}$, whence the result. □

Proposition 4.13.19 ([55, Proposition 3.8]) *Let G be a finite group and U a finitely generated $\mathcal{O}$-free $\mathcal{O}G$-module. Suppose that* $\mathrm{char}(K) = 0$. *We have $\underline{\mathrm{End}}_{\mathcal{O}G}(U) \cong \mathcal{O}/|G|\mathcal{O}$ if and only if $U^* \otimes_{\mathcal{O}} U \cong \mathcal{O} \oplus V \oplus X$ for some projective $\mathcal{O}G$-module X and some $\mathcal{O}G$-module V that has no nonzero projective direct summand and that satisfies $V^G = \{0\}$.*

Proof Suppose that $\underline{\mathrm{End}}_{\mathcal{O}G}(U) \cong \mathcal{O}/|G|\mathcal{O}$. Write $U^* \otimes_{\mathcal{O}} U = T \oplus X$, where X is a projective $\mathcal{O}G$-module and where T is an $\mathcal{O}G$-module that does not have a nonzero projective summand. Then $X^G = X_1^G$. Thus, by 2.13.11 and the assumptions, we have $\mathcal{O}/|G|\mathcal{O} \cong \underline{\mathrm{End}}_{\mathcal{O}G}(U) \cong T^G/T_1^G$. By 4.13.18, we have $T_1^G \subseteq J(\mathcal{O})T^G$, and hence $T^G \cong \mathcal{O}$. Let V be the annihilator of $\sum_{x\in G} x$ in T. Then V is an $\mathcal{O}$-pure submodule of T, and $T^G \cap V = \{0\}$. Thus $K \otimes_{\mathcal{O}} T = (K \otimes_{\mathcal{O}} T^G) \oplus (K \otimes_{\mathcal{O}} V)$. Identify T with its canonical image in $K \otimes_{\mathcal{O}} T$, and let t be a nonzero element in T^G such that $T^G = \mathcal{O}t$. Let $u \in T$. By the above, $u = \mu t + v$ for some unique $\mu \in K$ and $v \in K \otimes_{\mathcal{O}} V$. We have

$$\Big(\sum_{x\in G} x\Big)(u) = \Big(\sum_{x\in G} x\Big)(\mu t + v) = \mu|G|t$$

and this is an element in T^G. Thus $v = u - \mu|G|t$ belongs to $K \otimes_{\mathcal{O}} V \cap T = V$, where we use that V is $\mathcal{O}$-pure. This shows that $T = \mathcal{O} \oplus V$, which proves one implication. The converse is immediate. □

Corollary 4.13.20 *Let G be a finite group of order divisible by p. Suppose that* $\mathrm{char}(K) = 0$. *Let d be the positive integer such that π^d is equal to the* *p-part of $|G|$. Then d is the smallest positive integer such that π^d annihilates $\underline{\mathrm{Hom}}_{\mathcal{O}G}(U, V)$ for all finitely generated $\mathcal{O}G$-modules U and V.*

Proof Let U, V be finitely generated $\mathcal{O}G$-modules. By 4.13.19 we have $\underline{\mathrm{End}}_{\mathcal{O}G}(\mathcal{O}) \cong \mathcal{O}/|G|\mathcal{O} \cong \mathcal{O}/\pi^d\mathcal{O}$. In particular, $\pi^d\mathrm{Id}_{\mathcal{O}}$ factors through a projective $\mathcal{O}G$-module. Tensoring with U shows that $\pi^d\mathrm{Id}_U$ factors through a relatively $\mathcal{O}$-projective $\mathcal{O}G$-module. Thus, if $\varphi \in \mathrm{Hom}_{\mathcal{O}G}(U, V)$, then $\pi^d\varphi = \varphi \circ (\pi^d\mathrm{Id}_U)$ factors through a relatively $\mathcal{O}$-projective module, and hence π^d annihilates $\underline{\mathrm{Hom}}_{\mathcal{O}G}(U, V)$. □

Corollary 4.13.21 *Let P be a finite p-group. Suppose that* $\mathrm{char}(K) = 0$. *Then $|P|$ is the smallest positive integer that annihilates $\underline{\mathrm{Hom}}_{\mathcal{O}G}(U, V)$ for all finitely generated $\mathcal{O}P$-modules U and V.*

Proof This is an obvious consequence of 4.13.20. □

Corollary 4.13.22 *Let G be a finite group of order divisible by p, and let P be a p-subgroup of G. Suppose that* $\mathrm{char}(K) = 0$. *Let U, V be finitely generated $\mathcal{O}G$-modules. Supppose that U or V is relatively P-projective. Then $|P| \cdot \underline{\mathrm{Hom}}_{\mathcal{O}G}(U, V) = \{0\}$.*

Proof Suppose that U is relatively P-projective; that is, there is a finitely generated $\mathcal{O}P$-module W such that U is isomorphic to a direct summand of $\mathrm{Ind}_P^G(W)$. Thus $\underline{\mathrm{Hom}}_{\mathcal{O}G}(U, V)$ is a direct summand of $\underline{\mathrm{Hom}}_{\mathcal{O}G}(\mathrm{Ind}_P^G(W), V)$. By Frobenius' reciprocity, this is isomorphic to $\underline{\mathrm{Hom}}_{\mathcal{O}P}(W, \mathrm{Res}_P^G(V))$, hence annihilated by $|P|$. A similar argument shows the result if V is relatively P-projective. □

We conclude this section with a proof of the Auslander–Reiten conjecture for finite p-group algebras.

Theorem 4.13.23 ([55, Theorem 3.4]) *Suppose that k is a field of prime characteristic p. Let P be a nontrivial finite p-group and let A be a split finite-dimensional k-algebra such that there is a k-linear equivalence $\underline{\mathrm{mod}}(A) \cong \underline{\mathrm{mod}}(kP)$. Then A has a unique isomorphism class of nonprojective simple modules.*

Proof Since P is nontrivial, so is the category $\underline{\mathrm{mod}}(kP)$, and hence A has at least one isomorphism class of nonprojective simple modules. Arguing by contradiction, suppose that A has two nonisomorphic nonprojective simple modules S and T. Let U and V be kP-modules isomorphic to the images of S and T under an equivalence $\underline{\mathrm{mod}}(A) \cong \underline{\mathrm{mod}}(kP)$. Then $\underline{\mathrm{End}}_{kP}(U) \cong \underline{\mathrm{End}}_{kP}(V) \cong k$ and $\underline{\mathrm{Hom}}_{kP}(U, V) = \{0\} = \underline{\mathrm{Hom}}_{kP}(V, U)$. Thus, using 2.13.11, we have $U^* \otimes_k U \cong L \oplus X$ for some projective kP-module X and a kP-module L that has no nonzero projective direct summand and satisfies $L^P \cong k$. Similarly, $V^* \otimes_k V \cong M \oplus Y$ for some projective kP-module Y and a kP-module M that has no nonzero projective direct summand and satisfies $M^P \cong k$. Since L^P and M^P are the socles of L and M, respectively, and since kP is selfinjective, it follows that L and M are isomorphic to proper submodules of kP. In particular, $|P|$ does not divide the dimensions of either L or M, and hence $|P|$ does not divide the dimensions of either $U^* \otimes_k U$ or $V^* \otimes_k V$. Again by 2.13.11, the kP-modules $U^* \otimes_k V$ and $V^* \otimes_k U$ are projective, and hence their dimensions are divisible by $|P|$. Thus the dimension of $U^* \otimes_k U \otimes_k V^* \otimes_k V$ is divisible by $|P|^2$. But then at least one of $U^* \otimes_k U$ or $V^* \otimes_k V$ has dimension divisible by $|P|$. This contradiction proves the result. □

4.14 Stable equivalences of Morita type between $\mathcal{O}$-algebras

Let $\mathcal{O}$ be a complete local principal ideal domain with residue field k. We develop general properties of stable equivalences of Morita type between $\mathcal{O}$-algebras, with a focus on algebras that are $\mathcal{O}$-free of finite rank and whose reductions modulo $J(\mathcal{O})$ are selfinjective k-algebras; this includes symmetric $\mathcal{O}$-algebras.

Proposition 4.14.1 *Let A and B be $\mathcal{O}$-algebras that are finitely generated as $\mathcal{O}$-modules. Suppose that all simple left and right $k \otimes_{\mathcal{O}} A$-modules and that all simple left and right $k \otimes_{\mathcal{O}} B$-modules are nonprojective. Let M be an A-B-bimodule and N a B-A-bimodule inducing a stable equivalence of Morita type. Then M is a progenerator as a left A-module and as a right B-module, and N is a progenerator as a left B-module and as a right A-module.*

Proof One can prove this by observing that this stable equivalence of Morita type is a special case of a separable equivalence, as pointed out in 2.17.2, and then use 2.6.16. One can show this also directly. Let j be a primitive idempotent in B, and set $T = Bj/J(B)j$. Then T is a simple $k \otimes_{\mathcal{O}} B$-module, hence nonprojective as a $k \otimes_{\mathcal{O}} B$-module by the assumptions. Thus $M \otimes_B T$ is not projective as a $k \otimes_{\mathcal{O}} A$-module; in particular, $M \otimes_B T$ is nonzero. Up to isomorphism, T is the unique simple $k \otimes_{\mathcal{O}} B$-module such that $jT \neq \{0\}$. Note that $jT \cong jB \otimes_B T$. Since M is projective as a right B-module, it follows that jB is isomorphic to a direct summand of M as a right B-module. Thus every projective indecomposable B-module is isomorphic to a direct summand of M as a right B-module, and hence M is a progenerator as a right B-module. The rest follows similarly. □

Simple right $k \otimes_{\mathcal{O}} A$-modules are duals of simple left $k \otimes_{\mathcal{O}} A$-modules, and finitely generated injective modules are k-duals of finitely generated projective modules (cf. 2.9.2). Thus if $k \otimes_{\mathcal{O}} A$ is selfinjective and all simple left $k \otimes_{\mathcal{O}} A$-modules are nonprojective, then also all simple right $k \otimes_{\mathcal{O}} A$-modules are nonprojective. Stable equivalences of Morita type between indecomposable $\mathcal{O}$-algebras are essentially induced by indecomposable bimodules.

Theorem 4.14.2 ([55, 2.4]) *Let A, B be indecomposable $\mathcal{O}$-algebras that are finitely generated free as $\mathcal{O}$-modules. Suppose that A is not projective as an $A \otimes_{\mathcal{O}} A^{\mathrm{op}}$-module and that B is not projective as a $B \otimes_{\mathcal{O}} B^{\mathrm{op}}$-module. Let M be an A-B-bimodule and N a B-A-bimodule inducing a stable equivalence of Morita type between A and B. Then $M \cong M' \oplus M''$ for some indecomposable nonprojective $A \otimes_{\mathcal{O}} B^{\mathrm{op}}$-module M' and some projective $A \otimes_{\mathcal{O}} B^{\mathrm{op}}$-module M''. Moreover, if A and B are relatively $\mathcal{O}$-injective, then $k \otimes_{\mathcal{O}} M'$ remains indecomposable as a $k \otimes_{\mathcal{O}} (A \otimes_{\mathcal{O}} B^{\mathrm{op}})$-module.*

Proof Write $M \otimes_B N \cong A \oplus X$ and $N \otimes_A M \cong B \oplus Y$ for suitable projective bimodules X and Y. Since A is indecomposable as an A-A-bimodule, the Krull–Schmidt Theorem implies that there is an indecomposable direct summand M' of M such that A is isomorphic to a direct summand of $M' \otimes_B N$. Let M'' be a complement of M' in M; that is, $M = M' \oplus M''$. Since any summand of $M \otimes_B N$ other than A is projective, it follows that the bimodule $M'' \otimes_B N$ is projective. But then $M'' \otimes_B N \otimes_A M \cong M'' \oplus (M'' \otimes_B Y)$ is projective as an $A \otimes_{\mathcal{O}} B^{\mathrm{op}}$-module, which implies that M'' is projective, whence the statement on the structure of M. The same argument applied to $k \otimes_{\mathcal{O}} M'$ shows that $k \otimes_{\mathcal{O}} M'$ is a direct sum of an indecomposable nonprojective bimodule and a projective bimodule. It follows from 4.13.13 that $k \otimes_{\mathcal{O}} M'$ remains indecomposable. □

Theorem 4.14.2 implies in particular that when it comes to stable equivalences of Morita type between block algebras, it is no loss of generality to assume that the bimodules inducing a stable equivalence of Morita type are indecomposable. In order to detect whether a bimodule between $\mathcal{O}$-free $\mathcal{O}$-algebras induces a Morita equivalence or a stable equivalence of Morita type, it suffices to verify this at the level of k-algebras. To see this, we first describe the kernel of the canonical group homomorphism $\mathrm{Pic}(A) \to \mathrm{Pic}(k \otimes_{\mathcal{O}} A)$.

Lemma 4.14.3 *Let A be an $\mathcal{O}$-algebra that is free of finite rank as an $\mathcal{O}$-module. Let M be an $\mathcal{O}$-free A-A-bimodule. The following are equivalent.*

(i) We have an isomorphism of $k \otimes_{\mathcal{O}} A$-$k \otimes_{\mathcal{O}} A$-bimodules $k \otimes_{\mathcal{O}} M \cong k \otimes_{\mathcal{O}} A$.

(ii) We have an isomorphism of A-A-bimodules $M \cong A_\alpha$ for some automorphism α of A which induces the identity on $k \otimes_{\mathcal{O}} A$.

Proof Suppose that (i) holds. Since projective $k \otimes_{\mathcal{O}} A$-modules lift uniquely, up to isomorphism, to projective A-modules, it follows that M is isomorphic to A as a left A-module and as a right A-module. Morita's Theorem implies that $M \otimes_A -$ is an equivalence. It follows from 2.8.16 (v) that $M \cong A_\alpha$ for some $\alpha \in \mathrm{Aut}(A)$. Since $k \otimes_{\mathcal{O}} M \cong k \otimes_{\mathcal{O}} A$, it follows from 2.8.16 that α induces an inner automorphism of $k \otimes_{\mathcal{O}} A$. Any invertible element in $k \otimes_{\mathcal{O}} A$ lifts to an invertible element in A, and hence we may modify α by an inner automorphism in such a way that α induces the identity map on $k \otimes_{\mathcal{O}} A$. Thus (i) implies (ii), and the converse is trivial. □

Proposition 4.14.4 *Let A, B be $\mathcal{O}$-algebras that are free of finite rank as $\mathcal{O}$-modules, such that the k-algebras $k \otimes_{\mathcal{O}} A$ and $k \otimes_{\mathcal{O}} B$ are indecomposable nonsimple selfinjective with separable semisimple quotients. Let M be an A-B-bimodule and N a B-A-bimodule. The following are equivalent.*

(i) The bimodules $k \otimes_{\mathcal{O}} M$ and $k \otimes_{\mathcal{O}} N$ induce a stable equivalence of Morita type between $k \otimes_{\mathcal{O}} A$ and $k \otimes_{\mathcal{O}} B$.
(ii) There is an automorphism α of A that induces the identity on $k \otimes_{\mathcal{O}} A$ such that the bimodules M and N_α induce a stable equivalence of Morita type between A and B.

Proof Suppose that (i) holds. Then $(k \otimes_{\mathcal{O}} M) \otimes_{(k \otimes_{\mathcal{O}} B)} (k \otimes_{\mathcal{O}} N) \cong (k \otimes_{\mathcal{O}} A) \oplus \bar{X}$ for some projective $(k \otimes_{\mathcal{O}} A)$-$(k \otimes_{\mathcal{O}} A)$-bimodule $\bar{X}$. It follows from 4.14.3 and 4.13.13 that $M \otimes_B N \cong A_{\alpha^{-1}} \oplus X'$ for some projective A-A-bimodule X' and some automorphism α of A that induces the identity on $k \otimes_{\mathcal{O}} A$. Tensoring with $- \otimes_A A_\alpha$ yields an isomorphism $M \otimes_B N_\alpha \cong A \oplus X$ for some projective A-A-bimodule X. The same argument with N_α and M instead of M and N yields an isomorphism $N_\alpha \otimes_A M_\beta \cong B \oplus Y$ for some projective B-B-bimodule and some automorphism β that induces the identity on $k \otimes_{\mathcal{O}} B$. Tensoring this isomorphism on the left by $M \otimes_B -$ yields an isomorphism

$$(M \otimes_B N_\alpha) \otimes_A M_\beta \cong M \oplus M \otimes_B Y.$$

Since A is the unique nonprojective summand of $M \otimes_B N_\alpha$, it follows that M_β is the unique nonprojective summand, up to isomorphism, on the left side of this isomorphism. But then comparing this with the right side yields an isomorphism $M_\beta \cong M$, and hence $N_\alpha \otimes_A M \cong B \oplus Y$. Thus (i) implies (ii). The converse is trivial. □

In view of later applications, we formulate this in a slightly different way for symmetric algebras, with a different proof.

Proposition 4.14.5 *Let A, B be symmetric $\mathcal{O}$-algebras. Let M be an A-B-bimodule that is finitely generated projective as a left and right module. Then M and its $\mathcal{O}$-dual M^* induce a stable equivalence of Morita type (resp. Morita equivalence) between A and B if and only if $k \otimes_{\mathcal{O}} M$ and its k-dual $(k \otimes_{\mathcal{O}} M)^*$ induce a stable equivalence of Morita type (resp. Morita equivalence) between $k \otimes_{\mathcal{O}} A$ and $k \otimes_{\mathcal{O}} B$.*

Proof Set $\bar{A} = k \otimes_{\mathcal{O}} A$, $\bar{B} = k \otimes_{\mathcal{O}} B$, and $\bar{M} = k \otimes_{\mathcal{O}} M$. Suppose that $\bar{M}$ and its k-dual $\bar{M}^*$ induce a stable equivalence of Morita type. That is, after possibly adding a suitable projective summand to M (which may be needed as we have not excluded that A may have a nonzero projective summand as an A-A-bimodule), the adjunction map $\bar{M} \otimes_{\bar{B}} \bar{M}^* \to \bar{A}$ is surjective with a projective kernel $\bar{X}$. Nakayama's Lemma implies that the adjunction map $M \otimes_B M^* \to A$ is surjective. Thus its kernel X satisfies $\bar{X} \cong k \otimes_{\mathcal{O}} X$, and hence X is a projective A-A-bimodule, hence relatively $\mathcal{O}$-injective, and thus $M \otimes_B M^* \cong A \oplus X$. Using the same argument with the roles of A and B exchanged implies that M and M^* induce a stable equivalence of Morita type. The converse is trivial. Since $\bar{X}$ is zero if and only if X is zero, it follows that $\bar{M}$ and $\bar{M}^*$ induce a Morita equivalence if and only if M and M^* induce a Morita equivalence. □

It is not true in general that if an A-module U is indecomposable, then so is $k \otimes_{\mathcal{O}} U$. Under suitable hypotheses, the property of having an indecomposable reduction modulo $J(\mathcal{O})$ is invariant under stable equivalences of Morita type. Note that a nonprojective $\mathcal{O}$-free module is not relatively $\mathcal{O}$-projective.

Proposition 4.14.6 *Let A, B be $\mathcal{O}$-algebras that are finitely generated free as $\mathcal{O}$-modules such that $k \otimes_{\mathcal{O}} A$ and $k \otimes_{\mathcal{O}} B$ are selfinjective k-algebras. Let M be an A-B-bimodule and N a B-A-bimodule inducing a stable equivalence of Morita type between A and B. Let V be an $\mathcal{O}$-free finitely generated indecomposable nonprojective B-module. Let U be the up to isomorphism unique indecomposable nonprojective direct summand of $M \otimes_B V$. Then $k \otimes_{\mathcal{O}} U$ is indecomposable if and only if $k \otimes_{\mathcal{O}} V$ is indecomposable.*

Proof Since $M \otimes_B -$ induces a stable equivalence, it follows from 4.13.2 that $M \otimes_B V = U \oplus U'$ for some indecomposable nonprojective A-module U and a projective A-module U'. Suppose that $k \otimes_{\mathcal{O}} V$ is indecomposable. Then $\bar{V} = k \otimes_{\mathcal{O}} V$ is nonprojective, as otherwise the last statement in 4.13.13 would imply that V is projective since it is $\mathcal{O}$-free. Thus $M \otimes_B \bar{V}$ has a unique nonprojective indecomposable summand as a $k \otimes_{\mathcal{O}} A$-module. Since by 4.13.13 any nonzero projective summand of $\bar{U} = k \otimes_{\mathcal{O}} U$ as a $k \otimes_{\mathcal{O}} A$-module lifts to a nonzero projective summand of U as an A-module, it follows that $k \otimes_{\mathcal{O}} U$ is indecomposable. Exchanging the roles of A and B concludes the proof. □

The property of a module defined over k to be 'liftable' to an $\mathcal{O}$-free module is invariant under stable equivalences of Morita type, and then such lifts correspond to each other through the stable equivalence of Morita type.

Proposition 4.14.7 *Let A, B be $\mathcal{O}$-algebras that are finitely generated free as $\mathcal{O}$-modules such that $k \otimes_{\mathcal{O}} A$ and $k \otimes_{\mathcal{O}} B$ are selfinjective k-algebras. Let M be an A-B-bimodule and N a B-A-bimodule inducing a stable equivalence of Morita type between A and B. Let U be a finite-dimensional $k \otimes_{\mathcal{O}} A$-module and V a finite-dimensional $k \otimes_{\mathcal{O}} B$-module such that $M \otimes_B V \cong U \oplus P$ for some projective $k \otimes_{\mathcal{O}} A$-module P. If $\hat{V}$ is an $\mathcal{O}$-free B-module satisfying $k \otimes_{\mathcal{O}} \hat{V} \cong V$, then $M \otimes_B \hat{V}$ has, up to isomorphism, a unique summand $\hat{U}$ satisfying $k \otimes_{\mathcal{O}} \hat{U} \cong U$. The correspondence $\hat{V} \mapsto \hat{U}$ induces a bijection between the isomorphism classes of B-modules and A-modules that lift V and U, respectively.*

Proof This is an immediate consequence of 4.13.14. □

Proposition 4.14.8 (cf. [55, 2.3]) *Let A, B be finite-dimensional selfinjective k-algebras. Suppose that $A/J(A)$ and $B/J(B)$ are separable. Let M be a finitely generated A-B-bimodule. If $M\mathrm{soc}(B)$ has a nonzero projective direct summand as an A-module, then M has a nonzero projective direct summand as an $A \otimes_k B^{\mathrm{op}}$-module.*

Proof The algebra $A \otimes_k B^{\mathrm{op}}$ is again selfinjective and $\mathrm{soc}(A \otimes_k B^{\mathrm{op}}) = \mathrm{soc}(A) \otimes_k \mathrm{soc}(B^{\mathrm{op}})$ by the separability hypothesis. If $M\mathrm{soc}(B)$ has a nonzero projective direct summand as an A-module, then $\mathrm{soc}(A)M\mathrm{soc}(B) \neq \{0\}$, and hence M has a nonzero projective direct summand by 4.11.7 (iii). □

Proposition 4.14.9 (cf. [55, 2.3]) *Let A, B be finite-dimensional selfinjective k-algebras. Suppose that $A/J(A)$ and $B/J(B)$ are separable. Let M be an A-B-bimodule such that M has no nonzero projective direct summand as an A-B-bimodule and such that M is finitely generated projective as a right B-module. For any simple nonprojective B-module T the A-module $M \otimes_B T$ has no nonzero projective direct summand.*

Proof Since M has no nonzero projective summand as an A-B-bimodule, it follows from 4.14.8 that $M\mathrm{soc}(B)$ has no nonzero projective direct summand as an A-module. As M is projective as a right B-module we have $M\mathrm{soc}(B) \cong M \otimes_B \mathrm{soc}(B)$. In particular, $M \otimes_B \mathrm{soc}(B)$ has no nonzero projective direct summand. Since B is selfinjective, every simple B-module is isomorphic to a direct summand of $\mathrm{soc}(B)$, whence the result. □

Theorem 4.14.10 (cf. [55, 2.5]) *Let A, B be $\mathcal{O}$-algebras that are free of finite rank as $\mathcal{O}$-modules, such that $k \otimes_{\mathcal{O}} A$ and $k \otimes_{\mathcal{O}} B$ are indecomposable non-simple selfinjective with separable semisimple quotients. Let M be an A-B-bimodule and N a B-A-bimodule such that M and N induce a stable equivalence of Morita type between A and B. Suppose that the bimodules M and N are indecomposable. The following are equivalent.*

(i) *The bimodules M and N induce a Morita equivalence.*
(ii) *For any simple B-module T, the A-module $M \otimes_B T$ is simple.*

Proof The implication (i) $\Rightarrow$ (ii) is trivial. Suppose that (ii) holds. By 4.14.4 we may assume that $\mathcal{O} = k$. Write $M \otimes_B N \cong A \oplus X$ for some projective A-A-bimodule X and $N \otimes_A M \cong B \oplus Y$ for some projective B-B-bimodule Y. We need to show that X and Y are zero. It suffices to show that Y is zero; indeed, if $N \otimes_A M \cong B$, then $A \oplus X \cong M \otimes_B N \cong M \otimes_B (N \otimes_A M) \otimes_B N \cong (M \otimes_B N) \otimes_A (M \otimes_B N) \cong (A \oplus X) \otimes_A (A \oplus X) \cong A \oplus X \oplus X \oplus X \otimes_A X$, which forces $X = \{0\}$ by the Krull–Schmidt Theorem. Arguing by contradiction, assume that Y is nonzero. Let S be a simple A-module. By the assumptions, S is not projective, and hence $N \otimes_A S$ is not projective because $N \otimes_A -$ induces a stable equivalence. Let T be a simple B-submodule of a nonprojective indecomposable summand of $N \otimes_A S$. Then the inclusion $T \to N \otimes_A S$ does not factor through a projective B-module. Since $M \otimes_B -$ induces a stable equivalence, it follows that the corresponding map $M \otimes_B T \to S$ does not factor through a projective A-module. Thus $M \otimes_B T \cong S$, as $M \otimes_B T$ is simple by the assumption (ii). This shows that every simple A-module is isomorphic to $M \otimes_B T$ for some simple B-module T. It follows from 4.14.9 applied to N and a simple A-module S that $N \otimes_A S$ has no nonzero projective direct summand, and hence that $N \otimes_A S$ is indecomposable. Since there is a simple B-module T such that $S \cong M \otimes_B T$, it follows that $N \otimes_A S \cong N \otimes_A M \otimes_B T \cong T \oplus Y \otimes_B T$. The indecomposability of $N \otimes_A S$ implies that $N \otimes_A S \cong T$ is simple, and that $Y \otimes_B T = \{0\}$. The second condition, which holds for all simple B-modules T, implies that Y is zero. Thus (ii) implies (i). □

Remark 4.14.11 This theorem says that 'up to self Morita equivalences' a stable equivalence of Morita type between two indecomposable symmetric $\mathcal{O}$-algebras is determined by the images of the simple modules. Indeed, if M and N are two indecomposable nonprojective A-B-bimodules both inducing (with their duals) stable equivalences of Morita type between A and B in such a way that $M \otimes_B T \cong N \otimes_B T$ for any simple B-module T, then the B-B-bimodule $N^* \otimes_A M$ induces a stable equivalence on $\underline{\text{mod}}(B)$ preserving the simple B-modules, and hence the unique indecomposable nonprojective direct summand of $N^* \otimes_A M$ induces a self Morita equivalence on mod(B). This has led Okuyama to formulate a technique, now known as *Okuyama's method*, to construct derived equivalences from stable equivalences of Morita type. See the introduction of [80] for a brief description of Okuyama's method and references to further applications.

In many cases, a stable equivalence of Morita type can be played back to the restriction and induction functors between an algebra A and a subalgebra B. In what follows we use the subscript B to indicate the restriction to B of an A-module structure.

Proposition 4.14.12 *Let A be an $\mathcal{O}$-algebra that is finitely generated free as an $\mathcal{O}$-module. Let B be a subalgebra of A. The following are equivalent.*

(i) The algebra A is relatively B-separable and ${}_BA_B = B \oplus Y$ for some projective B-B-bimodule Y.
(ii) The bimodules A_B and ${}_BA$ induce a stable equivalence of Morita type between A and B.

Proof Suppose that (i) holds. We have ${}_BA \otimes_A A_B \cong {}_BA_B \cong B \oplus Y$ with Y projective by the assumptions. Since A is relatively B-separable, we have $A \otimes_B A \cong A \oplus X$ for some A-A-bimodule X. In order to prove (ii) it suffices to show that X is projective. Since A is relatively B-separable, it follows that X is a direct summand of $A \otimes_B X \otimes_B A$, and hence it suffices to show that ${}_BX_B$ is a projective B-B-bimodule. We have ${}_BA \otimes_B A_B \cong (B \oplus Y) \otimes_B (B \oplus Y) \cong B \oplus Y \oplus Y \oplus Y \otimes_B Y$. Since also ${}_BA \otimes_B A_B \cong {}_BA_B \oplus {}_BX_B \cong B \oplus Y \oplus {}_BX_B$, it follows from the Krull–Schmidt Theorem that ${}_BX_B \cong Y \oplus Y \otimes_B Y$, which is projective as Y is so. Thus (i) implies (ii), and the converse is obvious. □

It is shown in [27, 5.1] that a stable equivalence of Morita type between finite-dimensional algebras with separable semisimple quotients can always be played back to a stable equivalence that is induced by induction and restriction (by replacing one of the algebras by a suitable Morita equivalent algebra). Stable equivalences of Morita type preserve the absolute value of the determinant of the Cartan matrix. For A a finite-dimensional k-algebra, we denote by $R(A)$ the Grothendieck group of finite-dimensional A-modules, and by Pr(A) the subgroup of $R_k(A)$ generated by the images of the finite-dimensional projective A-modules; see 1.8.7 and 4.10.6.

Proposition 4.14.13 *Let A, B be finite-dimensional k-algebras with Cartan matrices C_A, C_B, respectively. Let M be an A-B-bimodule and N a B-A-bimodule inducing a stable equivalence of Morita type. Then $M \otimes_B -$ and*

$N \otimes_A -$ induce inverse group isomorphisms

$$R(A)/\mathrm{Pr}(A) \cong R(B)/\mathrm{Pr}(B);$$

in particular, if one of C_A, C_B is nonsingular, then so is the other, and we have $|\det(C_A)| = |\det(C_B)|$.

Proof The functor $M \otimes_B -$ from $\mathrm{mod}(B)$ to $\mathrm{mod}(A)$ is exact and preserves projectives, hence induces a group homomorphism $R(B) \to R(A)$, mapping $\mathrm{Pr}(B)$ to $\mathrm{Pr}(A)$. A similar statement holds for $N \otimes_A -$. Thus $M \otimes_B N \otimes_A -$ induces a group endomorphism of $R(A)$ preserving $\mathrm{Pr}(A)$. By the assumptions we have $M \otimes_B N \cong A \oplus X$ for some projective $A \otimes_k A^{\mathrm{op}}$-mdoule. The projectivity of X as a bimodule implies that $X \otimes_A -$ sends $R(A)$ to $\mathrm{Pr}(A)$, hence induces the zero map on $R(A)/\mathrm{Pr}(A)$. Thus the endomorphism induced by $M \otimes_B N \otimes_A -$ on $R(A)/\mathrm{Pr}(A)$ is equal to that induced by $A \otimes_A -$, hence the identity. The same argument with reversed roles shows that $N \otimes_A M \otimes_B -$ induces the identity on $R(B)/\mathrm{Pr}(B)$, whence the isomorphism $R(A)/\mathrm{Pr}(A) \cong R(B)/\mathrm{Pr}(B)$. The last statement follows from 4.10.6. □

Proposition 4.14.12 admits a generalisation for indecomposable symmetric $\mathcal{O}$-algebras. This is based on an argument due to Rickard, in the proof of [79, Theorem 2.1], playing off general properties of adjunction maps and the Krull–Schmidt Theorem.

Theorem 4.14.14 *Let A and B be indecomposable symmetric $\mathcal{O}$-algebras such that $k \otimes_{\mathcal{O}} A$ is not simple. Let M be an indecomposable A-B-bimodule such that M is finitely generated projective as a left A-module and as a right B-module. Suppose that $M \otimes_B M^* \cong A \oplus P$ for some projective $A \otimes_k A^{\mathrm{op}}$-module P. Then M and M^* induce a stable equivalence of Morita type between A and B.*

Proof Since A is nonprojective as an $A \otimes_{\mathcal{O}} A^{\mathrm{op}}$-module and isomorphic to a direct summand of $M \otimes_B M^*$, it follows that M is not projective as an $A \otimes_{\mathcal{O}} B^{\mathrm{op}}$-module. Denote by $\epsilon : B \to M^* \otimes_A M$ the adjunction unit of $M \otimes_B -$ being left adjoint to $M \otimes_A -$, as in Equation (2.12.1). Denote by $\eta : M^* \otimes_A M \to B$ the adjunction counit of $M^* \otimes_A -$ being left adjoint to $M^* \otimes_B -$, as in Equation (2.12.4). The main theorem on adjoint functors 2.3.5 implies that the map $\mathrm{Id}_X \otimes \epsilon : M \to M \otimes_B M^* \otimes_A M$ is split injective. Since $M \otimes_B M^*$ is isomorphic to A in the $\mathcal{O}$-stable category of A-A-bimodules, it follows that $\mathrm{Id}_M \otimes \epsilon$ is an isomorphism in the $\mathcal{O}$-stable category of A-B-bimodules. Similarly, $\mathrm{Id}_X \otimes \eta : M \otimes_B M^* \otimes_A M \to M$ is split surjective, hence an isomorphism, in the $\mathcal{O}$-stable category of A-B-bimodules. Thus $\mathrm{Id}_X \otimes (\eta \circ \epsilon)$ is an automorphism of M in the $\mathcal{O}$-stable bimodule category, hence also in the category of A-B-bimodules, as M is indecomposable nonprojective. In particular,

$\eta \circ \epsilon$ is not contained in the radical of $\mathrm{End}_{B\otimes_{\mathcal{O}}B^{\mathrm{op}}}(B)$. As B is indecomposable as a bimodule, this implies that $\eta \circ \epsilon$ is an automorphism of B. Thus ϵ is split injective and η is split surjective. In particular, we have $M^* \otimes_A M \cong B \oplus Q$ for some $B \otimes_{\mathcal{O}} B^{\mathrm{op}}$-module Q. We calculate the bimodule $M^* \otimes_A M \otimes_B M^* \otimes_A M$ in two ways. We have $(M^* \otimes_A M) \otimes_B (M^* \otimes_A M) \cong (B \oplus Q) \otimes_B (B \oplus Q) \cong B \oplus Q \oplus Q \oplus Q \otimes_B Q$. In particular, this bimodule has two copies of Q as a direct summand. We have $M^* \otimes_A (M \otimes_B M^*) \otimes_A M \cong M^* \otimes_A (A \oplus P) \otimes_A M \cong B \oplus Q \oplus M^* \otimes_A P \otimes_A M$. The Krull–Schmidt Theorem forces Q to be a direct summand of the projective summand $M^* \otimes_A P \otimes_A M$, hence Q is projective. This concludes the proof. □

Corollary 4.14.15 *Let A be a symmetric $\mathcal{O}$-algebra and let B be a symmetric $\mathcal{O}$-subalgebra of A. Suppose that $k \otimes_{\mathcal{O}} A$ and $k \otimes_{\mathcal{O}} B$ are indecomposable non-simple. If ${}_BA_B = B \oplus Y$ for some projective $B \otimes_{\mathcal{O}} B^{\mathrm{op}}$-subbimodule Y of ${}_BA_B$, then A_B and ${}_BA$ induce a stable equivalence of Morita type between A and B.*

Proof Set $M = {}_BA$. Since A is symmetric, it follows that $M^* \cong A_B$. We have $M \otimes_A M^* \cong {}_BA_B = B \oplus Y$. The result follows from 4.14.14, with the roles of A and B exchanged. □

Rickard's argument in the proof of 4.14.14 has an analogue for Rickard complexes.

Theorem 4.14.16 *Let A and B be indecomposable symmetric $\mathcal{O}$-algebras such that $k \otimes_{\mathcal{O}} A$ is not simple. Let X be an indecomposable bounded complex of A-B-bimodules whose terms are finitely generated projective as left A-modules and as right B-modules. Suppose that $X \otimes_B X^* \simeq A$. Then X is a Rickard complex of A-B-bimodules.*

Proof By the hypotheses, we have an isomorphism of complexes $X \otimes_B X^* \cong A \oplus P$ for some bounded contractible complex P. The rest of the proof follows that of 4.14.14 with stable categories replaced by homotopy categories and with projective bimodules replaced by contractible complexes of bimodules. Since all involved complexes are bounded and consist of finitely generated modules, the Krull–Schmidt Theorem applies as in the proof of 4.14.14. □

Remark 4.14.17 In 4.14.14 and in 4.14.16 it is crucial to assume that M and the terms of X are projective as one-sided modules. Without such a hypothesis, these results would not be true, not even for Morita equivalences. Indeed, if e is any idempotent in an algebra A, then $eA \otimes_A Ae \cong eAe$, but if AeA is a proper ideal in A, then $Ae \otimes_{eAe} eA$ is not isomorphic to A, and eA is not projective as a left eAe-module; see 2.8.7.

Remark 4.14.18 Stable equivalences of Morita type between finite-dimensional k-algebras have been investigated well beyond the selfinjective case. See for instance Dugas and Martinez-Villa [27], Liu and Xi [62], and Xi [94], as well as the references therein.

4.15 Stable equivalences of Morita type and automorphisms

A stable equivalence of Morita type between two algebras preserves the stable Picard groups, but it need not preserve the Picard groups. We will show that a stable equivalence of Morita type preserves certain subgroups of the Picard groups. Let $\mathcal{O}$ be a complete local principal ideal domain with residue field k. An algebra automorphism α of an $\mathcal{O}$-algebra A is said to stabilise the isomorphism classes of finitely generated $k \otimes_{\mathcal{O}} A$-modules if ${}_{\alpha}U \cong U$ for any finitely generated $k \otimes_{\mathcal{O}} A$-module U, viewed as an A-module via the canonical surjection $A \to k \otimes_{\mathcal{O}} A$. Note that any inner automorphism of A has this property. Thus the set of automorphisms that stabilise the isomorphism classes of finitely generated $k \otimes_{\mathcal{O}} A$-modules form a subgroup of $\mathrm{Aut}(A)$ which contains $\mathrm{Inn}(A)$. We denote this subgroup by $\mathrm{Aut}_0(A)$, and we denote its image in $\mathrm{Out}(A)$ by $\mathrm{Out}_0(A)$. Through the canonical map $\mathrm{Out}(A) \to \mathrm{Pic}(A)$ this can be identified with a subgroup of $\mathrm{Pic}(A)$. The set of automorphisms of A that induce the identity automorphism on $k \otimes_{\mathcal{O}} A$ is a subgroup of $\mathrm{Aut}(A)$, denoted by $\mathrm{Aut}_1(A)$. This is a subgroup of $\mathrm{Aut}_0(A)$. We denote by $\mathrm{Out}_1(A)$ the image of $\mathrm{Aut}_1(A)$ in $\mathrm{Out}(A)$; that is, $\mathrm{Out}_1(A) \cong \mathrm{Aut}_1(A)\mathrm{Inn}(A)$ $/\mathrm{Inn}(A)$.

Theorem 4.15.1 ([55, Theorem 4.2]) *Let A, B be $\mathcal{O}$-algebras that are free of finite rank as $\mathcal{O}$-modules, such that $k \otimes_{\mathcal{O}} A$ and $k \otimes_{\mathcal{O}} B$ are indecomposable nonsimple selfinjective with separable semisimple quotients. Let M be an A-B-bimodule and N a B-A-bimodule such that M and N induce a stable equivalence of Morita type between A and B. Suppose that the bimodules M and N are indecomposable.*

(i) Let $\alpha \in \mathrm{Aut}_0(A)$. Then there is $\beta \in \mathrm{Aut}_0(B)$ such that we have an isomorphism of A-B-bimodules ${}_{\alpha^{-1}}M \cong M_{\beta}$. In that case we also have an isomorphism of B-A-bimodules ${}_{\beta^{-1}}N \cong N_{\alpha}$.

(ii) The map sending α to β as in (i) induces a group isomorphism $\mathrm{Out}_0(A) \cong \mathrm{Out}_0(B)$.

(iii) The map sending α to β as in (i) induces a group isomorphism $\mathrm{Out}_1(A) \cong \mathrm{Out}_1(B)$.

Proof Let $\alpha \in \mathrm{Aut}(A)$. Suppose that α stabilises the isomorphism classes of all finitely generated $k \otimes_{\mathcal{O}} A$-modules. It follows that the self-stable equivalence induced by the B-B-bimodule $N \otimes_A ({}_{\alpha^{-1}}M)$ stabilises, up to projective summands, the isomorphism class of all finitely generated $k \otimes_{\mathcal{O}} B$-modules. By 4.14.2, the bimodule $N \otimes_A ({}_{\alpha^{-1}}M)$ has a unique nonprojective summand Z, up to isomorphism. It follows from 4.14.9 that $Z \otimes_B T$ is indecomposable for any simple B-module T. Thus in fact $Z \otimes_B T \cong T$ for any simple B-module T. It follows from 4.14.10 that the B-B-bimodule Z induces a Morita equivalence on B that stabilises the isomorphism classes of all $k \otimes_{\mathcal{O}} B$-modules. In particular, $Z \otimes_B B \cong B$ as left B-modules, and hence 2.8.16 (v) implies that $Z \cong B_\beta$ for some automorphism β of B that stabilises the isomorphism class of all finitely generated $k \otimes_{\mathcal{O}} B$-modules. Thus $M_\beta = M \otimes_B B_\beta$ is the unique nonprojective direct summand, up to isomorphism, of the A-B-bimodule

$$M \otimes_B N \otimes_A ({}_{\alpha^{-1}}M) \cong (A \oplus X) \otimes_A ({}_{\alpha^{-1}}M) \cong {}_{\alpha^{-1}}M \oplus X \otimes_A ({}_{\alpha^{-1}}M).$$

The last bimodule summand is projective because X is so, and hence $M_\beta \cong {}_{\alpha^{-1}}M$. Using 2.8.17, we have B-A-bimodule isomorphisms

$$\begin{aligned} N \otimes_A M_\beta \otimes_B N &\cong N \otimes_A ({}_{\alpha^{-1}}M) \otimes_B N \cong N \otimes_A ({}_{\alpha^{-1}}(A \oplus X)) \\ &\cong N_\alpha \oplus (N_\alpha \otimes_A X), \end{aligned}$$

and similarly, we have

$$\begin{aligned} N \otimes_A M_\beta \otimes_B N &\cong N \otimes_A M \otimes_B ({}_{\beta^{-1}}N) \cong (M \oplus Y) \otimes_B ({}_{\beta^{-1}}N) \\ &\cong {}_{\beta^{-1}}N \oplus (Y_\beta \otimes_B N). \end{aligned}$$

Comparing the nonprojective summands in these two expressions yields an isomorphism ${}_{\beta^{-1}}N \cong N_\alpha$ as stated in (i). We show next that the class of β in $\mathrm{Out}(B)$ depends only on that of $\alpha \in \mathrm{Out}(A)$. It suffices to show that if α is inner, then so is β. If α is inner, then so is α^{-1}, and hence ${}_{\alpha^{-1}}M \cong M$ as A-B-bimodules. Thus $M_\beta \cong M$ as A-B-bimodules. It follows that $N \otimes_A M_\beta \cong N \otimes_A M$ as B-B-bimodules. Comparing the nonprojective summands of these two bimodules yields $B_\beta \cong B$ as B-B-bimodules. Thus β is inner by 2.8.16 (i). This proves (i). Let α, α' be automorphisms of A stabilising the isomorphism classes of finitely generated $k \otimes_{\mathcal{O}} A$-modules. Then $\alpha' \circ \alpha$ stabilises the isomorphism classes of finitely generated $k \otimes_{\mathcal{O}} A$-modules. By (i), there are $\beta, \beta' \in \mathrm{Aut}(B)$ such that ${}_{\alpha^{-1}}M \cong M_\beta$ and ${}_{\alpha'^{-1}}M \cong M_{\beta'}$. Then ${}_{(\alpha' \circ \alpha)^{-1}}M = {}_{\alpha^{-1} \circ \alpha'^{-1}}M = {}_{\alpha'^{-1}}({}_{\alpha^{-1}}M) \cong {}_{\alpha'^{-1}}M_\beta \cong (M_{\beta'})_\beta = M_{\beta' \circ \beta}$. This shows that the correspondence sending α to β induces a group homomorphism $\mathrm{Out}_0(A) \to \mathrm{Out}_0(B)$. The same argument, using the isomorphism ${}_{\beta^{-1}}N \cong N_\alpha$, implies that the correspondence sending β

to α yields the inverse of this group homomorphism, whence (ii). For (iii), suppose that α induces the identity on $k \otimes_{\mathcal{O}} A$. Then $k \otimes_{\mathcal{O}} M_\beta \cong k \otimes_{\mathcal{O}} ({}_{\alpha^{-1}}M) \cong k \otimes_{\mathcal{O}} M$. As before, tensoring by $k \otimes_{\mathcal{O}} N \otimes_A -$ and comparing nonprojective summands shows that $k \otimes_{\mathcal{O}} B \cong k \otimes_{\mathcal{O}} B_\beta$, and hence 2.8.16 (i) implies that β induces an inner automorphism on $k \otimes_{\mathcal{O}} B$, given by conjugation with an element $d \in (k \otimes_{\mathcal{O}} B)^\times$. Since the canonical map $B^\times \to (k \otimes_{\mathcal{O}} B)^\times$ is surjective, it follows that d is the image of an element $c \in B^\times$. Thus, after modifying β with the inner automorphism given by conjugation with c^{-1}, it follows that β can be chosen to induce the identity map on $k \otimes_{\mathcal{O}} B$. This concludes the proof. □

It is possible to formulate Theorem 4.15.1 in a slightly more general way for certain subgroups of the Picard groups of A and B; see [55, Theorem 4.1]. In the situation of a stable equivalence of Morita type given by induction and restriction, the correspondence in 4.15.1 can be described as follows.

Proposition 4.15.2 *Let A be an $\mathcal{O}$-algebra that is finitely generated free as an $\mathcal{O}$-module. Let B be a subalgebra of A such that A_B and ${}_BA$ induce a stable equivalence of Morita type between A and B. Any automorphism $\beta \in \mathrm{Aut}_{\mathcal{O}}(B)$ extends to some automorphism $\alpha \in \mathrm{Aut}_{\mathcal{O}}(A)$. The class of α in $\mathrm{Out}_{\mathcal{O}}(A)$ is uniquely determined by the class of β in $\mathrm{Out}_{\mathcal{O}}(B)$, we have ${}_{\alpha^{-1}}A \cong A_\beta$ as A-B-bimodules, and we have $\alpha \in \mathrm{Aut}_1(A)\mathrm{Inn}(A)$ if and only if $\beta \in \mathrm{Aut}_1(B)\mathrm{Inn}(B)$.*

Proof Let $\beta \in \mathrm{Aut}_{\mathcal{O}}(B)$. By 4.15.1 there is $\alpha \in \mathrm{Aut}_{\mathcal{O}}(A)$ such that ${}_{\alpha^{-1}}A \cong A_\beta$ as A-B-bimodules. By 2.8.16 we have ${}_{\alpha^{-1}}A \cong A_\alpha$ as A-A-bimodules. Thus there is an isomorphism of A-B-bimodules $A_\alpha \cong A_\beta$. Any such isomorphism is in particular an automorphism of A as a left A-module, hence given by right multiplication with an element $c \in A^\times$. For this map to also be a homomorphism of right B-modules, we must have $\alpha(b)c = c\beta(b)$ for all $b \in B$. Therefore, after replacing α by its conjugate, sending $a \in A$ to $c^{-1}\alpha(b)c$, it follows that α extends β. The rest follows from 4.15.1. □

Exercise 4.15.3 Let A be an $\mathcal{O}$-algebra that is finitely generated free as an $\mathcal{O}$-module. Show that $\mathrm{Aut}_1(A)\mathrm{Inn}(A)$ is equal to the group of all automorphisms of A that induce an inner automorphism on $k \otimes_{\mathcal{O}} A$.

Exercise 4.15.4 Let A be an $\mathcal{O}$-algebra that is finitely generated free as an $\mathcal{O}$-module. Show that there are canonical group isomorphisms

$$\ker(\mathrm{Pic}(A) \to \mathrm{Pic}(k \otimes_{\mathcal{O}} A)) \cong \mathrm{Out}_1(A) \cong \ker(\mathrm{StPic}(A) \to \mathrm{StPic}(k \otimes_{\mathcal{O}} A)).$$

4.16 The decomposition matrix

Let $\mathcal{O}$ be a complete discrete valuation ring with quotient field K and residue field k. The decomposition matrix of an $\mathcal{O}$-algebra A relates simple $K \otimes_{\mathcal{O}} A$-modules to simple $k \otimes_{\mathcal{O}} A$-modules.

Definition 4.16.1 Let A be an $\mathcal{O}$-algebra that is finitely generated free as an $\mathcal{O}$-module. Suppose that $K \otimes_{\mathcal{O}} A$ is a direct product of matrix algebras over K and that $k \otimes_{\mathcal{O}} A$ is split. For any simple $K \otimes_{\mathcal{O}} A$-module X and any simple $k \otimes_{\mathcal{O}} A$-module S define a non negative integer d_S^X by

$$d_S^X = \dim_K(iX)$$

where i is a primitive idempotent in A such that $Ai/J(A)i \cong S$. The *decomposition matrix of* A is the matrix $D = (d_S^X)_{X,S}$, with X and S running over sets of representatives of the isomorphism classes of simple modules over $K \otimes_{\mathcal{O}} A$ and $k \otimes_{\mathcal{O}} A$, respectively.

The idempotent i is determined uniquely up to conjugacy in $A^\times$ by the simple $k \otimes_{\mathcal{O}} A$-module S, and thus this definition makes sense. Morita equivalent $\mathcal{O}$-algebras satisfying the hypotheses of this definition have the same decomposition matrices, because $iX \cong \mathrm{Hom}_{K\otimes_{\mathcal{O}}A}(K \otimes_{\mathcal{O}} Ai, X)$, and a Morita equivalence preserves projective indecomposable modules and extends to a Morita equivalence between the corresponding K-algebras, preserving simple modules over K. A derived equivalence does not in general preserve decomposition matrices. The term 'decomposition matrix' refers to the fact that the numbers d_S^X have another interpretation as the multiplicity of X in the scalar extension $K \otimes_{\mathcal{O}} Ai$ of the projective indecomposable A-module Ai, where X, S, i are as in the definition above.

Theorem 4.16.2 *Let A be an $\mathcal{O}$-algebra that is free of finite rank as an $\mathcal{O}$-module. Suppose that $K \otimes_{\mathcal{O}} A$ is a direct product of matrix algebras over K and that $k \otimes_{\mathcal{O}} A$ is split. Let X be a simple $K \otimes_{\mathcal{O}} A$-module, S a simple $k \otimes_{\mathcal{O}} A$-module and i a primitive idempotent in A such that $Ai/J(A)i \cong S$. Then d_S^X is the number of direct summands isomorphic to X in a decomposition of $K \otimes_{\mathcal{O}} Ai$ as a direct sum of simple $K \otimes_{\mathcal{O}} A$-modules.*

Proof Write $K \otimes_{\mathcal{O}} Ai = \oplus_{j\in J} Y_j$, where J is a finite set and Y_j a simple $K \otimes_{\mathcal{O}} A$-module, for $j \in J$. By 1.7.4 we have $\dim_K(iX) = \dim_K(\mathrm{Hom}_{K\otimes_{\mathcal{O}}A}(K \otimes_{\mathcal{O}} Ai, X)) = \sum_{j\in J} \dim_K(\mathrm{Hom}_{K\otimes_{\mathcal{O}}A}(Y_j, X))$. The result follows from Schur's Lemma 1.9.2. $\square$

This interpretation of decomposition numbers yields the following relation between decomposition numbers and the Cartan matrix:

Theorem 4.16.3 *Let A be an $\mathcal{O}$-algebra that is finitely generated free as an $\mathcal{O}$-module. Suppose that $K \otimes_\mathcal{O} A$ is a direct product of matrix algebras over K and that $k \otimes_\mathcal{O} A$ is split. Denote by D the composition matrix of A and by C the Cartan matrix of A. Then ${}^tD \cdot D = C$. In particular, the Cartan matrix of A is symmetric.*

Proof Let i, j be primitive idempotents in A, let S, T be simple $k \otimes_\mathcal{O} A$-modules such that $Ai/J(A)i \cong S$ and $Aj/J(A)j \cong T$. By 4.16.2 we can write

$$K \otimes_\mathcal{O} Ai \cong \oplus_X X^{d_S^X}$$

$$K \otimes_\mathcal{O} Aj \cong \oplus_X X^{d_T^X}$$

where X runs over a set of representatives of the isomorphism classes of simple $K \otimes_\mathcal{O} A$-modules. Thus the Cartan number $c_{i,j}$ is equal to $\dim_k(\bar{i}k \otimes_\mathcal{O} A\bar{j}) = \mathrm{rk}_\mathcal{O}(iAj) = \dim_K(iK \otimes_\mathcal{O} A)j) = \dim_K(\mathrm{Hom}_{K \otimes_\mathcal{O} A}(K \otimes_\mathcal{O} Ai, K \otimes_\mathcal{O} Aj)) = \sum_X d_S^X d_T^X$, where in the last equality we used Schur's Lemma 1.9.2. The result follows. □

There is another interpretation of decomposition numbers based on the following result, which does not require the completeness of $\mathcal{O}$ but only the fact that finitely generated torsion free $\mathcal{O}$-modules are free. This goes back to a technique in Thompson [91] which makes use of $\mathcal{O}$-pure submodules, and which has applications in the context of blocks with cyclic defect groups (cf. 11.10.4 below). An A-submodule V of an A-module U is $\mathcal{O}$-pure in U if it is a direct summand of U as an $\mathcal{O}$-module (but not necessarily as an A-module); see 4.2.6.

Theorem 4.16.4 *Let A be an $\mathcal{O}$-algebra and let V be an A-module that is finitely generated free as an $\mathcal{O}$-module. Suppose that $K \otimes_\mathcal{O} V = X \oplus X'$ for some $K \otimes_\mathcal{O} A$-submodules X, X' of $K \otimes_\mathcal{O} V$. Identify V with its image $1 \otimes V$ in $K \otimes_\mathcal{O} V$. Set $Y = X \cap V$ and $Y' = X \cap V'$. The following hold.*

(i) *The A-submodules Y and Y' of V are $\mathcal{O}$-pure in V, and we have $Y \cap Y' = \{0\}$.*
(ii) *The inclusions $Y \subseteq X$ and $Y' \subseteq X'$ induce isomorphisms of $K \otimes_\mathcal{O} A$-modules $K \otimes_\mathcal{O} Y \cong X$ and $K \otimes_\mathcal{O} Y' \cong X'$.*
(iii) *The A-submodule $Y \oplus Y'$ of V has the same $\mathcal{O}$-rank as V; equivalently, the quotient $V/(Y \oplus Y')$ is an $\mathcal{O}$-torsion A-module.*
(iv) *The character of V is equal to the sum of the characters of Y and Y'.*

Proof Let $v \in V$ and $\lambda \in \mathcal{O}$, $\lambda \neq 0$, such that $\lambda v \in Y = X \cap V$. Then $v = \lambda^{-1}\lambda v \in X \cap V = Y$, hence Y is $\mathcal{O}$-pure by 4.2.6. The same argument shows that Y' is $\mathcal{O}$-pure. Since $X \cap X' = \{0\}$ we have $Y \cap Y' = \{0\}$. This proves (i). The inclusion $Y \subseteq X$ induces a homomorphism of $K \otimes_{\mathcal{O}} A$-modules $\tau : K \otimes_{\mathcal{O}} Y \to X$ sending $\lambda \otimes y$ to λy, where $\lambda \in K$ and $y \in Y$. Since V has finite $\mathcal{O}$-rank and since K is the quotient field of $\mathcal{O}$, for any $x \in X$ there is a nonzero $\mu \in \mathcal{O}$ such that $y = \mu x \in Y$. Then α sends $\mu^{-1} \otimes y$ to x, hence τ is surjective. Let $\sum_{j \in J} \alpha_j \otimes y_j$ be an element in the kernel of τ, where J is a finite set, $\alpha_j \in K$ and $y_j \in Y$ for $j \in J$. In other words, $\sum_{j \in J} \alpha_j y_j = 0$. Let $\lambda \in \mathcal{O}$ such that $\lambda \neq 0$ and $\lambda\alpha_j \in \mathcal{O}$ for $j \in J$. Then

$$\sum_{j \in J} \lambda\alpha_j \otimes y_j = 1 \otimes \Big(\sum_{j \in J} \lambda\alpha_j\Big) y_j = 0.$$

Since $K \otimes_{\mathcal{O}} Y$ is torsion free as a $\mathcal{O}$-module this implies $\sum_{j \in J} \alpha_j \otimes y_j = 0$, showing that the map $K \otimes_{\mathcal{O}} Y \to X$ is also injective. This proves (ii) and implies further that the dimensions of $K \otimes_{\mathcal{O}} (Y \oplus Y')$ and of $K \otimes_{\mathcal{O}} V$ are equal. The statements (iii) and (iv) follow. □

Theorem 4.16.4 has an obvious generalisation to decompositions of $K \otimes_{\mathcal{O}} U$ as a direct sum of more than two summands, and is therefore particularly useful if $K \otimes_{\mathcal{O}} A$ is semisimple, as then $K \otimes_{\mathcal{O}} U$ is a direct sum of simple modules.

Theorem 4.16.5 *Let A be an $\mathcal{O}$-algebra that is finitely generated free as $\mathcal{O}$-module. Suppose that $K \otimes_{\mathcal{O}} A$ is semisimple. For any finitely generated $K \otimes_{\mathcal{O}} A$-module X there is an A-module Y that is finitely generated free as an $\mathcal{O}$-module such that $X \cong K \otimes_{\mathcal{O}} Y$.*

Proof Since $K \otimes_{\mathcal{O}} A$ is semisimple, X is a finite direct sum of simple $K \otimes_{\mathcal{O}} A$-modules, and hence we may assume that X is simple. Then X is isomorphic to a direct summand of $K \otimes_{\mathcal{O}} A$ as $K \otimes_{\mathcal{O}} A$-module. Thus we may assume that $K \otimes_{\mathcal{O}} A = X \oplus X'$ for some $K \otimes_{\mathcal{O}} A$-module X'. Set $Y = A \cap X$, where we identify A to its canonical image $1_K \otimes A$ in $K \otimes_{\mathcal{O}} A$. Since A is free of finite rank over $\mathcal{O}$, so is Y, and Y is an $\mathcal{O}$-pure submodule of A. The inclusion $Y \subseteq X$ induces a $(K \otimes_{\mathcal{O}} A)$-homomorphism mapping $\alpha \otimes y$ to αy, where $\alpha \in K$ and $y \in Y$. By 4.16.4 (ii) this is an isomorphism. □

The A-module Y is not uniquely determined by the $K \otimes_{\mathcal{O}} A$-module X, in general. If X is simple, then the arguments in the proof of the previous theorem imply that Y can be chosen in such a way that $k \otimes_{\mathcal{O}} Y$ has a unique maximal or unique minimal submodule.

Theorem 4.16.6 *Let A be an $\mathcal{O}$-algebra that is finitely generated free as $\mathcal{O}$-module. Suppose that $K \otimes_{\mathcal{O}} A$ is semisimple. Let X be a simple $K \otimes_{\mathcal{O}} A$-module.*

(i) There is an $\mathcal{O}$-free A-module Y such that $K \otimes_{\mathcal{O}} Y \cong X$ and such that $k \otimes_{\mathcal{O}} Y$ has a unique maximal submodule; in particular, $k \otimes_{\mathcal{O}} Y$ remains indecomposable.

(ii) There is an $\mathcal{O}$-free A-module Y such that $K \otimes_{\mathcal{O}} Y \cong X$ and such that $k \otimes_{\mathcal{O}} Y$ has a unique minimal submodule; in particular, $k \otimes_{\mathcal{O}} Y$ remains indecomposable.

Proof Let I be a primitive decomposition of 1 in A. Then $A = \oplus_{i \in I} Ai$ as a left A-module, hence $K \otimes_{\mathcal{O}} A = \oplus_{i \in I} K \otimes_{\mathcal{O}} Ai$ as a left $K \otimes_{\mathcal{O}} A$-module. Since X is simple there is a primitive idempotent i in A such that X is isomorphic to a direct summand of $K \otimes_{\mathcal{O}} Ai$. Thus we may assume that $K \otimes_{\mathcal{O}} Ai = X \oplus X'$ for some $K \otimes_{\mathcal{O}} A$-module X'. Set $Y' = X' \cap Ai$, where Ai is identified to its image $1 \otimes Ai$ in $K \otimes_{\mathcal{O}} Ai$. Thus Y' is an $\mathcal{O}$-pure submodule of Ai, and the quotient $Y = Ai/Y'$ satisfies $K \otimes_{\mathcal{O}} Y \cong X$. Moreover, since $k \otimes_{\mathcal{O}} Y$ is a quotient of the projective indecomposable $k \otimes_{\mathcal{O}} A$-module $k \otimes_{\mathcal{O}} Ai$, it follows from 4.5.3 (i) that $k \otimes_{\mathcal{O}} Y$ has a unique maximal submodule. This shows (i). Applying (i) to the simple right $K \otimes_{\mathcal{O}} A$-module X^* yields an $\mathcal{O}$-free right A-module Z such that $K \otimes_{\mathcal{O}} Z \cong X^*$ and such that $k \otimes_{\mathcal{O}} Z$ has a unique maximal submodule. Setting $Y = Z^*$, the $\mathcal{O}$-dual of Z, we get that $K \otimes_{\mathcal{O}} Y \cong X$ and $k \otimes_{\mathcal{O}} Y$ has a unique minimal submodule. □

Theorem 4.16.7 *Let A be an $\mathcal{O}$-algebra that is finitely generated free as an $\mathcal{O}$-module. Suppose that $K \otimes_{\mathcal{O}} A$ is a direct product of matrix algebras over K and that $k \otimes_{\mathcal{O}} A$ is split. Let X be a simple $K \otimes_{\mathcal{O}} A$-module, S a simple $k \otimes_{\mathcal{O}} A$-module and Y an $\mathcal{O}$-free A-module such that $X \cong K \otimes_{\mathcal{O}} Y$. Then d_S^X is equal to the number of composition factors isomorphic to S in a composition series of $k \otimes Y$.*

Proof We have $\dim_K(iX) = \mathrm{rk}_{\mathcal{O}}(iY) = \dim_k(\bar{i}\bar{Y})$, where $\bar{Y} = k \otimes_{\mathcal{O}} Y$ and $\bar{i}$ is the image of i in $\bar{A} = k \otimes_{\mathcal{O}} A$. By 4.10.3, this dimension is equal to the number of composition factors of $\bar{Y}$ isomorphic to S. □

As mentioned before, Y is not determined by X, but the argument in the proof of the previous result implies in particular that the composition series of Y are determined up to equivalence.

Corollary 4.16.8 *Let A be an $\mathcal{O}$-algebra that is finitely generated free as $\mathcal{O}$-module. Suppose that $K \otimes_{\mathcal{O}} A$ is a direct product of matrix algebras over K and that $k \otimes_{\mathcal{O}} A$ is split. If Y, Y' are finitely generated $\mathcal{O}$-free A-modules such* *that $K \otimes_{\mathcal{O}} Y \cong K \otimes_{\mathcal{O}} Y'$ then $k \otimes_{\mathcal{O}} Y$ and $k \otimes_{\mathcal{O}} Y'$ have equivalent composition series.*

Proof The hypotheses imply that for any primitive idempotent i in A we have $\dim_k(\bar{i}k \otimes_{\mathcal{O}} Y) = \mathrm{rk}_{\mathcal{O}}(iY) = \dim_K(iK \otimes_{\mathcal{O}} Y) = \dim_K(iK \otimes_{\mathcal{O}} Y') = \mathrm{rk}_{\mathcal{O}}(iY') = \dim_k(\bar{i}k \otimes_{\mathcal{O}} Y')$, where $\bar{i}$ is the image of A in $k \otimes_{\mathcal{O}} A$. The result follows from 4.10.3. □

Decomposition matrices and Cartan matrices decompose according to block decompositions.

Proposition 4.16.9 *Let A be an $\mathcal{O}$-algebra that is finitely generated free as an $\mathcal{O}$-module. Suppose that $K \otimes_{\mathcal{O}} A$ is a direct product of matrix algebras over K and that $k \otimes_{\mathcal{O}} A$ is split. Denote by D the decomposition matrix of A and by C the Cartan matrix of A. For a suitable order of the isomorphism classes of simple $K \otimes_{\mathcal{O}} A$-modules and simple $k \otimes_{\mathcal{O}} A$-modules, the decomposition matrix D of A is a block matrix with blocks D_b, and the Cartan matrix C of A is a block diagonal matrix with blocks C_b, where b runs over the blocks of A and where D_b and C_b denote the decomposition matrix and Cartan matrix of Ab, respectively.*

Proof Let U be an $\mathcal{O}$-free A-module such that $K \otimes_{\mathcal{O}} U$ is a simple $K \otimes_{\mathcal{O}} A$-module. Then U is indecomposable, hence belongs to a unique block b of A. Since b acts as identity on U, it also acts as identity on any composition factor S of $k \otimes_{\mathcal{O}} A$. Thus $d_S^X \neq 0$ forces X and S to belong to the same block. The result follows. □

4.17 The decomposition map

The material from the previous section can be expressed in terms of Grothendieck groups. Let $\mathcal{O}$ be a complete discrete valuation ring with quotient field K and residue field k. For A an $\mathcal{O}$-algebra that is $\mathcal{O}$-free of finite $\mathcal{O}$-rank, we set $R_K(A) = R(K \otimes_{\mathcal{O}} A)$ and $R_k(A) = R(k \otimes_{\mathcal{O}} A)$. That is, $R_K(A)$ and $R_k(A)$ are the Grothendieck groups of finitely generated modules over $K \otimes_{\mathcal{O}} A$ and $k \otimes_{\mathcal{O}} A$, respectively. We consider $R_K(A)$ with the scalar product $\langle -, - \rangle_A$ defined by

$$\langle [X], [X'] \rangle_A = \dim_K(\mathrm{Hom}_{K \otimes_{\mathcal{O}} A}(X, X'))$$

where X, X' are finitely generated $K \otimes_{\mathcal{O}} A$-modules and where $[X]$, $[X']$ are their images in $R_K(A)$. If no confusion arises, we write $\langle [X], [X'] \rangle$ instead of $\langle [X], [X'] \rangle_A$. If X, X' are nonisomorphic simple $K \otimes_{\mathcal{O}} A$-modules, then Schur's Lemma implies that $\langle [X], [X'] \rangle = 0$. Clearly $\langle [X], [X] \rangle = \dim_K(\mathrm{End}_K(X))$ is a positive integer. For an arbitrary nonzero element χ in $R_K(A)$, it follows from writing χ as a $\mathbb{Z}$-linear combination of isomorphism classes $[X]$ of simple $K \otimes_{\mathcal{O}} A$-modules X that $\langle \chi, \chi \rangle > 0$; in other words, this scalar product is positive definite. For U a subgroup of $R_K(A)$, we denote by $U^\perp$ the subgroup of all $\psi \in R_K(A)$ satisfying $\langle \chi, \psi \rangle = 0$ for all $\chi \in U$. Since $\langle -, - \rangle$ is positive definite, it follows that $U \cap U^\perp = \{0\}$, that the sum of the $\mathbb{Z}$-ranks of U and $U^\perp$ is equal to the $\mathbb{Z}$-rank of $R_K(A)$, and hence $U \oplus U^\perp$ is a subgroup of finite index in $R_K(A)$. We have $U \subseteq (U^\perp)^\perp$, but this inclusion need not be an equality.

We denote by $\mathrm{Pr}_{\mathcal{O}}(A)$ the subgroup of $R_K(A)$ generated by the isomorphism classes of the $K \otimes_{\mathcal{O}} A$-modules $[K \otimes_{\mathcal{O}} Ai]$, with i running over the primitive idempotents in A, and by $L^0(A)$ the subgroup of all $\chi \in R_K(A)$ satisfying $\langle \Psi, \chi \rangle = 0$ for all $\Psi \in \mathrm{Pr}_{\mathcal{O}}(A)$. With the notation above, we have $L^0(A) = \mathrm{Pr}_{\mathcal{O}}(A)^\perp$. We denote by $\mathrm{Pr}_k(A)$ the subgroup of $R_k(A)$ generated by the isomorphism classes of the $k \otimes_{\mathcal{O}} A$-modules $[k \otimes_{\mathcal{O}} Ai]$, with i running over the primitive idempotents in A.

Theorem 4.17.1 *Let A be an $\mathcal{O}$-algebra that is finitely generated free as an $\mathcal{O}$-module. Suppose that $K \otimes_{\mathcal{O}} A$ is a direct product of matrix algebras over K and that $k \otimes_{\mathcal{O}} A$ is split. There is a unique group homomorphism $d_A : R_K(A) \to R_k(A)$ such that $d_A([K \otimes_{\mathcal{O}} Y]) = [k \otimes_{\mathcal{O}} Y]$ for any finitely generated $\mathcal{O}$-free A-module Y. Moreover, the following hold.*

(i) We have $L^0(A) = \ker(d_A)$.
(ii) We have $\mathrm{Pr}_{\mathcal{O}}(A) \cap L^0(A) = \{0\}$ and $\mathrm{Pr}_{\mathcal{O}}(A) \oplus L^0(A)$ has finite index in $R_K(A)$.
(iii) The map d_A induces an isomorphism $\mathrm{Pr}_{\mathcal{O}}(A) \cong \mathrm{Pr}_k(A)$.
(iv) If the Cartan matrix C of A is non singular then $R_k(A)/\mathrm{Pr}_k(A)$ is a finite abelian group of order $|\det(C)|$.

Proof The existence and uniqueness of the map d_A as stated is an immediate consequence of 4.16.5 and 4.16.8. Any element η in $R_K(A)$ can be written in the form $[K \otimes_{\mathcal{O}} Y_1] - [K \otimes_{\mathcal{O}} Y_2]$ for some finitely generated $\mathcal{O}$-free A-modules Y_1 and Y_2. Let I be a set of representative of the conjugacy classes of primitive idempotents in A. For $i \in I$, denote by Φ_i the image of $K \otimes_{\mathcal{O}} Ai$ in $R_K(A)$. Since $\mathrm{Pr}_{\mathcal{O}}(A)$ is generated by the Φ_i, with $i \in I$, we get that $\eta \in L^0(A)$ if and only if $\langle \Phi_i, \eta \rangle_A = 0$ for all $i \in I$. This is equivalent to $\mathrm{rank}_{\mathcal{O}}(iY_1) = \mathrm{rank}_{\mathcal{O}}(iY_2)$, hence to $\dim_k(ik \otimes_{\mathcal{O}} Y_1) = \dim_k(ik \otimes_{\mathcal{O}} Y_2)$ for all $i \in I$. By 4.10.3 this is just a reformulation of the statement that η belongs to $\ker(d_A)$, proving (i). Since $\langle -, - \rangle$ is positive definite on $R_K(A)$, and since $L^0(A) = \mathrm{Pr}_{\mathcal{O}}(A)^\perp$, statement (ii) follows from the remarks preceding the Theorem. Since any projective $k \otimes_{\mathcal{O}} A$-module lifts to a projective A-module, the map d_A induces a surjective map

$\mathrm{Pr}_{\mathcal{O}}(A) \to \mathrm{Pr}_k(A)$. It follows from (ii) that is is an isomorphism, whence (iii). The last statement is from 4.10.6. □

Definition 4.17.2 With the notation and hypotheses of 4.17.1, the group homomorphism $d_A : R_K(A) \to R_k(A)$ is called the *decomposition map* of A.

It follows from 4.16.9 that the decomposition map d_A is the sum of the decomposition maps d_{Ab}, where b runs over the set of blocks of A. We will see in 5.14.1 that the decomposition map d_A is surjective if $A = \mathcal{O}G$ for some finite group G. The following is also known as Brauer's reciprocity:

Theorem 4.17.3 *Let A be an $\mathcal{O}$-algebra that is free of finite rank as an $\mathcal{O}$-module. Suppose that $K \otimes_{\mathcal{O}} A$ is split semisimple, that $k \otimes_{\mathcal{O}} A$ is split, and that the decomposition map $d_A : R_K(A) \to R_k(A)$ is surjective. Let I be a system of representatives of the conjugacy classes of primitive idempotents in A. For any $i \in I$ denote by Φ_i the image in $\mathrm{Pr}_{\mathcal{O}}(A)$ of $K \otimes_{\mathcal{O}} Ai$ and by φ_i the image of $Ai/J(A)i$ in $R_k(A)$. The scalar product $\langle -, - \rangle$ induces a scalar product $\langle -, - \rangle' : \mathrm{Pr}_{\mathcal{O}}(A) \times R_k(A) \to \mathbb{Z}$ with the following properties:*

(i) We have $\langle \Phi_i, \varphi_j \rangle' = \delta_{i,j}$ for any $i, j \in I$; in particular, the Cartan matrix of A is non singular.
(ii) We have $L^0(A)^{\perp} = \mathrm{Pr}_{\mathcal{O}}(A)$.

Proof The scalar product $\langle -, - \rangle$ on $R_K(A)$ restricts to a bilinear form, still denoted $\langle -, - \rangle : \mathrm{Pr}_{\mathcal{O}}(A) \times R_K(A) \to \mathbb{Z}$. This bilinear form vanishes on $\mathrm{Pr}_{\mathcal{O}}(A) \times L^0(A)$, hence, since d_A is surjective with kernel $L^0(A)$, induces a bilinear form $\langle -, - \rangle' : \mathrm{Pr}_{\mathcal{O}}(A) \times R_k(A) \to \mathbb{Z}$. Again since d_A is surjective, for any $i \in I$ and any $\chi \in \mathrm{Irr}_K(A)$ there are integers m_i^χ satisfying

$$\varphi_i = \sum_{\chi \in \mathrm{Irr}_K(A)} m_i^\chi \cdot d_A(\chi) = \sum_{j \in I} \sum_{\chi \in \mathrm{Irr}_K(A)} m_i^\chi d_j^\chi \varphi_j$$

which implies that $\sum_{\chi \in \mathrm{Irr}_K(A)} m_i^\chi d_j^\chi = \delta_{i,j}$ for all $i, j \in I$. We also have $\Phi_i = \sum_{\chi \in \mathrm{Irr}_K(A)} d_i^\chi \chi$ and hence

$$\langle \Phi_i, \varphi_j \rangle' = \langle \Phi_i, \sum_{\chi \in \mathrm{Irr}_K(A)} m_j^\chi \chi \rangle = \sum_{\chi \in \mathrm{Irr}_K(A)} d_i^\chi m_j^\chi = \delta_{i,j}.$$

This implies that the subgroup $\mathrm{Pr}_k(A)$ has the same rank as $R_k(A)$, and hence that the Cartan matrix of A is nonsingular, proving (i). The hypotheses imply that extending coefficients yields

$$\mathbb{Q} \otimes_{\mathbb{Z}} R_K(A) = (\mathbb{Q} \otimes_{\mathbb{Z}} L^0(A)) \oplus (\mathbb{Q} \otimes_{\mathbb{Z}} \mathrm{Pr}_{\mathcal{O}}(A)).$$

Let $\theta \in L^0(A)^\perp$. Then, in $\mathbb{Q} \otimes_{\mathbb{Z}} R_K(A)$, we have $1_K \otimes \theta = \sum_{i \in I} q_i \otimes \Phi_i$ for some $q_i \in \mathbb{Q}$. For any $i \in I$ we have

$$q_i = \langle \theta, \varphi_i \rangle' = \sum_{\chi \in \mathrm{Irr}_K(A)} m_i^\chi \langle \theta, \chi \rangle$$

which is an integer as $\theta \in R_K(A)$. This shows (ii). □

Decomposition maps commute with group homomorphisms induced by certain exact functors:

Proposition 4.17.4 *Let A, B be $\mathcal{O}$-algebras that are finitely generated free as $\mathcal{O}$-modules. Suppose that $K \otimes_{\mathcal{O}} A$, $K \otimes_{\mathcal{O}} B$ are direct products of matrix algebras over K and that $k \otimes_{\mathcal{O}} A$, $k \otimes_{\mathcal{O}} B$ are split. Let M be an A-B-bimodule that is finitely generated projective as a right B-module. The functor $M \otimes_B -$ induces group homomorphisms $\Phi : R_K(B) \to R_K(A)$ and $\bar{\Phi} : R_k(B) \to R_k(A)$ satisfying $d_A \circ \Phi = \bar{\Phi} \circ d_B$.*

Proof Let V be a finitely generated $\mathcal{O}$-free B-module. Since M is projective as a right B-module we have an obvious isomorphism of $k \otimes_{\mathcal{O}} A$-modules $k \otimes_{\mathcal{O}} (M \otimes_B V) \cong (k \otimes_{\mathcal{O}} M) \otimes_{k \otimes_{\mathcal{O}} B} (k \otimes_{\mathcal{O}} V)$. This is equivalent to the assertion $d_A(\Phi([V])) = \bar{\Phi}(d_B([V]))$, where $[V]$ is the image of $K \otimes_{\mathcal{O}} V$ in $R_K(B)$. The result follows. □

Remark 4.17.5 With the notation of 4.17.4, if M induces a Morita equivalence between A and B, then M induces *isomorphisms* $\Phi : R_K(B) \cong R_K(A)$ and $\bar{\Phi} : R_k(B) \cong R_k(A)$ satisfying $d_A \circ \Phi = \bar{\Phi} \circ d_B$.

Remark 4.17.6 The material of this section relates the Grothendieck groups of A over K and over k. If $A = \mathcal{O}G$ for some finite group G, then $R_K(A)$ can be identified with the group of K-valued generalised characters of G. We will see later that $R_k(A)$ has in that case an interpretation in terms of K-valued central functions on the set $G_{p'}$ of p-regular elements in G.

5

Group Algebras and Modules over p-Local Rings

Green's theory of vertices and sources seems to be the first systematic module theoretic approach to group algebras over local rings and fields with positive characteristic p, combining properties of algebras over p-local rings with properties of induction and restriction, developed in earlier chapters. Throughout this chapter, $\mathcal{O}$ is a complete discrete valuation ring with residue field $k = \mathcal{O}/J(\mathcal{O})$ of prime characteristic p and quotient field K. If not stated otherwise, we allow the case $k = \mathcal{O} = K$. From Section 5.13 onwards we assume that $\mathrm{char}(K) = 0$. Green's theory of vertices and sources and the Green correspondence are formulated for finitely generated modules, since they rely on the Krull–Schmidt Theorem. Almost all statements on finitely generated modules over a finite group algebra $\mathcal{O}G$ in the first two sections of this chapter hold verbatim for bounded complexes of finitely generated $\mathcal{O}G$-modules. For those statements for which finite generation of modules is essential, we use without further comment that all the standard constructions between modules of finite group algebras, including restriction, induction, tensor products, as well as the Brauer construction below, preserve finite generation.

5.1 Vertices and sources

Given a finite group G and an $\mathcal{O}G$-module (or complex of $\mathcal{O}G$-modules) M, recall that M is relatively Q-projective for some subgroup Q of G if M is isomorphic to a direct summand of $\mathrm{Ind}_Q^G(V)$ for some $\mathcal{O}Q$-module (or complex of $\mathcal{O}Q$-modules) V. By Higman's criterion 2.6.2, M is relatively Q-projective if and only if M is isomorphic to a direct summand of $\mathrm{Ind}_Q^G(\mathrm{Res}_Q^G(M))$. If no confusion arises, we suppress some brackets when composing induction and restriction functors.

Definition 5.1.1 Let G be a finite group and let M be a finitely generated indecomposable $\mathcal{O}G$-module. A subgroup Q of G is called a *vertex of M* if Q

is a minimal with the property that M is relatively Q-projective. If Q is a vertex of M, an *$\mathcal{O}Q$-source of M* is an indecomposable $\mathcal{O}Q$-module V such that M is isomorphic to a direct summand of $\mathrm{Ind}_Q^M(V)$.

Since we allow the case $\mathcal{O} = k$ this definition covers also modules of group algebras over a field. Again by Higman's criterion 2.6.2, a subgroup Q of G is a vertex of M if and only if Q is a minimal subgroup of G such that $\mathrm{Id}_M \in \mathrm{Tr}_Q^G(\mathrm{End}_{\mathcal{O}Q}(M))$. If a vertex Q of M is clear from the context, we will call V a source of M, rather than $\mathcal{O}Q$-source. The definition of vertices and sources would have made sense for arbitrary modules, but we will see in the Theorem 5.1.2 below that the condition on M being finitely generated indecomposable is crucial for these notions to have good properties, mostly because the Krull–Schmidt Theorem fails for infinitely generated modules. Theorem 5.1.2 states that every indecomposable $\mathcal{O}G$-module M has a vertex Q, every vertex is a p-group and for every such vertex there is an $\mathcal{O}Q$-source V. The Theorem tells us further where to look for sources, namely in the restriction to a vertex Q of M, and concludes that pairs (Q, V) consisting of a vertex and a source are unique up to conjugation by elements in G. Its proof combines relative projectivity, Mackey's formula and the Krull–Schmidt Theorem. The significance of this theorem is that it relates the module theory for finite groups over p-local rings to the module theory of p-groups.

Theorem 5.1.2 *Let G be a finite group and let M be a finitely generated indecomposable $\mathcal{O}G$-module.*

(i) M has a vertex, and every vertex of M is a p-subgroup of G.
(ii) For every vertex Q of M there is an $\mathcal{O}Q$-source V of M.
(iii) Given a vertex Q every $\mathcal{O}Q$-source V of M is isomorphic to a direct summand of $\mathrm{Res}_Q^G(M)$.
(iv) Given a vertex Q of M, an indecomposable direct summand V of $\mathrm{Res}_Q^G(M)$ is a source of M if and only if V has vertex Q.
(v) Given a vertex Q of M, an $\mathcal{O}Q$-source V, a p-subgroup R of G and an indecomposable $\mathcal{O}R$-module W such that W has R as vertex and is isomorphic to a direct summand of $\mathrm{Res}_R^G(W)$, there is $x \in G$ such that ${}^xR \subseteq Q$ and such that ${}^xW$ is isomorphic to a direct summand of $\mathrm{Res}_{{}^xR}^Q(V)$.
(vi) Given two vertices Q, R of M, an $\mathcal{O}Q$-source V and $\mathcal{O}R$-source W of M, there is $x \in G$ such that ${}^xR = Q$ and ${}^xW \cong V$.

In particular, the set of vertices of M is a conjugacy class of p-subgroups of G.

Proof The existence of a vertex Q is trivial: since G is finite, there is a minimal subgroup Q for which one can find an $\mathcal{O}Q$-module V such that M is a summand of $\mathrm{Ind}_Q^G(V)$. Write $V = \oplus_{1\leq i\leq n} V_i$, where the V_i are indecomposable $\mathcal{O}Q$-modules for $1 \leq i \leq n$. Thus M is a direct summand of $\oplus_{1\leq i\leq n} \mathrm{Ind}_Q^G(V_i)$. It follows from the Krull–Schmidt Theorem that in fact M is the summand of one of the $\mathrm{Ind}_Q^G(V_i)$, for some i. Thus we may replace V by V_i, or equivalently, we may assume that V is an $\mathcal{O}Q$-source of M. Suppose that Q is not a p-group. Let S be a Sylow p-subgroup of Q. Then $|Q : S|$ is prime to p, hence invertible in $\mathcal{O}$, and thus $\mathrm{Id}_V = \frac{1}{|Q:S|}\mathrm{Tr}_S^Q(\mathrm{Id}_V)$. By Higman's criterion, there is an $\mathcal{O}S$-module W such that V is a summand of $\mathrm{Ind}_S^Q(W)$. But then M is a summand of $\mathrm{Ind}_Q^G(\mathrm{Ind}_S^Q(W)) = \mathrm{Ind}_S^G(W)$, contradicting the minimality of Q. Thus Q is a p-group. This completes the proof of (i) and (ii). Again by Higman's criterion, if Q is a vertex of M then M is a summand of $\mathrm{Ind}_Q^G\mathrm{Res}_Q^G(M)$. Thus we may choose an indecomposable direct summand V of $\mathrm{Res}_Q^G(M)$ such that M is a summand of $\mathrm{Ind}_Q^G(V)$. This shows that at least some source V of M is a summand of $\mathrm{Res}_Q^G(M)$. Note that a source must have vertex Q, because otherwise Q would not be minimal with the property that M is relatively Q-projective. Let W be another $\mathcal{O}Q$-source of M. Then M is a summand of $\mathrm{Ind}_Q^G(W)$. Thus V is a summand of $\mathrm{Res}_Q^G(M)$, which is a summand of

$$\mathrm{Res}_Q^G\mathrm{Ind}_Q^G(W) = \oplus_{x\in[Q\backslash G/Q]} \mathrm{Ind}_{Q\cap {}^xQ}^Q \mathrm{Res}_{Q\cap {}^xQ}^{{}^xQ}({}^xW)$$

by Mackey's formula. Since V is indecomposable, V is a summand of $\mathrm{Ind}_{Q\cap {}^xQ}^Q \mathrm{Res}_{Q\cap {}^xQ}^{{}^xQ}({}^xW)$ for some $x \in G$. But V must have vertex Q, so this forces $Q = {}^xQ$, and hence $V \cong {}^xW$, as W is indecomposable. Since ${}^xM \cong M$ via the map sending $m \in M$ to xm, we get that W itself is a summand of M restricted to Q. This proves (iii) and (iv). Note that this also proves a special case of (v), namely when $Q = R$. Let now Q, V, R, W be as in (v). Then, by assumption, W is a summand of $\mathrm{Res}_R^G(M)$, and M is a summand of $\mathrm{Ind}_Q^G(V)$. Thus W is a summand of

$$\mathrm{Res}_R^G\mathrm{Ind}_Q^G(V) = \oplus_{y\in[R\backslash G/Q]}\mathrm{Ind}_{R\cap {}^yQ}^R \mathrm{Res}_{R\cap {}^yQ}^{{}^yQ}(V)$$

by Mackey's formula. Since W is indecomposable, the Krull–Schmidt Theorem implies that W is a summand of $\mathrm{Ind}_{R\cap {}^yQ}^R \mathrm{Res}_{R\cap {}^yQ}^{{}^yQ}({}^yV)$ for some $y \in G$. Since R is a vertex of W, it follows that $R \subseteq {}^yQ$ and hence that W is isomorphic to a direct summand of $\mathrm{Res}_R^{{}^yQ}({}^yV)$. Conjugating this back with $x = y^{-1}$ yields (v). In the situation of (vi), applying (v) twice implies that actually ${}^xR = Q$, and hence ${}^xW \cong V$ for some $x \in G$. This concludes the proof of (vi). □

Every indecomposable $\mathcal{O}G$-module M determines hence a pair (Q, V) consisting of a vertex Q and an $\mathcal{O}Q$-source V, uniquely up to G-conjugacy. In general, a vertex-source pair (Q, V) of an indecomposable $\mathcal{O}G$-module M does not determine the isomorphism class of M. One can show that given (Q, V)

there are at most finitely many isomorphism classes of indecomposable $\mathcal{O}G$-modules M having Q as a vertex and V as a source. These can be parametrised by a third invariant, the *multiplicity module*, defined by Puig [73]; see 5.7.6. Every indecomposable $\mathcal{O}Q$-module V with vertex Q arises as a source of some $\mathcal{O}G$-module; this follows from the fact that every indecomposable direct summand of $\mathrm{Ind}_Q^G(V)$ has a vertex contained in Q, and at least one such summand M has the property that V is a summand of $\mathrm{Res}_Q^G(M)$ because V is a summand of $\mathrm{Res}_Q^G(\mathrm{Ind}_Q^G(V))$. Not every indecomposable kQ-module V with vertex Q arises however as a source of a *simple* kG-module for some finite group G containing Q. The following conjecture of Feit, if true, would explain this:

Conjecture 5.1.3 (Feit, 1979) *Let Q be a finite p-group. There are only finitely many isomorphism classes of indecomposable kQ-modules V such that there exists a finite group G containing Q and a simple kG-module S with vertex Q and kQ-source V.*

Dade showed in [25] that this is true if one bounds the dimension of V: there are only finitely many isomorphism classes of indecomposable kQ-modules V of a fixed dimension such that there exists a finite group G containing Q and a simple kG-module S with vertex Q and kQ-source V. Vertices and sources of trivial and projective indecomposable modules are easy to determine:

Proposition 5.1.4 *Let G be a finite group.*

(i) The trivial $\mathcal{O}G$-module $\mathcal{O}$ and any nonzero quotient W of $\mathcal{O}$ has as a vertex any Sylow p-subgroup P of G, and as an $\mathcal{O}P$-source the trivial $\mathcal{O}P$-module or nonzero quotient W of $\mathcal{O}$, respectively, as a source.

(ii) A finitely generated indecomposable $\mathcal{O}G$-module U is projective if and only if U has the trivial group $\{1\}$ as a vertex and $\mathcal{O}$ as a source.

(iii) A finitely generated $\mathcal{O}G$-module U is projective if and only if $\mathrm{Res}_P^G(U)$ is projective for some Sylow p-subgroup P of G.

Proof Let P be a subgroup of G. Let $W = \mathcal{O}$ or $W = \mathcal{O}/\pi^n\mathcal{O}$ for some positive integer n. By Higman's criterion 2.6.2, the $\mathcal{O}G$-module W is relatively P-projective if and only if there is $\varphi \in \mathrm{End}_{\mathcal{O}P}(W)$ such that $\mathrm{Tr}_P^G(\varphi) = \mathrm{Id}_W$. Since the elements of G act as identity on W we have $\mathrm{Tr}_P^G(\varphi) = |G : P|\varphi$. Thus the equality $\mathrm{Tr}_P^G(\varphi) = \mathrm{Id}_W$ is equivalent to $|G : P|$ being invertible in $\mathcal{O}$ and $\varphi = \frac{1}{|G:P|}Id_W$. Since $|G : P|$ is invertible in $\mathcal{O}$ if and only if $|G : P|$ is prime to p, the minimal subgroups of G with this property are exactly the Sylow p-subgroups of G. Since an $\mathcal{O}P$-source of W is a direct summand of $\mathrm{Res}_P^G(W)$, the trivial $\mathcal{O}P$-module W is, up to isomorphism, the unique $\mathcal{O}P$-source of the trivial

$\mathcal{O}G$-module W. This shows (i). We have $\mathrm{Ind}_{\{1\}}^G(\mathcal{O}) = \mathcal{O}G \otimes_{\mathcal{O}} \mathcal{O} \cong \mathcal{O}G$. Thus a finitely generated indecomposable $\mathcal{O}G$-module U has vertex $\{1\}$ and trivial source if and only if U is isomorphic to a direct summand of $\mathcal{O}G$ as a left $\mathcal{O}G$-module, hence if and only if U is a projective indecomposable $\mathcal{O}G$-module. This shows (ii). For the last statement we may assume that U is indecomposable. If U is projective, then so is its restriction to any subgroup, by 2.1.4. If $\mathrm{Res}_P^G(U)$ is projective, where P is a Sylow p-subgroup, then U is $\mathcal{O}$-free and has vertex $\{1\}$, hence is projective by (ii). This completes the proof. □

Proposition 5.1.5 *Let G be a finite group and H a subgroup of G.*

(i) Let Q be a subgroup of H and V a finitely generated relatively Q-projective $\mathcal{O}H$-module. Then any indecomposable direct summand of the $\mathcal{O}G$-module $\mathrm{Ind}_H^G(V)$ has a vertex contained in Q.
(ii) Let P be a subgroup of G and U a finitely generated relatively P-projective $\mathcal{O}G$-module. Then any indecomposable direct summand of the $\mathcal{O}H$-module $\mathrm{Res}_H^G(U)$ has a vertex contained in $H \cap {}^xP$ for some $x \in G$.
(iii) Let U be a finitely generated indecomposable $\mathcal{O}G$-module such that H acts trivially on U. Then any Sylow p-subgroup Q of H is contained in a vertex of U.

Proof Since V is relatively Q-projective, there is an $\mathcal{O}Q$-module W such that V is isomorphic to a direct summand of $\mathrm{Ind}_Q^H(W)$. Thus $\mathrm{Ind}_H^G(V)$ is isomorphic to a direct summand of $\mathrm{Ind}_Q^G(W)$, whence (i). Since U is relatively P-projective, there is an $\mathcal{O}P$-module X such that U is isomorphic to a direct summand of $\mathrm{Ind}_P^G(X)$. Mackey's formula implies that $\mathrm{Res}_H^G(U)$ is isomorphic to a direct summand of $\mathrm{Res}_H^G(\mathrm{Ind}_P^G(X)) \cong \oplus_x \mathrm{Ind}_{H\cap {}^xP}^H(\mathrm{Res}_{H\cap {}^xP}^{{}^xP}(X))$, with x running over a set of representatives of $H\backslash G/P$ in G. Thus every indecomposable direct summand of $\mathrm{Res}_H^G(U)$ is isomorphic to a direct summand of $\mathrm{Ind}_{H\cap {}^xP}^H(\mathrm{Res}_{H\cap {}^xP}^{{}^xP}({}^xX))$ for some $x \in G$, whence (ii). If U is indecomposable and P a vertex of U, then $H \cap {}^xP$ contains a vertex of some summand of $\mathrm{Res}_H^G(U)$, for some $x \in G$. If H acts trivially on U, then every summand of $\mathrm{Res}_H^G(U)$ is trivial, hence has a Sylow p-subgroup of H as a vertex. Statement (iii) follows from 5.1.4. □

Proposition 5.1.6 *Let G be a finite group and H a subgroup of G. Let M be a finitely generated indecomposable $\mathcal{O}G$-module with a vertex Q and $\mathcal{O}Q$-source V. Suppose that $Q \subseteq H$. Then $\mathrm{Res}_H^G(M)$ has an indecomposable direct summand N with vertex Q and $\mathcal{O}Q$-source V.*

Proof By 5.1.2, V is isomorphic to a direct summand of $\mathrm{Res}^G_Q(M)$. Thus $\mathrm{Res}^G_H(M)$ has an indecomposable direct summand N such that V is isomorphic to a direct summand of $\mathrm{Res}^H_Q(N)$. This implies that Q is contained in a vertex of N. By 5.1.5 (ii), the vertices of N are contained in G-conjugates of Q. Thus Q is a vertex of N, and V a source of N. □

Proposition 5.1.7 *Let G be a finite group, H a subgroup of G, and let N be a finitely generated indecomposable $\mathcal{O}H$-module. Let (Q, V) be a vertex-source pair of N. Every indecomposable direct summand of $\mathrm{Ind}^G_H(N)$ has a vertex contained in Q, and there is an indecomposable direct summand M of $\mathrm{Ind}^G_H(N)$ such that (Q, V) is a vertex-source pair of M.*

Proof Since N is isomorphic to a direct summand of $\mathrm{Ind}^H_Q(V)$, it follows that $\mathrm{Ind}^G_H(N)$, and hence any direct summand of $\mathrm{Ind}^G_H(N)$, is isomorphic to a direct summand of $\mathrm{Ind}^G_Q(V)$. This shows that any indecomposable direct summand of $\mathrm{Ind}^G_H(N)$ has a vertex contained in Q. Since V is a direct summand of $\mathrm{Res}^H_Q(N)$ and N is a direct summand of $\mathrm{Res}^G_H(\mathrm{Ind}^G_H(N))$, it follows that there is an indecomposable direct summand M of $\mathrm{Ind}^G_H(N)$ such that V is isomorphic to a direct summand of $\mathrm{Res}^G_Q(M)$. It follows from 5.1.5 (ii) that Q is contained in a vertex of M. But Q also contains a vertex of M, hence is a vertex of M. It follows from 5.1.2 (iv) that V is a source of M. □

Proposition 5.1.8 *Let G be a finite group and Q a normal p-subgroup of G. Let R be a p-subgroup of G containing Q. Let M be a finitely generated indecomposable $\mathcal{O}G/Q$-module. Then M is relatively $\mathcal{O}R/Q$-projective as an $\mathcal{O}G/Q$-module if and only if M is relatively $\mathcal{O}R$-projective when viewed as an $\mathcal{O}G$-module via the canonical surjection $\mathcal{O}G \to \mathcal{O}G/Q$. In particular, R/Q is a vertex of M as an $\mathcal{O}G/Q$-module if and only if R is a vertex of M as an $\mathcal{O}G$-module. In that case the canonical surjection $R \to R/Q$ induces a bijection between the isomorphism classes of $\mathcal{O}R/Q$-sources and $\mathcal{O}R$-sources of M as a module over $\mathcal{O}G/Q$ and $\mathcal{O}G$, respectively.*

Proof The relative trace map $\mathrm{Tr}^{G/Q}_{R/Q}$ on $\mathrm{End}_{\mathcal{O}}(M)$, viewed as an interior G/Q-algebra is equal to the relative trace map Tr^G_Q on $\mathrm{End}_{\mathcal{O}}(M)$, viewed as an interior G-algebra via thc canonical surjection $G \to G/Q$. Thus the statements on relative projectivity and vertices follow immediately from Higman's criterion 2.6.2. First restricting M to R/Q and then inflating to R coincides with first inflating M to G and then restricting to R, so the statement on sources in trivial. □

Given a finite group G, the $\mathcal{O}$-dual $M^* = \mathrm{Hom}_{\mathcal{O}}(M, \mathcal{O})$ of a finitely generated $\mathcal{O}$-free $\mathcal{O}G$-module M is considered as a left $\mathcal{O}G$-module via $(x \cdot \mu)(m) =$

$\mu(x^{-1}m)$ for $m \in M$, $\mu \in M^*$ and $x \in G$. Taking $\mathcal{O}$-duals of finitely generated $\mathcal{O}$-free modules preserves the property of being indecomposable because applying duality twice is the identity functor on finitely generated $\mathcal{O}$-free modules. Vertices and sources are compatible with duality:

Proposition 5.1.9 *Let G be a finite group and M a finitely generated indecomposable $\mathcal{O}$-free $\mathcal{O}G$-module. Let Q be a vertex of M and V an $\mathcal{O}Q$-source of M. Then Q is a vertex of M^* and V^* is an $\mathcal{O}Q$-source of M^*.*

Proof The result follows from the obvious isomorphism $(\mathrm{Ind}_Q^G(V))^* \cong \mathrm{Ind}_Q^G(V^*)$. □

Vertices and sources are compatible with the Heller operator Ω_G and its inverse Σ_G on the $\mathcal{O}$-stable category $\underline{\mathrm{mod}}(\mathcal{O}G)$. If M is an indecomposable $\mathcal{O}G$-module, then $\Omega_G(M)$ is indecomposable in $\underline{\mathrm{mod}}(\mathcal{O}G)$; by choosing a minimal relatively $\mathcal{O}$-projective cover of M we may assume that $\Omega_G(M)$ is chosen indecomposable (and so the next statement makes sense).

Proposition 5.1.10 *Let G be a finite group and M a finitely generated indecomposable $\mathcal{O}G$-module with a vertex Q and $\mathcal{O}Q$-source V. Then, for any integer n, the $\mathcal{O}G$-module $\Omega_G^n(M)$ has vertex Q and $\mathcal{O}Q$-source $\Omega_Q^n(V)$.*

Proof Let $0 \longrightarrow \Omega_G(M) \longrightarrow Y \longrightarrow M \longrightarrow 0$ be an $\mathcal{O}$-split short exact sequence of $\mathcal{O}G$-modules with Y finitely generated relatively $\mathcal{O}$-projective. The restriction to $\mathcal{O}Q$ of this sequence is again $\mathcal{O}$-split, and the middle term remains relatively $\mathcal{O}$-projective. Since V is a direct summand of $\mathrm{Res}_Q^G(M)$, it follows that $\Omega_Q(V)$ is a direct summand of $\Omega_G(M)$. Therefore, $\Omega_G(M)$ has a vertex containing Q. Since Ω_G and Σ_G are inverse functors on $\underline{\mathrm{mod}}(\mathcal{O}G)$, it follows that Q is a vertex of $\Omega_G(M)$, and hence $\Omega_Q(V)$ is an $\mathcal{O}Q$-source of $\Omega_G(M)$. Iterating this argument and its dual using Σ_G implies the result. □

Given a finite group G and two $\mathcal{O}G$-modules U, U' we consider the tensor product $U \otimes_{\mathcal{O}} U'$ as an $\mathcal{O}G$-module with $x \in G$ acting diagonally by $x(u \otimes u') = xu \otimes xu'$ for all $u \in U$, $u' \in U'$.

Theorem 5.1.11 *Let G be a finite group and let U, U' be finitely generated indecomposable $\mathcal{O}G$-modules with vertices Q, Q', respectively. Then every indecomposable direct summand of $U \otimes_{\mathcal{O}} U'$ has a vertex contained in $Q \cap {}^x(Q')$ for some $x \in G$.*

Proof This follows from 2.4.9. Alternatively, by Higman's criterion there is $\varphi \in \mathrm{End}_{\mathcal{O}Q}(U)$ such that $\mathrm{Id}_U = \mathrm{Tr}_Q^G(\varphi)$. A trivial computation shows that

$\mathrm{Tr}_Q^G(\varphi \otimes \mathrm{Id}_{U'}) = \mathrm{Id}_U \otimes \mathrm{Id}_{U'}$. Thus every indecomposable direct summand V of $U \otimes_{\mathcal{O}} U'$ has a vertex R contained in Q. The same argument shows that V has a vertex R' contained in Q'. Since all vertices of V are conjugate, there is $x \in G$ such that $R = {}^x(R') \subseteq {}^x(Q')$, hence $R \subseteq Q \cap {}^x(Q')$. □

Corollary 5.1.12 *Let G be a finite group and let U, U' be finitely generated $\mathcal{O}$-free $\mathcal{O}G$-modules. If one of U, U' is projective then $U \otimes_{\mathcal{O}} U'$ and $\mathrm{Hom}_{\mathcal{O}}(U, U')$ are projective.*

Proof This follows from 2.4.10. Alternatively, if one of U, U' is projective, hence relatively 1-projective, then by 5.1.11, $U \otimes_{\mathcal{O}} U'$ is relatively 1-projective. Since U, U' are $\mathcal{O}$-free, so is $U \otimes_{\mathcal{O}} U'$, hence every indecomposable summand of $U \otimes_{\mathcal{O}} U'$ has $\mathcal{O}$ as a source, and hence is projective. The statement on $\mathrm{Hom}_{\mathcal{O}}(U, U')$ is a consequence of this together with the isomorphism $\mathrm{Hom}_{\mathcal{O}}(U, U') \cong U' \otimes_{\mathcal{O}} U^*$ from 2.9.4 □

Although 5.1.11 gives some information about vertices of summands of tensor products of $\mathcal{O}G$-modules, it says little about the structure of these summands. If a tensor product $U \otimes_{\mathcal{O}} V$ has a trivial direct summand, then since a trivial $\mathcal{O}G$-module has a Sylow p-subgroup P of G as a vertex, it follows from 5.1.11 that in that case both U, V have P as a vertex. The following result – for $\mathcal{O} = k$ also known as Benson's Lemma – is considerably more precise:

Theorem 5.1.13 ([7, 2.1], [1]) *Let G be a finite group and U, V finitely generated absolutely indecomposable $\mathcal{O}$-free $\mathcal{O}G$-modules. The following are equivalent.*

(i) The $\mathcal{O}G$-module $U \otimes_{\mathcal{O}} V$ has a trivial direct summand $\mathcal{O}$.
(ii) We have $V \cong U^*$ and the $\mathcal{O}$-rank of U is prime to p.

Moreover, if $\mathcal{O} = k$ is algebraically closed, then the multiplicity of k as a direct summand of $U \otimes_k U^*$ is at most 1.

Proof If (ii) holds then $\mathrm{tr}_U : \mathrm{End}_{\mathcal{O}}(U) \to \mathcal{O}$ is surjective, with kernel $\ker(tr_U)$. Thus $U \otimes_{\mathcal{O}} U^* \cong \mathrm{End}_{\mathcal{O}}(U) \cong \mathcal{O}\mathrm{Id}_U \oplus \ker(tr_U)$. This is a direct sum decomposition of $\mathcal{O}G$-modules, whence (i). Conversely, suppose that (i) holds. By 2.9.7, both $U \otimes_{\mathcal{O}} U^*$ and $V \otimes_{\mathcal{O}} V^*$ have a trivial direct summand. It follows from 2.7.2 that the trace map tr_U (which represents the adjunction counit of 2.9.5) is split surjective. Thus there is $\tau \in \mathrm{End}_{\mathcal{O}}(U)$ satisfying $\mathrm{tr}_U(\tau) = 1$. Since U is absolutely indecomposable, we may write $\tau = \lambda \mathrm{Id}_U + \rho$ for some $\lambda \in \mathcal{O}$ and some $\rho \in J(\mathrm{End}_{\mathcal{O}G}(U))$. Then $1 = \mathrm{tr}_U(\tau) = \lambda \mathrm{rk}_{\mathcal{O}}(U) + \mathrm{tr}_U(\rho)$. Since $\mathrm{tr}_U(\rho) \in J(\mathcal{O})$ we get that λ and $\mathrm{rk}_{\mathcal{O}}(U)$ are invertible in $\mathcal{O}$; in particular, the rank of U is prime to p. Consider next the commutative diagram

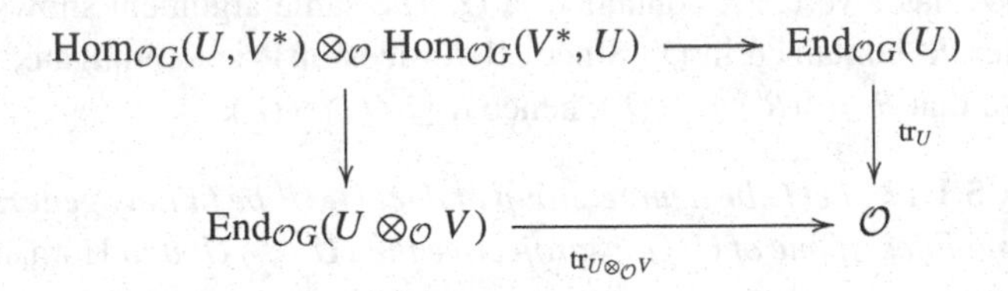

obtained from 2.10.3 by taking G-fixed points. Let T be a trivial direct summand of $U \otimes_{\mathcal{O}} V$ and let $\tau \in \mathrm{End}_{\mathcal{O}G}(U \otimes_{\mathcal{O}} V)$ be a projection of $U \otimes_{\mathcal{O}} V$ onto T. Clearly $\mathrm{tr}_{U\otimes_{\mathcal{O}}V}(\tau) = 1$. Through the isomorphism $\mathrm{End}_{\mathcal{O}}(U \otimes_{\mathcal{O}} V) \cong U^* \otimes_{\mathcal{O}} V^* \otimes_{\mathcal{O}} U \otimes_{\mathcal{O}} V$, the endomorphism τ corresponds to an element in $U^* \otimes_{\mathcal{O}} V^* \otimes_{\mathcal{O}} T$, hence an element of the form $w \otimes t$ for some nonzero $t \in T$ and some $w \in U^* \otimes_{\mathcal{O}} V^*$. Since τ and t are fixed by G, so is w. Under the isomorphism $U^* \otimes_{\mathcal{O}} V^* \cong \mathrm{Hom}_{\mathcal{O}}(U, V^*)$, the element w corresponds thus to an element $\alpha \in \mathrm{Hom}_{\mathcal{O}G}(U, V^*)$. Through the isomorphism $U \otimes_{\mathcal{O}} V \cong \mathrm{Hom}_{\mathcal{O}}(V^*, U)$, the element t corresponds to an element $\beta \in \mathrm{Hom}_{\mathcal{O}G}(V^*, U)$. Thus τ is the image of $\alpha \otimes \beta$ under the left vertical map in the above diagram. It follows that the map from the left upper corner to the right lower corner in this diagram is still surjective. In particular, the image of the upper horizontal map is not contained in $J(\mathrm{End}_{\mathcal{O}G}(U))$. Since U is absolutely indecomposable this forces $U \cong V^*$. If $\mathcal{O} = k$ is algebraically closed, the above argument also shows that every nonzero element in a trivial direct summand of $U \otimes_k U^*$ corresponds to an automorphism of U. Thus, if $k \oplus k$ were a summand of $U \otimes_k U^*$, we would get a 2-dimensional subspace of $\mathrm{End}_{kG}(U)$ all of whose nonzero elements are invertible. This is, however, impossible by 1.9.4. □

Combining 5.1.11, 5.1.4 and 5.1.13 shows in particular that an indecomposable $\mathcal{O}$-free $\mathcal{O}G$-module whose rank is prime to p must have a Sylow p-subgroup of G as a vertex. See 5.12.13 below for more precise statements.

Proposition 5.1.14 *Let G be a finite group, Q a subgroup and U an $\mathcal{O}G$-module. Let $\mathcal{O}'/\mathcal{O}$ be an extension of complete local Noetherian commutative rings. If U is relatively Q-projective then the $\mathcal{O}'G$-module $\mathcal{O}' \otimes_{\mathcal{O}} U$ is relatively Q-projective; in particular, if U is finitely generated indecomposable with Q as a vertex then any indecomposable direct summand of the $\mathcal{O}'G$-module $\mathcal{O}' \otimes_{\mathcal{O}} U$ has a vertex contained in Q.*

Proof By Higman's criterion, U is relatively Q-projective if and only if $\mathrm{Id}_U = \mathrm{Tr}_Q^G(\varphi)$ for some $\varphi \in \mathrm{End}_{\mathcal{O}Q}(U)$. In that case, $\mathrm{Id}_{\mathcal{O}'\otimes_{\mathcal{O}}U} = \mathrm{Id}_{\mathcal{O}'} \otimes \mathrm{Id}_U = \mathrm{Tr}_Q^G(\mathrm{Id}_{\mathcal{O}'} \otimes \varphi)$, hence $\mathcal{O}' \otimes_{\mathcal{O}} U$ is relatively Q-projective. The result follows. □

We have mentioned already that for G, H two finite groups, an $\mathcal{O}G$-$\mathcal{O}H$-bimodule M can be viewed as an $\mathcal{O}(G \times H)$-module, with $(x, y) \in G \times H$ acting on $m \in M$ by $(x, y) \cdot m = xmy^{-1}$, and vice versa, via the same formula, any $\mathcal{O}(G \times H)$-module can be viewed as an $\mathcal{O}G$-$\mathcal{O}H$-bimodule. Thus if M is indecomposable as a bimodule, it is indecomposable as a left $\mathcal{O}(G \times H)$-module, and hence gives rise to a vertex (a p-subgroup of $G \times H$) and a source. We extend in this way the terminology of vertices and sources to bimodules. For P a common subgroup of groups G, H, we denote by ΔP the 'diagonal' subgroup $\Delta P = \{(y, y)|y \in P\}$ of $G \times H$. Any $\mathcal{O}P$-module V can be viewed as an $\mathcal{O}\Delta P$-module via the obvious isomorphism $P \cong \Delta P$ sending $y \in P$ to (y, y). The following lemma will be useful for identifying vertices of bimodules reinterpreted as left modules in cases where the vertices are diagonal subgroups.

Lemma 5.1.15 *Let G, H be a finite group, P a common subgroup of G, H, and U an $\mathcal{O}P$-module.*

(i) We have an isomorphism of $\mathcal{O}G$-$\mathcal{O}H$-bimodules

$$\mathrm{Ind}_{\Delta P}^{G\times H}(U) \cong \mathcal{O}G \otimes_{\mathcal{O}P} (U \otimes_{\mathcal{O}} \mathcal{O}H)$$

sending $(g, h) \otimes u$ to $g \otimes (u \otimes h^{-1})$, where $g \in G$, $h \in H$ and $u \in U$. On the right side, $U \otimes_{\mathcal{O}} \mathcal{O}H$ is considered as an $\mathcal{O}P$-$\mathcal{O}H$-bimodule via $y(v \otimes h) = yv \otimes yh$ and $(u \otimes h)z = u \otimes hz$, for $v \in V$, $y \in P$, $h, z \in H$.

(ii) We have an isomorphism of $\mathcal{O}G$-$\mathcal{O}H$-bimodules

$$\mathrm{Ind}_{\Delta P}^{G\times H}(\mathcal{O}) \cong \mathcal{O}G \otimes_{\mathcal{O}P} \mathcal{O}H$$

sending $(g, h) \otimes \lambda$ to $g \otimes (\lambda \otimes h^{-1})$, where $g \in G$, $h \in H$ and $\lambda \in \mathcal{O}$.

Proof The first statement is a straightforward verification, and the second statement is the special case where $U = \mathcal{O}$. □

This Lemma is used to show the following bimodule version of Theorem 5.1.11.

Theorem 5.1.16 *Let G, H, L be finite groups, P a common subgroup of G and H, and Q a common subgroup of H and L. Let M be an indecomposable $\mathcal{O}G$-$\mathcal{O}H$-bimodule with a vertex $\mathcal{O}\Delta P$ and $\mathcal{O}\Delta P$-source U, and let N be an indecomposable $\mathcal{O}H$-$\mathcal{O}L$-bimodule with a vertex ΔQ and $\mathcal{O}\Delta Q$-source V. Then any indecomposable direct summand of the $\mathcal{O}G$-$\mathcal{O}L$-bimodule $M \otimes_{\mathcal{O}H} N$ has a vertex R and $\mathcal{O}R$-source W such that R is a subgroup of $\{(y, {}^{s^{-1}}y)|y \in P \cap {}^sQ\}$ for some $s \in H$, and W is a direct summand of the restriction to R of $U \otimes_{\mathcal{O}} {}^sV$.*

Proof Note that $\{(y, {}^{s^{-1}}y)|y \in P \cap {}^sQ\}$, where $s \in H$, is a subgroup of $P \times Q$, hence of $G \times L$ by the assumptions, so the statement makes sense. Since M is a summand of $\mathrm{Ind}_{\Delta P}^{G\times H}(U)$ and N a summand of $\mathrm{Ind}_{\Delta Q}^{H\times L}(V)$, it follows from 5.1.15

that an indecomposable direct summand X of $M \otimes \mathcal{O}HN$ is an indecomposable direct summand of

$$\mathcal{O}G \otimes_{\mathcal{O}P} (U \otimes_{\mathcal{O}} \mathcal{O}H) \otimes_{\mathcal{O}Q} \mathcal{O}H \otimes_{\mathcal{O}Q} (V \otimes_{\mathcal{O}} \mathcal{O}L)$$

hence of

$$\mathcal{O}G \otimes_{\mathcal{O}P} Y \otimes_{\mathcal{O}Q} \mathcal{O}H \otimes_{\mathcal{O}Q} (V \otimes_{\mathcal{O}} \mathcal{O}L)$$

for some indecomposable direct summand Y of the $\mathcal{O}P$-$\mathcal{O}Q$-bimodule $U \otimes_{\mathcal{O}} \mathcal{O}H$, where P acts on the left by $y(u \otimes h) = yu \otimes yh$ and Q acts on the right by $u \otimes h)z = u \otimes hz$, for $u \in U, y \in P, z \in Q$ and $h \in H$. Thus Y is isomorphic to a direct summand of

$$\begin{aligned} U \otimes_{\mathcal{O}} \mathcal{O}[PsQ] &\cong U \otimes_{\mathcal{O}} \mathcal{O}P \otimes_{\mathcal{O}(P \cap {}^sQ)} \mathcal{O}[sQ] \\ &\cong \mathcal{O}P \otimes_{\mathcal{O}(P \cap {}^sQ)} (U \otimes_{\mathcal{O}} \mathcal{O}[sQ]) \end{aligned}$$

for some element $s \in H$, where the first isomorphism is from 2.4.1 (ii) and the second isomorphism sends $u \otimes (y \otimes sz)$ to $y \otimes (u \otimes sz)$, for $u \in U, y \in P$ and $z \in Q$. Thus X is isomorphic to a direct summand of

$$\begin{aligned} &\mathcal{O}G \otimes_{\mathcal{O}(P \cap {}^sQ)} (U \otimes \mathcal{O}[sQ]) \otimes_{\mathcal{O}Q} \otimes (V \otimes_{\mathcal{O}} \mathcal{O}L) \\ &\cong \mathcal{O}G \otimes_{\mathcal{O}(P \cap {}^sQ)} (U \otimes {}^sV) \otimes_{\mathcal{O}} \mathcal{O}L. \end{aligned}$$

The result follows from 5.1.15, applied to the subgroup $P \cap {}^sQ$ of G, identified to a subgroup of L via the homomorphism induced by conjugation with s^{-1}. □

Remark 5.1.17 Theorem 5.1.16 is a special case of a more general formula for tensor products of induced bimodules due to Bouc [11]. The hypothesis that P is a common subgroup of G and H is somewhat artificial in that it amounts to choosing an identification of a subgroup of G with a subgroup of H. The conclusion of Theorem 5.1.16 illustrates that this may become an issue: the subgroup R is no longer a diagonal subgroup of $G \times L$ but a ‘twisted’ diagonal subgroup. In applications of this theorem further along we will usually have additional assumptions on fusion which will permit us to play back the situation to a case where R is again a diagonal subgroup. This issue will reappear in Chapter 9; see Remark 9.11.12.

We conclude this section with a consequence of 5.1.13 that shows that relative separability is preserved under certain Morita equivalences.

Proposition 5.1.18 *Let P be a finite group, A a relatively $\mathcal{O}P$-separable interior P algebra and V an $\mathcal{O}$-free $\mathcal{O}P$-module of finite rank n that is prime to p.*

Set $S = \mathrm{End}_{\mathcal{O}}(V)$, viewed as an interior P-algebra. Then the interior P-algebra $S \otimes_{\mathcal{O}} A$ is relatively $\mathcal{O}P$-separable.

Proof The interior P-algebra structure of $S \otimes_{\mathcal{O}} A$ is given diagonally. We need to show that $S \otimes_{\mathcal{O}} A$ is isomorphic to a direct summand of $(S \otimes_{\mathcal{O}} A) \otimes_{\mathcal{O}P} (S \otimes_{\mathcal{O}} A)$ as an $S \otimes_{\mathcal{O}} A$-$S \otimes_{\mathcal{O}} A$-bimodule. By the assumptions, A is isomorphic to a direct summand of $A \otimes_{\mathcal{O}P} A$ as an A-A-bimodule. Tensoring with S implies that $S \otimes_{\mathcal{O}} A$ is isomorphic to a direct summand of $(S \otimes_{\mathcal{O}} A) \otimes_{(S \otimes_{\mathcal{O}} \mathcal{O}P)} (S \otimes_{\mathcal{O}} A)$ as a bimodule. Thus it suffices to show that $S \otimes_{\mathcal{O}} \mathcal{O}P$ is isomorphic to a direct summand of $(S \otimes_{\mathcal{O}} \mathcal{O}P) \otimes_{\mathcal{O}P} (S \otimes_{\mathcal{O}} \mathcal{O}P)$ as a bimodule. Consider the canonical isomorphism $S \cong V \otimes_{\mathcal{O}} V^*$, where on the right side the S-S-bimodule structure is given by the canonical left and right S-module structure on V and V^*, respectively. Thus $S \otimes_{\mathcal{O}} \mathcal{O}P \cong (V \otimes_{\mathcal{O}} V^*) \otimes_{\mathcal{O}} \mathcal{O}P$. The right $\mathcal{O}P$-module structure given by the (diagonal) interior P-algebra structure of $S \otimes_{\mathcal{O}} \mathcal{O}P$ is given by the diagonal action of P on the right of V^* and $\mathcal{O}P$; the corresponding left action is given by the diagonal action of P on V and $\mathcal{O}P$. With these actions, we have

$$
\begin{aligned}
(S \otimes_{\mathcal{O}} \mathcal{O}P) \otimes_{\mathcal{O}P} (S \otimes_{\mathcal{O}} \mathcal{O}P) &\cong (V \otimes_{\mathcal{O}} V^* \otimes_{\mathcal{O}} \mathcal{O}P) \otimes_{\mathcal{O}P} (V \otimes_{\mathcal{O}} V^* \otimes_{\mathcal{O}} \mathcal{O}P) \\
&\cong V \otimes_{\mathcal{O}} (V^* \otimes_{\mathcal{O}} \mathcal{O}P) \otimes_{\mathcal{O}P} (\mathcal{O}P \otimes_{\mathcal{O}} V) \otimes V^*.
\end{aligned}
$$

The middle part of this expression is the $\mathcal{O}P$-$\mathcal{O}P$-bimodule $(V^* \otimes_{\mathcal{O}} \mathcal{O}P) \otimes_{\mathcal{O}P} (\mathcal{O}P \otimes_{\mathcal{O}} V)$, and it suffices to show that this bimodule has a direct summand isomorphic to $\mathcal{O}P$. Using the appropriate versions of 2.4.12 and 2.4.13 it follows that this bimodule is isomorphic to

$$
\mathrm{Ind}_{\Delta P}^{P \times P}(V) \otimes_{\mathcal{O}P} \mathrm{Ind}_{\Delta P}^{P \times P}(V^*) \cong \mathrm{Ind}_{\Delta P}^{P \times P}(V \otimes_{\mathcal{O}} V^*).
$$

Since V has rank prime to p, it follows from 5.1.13 that $V \otimes_{\mathcal{O}} V^*$ has a trivial direct summand, and hence the above bimodule has a direct summand $\mathrm{Ind}_{\Delta P}^{P \times P}(\mathcal{O}) \cong \mathcal{O}P$ as required. The result follows. □

Remark 5.1.19 Let P be a finite p-group, and let A be an interior P-algebra. Suppose that A is relatively $\mathcal{O}P$-separable. Then every A-module is relatively $\mathcal{O}P$-projective. Thus, for any indecomposable A-module M there is a minimal subgroup Q of P such that M is relatively $\mathcal{O}Q$-projective, and then there is an indecomposable direct summand V of $\mathrm{Res}_Q(M)$ such that M is isomorphic to a direct summand of $A \otimes_{\mathcal{O}Q} V$. We call Q a *vertex of M* and V an *$\mathcal{O}Q$-source*. The main drawback of allowing the use of this terminology in this generality is that vertex-source pairs need not be unique in any sense, and one needs to carefully consider this aspect depending on the circumstances.

Remark 5.1.20 We took care to formulate Higman's criterion 2.6.2 and the Krull–Schmidt Theorem 4.6.11 for bounded complexes of finitely generated modules over finite group algebras. Thus the definition of a vertex and a source of an indecomposable module extends to indecomposable bounded complexes of finitely generated modules. The results 5.1.2, 5.1.5, 5.1.6, 5.1.7, 5.1.8, 5.1.9, and 5.1.14 extend verbatim to bounded complexes of finitely generated modules.

5.2 The Green correspondence

The Green correspondence, due to J. A. Green, is a vertex and source preserving correspondence between indecomposable modules of a finite group G over a complete p-local ring and modules of normalisers of p-subgroups of G. Modules in this section are finitely generated.

Theorem 5.2.1 (Green Correspondence) *Let G be a finite group, let Q be a p-subgroup of G and let H be a subgroup of G containing $N_G(Q)$. There is a bijection between the sets of isomorphism classes of indecomposable $\mathcal{O}G$-modules with Q as a vertex and indecomposable $\mathcal{O}H$-modules with Q as a vertex given as follows.*

(i) If U is an indecomposable $\mathcal{O}G$-module having Q as a vertex, then there is, up to isomorphism, a unique indecomposable direct summand $f(U)$ of $\mathrm{Res}^G_H(U)$ having Q as a vertex. Moreover, every $\mathcal{O}Q$-source of $f(U)$ is a source of U, and every other direct summand of $\mathrm{Res}^G_H(U)$ has a vertex that is contained in ${}^xQ \cap H$ for some $x \in G \setminus H$ and that is not H-conjugate to Q.

(ii) If V is an indecomposable $\mathcal{O}H$-module having Q as a vertex, then there is, up to isomorphism, a unique indecomposable direct summand $g(V)$ of $\mathrm{Ind}^G_H(V)$ having Q as a vertex. Moreover, every $\mathcal{O}Q$-source of V is a source of $g(V)$, and every other indecomposable direct summand of $\mathrm{Ind}^G_H(V)$ has a vertex contained in $Q \cap {}^xQ$ for some $x \in G \setminus H$.

(iii) We have $g(f(U)) \cong U$ and $f(g(V)) \cong V$.

The $\mathcal{O}H$-module $f(U)$ in 5.2.1 is called the *Green correspondent of U with respect to Q*, or simply *Green correspondent*, if Q is clear from the context. Similarly, the $\mathcal{O}G$-module $g(V)$ is called the *Green correspondent of V with respect to Q*. The Green correspondence does not say anything about the structural connections between U and $f(U)$. For instance, if one of U, $f(U)$ is simple this does not imply that the other is simple as well. If Q and Q' are two vertices of U such that $N_G(Q)$ and $N_G(Q')$ are contained in H, then the Green correspondents $f(U)$ and $f'(U)$ with respect to Q and Q', respectively, need not be isomorphic because Q and Q' need no longer be conjugate in H. By contrast, if Q and Q' are two vertices of V such that $N_G(Q)$ and $N_G(Q')$ are contained in H, then the Green correspondents $g(U)$ and $g'(U)$ with respect to Q and Q', respectively, are isomorphic. Indeed, the vertices Q and Q' of V are H-conjugate, so also G-conjugate, and hence Q, Q' are both vertices of $g'(U)$. The uniqueness statement in 5.2.1 (i) implies that $g'(U) \cong g(U)$.

We formulate a technical part of the proof of the Green correspondence as a Lemma, for future reference.

Lemma 5.2.2 *Let G be a finite group, let Q be a p-subgroup of G and let H be a subgroup of G containing $N_G(Q)$. Let V be a relatively Q-projective indecomposable $\mathcal{O}H$-module.*

(i) We have $\mathrm{Res}^G_H \mathrm{Ind}^G_H(V) = V \oplus V'$ for some $\mathcal{O}H$-module V' such that every indecomposable summand of V' has a vertex contained in $H \cap {}^xQ$ for some $x \in G \setminus H$. In particular, no indecomposable direct summand of V' has Q as a vertex.

(ii) We have $\mathrm{Ind}^G_H(V) = U \oplus U'$ for some $\mathcal{O}G$-modules U, U' such that U is indecomposable, V is a summand of $\mathrm{Res}^G_H(U)$, and every indecomposable summand of U' has a vertex contained in $Q \cap {}^xQ$ for some $x \in G \setminus H$. In particular, no indecomposable direct summand of U' has Q as a vertex.

Proof By Mackey's formula we have

$$\mathrm{Res}^G_H \mathrm{Ind}^G_H(V) = \oplus_{x \in [H \backslash G/H]} \mathrm{Ind}^H_{H \cap {}^xH} \mathrm{Res}^{{}^xH}_{H \cap {}^xH}({}^xV) \cong V \oplus V'.$$

Now V is relatively Q-projective, and V is indecomposable. Thus V is a direct summand of $\mathrm{Ind}^H_Q(S)$ for some indecomposable $\mathcal{O}Q$-module S. Write

$$\mathrm{Ind}^H_Q(S) = V \oplus V_0$$

for some $\mathcal{O}H$-module V_0. As for V we may write

$$\mathrm{Res}^G_H \mathrm{Ind}^G_H(V_0) = V_0 \oplus V_0'$$

for some $\mathcal{O}H$-module V_0'. We compute $\mathrm{Res}^G_H \mathrm{Ind}^G_Q(S)$ in two different ways. On the one hand, the Mackey formula yields that

$$\mathrm{Res}^G_H \mathrm{Ind}^G_Q(S) = \oplus_{x \in [H \backslash G/Q]} \mathrm{Ind}^H_{H \cap {}^xQ} \mathrm{Res}^{xQ}_{H \cap {}^xQ}({}^xS) = \mathrm{Ind}^H_Q(S) \oplus V'',$$

where V'' has the property that all indecomposable summands have a vertex contained in $H \cap {}^xQ$ for some $x \in G \setminus H$. On the other hand,

$$\begin{aligned}\mathrm{Res}^G_H\mathrm{Ind}^G_S(S) &= \mathrm{Res}^G_H\mathrm{Ind}^G_H(V) \oplus \mathrm{Res}^G_H\mathrm{Ind}^G_H(V_0) = V \oplus V' \oplus V_0 \oplus V_0' \\ &= \mathrm{Ind}^H_Q(S) \oplus V' \oplus V_0'.\end{aligned}$$

The Krull–Schmidt Theorem implies that $V'' = V' \oplus V_0'$. In particular, all indecomposable summands of V' have a vertex contained in $H \cap {}^xQ$ for some $x \in G \setminus H$. Thus any indecomposable direct summand of V' has a vertex R contained in $H \cap {}^xQ$ for some $x \in G \setminus H$. If Q were a vertex of that summand as well, then ${}^hR = Q$ for some $h \in H$. But then $Q \leq H \cap {}^{hx}Q$, hence $Q \leq Q \cap {}^{hx}Q$, which is impossible since $hx \notin H$ does not normalise Q. This proves (i). For (ii), note that since V is a summand of $\mathrm{Res}^G_H(U)$ it follows from (i) that $\mathrm{Res}^G_H(U')$ is a summand of V', by the Krull–Schmidt Theorem. Every indecomposable summand of U' has a vertex contained in Q because U' is a summand of $\mathrm{Ind}^G_Q(S)$. Let U_1 be an indecomposable summand of U' and let $R \subseteq Q$ be a vertex of U_1. Let T be an $\mathcal{O}R$-source of U_1. Then T is a summand of $\mathrm{Res}^G_R(U_1)$, by 5.1.2. Thus, for some indecomposable summand V_1 of $\mathrm{Res}^G_H(U_1)$, the module T is a summand of $\mathrm{Res}^H_R(V_1)$. By (i), V_1 has a vertex contained in $H \cap {}^xQ$ for some $x \in G \setminus H$. Since all vertices of V_1 are H-conjugate, there is $y \in H$ and $x \in G \setminus H$ such that ${}^yR \subseteq H \cap {}^xQ$. Thus $R \subseteq H \cap {}^{y^{-1}x}Q$, hence $R \subseteq Q \cap {}^{y^{-1}x}Q$ as required. □

Proof of Theorem 5.2.1 Let S be an $\mathcal{O}Q$-source of U. Then U is a summand of $\mathrm{Ind}^G_Q(S) = \mathrm{Ind}^G_H\mathrm{Ind}^H_Q(S)$. Since U is indecomposable, there is an indecomposable summand V of $\mathrm{Ind}^H_Q(S)$ such that U is a summand of $\mathrm{Ind}^G_H(V)$. Write $\mathrm{Ind}^H_Q(S) = V \oplus V_0$ for some $\mathcal{O}H$-module V_0. Note that then Q is a vertex of V and S an $\mathcal{O}Q$-source of V. By Mackey's formula we can write

$$\mathrm{Res}^G_H\mathrm{Ind}^G_H(V) = V \oplus V'$$

for some $\mathcal{O}H$-module V'. We also have

$$\mathrm{Ind}^G_Q(S) = \mathrm{Ind}^G_H(V) \oplus \mathrm{Ind}^G_H(V_0).$$

Now $\mathrm{Res}^G_H(U)$ is a summand of $\mathrm{Res}^G_H\mathrm{Ind}^G_H(V)$, thus, by 5.2.2, $\mathrm{Res}^G_H(U)$ has at most one summand with vertex Q, namely V, and all other summands have a vertex contained in $H \cap {}^xQ$ for some $x \in G \setminus H$. Since U has vertex Q, which is contained in H, U is a summand of $\mathrm{Ind}^G_H\mathrm{Res}^G_H(U)$, by Higman's criterion. Since all vertices of U are G-conjugate, no vertex can be contained in $H \cap {}^xQ$ with $x \in G \setminus H$, and hence $\mathrm{Res}^G_H(U)$ must have V as a direct summand. Thus setting $f(U) = V$ yields the result. This proves (i). Note that this will also help to prove (iii), because by construction of V we have that U is a summand of $\mathrm{Ind}^G_H(V)$. By 5.2.2 there is an indecomposable summand U of $\mathrm{Ind}^G_H(V)$ such that V is a summand of $\mathrm{Res}^G_H(U)$, and then $\mathrm{Ind}^G_H(V) = U \oplus U'$, where U' is an $\mathcal{O}G$-module all of whose indecomposable summands have a vertex contained in $Q \cap {}^xQ$ for some $x \in G \setminus H$. Setting $g(V) = U$ proves (ii). Statement (iii) follows from the construction of the correspondences in (i) and (ii). □

The following Proposition from [47] adds some precisions regarding the Green correspondence and the passage to the normaliser of a vertex-source pair, rather than just the normaliser of a source. For G a finite group, H a subgroup of G, an $\mathcal{O}G$-module U and an indecomposable $\mathcal{O}H$-module V, the *multiplicity of V in a direct sum decomposition of* $\mathrm{Res}^G_H(U)$ is the number of direct sums isomorphic to V in a decomposition of $\mathrm{Res}^G_H(U)$ as a direct sum of indecomposable $\mathcal{O}H$-modules. By the Krull–Schmidt Theorem, this is well-defined.

Proposition 5.2.3 *Let G be a finite group and U an $\mathcal{O}$-free indecomposable $\mathcal{O}G$-module with vertex Q and $\mathcal{O}Q$-source Y. Set $N = N_G(Q)$ and $T = N_G(Q, Y) = \{x \in N | {}^xY \cong Y\}$. Denote by V an indecomposable direct summand of $\mathrm{Res}^G_N(U)$ with vertex Q and $\mathcal{O}Q$-source Y. Denote by m the multiplicity of V in a direct sum decomposition of $\mathrm{Res}^G_Q(U)$. The following hold.*

(i) The integer m is equal to the multiplicity of Y in a direct sum decomposition of $\mathrm{Res}^N_Q(V)$.
(ii) There is, up to isomorphism, a unique indecomposable direct summand W of $\mathrm{Res}^N_T(V)$ such that Y is isomorphic to a direct summand of $\mathrm{Res}^T_Q(W)$.
(iii) We have $\mathrm{Ind}^N_T(W) \cong V$ and $\mathrm{Res}^T_Q(W) \cong Y^m$.

Proof By the Green correspondence 5.2.1 we have $\mathrm{Res}^G_N(U) \cong V \oplus V'$ for some $\mathcal{O}N$-module V' which is a direct sum of indecomposable modules with vertices strictly contained in Q. In particular, Y is not isomorphic to a direct summand of $\mathrm{Res}^N_Q(V')$, whence (i). Again by the Green correspondence, V has Q as a vertex and Y as a source, so V is isomorphic to a direct summand of $\mathrm{Ind}^N_Q(Y) \cong \mathrm{Ind}^N_T(\mathrm{Ind}^T_Q(Y))$. Since V is indecomposable, it follows that there is an indecomposable direct summand W of $\mathrm{Ind}^T_Q(Y)$ such that V is isomorphic to a direct summand of $\mathrm{Ind}^N_T(W)$. Since T stabilises the isomorphism class of Y, it follows that $\mathrm{Res}^T_Q(W) \cong Y^n$ for some positive integer n. Since the elements in $N \setminus T$ do not stabilise Y, it follows that W is the unique indecomposable direct summand of $\mathrm{Res}^N_T(\mathrm{Ind}^N_T(W))$ such that Y is isomorphic to a direct summand of $\mathrm{Res}^T_Q(W)$. But then W must be isomorphic to a direct summand of $\mathrm{Res}^N_T(V)$, because Y is an $\mathcal{O}Q$-source of V. Note that this shows $n = m$. Since V is an $\mathcal{O}N$-module, its restriction to $\mathcal{O}Q$ has all conjugates ${}^xY$ as a direct summand with the same multiplicity as Y, so its $\mathcal{O}$-rank is $m \cdot \mathrm{rk}_{\mathcal{O}}(V) \cdot |N : T|$, which is equal to the $\mathcal{O}$-rank of $\mathrm{Ind}_T^N(W)$. Thus $V \cong \mathrm{Ind}_T^N(W)$, which completes the proof. □

The passage from N to T in the above Proposition is a special case of the passage from normalisers of p-groups to normalisers of *pointed groups*; see 5.5.1 below. The Green correspondence is particularly useful in situations where the intersections ${}^xQ \cap H$ and $Q \cap {}^xQ$ arising in 5.2.1 are trivial.

Corollary 5.2.4 *Let G be a finite group, P a p-subgroup and H a subgroup of G containing P such that $P \cap {}^xP = \{1\}$ for all $x \in G \setminus H$. Let U be an indecomposable $\mathcal{O}G$-module having a nontrivial vertex contained in P, and let V be an indecomposable $\mathcal{O}H$-module having a nontrivial vertex contained in P. Then $N_G(Q) \leq H$ for any nontrivial subgroup Q of P, and the following hold.*

(i) We have $\mathrm{Res}_H^G(U) \cong f(U) \oplus Y \oplus Z$ for some relatively $\mathcal{O}$-projective $\mathcal{O}H$-module Y and an $\mathcal{O}H$-module Z which is a direct sum of indecomposable $\mathcal{O}H$-modules having no vertex contained in P.
(ii) We have $\mathrm{Ind}_G^G(V) \cong g(V) \oplus X$ for some relatively $\mathcal{O}$-projective $\mathcal{O}G$-module X.

Proof Denote by Q a vertex of U such that $Q \subseteq P$. Since Q is nontrivial, it follows from the assumptions on H that $N_G(Q) \subseteq H$. Thus, by 5.2.1, the module $\mathrm{Res}_H^G(U)$ has a direct summand isomorphic to the Green correspondent $f(U)$ with respect to H. Let W be an indecomposable direct summand of $\mathrm{Res}_H^G(U)$ not isomorphic to $f(U)$. If W has a trivial vertex, then W is relatively $\mathcal{O}$-projective. Suppose that W has a nontrivial vertex R. By 5.2.1 (i), the subgroup R is contained in ${}^xQ \cap H$ for some $x \in G \setminus H$. Since ${}^xQ \cap P = \{1\}$, it follows that R is not contained in P. Statement (i) follows. The assumptions on H imply that ${}^xQ \cap Q = \{1\}$ for any subgroup Q of P, and hence (ii) follows from 5.2.1 (ii). □

This can be used to show a more precise result: if a Sylow p-subgroup of G has the trivial intersection property as in 5.2.4, then the restriction and induction functors yield a stable equivalence of Morita type.

Proposition 5.2.5 *Let G be a finite group, P a Sylow p-subgroup and H a subgroup of G containing P such that $P \cap {}^xP = \{1\}$ for all $x \in G \setminus H$. Then the functors Ind_H^G and Res_H^G induce an $\mathcal{O}$-stable equivalence between the $\mathcal{O}$-stable categories $\underline{\mathrm{Mod}}(\mathcal{O}G) \cong \underline{\mathrm{Mod}}(\mathcal{O}H)$. More precisely, the bimodule $M = \mathcal{O}G$, viewed as an $\mathcal{O}G$-$\mathcal{O}H$-bimodule, and its dual $M^* \cong \mathcal{O}G$ viewed as an $\mathcal{O}H$-$\mathcal{O}G$-bimodule, induce a stable equivalence of Morita type between $\mathcal{O}G$ and $\mathcal{O}H$.*

Proof Since $M \otimes_{\mathcal{O}H} -$ is isomorphic to the induction functor Ind_H^G and $M^* \otimes_{\mathcal{O}G} -$ is isomorphic to the restriction functor Res_H^G, the first statement follows from the second. We give a direct proof of the first statement for the sake of illustrating the use of the Green correspondence. Every p-subgroup of H is H-conjugate to a subgroup of the Sylow p-subgroup P of H. It follows from 5.2.4 that for any indecomposable $\mathcal{O}G$-module U with a nontrivial vertex we have $\mathrm{Res}_H^G(U) \cong f(U) \oplus Y$ for some projective $\mathcal{O}H$-module Y. Together with 5.2.4 (ii) and 5.2.1 (iii) it follows that the functors Res_H^G and Ind_H^G induce an $\mathcal{O}$-stable equivalence. To see that this is actually a stable equivalence of Morita type, consider $M^* \otimes_{\mathcal{O}G} M \cong \mathcal{O}G$, viewed as an $\mathcal{O}H$-$\mathcal{O}H$-bimodule. This is isomorphic to $\mathcal{O}H \oplus \mathcal{O}[G \setminus H]$. The action by $P \times P$ on $G \setminus H$ given by left and right multiplication with elements in P is free. Indeed if $u, v \in P$ and $x \in G \setminus H$, then the equality $uxv^{-1} = x$ implies $x^{-1}ux = v$, forcing $u = v = 1$ because of the assumption $P \cap {}^xP = \{1\}$. Thus $\mathcal{O}[G \setminus H]$ is projective as an $\mathcal{O}(P \times P)$-module. Since $P \times P$ is a Sylow p-subgroup of $H \times H$, it follows from 5.1.4 (iii) that the indecomposable direct summands of $\mathcal{O}[G \setminus H]$ as an $\mathcal{O}(H \times H)$-module are projective. Thus $\mathcal{O}[G \setminus H]$ is a projective $\mathcal{O}H$-$\mathcal{O}H$-bimodule. Consider next $M \otimes_{\mathcal{O}H} M^* \cong \mathcal{O}G \otimes_{\mathcal{O}H} \mathcal{O}G$. By the previous argument, we have an equality of $\mathcal{O}H$-$\mathcal{O}H$-bimodules $\mathcal{O}G = \mathcal{O}H \oplus Y$ for some projective $\mathcal{O}H$-$\mathcal{O}H$-bimodule Y. Thus, as an $\mathcal{O}H$-$\mathcal{O}H$-bimodule, we have $\mathcal{O}G \otimes_{\mathcal{O}H} \mathcal{O}G \cong \mathcal{O}H \oplus Y \oplus Y \oplus Y \otimes_{\mathcal{O}H} Y$, and clearly all but possibly the first summand are projective as $\mathcal{O}H$-$\mathcal{O}H$-bimodules. Now by 2.6.9, as an $\mathcal{O}G$-$\mathcal{O}G$-bimodule $\mathcal{O}G \otimes_{\mathcal{O}H} \mathcal{O}G$ has a summand isomorphic to $\mathcal{O}G$. Since $\mathcal{O}H$ is a summand of $\mathcal{O}G$ as an $\mathcal{O}H$-$\mathcal{O}H$-bimodule, it follows that every summand of the $\mathcal{O}G$-$\mathcal{O}G$-bimodule $\mathcal{O}G \otimes_{\mathcal{O}H} \mathcal{O}G$ other than $\mathcal{O}G$ is projective as an $\mathcal{O}H$-$\mathcal{O}H$-bimodule, hence as an $\mathcal{O}G$-$\mathcal{O}G$-bimodule by 5.1.4 (iii). □

The uniqueness properties of the Green correspondence imply the following transitivity result.

Proposition 5.2.6 *Let G be a finite group, H, L subgroups of G such that $L \leq H$, and let Q be a p-subgroup of L such that $N_G(Q) \leq L$. Let U, V, W be indecomposable module over $\mathcal{O}G$, $\mathcal{O}H$, $\mathcal{O}L$, respectively, with Q as a vertex. If two of these modules are Green correspondents of the third, then any two of these three modules are Green correspondents.*

Proof Suppose that V and W are Green correspondents of U. Then W is direct summand of $\mathrm{Res}_L^G(U)$, and V is a direct summand of $\mathrm{Res}_N^G(U)$. Thus there is an indecomposable direct summand V' of U such that W is a summand of $\mathrm{Res}_L^H(V')$. By 5.1.5 a vertex R of V' is contained in $H \cap xQx^{-1}$ for some $x \in Q$. By 5.1.5 again, W has a vertex contained in $L \cap yxQx^{-1}y$ for some $y \in H$. Since

W has Q as a vertex, it follows that $yxQx^{-1}y^{-1}$ is contained in L and conjugate to Q. Thus Q is a vertex of V', and hence V' is a Green correspondent of U. The uniqueness of the Green correspondence implies that $V' \cong V$. Thus W is a Green correspondent of V. Suppose that U and W are Green correspondents of V. Then V is a summand of $\mathrm{Res}^G_H(U)$ and W is a summand of $\mathrm{Res}^H_L(V)$. The transitivity of the restriction functor implies that W is a summand of $\mathrm{Res}^G_L(U)$. Suppose finally that V and U are Green correspondents of W. Then W is a summand of $\mathrm{Res}^G_L(U)$. Thus there is an indecomposable direct summand V' of $\mathrm{Res}^G_H(U)$ such that W is a summand of $\mathrm{Res}^H_L(V')$. Since Q is a vertex of W it follows that V' has a vertex containing Q. Since Q is a vertex of U it follows that V' has a vertex contained in a conjugate of Q, and hence Q is a vertex of V'. The uniqueness of the Green correspondence implies that $V' \cong V$. The result follows. □

Examples 5.2.7 The situation of 5.2.4 arises if P is cyclic or a generalised quaternion 2-group.

(1) Let p be a prime and let

$$G = SL_2(p) = \left\{ \begin{pmatrix} a & b \\ c & d \end{pmatrix} \mid a, b, c, d \in \mathbb{F}_p, ad - bc = 1 \right\}$$

be the special linear group on $\mathbb{F}_p^2$. Then $|G| = p(p-1)(p+1)$, and hence the group

$$P = \left\{ \begin{pmatrix} 1 & b \\ 0 & 1 \end{pmatrix} \mid b \in \mathbb{F}_p \right\}$$

is a Sylow p-subgroup of G. Elementary matrix calculations show that

$$N_G(P) = \left\{ \begin{pmatrix} a & b \\ 0 & a^{-1} \end{pmatrix} \mid a \in \mathbb{F}_p{}^{\times}, b \in \mathbb{F}_p \right\} = P \rtimes E,$$

where

$$E = \left\{ \begin{pmatrix} a & 0 \\ 0 & a^{-1} \end{pmatrix} \mid a \in \mathbb{F}_p{}^{\times} \right\}.$$

Since P is cyclic of order p, for every $x \in G \setminus N_G(P)$ we get that $N_G(P) \cap {}^xP = 1 = P \cap {}^xP$. Thus 5.2.5 implies that for any indecomposable nonprojective $\mathcal{O}$-free $\mathcal{O}G$-module U, the group P is a vertex of U and we have

$$\mathrm{Res}^G_{N_G(P)}(U) = f(U) \oplus X$$

for some projective $\mathcal{O}N_G(P)$-module X. Similarly, for every indecomposable nonprojective $\mathcal{O}$-free $\mathcal{O}N_G(P)$-module V, the group P is a vertex of V and we have

$$\mathrm{Ind}^G_{N_G(P)}(V) = g(V) \oplus Y$$

for some projective $\mathcal{O}G$-module Y. It is easy to determine the set $\mathrm{Irr}_K(N_G(P))$, where the quotient field K of $\mathcal{O}$ is assumed to be large enough, and from this information one can then in fact determine the set $\mathrm{Irr}_K(G)$.

(2) Suppose $p = 2$. Let G be a finite group having a quaternion Sylow 2-subgroup

$$Q = \langle x, y \mid x^{2^{n-1}} = 1 = y^4, x^{2^{n-2}} = y^2, yxy^{-1} = x^{-1} \rangle$$

of order 2^n for some integer $n \geq 3$. The element $z = y^2 = x^{2^{n-2}}$ is the unique involution of Q, and $Z(Q) = \{1, z\}$ has order 2. Set $H = C_G(z)$. Then z is the unique involution in H. Indeed, Q is contained in H because z is central in Q, hence Q is a Sylow 2-subgroup of H. If z' is an involution in H, then there is $y \in H$ such that ${}^y(z') \in Q$. Since z is the unique involution in Q this implies that ${}^y(z') = z$. But z is central in H and therefore $z' = z$. It follows that for any $x \in G \setminus H$ we have $H \cap {}^xQ = Q \cap {}^xQ = 1$. The uniqueness of the involution z implies that z is contained in any nontrivial subgroup R of Q. Thus z is the unique involution of any nontrivial subgroup R of Q, and hence $N_G(R) \subseteq H$. By 5.2.5, the Green correspondence implies that there is a bijection between the sets of isomorphism classes of indecomposable nonprojective $\mathcal{O}$-free modules over $\mathcal{O}G$ and over $\mathcal{O}H$.

(3) Suppose that $p = 2$. The alternating group A_4 has a Klein four subgroup P generated by the involutions $(12)(34)$ and $(13)(24)$ as a Sylow 2-subgroup. By considering A_4 as a subgroup of A_5, the group P remains a Sylow 2-subgroup of A_5. An easy verification shows that ${}^xP \cap P = \{1\}$ for $x \in A_5 \setminus A_4$. It follows from 5.2.5 that the functors $\mathrm{Res}^{A_5}_{A_4}$ and $\mathrm{Ind}^{A_5}_{A_4}$ induce a stable equivalence $\underline{\mathrm{Mod}}(\mathcal{O}A_5) \cong \underline{\mathrm{Mod}}(\mathcal{O}A_4)$.

Remark 5.2.8 As with many statements in the previous section, the Green correspondence 5.2.1 and the subsequent results 5.2.3, 5.2.4, and 5.2.6 can all be formulated more generally for bounded complexes of finitely generated modules. In fact, via the standard translation of direct summands of a module or a bounded complex of modules to points on its endomorphism algebra, the Green correspondence can be formulated more generally as a bijection of certain sets of points on G-algebras. This generalisation of the Green correspondence is called *Puig correspondence*. See [90, §19, §20] for more details.

5.3 Group actions on matrix algebras

Let V be a free $\mathcal{O}$-module of finite rank n and let G be a finite group acting on the matrix algebra $S = \mathrm{End}_{\mathcal{O}}(V)$. By the Skolem–Noether Theorem 4.1.5, any algebra automorphism of S is inner, and hence for any $x \in G$ there is an element $s_x \in S^\times$ such that x acts as conjugation by s_x on S; that is, such that ${}^x t = s_x t (s_x)^{-1}$ for all $t \in S$. Since $Z(S) \cong \mathcal{O}$, the elements s_x are unique up to nonzero scalars. For any $x, y \in G$ the elements $s_x s_y$ and s_{xy} both act as xy on S, and hence we have

$$s_x s_y = \alpha(x, y) s_{xy}$$

for some scalar $\alpha(x, y) \in \mathcal{O}^\times$. In other words, x acts on V as the linear map s_x, but this assignment falls short of defining an $\mathcal{O}G$-module structure of V because of the scalars $\alpha(x, y)$. This is traditionally called a 'projective representation', with the term 'projective' here borrowed from 'projective varieties', but we will not use this terminology in order to avoid confusion with the notions of projective modules and, more generally, projective objects in categories. The map $\alpha : G \times G \to \mathcal{O}^\times$ defined above is a 2-cocycle of G with coefficients in $\mathcal{O}^\times$. This is a routine verification: if x, y, $z \in G$ then $(s_x s_y) s_z = \alpha(x, y) s_{xy} s_z = \alpha(x, y)\alpha(xy, z) s_{xyz}$, but this is also equal to $s_x(s_y s_z) = \alpha(y, z) s_x s_{yz} = \alpha(y, z)\alpha(x, yz) s_{xyz}$, whence the 2-cocycle identity $\alpha(x, y)\alpha(xy, z) = \alpha(y, z)\alpha(x, yz)$. A different choice for the s_x leads to a 2-cocycle α' representing the same class in $H^2(G; \mathcal{O}^\times)$; indeed, if s'_x acts as x on S, then $s'_x = \gamma(x) s_x$ for some $\gamma(x) \in \mathcal{O}^\times$, and then a short calculation shows that $s'_x s'_y = \alpha'(x, y) s'_{xy}$, where $\alpha'(x, y) = \alpha(x, y)\gamma(x)\gamma(y)\gamma(xy)^{-1}$. The map sending $x \in G$ to s_x induces an algebra homomorphism

$$\mathcal{O}_\alpha G \to S = \mathrm{End}_{\mathcal{O}}(V),$$

implying that V becomes an $\mathcal{O}_\alpha G$-module. Thus the action of G on S 'lifts' to a group homomorphism $G \to S^\times$, or equivalently, to a module structure of $\mathcal{O}G$ on V, if and only if α represents the trivial class in $H^2(G; \mathcal{O}^\times)$, so if and only if the s_x can be chosen in such a way that α is the constant map on $G \times G$ taking the value 1. In that case the group homomorphism $G \to S^\times$ need not be unique, but if $\sigma, \sigma' : G \to S^\times$ are two group homomorphisms lifting the same action of G on S, then $\sigma'(x) = \beta(x)\sigma(x)$ for all $x \in G$, for some group homomorphism $\beta : G \to \mathcal{O}^\times$. If $\mathcal{O} = k$ and if G is a finite p-group, then the action of P on a matrix algebra lifts uniquely.

Proposition 5.3.1 *Suppose that k is perfect. Let n be a positive integer, V an n-dimensional k-vector space and P a finite p-group acting on the matrix algebra $S = \mathrm{End}_k(V)$. There is a unique group homomorphism $P \to S^\times$ which lifts the action of P on S; equivalently, V can be endowed with a unique kP-module structure, up to isomorphism, such that $S \cong \mathrm{End}_k(V)$ as P-algebras.*

Proof By 1.2.9 the group $H^2(P; k^\times)$ is trivial, which means that there exists a group homomorphism $P \to S^\times$ lifting the action of P on S. The uniqueness follows from the fact that there is no nontrivial group homomorphism from P to $k^\times$, or equivalently, $H^1(P; k^\times)$ is trivial. □

In general, a 2-cocycle α with coefficients in $\mathcal{O}^\times$ determines a central extension $\hat{G}$ of G by $\mathcal{O}^\times$. If α is determined by the action of G on a matrix algebra S, then the group $\hat{G}$ can be explicitly constructed as the subgroup of $G \times S^\times$ consisting of all pairs $(x, s) \in G \times S^\times$ such that ${}^x t = sts^{-1}$ for all $t \in S$. Equivalently, $\hat{G}$ is the pullback of the two obvious group homomorphisms $G \to \mathrm{Aut}(S) \cong S^\times/\mathcal{O}^\times$ and $S^\times \to S^\times/\mathcal{O}^\times$. We have a short exact sequence of groups

$$1 \longrightarrow \mathcal{O}^\times \xrightarrow{\ \theta\ } \hat{G} \xrightarrow{\ \pi\ } G \longrightarrow 1$$

where $\theta(\lambda) = (1_G, \lambda 1_S)$ and $\pi(x, s) = x$ for all $\lambda \in \mathcal{O}^\times$ and $(x, s) \in \hat{G}$. Up to isomorphism, this central extension depends only on the class of α in $H^2(G; \mathcal{O}^\times)$. Note that so far, the only hypothesis on the ring $\mathcal{O}$ we have used is that $\mathcal{O}$ is commutative local, in order to ensure that we can apply the Skolem–Noether Theorem. In order to understand the group $H^2(G; \mathcal{O}^\times)$, a slightly more precise hypothesis on $\mathcal{O}$ will be helpful: since k is algebraically closed it follows from 4.3.9 that we have a canonical group isomorphism $\mathcal{O}^\times = k^\times \times (1 + J(\mathcal{O}))$, implying that $H^2(G; \mathcal{O}^\times) \cong H^2(G; k^\times) \times H^2(G; 1 + J(\mathcal{O}))$. In particular, any element α in $H^2(G; k^\times)$ can be viewed as an element in $H^2(G; \mathcal{O}^\times)$, hence gives rise to a twisted group algebra $\mathcal{O}_\alpha G$ over $\mathcal{O}$. This yields the following immediate generalisation of 1.2.18.

Proposition 5.3.2 *Suppose that k is algebraically closed. Let G be a finite group and $\alpha \in H^2(G; k^\times)$. There is a central group extension*

$$1 \longrightarrow Z \xrightarrow{\ \iota\ } G' \xrightarrow{\ \pi\ } G \longrightarrow 1$$

for some finite subgroup Z of $k^\times$ of order dividing $|G|$ such that π induces an $\mathcal{O}$-algebra isomorphism

$$\mathcal{O}G'e \cong \mathcal{O}_\alpha G,$$

where e is the central idempotent in $\mathcal{O}G'$ defined by $e = \frac{1}{|Z|}\sum_{z\in Z} \iota(z)$.

Proof By 1.2.9, the class α has a representative in $Z^2(G; k^\times)$, abusively still denoted by α, with values in the subgroup Z of $|G|$-th roots of unity in $k^\times$. This gives rise to a central extension G' of G by Z, and exactly as in the proof of 1.2.18 one shows that this extension has the required properties. □

The following result shows that a class $\alpha \in H^2(G; \mathcal{O}^\times)$ determined by the action of a finite group G acting on a matrix algebra of rank prime to p belongs to the subgroup $H^2(G; k^\times)$.

Proposition 5.3.3 *Suppose that k is algebraically closed. Let n be a positive integer not divisible by p, let V be a free $\mathcal{O}$-module of rank n and let G be a finite group acting on the matrix algebra $S = \mathrm{End}_{\mathcal{O}}(V)$. Denote by $\hat{G}$ the subgroup of all $(x, s) \in G \times S^\times$ satisfying ${}^x t = sts^{-1}$ for all $t \in S$.*

(i) For any $x \in G$ there is an element $s_x \in S^\times$ such that $(x, s_x) \in \hat{G}$ and such that $\det(s_x) = 1$. The order of s_x in $S^\times$ is finite and divides nm, where m is the order of x.
(ii) If s'_x is another element in $S^\times$ such that $(x, s'_x) \in \hat{G}$ and such that $\det(s'_x) = 1$, then there is an n-th root of unity ζ such that $s'_x = \zeta s_x$.
(iii) For any $x \in G$ choose $s_x \in S^\times$ such that $(x, s_x) \in \hat{G}$ and such that $\det(s_x) = 1$. For $x, y \in G$, let $\alpha(x, y) \in \mathcal{O}^\times$ such that $s_x s_y = \alpha(x, y) s_{xy}$. Then $\alpha(x, y)^n = 1$. In particular, we have $\alpha \in H^2(G; k^\times)$, where we identify $k^\times$ with its canonical preimage in $\mathcal{O}^\times$.

Proof Let $s \in S^\times$ such that $(x, s) \in G$. Since n is coprime to p and since k is algebraically closed, there is $\tau \in \mathcal{O}^\times$ such that $\tau^n = \det(s)$. Then $s_x = \tau^{-1}s$ has determinant 1. Since $(s_x)^m$ acts as the identity on S and has determinant 1, we have $(s_x)^m = \zeta 1_S$ for some n-th root of unity ζ. This shows (i). Similarly, if s'_x is another element with determinant 1 and acting as x on S, then $s'_x s_x^{-1}$ has determinant 1 and acts as identity on S, hence is equal to $\zeta 1_S$ for some n-th root of unity ζ. This shows (ii), and (iii) follows from (ii) because $s_x s_y$ and s_{xy} are two elements in $S^\times$ acting both as xy on S and having both determinant 1. □

Statement (iii) can be interpreted as saying that the class of α in $H^2(G; \mathcal{O}^\times)$ is contained in the image of the map $H^2(G; \mu_n) \to H^2(G; \mathcal{O}^\times)$ induced by the inclusion of the subgroup of n-th roots of unity μ_n in $\mathcal{O}^\times$.

Corollary 5.3.4 *Suppose that k is algebraically closed. Let n be a positive integer not divisible by p, let V be a free $\mathcal{O}$-module of rank n, and let P be a finite p-group acting on the matrix algebra $S = \mathrm{End}_{\mathcal{O}}(V)$. There is a unique group homomorphism $\sigma : P \to S^\times$ lifting the action of P on S such that $\det(\sigma(x)) = 1$ for all $x \in P$. In particular, the central extension $\hat{P}$ determined by the action of P on S splits canonically as a direct product $\hat{P} \cong P \times \mathcal{O}^\times$.*

Proof Denote by μ_n the group of n-th roots of unity in $\mathcal{O}^\times$. This is a cyclic group of order prime to p. The existence of a group homomorphism σ with the properties as stated follows from 5.3.3 (iii) and the fact that the group $H^2(P; \mu_n)$ is trivial, by 1.2.9. Any two different homomorphisms satisfying the conclusions would 'differ' by a homomorphism from P to μ_n, but since μ_n has order prime to p this forces them to be equal. □

In order to calculate the 2-cocycle determined by a group action on a matrix algebra, one can always assume that the the matrix algebra is primitive for the group action.

Proposition 5.3.5 *Let n be a positive integer, V a free $\mathcal{O}$-module of rank n, and G a finite group acting on the matrix algebra $S = \mathrm{End}_{\mathcal{O}}(V)$. Let i be an idempotent in S^G. Then G acts on the matrix algebra iSi, and the central extension of G determined by the actions of G on S and on iSi are isomorphic.*

Proof For $x \in G$ choose s_x in $S^\times$ such that the action of x on S coincides with the conjugation action by s_x. Then s_x commutes with i since G fixes i, hence $s_x i$ is invertible in iSi and induces the same action as x on iSi. For $x, y \in G$, let $\alpha(x, y) \in \mathcal{O}^\times$ such that $s_x s_y = s_{xy}$. By the above, we have $s_x i s_y i = \alpha(x, y) s_{xy} i$, hence the elements $s_x i$ yield the same 2-cocycle α. □

Corollary 5.3.6 *Suppose that k is algebraically closed. Let n be a positive integer, V a free $\mathcal{O}$-module of rank n, and P a finite p-group acting on the matrix algebra $S = \mathrm{End}_{\mathcal{O}}(V)$. If there is an idempotent $i \in S^P$ such that the rank of iSi is prime to p, then the action of P on S lifts to a unique group homomorphism $\sigma : P \to S^\times$ satisfying $\det(\sigma(x)i) = 1$ for all $x \in P$.*

Proof By 5.3.4, the action of P on iSi lifts to a group homomorphism $P \to (iSi)^\times$, or equivalently, a 2-cocycle determined by this action represents the trivial class in $H^2(P; \mathcal{O}^\times)$. It follows from 5.3.5 that a 2-cocycle determined by the action of P on S represents the trivial class as well. Equivalently, the action of P on S lifts to a group homomorphism $P \to S^\times$. Since this group homomorphism is unique up to a group homomorphism $P \to \mathcal{O}^\times$, it follows from 5.3.4 that σ is uniquely determined by the additional condition $\det(\sigma(x)i) = 1$ for all $x \in P$. □

Proposition 5.3.7 *Let G be a finite group, H a subgroup of G and $\alpha \in Z^2(G; \mathcal{O}^\times)$ such that $\alpha(x, y) = 1$ if $x, y \in G$ such that at least one of x, y is in H. Then $\mathcal{O}H$ is a subalgebra of $\mathcal{O}_\alpha G$, and we have an isomorphism $\mathcal{O}_\alpha G \cong \mathcal{O}G$ as $\mathcal{O}H$-$\mathcal{O}H$-bimodules.*

Proof Since the restriction of α to $H \times H$ is constant 1, it follows that $\mathcal{O}H$ is a subalgebra of $\mathcal{O}_\alpha G$. It suffices to show that for any $x \in G$ the H-H-orbit $H \cdot x \cdot H$ in $\mathcal{O}_\alpha G$ is isomorphic, as an H-H-biset, to the double coset HxH in G. For that we need to show that if $xy = zx$ for some $y, z \in H$, then also $x \cdot y = z \cdot x$. This is however clear by the assumptions on α. □

Corollary 5.3.8 *Let P be a finite p-group, E a subgroup of* Aut(P) *and let $\alpha \in Z^2(E; \mathcal{O}^\times)$. Then α extends to $P \rtimes E$, and we have an isomorphism of $\mathcal{O}P$-$\mathcal{O}P$-bimodules $\mathcal{O}_\alpha(P \rtimes E) \cong \mathcal{O}(P \rtimes E)$.*

Proof By 1.2.14 we may assume that α is normalised; that is, $\alpha(1, y) = \alpha(y, 1) = 1$ for all $y \in E$. Inflating α to a map on $P \rtimes E$ via the canonical map $P \rtimes E \to E$ yields a 2-cocycle for $P \rtimes E$ satisfying $\alpha(y, z) = 1$ if one of y, z is in P. Thus 5.3.7 yields the result. □

Corollary 5.3.9 *Suppose that k is algebraically closed. Let G be a finite group, P a p-subgroup of G, and $\alpha \in H^2(G; k^\times)$. Then $\mathcal{O}P$ is a subalgebra of $\mathcal{O}_\alpha G$, and we have an isomorphism $\mathcal{O}_\alpha G \cong \mathcal{O}G$ as $\mathcal{O}P$-$\mathcal{O}P$-bimodules.*

Proof By 5.3.2, there is a central extension

$$1 \longrightarrow Z \xrightarrow{\iota} G' \xrightarrow{\pi} G \longrightarrow 1$$

for some finite subgroup Z of $k^\times$ of order dividing $|G|$ such that π induces an $\mathcal{O}$-algebra isomorphism $\mathcal{O}G'e \cong \mathcal{O}_\alpha G$, where $e = \frac{1}{|Z|}\sum_{z \in Z} \iota(z)$. Since Z is a central subgroup of order prime to p, it follows that there is a unique p-subgroup P' of G' such that π maps P' isomorphically onto P. Let $u', v' \in P'$ and $x \in G'$. Denote by u, v, x the images of u', v', x' in G, respectively. If $ux = xv$ in G, then $u'x' = x'v'z$ in G' for some $z \in Z$. Thus $v'z$ is conjugate to u' in G', and hence $v'z$ is a p-element. This forces $z = 1$. This shows that the P'-P'-orbit of $x' \in G'$ has the same cardinal as the P-P-orbit of $x \in G$. Thus there is a P'-P'-stable subset X of G' that is mapped by π bijectively onto G, and via the canonical isomorphism $P \cong P'$, this is an isomorphism of P-P-bisets. For $x \in G$, denote by $\hat{x}$ the unique element in X such that $\pi(\hat{x}) = x$. With respect to this choice of representatives of G in G', the 2-cocycle, still denoted α, satisfying $\hat{x}\hat{y} = \alpha(x, y)\widehat{xy}$ has now the property that $\alpha(x, y) = 1$ whenever one of x, y is in P. The result follows from 5.3.7. □

Corollary 5.3.10 *Suppose that k is algebraically closed. Let n be a positive integer not divisible by p and G a finite group acting on the matrix algebra $S = M_n(\mathcal{O})$. Denote by α the class in $H^2(G; \mathcal{O}^\times)$ determined by the action of G on S. Let P be a p-subgroup of G. Then $\mathcal{O}_\alpha G \cong \mathcal{O}G$ as $\mathcal{O}P$-$\mathcal{O}P$-bimodules.*

Proof By 5.3.3 the class α belongs to $H^2(G; k^\times)$. The result follows from 5.3.9. □

Remark 5.3.11 We assumed in the statements 5.3.2, 5.3.3, 5.3.6, 5.3.9, 5.3.10 that k is algebraically closed. One can prove these statements under the weaker hypothesis that k is perfect, modulo replacing k by a suitable finite extension in the conclusions of these statements, along the lines of the earlier Remark 1.2.19 on this theme.

One of the standard applications of group actions on matrix algebras arises in the context of G-stable simple quotients of kN, where N is a normal subgroup of a finite group G. The following result holds without any restriction on the characteristic of k.

Theorem 5.3.12 *Let G be a finite group, N a normal subgroup of G, S a finite-dimensional matrix algebra over k and $\pi : kN \to S$ a surjective algebra homomorphism such that* $\ker(\pi)$ *is G-stable. Denote by $\rho : H^2(G/N; k^\times) \to H^2(G; k^\times)$ the map induced by the canonical surjection $G \to G/N$. The conjugation action of G on kN induces an action of G on S. Denote by $\alpha \in H^2(G; k^\times)$ the class determined by this action.*

(i) There exist elements $s_x \in S^\times$, where $x \in G$, such that for any $x \in G$, any $s \in S$ and any $n \in N$ we have ${}^x s = s_x s (s_x)^{-1}$ and $s_{xn} = s_x \pi(n)$.
(ii) There is a canonical choice of a class $\beta \in H^2(G/N; k^\times)$ such that $\rho(\beta) = \alpha$. In particular, α has a representative in $Z^2(G; k^\times)$, again denoted by α, such that $\alpha(xm, yn) = \alpha(x, y)$ for all $x, y \in G$ and all $m, n \in N$.
(iii) For any representative in $Z^2(G/N; k^\times)$ of β, again denoted by β, there is a k-algebra isomorphism

$$kG/\ker(\pi)kG \cong S \otimes_k k_{\beta^{-1}} G/N$$

which sends $x + \ker(\pi)$ to $s_x \otimes \bar{x}$, where $x \in G$ and $\bar{x}$ is the image of xN in $k_{\beta^{-1}} G/N$.

Proof Since $\ker(\pi)$ is assumed to be G-stable, it is clear that G acts on the quotient $kN/\ker(\pi) \cong S$. Choose a system of representatives $\mathcal{R}$ of G/N in G. For $x \in \mathcal{R}$ choose $s_x \in S^\times$ such that x and s_x act in the same way on S. For $x \in \mathcal{R}$ and $n \in N$ set $s_{xn} = s_x \pi(n)$. This defines s_x for all $x \in G$. By construction, s_x acts as x on S for all $x \in G$, and we have $s_{xn} = s_x \pi(n)$ for all $x \in G$ and all $n \in N$. This shows (i). Define $\alpha \in Z^2(G; k^\times)$ by $s_x s_y = \alpha(x, y) s_{xy}$ for all $x, y \in G$. Then α represents the class determined by the action of G on S. A short calculation shows that this particular choice of the s_x implies that

$\alpha(x, y) = \alpha(xm, yn)$ for all $x, y \in G$ and all $m, n \in N$. It follows that setting $\beta(xN, yN) = \alpha(x, y)$ yields a class $\beta \in H^2(G/N; k^\times)$ such that $\rho(\beta) = \alpha$. In order to show that the class β is canonically determined, we need to show that β is independent of the choice of the s_x satisfying (i). For $x \in G$, let $t_x \in S^\times$ be another element acting as x on S, such that $t_{xn} = t_x\pi(n)$ for all $n \in N$. Thus $t_x = \lambda(x)s_x$ for some $\lambda(x) \in k^\times$, satisfying $\lambda(xn) = \lambda(x)$ for all $x \in G$ and all $n \in N$. Let $\alpha' \in Z^2(G; k^\times)$ such that $t_xt_y = \alpha'(x, y)t_{xy}$ for all $x, y \in G$. As before, we have $\alpha' = \rho(\beta')$, where $\beta'(xN, yN) = \alpha'(x, y)$ for $x, y \in G$. Combining the previous equalities yields $\lambda(x)\lambda(y)\alpha(x, y) = \lambda(xy)\alpha'(x, y)$. Since all elements in this equation depend only on the images of x, y in G/N, it follows that β and β' determine the same class in $H^2(G/N; k^\times)$. This shows (ii). Since $\ker(\pi)$ is G-stable, we have $\ker(\pi)kG = kG\ker(\pi)$, and hence $\ker(\pi)kG$ is an ideal in kG. We check first that the linear map $\tau : kG \to S \otimes_k k_{\beta^{-1}}G/N$ sending $x \in G$ to $s_x \otimes \bar{x}$ is an algebra homomorphism. Let $x, y \in G$. Then $\tau(x)\tau(y) = s_xs_y \otimes \bar{x} \cdot \bar{y} = \beta(xN, yN)s_{xy} \otimes \beta^{-1}(xN, yN) = s_{xy} \otimes \overline{xy}$. Thus τ is an algebra homomorphism. By construction, τ maps kN onto $S \otimes \bar{1}$, hence kNx onto $S \otimes \bar{x}$, where $x \in G$. Thus τ is surjective. The kernel of τ contains $\ker(\pi)$, hence also the ideal $\ker(\pi)kG$. Since kG is free of rank $|G/N|$ as a left kN-module, it follows that both $kG/\ker(\pi)kG$ and $S \otimes_k k_{\beta^{-1}}G/N$ have dimension $\dim_k(S) \cdot |G/N|$, whence the isomorphism in (iii). □

As mentioned above, the previous result holds regardless of the characteristic of k, and can be used to prove the following fact on the extendibility of simple modules and irreducible characters alike.

Corollary 5.3.13 *Let G be a finite group and N a normal subgroup. Suppose that k is large enough. Let V be a G-stable simple kN-module. Suppose that the canonical class in $H^2(G/N; k^\times)$ determined by the action of G on $\mathrm{End}_k(V)$ is trivial. Then there exists a simple kG-module U such that $\mathrm{Res}^G_N(U) \cong V$. If G/N is abelian and if the characteristic of k is coprime to $|G/N|$, then there exist exactly $|G/N|$ isomorphism classes of simple kG-modules whose restrictions to kN are isomorphic to V.*

Proof Denote by $\pi : kN \to \mathrm{End}_k(V)$ the structural homomorphism. The assumption that V is G-stable is equivalent to requiring that $\ker(\pi)$ is G-stable in kN. If U is a kG-module that extends V, then U is annihilated by $\ker(\pi)$, hence by the ideal $\ker(\pi)kG$ generated by $\ker(\pi)$ in kG. The hypotheses imply, together with 5.3.12 (iii), that we have an algebra isomorphism

$$kG/\ker(\pi)kG \cong \mathrm{End}_k(V) \otimes_k kG/N.$$

If W is a 1-dimensional kG/N-module, then $V \otimes_k W$ corresponds through this isomorphism to a simple kG-module extending V. If G/N is abelian and the characteristic of k is coprime to $|G/N|$, then kG/N is a direct product of $|G/N|$ copies of k, so yields exactly $|G/N|$ pairwise nonisomorphic extensions of V. □

By 1.2.10, the vanishing hypothesis of $H^2(G/N; k^\times)$ in 5.3.13 is in particular satisfied if G/N is cyclic and k is an algebraically closed field. There are more general results on extensions of module structures; see 5.7.11 below.

Remark 5.3.14 With the notation of 5.3.12, the map $\rho : H^2(G/N; k^\times) \to H^2(G; k^\times)$ need not be injective. In other words, 5.3.12 does *not* assert that there is a unique $\beta \in H^2(G/N; k^\times)$ satisfying $\rho(\beta) = \alpha$. It merely says that the circumstances in which α arises lead to a canonical choice of such a β. In particular, it is possible that α is the trivial class while β is nontrivial; in that case, the simple kN-module corresponding to S does not extend to a simple kG-module, despite the fact that the action of G on S lifts to a group homomorphism $G \to S^\times$; see the exercise 5.3.15 below. In terms of central extensions, this can be expressed as follows. Let $\hat{G}$ be the subgroup of $S^\times \times G$ consisting of all pairs (s, x) such that s and x act in the same way on S. Denote by $\hat{N}$ the inverse image of N in $\hat{G}$ under the canonical map $\hat{G} \to G$ given by the projection $(s, x) \to x$. Then $\hat{N}$ splits canonically (but not uniquely) as $\hat{N} \cong k^\times \times N$ via the homomorphism $\tau : N \to \hat{N}$ sending $n \in N$ to $(\pi(n), n)$. The image of τ is a normal subgroup of $\hat{G}$, isomorphic to N, and the quotient $\hat{G}/\mathrm{Im}(\tau)$ is a central extension of G/N by $k^\times$ corresponding to the class β. Any group homomorphism $\nu : N \to k^\times$ yields a different splitting τ', sending $n \in N$ to $(\nu(n)\pi(n), n)$, and the quotient $\hat{G}/\mathrm{Im}(\tau')$ is again a central extension of G/N, corresponding to a class $\beta' \in H^2(G/N; k^\times)$ still satisfying $\rho(\beta') = \alpha$, but β' need not be equal to β. In fact, one can show that the classes β and β' are equal if and only if ν extends to a group homomorphism from G to $k^\times$. This reflects the fact that there is an exact sequence in low degree group cohomology $H^1(G; k^\times) \to H^1(N; k^\times) \to H^2(G/N; k^\times) \to H^2(G; k^\times)$, obtained from the Lyndon–Hochschild–Serre spectral sequence, together with the identifications $H^1(G; k^\times) = \mathrm{Hom}(G; k^\times)$ and $H^1(N; k^\times) = \mathrm{Hom}(N; k^\times)$. See [93, 6.8.3].

Theorem 5.3.12 states that the obstruction to extending the module structure of a simple kN-module to kG lies in the group $H^2(G/N; k^\times)$, and 5.3.13 parametrises all possible extensions. This is true in greater generality; see for instance [88].

Exercise 5.3.15 Let Z be the cyclic subgroup of order 2 of the quaternion group Q_8 of order 8, and let η be the nontrivial complex valued linear character of Z.

(1) Show that η is Q_8-stable but does not extend to a character of Q_8.
(2) Show that the element α in $H^2(Q_8; \mathbb{C}^\times)$ determined by the action of Q_8 on $\mathbb{C}Ze(\eta)$ is trivial. (In fact, one can show that $H^2(Q_8; \mathbb{C}^\times)$ is trivial.)
(3) Show that the canonical class in $H^2(Q_8/Z; \mathbb{C}^\times)$ determined by the action of Q_8 on $\mathbb{C}Ze(\eta)$ is nontrivial.

A *character triple* is a triple (G, N, η) consisting of a finite group G, a normal subgroup N of G, and a G-stable complex valued irreducible character η of N. Character triples are a fundamental concept for Clifford theoretic reductions in finite group representation theory.

5.4 The Brauer construction

Given a finite group G acting on an algebra A and a subgroup H of G, we denote as in §2.5 by A^G the G-fixed point subalgebra of A and by A^G_H the image of the relative trace map Tr^G_H sending $a \in A^H$ to $\sum_{x\in[G/H]} {}^x a \in A^G$; this is an ideal in A^G. The canonical k-linear projection $kG \to kH$ sending $\sum_{x\in G} \lambda_x x$ to $\sum_{x\in H} \lambda_x x$ is not an algebra homomorphism in general. Linear projections do yield algebra homomorphisms upon restriction to suitable subalgebras:

Theorem 5.4.1 *Let G be a finite group and let P be a p-subgroup of G. The canonical k-linear projection $kG \to kC_G(P)$ induces a split surjective homomorphism of $N_G(P)$-algebras over k,*

$$\mathrm{Br}_P : (kG)^P \to kC_G(P).$$

We have $\ker(\mathrm{Br}_P) = \sum_{Q<P}(kG)^P_Q$*, where in the sum Q runs over the proper subgroups of P.*

Proof If $a \in (kG)^P$ and $x \in N_G(P)$ then ${}^x a \in (kG)^P$; in other words, $(kG)^P$ is an $N_G(P)$-algebra. Similarly, $kC_G(P)$ is an $N_G(P)$-algebra. The map Br_P clearly commutes with the action of $N_G(P)$. The algebra $(kG)^P$ has as k-basis the set of P-conjugacy class sums of elements in G. For an element $x \in G$ and $u, v \in P$ the conjugates uxu^{-1} and vxv^{-1} are equal if and only if $v^{-1}u \in C_P(x)$, or equivalently, if and only if $uQ = v$, where $Q = C_P(x)$. Thus the P-conjugacy class sum of $x \in G$ is $\mathrm{Tr}^P_Q(x)$, where $Q = C_P(x)$. Note that $Q = P$ if and only if $x \in C_G(P)$. Note also that if $x \in C_G(P)$ and Q is a proper subgroup of P then $\mathrm{Tr}^P_Q(x) = |P : Q|x = 0$. This means that $\ker(\mathrm{Br}_P)$ is the k-vector space $\sum_{Q<P}(kG)^P_Q$. By 2.5.8, this space is an ideal in $(kG)^P$. Thus Br_P is an algebra homomorphism. Since $kC_G(P) \subseteq (kG)^P$, the homomorphism Br_P is split, having as section the inclusion homomorphism $kC_G(P) \subseteq (kG)^P$. $\square$

If P is trivial, then the indexing set of proper subgroups of P in the statement of 5.4.1 is empty; we adopt the standard convention that a sum taken over an empty indexing set is zero. One can precompose Br_P with the canonical homomorphism $(\mathcal{O}G)^P \to (kG)^P$, which yields a surjective algebra homomorphism, usually denoted again $\mathrm{Br}_P : (\mathcal{O}G)^P \to kC_G(P)$. This is a homomorphism of $\mathcal{O}$-algebras, where $kC_G(P)$ is viewed as $\mathcal{O}$-algebra via the canonical map $\mathcal{O} \to k$, and this homomorphism is no longer split. The proof of 5.4.1 suggests how to define a Brauer homomorphism more generally for any G-algebra.

Definition 5.4.2 Let G be a finite group, let A be a G-algebra over $\mathcal{O}$, and let P be a p-subgroup of G. We set

$$A(P) = A^P / \Big(\sum_{Q<P} A_Q^P + J(\mathcal{O})A^P \Big)$$

and denote by $\mathrm{Br}_P^A : A^P \to A(P)$ the canonical surjective algebra homomorphism, called *Brauer homomorphism* with respect to P.

The subalgebra A^P of A is an $N_G(P)$-algebra, and $A(P)$ inherits an $N_G(P)$-algebra structure because the ideal $\ker(\mathrm{Br}_P^A)$ is stable under the action of $N_G(P)$. In this way, Br_P^A is a surjective homomorphism of $N_G(P)$-algebras. Since P acts trivially on A^P and on $A(P)$, the action of $N_G(P)$ induces an action of $N_G(P)/P$ on A^P and on $A(P)$. If $A = \mathcal{O}G$ we identify $(\mathcal{O}G)(P)$ and $kC_G(P)$, using 5.4.1. In general $A(P)$ can be zero. If $f : A \to B$ is a homomorphism of G-algebras, then for any p-subgroup P of G the restriction of f to A^P induces an algebra homomorphism $f^P : A^P \to B^P$ which sends $\ker(\mathrm{Br}_P^A)$ to $\ker(\mathrm{Br}_P^B)$, and hence induces a unique algebra homomorphism $f(P) : A(P) \to B(P)$ satisfying $\mathrm{Br}_P^B \circ f^P = \mathrm{Br}_P^A \circ f(P)$. In this way the Brauer construction becomes a functor from G-algebras to $N_G(P)/P$-algebras. In order to determine the image of a fixed point algebra A^G in the Brauer quotient $A(P)$ we will require the following result.

Proposition 5.4.3 *Let G be a finite group, let P be a p-subgroup of G, and let A be a G-algebra over $\mathcal{O}$. Suppose that $1 \in A_P^G$. Then $A^G = A_P^G$ and*

$$\ker(\mathrm{Br}_P^A) \cap A^G = \sum_{Q<P} A_Q^G + J(\mathcal{O})A^G.$$

Proof Let $c \in A^P$ such that $\mathrm{Tr}_P^G(c) = 1$. For any $z \in A^G$, we have $z = 1 \cdot z = \mathrm{Tr}_P^G(c)z = \mathrm{Tr}_P^G(cz) \in A_P^G$. This proves the equality $A^G = A_P^G$. Let $u \in \ker(\mathrm{Br}_P^A) \cap A_P^G$. Write $u = \sum_{Q<P} \mathrm{Tr}_Q^P(a_Q) + \lambda v$ for some $\lambda \in J(\mathcal{O})$, $v \in A^P$ and $a_Q \in A^Q$ for any proper subgroup Q of P. Using that u commutes with the action

of G we have

$$\begin{aligned} u &= 1 \cdot u = \mathrm{Tr}_P^G(c)u = \mathrm{Tr}_P^G(uc) \\ &= \mathrm{Tr}_G^P\Big(\sum_{Q<P} \mathrm{Tr}_Q^P(a_Q c) + \lambda vc\Big) = \sum_{Q<P} \mathrm{Tr}_Q^G(a_Q c) + \lambda \mathrm{Tr}_P^G(vc)\Big). \end{aligned}$$

This expression belongs to $\sum_{Q<P} A_Q^G + J(\mathcal{O})A^G$. Conversely, we have $J(\mathcal{O})A^G \subseteq J(\mathcal{O})A^P \cap A^G$, and if Q is a proper subgroup of P then for any $d \in A^Q$ we have $\mathrm{Tr}_Q^G(d) = \sum_{x\in[P\backslash G/Q]} \mathrm{Tr}_{P\cap {}^xQ}^P({}^xc)$, which is contained in $\ker(\mathrm{Br}_P^A)$ because $P \cap {}^xQ$ is a proper subgroup of P for any $x \in G$. The result follows. □

Proposition 5.4.4 *Let G be a finite group and P a normal p-subgroup of G. Then* $\ker(\mathrm{Br}_P) \subseteq J((\mathcal{O}G)^P)$.

Proof Assume first that $\mathcal{O} = k$. It suffices to show that the ideal $\ker(\mathrm{Br}_P)$ is nilpotent. Let S be a simple kG-module. Then $\mathrm{Res}_P^G(S)$ is semisimple, by Clifford's Theorem 1.9.9, because P is normal in G. The trivial kP-module is up to isomorphism the unique simple kP-module. Thus all elements of P act as the identity on S. Let Q be a proper subgroup of P and let $a \in (kG)^Q$. Then, for any $s \in S$, we have $\mathrm{Tr}_Q^P(a)s = \sum_{y\in[P/Q]} yay^{-1}s = \sum_{y\in[P/Q]} yas = [P:Q]as = 0$. Thus $\ker(\mathrm{Br}_P) \subseteq J(kG)$. But $J(kG)$ is a nilpotent ideal in kG, and hence $\ker(\mathrm{Br}_P)$ is a nilpotent ideal in $(kG)^P$, hence contained in $J((kG)^P)$. For general $\mathcal{O}$ the result follows from the fact that $J((\mathcal{O}G)^P)$ is the inverse image in $(\mathcal{O}G)^P$ of $J((kG)^P)$. □

Proposition 5.4.5 *Let G be a finite group, let A be a G-algebra over $\mathcal{O}$ that is finitely generated as $\mathcal{O}$-module and let P be a p-subgroup of G. Suppose that $A(P) \neq \{0\}$. The Brauer homomorphism* $\mathrm{Br}_P^A : A^P \to A(P)$ *is a surjective $N_G(P)$-algebra homomorphism, and for every $a \in A^P$ we have*

$$\mathrm{Br}_P^A(\mathrm{Tr}_P^G(a)) = \mathrm{Tr}_P^{N_G(P)}(\mathrm{Br}_P^A(a));$$

in particular, Br_P^G *maps* A_P^G *onto* $A(P)_P^{N_G(P)}$.

Proof Since P is normal in $N_G(P)$, the action of $N_G(P)$ on A induces an action of $N_G(P)$ on A^P, and this action preserves $\ker(\mathrm{Br}_P^A)$. By Mackey's formula 2.5.5 for relative traces we have $\mathrm{Tr}_P^G(a) = \sum_{x\in[P\backslash G/P]} \mathrm{Tr}_{P\cap {}^xP}^P({}^xa)$. If $P \cap {}^xP$ is a proper subgroup of P then the summand $\mathrm{Tr}_{P\cap {}^xP}^P({}^xa)$ is contained in $\ker(\mathrm{Br}_P^A)$. If $P \cap {}^xP = P$ then $x \in N_G(P)$ and $PxP = xP$. Thus $\mathrm{Br}_P^A(\mathrm{Tr}_P^G(a)) = \sum_{x\in[N_G(P)/P]} {}^xa = \mathrm{Tr}_P^{N_G(P)}(a)$. The rest is clear. □

Setting $\bar{N} = N_G(P)/P$ and using that P acts trivially on $A(P)$, this Proposition can be reformulated as follows. We have $\mathrm{Br}_P^A(\mathrm{Tr}_P^G(a)) = \mathrm{Tr}_1^{\bar{N}}(\mathrm{Br}_P^A(a))$; in particular, Br_P^A maps A_P^G onto $A(P)_1^{\bar{N}}$. The version 4.7.5 of the Lifting Theorem for idempotents implies that primitive idempotents in A_P^G not contained in $\ker(\mathrm{Br}_P^A)$ correspond bijectively to primitive idempotents in $A(P)_P^{N_G(P)}$. This observation will be used later in the proof of Brauer's first main theorem. For group algebras 5.4.3 can be made more precise, the reason being that group algebras have permutation bases:

Proposition 5.4.6 *Let G be a finite group, H a subgroup of G and P a p-subgroup of H. The following hold.*

(i) $(\mathcal{O}G)_P^H = \mathcal{O}C_G(P)_P^H + \sum_{Q;Q<P}(\mathcal{O}G)_Q^H$.

(ii) The trace map $\mathrm{Tr}_{N_G(P)}^H$ *induces a linear isomorphism* $(kC_G(P))_P^{N_G(P)} \cong (kC_G(P))_P^H$ *with inverse induced by* Br_P.

(iii) $\ker(\mathrm{Br}_P) \cap (\mathcal{O}G)_P^H = \sum_{Q<P}(\mathcal{O}G)_Q^H + J(\mathcal{O})(\mathcal{O}G)^H$.

(iv) $(kG)^H = \sum_Q(\mathcal{O}C_G(Q))_Q^H$, *where Q runs over the p-subgroups of H.*

Proof The right side in statement (i) is trivially contained in the left side. For the reverse inclusion, observe that every element in $(\mathcal{O}G)_P^H$ is a linear combination of elements of the form $\mathrm{Tr}_P^G(c)$, where c is a P-conjugacy class sum of an element y in G. Thus $c = \mathrm{Tr}_Q^P(y)$, where $Q = C_P(y)$. If $Q = P$, then $c = y \in C_G(P)$, and hence $\mathrm{Tr}_P^G(c)$ belongs to the right side. If $Q \neq P$, then $\mathrm{Tr}_P^G(c) = \mathrm{Tr}_Q^G(y)$ belongs to the right side as well, proving (i). For (ii), the trace map $\mathrm{Tr}_{N_H(P)}^H$ maps $(kC_G(P))_P^{N_G(P)}$ onto $(kC_G(P))_P^H$. Since $\mathrm{Br}_P(c) = c$ for any $c \in kC_G(P)$, it follows from 5.4.5 that Br_P induces an inverse of this map as stated. The right side in statement (iii) is obviously contained in the left side. In order to prove the converse we may assume that $\mathcal{O} = k$. Using (i), it suffices to show that $\ker(\mathrm{Br}_P) \cap (kC_G(P))_P^H$ is zero. Let $c \in kC_G(P)$. Suppose that $\mathrm{Br}_P(\mathrm{Tr}_P^H(c)) = 0$. By (ii) we have $\mathrm{Tr}_P^H(c) = 0$, which shows (iii). For (iv), let P be a Sylow p-subgroup of H. Then $(\mathcal{O}G)^H = (\mathcal{O}G)_P^H$. Statement (iv) follows from (i) and an easy induction. □

Proposition 5.4.7 *Let G be a finite group. Let $\mathcal{R}$ be a set of representatives of the conjugacy classes of p-subgroups of G. We have vector space decomposition*

$$Z(kG) = \oplus_{Q\in\mathcal{R}} (kC_G(Q))_Q^G.$$

In particular, we have

$$\dim_k(Z(kG)) = \sum_{Q\in\mathcal{R}} \dim_k((kC_G(Q))_Q^{N_G(Q)}).$$

Proof Note that $(kC_G(Q))_Q^G = (kC_G(R))_R^G$ whenever Q and R are G-conjugate subgroups of G. Therefore, applying 5.4.6 (iv) with $H = G$ and k instead of $\mathcal{O}$ yields $Z(kG) = (kG)^G = \sum_{Q\in\mathcal{R}}(kC_G(Q))_Q^G$. We need to show that this sum is direct; we do this by comparing dimensions. Let $c \in C_G(Q)$. Then $\mathrm{Tr}_Q^G(c)$ is the conjugacy class sum of c multiplied by the index $|C_G(c) : Q|$. Thus this is zero unless Q is a Sylow p-subgroup of $C_G(c)$. For $x \in G$, after possibly replacing x by a conjugate, we may assume that $C_G(x)$ has a Sylow p-subgroup Q belonging to the set $\mathcal{R}$. For each Q in $\mathcal{R}$, it follows from 5.4.6 (ii) that $(kC_G(Q))_Q^G$ has the same dimension as $(kC_G(Q))_Q^{N_G(Q)}$, and by the above, this dimension is the number of $N_G(Q)$-conjugacy classes of elements c in $C_G(Q)$ such that Q is a Sylow p-subgroup of c. Thus it suffices to show that if x, y are two elements in $C_G(Q)$ such that Q is a Sylow p-subgroup of $C_G(x)$ and $C_G(y)$, then x, y are G-conjugate if and only is they are $N_G(Q)$-conjugate. If $y = gxg^{-1}$ for some $g \in G$, then Q and gQg^{-1} are both Sylow p-subgroups of $C_G(y)$, and hence $gQg^{-1} = hQh^{-1}$ for some $h \in C_G(y)$. Then $h^{-1}g \in N_G(Q)$, and $h^{-1}gxg^{-1}h = h^{-1}yh = y$. The result follows. □

Proposition 5.4.8 *Let G be a finite group, let A be a G-algebra over $\mathcal{O}$ that is finitely generated as an $\mathcal{O}$-module and let P be a p-subgroup of G. Suppose that $A(P) \neq \{0\}$. The Brauer homomorphism Br_P^A maps $Z(A)^P$ to $Z(A(P))$.*

Proof The statement makes sense because the action of G on A stabilises $Z(A)$. Let $z \in Z(A)^P$, and let $a \in A^P$. Then $\mathrm{Br}_P(a)\mathrm{Br}_P(z) = \mathrm{Br}_P(az) = \mathrm{Br}_P(za) = \mathrm{Br}_P(z)\mathrm{Br}_P(a)$, whence the result. □

Example 5.4.9 Every $\mathcal{O}G$-module M gives rise to a G-algebra, namely $\mathrm{End}_{\mathcal{O}}(M)$, with $x \in G$ acting on $\varphi \in \mathrm{End}_{\mathcal{O}}(M)$ by $({}^x\varphi)(m) = x\varphi(x^{-1}m)$ for all $m \in M$. If P is a p-subgroup of G, the Brauer construction 5.4.2 reads

$$(\mathrm{End}_{\mathcal{O}}(M))(P) = \mathrm{End}_{\mathcal{O}P}(M)/\Big(\sum_{Q<P}\mathrm{Tr}_Q^P(\mathrm{End}_{\mathcal{O}Q}(M)) + J(\mathcal{O})\mathrm{End}_{\mathcal{O}P}(M)\Big).$$

Higman's criterion states that M is relatively P-projective if and only if $\mathrm{Id}_M \in \mathrm{Tr}_P^G(\mathrm{End}_{\mathcal{O}P}(M))$.

The Brauer construction as defined in 5.4.2 uses only the action of G on A. Thus we can extend this construction to modules over group algebras as follows:

Definition 5.4.10 Let G be a finite group, let P be a p-subgroup of G and let M be an $\mathcal{O}G$-module. We set

$$M(P) = M^P / \Big(\sum_{Q<P} M_Q^P + J(\mathcal{O})M^P \Big)$$

and denote by $\mathrm{Br}_P^M : M^P \to M(P)$ the canonical surjective map. We consider M^P as an $\mathcal{O}N_G(P)$-module and $M(P)$ as a $kN_G(P)$-module or as a $kN_G(P)/P$-module.

If $\alpha : M \to N$ is a homomorphism of $\mathcal{O}G$-modules, then α maps M^P to N^P and M_Q^P to N_Q^P for any subgroup Q of P. Thus α induces a homomorphism of $kN_G(P)$-modules $M(P) \to N(P)$. In this way, the Brauer construction defines a covariant functor from the category of $\mathcal{O}G$-modules to the category of $kN_G(P)$-modules, or also the category of $kN_G(P)/P$-modules – whichever point of view is more convenient. If Q is a normal subgroup of P then P acts on $M(Q)$, and hence the Brauer construction $(M(Q))(P)$ is defined. In general, this need not be isomorphic to $M(P)$.

Proposition 5.4.11 *Let G be a finite group, let M be an $\mathcal{O}G$-module, and let P be a p-subgroup of G. The Brauer homomorphism $\mathrm{Br}_P^M : M^P \to M(P)$ is a surjective homomorphism of $\mathcal{O}N_G(P)$-modules, and for every $m \in M^P$ we have*

$$\mathrm{Br}_P^M(\mathrm{Tr}_P^G(m)) = \mathrm{Tr}_P^{N_G(P)}(\mathrm{Br}_P^A(m));$$

in particular, Br_P^M maps M_P^G onto $M(P)_P^{N_G(P)}$.

Proof The proof is identical to that of 5.4.5 with M instead of A. □

For the purpose of calculating character values, the following observation is useful.

Proposition 5.4.12 *Let G be a finite group and A an interior G-algebra over $\mathcal{O}$. Let P be a p-subgroup of G. We have $\ker(\mathrm{Br}_P^A) \subseteq J(\mathcal{O})A + [A, A]$. In particular, if $\chi : A \to \mathcal{O}$ is a central function, then $\chi(\ker(\mathrm{Br}_P^A)) \subseteq J(\mathcal{O})$.*

Proof Identify the elements of G to their images in $A^\times$. Let $a \in A$ and $x \in G$. We have $xax^{-1} - a = [xa, x^{-1}] \in [A, A]$. Thus, for Q a proper subgroup of P and $a \in A^Q$ we have $\mathrm{Tr}_Q^P(a) - |P : Q|a \in [A, A]$. Since $|P : Q| \in J(\mathcal{O})$, the first statement follows. Since a central function on A annihilates $[A, A]$ this implies the second statement. □

Remark 5.4.13 Let G be a finite group and A an interior G-algebra with structural homomorphism $\sigma : G \to A^\times$. Then A has an $\mathcal{O}G$-$\mathcal{O}G$-bimodule structure via σ, or equivalently, A can be considered as an $\mathcal{O}(G \times G)$-module, with

$(x, y) \in G \times G$ acting on $a \in A$ via $(x, y) \cdot a = \sigma(x)a\sigma(y^{-1})$. For H a subgroup of G, the H-fixed points A^H in A are actually the fixed points with respect to the action of the diagonal subgroup $\Delta H = \{(y, y)|y \in H\}$ of $G \times G$, so strictly speaking, we should write $A^{\Delta H}$ instead of A^H if A is an interior G-algebra. As long as there is no confusion, we will write A^H for the sake of simplicity, but there will be a point where the extra notational precision is needed when dealing with fusion and compatibility of the Brauer construction with tensor products.

5.5 Pointed groups on G-algebras

The terminology and results in this section go back to work of Puig [71]. The underlying philosophy is to investigate the structure of a G-algebra A by a systematic use of conjugacy classes of primitive idempotents in fixed point subalgebras A^H, for H a subgroup of the finite group G. All algebras over $\mathcal{O}$ in this section are assumed to be finitely generated as $\mathcal{O}$-modules.

Definition 5.5.1 Let G be a finite group and let A be a G-algebra over $\mathcal{O}$. Let H be a subgroup of G. A *point of H on A* is an $(A^H)^\times$-conjugacy class β of primitive idempotents in A^H; that is β is a point of the H-fixed point algebra A^H. We call H_β a *pointed group on A*. If $H = \langle y \rangle$ is cyclic with generator y, we call y_β a *pointed element on A*. The *multiplicity of a point β on A of a subgroup H of G* is the integer $m_\beta = |\beta \cap J|$, where J is a primitive decomposition of 1 in A^H.

Since primitive decompositions of 1 in A^H are conjugate by an element in $(A^H)^\times$, it follows that the integer m_β does not depend on the choice of J.

Example 5.5.2 Let G be a finite group, let M be a finitely generated $\mathcal{O}G$-module and let H be a subgroup of G. A primitive idempotent in $(\mathrm{End}_\mathcal{O}(M))^H = \mathrm{End}_{\mathcal{O}H}(M)$ is a projection of $\mathrm{Res}^G_H(M)$ onto an indecomposable direct summand of $\mathrm{Res}^G_H(M)$. Thus a point β of H on $\mathrm{End}_\mathcal{O}(M)$ corresponds to an isomorphism class of an indecomposable direct summand V of $\mathrm{Res}^G_H(M)$. The multiplicity of β is the number of summands isomorphic to V in a decomposition of $\mathrm{Res}^G_H(M)$ as a direct sum of indecomposable $\mathcal{O}H$-modules.

Definition 5.5.3 Let G be a finite groups and let A be a G-algebra over $\mathcal{O}$. Let P be a p-subgroup of G. A *local point of P on A* is a point γ of P on A such that $\mathrm{Br}_P(\gamma) \neq 0$, where $\mathrm{Br}_P : A^P \to A(P)$ is the Brauer homomorphism. If $P = \langle u \rangle$ is cyclic with generator u, we call the pair u_γ a *local pointed element on A*.

The local pointed groups on A are a G-subset of the set of all pointed groups on A. The local pointed elements on A form a G-subset of the set of all pointed elements on A.

Example 5.5.4 Let G be a finite group, M an $\mathcal{O}G$-module, P a subgroup of G and γ a local point of P on $\mathrm{End}_{\mathcal{O}}(M)$. That is, for $i \in \gamma$, the space $i(M)$ is an indecomposable direct factor of $\mathrm{Res}^G_P(M)$. Since $\mathrm{Br}_P(i) \neq 0$ it follows from Higman's criterion that $i(M)$ has P as a vertex P. In particular, if M is indecomposable and P is a vertex of M then the local points of P on $\mathrm{End}_{\mathcal{O}}(M)$ correspond exactly to the sources of M. This shows also that all local points of P on $\mathrm{End}_{\mathcal{O}}(M)$ are $N_G(P)$-conjugate, a fact that we are going to prove in greater generality below.

Theorem 5.5.5 *Let G be a finite group, let A be a G-algebra, and let α be a point of G on A. Let P be a subgroup of G such that $\alpha \in A^G_P$ and let Q be a p-subgroup of G such that $\mathrm{Br}_Q(\alpha) \neq 0$. Then Q is G-conjugate to a subgroup of P.*

Proof Let $e \in \alpha$. Let P be a subgroup of G such that $e \in A^G_P$ and let Q be a p-subgroup of G such that $\mathrm{Br}_Q(e) \neq 0$. Write $e = \mathrm{Tr}^G_P(c)$ for some $c \in A^P$. The Mackey formula yields

$$e = \sum_{y \in [Q\backslash G/P]} \mathrm{Tr}^Q_{Q\cap {}^yP}({}^yc).$$

The condition $\mathrm{Br}_Q(e) \neq 0$ implies that there is $y \in G$ such that $Q \subseteq {}^yP$. □

Theorem 5.5.6 *Let G be a finite group, let A be a G-algebra, and let α be a point of G on A. Let P be a minimal subgroup of G such that $\alpha \subseteq A^G_P$. The following hold.*

(i) The group P is a p-subgroup of G.
(ii) We have $\mathrm{Br}_P(\alpha) \neq 0$.
(iii) If H is a subgroup of G such that $\alpha \in A^G_H$, then H contains a G-conjugate of P.

Proof Let S be a Sylow p-subgroup of P. Then $|P : S|$ is invertible in $\mathcal{O}$, hence, for any $c \in A^P$ we have $c = \frac{1}{|P:S|}\mathrm{Tr}^P_S(c)$ and therefore $\mathrm{Tr}^G_P(c) = \mathrm{Tr}^G_S(\frac{1}{|P:S|}c)$. The minimality of P with the property $1_A \in A^G_P$ forces $S = P$, thus P is a p-subgroup of G. This proves (i). Let $e \in \alpha$. By 5.4.3 we have $\ker(\mathrm{Br}_P) \cap A^G = \sum_{Q<P} A^G_P + J(\mathcal{O})A^G$. Note that $J(\mathcal{O})A^G$ contains no idempotent. Thus if $e \in \ker(\mathrm{Br}_P)$, then Rosenberg's Lemma 4.4.8 implies that $e \in A^G_Q$ for some proper subgroup A of P, contradicting the choice of P. This proves $\mathrm{Br}_P(e) \neq 0$. But

then P is conjugate to a subgroup of any other subgroup H of G satisfying $\alpha \subseteq A_H^G$ by 5.5.5. □

Theorem 5.5.7 *Let G be a finite group, let A be a G-algebra over $\mathcal{O}$, and let α be a point of G on A. Let P be a subgroup of G. The following are equivalent:*

(i) P is a minimal subgroup of G such that $\alpha \subseteq A_P^G$.
(ii) P is a maximal p-subgroup of G such that $\mathrm{Br}_P(\alpha) \neq 0$.
(iii) P is a p-subgroup of G such that $\alpha \subseteq A_P^G$ and such that $\mathrm{Br}_P(\alpha) \neq 0$.

Proof Suppose that (i) holds. Then $\mathrm{Br}_P(\alpha) \neq 0$ by 5.5.6. By 5.5.5, P contains a G-conjugate of any other p-subgroup Q satisfying $\mathrm{Br}_Q(\alpha) \neq 0$, and hence P is maximal with this property. This shows that (i) implies (ii). Suppose conversely that (ii) holds. Let R be a minimal subgroup of G such that $\alpha \subseteq A_R^G$. Then P is conjugate to a subgroup of R by 5.5.5, so we may assume that $P \subseteq R$. But $\mathrm{Br}_R(\alpha) \neq 0$ by 5.5.6, so the maximality of P implies $P = R$, whence the equivalence of (i) and (ii). This shows also that any one of the statements (i) or (ii) implies (iii). Suppose that (iii) holds. Let Q be a minimal subgroup of P such that $\alpha \subseteq A_Q^G$. Then Q is maximal such that $\mathrm{Br}_Q(1_A) \neq 0$. Since $\mathrm{Br}_P(1_A) \neq 0$ this forces $Q = P$. The proof is complete. □

Theorem 5.5.8 *Let G be a finite group, and let A be a primitive G-algebra over $\mathcal{O}$. Let H be a normal subgroup of G such that G/H is a p'-group. Then G acts transitively on the set of points of H on A.*

Proof Let β be a point of H on A. Let I be a primitive decomposition of 1_A in A^H. Denote by J the subset of all $j \in I$ such that $j \in {}^x\beta$ for some $x \in G$ and set $e = \sum_{j \in J} j$. Denote by $s : A^H \to A^H/J(A^H)$ the canonical surjection. Since $x \in G$ permutes the G-conjugates of β, the idempotent $s(e)$ is the sum of the unit elements of the simple factors of the semisimple algebra $A^H/J(A^H)$ corresponding to the G-conjugates of β. Thus $s(e)$ is a G-stable and central idempotent in $A^H/J(A^H)$. Since G/H is a p'-group, we have $A^G = A_H^G$, and hence s maps A^G onto $(A^H/J(A^H))^G$. Thus $s(e)$ lifts to an idempotent f in A^G. However, A^G is local, hence $f = 1$, which forces $e = 1$. The result follows. □

Corollary 5.5.9 *Let G be a finite group and A a primitive G-algebra over $\mathcal{O}$. Suppose that G has a normal Sylow p-subgroup P and that $A(P) \neq 0$. Then $\ker(\mathrm{Br}_P^A) \subseteq J(A^P)$, or equivalently, every point of P on A is local. In particular, the algebra A^P is split if and only if $A(P)$ is split.*

Proof Since $A(P) \neq 0$ there is at least one local point of P on A. By the assumptions and 5.5.8, the group G acts transitively on the set of points of P on A.

Since a G-conjugate of a local point of P on A remains local, every point of P on A is local, or equivalently, $\ker(Br_P^A)$ contains no idempotent, whence $\ker(\mathrm{Br}_P^A) \subseteq J(A^P)$. In particular, A^P and $A(P)$ have isomorphic semisimple quotients, implying the last statement. □

Corollary 5.5.10 *Let G be a finite group having a normal Sylow p-subgroup, and let U be a finitely generated indecomposable $\mathcal{O}G$-module having P as a vertex. Then every indecomposable direct summand of $\mathrm{Res}^G_P(U)$ has P as a vertex.*

Proof The points of P on $\mathrm{End}_{\mathcal{O}}(U)$ correspond to isomorphism classes of indecomposable direct summands of $\mathrm{Res}^G_P(U)$. The local points of P on $\mathrm{End}_{\mathcal{O}}(U)$ correspond to isomorphism classes of indecomposable direct summands of $\mathrm{Res}^G_P(U)$ having P as a vertex. By 5.5.9, every point of P on $\mathrm{End}_{\mathcal{O}}(U)$ is local, whence the result. □

Corollary 5.5.11 *Let G be a finite group and A a primitive interior G-algebra over $\mathcal{O}$. Suppose that $G = PC_G(P)$ for some Sylow p-subgroup P of G. Then P has a unique point on A.*

Proof The image of $C_G(P)$ in $A^\times$ under the structural homomorphism $G \to A^\times$ belongs to A^P; thus $C_G(P)$ stabilises every point of P on A. Since P stabilises every point of P on A, it follows that $G = PC_G(P)$ stabilises every point of P on A. By 5.5.8, the group G acts transitively on the set of points of P on A. The result follows. □

The set of pointed groups on a G-algebra A is a G-set, with G acting by conjugation. More explicitly, for H_β a pointed group on A and $x \in G$ we define ${}^xH_\beta = {}^xH_{({}^x\beta)}$, where ${}^xH = xHx^{-1}$ and ${}^x\beta = x\beta x^{-1}$. This makes sense as the action of x on A sends A^H to $A^{{}^xH}$. The set of local pointed groups on A is a G-subset of the G-set of pointed groups. Similarly, the pointed elements on A form a G-set via ${}^x(y_\beta) = {}^xy_{{}^x\beta}$. Note that for two pointed elements y_β, $y'_{\beta'}$ it is possible that the pointed groups $\langle y\rangle_\beta$ and $\langle y'\rangle_{\beta'}$ are G-conjugate while the pointed elements are not G-conjugate. The G-set of pointed groups on A admits a partial order.

Definition 5.5.12 Let G be a finite group, let A be a G algebra over $\mathcal{O}$, let P_γ, Q_δ be pointed groups on A, and let u_ϵ be a pointed element on A. We say that Q_δ is contained in P_γ or that P_γ contains Q_δ and write $Q_\delta \leq P_\gamma$ if $Q \subseteq P$ and if there are $i \in \gamma$, $j \in \delta$ such that $ij = j = ji$. We say that u_ϵ is contained in P_γ and write $u_\epsilon \in P_\gamma$ if $\langle u\rangle_\epsilon \leq P_\gamma$. The *relative multiplicity of δ in γ* is the nonnegative integer $m_\delta^\gamma = |\delta \cap J|$, where J is a primitive decomposition in A^Q of an element $i \in \gamma$.

With the notation of the above definition, if $Q \subseteq P$ then $A^P \subseteq A^Q$. The primitive idempotent $i \in A^P$ need no longer be primitive in the potentially bigger algebra A^Q. The condition $ji = ij = j$ means that j appears in some primitive decomposition of i in A^Q. If this holds then in fact for *any* $i \in \gamma$ there exists some $j \in \delta$ such that $ij = ji = i$. In this way, the set of pointed groups on A becomes a G-poset, and the set of local pointed groups on A is a G-subposet. If $Q \subseteq P$, then the relative multiplicity m_δ^γ is nonzero if and only if $Q_\delta \leq P_\gamma$. Since primitive decompositions of an idempotent are all conjugate, the integer m_δ^γ does not depend on the choice of i in γ or of the decomposition J of i in A^Q.

Example 5.5.13 Let G be a finite group, let M be a finitely generated $\mathcal{O}G$-module and let P_γ, Q_δ be pointed groups on $\mathrm{End}_\mathcal{O}(M)$. Thus, if we choose $i \in \gamma$ and $j \in \delta$ then $i \in \mathrm{End}_{\mathcal{O}P}(M)$ projects M onto an indecomposable direct summand $i(M)$ of $\mathrm{Res}_P^G(M)$. Similarly, j projects M onto an indecomposable direct summand $j(M)$ of $\mathrm{Res}_Q^G(M)$. The inclusion $Q_\delta \leq P_\gamma$ is equivalent to $j(M)$ being isomorphic to a direct summand of $\mathrm{Res}_Q^P(i(M))$.

The following result generalises the last statement in Corollary 4.6.9.

Proposition 5.5.14 *Let G be a finite group and let A be a G-algebra over $\mathcal{O}$. Let P_γ and Q_δ be pointed groups on A such that $Q \subseteq P$. Let $i \in \gamma$ and $j \in \delta$. We have $Q_\delta \leq P_\gamma$ if and only if $jA^Q i \not\subseteq J(A^Q)$.*

Proof We have $jA^Q i \not\subseteq J(A^Q)$ if and only if the image of $jA^Q i$ in $A^Q/J(A^Q)$ is non zero. The image of j in $A^Q/J(A^Q)$ is a primitive idempotent, hence contained in a unique simple factor B of $A^Q/J(A^Q)$. Therefore the image of $jA^Q i$ in $A^Q/J(A^Q)$ is contained in B, and this is non zero if and only if in a primitive decomposition of i in A^Q appears an idempotent whose image is in B, hence which is in the same point δ as j. □

Given a point α of a finite group G on a G-algebra A, we observed in the last section that the maximal p-subgroups P of G satisfying $\mathrm{Br}_P(\alpha) \neq 0$ coincide with the minimal subgroups P of G satisfying $\alpha \subseteq A_P^G$. The condition $\mathrm{Br}_P(\alpha) \neq 0$ means that for some (and hence any) element e in the point α we have $\mathrm{Br}_P(e) \neq 0$, so in a primitive decomposition of e in A^P there must occur a primitive idempotent $i \in A^P$ with the property $\mathrm{Br}_P(i) \neq 0$. This means exactly that i belongs to a local point γ of P on A. The corresponding local pointed group P_γ satisfies obviously $P_\gamma \leq G_\alpha$. The maximality of P with the property $\mathrm{Br}_P(\alpha) \neq 0$ implies that P_γ is a maximal local pointed group on A with the property $P_\gamma \leq G_\alpha$.

Definition 5.5.15 Let G be a finite group, A a G-algebra and α a point of G on A. A *defect pointed group of* G_α is a maximal local pointed group P_γ on A satisfying $P_\gamma \leq G_\alpha$.

The following result implies in particular that all defect pointed groups on A of G_α are G-conjugate, and any defect pointed group P_γ of G_α has the property that $\alpha \subseteq \mathrm{Tr}_P^G(A^P\gamma A^P)$; that is, not only is any $e \in \alpha$ of the form $e = \mathrm{Tr}_P^G(c)$ for some $c \in A^P$, but the element c can actually be chosen in the two-sided ideal $A^P\gamma A^P$ generated by the local point γ in A^P.

Theorem 5.5.16 *Let G be a finite group, A a G-algebra, and let α be a point of G on A. Let P be a subgroup of G that is minimal such that $\alpha \subseteq A_P^G$. Then there is a local point γ of P on A such that $P_\gamma \leq G_\alpha$, and moreover the following hold.*

(i) *We have $\alpha \subseteq \mathrm{Tr}_P^G(A^P\gamma A^P)$.*
(ii) *For any $i \in \gamma$ and $e \in \alpha$, there are $c, d \in (eAe)^P$ such that $e = \mathrm{Tr}_P^G(cid)$.*
(iii) *For any local pointed group Q_δ on A such that $Q_\delta \leq G_\alpha$ there exists an element $x \in G$ such that ${}^x(Q_\delta) \leq P_\gamma$. In particular, P_γ is a defect pointed group of G_α.*
(iv) *The defect pointed groups of G_α are G-conjugate.*
(v) *The local points γ' of P on A satisfying $P_{\gamma'} \leq G_\alpha$ are $N_G(P)$-conjugate.*

Proof We may replace A by eAe for some $e \in \alpha$; in other words, we may assume that $\alpha = \{1_A\}$. By 5.5.6 we have $\mathrm{Br}_P(1_A) \neq 0$, so the existence of a local point of P on A is clear. Decompose 1_A as a sum of primitive orthogonal idempotents in A^P. Then A^P is a sum of two-sided ideals of the form A^PiA^P, with i running over a set of primitive idempotents. Hence 1_A is in the sum of ideals $\mathrm{Tr}_P^G(A^PiA^P)$, with i running over a set of primitive idempotents in A^P, and hence, by Rosenberg's Lemma again, we have $1_A \in \mathrm{Tr}_P^G(A^PiA^P)$ for some primitive idempotent $i \in A^P$. Then $\mathrm{Br}_P(i) \neq 0$ because if $\mathrm{Br}_P(i) = 0$ then $i \in A_Q^P$ for some proper subgroup Q, which would imply that $1_A \in \mathrm{Tr}_P^G(A^PiA^P) \subseteq A_Q^G$, contradicting the minimality of P with this property. Thus i belongs to a local point γ of P on A. This proves (i) for this particular choice of γ, but we will see that this holds then for any local point of P on A. In order to see this, we first prove (iii). Let Q_δ be a local pointed group on A and let $j \in \delta$. Note that $A^P\gamma A^P = A^PiA^P$ for any $i \in \gamma$. Write

$$1_A = \sum_{s \in S} \mathrm{Tr}_P^G(a_s i b_s)$$

for some finite indexing set S and elements $a_s, b_s \in A^P$, where $s \in S$. Multiplying this expression by j and applying Mackey's formula yields

$$j = \sum_{s\in S} \sum_{y\in[Q\backslash G/P]} \mathrm{Tr}^Q_{Q\cap {}^yP}(j({}^y(a_s i b_s))).$$

Since $\mathrm{Br}_Q(j) \neq 0$ there is $y \in G$ such that $Q \subseteq {}^yP$ and such that $\mathrm{Br}_Q(j({}^ya_s)({}^yi)({}^yb_s)) \neq 0$, hence $jA^Q({}^yi) \not\subseteq J(A^Q)$ because $\mathrm{Br}_Q(j)$ is a non zero idempotent, hence not contained in $J(A(Q))$. By 5.5.14 this implies that $Q_\delta \leq {}^y(P_\gamma)$, whence (ii). This implies in particular that all defect pointed groups of G_α are G-conjugate, whence (iv) and (v). It also implies that (i) holds for *any* local point of P on A, because the local points of P on A are all G-conjugate and the statement in (i) is invariant under conjugation with elements in G. In order to prove (ii), observe that not all terms of the sum $1_A = \sum_{s\in S} \mathrm{Tr}^G_P(a_s i b_s)$ are in $J(A^G)$. Since A^G is local, at least one these summands is invertible in A^G. Thus there are elements $c, d \in A^P$ such that $u = \mathrm{Tr}^G_P(cid)$ is invertible in A^G. Then $1_A = u^{-1}\mathrm{Tr}^G_P(cid) = \mathrm{Tr}^G_P(u^{-1}cid)$, so after replacing c by $u^{-1}c$, we get the formula in (ii). □

The following theorem, known as the Burry–Carlson–Puig Theorem, was proved by Puig in [71] and for modules over group algebras independently by Burry and Carlson in [17].

Theorem 5.5.17 ([71, 1.4]) *Let G be a finite group, A a G algebra, P_γ a local pointed group on A, H_β a pointed group on A and α a point of G on A. Suppose that $P_\gamma \leq H_\beta \leq G_\alpha$ and that $N_G(P_\gamma) \leq H$. Then P_γ is a defect pointed group of H_β if and only if P_γ is a defect pointed group of G_α.*

Proof Suppose that P_γ is a defect pointed group of H_β. After replacing A by iAi for some $i \in \alpha$, we may assume that 1_A is primitive in A^G. Let $j \in \beta$. By 5.5.16 (i) there is $c \in A^P\gamma A^P$ such that $j = \mathrm{Tr}^H_P(c)$. By 2.5.5 we have $\mathrm{Tr}^G_H(j) = \mathrm{Tr}^G_P(c) = \sum_{x\in[P\backslash G/P]} \mathrm{Tr}^P_{P\cap {}^xP}({}^xc)$. Denote by $S(\gamma)$ the unique simple quotient of A^P such that the canonical image of γ in $S(\gamma)$ is nonzero. Denote by $s : A^P \to S(\gamma)$ the canonical surjection. Note that $\ker(\mathrm{Br}_P) \subseteq \ker(s)$ as the point γ is local. Thus, if x is not contained in $N_G(P)$, then the summand $\mathrm{Tr}^P_{P\cap {}^xP}({}^xc)$ is contained in $\ker(s)$. If $x \in N_G(P)$ but $x \notin N_G(P_\gamma)$, then ${}^xc$ belongs to $A^P({}^x\gamma)A^P$, which is also contained in $\ker(s)$ as ${}^x\gamma \neq \gamma$ in that case. It follows that $\mathrm{Tr}^P_{P\cap {}^xP}({}^xc)$ is contained in $\ker(s)$ for all $x \in G \setminus H$. Therefore we have $s(\mathrm{Tr}^G_H(j)) = s(\mathrm{Tr}^G_P(c)) = s(\mathrm{Tr}^H_P(c)) = s(j)$. Note that $s(j) \neq 0$ since $P_\gamma \leq H_\beta$ and $s(\gamma) \neq 0$. Thus $s(j)$ is an idempotent in $S(\gamma)$. This means that under the canonical map $A^G \to A^P \to S(\gamma)$ the image of the ideal $\mathrm{Tr}^G_P(A^P\gamma A^P)$ of A^G contains an idempotent. In particular, this ideal is not contained in

$J(A^G)$. Since 1_A is primitive in A^G, it follows that A^G is local, and hence that $A^G = \mathrm{Tr}_P^G(A^P\gamma A^P)$. This implies that P_γ is a defect pointed group of G_α. The converse is trivial. □

Corollary 5.5.18 *Let G be a finite group, A a G-algebra and P_γ, Q_δ local pointed groups on A such that $Q_\delta < P_\gamma$. Then there is a local pointed group R_ϵ on A such that $Q_\delta < R_\epsilon \leq P_\gamma$ and such that $R \leq N_P(Q_\delta)$. In particular, $Q < N_P(Q_\delta)$.*

Proof Since Q_δ is not maximal local in P_γ this follows from applying 5.5.17 with P_γ, $N_P(Q_\delta)$, Q_δ instead of G_α, H, P_γ, respectively. □

Corollary 5.5.19 *Let G be a finite group, P a p-subgroup, U a finitely generated indecomposable $\mathcal{O}G$-module, W an indecomposable direct summand of $\mathrm{Res}_P^G(U)$ having P as a vertex, and H a subgroup of G containing the stabiliser of W in $N_G(P)$. Let V be an indecomposable direct summand of $\mathrm{Res}_H^G(U)$ such that W is isomorphic to a direct summand of $\mathrm{Res}_P^H(V)$. Then (P, W) is a vertex-source pair for the $\mathcal{O}G$-module U if and only if (P, W) is a vertex source pair for the $\mathcal{O}H$-module V.*

Proof This is the special case of 5.5.17 applied to the G-algebra $A = \mathrm{End}_{\mathcal{O}}(U)$, the point $\alpha = \{\mathrm{Id}_U\}$ of G on A, the point β of H on A determined by a projection of $\mathrm{Res}_H^G(U)$ onto V and the point γ of P on A determined by a projection of $\mathrm{Res}_P^G(U)$ onto a direct summand of $\mathrm{Res}_P^H(V)$ isomorphic to W. □

Corollary 5.5.20 ([17, Theorem 5]) *Let G be a finite group, P a p-subgroup, and let U be a finitely generated indecomposable $\mathcal{O}G$-module. Let H be a subgroup of G containing $N_G(P)$. Then P is a vertex of U if and only if $\mathrm{Res}_H^G(U)$ has an indecomposable direct summand V with P as a vertex. Moreover, in that case, V is the Green correspondent of U, and any $\mathcal{O}P$-source of V is a source of U.*

As a consequence, direct summands with vertex P of an $\mathcal{O}G$-module U are completely determined by direct summands with vertex P of $\mathrm{Res}_{N_G(P)}^G(U)$.

Corollary 5.5.21 *Let G be a finite group, P a p-subgroup of G, and let U, U' be finitely generated $\mathcal{O}G$-modules. Set $N = N_G(P)$. The following are equivalent.*

(i) *The direct sum of all indecomposable direct summands with vertex P in a direct sum decomposition of U is isomorphic to the direct sum of all indecomposable direct summands with vertex P in a direct sum decomposition of U'.*

(ii) The direct sum of all indecomposable direct summands with vertex P in a direct sum decomposition of $\mathrm{Res}^G_N(U)$ *is isomorphic to the direct sum of all indecomposable direct summands with vertex P in a direct sum decomposition of* $\mathrm{Res}^G_N(U')$.

Proof If V is an indecomposable direct summand of U with vertex P, then the Green correspondent of V is the unique (up to isomorphism) indecomposable direct summand of $\mathrm{Res}^G_{N_G(P)}(U)$ with vertex P. If V does not have P as a vertex, then 5.5.20 implies that $\mathrm{Res}^G_{N_G(P)}(V)$ has no indecomposable direct summand with with vertex P. The equivalence between (i) and (ii) follows from the Green correspondence. □

Remark 5.5.22 Let G be a finite group, A a G-algebra, and Q_δ a local pointed group on A. Suppose that A^Q is split. Denote by $A(\delta)$ the unique simple quotient of A^Q such that the image of δ under the canonical surjection $s : A^Q \to A(\delta)$ is nonzero. Note that s factors through the Brauer homomorphism $\mathrm{Br}_Q : A^Q \to A(Q)$ because δ is local. Then $s(\delta)$ is the unique point of the matrix algebra $A(\delta)$. The group $N_G(Q_\delta)$ acts on A^Q and stabilises δ, hence $A(\delta)$, and thus induces an action of $N_G(Q_\delta)$ on $A(\delta)$. This action determines a class $\alpha \in H^2(N_G(Q_\delta); k^\times)$. A variation of the proof of 5.3.12 shows that if the action of $C_G(Q)$ on A is induced by an $N_G(Q_\delta)$-invariant group homomorphism $\sigma : C_G(Q) \to A^\times$, then this class is in the image of the map $H^2(N_G(Q_\delta)/C_G(Q); k^\times) \to H^2(N_G(Q_\delta); k^\times)$ induced by the canonical group homomorphism $N_G(Q_\delta) \to N_G(Q_\delta)/C_G(Q)$, and there is a canonical choice for a preimage of α. To see this, note that σ sends $C_G(Q)$ to $(A^Q)^\times$, and hence any element $z \in C_G(Q)$ acts on $A(\delta)$ as conjugation by $s(\sigma(z))$. Let $\mathcal{R}$ be a set of representatives of $N_G(Q_\delta)/C_G(Q)$ in $N_G(Q_\delta)$. For any $x \in R$ choose $t_x \in A(\delta)^\times$ such that t_x acts as x on $A(\delta)$, and set $t_{xy} = t_x s(\sigma(z))$ for all $x \in \mathcal{R}$ and all $z \in C_G(Q)$. Thus t_x acts as x on $A(\delta)$ for all $x \in G$, and we have $t_x t_y = \alpha(x, y)t_{xy}$ for some $\alpha \in k^\times$ whose value depends only on the classes $xC_G(Q)$ and $yC_G(Q)$, for all $x, y \in G$.

5.6 Defect groups and source algebras of primitive *G*-algebras

All $\mathcal{O}$-algebras in this section are assumed to be finitely generated as $\mathcal{O}$-modules.

Definition 5.6.1 Let G be a finite group. A G-algebra A over $\mathcal{O}$ is called *primitive* if 1_A is a primitive idempotent in A^G.

Example 5.6.2 Let G be a finite group and let M be a finitely generated $\mathcal{O}G$-module. Then $\mathrm{End}_{\mathcal{O}}(M)$ is a primitive G-algebra if and only if M is indecomposable. Indeed, any idempotent in $(\mathrm{End}_{\mathcal{O}}(M))^G = \mathrm{End}_{\mathcal{O}G}(M)$ is a projection of M onto a direct summand of M.

Definition 5.6.3 Let G be a finite group and let A be a primitive G-algebra over $\mathcal{O}$ that is finitely generated as $\mathcal{O}$-module. A *defect group of A* is a minimal subgroup P of G such that $1_A \in A_P^G$, or equivalently, such that $1_A = \mathrm{Tr}_P^G(c)$ for some $c \in A^P$.

Example 5.6.4 Let G be a finite group and let M be a finitely generated indecomposable $\mathcal{O}G$-module. By Higman's criterion 2.6.2, a subgroup P of G is a defect group of the primitive G-algebra $\mathrm{End}_{\mathcal{O}}(M)$ if and only if P is a vertex of M.

We restate some of the results of the previous section for primitive G-algebras.

Theorem 5.6.5 *Let G be a finite group and let A be a primitive G-algebra over $\mathcal{O}$. Let P be a defect group of A.*

(i) The group P is a p-subgroup of G.
(ii) For any subgroup H of G such that $1_A \in A_H^G$ there is an element $x \in G$ such that $P \subseteq {}^xH$.
(iii) The defect groups of A form a G-conjugacy class of p-subgroups of G.

Proof The statements (i) and (ii) are specialisations to primitive G-algebras of statements in 5.5.6. Let Q be another defect group of b. By (ii) applied twice we have $P \subseteq {}^yQ$ and $Q \subseteq {}^zP$ for some $y, z \in G$ and hence P and Q are G-conjugate. Conversely, if ${}^yQ = P$ for some $y \in G$ then $1_A = \mathrm{Tr}_P^G(c) = \mathrm{Tr}_Q^G({}^xc)$, hence any conjugate of a defect group is a defect group. This proves (iii). □

Theorem 5.6.6 *Let G be a finite group, A a primitive G-algebra over $\mathcal{O}$ and P be a defect group of A.*

(i) We have $\mathrm{Br}_P(1_A) \neq 0$.
(ii) For any p-subgroup Q such that $\mathrm{Br}_Q(1_A) \neq 0$ there is $x \in G$ such that $Q \subseteq {}^xP$.

Proof Statement (i) is a special case of 5.5.6, and (ii) is a special case of 5.5.5. □

Theorem 5.6.7 *Let G be a finite group, A a primitive G-algebra over $\mathcal{O}$ and P a subgroup of G. The following are equivalent.*

(i) *P is a defect group of A.*
(ii) *P is a maximal p-subgroup of G such that $A(P) \neq \{0\}$.*
(iii) *P is a p-subgroup of G satisfying $1_A \in A^G_P$ and $\mathrm{Br}_P(1_A) \neq 0$.*

Proof Note that $A(P) \neq \{0\}$ is equivalent to $\mathrm{Br}_P(1_A) \neq 0$. Thus this theorem is a special case of 5.5.7. □

Corollary 5.6.8 *Let G be a finite group, A a primitive G-algebra, and P_γ a defect pointed group of $G_{\{1_A\}}$. Then P is a defect group of A, and we have $1_A \subseteq \mathrm{Tr}^G_P(A^P\gamma A^P)$.*

Proof This is an immediate consequence of the above, or also of 5.5.16. □

As an application, we get the following characterisation of vertices of indecomposable modules:

Theorem 5.6.9 *Let G be a finite group, let P be a subgroup of G and let M be a finitely generated indecomposable $\mathcal{O}G$-module. The following are equivalent:*

(i) *P is a vertex of M.*
(ii) *P is a maximal p-subgroup of G with the property $(\mathrm{End}_\mathcal{O}(M))(P) \neq 0$.*

Proof By Higman's criterion 2.6.2, P is a vertex of M if and only if P is a defect group of the primitive G-algebra $\mathrm{End}_\mathcal{O}(M)$. The equivalence of (i) and (ii) is therefore a special case of the equivalence of the statements (i) and (ii) in 5.6.7.
□

A G-algebra A is in particular an $\mathcal{O}G$-module, and by our general assumption, A is in fact a finitely generated $\mathcal{O}G$-module. As another application of Higman's criterion we obtain the following relative projectivity statements for G-algebras and interior G-algebras.

Proposition 5.6.10 *Let G be a finite group, A a primitive G-algebra, and let P be a defect group of A. Then, as an $\mathcal{O}G$-module, A is relatively $\mathcal{O}P$-projective.*

Proof For $c \in A$, denote by γ_c the $\mathcal{O}$-linear endomorphism of A sending $a \in A$ to ca. The action of G on A induces an action on $\mathrm{End}_\mathcal{O}(A)$ such that $({}^x(\gamma_c))(a) = {}^x(\gamma_c({}^{x^{-1}}a)) = {}^x(c \cdot ({}^{x^{-1}}a)) = ({}^xc) \cdot a = \gamma_{{}^xc}(a)$. It follows that if $c \in A^P$, then $\gamma_c \in \mathrm{End}_{\mathcal{O}P}(A)$, and we have $\mathrm{Tr}^G_P(\gamma_c) = \gamma_{\mathrm{Tr}^G_P(c)}$, where the first trace map is understood in $\mathrm{End}_\mathcal{O}(A)$ and the second in A. Since P is a defect group of A, there is $c \in A^P$ such that $\mathrm{Tr}^G_P(c) = 1_A$, hence such that $\mathrm{Tr}^G_P(\gamma_c) = \mathrm{Id}_A$. Higman's criterion 2.6.2 implies that A is relatively $\mathcal{O}P$-projective as an $\mathcal{O}G$-module.
□

Proposition 5.6.11 *Let G be a finite group, A a primitive interior G-algebra with structural homomorphism $\sigma : G \to A^\times$, and let P be a defect group of A. Then, as an $\mathcal{O}G$-A-bimodule, A is relatively $\mathcal{O}P$-projective. More precisely, the canonical map $\mathcal{O}G \otimes_{\mathcal{O}P} A \to A$ sending $x \otimes a$ to $\sigma(x)a$ splits as a homomorphism of $\mathcal{O}G$-A-bimodules.*

Proof This is a variation of the proof of 5.6.10. The map γ_c sending $a \in A^{\Delta P}$ to ca is an $\mathcal{O}P$-A-bimodule endomorphism of A. If c is chosen such that $\mathrm{Tr}_P^G(c) = 1_A$, then the same reasoning as in 5.6.10 yields $\mathrm{Tr}_P^G(\gamma_c) = \mathrm{Id}_A$. Higman's criterion 2.6.2 implies the result. □

Definition 5.6.12 Let G be a finite group, A a primitive G-algebra over $\mathcal{O}$ and P a defect group of A. Let i be a primitive idempotent in A^P such that $\mathrm{Br}_P(i) \neq 0$. The primitive P-algebra iAi is called a *source algebra of* A, and the idempotent i is called a *source idempotent of* A.

Equivalently, if P_γ is a defect pointed group of the primitive G-algebra, then, for any $i \in \gamma$, the algebra iAi is a source algebra of A. If j is another element of γ, then $j = cic^{-1}$ for some $c \in (A^{\Delta P})^\times$, and hence the map sending $a \in iAi$ to cac^{-1} is an isomorphism of P-algebras $iAi \cong jAj$. By 5.5.16, the group $N_G(P)$ acts transitively on the set of local points of P on A, and hence all source algebras of A are isomorphic as $\mathcal{O}$-algebras, with the P-algebra structure possibly 'twisted' by an automorphism of P given by conjugation with an element in $N_G(P)$. That is, the source algebras of A are isomorphic as P-algebras 'up to twisting by automorphisms of P'.

Theorem 5.6.13 (cf. [71, 3.5]) *Let G be a finite group, A a primitive interior G-algebra, P a defect group of A and i an idempotent in A^P satisfying $\mathrm{Br}_P(i) \neq 0$.*

(i) There are c, $d \in A^P$ such that $1_A = \mathrm{Tr}_P^G(cid)$. In particular, we have $A = AiA$.
(ii) Then the algebras A and iAi are Morita equivalent via the bimodules Ai and iA. In particular, A is Morita equivalent to any of its source algebras.

Proof Denote by $\sigma : G \to A^\times$ the structural homomorphism. Any idempotent in a primitive decomposition of i in A^P is contained in the ideal $A^P iA^P$, so we may assume that i is primitive in A^P. By 5.5.16 there are $c, d \in A^P$ such that

$$1_A = \mathrm{Tr}_P^G(cid) = \sum_{x\in[G/P]} \sigma(x)cid\sigma(x^{-1})$$

which shows (i). Statement (ii) follows from 2.8.7. □

Exercise 5.6.14 The purpose of this exercise is to show that 5.6.13 need not hold for G-algebras that are not interior G-algebras. Consider the group $G = \{1, t\}$ of order 2 and the G-algebra $A = \mathcal{O} \times \mathcal{O}$ with t acting on $(\lambda, \mu) \in A$ by ${}^t(\lambda, \mu) = (\mu, \lambda)$. Show that A is a primitive G-algebra, having the trivial subgroup of G as a defect group, and a source algebra isomorphic to $\mathcal{O}$. Deduce that A is not Morita equivalent to its source algebras.

There is a way to reconstruct a primitive interior G-algebra from its source algebras.

Theorem 5.6.15 (cf. [71, Theorem 3.4]) *Let G be a finite group, A a primitive interior G-algebra, P a defect group of A and $i \in A^{\Delta P}$ a source idempotent. Set $B = iAi$.*

(i) As an $\mathcal{O}G$-B-bimodule, Ai is isomorphic to a direct summand of $\mathcal{O}G \otimes_{\mathcal{O}P} B$.
(ii) There is a primitive idempotent e in $\mathrm{End}_{\mathcal{O}G \otimes_{\mathcal{O}P} B^{\mathrm{op}}}(\mathcal{O}G \otimes_{\mathcal{O}P} B)$ such that we have an isomorphism of interior G-algebra

$$A \cong e \cdot \mathrm{End}_{B^{\mathrm{op}}}(\mathcal{O}G \otimes_{\mathcal{O}P} B) \cdot e.$$

Proof We show first that (i) implies (ii). We have an obvious isomorphism $A \cong \mathrm{End}_{A^{\mathrm{op}}}(A)$. Since A and B are Morita equivalent via the A-B-bimodule Ai, it follows that $A \cong \mathrm{End}_{B^{\mathrm{op}}}(Ai)$. Thus if (i) holds, then taking for e a projection of $\mathcal{O}G \otimes_{\mathcal{O}P} B$ onto a summand isomorphic to Ai yields the isomorphism in (ii). Using again the Morita equivalence between A and B, it suffices to show that as an $\mathcal{O}G$-A-bimodule, A is isomorphic to a direct summand of $\mathcal{O}G \otimes_{\mathcal{O}P} iA$. There are two ways to do this. From 5.6.11 we get that A is isomorphic to a direct summand of $\mathcal{O}G \otimes_{\mathcal{O}P} A$. Since A is a primitive interior G-algebra, it is indecomposable as an $\mathcal{O}G$-A-module, and hence A is isomorphic to a direct summand of $\mathcal{O}G \otimes_{\mathcal{O}P} jA$ for some primitive idempotent $j \in A^{\Delta P}$. Then $\mathrm{Br}_{\delta P}(j) \neq 0$ because $A(\Delta P) \neq 0$. This shows that j belongs to a local point of P on A. By 5.5.16, the local points of P on A form a single $N_G(P)$-orbit for the conjugation action, and hence we may take $j = i$. The second proof uses the fact from 5.5.16 that $1_A = \mathrm{Tr}_P^G(cid)$ for some $c, d \in A^{\Delta P}$. A straightforward verification shows that the canonical $\mathcal{O}G$-A-bimodule homomorphism $\mathcal{O}G \otimes_{\mathcal{O}P} iA \to A$ has as a section the map sending $a \in A$ to $\sum_{x \in [G/P]} x \otimes cix^{-1}a$. □

Remark 5.6.16 The algebra $\mathrm{End}_{B^{\mathrm{op}}}(\mathcal{O}G \otimes_{\mathcal{O}P} B)$ is isomorphic, as an interior G-algebra, to what is called the *induced algebra* $\mathrm{Ind}_P^G(B)$ in [71, 3.3]. The next proposition justifies this terminology in that it establishes a link with the usual induction of modules. See [58] for more details and references.

Proposition 5.6.17 *Let G be a finite group, H a subgroup of G, and V a finitely generated $\mathcal{O}H$-module. Set $S = \mathrm{End}_{\mathcal{O}}(V)$. We have an isomorphism of interior G-algebras*

$$\mathrm{End}_{\mathcal{O}}(\mathrm{Ind}_H^G(V)) \cong \mathrm{End}_{S^{\mathrm{op}}}(\mathcal{O}G \otimes_{\mathcal{O}H} S).$$

Proof Consider the $\mathcal{O}$-module decomposition $\mathrm{Ind}_H^G(V) = \oplus_{x\in[G/H]}\, x \otimes V$. Through this decomposition, every element in $\mathrm{End}_{\mathcal{O}}(\mathrm{Ind}_H^G(V))$ corresponds to a matrix of $\mathcal{O}$-linear maps $x \otimes V \to y \otimes V$, where $x, y \in [G/H]$, hence to a matrix indexed by $x, y \in [G/H]$ with entries in S. Consider next the decomposition as right S-modules $\mathcal{O}G \otimes_{\mathcal{O}H} S = \oplus_{x\in[G/H]}\, x \otimes S$. Any S^{op}-endomorphism of S is given by left multiplication with an element in S. Thus, through this decomposition, any element in $\mathrm{End}_{S^{\mathrm{op}}}(\mathcal{O}G \otimes_{\mathcal{O}H} S)$ corresponds again to a matrix of elements in S, indexed by $x, y \in [G/H]$. One verifies that this is an isomorphism of interior G-algebras. □

5.7 Multiplicity modules

The isomorphism class of an indecomposable $\mathcal{O}G$-module U is not determined uniquely by a vertex-source pair, but it is determined up to finitely many possibilities. Those finitely many possibilities are parametrised by the *multiplicity module* of U. Multiplicity modules are crucial in work of Knörr on vertices of modules with irreducible characters; see Section 10.3. We will be mainly interested in multiplicity modules of indecomposable $\mathcal{O}G$-modules and kG-modules, but it takes no extra effort to introduce this notion for arbitrary interior algebras. In this generality, this concept is due to Puig. In this section, algebras over $\mathcal{O}$ are finitely generated as $\mathcal{O}$-modules, and modules over $\mathcal{O}$-algebras are finitely generated.

Definition 5.7.1 Let G be a finite group, A an interior G-algebra and P_γ a local pointed group on A. Denote by $S(\gamma)$ the unique simple quotient of A^P such that the image of γ in $S(\gamma)$ is nonzero. Suppose that $S(\gamma)$ is split (that is, a matrix algebra). The group $\bar{N} = N_G(P_\gamma)/P$ acts on $S(\gamma)$; denote by $\alpha \in H^2(\bar{N}; k^\times)$ the class determined by this action. Let M_γ be a $k_\alpha\bar{N}$-module such that $S(\gamma) \cong \mathrm{End}_k(M_\gamma)$ as $\bar{N}$-algebras. Then M_γ is called a *multiplicity module of P_γ on A.*

This applies in particular to $A = \mathrm{End}_{\mathcal{O}}(U)$ for an indecomposable $\mathcal{O}G$-module U. In that case, P is a vertex of U and for $i \in \gamma$, the $\mathcal{O}P$-module $i(U)$ is a source of U, and we will call M_γ a *multiplicity module of U*, assuming a choice of a vertex-source pair. The hypothesis on $S(\gamma)$ being a matrix algebra amounts to requiring that the source $i(U)$ of U is absolutely indecomposable.

Remark 5.7.2 The multiplicity module M_γ is well-defined up to a linear character $N_G(P_\gamma)/P \to k^\times$, since any two 'lifts' of the action of $\bar{N}$ on $S(\gamma)$ to a module structure on M_γ will differ by such a linear character. One can be slightly more precise regarding the uniqueness of M_γ by making use of the fact that $S(\gamma)$ is an interior $C_G(P)$-algebra. More precisely, since P_γ is a local pointed group on A, it follows that the image of γ in $A(P)$ is nonzero, and hence $S(\gamma)$ is a quotient of $A(P)$. The action of $N_G(P_\gamma)$ on $A(P)$ stabilises the image of γ and hence induces an action on the matrix algebra $S(\gamma)$. Moreover, P acts trivially on A^P, hence trivially on $A(P)$ and on $S(\gamma)$, so this induces an action of $\bar{N} = N_G(P_\gamma)/P$ on $S(\gamma)$, giving rise to a 2-cocycle α as above. Since A is an interior G-algebra, it follows that A^P is an interior $C_G(P)$-algebra, and hence $S(\gamma)$ is an interior $C_G(P)$-algebra. The action of $N_G(P_\gamma)$ extends the interior $C_G(P)$-algebra structure of $S(\gamma)$. The image of $P \cap C_G(P) = Z(P)$ in A^P is contained in $Z(A^P)$, hence its image in $S(\gamma)$ is contained in $Z(S(\gamma)) \cong k$. Since $\mathrm{char}(k) = p$, it follows that the image of $Z(P)$ in $S(\gamma)$ is trivial. Thus the interior $C_G(P)$-algebra structure of $S(\gamma)$ induces an interior $PC_G(P)/P$-algebra structure on $S(\gamma)$, and the action of $N_G(P_\gamma)/P$ on $S(\gamma)$ extends the interior $PC_G(P)/P$-algebra structure of $S(\gamma)$. Thus M_γ is in fact uniquely determined up to a linear character of $N_G(P_\gamma)/PC_G(P)$.

In what follows, whenever we use multiplicity modules, we assume implicitly that k is large enough for the simple quotient $S(\gamma)$ of $A(P)$. The following observation explains the terminology:

Proposition 5.7.3 *Let G be a finite group, A an interior G-algebra and P_γ a local pointed group on A with multiplicity module M_γ. Then $\dim_k(M_\gamma) = m_\gamma$, the multiplicity of γ on A.*

Proof Let I be a primitive decomposition of 1 in A^P. Denote by $s_\gamma : A^P \to S(\gamma)$ the canonical map. For $j \in I$ we have $s_\gamma(j) = 0$ if and only if j does not belong to γ. If $j \in \gamma$, then $s_\gamma(j)$ is a primitive idempotent in $S(\gamma)$. Thus $s_\gamma(\gamma \cap I)$ is a primitive decomposition of 1 in the matrix algebra $S(\gamma) = \mathrm{End}_k(M_\gamma)$, hence $\dim_k(M_\gamma) = |\gamma \cap I|$, which by definition is the multiplicity of γ on A. □

Example 5.7.4 Let G be a finite group and U an indecomposable $\mathcal{O}G$-module. Set $A = \mathrm{End}_\mathcal{O}(U)$. Let P_γ be a defect pointed group of A with multiplicity module M_γ. Then P is a vertex of U and $Y = i(U)$ is an $\mathcal{O}P$-source of U. Suppose that Y is absolutely indecomposable. Then the multiplicity $m_\gamma = \dim_k(M_\gamma)$ of

γ is the multiplicity of the source Y in a decomposition of $\mathrm{Res}^G_P(U)$ as a direct sum of indecomposable $\mathcal{O}P$-modules. We say in this situation that *M_γ is the multiplicity module of U with respect to the source Y*. The next result implies that M_γ is a projective indecomposable module over a twisted group algebra $k_\alpha T/P$, where T is the stabiliser of the isomorphism class of Y in $N_G(P)$, and where α is a class in $H^2(T/Q; k^\times)$.

Theorem 5.7.5 *Let G be a finite group, A a primitive interior G-algebra and P_γ a local pointed group on A. Denote by $S(\gamma)$ the simple quotient of A^P determined by γ, and set $\bar{N} = N_G(P_\gamma)/P$. Suppose that $S(\gamma)$ is a matrix algebra over k, and denote by α the class in $H^2(\bar{N}; k^\times)$ determined by the action of $\bar{N}$ on $S(\gamma)$. Let M_γ be a $k_\alpha\bar{N}$-module satisfying $S(\gamma) \cong \mathrm{End}_k(M_\gamma)$ as $\bar{N}$-algebras.*

(i) Let $s_\gamma : A^P \to S(\gamma)$ be the canonical surjection. For any $a \in A^P\gamma A^P$ we have $s_\gamma(\mathrm{Tr}^G_P(a)) = \mathrm{Tr}^{\bar{N}}_1(s_\gamma(a))$. In particular, we have $s_\gamma(\mathrm{Tr}^G_P(A^P\gamma A^P)) = \mathrm{Tr}^{\bar{N}}_1(S(\gamma))$.

(ii) The local pointed group P_γ is a defect pointed group on A if and only if the multiplicity module M_γ has a nonzero projective direct summand as a $k_\alpha\bar{N}$-module.

(iii) If P_γ is a defect pointed group on A, then M_γ is a projective indecomposable $k_\alpha\bar{N}$-module, and the canonical map $s_\gamma : A^P \to S(\gamma)$ induces a surjective algebra homomorphism $A^G \to \mathrm{End}_{k_\alpha\bar{N}}(M_\gamma)$ with kernel containing $A^G \cap \ker(\mathrm{Br}^A_P)$.

Proof The image of γ in $A(P)$ is nonzero as γ is a local point, and hence the matrix algebra $S(\gamma)$ is a quotient of $A(P)$. It follows that $\ker(\mathrm{Br}^A_P) \subseteq \ker(s_\gamma)$. Thus s_γ factors as $s_\gamma = t_\gamma \circ \mathrm{Br}^A_P$ for some surjective algebra homomorphism $t_\gamma : A(P) \to S(\gamma)$. The surjective algebra homomorphism s_γ maps the ideal $A^P\gamma A^P$ to a nonzero ideal in $S(\gamma)$, hence onto $S(\gamma)$ as $S(\gamma)$ is a simple algebra. Let $a \in A^P\gamma A^P$. Then $\mathrm{Br}^A_P(\mathrm{Tr}^G_P(a)) = \mathrm{Tr}^{N_G(P)}_P(\mathrm{Br}^A_P(a)) = \sum_{x\in[N_G(P)/P]} \mathrm{Br}^A_P(xax^{-1})$. The summands for which $x \in N_G(P)$ is not in $N_G(P_\gamma)$ are mapped to zero under t_γ, because in that case $xax^{-1} \in A^{Px}\gamma A^P$ and ${}^x\gamma$ is a local point of P on A which is different from γ, hence is mapped to zero under s_γ. Thus $s_\gamma(\mathrm{Tr}^G_P(a)) = \mathrm{Tr}^{N_G(P_\gamma)}_P(s_\gamma(a)) = \mathrm{Tr}^{\bar{N}}_1(s_\gamma(a))$. This proves (i). If P_γ is a defect pointed group on A, then $1_A \in \mathrm{Tr}^G_P(A^P\gamma A^P)$. Thus $1_{S(\gamma)} \in (S(\gamma))^{\bar{N}}_1$, or equivalently, $(S(\gamma))^{\bar{N}}_1 = (S(\gamma))^{\bar{N}} = \mathrm{End}_{k_\alpha\bar{N}}(M_\gamma)$. This shows that s_γ induces a surjective algebra homomorphism $A^G \to \mathrm{End}_{k_\alpha\bar{N}}(M_\gamma)$ with kernel containing $A^G \cap \ker(\mathrm{Br}^A_P)$. The version 2.6.19 of Higman's criterion for twisted group algebras implies that M_γ is projective as a $k_\alpha\bar{N}$-module. Since s_γ maps the local algebra A^G onto $S(\gamma)^{\bar{N}}$, it follows that $S(\gamma)^{\bar{N}}$ is local, which implies that M_γ is indecomposable. This shows (iii) and the forward direction in (ii). For the converse in (ii), if M_γ has a nonzero projective summand, then $S(\gamma)_1^{\bar{N}}$ contains an idempotent. Thus there is $a \in \mathrm{Tr}_P^G(A^P\gamma A^P)$ such that $s_\gamma(a)$ is an idempotent in $S(\gamma)$. In particular, a does not belong to the radical of A^G. Since A^G is local this forces that a is invertible in A^G and hence that $A^G = \mathrm{Tr}_P^G(A^P\gamma A^P)$. Thus P_γ is a defect pointed group on A. □

The following result show that a vertex, source, and multiplicity module determine the isomorphism class of an indecomposable $\mathcal{O}G$-module.

Proposition 5.7.6 *Let G be a finite group, P a p-subgroup and V an absolutely indecomposable $\mathcal{O}P$-module having P as a vertex. Set $A = \mathrm{End}_{\mathcal{O}}(\mathrm{Ind}_P^G(V))$, denote by γ the local point of P on A determined by V and by α the class in $H^2(\bar{N}; k^\times)$ given by the action of the group $\bar{N} = N_G(P_\gamma)/P$ on the simple quotient $S(\gamma)$ of A^P determined by γ. Write $S(\gamma) = \mathrm{End}_k(V_\gamma)$ for some $k_\alpha\bar{N}$-module V_γ. Any indecomposable direct summand U of $\mathrm{Ind}_P^G(V)$ with P as a vertex has a multiplicity module that is isomorphic to an indecomposable direct summand of V_γ, and this correspondence induces a bijection between the isomorphism classes of indecomposable direct summands of $\mathrm{Ind}_P^G(V)$ having P as a vertex and V as a source and the isomorphism classes of projective indecomposable $k_\alpha\bar{N}$-modules.*

Proof Let U be an indecomposable direct summand of $\mathrm{Ind}_P^G(V)$. Denote by $e \in A^G$ an idempotent corresponding to a projection of $\mathrm{Ind}_P^G(V)$ onto U. Then $\mathrm{End}_{\mathcal{O}}(U) \cong eAe$. If U has P as a vertex and V as a source, then the image of e in $S(\gamma)$ is a nonzero idempotent $\bar{e}$ in $S(\gamma)^{\bar{N}}$. Thus, by 5.3.5 the classes determined by the actions of $\bar{N}$ on $S(\gamma)$ and on $\bar{e}S(\gamma)\bar{e}$ coincide. This shows that the multiplicity modules of different direct summands of $\mathrm{Ind}_P^G(V)$ are all defined over the same twisted group algebra $k_\alpha\bar{N}$, so long as they have P as a vertex and V as a source. The rest follows from 5.7.5. □

One can push this further to obtain a parametrisation of interior P-algebras in terms of suitable generalisations of vertices, sources and multiplicity modules; see [89] and [74]. See also [90, §13, §14, §26] for a broader exposition of this material. The next observation is a sufficient criterion for multiplicity modules to be simple.

Proposition 5.7.7 ([71, 1.6]) *Let G be a finite group, A a primitive interior G-algebra, P_γ a defect pointed group on A. Denote by $S(\gamma)$ the simple quotient of A^P determined by γ, and set $\bar{N} = N_G(P_\gamma)/P$. Suppose that $S(\gamma)$ is a matrix algebra over k, and denote by α the class in $H^2(\bar{N}; k^\times)$ determined by the action of $\bar{N}$ on $S(\gamma)$. Let M be a $k_\alpha\bar{N}$-module satisfying $S(\gamma) \cong \mathrm{End}_k(M)$*

as $\bar{N}$-algebras. If $A^G = \mathcal{O} \cdot 1 + (A^G \cap \ker(\mathrm{Br}_P^A))$, then M is absolutely simple projective.

Proof By 5.7.5 (i), the $k_\alpha\bar{N}$-module M is projective indecomposable. Since the algebra $k_\alpha\bar{N}$ is symmetric, it follows that the top and bottom composition factors of M are isomorphic. In particular, there is a nonzero endomorphism of M sending M onto $\mathrm{soc}(M)$. If $A^G = \mathcal{O} \cdot 1 + (A^G \cap \ker(\mathrm{Br}_P^A))$, then 5.7.5 (iii) implies that $\mathrm{End}_{k_\alpha\bar{N}}(M) \cong k$. Thus any nonzero endomorphism of M is an automorphism, and hence $M = \mathrm{soc}(M)$ is absolutely simple. □

Corollary 5.7.8 *Let G be a finite group, U an indecomposable $\mathcal{O}G$-module with a nontrivial vertex P, an absolutely indecomposable source V and multiplicity $k_\alpha\bar{N}$-module M, where $\bar{N} = N_G(P, V)/P$ and $\alpha \in H^2(\bar{N}, k^\times)$. If $\underline{\mathrm{End}}_{\mathcal{O}G}(U) \cong \mathcal{O}/J(\mathcal{O})^m$ for some positive integer m, then M is a simple projective $k\bar{N}$-module.*

Proof Since P is nontrivial, we have $(\mathrm{End}_{\mathcal{O}}(U))_1^G \subseteq \ker(\mathrm{Br}_P)$. The result is a special case of 5.7.7. □

Corollary 5.7.9 *Let G be a finite group. Suppose that K is a splitting field for G of characteristic zero.*

(i) If U is a simple kG-module, then U has a simple multiplicity module.

(ii) If X is an $\mathcal{O}$-free $\mathcal{O}G$-module with an irreducible character, then X has a simple multiplicity module.

Proof If U is a simple kG-module, then $\mathrm{End}_{kG}(U) \cong k$. If X has an irreducible character, then $\mathrm{End}_{\mathcal{O}G}(X) \cong \mathcal{O}$, hence $\underline{\mathrm{End}}_{\mathcal{O}G}(X) \cong \mathcal{O}/J(\mathcal{O})^m$ for some positive integer m. Thus both statements follow from 5.7.8. □

Remark 5.7.10 We will see later some implications for an interior algebra to have a simple multiplicity module. If a primitive interior G-algebra A with defect pointed group P_γ has a simple multiplicity module M_γ over the twisted group algebra $k_\alpha\bar{N}$ as above, then by 6.6.4, we have $O_p(\bar{N}) = \{1\}$, and we will show in 10.3.1 that P is *centric* in a fusion system of the unique block of $\mathcal{O}G$ to which A belongs.

As an application we prove a result on the parametrisation of extensions of a module of a normal subgroup.

Theorem 5.7.11 ([88]) *Let G be a finite group and N a normal subgroup of index prime to p. Suppose that k is algebraically closed. Let V be a finitely generated indecomposable $\mathcal{O}N$-module, and let U, U' be $\mathcal{O}G$-modules such that $\mathrm{Res}_N^G(U) \cong \mathrm{Res}_N^G(U') \cong V$. Then there is an $\mathcal{O}$-free $\mathcal{O}G/N$-module T of*

$\mathcal{O}$-rank 1 such that $U' \cong T \otimes_{\mathcal{O}} U$, where T is considered an an $\mathcal{O}G$-module via the canonical surjection $G \to G/N$.

Proof Denote by $\rho : G \to \mathrm{Aut}(U)$ and $\sigma : G \to \mathrm{Aut}(U')$ the structural homomorphisms. Since U, U' extend the same module V, the maps ρ and σ coincide on N. For $x \in G$, set $\tau(x) = \rho(x)\sigma(x^{-1})$. We verify that $\tau(x) \in \mathrm{End}_{\mathcal{O}N}(V)$. Let $y \in N$ and $x \in G$. Then $xyx^{-1} \in N$. Thus $\rho(y) = \sigma(y)$ and $\rho(xyx^{-1}) = \sigma(xyx^{-1})$. We therefore have $\tau(x)\rho(y) = \rho(x)\sigma(x^{-1})\sigma(y) = \rho(x)\sigma(x^{-1}y) = \rho(x)\sigma(x^{-1}yx)\sigma(x^{-1}) = \rho(x)\rho(x^{-1}yx)\sigma(x^{-1}) = \rho(y)\rho(x)\sigma(x^{-1}) = \rho(y)\tau(x)$. This shows that $\tau(x)$ is an invertible element in the algebra $E = \mathrm{End}_{\mathcal{O}N}(V)$. The element $\tau(x) \in E^{\times}$ depends clearly only on the image of x in G/N. Thus τ induces a map $\bar{\tau} : G/N \to E^{\times}$. An easy verification shows that for any x, $y \in G$, we have $\tau(xy) = \tau(x)({}^{y}\tau(y))$, and hence $\bar{\tau} \in H^1(G/N; E^{\times})$. Since V is indecomposable and k is algebraically closed, we have $E^{\times} = k^{\times} \times (1 + J(E))$, where we identify $k^{\times}$ with its canonical preimage in $\mathcal{O}^{\times}$. Note that the group $1 + J(E)$ need not be abelian, since E need not be commutative.

One way to conclude the proof is to show directly that $H^1(G/N; 1 + J(E))$ is zero, using that G/N has order prime to p. This can be done by filtering the group $1 + J(E)$ by the normal subgroups $1 + J(E)^n$ and then using that $1 + J(E)^n/1 + J(E)^{n+1}$ is abelian of exponent p; see [88]. Having shown that $H^1(G/N; 1 + J(E))$ is trivial, it follows that the canonical map $E^{\times} \to k^{\times}$ induces an isomorphism $H^1(G/N; E^{\times}) \cong H^1(G/N; k^{\times}) = \mathrm{Hom}(G/N; k^{\times})$. Any such homomorphism determines, via the identification of $k^{\times}$ with its canonical inverse image in $\mathcal{O}^{\times}$ a unique $\mathcal{O}G/N$-module T of $\mathcal{O}$-rank 1. The result follows.

Another way to conclude the proof starts with the observation that if V is simple, then $E = k$, so $1 + J(E)$ is trivial, and hence $H^1(G/N; E^{\times}) = \mathrm{Hom}(G/N, k^{\times})$, which concludes the proof as before. Thus the theorem holds if V is simple. Suppose next that V is a projective indecomposable $\mathcal{O}N$-module. Since G/N is a p'-group, we have $J(\mathcal{O}G) = J(\mathcal{O}N)\mathcal{O}G$ by 4.1.9. It follows that any extension of V to G remains a projective indecomposable $\mathcal{O}G$-module, and that the radical of V as an $\mathcal{O}N$-module is also the radical of any such extension as an $\mathcal{O}G$-module. Therefore any such extension is determined by the corresponding extension of the simple $\mathcal{O}N$-module $V/J(\mathcal{O}N)V$, and these are parametrised by $\mathrm{Hom}(G/N; k^{\times})$, so the same is true for the extensions of V. Note that this argument for extending projective indecomposable modules remains valid for twisted group algebras. We play the general case back to this via multiplicity modules. If (Q, Y) is a vertex-source pair of V, then (Q, V) is a vertex-source pair of the extended module U, since G/N is a p'-group. A Frattini argument yields $G = NN_G(Q, Y)$. Set $A = \mathrm{End}_{\mathcal{O}}(\mathrm{Ind}_Q^G(Y))$.

The module $\mathrm{Ind}_Q^N(Y)$ is a direct summand of $\mathrm{Res}_N^G(\mathrm{Ind}_Q^G(Y))$; denote by $f \in A^N = \mathrm{End}_{\mathcal{O}N}(\mathrm{Ind}_Q^G(Y))$ a projection onto this direct summand. The 2-class of the action of $N_N(Q, Y)/Q$ on the simple quotient of A^Q corresponding to Y is equal to that on the corresponding simple quotient of fA^Qf, and hence this is the restriction of the 2-class determined by the action of $N_G(Q, Y)/Q$ on the relevant simple quotient of A^Q. Thus the algebra $k_\alpha N_N(Q, Y)/Q$ over which the multiplicity module of V is defined is a subalgebra of the algebra $k_\alpha N_H(Q, Y)/Q$ over which the multiplicity module of any extension of U is defined. The dimension of the multiplicity module of V and of U coincide since they are both equal to the multiplicity of Y in decompositions of $\mathrm{Res}_Q^N(V) = \mathrm{Res}_Q^G(U)$. Thus the extensions of V are parametrised by the extensions of the multiplicity module of V from $k_\alpha N_N(Q, Y)/Q$ to $k_\alpha N_G(Q, Y)/Q$. Since multiplicity modules are projective indecomposable, this is the situation dealt with previously. □

5.8 The Brauer construction applied to permutation modules

Let G be a finite group, P a p-subgroup of G and M a kG-module. Applying the Brauer construction to $\mathrm{End}_k(M)$ yields the $N_G(P)$-algebra $(\mathrm{End}_k(M))(P)$. Applying first the Brauer construction to M, yields a $kN_G(P)$-module $M(P)$, and hence its linear endomorphism algebra $\mathrm{End}_k(M(P))$ is also an $N_G(P)$-algebra. These two $N_G(P)$-algebras are not isomorphic, in general, but they *are* isomorphic if $\mathrm{Res}_P^G(M)$ is a permutation kP-module. In order to show this we first prove a result on how one computes $M(P)$ in case $\mathrm{Res}_P^G(M)$ is a permutation module.

Proposition 5.8.1 *Let G be a finite group, let P be a p-subgroup of G and let M be an $\mathcal{O}$-free $\mathcal{O}G$-module such that $\mathrm{Res}_P^G(M)$ is a permutation kP-module. Let X be a P-stable $\mathcal{O}$-basis of M and denote by X^P the set of fixed points of P in X. The image $\mathrm{Br}_P(X^P)$ of X^P in $M(P)$ is a k-basis of $M(P)$. In particular, we have $\mathrm{rk}_\mathcal{O}(M) \equiv \dim_k(M(P)) \pmod{p}$, and if P has no fixed point in X then $M(P) = \{0\}$.*

Proof This is essentially the same proof as in 5.4.1. The $\mathcal{O}$-space of P-fixed points M^P has as an $\mathcal{O}$-basis the set of P-orbit sums in X. For $x \in X$, the P-orbit sum of x is equal to $\mathrm{Tr}_Q^P(x)$, where Q is the stabiliser of x in P. Since $x \in X^P$ if and only if $Q = P$ it follows that $\mathrm{Br}_P(X^P)$ generates $M(P)$ as a k-vector space. The nontrivial P-orbit sums in X span $\ker(\mathrm{Br}_P)$, thus no nontrivial linear combination of elements of X^P lies in $\ker(\mathrm{Br}_P)$. This shows that the image of X^P in $M(P)$ is a basis. Since $X \setminus X^P$ consists of the nontrivial P-orbits, we have $|X| \equiv |X^P| \pmod{p}$, whence the result. □

Corollary 5.8.2 *Let P be a finite p-group and M a finitely generated permutation $\mathcal{O}P$-module. The following are equivalent.*

(i) M is a projective $\mathcal{O}P$-module.
(ii) $M^P = M_1^P$.
(iii) $M(Q) = \{0\}$ for any nontrivial subgroup Q of P.

Proof Let X be a P-stable $\mathcal{O}$-basis of M. The $\mathcal{O}P$-module M is projective if and only if P acts freely on X. Equivalently, the P-orbit sum of any element $x \in X$ is $\mathrm{Tr}_1^P(x)$. Since the P-orbit sums of X span M^P, the equivalence of (i) and (ii) follows. Moreover, P acts freely on X if and only if X^Q is empty for any nontrivial subgroup Q of P. The equivalence of (i) and (iii) follows from 5.8.1. □

Corollary 5.8.3 *Let P be a finite p-group, M a finitely generated permutation $\mathcal{O}P$-module, and Q a proper subgroup of P. Set $R = N_P(Q)$. We have*

$$\mathrm{Br}_Q^M(M_Q^P) = M(Q)_Q^R = \cap_{Q<S\leq R} \ker(\mathrm{Br}_S^{M(Q)}),$$

the intersection taken over all subgroups S of R that properly contain Q.

Proof The first equality is a special case of 5.4.11. For the second equality, note that Q acts trivially on $M(Q)$, and hence $M(Q)_Q^R = M(Q)_1^{Q/R}$. It follows from 5.8.2 that this space is equal to the intersection of $\ker(\mathrm{Br}_{S/Q}^{M(Q)}) = \ker(\mathrm{Br}_S^{M(Q)})$, with S running over the subgroups of R that properly contain Q. □

Corollary 5.8.4 *Let G be a finite group and let P be a p-subgroup of G. We have $(\mathrm{Ind}_P^G(\mathcal{O}))(P) \cong \mathrm{Ind}_P^{N_G(P)}(k)$.*

Proof We have $\mathrm{Ind}_P^G(\mathcal{O}) \cong \mathcal{O}G/P$, and the cosets xP, with $x \in G$, form a permutation basis of this module. Such a coset is a P-fixed point if and only if $uxP = xP$ for all $u \in P$, hence if and only if $x^{-1}ux \in P$ for all $u \in P$, which in turn is equivalent to $x \in N_G(P)$. The result follows from 5.8.1. □

Proposition 5.8.5 *Let G be a finite group, P a p-subgroup of G and M a finitely generated kG-module such that $\mathrm{Res}_P^G(M)$ is a permutation kP-module. Let Q be a normal subgroup Q of P. The Brauer homomorphism Br_Q maps M^P onto $M(Q)^P$ and induces an isomorphism of $kN_{N_G(Q)}(P)$-modules $(M(Q))(P) \cong M(P)$. In particular, if $M(P) \neq \{0\}$, then $M(R) \neq \{0\}$ for any subgroup R of P.*

Proof Let X be a P-stable k-basis of M. Since Q is normal in P, it follows that the action of P on X stabilises the subset X^Q of Q-fixed points in X. It follows from 5.8.1 that $M(Q)^P$ is spanned by the P-orbit sums of the image of X^Q in $M(Q)$, hence Br_Q maps M^P onto $M(Q)^P$. Moreover, we have

$(X^Q)^P = X^P$, and hence again by 5.8.1 we have an isomorphism $(M(Q))(P) \cong M(P)$ as stated. In particular, if $M(P) \neq \{0\}$ then $M(Q) \neq \{0\}$. Since any subgroup of P is subnormal in P, the last statement follows inductively. □

Proposition 5.8.6 ([13, (3.3)]) *Let G be a finite group, P a p-subgroup of G and M a finitely generated kG-module such that $\mathrm{Res}^G_P(M)$ is a permutation kP-module. Every $\varphi \in \mathrm{End}_{kP}(M)$ induces an endomorphism $\bar{\varphi} \in \mathrm{End}_k(M(P))$, and this induces an isomorphism of $N_G(P)$-algebras $(\mathrm{End}_k(M))(P) \cong \mathrm{End}_k(M(P))$.*

Proof Every kP-endomorphism of M preserves M^P and $\ker(\mathrm{Br}^M_P)$, hence induces an endomorphism on $M(P)$. Let X be a P-stable k-basis of M. For $x, y \in X$ denote by $\epsilon_{x,y}$ the unique k-linear endomorphism of M which sends x to y and $x' \in X$ different from x to zero. The set $E = \{\epsilon_{x,y} | x, y \in X\}$ is clearly a P-stable k-basis of $\mathrm{End}_k(M)$. The set of P-fixed points in E is equal to $E^P = \{\epsilon_{x,y} | x, y \in X^P\}$. Together with 5.8.1 this shows that the dimension of $(\mathrm{End}_k(M))(P)$ is equal to $|X^P|^2$, which is also equal to the dimension of $\mathrm{End}_k(M(P))$ since $\dim_k(M(P)) = |X^P|$ by 5.8.1. Since $\mathrm{Br}_P(X^P)$ is a k-basis of $M(P)$, the image of E^P in $\mathrm{End}_k(M(P))$ is a k-basis of $\mathrm{End}_k(M(P))$, which implies the result. □

With the notation of 5.8.6, if $\varphi \in \mathrm{End}_{kP}(M)$ and if we identify $\mathrm{Br}_P(\varphi)$ with its image in $\mathrm{End}_k(M(P))$, then the proof of 5.8.6 shows that for any $m \in M^P$ we have $\mathrm{Br}_P(\varphi)(\mathrm{Br}_P(m)) = \mathrm{Br}_P(\varphi(m))$. The next result is also based on arguments from [13].

Proposition 5.8.7 *Let G be a finite group, P a p-subgroup of G and M a finitely generated kG-module such that $\mathrm{Res}^G_P(M)$ is a permutation kP-module. Then there is an isomorphism of $kN_G(P)$-modules $\mathrm{Res}^G_{N_G(P)}(M) \cong M(P) \oplus N$, where N is a $kN_G(P)$-module satisfying $N(P) = \{0\}$.*

Proof Let U be an indecomposable direct summand of $\mathrm{Res}^G_{N_G(P)}(M)$. We have to show that either $U \cong U(P)$ or that $U(P) = \{0\}$. Let R be a vertex of U. Then U is isomorphic to a direct summand of $\mathrm{Ind}^{N_G(P)}_R(k)$. By Mackey's formula, $\mathrm{Res}^{N_G(P)}_P(U)$ is isomorphic to a direct summand of

$$\oplus_x \mathrm{Ind}^P_{P \cap {}^xR}(k)$$

where x runs over a set of representatives of the double cosets $P \backslash N_G(P)/R$ in $N_G(P)$. If $P \leq R$, then $P \leq {}^xR$ for all $x \in N_G(P)$ since P is normal in $N_G(P)$. Thus, in that case, P acts trivially on U, or equivalently, $U \cong U(P)$. If $P \nleq R$ then $P \nleq {}^xR$ for all x as before, hence $U(P) = \{0\}$ in that case. The result follows. □

In the following result on the Brauer construction applied to permutation *bimodules* we need the extra notational precision mentioned in 5.4.13, using the diagonal subgroup $\Delta P = \{(u, u)|u \in P\}$ of $P \times P$, where P is a finite group P.

Lemma 5.8.8 ([57] 6.1 (iii)) *Let P be a finite p-group and let M be a finitely generated indecomposable $\mathcal{O}P$-$\mathcal{O}P$-bimodule having a $P \times P$-stable $\mathcal{O}$-basis such that M is projective as a left and right $\mathcal{O}P$-module. Then $M(\Delta P)$ is non zero if and only if $M \cong \mathcal{O}P$ as $\mathcal{O}P$-$\mathcal{O}P$-bimodule. In that case, $M(\Delta P) \cong kZ(P)$.*

Proof Let X be a $P \times P$-stable $\mathcal{O}$-basis of M. By the assumptions on M, the group P acts freely on the left and on the right of X. Suppose that $M(\Delta P) \neq 0$. By 5.8.1 the set X has a fixed point with respect to the action of ΔP on X; that is, there is $x \in X$ such that $(u, u) \cdot x = x$ for all $u \in P$, or equivalently, such that $ux = xu$ for all $u \in P$. This means that the subspace $\mathcal{O}[Px]$ of M is a direct summand of M as an $\mathcal{O}(P \times P)$-module, having as a complement the space $\mathcal{O}[X \setminus Px]$. Since M was assumed indecomposable this implies $M = \mathcal{O}[Px]$. Since moreover P acts freely on the left and on the right of X, the map sending u to ux induces an isomorphism of $\mathcal{O}P$-$\mathcal{O}P$-bimodules $\mathcal{O}P \cong \mathcal{O}[Px]$, hence $M \cong \mathcal{O}P$. Conversely, if $M \cong \mathcal{O}P$ then $M(\Delta P) \cong kZ(P)$ because the elements in $Z(P)$ are exactly the ΔP-fixed points in the set P. □

Lemma 5.8.9 *Let G be a finite group. Consider $\mathcal{O}G$ as an $\mathcal{O}(G \times G)$-module with $(x, y) \in G \times G$ acting by left multiplication with x and by right multiplication with y^{-1}. Let u, v be p-elements in G and denote by R the cyclic subgroup of $G \times G$ generated by (u, v). If $u = v$, then $(\mathcal{O}G)(R) = kC_G(u)$. If u, v are not conjugate in G then $(\mathcal{O}G)(R) = \{0\}$.*

Proof If $u = v$, then the conclusion $(\mathcal{O}G)(R) = kC_G(u)$ follows from 5.4.1. In general, by 5.8.1, the dimension of $(\mathcal{O}G)(R)$ is equal to the number of elements $x \in G$ that are fixed under R, or equivalently, that satisfy $uxv^{-1} = x$, or equivalently, $u = xvx^{-1}$. If u, v are not conjugate, then there is no such element, whence the result. □

The Brauer construction applied to permutation modules is compatible with taking tensor products:

Proposition 5.8.10 *Let G be a finite group and let M, N be finitely generated kG-modules. Let P be a p-subgroup of G such that $\mathrm{Res}^G_P(M)$ is a permutation kP-module. Then the inclusion $M^P \otimes_k N^P \subseteq (M \otimes_k N)^P$ induces an isomorphism of $kN_G(P)$-modules $M(P) \otimes_k N(P) \cong (M \otimes_k N)(P)$.*

Proof The inclusion $M^P \otimes_k N^P \subseteq (M \otimes_k N)^P$ induces a $kN_G(P)$-homomorphism $M(P) \otimes_k N(P) \to (M \otimes_k N)(P)$ which is natural in M and N. Thus it suffices to show the statement for $G = P$ and M indecomposable. By the assumptions on M, this implies that $M = \mathrm{Ind}_Q^P(k)$ for some subgroup Q of P. If $Q = P$, then $M = k$, and hence both sides are canonically isomorphic to $N(P)$. Suppose that Q is a proper subgroup of P. Then $M^P = M_Q^P$ by 2.5.7, so $M(P) \otimes_k N(P)$ is zero. Similarly, $M \otimes_k N \cong \mathrm{Ind}_Q^P(N)$, and hence $(M \otimes_k N)(P)$ is also zero. □

If both $\mathrm{Res}_P^G(M)$ and $\mathrm{Res}_P^G(N)$ are finitely generated permutation kP-modules, then one can also argue as follows. Let X, Y be P-stable k-bases of M, N, respectively. Then the image $X \otimes Y$ in $M \otimes_k N$ is a P-stable k-basis of $M \otimes_k N$. Given $x \in X$, $y \in Y$, the element $x \otimes y$ belongs to $(X \otimes Y)^P$ if and only if $x \in M^P$ and $y \in N^P$. The conclusion in Proposition 5.8.10 follows in that case from 5.8.1. Using the isomorphisms $\mathrm{End}_k(M) \cong M \otimes_k M^*$ and $\mathrm{End}_k(M(P)) \cong M(P) \otimes_k M(P)^*$ from 2.9.4 this yields another proof of the isomorphism $(\mathrm{End}_{\mathcal{O}}(M))(P) \cong \mathrm{End}_k(M(P))$ from 5.8.6.

Proposition 5.8.11 *Let G be a finite group, P a p-subgroup of G, let M, N be $\mathcal{O}G$-modules such that $\mathrm{Res}_P^G(M)$, $\mathrm{Res}_P^G(N)$ are permutation $\mathcal{O}P$-modules, and let $\varphi : M \to N$ be a homomorphism of $\mathcal{O}G$-modules. Suppose that N is finitely generated and relatively $\mathcal{O}P$-projective. The following are equivalent.*

(i) *The $\mathcal{O}G$-homomorphism φ is split surjective.*
(ii) *We have $\varphi(M^Q) = N^Q$ for any subgroup Q of P.*
(iii) *We have $\varphi(M_Q^P) = N_Q^P$ for any subgroup Q of P.*
(iv) *For any subgroup Q, the homomorphism $\varphi(Q) : M(Q) \to N(Q)$ induced by φ is surjective.*
(v) *For any P-stable $\mathcal{O}$-basis X of M there is a P-stable subset Y of X such that $|Y| = |\varphi(Y)|$ and such that $\varphi(Y)$ is a P-stable $\mathcal{O}$-basis of N.*

Proof The implications (i) ⇒ (ii) ⇒ (iii) ⇒ (iv) are trivial. If (v) holds, then φ is split surjective as a homomorphism of $\mathcal{O}P$-modules. Since N is assumed to be relatively $\mathcal{O}P$-projective, it follows that φ is split surjective as a homomorphism of $\mathcal{O}G$-modules. Thus (v) implies (i). Suppose that (iv) holds. Let Q be a subgroup of P. Since $\varphi(Q)$ is surjective, we have $N^Q = \varphi(M^Q) + \sum_{R<Q} N_R^Q + J(\mathcal{O})N^Q$. Nakayama's Lemma implies that $N^Q = \varphi(M^Q) + \sum_{R<Q} N_R^Q$. Arguing by induction over the order of Q, we may assume that $N_R^Q = \varphi(M_R^Q)$, and hence $N^Q = \varphi(M^Q)$. This shows that (iv) implies (ii). It remains to show that (iii) implies (v). Suppose that (iii) holds. In order to show (v) we may assume $\mathcal{O} = k$. We also may assume that $G = P$. We argue by induction over $\dim_k(N)$.

Let Q be a minimal subgroup of P such that N has a direct summand isomorphic to kP/Q. Write $N = U \oplus V$, where U is isomorphic to a direct sum of copies of kP/Q and where V has no direct summand isomorphic to kP/Q. Note that $U^P = U^P_Q$ and $U^P_R = \{0\}$ for any proper subgroup R of Q. Consider the kP-homomorphism $\varphi_U : M \to U$ obtained from composing $\varphi : M \to N$ with the projection $N \to U$ having V as kernel. We still have $\varphi_U(M^P_R) = U^P_R$ for all subgroups R of P. Thus $\varphi_U(M^P_Q) = U^P_Q = U^P$ and $\varphi_U(M^P_R) = U^P_R = \{0\}$ for any proper subgroup R of Q. Since M^Q is spanned by Q-orbit sums of elements of X and nontrivial orbit sums belong to $\ker(\mathrm{Br}_Q) = \sum_{R<Q} M^Q_R$, it follows that there is a subset X_U of X^Q such that $\varphi(\mathrm{Tr}^P_Q(X_U))$ is a k-basis of U^P. By deleting superfluous elements we may assume that $|X_U| = \dim_k(U^P)$. Set $Y_U = P \cdot X_U$. We are going to show that $|Y_U| = |\varphi_U(X_U)|$ and that $\varphi_U(Y_U)$ is a P-stable k-basis of U. Since Q acts trivially on X_U, we have $|Y_U| \le |P : Q| \cdot |X_U| = |P : Q|\dim_k(U^P) = \dim_k(U)$, where the last equality holds because U is a direct sum of copies of kP/Q. It suffices therefore to show that $\varphi_U(Y_U)$ spans U as a k-vector space. The space spanned by $\varphi_U(Y_U)$ is equal to $\sum_{x\in X_U} kP\varphi_U(x)$. every summand $kP\varphi_U(x)$ is isomorphic to kP/Q; this is because $\varphi_U(x) \in U^Q$ and $\mathrm{Tr}^P_Q(\varphi_U(x)) \ne 0$. Thus the sum $\sum_{x\in X_U} kP\varphi_U(x)$ is a direct sum, because every summand has a simple socle spanned by $\mathrm{Tr}^P_Q(\varphi_U(x))$, and these elements are linearly independent. Thus the dimension of this sum is $|P : Q| \cdot |X_U| = \dim_k(U)$ as required. Let now X' be a complement of X_U in X. Decompose $M = M_U \oplus M'$, where M_U has X_U as a k-basis and M' has X' as a k-basis. Denote by $\varphi_V : M \to V$ the composition of φ followed by the projection $M \to V$ with kernel U. By the minimal choice of Q we have $\varphi_V((M')^P_Q) \subseteq V^P_Q = \{0\}$. Since $(M_U)^P = (M_U)^P_Q$, it follows that $\varphi_V((M')^P_R) = V^P_R$ for any subgroup R of P because $\varphi_V((M_U)^P) = \{0\}$. By induction, there is a P-stable subset X_V in X' such that $|X_V| = |\varphi_V(X_V)|$ and such that $\varphi_V(X_V)$ is a k-basis of V. Set $Y = X_U \cup X_V$. We claim that $|Y| = |\varphi(Y)|$ and that $\varphi(Y)$ is a basis of N. Denote by U' and V' the submodules of N that are spanned by $\varphi(X_U)$ and $\varphi(X_V)$, respectively. Since $U' \cong U$ we have $(U')^P = (U')^P_Q \subseteq U^P_Q \oplus V^P_Q$, and since $V^P_Q = \{0\}$ by the choice of Q we get that $(U')^P = U^P$. Now $V' \cong V$, and $\varphi_V(V') = V$. Thus $\{0\} = V' \cap \ker(\varphi_V) = V' \cap U$, hence also $V' \cap U' = \{0\}$ because any simple submodule of this intersection would belong to $(U')^P$, hence to U. It follows that the sum of U' and V' is a direct sum, and comparing dimensions yields $U' \oplus V' = N$, which completes the proof. □

We conclude this section with two applications of Proposition 5.8.11.

Proposition 5.8.12 *Let G be a finite group, P a p-subgroup, let M, N be finitely generated relatively $\mathcal{O}P$-projective $\mathcal{O}G$-modules such that* $\mathrm{Res}^G_P(M)$, $\mathrm{Res}^G_P(N)$

are permutation $\mathcal{O}P$-modules and let $\varphi : M \to N$ be a homomorphism of $\mathcal{O}G$-modules. Then φ induces an isomorphism in the relatively $\mathcal{O}$-stable category $\underline{\mathrm{mod}}(\mathcal{O}G)$ if and only if the induced $kC_G(Q)$-homomorphism $\varphi(Q) : M(Q) \to N(Q)$ is an isomorphism for every nontrivial subgroup Q of P. In particular, if φ is surjective and $\varphi(Q)$ is an isomorphism for any nontrivial subgroup Q of P, then φ is split surjective with projective kernel.

Proof Adding a projective cover of N to the module M does not affect its image in the stable category, and the Brauer construction with respect to a nontrivial subgroup annihilates every projective module. Thus we may assume that φ is surjective. Suppose that φ induces an isomorphism in the stable category. Then $\ker(\varphi)$ is projective, hence a direct summand since a projective $\mathcal{O}G$-module is relatively $\mathcal{O}$-injective, and the involved modules are all $\mathcal{O}$-free. Thus, for Q a nontrivial subgroup of P we get $\ker(\varphi)(Q) = \{0\}$ and hence $M(Q) \cong N(Q)$. Suppose conversely that $\varphi(Q)$ is an isomorphism for all nontrivial subgroups Q of P. Since φ is assumed to be surjective, it follows from 5.8.11 that φ is split surjective. Moreover, since $\varphi(Q)$ is an isomorphism for any nontrivial subgroup Q of P we get that $\ker(\varphi)(Q) = \{0\}$ for any nontrivial subgroup Q of P. Thus $\ker(\varphi)$ is projective by 5.8.2. □

Proposition 5.8.13 *Let P be a finite p-group and A a P-algebra over $\mathcal{O}$ such that A has a finite P-stable $\mathcal{O}$-basis. Suppose that k has at least three distinct elements. Then A has a P-stable $\mathcal{O}$-basis contained in $A^\times$.*

Proof Consider $A^\times$ as a P-set, and set $M = \mathcal{O}[A^\times]$. Thus M is a permutation $\mathcal{O}P$-module with P-stable basis $A^\times$. Note that M^Q contains the image of $(A^\times)^Q$ in M. The inclusion $A^\times \subseteq A$ induces an $\mathcal{O}$-linear map $f : M \to A$, which is clearly an $\mathcal{O}P$-module homomorphism. For any subgroup Q of P we have $(A^Q)^\times = (A^\times)^Q$. Moreover, since A has a P-stable basis, it follows that $(k \otimes_\mathcal{O} A^Q) = (k \otimes_\mathcal{O} A)^Q$. Thus an element in A^Q is invertible if and only if its image in $(k \otimes_\mathcal{O} A)^Q$ is invertible. It follows from 1.13.7 that the set $(A^\times)^Q$ spans A^Q as an $\mathcal{O}$-module. Thus $f(M^Q) = A^Q$ for all subgroups Q of P, and hence 5.8.11 implies that there is a subset Y in $A^\times$ such that Y is a P-stable $\mathcal{O}$-basis of A. □

5.9 Brauer pairs and p-permutation G-algebras

Algebras over $\mathcal{O}$ are finitely generated as $\mathcal{O}$-modules in this section. If P_γ is a local pointed group on a G-algebra A, then γ is a conjugacy class of primitive idempotents in A^P which is mapped by the Brauer homomorphism Br_P to a conjugacy class of primitive idempotents in $A(P)$. Thus P_γ determines a unique block e_γ of $A(P)$ satisfying $\mathrm{Br}_P(\gamma) \subseteq A(P)e_\gamma$. This motivates the following definition.

Definition 5.9.1 Let G be a finite group, and let A be a G-algebra over $\mathcal{O}$. A (G, A)-*Brauer pair* is a pair (P, e) consisting of a p-subgroup P of G such that $A(P) \neq \{0\}$ and a block e of $A(P)$.

We will see in this section that if A has a p-permutation basis, then the partial order of the set of local pointed groups induces a partial order on the set of (G, A)-Brauer pairs. Note that if G is a finite group, A a primitive G-algebra, and (P, e) a (G, A)-Brauer pair, then $A(P)$ is nonzero, and hence P is contained in a defect group of A by 5.6.7.

Proposition 5.9.2 *Let G be a finite group and P be a p-subgroup of G. Let A, B be G-algebras over $\mathcal{O}$ having finite P-stable $\mathcal{O}$-bases. The inclusion $A^P \otimes B^P \subseteq (A \otimes_\mathcal{O} B)^P$ induces an isomorphism of $N_G(P)$-algebras $A(P) \otimes_k B(P) \cong (A \otimes_\mathcal{O} B)(P)$.*

Proof By 5.8.10, we have a $kN_G(P)$-module isomorphism $A(P) \otimes_k B(P) \cong (A \otimes_\mathcal{O} B)(P)$. This is clearly an algebra isomorphism. □

In particular, there is a bijection between the set of local points of P on $A \otimes_\mathcal{O} B$ and the set of pairs consisting of a local point of P on A and on B. If in addition the algebras $A(P)$, $B(P)$ are split, then under this bijection, the multiplicity of a local point of P on $A \otimes_\mathcal{O} B$ is the product of the multiplicities of the corresponding local points of P on A and on B. We spell this out in detail in the following special case.

Proposition 5.9.3 *Let P be a finite p-group, and let A, B be primitive P-algebras over $\mathcal{O}$ that have finite P-stable $\mathcal{O}$-bases. Suppose that the k-algebras $A(P)$, $B(P)$ are nonzero and split. Then $1_{A\otimes_\mathcal{O}B} = e + e'$ for some primitive idempotent $e \in (A \otimes_\mathcal{O} B)^P$ satisfying $\mathrm{Br}_P(e) \neq 0$ and some idempotent $e' \in (A \otimes_\mathcal{O} B)^P$ satisfying $\mathrm{Br}_P(e') = 0$. Equivalently, P has a unique local point on $A \otimes_\mathcal{O} B$, and the multiplicity of this local point is* 1.

Proof Since 1_A is primitive in A^P, it follows that $A(P)$ is local. Similarly, $B(P)$ is local. Since $A(P)$, $B(P)$ are split by the assumptions, it follows that $A(P) \otimes_k B(P)$ is split local. By 5.9.2, we have $A(P) \otimes_k B(P) \cong (A \otimes_\mathcal{O} B)(P)$. The unit element of $(A \otimes_\mathcal{O} B)(P)$ lifts therefore to a primitive idempotent e in $(A \otimes_\mathcal{O} B)^P$ through the Brauer homomorphism Br_P on $(A \otimes_\mathcal{O} B)^P$ with the properties as stated. □

Proposition 5.9.4 *Let G be a finite group having a normal Sylow p-subgroup P. Let A, B be G-algebras over $\mathcal{O}$ having finite P-stable $\mathcal{O}$-bases. Suppose that 1_A is primitive in A^P, that 1_B is primitive in B^P, and that the k-algebras $A(P)$ and $B(P)$ are nonzero and split. Let e be a primitive idempotent in $(A \otimes_{\mathcal{O}} B)^G$ satisfying $\mathrm{Br}_P(e) \neq 0$. Then e is primitive in $(A \otimes_{\mathcal{O}} B)^P$.*

Proof Since $\mathrm{Br}_P(e) \neq 0$, it follows from 5.5.9 that every point of P on $e(A \otimes_{\mathcal{O}} B)e$ is local. Thus, if f is a primitive idempotent in $(e(A \otimes_{\mathcal{O}} B)e)^P$, then $\mathrm{Br}_P(f) \neq 0$. By 5.9.3, we have $\mathrm{Br}_P(1 - f) = 0$, hence $\mathrm{Br}_P(e - f) = 0$, which forces $e - f = 0$, hence $e = f$. □

An important feature of a G-algebra with a p-permutation basis is that the partial order on its local pointed groups induces a partial order on its Brauer pairs. This is due to Broué and Puig. We start with a special case.

Proposition 5.9.5 *Let P be a finite p-group and A a P-algebra having a finite P-stable $\mathcal{O}$-basis. Suppose that every point of P on A is local and that the k-algebra $A(P)$ is indecomposable. Then, for any subgroup Q of P, the k-algebra $A(Q)$ is indecomposable.*

Proof Let Q be a subgroup of P. We argue by induction over $|P : Q|$. The conclusion holds for $Q = P$ by the assumptions. Let Q be a proper subgroup of P. Then $R = N_P(Q)$ contains Q properly. By 5.8.5, the algebra $A(Q)$ is nonzero because $A(P)$ is nonzero. The group R acts on $A(Q)$, hence on the set of block idempotents of $A(Q)$. Since $A(Q)(R) \cong A(R)$ is nonzero and indecomposable, R must fix at least one block idempotent $f \in Z(A(Q))$ of $A(Q)$ such that $\mathrm{Br}_R(f) \neq 0$. Set $f' = 1_{A(Q)} - f$. Thus $A(Q) = A(Q)f \times A(Q)f'$, hence for any subgroup S of R containing Q, we have $A(R) \cong A(Q)(S) \cong (A(Q)f)(S) \times (A(Q)f')(S)$, and $(A(Q)f)(S) \neq \{0\}$. If S contains Q properly, then $A(S)$ is indecomposable by induction, hence $\mathrm{Br}_S(f') = 0$ for any S such that $Q < S \leq R$. It follows from 5.8.3 that $f' \in A(Q)^R_Q$. But by 5.4.5 we have $A(Q)^R_Q = \mathrm{Br}_P(A^P_Q)$. Since every point of P on A is local, we have $A^P_Q \subseteq J(A^P)$, and hence $A(Q)^R_Q$ contains no idempotent. This shows that $f' = 0$, and hence $f = 1_{A(Q)}$ is the unique block idempotent of $A(Q)$. □

Theorem 5.9.6 ([15, Theorem 1.8]) *Let P be a finite p-group and A a P-algebra having a P-stable $\mathcal{O}$-basis such that $A(P) \neq \{0\}$. Let e be a block of $A(P)$, and let Q be a subgroup of P.*

(i) There is a unique block f of $A(Q)$ such that there exists a primitive idempotent $i \in A^P$ satisfying $\mathrm{Br}_P(i)e \neq 0$ and $\mathrm{Br}_Q(i)f \neq 0$.
(ii) The block f of $A(Q)$ has the property that for any primtive idempotent $j \in A^P$ satisfying $\mathrm{Br}_P(j)e \neq 0$ we have $0 \neq \mathrm{Br}_Q(j) \in A(Q)f$.

Note that if $\mathrm{Br}_P(i)e \neq 0$, then $\mathrm{Br}_P(i)$ is a primitive idempotent in $A(P)$, and hence $\mathrm{Br}_P(i) \in A(P)e$. By contrast, i need not be primitive in A^Q, hence $\mathrm{Br}_Q(i)$ need not be primitive in $A(Q)$, and hence it is a nontrivial observation that $\mathrm{Br}_Q(i)$ is still contained in a single block of $A(Q)$. The second statement in this theorem is equivalent to the first, because the uniqueness part of the first statement ensures that $\mathrm{Br}_Q(i) \in A(Q)f$ and that this does not depend on the choice of i satisfying $\mathrm{Br}_P(i)e \neq 0$.

Proof of Theorem 5.9.6 We have $\mathrm{Br}_P(i)e \neq 0$ if and only if i belongs to a local point of P on A satisfying $\mathrm{Br}_P(\gamma) \subseteq A(P)e$. Since $\mathrm{Br}_P : A^P \to A(P)$ it follows from standard lifting theorems for idempotents that there is an idempotent $\hat{e}$ in A^P such that $\mathrm{Br}_P(\hat{e}) = e$. We now replace $\hat{e}$ by an idempotent e' by 'discarding' all nonlocal points from $\hat{e}$. More precisely, let I be a primitive decomposition of $\hat{e}$ in A^P. Let J be the subset of all $i \in I$ such that $\mathrm{Br}_P(i) \neq 0$, and set $e' = \sum_{i \in J} i$. Since Br_P maps all elements in $I \setminus J$ to zero, it follows that $\mathrm{Br}_P(\hat{e} - e') = 0$, and hence the idempotent e' in A^P still satisfies $\mathrm{Br}_P(e') = e$. By the construction of e', every point of P on $e'Ae'$ is local, every point of $A(P)e$ has a representative that lifts to an element in J, and $(e'Ae')(P) \cong A(P)e$ is an indecomposable algebra because e is a block of $A(P)$. By 5.9.5, the algebra $(e'Ae')(Q) = \mathrm{Br}_Q(e')A(Q)\mathrm{Br}_Q(e')$ is indecomposable. This means that the idempotent $\mathrm{Br}_Q(e')$ is contained in $A(Q)f$ for a uniquely determined block f of $A(Q)$. By the construction of e' again, it follows that if i is *any* primitive idempotent in A^P satisfying $\mathrm{Br}_P(i)e \neq 0$, then $0 \neq \mathrm{Br}_Q(i) \in A(Q)f$. Both statements follow. □

Definition 5.9.7 Let G be a finite group, and let A be a G-algebra over $\mathcal{O}$. Let (P, e), (Q, f) be (G, A)-Brauer pairs. We write $(Q, f) \leq (P, e)$ and say that (Q, f) *is contained in* (P, e) or that (P, e) *contains* (Q, f) if there is a primitive idempotent $i \in A^P$ satisfying $\mathrm{Br}_P(i)e \neq 0$ and $\mathrm{Br}_Q(i)f \neq 0$.

With this definition, the previous theorem can be reformulated as follows.

Theorem 5.9.8 ([15, Theorem 1.8]) *Let G be a finite group, and let A be a G-algebra over $\mathcal{O}$. Let (P, e) be a (G, A)-Brauer pair and Q a subgroup of P. Suppose that A has a P-stable $\mathcal{O}$-basis. There is a unique block f of $A(Q)$ such that $(Q, f) \leq (P, e)$.*

Proof This is equivalent to the first statement in 5.9.6. □

If Q is normal in P, the inclusion of Brauer pairs can be characterised as follows.

Proposition 5.9.9 *Let G be a finite group, and let A be a G-algebra over $\mathcal{O}$. Let (P, e) be a (G, A)-Brauer pair and Q a normal subgroup of P. Suppose that A has a P-stable $\mathcal{O}$-basis. The unique block f of $A(Q)$ such that $(Q, f) \leq (P, e)$ is the unique P-stable block of $A(Q)$ satisfying $\mathrm{Br}_P(f)e = e$.*

Proof Since Q is normal in P, P acts on A^Q, inducing an action on $A(Q)$. The uniqueness of f satisfying $(Q, f) \leq (P, e)$ implies that f is P-stable. Let i be a primitive idempotent in A^P such that $\mathrm{Br}_P(i)e \neq 0$ and such that $\mathrm{Br}_Q(i)f \neq 0$. The uniqueness of f implies that $\mathrm{Br}_Q(i)f = \mathrm{Br}_Q(i) \neq 0$. Since $\mathrm{Br}_P(\mathrm{Br}_Q(i)) = \mathrm{Br}_P(i) \neq 0$, this implies in turn that $\mathrm{Br}_P(f) \neq 0$. By 5.4.8, the map Br_P on $A(Q)$ sends $Z(A(Q))^P$ to $Z(A(P))$, so $\mathrm{Br}_P(f)$ is a sum of blocks of $A(P)$. Since $\mathrm{Br}_P(i)e = \mathrm{Br}_P(i)$, the block e must occur in $\mathrm{Br}_P(f)$, hence $\mathrm{Br}_P(f)e = e$. Any other P-stable block f' of $A(Q)$ satisfies $f'f = 0$, hence $\mathrm{Br}_P(f')\mathrm{Br}_P(f) = 0$, and in particular $\mathrm{Br}_P(f')e = 0$. The result follows. □

The uniqueness property of the inclusion of Brauer pairs implies that if A has a stable basis for any p-subgroup of G, then the set of (G, A)-Brauer pairs is partially ordered. This partial order is induced by that of the set of local pointed groups on A:

Proposition 5.9.10 *Let G be a finite group, A a G-algebra over $\mathcal{O}$. Let P_γ, Q_δ be local pointed groups on A. Denote by e the unique block of $A(P)$ satisfying $\mathrm{Br}_P(\gamma) \subseteq A(P)e$ and by f the unique block of $A(Q)$ satisfying $\mathrm{Br}_Q(\delta) \subseteq A(Q)f$. Suppose that A has a P-stable $\mathcal{O}$-basis. If $Q_\delta \leq P_\gamma$, then $(Q, f) \leq (P, e)$.*

Proof Suppose that $Q_\delta \leq P_\gamma$; that is, there are $i \in \gamma$ and $j \in \delta$ such that $ij = j = ji$. We have $\mathrm{Br}_P(i) \in A(P)e$ and $\mathrm{Br}_Q(j) \in A(Q)f$. In order to show that f is the unique block of $A(Q)$ such that $(Q, f) \leq (P, e)$ we need to show that $\mathrm{Br}_Q(i)f \neq 0$. If $\mathrm{Br}_Q(i)f = 0$, then $0 = \mathrm{Br}_Q(j)\mathrm{Br}_Q(i)f = \mathrm{Br}_Q(ji)f = \mathrm{Br}_Q(j)f = \mathrm{Br}_Q(j)$, contradicting the fact that j belongs to a local point of Q on A. □

The partially ordered set of (G, A)-Brauer pairs is a G-poset: if (P, e) is a (G, A)-Brauer pair and $x \in G$, then the action of x on G induces an isomorphism $A^P \cong A^{^xP}$ sending A^P_Q to $A^{^xP}_{^xQ}$, hence inducing an isomorphism $A(P) \cong A(^xP)$. This isomorphism induces a bijection between the sets of blocks of $A(P)$ and of $A(^xP)$. We denote by ^xe the block of $A(^xP)$ corresponding to the block e of $A(P)$ under this bijection. Then $^x(P, e) = (^xP, {}^xe)$ is a (G, A)-Brauer pair. An easy verification shows that this action of G on the set of (G, A)-Brauer pairs preserves the inclusion as defined in 5.9.7. We have the following analogue of 5.5.16.

Theorem 5.9.11 *Let G be a finite group, A a primitive G-algebra over $\mathcal{O}$, and suppose that for any p-subgroup P of G, the algebra A has a P-stable $\mathcal{O}$-basis. Let (P, e) be a maximal (G, A)-Brauer pair. Then P is a defect group of A, and for any (G, A)-Brauer pair (Q, f) there is an element $x \in G$ such that ${}^x(Q, f) \leq (P, e)$. In particular, all maximal (G, A)-Brauer pairs are G-conjugate.*

Proof This follows from the analogous statement 5.5.16 applied to $G_{\{1_A\}}$, in conjunction with 5.9.10. □

It is sometimes useful to construct permutation bases containing the unit element of the algebra; this can be done in the following situations:

Lemma 5.9.12 *Let P be a finite p-group and A a primitive P-algebra having a finite P-stable $\mathcal{O}$-basis. Suppose that $A(P) \neq \{0\}$. Then there is a P-stable $\mathcal{O}$-basis of A containing 1_A.*

Proof Let X be a P-stable $\mathcal{O}$-basis of A. Since A is a primitive P-algebra, it follows that A^P is local. The image of X^P in $A(P)$ is a k-basis, and the algebra $A(P)$ is local. Thus there is $z \in X^P$ whose image in $A(P)$ is not contained in $J(A(P))$, and hence z is not contained in $J(A^P)$. Since A^P is local, this forces that z is invertible in A^P. Then $z^{-1}X$ is a P-stable $\mathcal{O}$-basis of A containing 1_A. □

Lemma 5.9.13 (cf. [72, 3.4]) *Let P be a finite p-group and A a primitive interior P-algebra over $\mathcal{O}$ with structural homomorphism $\sigma : P \to A^\times$. Suppose that $A(\Delta P) \neq \{0\}$ and that A has a finite $\mathcal{O}$-basis X that is P-P-stable; that is, which satisfies $\sigma(u)X = X = X\sigma(u)$ for all $u \in P$. Then A has a P-P-stable $\mathcal{O}$-basis containing 1_A.*

Proof This is proved using the same arguments as in the proof of 5.9.12. □

The theme of bases in an interior P-algebra A that are stable with respect to the conjugation action by P or with respect to the action of $P \times P$ on A is crucial for many important properties of finite group algebras, their block algebras and source algebras. For A an interior P-algebra with structural homomorphism $\sigma P \to A^\times$, we consider $A \otimes_{\mathcal{O}} \mathcal{O}P$ as an interior P-algebra with the diagonal structural homomorphism sending $u \in P$ to $\sigma(u) \otimes u$.

Lemma 5.9.14 *Let P be a finite p-group and let A be an interior P-algebra over $\mathcal{O}$ having a P-stable basis X with respect to the conjugation action of P on A. Then the interior P-algebra $A \otimes_{\mathcal{O}} \mathcal{O}P$ has a $P \times P$-stable $\mathcal{O}$-basis Y. Moreover, if X is contained in $A^\times$, then Y can be chosen in the subgroup $A^\times \otimes P$ of $(A \otimes_{\mathcal{O}} \mathcal{O}P)^\times$.*

Proof Denote by $\sigma : P \to A^\times$ the structural homomorphism of A. The set $Y = \{\sigma(u)x \otimes u | x \in X, u \in P\}$ is a $P \times P$-stable basis of $A \otimes_{\mathcal{O}} \mathcal{O}P$. Indeed, Y is invariant under left multiplication by $\sigma(v) \otimes v$, where $v \in P$. Using that ${}^v x \in X$ for any $x \in X$, it follows that $\sigma(u)x\sigma(v) \otimes uv = \sigma(uv)({}^v x) \otimes uv$ for all $u \in P$, and hence Y is also invariant under right multiplication by $\sigma(v) \otimes v$. If all elements of X are invertible, then the set Y lies in the subgroup $A^\times \otimes P$ as stated. □

The existence of Y in the above lemma could also have been proved using earlier results: by 2.4.5 we have an isomorphism as $\mathcal{O}(P \times P)$-modules $\mathcal{O}P \cong \mathrm{Ind}_{\Delta P}^{P \times P}(\mathcal{O})$. By 2.4.8, we have an isomorphism of $\mathcal{O}(P \times P)$-modules $A \otimes_{\mathcal{O}} \mathcal{O}P \cong \mathrm{Ind}_{\Delta P}^{P \times P}(\mathrm{Res}_{\Delta P}^{P \times P}(A))$. By the assumptions, A is a permutation $\mathcal{O}\Delta P$-module. Thus A is a permutation $\mathcal{O}(P \times P)$-module.

5.10 Trivial source modules

Modules in this section are finitely generated.

Definition 5.10.1 Let G be a finite group. An indecomposable $\mathcal{O}G$-module M is called a *trivial source module* if for some vertex Q of M the trivial $\mathcal{O}Q$-module is a source of M.

The trivial $\mathcal{O}G$-module $\mathcal{O}$ is a particular case of trivial source modules, and by 5.1.4 it has the Sylow p-subgroups of G as vertices. If an indecomposable $\mathcal{O}G$-module M has a trivial source for some vertex, then M has a trivial source for any vertex, because by 5.1.2, pairs consisting of a vertex and a source are permuted transitively, up to isomorphism, by the conjugation action of G. If M is a trivial source $\mathcal{O}G$-module, then every summand of the kG-module $\bar{M} = M/J(\mathcal{O})M$ is a trivial source kG-module. In fact, we will see below that $\bar{M}$ remains indecomposable. If M, N are $\mathcal{O}G$-modules, then every $\mathcal{O}G$-homomorphism $\varphi : M \to N$ maps $J(\mathcal{O})M$ to $J(\mathcal{O})N$, hence induces a kG-homomorphism $\bar{\varphi} : \bar{M} \to \bar{N}$ such that $\bar{\varphi}(m + J(\mathcal{O})M) = \varphi(m) + J(\mathcal{O})N$, where $m \in M$. We keep this notation in the following theorem, which characterises trivial source modules, and which implies in particular that isomorphism classes of trivial source modules over $\mathcal{O}G$ and over kG correspond bijectively to each other. This is a very special feature of trivial source modules – not every kG-modules 'lifts' to an $\mathcal{O}$-free $\mathcal{O}G$-module.

Theorem 5.10.2 *Let G be a finite group.*

(i) An indecomposable $\mathcal{O}G$-module M is a trivial source module if and only if M is a direct summand of a permutation module.

(ii) If M, N are direct sums of trivial source modules then the canonical map $\mathrm{Hom}_{\mathcal{O}G}(M,N) \to \mathrm{Hom}_{kG}(\bar{M},\bar{N})$ *is surjective.*

(iii) If M is a trivial source $\mathcal{O}G$-module with vertex Q then $\bar{M}$ is a trivial source kG-module with vertex Q; in particular, $\bar{M}$ remains indecomposable.

(iv) For any trivial source kG-module U there is, up to isomorphism, a unique trivial source $\mathcal{O}G$-module M such that $\bar{M} \cong U$.

Proof For the proof of (i), suppose first that M is a trivial source module. This means that if Q is vertex of M then M is a direct summand of $\mathrm{Ind}_Q^G(\mathcal{O})$. The latter is isomorphic to the permutation module $\mathcal{O}[G/Q]$, with G acting on the set G/Q by left multiplication. Suppose conversely that M is a direct summand of a permutation module. Since M is indecomposable, the Krull–Schmidt Theorem implies that M is a summand of $\mathrm{Ind}_H^G(\mathcal{O})$ for some subgroup H of G. Let Q be a vertex of M and let V be an $\mathcal{O}Q$-source of M. Then, by 5.1.2 the $\mathcal{O}Q$-module V is a summand of

$$\mathrm{Res}_Q^G\mathrm{Ind}_H^G(\mathcal{O}) = \oplus_{x\in[Q\backslash G/H]}\,\mathrm{Ind}_{Q\cap{}^xH}^Q\mathrm{Res}_{Q\cap{}^xH}^{{}^xH}({}^x\mathcal{O}) = \oplus_{x\in[Q\backslash G/H]}\,\mathrm{Ind}_{Q\cap{}^xH}^Q(\mathcal{O}),$$

where we used again the Mackey formula. Since V is indecomposable, V is a summand of $\mathrm{Ind}_{Q\cap{}^xH}^Q(\mathcal{O})$ for some $x \in G$, and since Q is a vertex of the source V it follows that $Q \subseteq {}^xH$, thus V is a summand of $\mathcal{O}$, hence isomorphic to $\mathcal{O}$. Therefore, M is a trivial source module. In order to show (ii), we show this first if $M = \mathrm{Ind}_Q^G(\mathcal{O})$ and $N = \mathrm{Ind}_R^G(\mathcal{O})$ for some subgroups Q, R in G. Using Frobenius' reciprocity, Mackey's formula and Frobenius' reciprocity again we get $\mathrm{Hom}_{\mathcal{O}G}(M,N) \cong \mathrm{Hom}_{\mathcal{O}R}(\mathrm{Res}_R^G\mathrm{Ind}_Q^G(\mathcal{O}),\mathcal{O}) = \oplus_{x\in[R\backslash G/Q]}\,\mathrm{Hom}_{\mathcal{O}R}(\mathrm{Ind}_{R\cap{}^xQ}^R(\mathcal{O}),\ \mathcal{O}) \cong \oplus_{x\in[R\backslash G/Q]}\mathrm{Hom}_{\mathcal{O}(R\cap{}^xQ)}(\mathcal{O},\ \mathcal{O}) \cong \oplus_{x\in[R\backslash G/Q]}\mathcal{O}$. Thus the $\mathcal{O}$-rank of $\mathrm{Hom}_{\mathcal{O}G}(M,N)$ is $|R\backslash G/Q|$; applied to $\mathcal{O} = k$, this argument shows that this is also the k-dimension of $\mathrm{Hom}_{kG}(\bar{M},\bar{N})$. Thus statement (ii) holds in this case. Then statement (ii) holds obviously for all direct sums and summands of modules of this form, and by (i), those are exactly the direct sums of trivial source modules. Let M be a trivial source $\mathcal{O}Q$-module with vertex Q. By (ii), $\mathrm{End}_{kG}(\bar{M})$ is a quotient of $\mathrm{End}_{\mathcal{O}G}(M)$, hence also local by 4.4.5, and thus $\bar{M}$ is indecomposable. By Higman's criterion 2.6.2, there is an $\mathcal{O}Q$-endomorphism φ of M such that $\mathrm{Id}_M = \mathrm{Tr}_Q^G(\varphi)$, the trace taken in $\mathrm{End}_{\mathcal{O}}(M)$, and Q is minimal with this property. The induced kQ-endomorphism $\bar{\varphi}$ of $\bar{M}$ satisfies $\mathrm{Id}_{\bar{M}} = \mathrm{Tr}_Q^G(\bar{\varphi})$, the trace here taken in $\mathrm{End}_k(\bar{M})$. Thus Q contains a vertex R of $\bar{M}$. Since M is a trivial source module, the trivial $\mathcal{O}Q$-module $\mathcal{O}$ is isomorphic to a direct summand of $\mathrm{Res}_Q^G(M)$. But then the trivial kQ-module k, which has Q as a vertex by 5.1.4 (i), is isomorphic to a direct summand of $\bar{M}$, and hence Q is contained in a vertex of $\bar{M}$. This shows (iii). Let U be a trivial source kG-module. Denote by Q a vertex of U; thus U is isomorphic to an indecomposable direct summand of $\mathrm{Ind}_Q^G(k)$. If M is a trivial source $\mathcal{O}G$-module such that $\bar{M} \cong U$ then M has vertex Q as well, by (iii); thus M is isomorphic to a direct summand of $\mathrm{Ind}_Q^G(\mathcal{O})$. Let $\tau \in \mathrm{End}_{kG}(\mathrm{Ind}_Q^G(k))$ be an idempotent such that $U = \tau(\mathrm{Ind}_Q^G(k))$. By 4.6.10, τ is unique up to conjugacy in $\mathrm{End}_{kG}(\mathrm{Ind}_Q^G(k))$. By (ii) the canonical map $\mathrm{End}_{\mathcal{O}G}(\mathrm{Ind}_Q^G(\mathcal{O})) \to \mathrm{End}_{kG}(\mathrm{Ind}_Q^G(k))$ is surjective. Thus, by the Lifting Theorem for idempotents 4.7.1, there is an idempotent $\pi \in \mathrm{End}_{\mathcal{O}G}(\mathrm{Ind}_Q^G(\mathcal{O}))$, unique up to conjugacy, such that τ is the canonical image of π. But then $M = \pi(\mathrm{Ind}_Q^G(\mathcal{O}))$ is a direct summand of $\mathrm{Ind}_Q^G(\mathcal{O})$ satisfying $\bar{M} = U$, and M is unique up to isomorphism. □

The uniqueness part of the lifting property (iv) holds only within the class of trivial source modules: given a finite p-group P, any rank one $\mathcal{O}P$-module lifts the trivial kP-module (but only the trivial $\mathcal{O}P$-module is a trivial source module amongst the rank one $\mathcal{O}P$-modules). When applied to p-permutation modules, the Brauer construction is particularly well-behaved. It can be used to give alternative descriptions of vertices of trivial source modules, or of the Green correspondence for trivial source modules.

Proposition 5.10.3 *Let G be a finite group and let M be a trivial source $\mathcal{O}G$-module. Let P be a p-subgroup of G. We have $M(P) \neq \{0\}$ if and only if P is contained in a vertex of M.*

Proof By 5.6.9 we have $(\mathrm{End}_{\mathcal{O}}(M))(P) \neq 0$ if and only if P is contained in a vertex of M. By 5.8.6 we have $(\mathrm{End}_{\mathcal{O}}(M))(P) \cong \mathrm{End}_k(M(P))$. The result follows. □

Proposition 5.10.4 *Let G be a finite group, Q a normal p-subgroup of G and M a trivial source $\mathcal{O}G$-module. If Q is contained in a vertex of M then $M = M^Q$; that is, Q acts trivially on M.*

Proof Let P be a vertex of M such that $Q \subseteq P$. Then M is isomorphic to a direct summand of $\mathrm{Ind}_P^G(\mathcal{O}) \cong \mathcal{O}G/P$. Since Q is normal in G we have $yxP = xP$ for all $x \in G$ and all $y \in Q$, which shows that Q acts trivially on $\mathrm{Ind}_P^G(\mathcal{O})$, hence on M. □

Theorem 5.10.5 ([13, (3,4)]) *Let G be a finite group, let P be a p-subgroup of G. Let M be a trivial source kG-module with vertex P. Then the $kN_G(P)$-module $M(P)$ is the Green correspondent of M. Viewed as a $kN_G(P)/P$-module, $M(P)$ is projective indecomposable, and the correspondence sending M to $M(P)$ induces a bijection between the isomorphism classes of trivial source modules with vertex P and projective indecomposable $kN_G(P)/P$-modules. Moreover, $M(P)$ is the multiplicity module of M.*

Proof By the assumptions, M is a direct summand of $\mathrm{Ind}_P^G(k)$. The Brauer construction is additive and hence $M(P)$ is a direct summand of $(\mathrm{Ind}_P^G(\mathcal{O})) \cong \mathrm{Ind}_P^{N_G(P)}(k)$, where the isomorphism is from 5.8.4. By 5.8.1, $M(P)$ is a direct summand of $\mathrm{Res}^G_{N_G(P)}(M)$. Since M and $M(P)$ are both relatively P-projective, Higman's criterion implies that $\mathrm{End}_{kG}(M) = (\mathrm{End}_k(M))_P^G$ and $\mathrm{End}_{kN_G(P)}(M(P)) = (\mathrm{End}_k(M(P))_P^{N_G(P)}$. Using 5.4.5 and 5.8.6 we get $\mathrm{Br}_P(\mathrm{End}_{\mathcal{O}G}(M)) = (\mathrm{End}_k(M(P))_P^{N_G(P)} = \mathrm{End}_{kN_G(P))}(M(P))$. In particular, $\mathrm{End}_{kN_G(P)}(M(P))$ is a quotient of the local algebra $\mathrm{End}_{\mathcal{O}G}(M)$, hence itself local and thus $M(P)$ is indecomposable as a $kN_G(P)$-module. Since P acts trivially on $M(P)$ any vertex of $M(P)$ contains P. If R is a vertex of $M(P)$ then $M(R) \cong M(P)(R) \neq 0$, hence R is contained in a vertex of M by 5.10.3, which forces $R = P$. Thus $M(P)$ is the Green correspondent of M. Again since P acts trivially on $M(P)$, we can consider $M(P)$ as $kN_G(P)/P$-module. Clearly $M(P)$ remains indecomposable as $kN_G(P)/P$-module, and the vertex is now trivial, hence $M(P)$ is projective by Higman's criterion. The uniqueness properties of the Green correspondence imply that the correspondence $M \mapsto M(P)$ is a bijection between the isomorphism classes as stated.

The projection of M onto a trivial summand of $\mathrm{Res}_P^G(M)$ is an idempotent in $\mathrm{End}_{\mathcal{O}P}(M)$ belonging to a local point of P on $\mathrm{End}_{\mathcal{O}}(M)$ corresponding to the local point denoted by γ in the Definition 5.7.1, and the isomorphism $\mathrm{End}_{\mathcal{O}}(M)(P) \cong \mathrm{End}_k(M(P))$ from 5.8.6 shows that $\mathrm{End}_k(M(P))$ corresponds to the simple quotient denoted $S(\gamma)$ in 5.7.1. Since the trivial $\mathcal{O}P$-module is stable under any automorphism of P, the group $N_G(P_\gamma)$ in 5.7.1 is $N_G(P)$. Set $\bar{N} = N_G(P)/P$. The fact that $M(P)$ is a $k\bar{N}$-module means that the action of $\bar{N}$ on $\mathrm{End}_{\mathcal{O}}(M)(P)$ lifts canonically, and hence the 2-cocycle appearing in 5.7.1 is trivial. Thus $M(P)$ is the multiplicity module of M. □

The following result, due to Okuyama, shows that simple modules with a trivial source have simple Green correspondents.

Proposition 5.10.6 ([67, 2.2]) *Let G be a finite group, and let S be an absolutely simple kG-module. Let P be a vertex of S. Suppose that S has a trivial source. Then the indecomposable $kN_G(P)$-module $f(S)$ with vertex P corresponding to S via the Green correspondence is absolutely simple.*

Proof It follows from 5.10.5 that the Green correspondent of S is $S(P)$. As a $kN_G(P)/P$-module, this is the multiplicity module of S, hence absolutely simple by by 5.7.8. □

We will later need the following criteria for a homomorphism starting from a trivial source module to be split injective.

Proposition 5.10.7 ([61, 6.1]) *Let G be a finite group, P a p-subgroup, U an indecomposable $\mathcal{O}G$-module with vertex P and trivial source, and let M be an $\mathcal{O}G$-module such that $\mathrm{Res}^G_P(M)$ is a permutation $\mathcal{O}P$-module. Set $N = N_G(P)/P$. Let $\alpha : U \to M$ be a homomorphism of $\mathcal{O}G$-modules. The following are equivalent.*

(i) *The $\mathcal{O}G$-homomorphism $\alpha : U \to M$ is split injective.*
(ii) *The kN-homomorphism $\alpha(P) : U(P) \to M(P)$ is injective.*

Proof The implication (i) ⇒ (ii) is trivial. Suppose that (ii) holds. Then $\alpha(P) : U(P) \to M(P)$ is split injective as a kN-homomorphism because $U(P)$ is projective, hence injective, as a kN-module. Using that $\mathrm{soc}(U(P))$ is simple it follows that M has an indecomposable direct summand M' such that the induced map $\beta(P) : U(P) \to M'(P)$ is still split injective, where β is the composition of α followed by the projection from M onto M'. The Brauer homomorphism applied to the algebra $\mathrm{End}_{\mathcal{O}}(M')$ maps $\mathrm{End}_{\mathcal{O}}(M')^G_P$ onto $(\mathrm{End}_k(M'))(P)^N_1 \cong \mathrm{End}_k(M'(P))^N_1$ (cf. 5.4.5). The summand of $M'(P)$ isomorphic to $U(P)$ corresponds to a primitive idempotent in $\mathrm{End}_k(M'(P))^N_1$, hence lifts to a primitive idempotent in $\mathrm{End}_{\mathcal{O}}(M')^G_P$. Since M' is indecomposable, this idempotent is $\mathrm{Id}_{M'}$, and hence, by Higman's criterion, M' has P as a vertex. But then M' has a trivial source, and so $M'(P)$ is indecomposable as a kN-module, hence isomorphic to $U(P)$. By the Green correspondence this implies $U \cong M'$. Composing β with the inverse of this isomorphism yields an endomorphism γ of U that induces an automorphism on $U(P)$. Since $\mathrm{End}_{\mathcal{O}G}(U)$ is local, this implies that γ is an automorphism of U, and hence that $\beta : U \to M'$ is an isomorphism. Thus α is split injective, whence the implication (ii) ⇒ (i). □

Proposition 5.10.8 ([61, 6.2]) *Let G be a finite group, P a p-subgroup, U an indecomposable $\mathcal{O}G$-module with vertex P and trivial source, and let M be an $\mathcal{O}G$-module such that $\mathrm{Res}^G_P(M)$ is a permutation $\mathcal{O}P$-module. Set $N = N_G(P)/P$. Suppose that N has a normal p-subgroup Z such that the kN-module $k \otimes_{kZ} U(P)$ is simple and such that $M(P)$ is projective as a kZ-module. Let $\alpha : U \to M$ be a homomorphism of $\mathcal{O}G$-modules. The following are equivalent.*

(i) *The $\mathcal{O}G$-homomorphism $\alpha : U \to M$ is split injective.*
(ii) *There is a nonzero direct summand W of $\mathrm{Res}^N_Z(U(P))$ such that the kZ-homomorphism $\alpha(P)|_W : W \to \mathrm{Res}^N_Z(M(P))$ is injective.*

Proof The implications (i) ⇒ (ii) is trivial. Suppose that (ii) holds. Since $U(P)$ is projective indecomposable as a kN-module, it has a simple socle and a simple top, and these are isomorphic. The restriction of $U(P)$ to kZ remains projective, and hence the module $k \otimes_{kZ} U(P)$ has the same dimension as the submodule $U(P)^Z$ of Z-fixed points in $U(P)$. Since Z is normal in N, it follows that $U(P)^Z$ is a kN-submodule of $U(P)$, hence that $U(P)^Z$ contains the simple socle of $U(P)$. Since $k \otimes_{kZ} U(P)$ is assumed to be simple, hence isomorphic to the top and bottom composition factor of $U(P)$, it follows that $U(P)^Z$ is equal to the socle $\mathrm{soc}(U(P))$ of $U(P)$ as a kN-module. By the assumption (ii), the kernel of the map $U(P) \to M(P)$ does not contain $U(P)^Z$. Since the socle of $U(P)$ as a kN-module is simple, this implies that $\alpha(P) : U(P) \to M(P)$ is injective, and hence that α is split injective by 5.10.7. This shows the implication (ii) $\Rightarrow$ (i). $\square$

5.11 p-permutation modules

It is a special feature of p-groups that transitive permutation modules over p-groups are indecomposable. This observation yields a different characterisation of trivial source modules. Modules in this section are finitely generated.

Definition 5.11.1 Let G be a finite group. An $\mathcal{O}G$-module M is called a *p-permutation module* if, for some Sylow p-subgroup P of G, the restriction $\mathrm{Res}^G_P(M)$ is a permutation $\mathcal{O}P$-module, or equivalently, if M is $\mathcal{O}$-free having an $\mathcal{O}$-basis X which is permuted by the action of elements of P on M.

Since all Sylow p-subgroups in G are conjugate, one easily sees that M is a p-permutation module if and only if for any p-subgroup Q of G the restriction $\mathrm{Res}^G_Q(M)$ is a permutation $\mathcal{O}Q$-module. The next result shows that for indecomposable modules the concepts "trivial source" and "p-permutation" coincide.

Theorem 5.11.2 *Let G be a finite group.*

(i) A finitely generated $\mathcal{O}G$-module M is a p-permutation module if and only if M is a direct sum of trivial source $\mathcal{O}G$-modules.
(ii) Every direct summand of a p-permutation module M is again a p-permutation module.
(iii) For any p-permutation kG-module N there is, up to isomorphism, a unique p-permutation $\mathcal{O}G$-module M satisfying $k \otimes_{\mathcal{O}} M \cong N$.
(iv) For any two p-permutation $\mathcal{O}G$-modules M, M', the canonical map from $\mathrm{Hom}_{\mathcal{O}G}(M, M')$ *to* $\mathrm{Hom}_{kG}(k \otimes_{\mathcal{O}} M, k \otimes_{\mathcal{O}} M')$ *is surjective.*

Proof The main point of the proof is that a direct summand of a p-permutation module is again a p-permutation module. Let P be a p-subgroup of G. Suppose that M is a p-permutation module. Then $\mathrm{Res}^G_P(M)$ is a permutation $\mathcal{O}P$-module; that is, $\mathrm{Res}^G_P(M)$ is a direct sum of transitive $\mathcal{O}P$-permutation modules, and every transitive $\mathcal{O}P$-permutation module is isomorphic to $\mathrm{Ind}_Q^P(\mathcal{O}) \cong \mathcal{O}P/Q$ for some subgroup Q of P. Any of the transitive $\mathcal{O}P$-permutation modules is indecomposable, by 1.11.4. This implies that if M' is a direct summand of M as an $\mathcal{O}G$-module, then $\mathrm{Res}_P^G(M')$ is also a direct sum of transitive $\mathcal{O}P$-permutation modules, where we use the Krull–Schmidt Theorem. In particular, every direct summand of M is again a p-permutation module. Moreover, since $\mathrm{Ind}_Q^P(\mathcal{O})$ has Q as a vertex, the only way for $\mathrm{Res}_P^G(M)$ to have an indecomposable direct summand of vertex P is to have the trivial $\mathcal{O}P$-module as a summand. Thus every indecomposable summand of M is a trivial source module. Conversely, if M is a trivial source module, then M is a direct summand of a permutation module $\mathrm{Ind}_P^G(\mathcal{O})$ for some P, and since a permutation module is obviously a p-permutation module, so is M by the first part of the argument. The statements (iii) and (iv) follow from 5.10.2. □

Proposition 5.11.3 *Let G be a finite group and M a p-permutation $\mathcal{O}G$-module. For any p-subgroup P of G, $M(P)$ is a p-permutation $kN_G(P)$-module on which P acts trivially.*

Proof The finite generation passes obviously from M to $M(P)$, and clearly P acts trivially on $M(P)$. Let R be a p-subgroup of $N_G(P)$ containing P. Then M has an R-stable basis X. By 5.8.1, the image of X^P in $M(P)$ is a k-basis of $M(P)$. Since R normalises P and stabilises X, it also stabilises X^P, whence the result. □

Combining some of the previous results we reformulate 5.8.7 for p-permutation modules over the ring $\mathcal{O}$ as follows.

Proposition 5.11.4 *Let G be a finite group, P a p-subgroup of G and M a p-permutation $\mathcal{O}G$-module. There is an isomorphism of $\mathcal{O}N_G(P)$-modules*

$$\mathrm{Res}^G_{N_G(P)}(M) \cong M' \oplus M''$$

such that P acts trivially on M' and such that $M''(P) = \{0\}$. In particular, $M(P) \cong M'(P) \cong k \otimes_{\mathcal{O}} M'$ as $kN_G(P)$-modules, every indecomposable direct summand of M' has a vertex containing P, and no indecomposable direct summand of M'' has a vertex containing P.

Proof In a decomposition of $\mathrm{Res}^G_{N_G(P)}(M)$ as a direct sum of indecomposable $\mathcal{O}N_G(P)$-modules, let M' be the sum of all summands with a vertex containing P and M'' the sum of all summands with no vertex containing P. By 5.11.2, any p-permutation $kN_G(P)$-module lifts uniquely, up to isomorphism, to a p-permutation $\mathcal{O}N_G(P)$-module. It follows from 5.10.2 that $M''(P) = \{0\}$, hence $M(P) \cong M'(P)$. It follows from 5.10.4 that P acts trivially on M', and hence $M'(P) \cong k \otimes_{\mathcal{O}} M'$. The last statement on vertices of indecomposable summands follows from 5.10.3. □

Thus, using 5.5.21, the $kN_G(P)$-module $M(P)$ determines the direct sum of summands of M with vertex P. This implies the following result; we include an alternative proof which does not make use of 5.5.21.

Proposition 5.11.5 *Let G be a finite group and let M, N be p-permutation $\mathcal{O}G$-modules. Suppose that M and N have no nonzero projective direct summands. We have $M \cong N$ as $\mathcal{O}G$-modules if and only if for any nontrivial p-subgroup P of G we have $M(P) \cong N(P)$ as $kN_G(P)$-modules.*

Proof Using the Krull–Schmidt Theorem, we may cancel isomorphic summands from M and N; that is, we may assume that M and N have no isomorphic indecomposable direct summands. Choose P maximal subject to $M(P) \neq \{0\}$. Since no nonzero summand of M is projective, clearly P is nontrivial. The maximality of P implies that M and N have no indecomposable direct summands with a vertex strictly containing P. Thus 5.10.5 implies that $M(P)$ is the direct sum of the Green correspondents of indecomposable direct summands of M with vertex P. Similarly for N. It follows that M and N have isomorphic summands with vertex P, contradicting the current assumption. The result follows. □

Any $\mathcal{O}G$-$\mathcal{O}G$-bimodule M can be viewed as $\mathcal{O}(G \times G)$-module by defining the action of $(x, y) \in G \times G$ on $m \in M$ via the formula

$$(x, y) \cdot m = xmy^{-1}.$$

In this way, an indecomposable $\mathcal{O}G$-$\mathcal{O}G$-bimodule becomes an indecomposable $\mathcal{O}(G \times G)$-module, and hence has a vertex and a source. The group algebra $\mathcal{O}G$ itself becomes a permutation $\mathcal{O}(G \times G)$-module in this way, and the following result elaborates on the question as to how vertices of indecomposable summands of $\mathcal{O}G$ as $\mathcal{O}(G \times G)$-module look like. The isomorphism in the following result has already been noted in 2.4.5, but we take the opportunity to describe this in more detail.

Proposition 5.11.6 *Let G be a finite group and H a subgroup of G. Set $\Delta G = \{(x, x) | x \in G\}$.*

(i) There is an isomorphism of $\mathcal{O}(G \times G)$-modules $\mathcal{O}G \cong \mathrm{Ind}_{\Delta G}^{G \times G}(\mathcal{O})$ mapping $x \in G$ to $(x, 1) \otimes 1$. Its inverse maps $(x, y) \otimes \lambda$ to λxy^{-1}, where $x, y \in G$ and $\lambda \in \mathcal{O}$.

(ii) Every indecomposable summand of $\mathcal{O}G$ as $\mathcal{O}(G \times G)$-module has a vertex contained in ΔG, and the trivial module as source.

(iii) Every indecomposable direct summand of $\mathcal{O}[G \setminus H]$ as an $\mathcal{O}(H \times H)$-module has a vertex of the form $\Delta_x Q = \{(u, x^{-1}ux) | u \in Q\}$ for some p-subgroup Q of $H \cap {}^xH$ and some $x \in G \setminus H$.

Proof The map sending $x \in G$ to $(x, 1) \otimes 1$ in $\mathrm{Ind}_{\Delta G}^{G \times G}(\mathcal{O}) = \mathcal{O}(G \times G) \otimes_{\mathcal{O}\Delta G} \mathcal{O}$ is obviously a homomorphism of left $\mathcal{O}G$-modules. We have $(x, 1) \otimes 1 = (1, x^{-1})(x, x) \otimes 1 = (1, x^{-1}) \otimes (x, x)1 = (1, x^{-1}) \otimes 1$. One checks that this means that this map is also a homomorphism of right $\mathcal{O}G$-modules. To see that the map sending $(x, y) \otimes \lambda$ to λxy^{-1} is well-defined, let $x, y, z \in G$ and $\lambda \in \mathcal{O}$. The image of $(xz, yz) \otimes \lambda$ is $xzz^{-1}y^{-1} \otimes \lambda = xy^{-1} \otimes \lambda$ as required. These two maps are inverse to each other: $x \in G$ is mapped to $(x, 1) \otimes 1$, which is mapped back to x, and $(x, y) \otimes 1$ is mapped to xy^{-1}, which in turn goes to $(xy^{-1}, 1) \otimes 1 = (xy^{-1}, 1) \otimes (y, y)1 = (xy^{-1}, 1)(y, y) \otimes 1 = (x, y) \otimes 1$. This proves (i), and (ii) is an immediate consequence of (i). Statement (iii) follows from 2.4.6. □

Proposition 5.11.6 shows that indecomposable summands of the $\mathcal{O}(G \times G)$-module $\mathcal{O}G$ are special cases of trivial source bimodules with a diagonal vertex. This is a key theme for the interplay of various equivalences and the local structure of blocks, which will be developed in greater detail in Section 9.4; at this point, we only mention some easy facts: a bimodule with a diagonal vertex is perfect, and a perfect p-permutation bimodule has a diagonal vertex in a sense specified below.

Proposition 5.11.7 *Let G, H be finite groups and P a common p-subgroup of G and of H. Set $\Delta P = \{(u, u) | u \in P\} \subseteq G \times H$. Let M be a relatively ΔP-projective p-permutation $\mathcal{O}G$-$\mathcal{O}H$-bimodule. For any subgroup Q of P, $M(\Delta Q)$ is a p-permutation $kC_G(Q)$-$kC_H(Q)$-bimodule which is projective as a left $kC_G(Q)$-module and as a right $kC_H(Q)$-module.*

Proof By the assumptions on M, ΔP contains a vertex of any indecomposable direct summand of M. Thus $G \times 1$ and $1 \times H$ have trivial intersections with any vertex of any indecomposable direct summand of M, and hence the restrictions to $G \times 1$ and $1 \times H$ of M are projective, by Mackey's formula. Since $C_G(Q) \times C_H(Q)$ is contained in $N_{\Delta Q}(G \times H)$, it follows from 5.11.3 that $M(\Delta Q)$ is a p-permutation $kC_G(Q)$-$kC_H(Q)$-bimodule. It follows from 5.8.7 that $M(\Delta Q)$ is isomorphic to a direct summand of $k \otimes_{\mathcal{O}} M$ restricted to $C_G(Q) \times C_H(Q)$, hence projective as a left and right module, since this is true for $k \otimes_{\mathcal{O}} M$ by the first argument. □

Proposition 5.11.8 *Let G, H be finite groups and let M be an indecomposable $\mathcal{O}G$-$\mathcal{O}H$-bimodule such that M is projective as a left $\mathcal{O}G$-module and as a right*

$\mathcal{O}H$-module. Let R be a vertex of M as an $\mathcal{O}(G \times H)$-module, and suppose that a source of M has $\mathcal{O}$-rank prime to p. Then $R = \{(u, \varphi(u)) | u \in Q\}$ for some subgroup Q of G and some injective group homomorphism $\varphi : Q \to H$.

Proof Since M is projective as a left $\mathcal{O}G$-module, the restriction of M to $(G \times 1) \cap R$ is projective. Thus the restriction of an $\mathcal{O}R$-source W of M to $(G \times 1) \cap R$ is projective. Since the $\mathcal{O}$-rank of W is prime to p, this forces $(G \times 1) \cap R = \{1\}$. Similarly, $(1 \times H) \cap R = \{1\}$. Thus the inclusion $R \subseteq G \times H$ followed by any of the projections from $G \times H$ to G or H yields injective group homomorphisms $\psi : R \to G$ and $\tau : R \to H$. It follows that $Q = \psi(R)$ and $\varphi = \tau \circ \psi^{-1}|_Q$ satisfy the conclusion. □

Note that 5.11.8 applies in particular if M is a trivial source bimodule.

Proposition 5.11.9 *Let G be a finite group and X a bounded complex consisting of finitely generated p-permutation $\mathcal{O}G$-modules. Then X is contractible if and only if $k \otimes_{\mathcal{O}} X$ is contractible.*

Proof Suppose that $\bar{X} = k \otimes_{\mathcal{O}} X$ is contractible. Denote by δ and $\bar{\delta}$ the differentials of X and $\bar{X}$, respectively. There is a homotopy $\bar{h}$ of $\bar{X}$ such that $\bar{h} \circ \bar{\delta} + \bar{\delta} \circ \bar{h} = \mathrm{Id}_{\bar{X}}$. By 5.11.2, the homotopy $\bar{h}$ lifts to a homotopy h on X. Then $\varphi = \delta \circ h + h \circ \delta$ is a chain map on X which lifts the identity on $\bar{X}$. Nakayama's Lemma implies that φ is an automorphism of X. Since also $\varphi \simeq 0$ we get that X is contractible. The converse is trivial. □

Theorem 5.11.10 (Bouc [10, 7.9]) *Let G be a finite group, P a p-subgroup and X a bounded complex of $\mathcal{O}G$-modules consisting of finitely generated relatively $\mathcal{O}P$-projective $\mathcal{O}G$-modules whose restrictions to $\mathcal{O}P$ are permutation $\mathcal{O}P$-modules.*

(i) The complex X is contractible if and only if the complex of $kC_G(Q)$-modules $X(Q)$ is exact for all subgroups Q of P.
(ii) The complex X is homotopy equivalent to a bounded complex of projective $\mathcal{O}G$-modules if and only if the complex of $kC_G(Q)$-modules $X(Q)$ is acyclic for all nontrivial subgroups Q of P.

Proof If X is contractible, then so is $X(Q)$ for any subgroup Q of P by the functoriality of the Brauer construction. The converse in (i) is a special case of (ii): if $X(1) \cong k \otimes_{\mathcal{O}} X$ is acyclic and homotopy equivalent to a bounded complex of projective $\mathcal{O}G$-modules, then $k \otimes_{\mathcal{O}} X$ is contractible by 1.18.8, hence X is contractible by 5.11.9. In order to prove (ii) we use the mapping cone construction. Let n be the largest integer such that $X_n \neq \{0\}$. Construct a complex Y, which is equal to X in all degrees except in degree $n + 1$, where we add a projective cover W of X_n and a surjective map $W \to X_{n+1}$ to the differential. The complex Y has X as a subcomplex, and the quotient Y/X is isomorphic to W concentrated in degree $n+1$. The last nonzero map γ from $Y_{n+1} = X_{n+1} \oplus W$ to $Y_n = X_n$ is now surjective, and remains surjective on applying the Brauer construction with respect to any nontrivial subgroup of P. Thus, by 5.8.11, this map is split surjective. This implies that $Y \cong Y' \oplus Z$ for some contractible complex Z having exactly two nonzero terms both isomorphic to X_n, and a complex Y' having fewer nonzero terms than X. But then X is homotopy equivalent to a shift of the mapping cone of the map $Y' \to W$. Note that $X(Q) \cong Y(Q) \simeq Y'(Q)$ is acyclic for all nontrivial subgroups Q of P. Arguing by induction over the length of X it follows that Y' is homotopy equivalent to a bounded complex Y'' of projective $\mathcal{O}G$-modules – but then so is the mapping cone of the chain map $Y'' \simeq Y' \to W$, and thanks to the invariance of the mapping cone construction under homotopy equivalences this is still homotopy equivalent to a shift of X. $\square$

We conclude this section with a characterisation of p-permutation modules which is a variation of a theorem of Conlon. The proof requires the fact, proved in 5.14.2 below, that group algebras have nonsingular Cartan matrices.

Proposition 5.11.11 *Let G be a finite group and let M, N be p-permutation $\mathcal{O}G$-modules. Suppose that k is large enough for all subgroups of G. The following are equivalent:*

(i) $M \cong N$.
(ii) For any p-subgroup Q of G and any primitive idempotent j in $kN_G(Q)/Q$ we have $\dim_k(jM(Q)) = \dim_k(jN(Q))$.

The proof uses induction over the vertices of indecomposable direct summands of M, N, in conjunction with the following lemma which handles the case where M, N are projective:

Lemma 5.11.12 *Let A be a split symmetric k-algebra such that the determinant of the Cartan matrix C of A is non zero. Let M, N be projective A-modules. We have $M \cong N$ if and only if* $\dim_k(jM) = \dim_k(jN)$ *for any primitive idempotent j in A.*

Proof Let I be a set of representatives of the points of A. Then C is the matrix $(\dim_k(iAj))_{i,j\in I}$. Suppose that $\dim_k(jM) = \dim_k(jN)$ for any $j \in I$. In order to show that $M \cong N$ we may cancel isomorphic summands and hence assume that M and N have no isomorphic non zero indecomposable direct summand. Since M, N are projective, this means that we may write I as disjoint union $I = J \cup J'$ such that $M \cong \oplus_{j\in J}(Aj)^{m_j}$ and $N \cong \oplus_{j'\in J'}(Aj)^{m_{j'}}$, with suitable non negative integers m_j for $j \in I$. For any primitive idempotent $i \in I$ we get that

$$\dim_k(iM) = \sum_{j \in J} m_j \dim_k(iAj) = \sum_{j' \in J'} m_{j'} \dim_k(iAj') = \dim_k(iN).$$

Thus the column vector consisting of the integers m_j with $j \in J$ and $-m_{j'}$ with $j' \in J'$, multiplied with the Cartan matrix, yields zero. By the assumption in the Cartan matrix having determinant non zero this forces all m_j to be zero, where $j \in I$. Thus $M \cong N$. The converse is trivial. □

Proof of 5.11.11 The implication (i) $\Rightarrow$ (ii) is trivial. Suppose that (ii) holds. After cancelling isomorphic summands we may assume that M, N have no nonzero isomorphic summands. Choose Q maximal such that $M(Q) \neq \{0\}$. Then, by 5.10.5, $M(Q)$ and $N(Q)$ are both projective $kN_G(Q)/Q$-modules, which are, when viewed as $kN_G(Q)$-modules, the direct sums of the Green correspondents of indecomposable direct summands of M, N having Q as vertex. By 5.14.2 below, finite group algebras have non singular Cartan matrices. Thus, by the assumptions and 5.11.12, the $kN_G(Q)/Q$-modules $M(Q)$ and $N(Q)$ are isomorphic. But then the Green correspondence implies that the indecomposable summands of M, N with vertex Q are isomorphic. The result follows. □

5.12 Green's Indecomposability Theorem

If P is a finite p-group and Q any subgroup of P, then $\mathrm{Ind}_Q^P(\mathcal{O})$ is an indecomposable module. Green's Indecomposability Theorem is a far reaching generalisation of that fact. Before we state it, we start with a simple observation. Modules in this section are finitely generated.

Theorem 5.12.1 *Let G be a finite group, H a normal subgroup of G and V an indecomposable $\mathcal{O}H$-module. Suppose that ${}^xV \not\cong V$ for any $x \in G \setminus H$. Then the $\mathcal{O}G$-module $\mathrm{Ind}_H^G(V)$ is indecomposable.*

Proof Let U be an indecomposable direct summand of $\mathrm{Ind}_H^G(V)$. By Mackey's formula 2.4.1 we have

$$\mathrm{Res}_H^G(\mathrm{Ind}_H^G(V)) = \oplus_{x \in [G/H]} {}^xV.$$

For any $x \in G$ the $\mathcal{O}H$-module ${}^xV$ remains obviously indecomposable. Thus, by the Krull–Schmidt Theorem 4.6.7, $\mathrm{Res}_H^G(U)$ has a summand isomorphic to ${}^xV$ for some $x \in G$. But since $\mathrm{Res}_H^G(U)$ is the restriction to $\mathcal{O}H$ of an $\mathcal{O}G$-module, it follows that $\mathrm{Res}_H^G(U)$ has a direct summand isomorphic to ${}^xV$ for any $x \in G$. Since the ${}^xV$, with $x \in [G/H]$, are pairwise non isomorphic, it follows that in fact the direct sum $\oplus_{x\in[G/H]}{}^xV$ is isomorphic to a direct summand of $\mathrm{Res}^G_H(U)$. But this forces clearly that $U \cong \mathrm{Ind}^G_H(V)$. □

Definition 5.12.2 Let A be an $\mathcal{O}$-algebra that is finitely generated as an $\mathcal{O}$-module. A finitely generated A-module U is called *absolutely indecomposable* if $\mathrm{End}_A(U)$ is split local; that is, if $\mathrm{End}_A(U)/J(\mathrm{End}_A(U)) \cong k$.

For any finite group G, the trivial $\mathcal{O}G$-module $\mathcal{O}$ is absolutely indecomposable. If $\mathcal{O}'$ is any local commutative flat ring extension of $\mathcal{O}$ such that $J(\mathcal{O}) \subseteq J(\mathcal{O}')$, then for any finitely generated absolutely indecomposable $\mathcal{O}$-free A-module U the $\mathcal{O}' \otimes_{\mathcal{O}} A$-module $\mathcal{O}' \otimes_{\mathcal{O}} U$ is indecomposable. Indeed, by 1.12.10, the hypotheses imply that $\mathrm{End}_{\mathcal{O}'\otimes_{\mathcal{O}}A}(\mathcal{O}' \otimes_{\mathcal{O}} U) \cong \mathcal{O}' \otimes_{\mathcal{O}} \mathrm{End}_A(U)$. Since $J(\mathcal{O}') \otimes_{\mathcal{O}} \mathrm{End}_A(U)$ and $\mathcal{O}' \otimes_{\mathcal{O}} J(\mathrm{End}_A(U))$ are both contained in the radical of this algebra, it follows that the quotient of this algebra by its radical is isomorphic to $k' = \mathcal{O}'/J(\mathcal{O}')$. There are, however, modules that remain indecomposable under any finite ring extension without being absolutely indecomposable in the sense of the above definition; see [39, 1.4].

A subgroup H of a group G is called *subnormal* if there is a finite sequence of subgroups $H = H_0 \subseteq H_1 \subseteq \cdots \subseteq H_n = G$ such that H_i is normal in H_{i+1} for $0 \le i \le n-1$.

Theorem 5.12.3 (Green's Indecomposability Theorem) *Let G be a finite group, H a subnormal subgroup of p-power index in G and V an absolutely indecomposable $\mathcal{O}H$-module. Then the $\mathcal{O}G$-module $\mathrm{Ind}^G_H(V)$ is absolutely indecomposable.*

Proof Arguing by induction over $|G : H|$ we may assume that H is normal in G of index p. Set $P = G/H$. We will denote abusively elements in P and representatives in G by the same letters. If ${}^xV \not\cong V$ for some (and hence any) $x \in G \setminus H$, theorem 5.12.1 implies the result. Assume that ${}^xV \cong V$ for all $x \in G$. Mackey's formula 2.4.1

$$\mathrm{Res}^G_H(\mathrm{Ind}^G_H(V)) = \oplus_{x\in P}\, {}^xV$$

and Frobenius' reciprocity

$$E = \mathrm{End}_{\mathcal{O}G}(\mathrm{Ind}^G_H(V)) \cong \mathrm{Hom}_{\mathcal{O}H}(V, \mathrm{Res}^G_H(\mathrm{Ind}^G_H(V))$$

imply that an $\mathcal{O}G$-endomorphism of $\mathrm{Ind}^G_H(V)$ is completely determined by its restriction to the summand V. Thus, setting $E_x = \{\varphi \in \mathrm{End}_{\mathcal{O}G}(\mathrm{Ind}^G_H(V)) \mid \varphi(V) \subseteq {}^xV\}$ we get that

$$E = \oplus_{x\in P} E_x$$

and $E_x \circ E_y \subseteq E_{xy}$ for all $x, y \in P$. In fact, choosing an isomorphism $V \cong {}^xV$ shows that every component E_x contains an invertible element φ_x of E, and composition with φ_x induces an isomorphism of E_1-modules $E_1 \cong E_x$. The elements $\varphi_x \circ \varphi_y$ and φ_{xy} differ by some automorphism in E_1, for any $x, y \in P$. Note that $E_1 \cong \mathrm{End}_{\mathcal{O}H}(V)$. Since V is absolutely indecomposable, we have $E_1/J(E_1) \cong k$. Furthermore, we have $J(E_1)E_x = E_xJ(E_1)$, and $E_x/J(E_1)E_x \cong k$ for any $x \in P$ because E_x contains an invertible element. Hence $J(E_1)E$ is an ideal in E contained in $J(E)$, and we have

$$E/J(E_1)E = \oplus_{x \in P} k\bar{\varphi}_x$$

where $\bar{\varphi}_x$ is the image in $E/J(E_1)E$ of φ. By the above, we have $\bar{\varphi}_x \circ \bar{\varphi}_y = \alpha(x, y)\bar{\varphi}_{xy}$ for some $\alpha(x, y) \in k^\times$ and $x, y \in P$. Thus $E/J(E_1)E$ is isomorphic to a twisted group algebra $k_\alpha P$ of P by a 2-cocycle $\alpha \in Z^2(P; k^\times)$. But $H^2(P; k^\times) = \{0\}$ and therefore $E/J(E_1)E \cong kP$. Since kP is split local, it follows that E is split local, and hence $\mathrm{Ind}_H^G(V)$ is absolutely indecomposable. □

Corollary 5.12.4 *Let G be a finite group, H a subnormal subgroup of p-power index in G and U a relatively H-projective $\mathcal{O}G$-module such that all indecomposable direct summands of $\mathrm{Res}_H^G(U)$ are absolutely indecomposable. Then $U \cong \mathrm{Ind}_H^G(V)$ for some direct summand V of $\mathrm{Res}_H^G(V)$. In particular, if U is a projective $\mathcal{O}G$-module and k is a splitting field for kH, then $U \cong \mathrm{Ind}_H^G(V)$ for some projective $\mathcal{O}H$-module V.*

Proof We may assume that U is indecomposable. Since U is relatively N-projective, Higman's criterion 2.6.2 implies that there is an indecomposable direct summand V of $\mathrm{Res}_H^G(U)$ such that U is isomorphic to a direct summand of $\mathrm{Ind}_H^G(V)$. By the assumptions, V is absolutely indecomposable, and hence 5.12.3 implies that $\mathrm{Ind}_H^G(V)$ is indecomposable. Thus $U \cong \mathrm{Ind}_H^G(V)$. If k is large enough for kH, then the projective indecomposable $\mathcal{O}H$-modules are absolutely indecomposable, and hence the second statement follows from the first. □

The statement on projective indecomposable modules in the previous result has the following variation in terms of primitive idempotents.

Corollary 5.12.5 *Let G be a finite group, H a subnormal subgroup of p-power index in G and i a primitive idempotent in $\mathcal{O}H$. Suppose that $i\mathcal{O}Hi$ is split local. Then i is primitive in $\mathcal{O}G$ and $i\mathcal{O}Gi$ is split local.*

Proof The hypotheses imply that $\mathcal{O}Hi$ is a projective and absolutely indecomposable $\mathcal{O}H$-module. Thus, by 5.12.3, the $\mathcal{O}G$-module $\mathrm{Ind}_H^G(\mathcal{O}Hi) \cong \mathcal{O}Gi$ is absolutely indecomposable, whence the result. □

Combined with Clifford's Theorem 1.9.9, this yields the following result.

Proposition 5.12.6 *Let G be a finite group and N a normal subgroup of index p. Suppose that k is a splitting field for all subgroups of G. Let S be a simple kG-module. Then either $\mathrm{Res}^G_N(S)$ is simple, or $S \cong \mathrm{Ind}^G_N(T)$ for some simple kN-module T.*

Proof Let S be a simple kG-module. By Clifford's Theorem 1.9.9, the kN-module $\mathrm{Res}^G_N(S)$ is semisimple, and the isotypic components of $\mathrm{Res}^G_N(S)$ are permuted transitively by G. We will first show that $\mathrm{Res}^G_N(S)$ is a direct sum of pairwise nonisomorphic simple modules. By 4.10.3, applied to the kN-module $\mathrm{Res}^G_N(S)$, it suffices to show that $\dim_k(iS)$ is at most 1 for any primitive idempotent i in kN. Since i remains primitive in kG by 5.12.5, this is an immediate consequence of 4.10.3 again, applied to the simple kG-module S. Thus $\mathrm{Res}^G_N(S)$ is a direct sum of pairwise nonisomorphic G-conjugates of a simple submodule T of $\mathrm{Res}^G_N(S)$. Since G/N has order p, it follows that either $\mathrm{Res}^G_N(S) \cong T$ is simple, or that $\mathrm{Res}^G_N(S)$ is a direct sum of p conjugates of T. In the first case there is nothing further to prove. In the second case, S and $\mathrm{Ind}^G_N(T)$ have the same dimension. Frobenius' reciprocity applied to the inclusion map $T \to \mathrm{Res}^G_N(S)$ yields a nonzero map $\mathrm{Ind}^G_N(T) \to S$. Since S is simple, this map is surjective. Since both sides have the same dimension, it follows that this map is an isomorphism, whence the result. □

The following lemma relates pointed groups on a group algebra to bimodule summands of the group algebra.

Lemma 5.12.7 *Let G be a finite group, H a subgroup of G and i an idempotent in $(\mathcal{O}G)^H$. Consider $\mathcal{O}G$ as an $\mathcal{O}(G \times H)$-module with $(x, y) \in G \times H$ acting on $a \in \mathcal{O}G$ by $(x, y) \cdot a = xay^{-1}$. Then the following hold.*

(i) Any direct summand of $\mathcal{O}G$ as an $\mathcal{O}(G \times H)$-module is equal to $\mathcal{O}Gi$ for some idempotent $i \in (\mathcal{O}G)^H$.

(ii) For any idempotent $i \in (\mathcal{O}G)^H$ we have an $\mathcal{O}$-algebra isomorphism

$$(i\mathcal{O}Gi)^H \cong \mathrm{End}_{\mathcal{O}(G\times H)}(\mathcal{O}Gi)^{\mathrm{op}}$$

sending $z \in (i\mathcal{O}Gi)^H$ to right multiplication by z on $\mathcal{O}Gi$.

(iii) For any idempotent $i \in (\mathcal{O}G)^H$, the $\mathcal{O}(G \times H)$-module $\mathcal{O}Gi$ is indecomposable if and only if i is primitive in $(\mathcal{O}G)^H$.

Proof This is a variation of 1.7.4. Write $\mathcal{O}G = U \oplus V$ as direct sum of $\mathcal{O}(G \times H)$-submodules U and V of $\mathcal{O}G$. Thus there are unique elements $i \in U$ and $j \in V$ such that $1 = i + j$. Since U and V are left $\mathcal{O}G$-modules, we have $i = i \cdot 1 = i^2 + ij$, with $i^2 \in U$ and $ij \in V$. But $i \in U$, and thus $i^2 = i$ and $ij = 0$.

Similarly, $j^2 = j$ and $ji = 0$. Thus i and j are orthogonal idempotents, and hence $\mathcal{O}G = \mathcal{O}Gi \oplus \mathcal{O}Gj$. Since $\mathcal{O}Gi \subseteq U$ and $\mathcal{O}Gj \subseteq V$ we get $\mathcal{O}Gi = U$ and $\mathcal{O}Gj = v$. Conjugation by any element in H leaves U and V invariant. Since i and j are uniquely determined as the components of 1 in U and V, respectively, it follows that i and j are in $(\mathcal{O}G)^H$. This proves (i). Given an idempotent $i \in (\mathcal{O}G)^H$, any endomorphism of $\mathcal{O}Gi$ as $\mathcal{O}(G \times H)$-module is in particular an endomorphism of $\mathcal{O}Gi$ as left $\mathcal{O}G$-module, hence induced by right multiplication with an element $z \in i\mathcal{O}Gi$. Since this is also an endomorphism of $\mathcal{O}Gi$ as right $\mathcal{O}H$-module we have $z \in (i\mathcal{O}Gi)^H$. Statement (ii) follows, and (iii) is an easy consequence of (ii). □

Theorem 5.12.8 *Let G be a finite group, Q a p-subgroup of G and H a p'-subgroup of $C_G(Q)$. Let i be a primitive idempotent in $(\mathcal{O}G)^{QH}$.*

(i) If R is a subgroup of Q such that $i \in (\mathcal{O}G)^{QH}_{RH}$ then there is a primitive idempotent j in $(i\mathcal{O}Gi)^{RH}$ such that $i = \mathrm{Tr}_R^Q(j)$ and such that $({}^u j)j = 0 = j({}^u j)$ for $u \in Q \setminus R$; in particular, $\mathcal{O}Gi \cong \mathrm{Ind}_{G\times RH}^{G\times QH}(\mathcal{O}Gj)$.

(ii) If R is a minimal subgroup of Q such that $i \in (\mathcal{O}G)^{QH}_{RH}$ and j a primitive idempotent in $(i\mathcal{O}Gi)^{RH}$ such that $\mathcal{O}Gi \cong \mathrm{Ind}_{G\times RH}^{G\times QH}(\mathcal{O}Gj)$ then $\mathrm{Br}_R(j) \neq 0$.

(iii) A subgroup R of Q is minimal with the property $i \in (\mathcal{O}G)^{QH}_{RH}$ if and only if R is maximal with the property $\mathrm{Br}_R(i) \neq 0$.

Proof In order to prove (i) we may extend, if necessary, the ring $\mathcal{O}$ in such a way that indecomposable direct summands of $\mathcal{O}G$ as modules over subgroups of $G \times G$ are absolutely indecomposable. Write $i = \mathrm{Tr}_{RH}^{QH}(z)$ for some $z \in (\mathcal{O}G)^{RH}$. Since i is an idempotent, by multiplying this trace on the left and on the right by i we may in fact choose z in $(i\mathcal{O}Gi)^{RH}$. Then, in particular, z commutes with i. Set $\varphi(a) = az$ for all $a \in \mathcal{O}Gi$. Since z commutes with i and RH, the map φ is an $\mathcal{O}(G \times RH)$-endomorphism of $\mathcal{O}Gi$. An easy calculation shows that $\mathrm{Tr}_{G\times RH}^{G\times QH}(\varphi)$ is equal to right multiplication by i, hence the identity on $\mathcal{O}Gi$. By Higman's criterion 2.6.2 , this shows that $\mathcal{O}Gi$ is relatively $G \times RH$-projective. Thus there is an indecomposable direct summand X of $\mathcal{O}Gi$ as $\mathcal{O}(G \times RH)$-module such that $\mathcal{O}Gi$ is isomorphic to a direct summand of $\mathrm{Ind}_{G\times RH}^{G\times QH}(X)$. By 5.12.7, the direct summand X is of the form $\mathcal{O}Gj$ for some primitive idempotent $j \in (i\mathcal{O}Gi)^{RH}$. Since $G \times RH$ is normal of p-power index in $G \times QR$, the $\mathcal{O}(G \times QH)$-module $\mathrm{Ind}_{G\times RH}^{G\times QH}(\mathcal{O}Gj)$ is indecomposable, hence isomorphic to $\mathcal{O}Gi$. In particular, we get that

$$\mathrm{End}_{\mathcal{O}(G\times 1)}(\mathrm{Ind}_{G\times RH}^{G\times QH}(\mathcal{O}Gj)) \cong \mathrm{End}_{\mathcal{O}(G\times 1)}(\mathcal{O}Gi) \cong (i\mathcal{O}Gi)^{\mathrm{op}}.$$

This is an isomorphism of interior QH-algebras. The projection π of $\mathrm{Ind}_{G\times RH}^{G\times QH}(\mathcal{O}Gj)$ onto the summand $1\otimes\mathcal{O}Gj$ is an idempotent in $\mathrm{End}_{\mathcal{O}(G\times RH)}(\mathrm{Ind}_{G\times RH}^{G\times QH}(\mathcal{O}Gj))$ whose different Q-conjugates are pairwise orthogonal and whose trace $\mathrm{Tr}_Q^R(\pi)$ is the identity on $\mathrm{Ind}_{G\times RH}^{G\times QH}(\mathcal{O}Gj)$ Thus the image of π in $(i\mathcal{O}Gi)^{op}$ yields an idempotent in $(i\mathcal{O}Gi)^{RH}$ with the required properties. This proves (i). If $\mathrm{Br}_R(j)=0$ then $j\in\ker(\mathrm{Br}_R)\cap(\mathcal{O}G)^{RH}$, and since H is a p'-group, we have $(\mathcal{O}G)^{RH}=(\mathcal{O}G)_R^{RH}$. By 5.4.6, we have $\ker(\mathrm{Br}_R)\cap(\mathcal{O}G)^{RH}=\sum_{S<R}(\mathcal{O}G)_{SH}^{RH}+J(\mathcal{O})(\mathcal{O}G)^{RH}$, where S runs over the proper subgroups of R. Rosenberg's Lemma 4.4.8, implies that there is a proper subgroup S of R such that $j\in(\mathcal{O}G)_{SH}^{RH}$. Applying (i) to R, S and j instead of Q, R and i, respectively, shows that R is not minimal with the property as stated. This shows (ii). Since Q and H commute, we can view $i(\mathcal{O}G)^H i$ as an Q-algebra. Since i is primitive in $(\mathcal{O}G)^{QH}$ the algebra $i(\mathcal{O}G)^H i$ is in fact a primitive Q-algebra. Statement (iii) is then a particular case of the characterisation of defect groups in 5.6.7. □

We need the following variation of 5.12.7.

Lemma 5.12.9 *Let G be a finite group, N a normal subgroup of G and P a subgroup of G. Let i be an idempotent in $(\mathcal{O}N)^P$. Then $H=(N\times 1)\Delta P$ is a subgroup of $G\times P$, and the following hold.*

(i) The action of $G\times P$ on $\mathcal{O}Gi$ restricts to an action of H on $\mathcal{O}Ni$.
(ii) We have an $\mathcal{O}$-algebra isomorphism $(i\mathcal{O}Ni)^P\cong\mathrm{End}_{\mathcal{O}H}(\mathcal{O}Ni)^{\mathrm{op}}$. In particular, $\mathcal{O}Ni$ is indecomposable as an $\mathcal{O}H$-module if and only if i is primitive in $(\mathcal{O}N)^P$.
(iii) We have an isomorphism of $\mathcal{O}(G\times P)$-modules $\mathrm{Ind}_H^{G\times P}(\mathcal{O}Ni)\cong\mathcal{O}Gi$.

Proof Since N is normal in G, it follows that H is a subgroup of $G\times P$. The module structure of $\mathcal{O}H$ on $\mathcal{O}Ni$ is obtained by the left action of N on $\mathcal{O}Ni$ and the conjugation action on $\mathcal{O}Ni$; this makes sense as P normalises N and fixes i. This proves (i). Any $\mathcal{O}N$-endomorphism of $\mathcal{O}Ni$ is given by right multiplication with an element $c\in i\mathcal{O}Ni$. Such an endomorphism commutes with the action of ΔP if and only if $c\in(i\mathcal{O}Ni)^P$, whence (ii). For (iii) one verifies that the map $(x,u)\otimes w\mapsto xwu^{-1}$ is an isomorphism as stated, where $x\in G$, $u\in P$, and $w\in\mathcal{O}Ni$. That is, one checks that this map is $\mathcal{O}H$-balanced, surjective, and that it is an $\mathcal{O}(G\times P)$-module homomorphism. Since both sides have the same $\mathcal{O}$-rank, the result follows. □

Proposition 5.12.10 *Let G be a finite group and N a normal subgroup of G such that G/N is a p-group. Suppose that k is a splitting field for all subgroups* *of G. Let P be a p-subgroup of G. Any primitive idempotent in $(\mathcal{O}N)^P$ remains primitive in $(\mathcal{O}G)^P$.*

Proof Let i be a primitive idempotent in $(\mathcal{O}N)^P$. By 5.12.9 (i), $\mathcal{O}Ni$ is indecomposable as a module over $\mathcal{O}(N \times 1)\Delta P$, and in fact absolutely indecomposable by the hypothesis on k. Since $N \times 1$ is normal of p-power index in $G \times P$, it follows that $(N \times 1)\Delta P$ is subnormal of p-power index in $G \times P$. By 5.12.9 (iii), we have $\mathcal{O}Gi \cong \mathrm{Ind}_{(N\times 1)\Delta P}^{G\times P}(\mathcal{O}Ni)$, and this module is indecomposable, by Green's Indecomposability Theorem. Thus i is primitive in $(\mathcal{O}G)^P$ by 5.12.7 (iii). □

The above proposition is a special case of what is called a *semicovering* in [52]. We will need this in the following form in the context of extensions of nilpotent blocks.

Corollary 5.12.11 *Let G be a finite group and N a normal subgroup of G such that G/N is a p-group. Suppose that k is a splitting field for all subgroups of G. Let P be a p-subgroup of G and i an idempotent in $(\mathcal{O}N)^P$. Let Q be a subgroup of P. Any primitive idempotent in $(i\mathcal{O}Ni)^Q$ remains primitive in $(i\mathcal{O}Gi)^Q$.*

Proof Let j be a primitive idempotent in $(i\mathcal{O}Ni)^Q$. Then j is primitive in $(\mathcal{O}N)^Q$, hence in $(\mathcal{O}G)^Q$ by 5.12.10, and thus also in $(i\mathcal{O}Gi)^Q$. □

Proposition 5.12.12 *Let P be a finite p-group, set $\Delta P = \{(u, u) | u \in P\}$, and let W be an $\mathcal{O}\Delta P$-module such that for every subgroup R of P every indecomposable direct summand of $\mathrm{Res}_{\Delta R}^{\Delta P}(W)$ is absolutely indecomposable. Then the following are equivalent.*

(i) The $\mathcal{O}(P \times P)$-module $\mathcal{O}P$ is isomorphic to a direct summand of $\mathrm{Ind}_{\Delta P}^{P\times P}(W)$.
(ii) The trivial $\mathcal{O}\Delta P$-module $\mathcal{O}$ is isomorphic to a direct summand of W.

Proof Suppose that (ii) holds. Since $\mathrm{Ind}_{\Delta P}^{P\times P}$ is an exact $\mathcal{O}$-linear functor, it follows that $\mathrm{Ind}_{\Delta P}^{P\times P}(\mathcal{O})$ is isomorphic to a direct summand of $\mathrm{Ind}_{\Delta P}^{P\times P}(W)$. By 2.4.5 we have $\mathrm{Ind}_{\Delta P}^{P\times P}(\mathcal{O}) \cong \mathcal{O}P$, and hence (ii) implies (i). Suppose conversely that (i) holds. Let $\mathcal{R}$ be a subset of P such that $1 \times \mathcal{R}$ is a set of representatives of the ΔP-ΔP-double cosets in $P \times P$. Mackey's formula yields

$$\mathrm{Res}_{\Delta P}^{P\times P}\mathrm{Ind}_{\Delta P}^{P\times P}(W) \cong \oplus_{y\in\mathcal{R}} \mathrm{Ind}_{\Delta P\cap^{(1,y)}\Delta P}^{\Delta P}\mathrm{Res}_{\Delta P\cap^{(1,y)}\Delta P}^{(1,y)\Delta P}({}^{(1,y)}W).$$

Since $\mathcal{O}$ is a direct summand of $\mathrm{Res}_{\Delta P}^{P\times P}(\mathcal{O}P)$, it follows that $\mathcal{O}$ is isomorphic to a direct summand of one of the summands on the right side. If y does not centralise P, then $\Delta P \cap {}^{(1,y)}\Delta P$ is a proper subgroup of ΔP, and hence, by Green's Indecomposability Theorem 5.12.3, every indecomposable direct summand of the summand indexed by such an element y has a rank divisible by p. Thus $\mathcal{O}$ is isomorphic to a direct summand on the right side indexed by some element $y \in Z(P)$. The summands on the right side indexed by elements $y \in Z(P)$ are all isomorphic to W, and thus $\mathcal{O}$ is isomorphic to a direct summand of W. This shows that (i) implies (ii). □

The fact that a character induced from a normal subgroup vanishes outside that normal subgroup by 3.1.24, together with Green's Indecomposability Theorem yields the following vanishing criterion for characters of modules. By 1.15.6 any element x in a finite group G can be uniquely written in the form $x = us$ such that u is a p-element, s is a p'-element and such that $us = su$.

Theorem 5.12.13 *Let G be a finite group and U an indecomposable $\mathcal{O}G$-module that is free of finite rank as an $\mathcal{O}$-module. Let Q be a vertex of U. Denote by $\chi : G \to \mathcal{O}$ the character of U.*

(i) For any element $x \in G$ whose p-part is not conjugate to an element in Q we have $\chi(x) = 0$.
(ii) The $\mathcal{O}$-rank of U is divisible by the p-part of the index $|G : Q|$.

Proof If $\mathcal{O}'$ is an extension of $\mathcal{O}$ as a complete local Noetherian ring, then every indecomposable summand of the $\mathcal{O}'G$-module $\mathcal{O}' \otimes_{\mathcal{O}} U$ has a vertex contained in Q. We may therefore assume that the indecomposable modules in the proof below are absolutely indecomposable. Write $x = us$, where u is the p-part of x and s is the p'-part of x. Set $H = \langle x \rangle$. Let W be an indecomposable direct summand of $\mathrm{Res}^G_H(U)$. Since U is a summand of $\mathrm{Ind}^G_Q(V)$ for some $\mathcal{O}Q$-module V, Mackey's formula implies that W is a summand of a module of the form $\mathrm{Ind}^H_{H \cap yQy^{-1}}(X)$ for some $y \in G$ and some indecomposable $\mathcal{O}(H \cap yQy^{-1})$-module X. By the assumptions $H \cap yQy^{-1}$ is a proper subgroup of $\langle u \rangle$. Set $L = \langle s \rangle (H \cap yQy^{-1})$. Then L is normal of p-power index in H and W is a summand of $\mathrm{Ind}^H_L(Y)$ for some indecomposable $\mathcal{O}L$-module Y. Green's Indecomposability Theorem implies that actually $W \cong \mathrm{Ind}^H_L(Y)$. Since us is not contained in L, the character of W vanishes at us by 3.1.24. This holds for all indecomposable direct summands of $\mathrm{Res}^G_H(U)$, and hence the character of U vanishes at us as claimed in (i). Let S be a Sylow p-subgroup of G containing Q. By Mackey's formula, $\mathrm{Res}^G_S \mathrm{Ind}^G_Q(V)$ is a direct sum of modules of the form $\mathrm{Ind}^S_{S \cap {}^xQ}({}^xV)$, with x running over a suitable subset of G. If T is an indecomposable direct summand of ${}^xV$ as an $\mathcal{O}(S \cap {}^xQ)$-module then $\mathrm{Ind}^S_{S \cap {}^xQ}(T)$ is indecomposable, by Green's Indecomposability Theorem 5.12.3; in particular, its $\mathcal{O}$-rank is divisible by $|S : Q|$. The Krull–Schmidt Theorem 4.6.7 implies that $\mathrm{Res}^G_S(U)$ is a direct sum of modules of the form $\mathrm{Ind}^S_{S \cap {}^xQ}(T_x)$, with T_x an

$\mathcal{O}(S \cap {}^xQ)$-module and x running over a subset of G. Thus the $\mathcal{O}$-rank of U is divisible by $|S : Q|$, whence the result. □

Corollary 5.12.14 *Let G be a finite group and U an indecomposable projective $\mathcal{O}G$-module. Denote by $\chi : G \to \mathcal{O}$ the character of U. If x is an element in G whose order is divisible by p then $\chi(x) = 0$.*

Proof Since U is indecomposable projective its vertex is the trivial group and hence the statement follows from 5.12.13. □

Combining Theorem 5.12.13 and Proposition 5.2.3 yields the following result.

Theorem 5.12.15 *Let G be a finite group and U a finitely generated $\mathcal{O}$-free indecomposable $\mathcal{O}G$-module with vertex Q. Set $N = N_G(Q)$. Let V be an indecomposable $\mathcal{O}N$-module with vertex Q that is a Green correspondent of U. Let Y be an $\mathcal{O}Q$-source of V, and suppose that Y is absolutely indecomposable. Let T be the stabiliser of Y in N, and let W be the summand of $\mathrm{Res}^N_T(V)$ with the property that Y is isomorphic to a summand of $\mathrm{Res}^T_Q(W)$. Denote by M the multiplicity module of U with respect to the source Y. The following are equivalent.*

(i) We have $\mathrm{rk}_{\mathcal{O}}(U)_p = |G : Q|$.
(ii) We have $\mathrm{rk}_{\mathcal{O}}(V)_p = |N : Q|$.
(iii) We have $\mathrm{rk}_{\mathcal{O}}(W)_p = |T : Q|$.
(iv) $\mathrm{rk}_{\mathcal{O}}(Y)$ is prime to p, and $\dim_k(M)_p = |T : Q|_p$.

Proof The hypothesis that Y is absolutely indecomposable ensures that the multiplicity module of U with respect to Y is defined. By the Green correspondence, we have $\mathrm{Ind}^G_N(V) \cong U \oplus U'$, and every indecomposable direct summand of U' has a vertex strictly smaller than Q, and hence $\mathrm{rk}_{\mathcal{O}}(U')$ is divisible by $p|G : N|_p$ by 5.12.13. Since $\mathrm{rk}_{\mathcal{O}}(\mathrm{Ind}^G_N(V)) = |G : N| \cdot \mathrm{rk}_{\mathcal{O}}(V)$, it follows that (i) and (ii) are equivalent. By 5.2.3 we have $V \cong \mathrm{Ind}^N_T(W)$, which implies the equivalence of (ii) and (iii). Again by 5.2.3, we have $\mathrm{rk}_{\mathcal{O}}(W) = m \cdot \mathrm{rk}_{\mathcal{O}}(Y)$, and m is the multiplicity of Y as a direct summand in a decomposition of $\mathrm{Res}^T_Q(W)$. By 5.7.3, we have $m = \dim_k(M)$. Since M is a projective module over a twisted group algebra $k_\alpha T/Q$, for some $\alpha \in H^2(T/Q; k^\times)$, it follows that $|T : Q|_p$ divides m. The equivalence between (iii) and (iv) follows. □

Theorem 5.12.16 ([15], 2.2) *Let G be a finite group, u a p-element in G and H a p'-subgroup of $C_G(u)$. Set $Q = \langle u \rangle$. Let i, j be idempotents in $(\mathcal{O}G)^{QH}$ such that $\mathrm{Br}_Q(i) = \mathrm{Br}_Q(j)$. Then, for any character χ of a finitely generated*

$\mathcal{O}$-free $\mathcal{O}G$-module U and any $s \in H$ we have $\chi(usi) = \chi(usj)$. In particular, if $i \in \ker(\mathrm{Br}_Q)$, then $\chi(usi) = 0$.

Proof View iU as $\mathcal{O}QH$-module. The value $\chi(usi)$ is then the character of the $\mathcal{O}QH$-module iU evaluated at us. We prove first the second statement; that is, we assume that $\mathrm{Br}_Q(i) = 0$. We may assume that i is primitive in $(\mathcal{O}G)^{QH}$. It follows from 5.12.8 (ii) that there is a proper subgroup subgroup R of Q such that $i \in (\mathcal{O}G)^{QH}_{RH}$. Thus, as an $\mathcal{O}QH$-module, iU is relatively $\mathcal{O}R$-projective, by the corollaries to Higman's criterion 2.6.2. Since $u \notin R$, the Vanishing Theorem 5.12.13 implies that $\chi(usi) = 0$. For the first statement, we may therefore assume that primitive decompositions of i, j in $(\mathcal{O}G)^{QH}$ contain no elements in $\ker(\mathrm{Br}_Q)$. Since H is a p'-group, the Brauer homomorphism induces a surjective map $(\mathcal{O}G)^{QH} \to kC_G(Q)^H$. Thus Br_Q maps primitive decompositions of i, j bijectively onto primitive decompositions of $\mathrm{Br}_Q(i) = \mathrm{Br}_Q(j)$ in $kC_G(Q)^H$. These are conjugate in $kC_G(Q)^H$ by the Krull–Schmidt Theorem 4.6.2. Since any invertible element in $kC_G(Q)^H$ lifts to an invertible element in $(\mathcal{O}G)^{QH}$, we may assume that i and j have primitive decompositions whose images in $kC_G(Q)^H$ are equal. Thus we may assume that i, j are primitive. But then i, j are conjugate, whence the result. □

Proposition 5.12.17 ([53, Proposition 1.1]) *Let G be a finite group, N a normal subgroup and suppose that K is a splitting field of characteristic zero for all subgroups of G. Denote by Λ a set of representatives of the G-orbits in $\mathrm{Irr}_K(N)$ with respect to the conjugation action by G. Let U be a finitely generated $\mathcal{O}G$-module that is free as an $\mathcal{O}$-module. For any $\lambda \in \Lambda$ there is a unique $\mathcal{O}$-pure $\mathcal{O}G$-submodule U_λ of U with the following properties.*

(i) Denoting by χ_λ the character of U_λ, all irreducible constituents of $\mathrm{Res}^G_N(\chi_\lambda)$ are G-conjugates of λ.
(ii) For any different λ, μ in Λ we have $\langle \chi_\lambda, \chi_\mu \rangle = 0$.
(iii) The character χ_U of U is equal to the character $\sum_{\lambda\in\Lambda} \chi_\lambda$ of the submodule $\oplus_{\lambda\in\Lambda} U_\lambda$ of U.

Moreover, if U is relatively $\mathcal{O}N$-projective, then U_λ is relatively $\mathcal{O}N$-projective, for any λ in Λ.

Proof We decompose $\mathrm{Res}^G_N(K \otimes_\mathcal{O} U) = \oplus_{\lambda\in\Lambda} Y_\lambda$, where Y_λ is the sum of the isotypic components of the simple KN-submodules whose characters are the G-conjugates of λ. This decomposition is unique by 1.9.8, and hence this is a decomposition of $K \otimes_\mathcal{O} U$ as a direct sum of KG-modules. Set $U_\lambda = Y_\lambda \cap U$, with U identified to its canonical image $1 \otimes U$ in $K \otimes_\mathcal{O} U$. By 4.16.4, U_λ is an $\mathcal{O}$-pure submodule of U satisfying (i), (ii), and (iii). The uniqueness of the U_λ follows from that of the Y_λ. Suppose that U is relatively N-projective. By Higman's criterion 2.6.2, this is equivalent to the existence of an $\mathcal{O}N$-endomorphism τ of U satisfying $\mathrm{Tr}_N^G(\tau) = \mathrm{Id}_U$. Extend τ to $K \otimes_{\mathcal{O}} U$. Then τ preserves the sums of isotypic components Y_λ, hence τ induces $\mathcal{O}N$-endomorphisms of U_λ. Since U_λ is an $\mathcal{O}G$-submodule of U we get that $\mathrm{Tr}_N^G(\tau|_{U_\lambda}) = \mathrm{Id}_{U_\lambda}$, whence that U_λ is relatively N-projective as well. □

It follows from 4.13.13 that a finitely generated $\mathcal{O}$-free $\mathcal{O}G$-module U is projective if and only if the kG-module $k \otimes_{\mathcal{O}} U$ is projective. One implication holds for modules with arbitrary vertices: Higman's criterion implies that if U is relatively H-projective, then $k \otimes_{\mathcal{O}} U$ is relatively H-projective. Using Green's Indecomposability Theorem, the following example illustrates that the converse need not hold for modules with nontrivial vertices.

Example 5.12.18 (Feit [29, p. 111]) Let G be a dihedral group of order 16, with generators x and t of order 8 and 2, respectively, such that $txt = x^{-1}$. Let H be the normal dihedral subgroup of order 8 generated by x^2 and t. Let $K = \mathbb{Q}_2(\zeta)$, where $\mathbb{Q}_2$ is the 2-adic completion of $\mathbb{Q}$ and ζ is a primitive 8-th root of unity. Denote by $\mathcal{O}$ the ring of algebraic integers in K. Define an action of G on $V = \mathcal{O}^2$ by setting

$$\rho(x) = \begin{pmatrix} \zeta & 1 \\ 0 & \zeta^{-1} \end{pmatrix}, \quad \rho(t) = \begin{pmatrix} 1 & 0 \\ \zeta^{-1} - \zeta & -1 \end{pmatrix}.$$

One verifies that this extends to a group homomorphism $\rho : G \to \mathrm{GL}(V)$. The description of the characters of dihedral groups implies that the character of $\mathrm{Res}_H^G(V)$ remains absolutely irreducible; that is, $\mathrm{End}_{\mathcal{O}H}(\mathrm{Res}_H^G(V)) \cong \mathcal{O}$. Green's Indecomposability Theorem implies that $\mathrm{Ind}_H^G(\mathrm{Res}_H^G(V))$ is indecomposable, hence cannot have V as a direct summand. This shows that V is not relatively H-projective. By contrast, the reduction $\bar{V}$ of V modulo the ideal generated by $\pi = 1 - \zeta$ has the property that t acts as identity on $\bar{V}$ and x acts as

$$\begin{pmatrix} 1 & 1 \\ 0 & 1 \end{pmatrix}.$$

This implies that $\bar{V} \cong \mathrm{Ind}_H^G(W)$, where W is the subspace of $\bar{V}$ fixed by both x and t. In particular, $\bar{V}$ is relatively H-projective.

The fact observed in 5.12.8 that under certain circumstances an idempotent can be written as a trace of an idempotent whose conjugates are pairwise orthogonal holds in much greater generality, and can be used to prove a generalisation of Green's Indecomposability Theorem, due to Puig [70]. We start with a result due to Thévenaz.

Proposition 5.12.19 (cf. [88, (22.1)]) *Let G be a finite group, A a G-algebra over $\mathcal{O}$ that is finitely generated as an $\mathcal{O}$-module, and let I be a G-stable ideal in A contained in $J(A)$. Let f be an idempotent in A/I such that $1_{A/I} = \mathrm{Tr}_1^G(f)$ and such that ${}^x f \cdot f = 0$ for $x \in G \setminus \{1\}$. Then there exists an idempotent e in A whose image in A/I is equal to f such that $1_A = \mathrm{Tr}_1^G(e)$ and such that ${}^x e \cdot e = 0$ for $x \in G \setminus \{1\}$.*

Proof Since I is contained in $J(A)$ there is an idempotent e' whose image in A/I is equal to f. We have $1 - \mathrm{Tr}_1^G(e') \in I \cap A^G \subseteq J(A) \cap A^G \subseteq J(A^G)$. Thus $u = \mathrm{Tr}_1^G(e')$ is invertible in A^G, and we have $1 = u^{-1}\mathrm{Tr}_1^G(e') = \mathrm{Tr}_1^G(u^{-1}e')$. It follows that $A = \sum_{x\in G} Au^{-1x}e' = \sum_{x\in G} A^x e'$. Since the idempotents ${}^x f$ are pairwise orthogonal in A/I, we have $A/I = \oplus_{x\in G} A/I^x f$. A projective module is determined up to isomorphism by its semisimple quotient, and hence $A \cong \oplus_{x\in G} A^x e'$. It follows that the previous sum $A = \sum_{x\in G} A^x e'$ is a direct sum. Thus we get that $A = \oplus_{x\in G} A^x e'$, and the summands in this decomposition are permuted by G. It follows from 1.7.5 that there is a unique (and hence G-stable) set of pairwise orthogonal idempotents e_x such that $A^x e' = Ae_x$, for $x \in G$. This uniqueness implies that $e = e_1$ lifts f and that ${}^x e_1 = e_x$ is orthogonal to e_1 for $x \in G \setminus \{1\}$. The result follows. □

Theorem 5.12.20 ([70]) *Let P be a finite p-group and A a P-algebra over $\mathcal{O}$ that is finitely generated as an $\mathcal{O}$-module. Suppose that k is large enough. Let Q be a subgroup of P. Let i be a primitive idempotent in A^P such that $i \in A_Q^P$. Then there exists a primitive idempotent j in $iA^Q i$ such that $i = \mathrm{Tr}_Q^P(j)$ and such that ${}^x j \cdot j = 0$ for $x \in P \setminus Q$.*

Proof We argue by induction over $|P : Q|$. We may replace A by iAi, or equivalently, we may assume that $i = 1$, which means that A is a primitive P-algebra. If $P = Q$, then there is nothing to prove. Suppose that Q is a proper subgroup of P. Let R be a maximal, hence normal, subgroup of P containing Q. If $R > Q$, then by induction we have $1 = \mathrm{Tr}_R^P(j')$ for some primitive idempotent $j' \in A^R$ such that the different P-conjugates of j' are pairwise orthogonal, and we also have $j' = \mathrm{Tr}_Q^R(j)$ for some primitive idempotent $j \in j'A^R j'$ whose different R-conjugates are pairwise orthogonal. But then $1 = \mathrm{Tr}_Q^P(j)$, and the different P-conjugates of j are pairwise orthogonal. Thus we may assume that Q is a maximal, and hence normal subgroup of P. Then A^Q becomes a P-algebra on which Q acts trivially. Thus we may replace P and Q by P/Q and $\{1\}$, respectively. Equivalently, we may assume that $Q = \{1\}$. It follows from 5.12.19 that we may replace A by its semisimple quotient $A/J(A)$, or equivalently, we may assume that A is a direct product $A = \prod_{t=1}^n A_t$ of matrix algebras A_t. The action of P permutes these factors. Denote by e_t the unit element of A_t. The idempotents

e_t are pairwise orthogonal, and they are permuted by P. The P-orbit sum of any one of the e_t yields an idempotent in A^P. Since 1 is the only idempotent in A^P this implies that P permutes the matrix factors A_t transitively. Let S be the stabiliser in P of $A_1 = Ae_1$. Then $1 = \mathrm{Tr}_S^P(e_1)$, and the different P-conjugates of e_1 are pairwise orthogonal. Thus we may replace P by S and 1 by e_1. Equivalently, we may assume that A is a matrix algebra on which P acts such that $1 \in A_1^P$. Writing $A = \mathrm{End}_k(V)$ for some vector space V, it follows from 5.3.1 that the action of P on A lifts to a kP-module structure on V. Since $1 \in A_1^P$, or equivalently, $\mathrm{Id}_V \in (\mathrm{End}_k(V))_1^P$, it follows from Higman's criterion 2.6.2 that V is a projective kP-module. It is also indecomposable, as A is a primitive P-algebra. Thus $V \cong kP$ as a kP-module. Denote by j the linear endomorphism of kP that sends 1 to 1 and $y \in P \setminus 1$ to 0. That is, j is a linear projection onto a 1-dimensional subspace of kP. Thus j is a primitive idempotent in $\mathrm{End}_k(kP)$. An easy verification shows that the P-conjugates of j are the canonical linear projection onto the 1-dimensional spaces ky, where $y \in P$, and hence they are pairwise orthogonal and satisfy $\mathrm{Tr}_1^P(j) = \mathrm{Id}_{kP}$ as required. □

Corollary 5.12.21 *Let P be a finite p-group and A a P-algebra over $\mathcal{O}$ that is finitely generated as an $\mathcal{O}$-module. Suppose that k is large enough. Let γ be a point of P on A, and let Q_δ be a defect local pointed group of P_γ. For every $i \in \gamma$ there is $j \in \delta$ such that $i = \mathrm{Tr}_Q^P(j)$ and such that the different P-conjugates of j are pairwise orthogonal.*

Proof Let $i \in \gamma$. By the assumptions we have $i \in A_Q^P$. By 5.12.20 there is a primitive idempotent $j \in A^Q$ such that $i = \mathrm{Tr}_Q^P(j)$ and such that the different P-conjugates of j are pairwise orthogonal. The minimality of Q subject to $i \in A_Q^P$ implies that j belongs to a local point ϵ of Q on A. Since the defect pointed groups of P_γ are all P-conjugate, it follows that we may assume that $\delta = \epsilon$, whence the result. □

Corollary 5.12.22 *Let G be a finite group, N a normal subgroup of p-power index and H a subgroup of G containing N. Let A be a G-algebra over $\mathcal{O}$ that is finitely generated as an $\mathcal{O}$-module. Let b be a primitive idempotent in A^G that is contained in A_H^G. Then there is a primitive idempotent c in A^H such that $\mathrm{Tr}_H^G(c) = b$ and such that ${}^x c \cdot c = 0$ for $x \in G \setminus H$.*

Proof Setting $P = G/N$, this follows from applying 5.12.20 to the P-algebra A^H and the subgroup $Q = H/N$ of P. □

Corollary 5.12.23 *Let G be a finite group and N a normal subgroup of p-power index. Let A be a G-algebra over $\mathcal{O}$ that is finitely generated as an $\mathcal{O}$-module. Let c be a primitive idempotent in A^N such that the different*

G-conjugates of c are pairwise orthogonal. Let H be the stabiliser of c in G. Then $\mathrm{Tr}_H^G(c)$ is primitive in A^G.

Proof Since H stabilises c, we have $c \in A^H$, and since A^H is a subalgebra of A^N, the idempotent c is primitive in A^H. Thus we may assume that H is a proper subgroup of G. Let M be a maximal subgroup of G containing H. Since G/N is a p-group it follows that M is normal of index p in G. Arguing by induction over $|G : H|$, we may assume that $d = \mathrm{Tr}_H^M(c)$ is primitive in A^M. We claim that M is the stabiliser of d. Indeed, let $x \in G$ such that ${}^x d = d$. Then the action of x permutes the M-conjugates of c. Thus there is $y \in M$ such that ${}^x d = {}^y d$. Then $y^{-1}x$ stabilises c, hence $y^{-1}x \in H$. Since $y \in M$ it follows that $x \in M$. The different G-conjugates of d are pairwise orthogonal. Thus we may replace H by M; equivalently, we may assume that H is a normal subgroup of index p in G. Let i be a primitive idempotent in $bA^G b$. Then $i = ib \in A_H^G$, hence by 5.12.22 we have $i = \mathrm{Tr}_H^G(j)$ for some primitive idempotent $j \in iA^H i$ such that ${}^x j \cdot j = 0$ for $x \in G \setminus H$. Since H is normal in G it follows that the G-conjugates of c and of j are primitive idempotents in A^H. Thus b and i have primitive decompositions in A^H of the same cardinality $|G : H|$. Since $bi = i = ib$ this forces $i = b$. □

Green's Indecomposability Theorem 5.12.3 is the special case of 5.12.23 applied to the interior G-algebra $\mathrm{End}_{\mathcal{O}}(\mathrm{Ind}_N^G(V))$ and the projection of $\mathrm{Res}_N^G(\mathrm{Ind}_H^G(V))$ onto $1 \otimes V$ instead of A and c, respectively.

5.13 Brauer characters

Suppose that $\mathrm{char}(K) = 0$ and that k is perfect. The purpose of this section is to reformulate the results on Grothendieck groups in the sections 4.16 and 4.17 in terms of character groups. Let G be a finite group. We denote by $R_K(G) = R(KG)$ and $R_k(G) = R(kG)$ the Grothendieck groups of finitely generated modules over KG and kG, respectively. The first orthogonality relations imply that $R_K(G)$ can be identified with $\mathbb{Z}\mathrm{Irr}_K(G)$. Characters over k do not carry enough information for an analogous statement involving $R_k(G)$; for instance, if G is a finite group whose order is divisible by p, then the character of the regular kG-module is zero. To circumvent this issue, we observe that if H is a p'-subgroup of G and M a finitely generated kG-module, then $\mathrm{Res}_H^G(M)$ is a projective kH-module. Thus, by 4.5.10, there is, up to isomorphism, a unique projective $\mathcal{O}H$-module U such that $k \otimes_{\mathcal{O}} U \cong \mathrm{Res}_H^G(M)$ as kH-modules. Therefore, instead of considering the trace of an element $x \in H$ on M, we consider the trace of x on U, which is an element in $\mathcal{O}$, and uniquely determined by M. Applying this reasoning to the identity element shows in particular that now the dimension of M can be recovered as the trace of 1_G on U. Any p'-element x of G generates a p'-subgroup $\langle x\rangle$, and so we may use this idea to associate an element in $\mathcal{O}$ with every p'-element in G. We denote by $G_{p'}$ the set of all elements of G of order prime to p. We have $1_G \in G_{p'}$ and $G_{p'}$ is the union of the conjugacy classes of p'-elements in G.

Definition 5.13.1 Let G be a finite group and M a finite-dimensional kG-module. The *Brauer character of M* is the map $\varphi_M : G_{p'} \to \mathcal{O}$ defined as follows. Let $x \in G_{p'}$ and let U_x be a projective $\mathcal{O}\langle x\rangle$-module such that $k \otimes_{\mathcal{O}} U_x \cong \mathrm{Res}^G_{\langle x\rangle}(M)$. We set $\varphi_M(x) = \chi_{U_x}(x)$, where χ_{U_x} is the character of U_x. A map $\varphi : G_{p'} \to \mathcal{O}$ is called a *Brauer character of kG* if there exists a finite-dimensional kG-module M such that $\varphi = \varphi_M$.

A trivial verification shows that isomorphic kG-modules have the same Brauer characters. Alternatively, one can define Brauer characters by observing that a p'-element x acts on a kG-module M with p'-roots of unity as eigenvalues. These lift uniquely to $\mathcal{O}$ by 4.3.7, and their sum is then defined to be the value of the Brauer character of M at x. By Remark 4.7.4, lifting p'-roots of unity is essentially equivalent to lifting projective modules over cyclic p'-groups, and so one easily sees that these definitions coincides. As in the case of ordinary characters we use simple modules to define the notion of irreducible Brauer characters:

Definition 5.13.2 Let G be a finite group. A Brauer character φ of kG is called *irreducible* if there is a simple kG-module S such that $\varphi = \varphi_S$. We denote by $\mathrm{IBr}_k(G)$ the set of irreducible Brauer characters of kG, and by $\mathbb{Z}\mathrm{IBr}_k(G)$ the group of $\mathcal{O}$-valued functions on $G_{p'}$ generated by $\mathrm{IBr}_k(G)$.

Remark 5.13.3 Since p'-roots of unity in $\mathcal{O}$ are contained in the ring of Witt vectors W over k, the definition of Brauer characters does not depend on the choice of the ring $\mathcal{O}$ with residue field k, justifying the notation $\mathrm{IBr}_k(G)$ rather than $\mathrm{IBr}_{\mathcal{O}}(G)$. The uniqueness properties of W imply that any field automorphism of k lifts uniquely to a ring automorphism of W, and that any ring automorphism of W sends a p'-root of unity ζ in W to ζ^q, where q is some power of p depending on the order of ζ. The field K has, in general, automorphisms that do not have this property, and hence the composition of a Brauer character with an automorphism of K need not be the Brauer character of a module.

If K and k are both splitting fields for a finite group G, then KG and $kG/J(kG)$ are direct products of matrix algebras over K and k, respectively. Thus the algebra $A = \mathcal{O}G$ satisfies the hypotheses needed for the results in the sections 4.10, 4.16 and 4.17. Replacing the set of isomorphism classes of simple KG-modules by $\mathrm{Irr}_K(G)$ and the set of isomorphism classes of simple kG-modules by $\mathrm{IBr}_k(G)$ leads to statements in terms of characters and Brauer characters that are equivalent to the results in those sections. Since we have some extra information about the number of isomorphism classes of simple KG-modules and of simple kG-modules in terms of numbers of conjugacy classes (of p'-elements), these reformulations will show in particular that the Cartan matrix of kG is non singular. We start with some observations on Brauer characters that are inherited from ordinary characters:

Proposition 5.13.4 *Let G be a finite group and M a finite-dimensional kG-module.*

(i) $\varphi_M(1_G) = \dim_k(M)$.
(ii) $\varphi_M(yxy^{-1}) = \varphi_M(x)$ *for all* $x \in G_{p'}$ *and* $y \in G$.

Proof Statement (i) follows from 3.1.3 (i). For (ii), let U and V be projective modules over $\mathcal{O}\langle x\rangle$, $\mathcal{O}\langle yxy^{-1}\rangle$ such that $k \otimes_{\mathcal{O}} U \cong \mathrm{Res}^G_{\langle x\rangle}(M)$ and $k \otimes_{\mathcal{O}} V \cong \mathrm{Res}^G_{\langle yxy^{-1}\rangle}(M)$, respectively. Denote by $\varphi : \langle x\rangle \cong \langle yxy^{-1}\rangle$ the group isomorphism sending x to yxy^{-1}. The map sending $m \in M$ to $y^{-1}m$ is an isomorphism $\mathrm{Res}^G_{\langle x\rangle}(M) \cong \mathrm{Res}_\varphi(\mathrm{Res}^G_{\langle yxy^{-1}\rangle}(M))$. Since U, V are projective, this lifts to an isomorphism $U \cong \mathrm{Res}_\varphi(V)$. Through this isomorphism, the trace of x on U corresponds to the trace of $\varphi(x) = yxy^{-1}$ on V, hence both are equal. □

This motivates the following terminology:

Definition 5.13.5 Let G be a finite group. A function $\psi : G_{p'} \to K$ with the property $\psi(x) = \psi(yxy^{-1})$ for all $x \in G_{p'}$ and all $y \in G$ is called a *p-regular class function on* G. The set of all p-regular class functions from G to K is denoted by $\mathrm{Cl}_K(G_{p'})$. We denote by $\mathrm{Res}^G_{G_{p'}} : \mathrm{Cl}_K(G) \to \mathrm{Cl}_K(G_{p'})$ the obvious restriction map.

In other words, $\mathbb{Z}\mathrm{IBr}_k(G)$ is the subgroup of $\mathrm{Cl}_K(G_{p'})$ generated by $\mathrm{IBr}_k(G)$. If a kG-module M can be lifted to an $\mathcal{O}$-free $\mathcal{O}G$-module Y then the Brauer character of M is simply the restriction to $G_{p'}$ of the character of Y. We state this for future reference:

Proposition 5.13.6 *Let G be a finite group and M a finite-dimensional kG-module. Suppose there is an $\mathcal{O}$-free $\mathcal{O}G$-module Y such that $k \otimes_{\mathcal{O}} Y \cong M$. Denote by φ the Brauer character of M and by χ the character of Y. Then* $\varphi = \mathrm{Res}^G_{G_{p'}}(\chi)$.

Proof Trivial. □

Proposition 5.13.7 *Let G be a finite group, $u \in G$ a p-element and set $Q = \langle u\rangle$. Let Y be a finitely generated p-permutation $\mathcal{O}G$-module. Denote by χ the* *character of Y and by φ the Brauer character of the $kC_G(u)$-module $Y(Q)$. For any p'-element s in $C_G(u)$ we have $\chi(us) = \varphi(s)$.*

Proof Write $\mathrm{Res}^G_{C_G(u)}(Y) = Y' \oplus Y''$ as a direct sum of $\mathcal{O}C_G(u)$-modules such that every indecomposable summand of Y' has a vertex containing Q and no indecomposable direct summand of Y'' has a vertex containing Q. By 5.12.13, the character of Y'' vanishes at us, hence $\chi(us)$ is equal to the character of us on Y'. By 5.11.4, u acts trivially on Y', and hence this value is equal to the character of Y' evaluated at s. By 5.13.6, this is further equal to the Brauer character of $k \otimes_{\mathcal{O}} Y'$ evaluated at s. By 5.11.4 again, we have $k \otimes_{\mathcal{O}} Y' \cong Y(Q)$, whence the result. □

As in the case of ordinary characters, the Brauer character of a module depends only on its composition series, hence on its image in the Grothendieck group $R_k(G)$:

Proposition 5.13.8 *Let G be a finite group and let M be a submodule of a finite-dimensional kG-module N. For any $x \in G_{p'}$ we have $\varphi_N(x) = \varphi_M(x) + \varphi_{N/M}(x)$.*

Proof Let $x \in G_{p'}$. Let V and W be projective $\mathcal{O}\langle x\rangle$-modules such that $k \otimes_{\mathcal{O}} V \cong \mathrm{Res}^G_{\langle x\rangle}(N)$ and $k \otimes_{\mathcal{O}} W \cong \mathrm{Res}^G_{\langle x\rangle}(N/M)$, respectively. Since V is projective, the canonical surjective map $\mu : N \to N/M$ lifts to a map $\beta : V \to W$. By Nakayama's Lemma 1.10.4, the map β is surjective. Since W is projective, β is split surjective, and $U = \ker(\beta)$ is a projective $\mathcal{O}\langle x\rangle$-module satisfying $k \otimes_{\mathcal{O}} U \cong \mathrm{Res}^G_{\langle x\rangle}(M)$. By 3.1.9 we get that $\chi_V = \chi_U + \chi_W$, whence the result. □

Corollary 5.13.9 *Let G be a finite group and M a finite-dimensional kG-module. For any simple kG-module S denote by d^M_S the number of composition factors isomorphic to S in a composition series of M. Then $\varphi_M = \sum_S d^M_S \varphi_S$, where S runs over a set of representative of the isomorphism classes of simple kG-modules.*

Proof Arguing by induction over the composition length of M this follows from 5.13.8. □

Corollary 5.13.10 *Let G be a finite group and Y a finitely generated $\mathcal{O}$-free $\mathcal{O}G$-module with character χ. For any simple kG-module S denote by d^Y_S the number of composition factors isomorphic to S in a composition series of $k \otimes_{\mathcal{O}} Y$. Then $\mathrm{Res}^G_{G_{p'}}(\chi) = \sum_S d^Y_S \varphi_S$, where S runs over a set of representative of the isomorphism classes of simple kG-modules.*

Proof This follows from 5.13.6 and 5.13.9. □

Proposition 5.13.11 *Let G be a finite group and M a finite-dimensional kG-module. For any $x \in G_{p'}$ we have $\varphi_{M^*}(x) = \varphi_M(x^{-1})$.*

Proof Let $x \in G_{p'}$ and let U be a projective $\mathcal{O}\langle x\rangle$-module such that $k \otimes_{\mathcal{O}} U \cong \mathrm{Res}^G_{\langle x\rangle}(M)$. Then the $\mathcal{O}$-dual U^* is again a projective $\mathcal{O}\langle x\rangle$-module such that $k \otimes_{\mathcal{O}} U^* \cong \mathrm{Res}^G_{\langle x\rangle}(M^*)$. The result follows from 3.2.2. □

Proposition 5.13.12 *Let G be a finite group and let M, N be finite-dimensional kG-modules. For any $x \in G_{p'}$ we have $\varphi_{M\otimes_k N}(x) = \varphi_M(x)\varphi_N(x)$.*

Proof Let $x \in G_{p'}$. Let U and V be projective $\mathcal{O}\langle x\rangle$-modules such that $k \otimes_{\mathcal{O}} U \cong \mathrm{Res}^G_{\langle x\rangle}(M)$ and $k \otimes_{\mathcal{O}} V \cong \mathrm{Res}^G_{\langle x\rangle}(N)$, respectively. Then $U \otimes_{\mathcal{O}} V$ is a projective $\mathcal{O}\langle x\rangle$-modules such that $k \otimes_{\mathcal{O}} (U \otimes_{\mathcal{O}} V) \cong \mathrm{Res}^G_{\langle x\rangle}(M \otimes_k N)$ and hence the result follows from 3.1.11. □

Combining the above results will show that the irreducible Brauer characters of a finite group G yield a basis of the space of K-valued class functions on $G_{p'}$, provided that k is large enough.

Theorem 5.13.13 *Let G be a finite group. Suppose that K and k are both splitting fields for G.*

(i) The set $\mathrm{IBr}_k(G)$ is a K-basis of the space $\mathcal{L}_K(G_{p'})$.
(ii) The map sending a finitely generated kG-module M to its Brauer character φ_M induces a group isomorphism $R_k(G) \cong \mathbb{Z}\mathrm{IBr}_k(G)$.
(iii) The Brauer characters of two nonisomorphic simple kG-modules are different.

Proof Let ℓ be the number of conjugacy classes of p'-elements in G. By 1.15.3 this is also the number of isomorphism classes of simple kG-modules. Thus the set $\mathrm{IBr}_k(G)$ has at most ℓ different elements. In order to prove (i) it suffices to show that $\mathrm{IBr}_k(G)$ spans $\mathrm{Cl}_K(G_{p'})$ as K-vector space. Since $\mathrm{Irr}_K(G)$ is a K-basis of $\mathrm{Cl}_K(G)$ by 3.3.6 every element in $\mathrm{Cl}_K(G_{p'})$ is the restriction to $G_{p'}$ of some K-linear combination $\sum_{\chi\in\mathrm{Irr}_K(G)} a_\chi \chi$. Thus it suffices to show that for each $\chi \in \mathrm{Irr}_K(G)$ its restriction to $G_{p'}$ lies in the span of $\mathrm{IBr}_k(G)$. For each $\chi \in \mathrm{Irr}_K(G)$ there is, by 4.16.5, an $\mathcal{O}$-free $\mathcal{O}G$-module Y with character χ. By 5.13.10 we have $\mathrm{Res}^G_{G_{p'}}(\chi) = \sum_S d^Y_S \varphi_S$, where S runs over a set of representatives of the isomorphism classes of simple kG-modules and where d^Y_S is the number of composition factors of $k \otimes_{\mathcal{O}} Y$ isomorphic to S in a composition series. This proves (i). The statements (ii) and (iii) are immediate consequences. □

5.14 The decomposition map for group algebras

Suppose that $\mathrm{char}(K) = 0$ and that k is perfect. Let G be a finite group. The decomposition map $d_{\mathcal{O}G} : R_K(G) \to R_k(G)$ defined in 4.17.1 induces, via the canonical isomorphisms $\mathbb{Z}\mathrm{Irr}_K(G) \cong R_K(G)$ and $\mathbb{Z}\mathrm{IBr}_k(G) \cong R_k(G)$ a group homomorphism $d_G : \mathbb{Z}\mathrm{Irr}_K(G) \to \mathbb{Z}\mathrm{IBr}_k(G)$. By 5.13.6, the map d_G sends χ to the restriction $\mathrm{Res}^G_{G_{p'}}(\chi)$ of χ to the set of p'-elements in G. The map d_G is called again *decomposition map of G*. Thanks to the interpretation of d_G as restriction from G to $G_{p'}$, it is only a minor abuse to extend the use of the notation $d_G(\chi) = \mathrm{Res}^G_{G_{p'}}(\chi)$ to arbitrary K-valued class functions $\chi \in \mathrm{Cl}_K(G)$. As a map from $\mathrm{Cl}_K(G)$ to $\mathrm{Cl}_K(G_{p'})$, the decomposition map is trivially surjective, with canonical section sending a class function χ' on $G_{p'}$ to the class function χ on G obtained by extending χ' by zero on all p-singular elements. The following result, due to Brauer, states the remarkable fact that d_G remains surjective as a map between generalised characters and Brauer characters, respectively:

Theorem 5.14.1 *Let G be a finite group. If both K and k are splitting fields for all subgroups of G, then the decomposition map* $d_G : \mathbb{Z}\mathrm{Irr}_K(G) \to \mathbb{Z}\mathrm{IBr}_k(G)$ *is surjective.*

Proof The proof of the surjectivity of d_G uses Brauer's Induction Theorem 3.7.6. We observe first that d_G is surjective if G is a p-group, because in that case the unique simple kG-module lifts to the $\mathcal{O}G$-module $\mathcal{O}$. The surjectivity holds also if G is a p'-group, because in that case any kG-module is projective, hence lifts to an $\mathcal{O}$-free $\mathcal{O}G$-module. Thus d_G is surjective for any nilpotent group, in particular, for any elementary group. Since the decomposition maps for subgroups of G commute with induction by 4.17.4, we get the surjectivity in general from 3.7.6. □

This result remains true if K is not a splitting field; the proof requires a generalisation of Brauer's Induction Theorem taking Galois groups into account; see [85, Ch. 16, 17]. Since $|\mathrm{IBr}(G)|$ is the number of p'-conjugacy classes in G and $|\mathrm{Irr}(G)|$ the number of all conjugacy classes of G, it follows that d_G is an isomorphism if and only if G is a p'-group. By theorem 5.13.13 there are unique integers d^χ_φ satisfying

$$\mathrm{Res}^G_{G_{p'}}(\chi) = \sum_{\varphi \in \mathrm{IBr}_k(G)} d^\chi_\varphi \varphi.$$

The matrix $D = (d^\chi_\varphi)_{\chi,\varphi}$ with $\chi \in \mathrm{Irr}_K(G)$ and $\varphi \in \mathrm{IBr}_k(G)$ is the decomposition matrix of the map d_G with respect to the obvious bases, as defined in 4.16.1 in the slightly more general context of an $\mathcal{O}$-algebra A that is free of finite rank over $\mathcal{O}$ such that $K \otimes_{\mathcal{O}} A$ is split semisimple and $k \otimes_{\mathcal{O}} A$ is split. More precisely, if X is a simple KG-module with character χ and S a simple kG-module with Brauer character φ then $d^{\chi}_{\varphi} = d^{X}_{S}$. The notation on the right-hand side is as in 4.16.1; that is, d^{X}_{S} is the number of composition factors of $k \otimes_{\mathcal{O}} Y$ isomorphic to S for some $\mathcal{O}$-free $\mathcal{O}G$-module Y satisfying $K \otimes_{\mathcal{O}} Y \cong X$. In particular, we have:

Theorem 5.14.2 *Let G be a finite group. Suppose that K and k are splitting fields for G. Denote by C the Cartan matrix of kG and by D the decomposition matrix of $\mathcal{O}G$.*

(i) ${}^tD \cdot D = C$.
(ii) C is symmetric and positive definite.
(iii) $\det(C) > 0$.

Proof Statement (i) is a special case of 4.16.3. Statement (ii) follows from (i) because the rank of D is equal to $|\mathrm{IBr}_k(G)|$ by 5.13.13. (The symmetry of C follows also from the fact that kG is a symmetric algebra.) The last statement is an immediate consequence of (ii). ☐

One can be more precise: the determinant of the Cartan matrix of kG is in fact a power of p; see 6.5.15 below. A simple kG-module determines a unique projective indecomposable $\mathcal{O}G$-module – namely its projective cover viewed as an $\mathcal{O}G$-module – and this in turn determines a character. We use the following notation.

Definition 5.14.3 Let G be a finite group. For any simple kG-module S we denote by Φ_S the character of a projective cover of S as an $\mathcal{O}G$-module and we denote by $\mathrm{IPr}_{\mathcal{O}}(G)$ the set of characters of projective indecomposable $\mathcal{O}G$-modules.

Since every projective indecomposable $\mathcal{O}G$-module has a unique simple quotient it follows that every character in $\mathrm{IPr}_{\mathcal{O}}(G)$ is equal to Φ_S for some simple kG-module S. The interpretation of decomposition numbers in 4.16.2 yields the following statement:

Theorem 5.14.4 *Let G be a finite group. Suppose that K and k are splitting fields for G. Let S be a simple kG-module, set $\varphi = \varphi_S$ and $\Phi = \Phi_S$. Then*

$$\Phi = \sum_{\chi \in \mathrm{Irr}_K(G)} d^{\chi}_{\varphi}\chi$$

and for any $\chi \in \mathrm{Irr}_K(G)$ we have $d^{\chi}_{\varphi} = \langle \Phi, \chi \rangle_G$.

Proof The first equality is a special case of 4.16.2 The second equality follows from the first and the orthogonality relations. □

There is another way to describe the decomposition numbers d_φ^χ in terms of a character value:

Theorem 5.14.5 *Let G be a finite group. Suppose that K and k are splitting fields for G. Let S be a simple kG-module and $\chi \in \mathrm{Irr}_K(G)$. Set $\varphi = \varphi_S$ and let i be a primitive idempotent on $\mathcal{O}G$ such that $\mathcal{O}Gi$ is a projective cover of S. Then $d_\varphi^\chi = \chi(i)$.*

Proof With the notation of 4.16.1 we have $d_\varphi^\chi = d_S^X = \dim_K(iX)$, where X is a simple KG-module with character χ. Since i acts as identity on iX and annihilates the complement $(1-i)X$ we get that $\dim_K(iX) = \chi(i)$. □

Corollary 5.14.6 *Let G be a finite group, i an idempotent in $\mathcal{O}G$ and Φ the character of $\mathcal{O}Gi$. We have*

$$\Phi = \sum_{\chi \in \mathrm{Irr}_K(G)} \chi(i)\chi.$$

Proof If i is primitive the result follows from the two previous theorems; the general case is obtained by considering a primitive decomposition of i. □

We will show that for non isomorphic simple kG-modules S, T, the associated characters Φ_S, Φ_T of their projective covers are different, as a consequence of an orthogonality relation involving the irreducible Brauer characters of G. Since Brauer characters are not defined on p-singular group elements, we cannot directly consider the scalar product between Brauer characters and ordinary characters, which is why we introduce the following notation.

Definition 5.14.7 Let G be a finite group. For any two K-valued functions α, β defined on a subset of G containing $G_{p'}$ we set

$$\langle \alpha, \beta \rangle'_G = \frac{1}{|G|} \sum_{x \in G_{p'}} \alpha(x)\beta(x^{-1}).$$

A class function defined on $G_{p'}$ extends to a class function on G by setting it to be zero on $G \setminus G_{p'}$. Thus the definition 3.1.19 and basic results on the induction and restriction for class functions extend to class functions on $G_{p'}$ and to Brauer characters. Brauer's reciprocity takes the following form in terms of characters and Brauer characters:

Theorem 5.14.8 *Let G be a finite group. Suppose that K and k are splitting fields for G. Let S, T be simple kG-modules. We have $\langle \Phi_S, \varphi_T \rangle'_G = 1$ if $S \cong T$*

and $\langle \Phi_S, \varphi_T \rangle'_G = 0$ *otherwise. In particular, the set* $\mathrm{IPr}_{\mathcal{O}}(G)$ *is a* K*-basis of the subspace of* $\mathrm{Cl}_K(G)$ *of class functions that vanish outside* $G_{p'}$.

The proof uses a technical argument which we separate for future reference:

Lemma 5.14.9 *Let* G *be a finite group. Suppose that* K *and* k *are splitting fields for* G.

(i) There are unique coefficients $a_\varphi^\chi \in K$*, where* $\chi \in \mathrm{Irr}_K(G)$ *and* $\varphi \in \mathrm{IBr}_k(G)$*, such that for any* $\varphi \in \mathrm{IBr}_k(G)$ *the class function*

$$\hat{\varphi} = \sum_{\chi \in \mathrm{Irr}_K(G)} a_\varphi^\chi \chi$$

coincides with φ *on* $G_{p'}$ *and is zero on* $G \setminus G_{p'}$.

(ii) For any two $\varphi, \psi \in \mathrm{IBr}_k(G)$ *we have*

$$\sum_{\chi \in \mathrm{Irr}_K(G)} a_\varphi^\chi d_\psi^\chi = \delta_{\varphi,\psi}.$$

(iii) For any $\varphi \in \mathrm{IBr}_k(G)$ *and any primitive idempotent* j *in* $\mathcal{O}G$ *we have* $\hat{\varphi}(j) = 1$ *if* $\mathcal{O}Gj$ *is a projective cover of a simple* kG*-module with Brauer character* φ*, and* $\hat{\varphi}(j) = 0$ *otherwise.*

Proof The existence and uniqueness of the a_φ^χ as in (i) is a trivial consequence of the fact that $\mathrm{Irr}_K(G)$ is a K-basis of $\mathrm{Cl}_K(G)$. Using the definition of the decomposition numbers d_φ^χ we get that

$$\varphi = \sum_{\chi \in \mathrm{Irr}_K(G)} a_\varphi^\chi \mathrm{Res}^G_{G_{p'}}(\chi) = \sum_{\chi \in \mathrm{Irr}_K(G)} \sum_{\psi \in \mathrm{IBr}_k(G)} a_\varphi^\chi d_\psi^\chi \psi$$
$$= \sum_{\psi \in \mathrm{IBr}_k(G)} \Big(\sum_{\chi \in \mathrm{Irr}_K(G)} a_\varphi^\chi d_\psi^\chi \Big) \psi$$

and now the linear independence of $\mathrm{IBr}_k(G)$ implies statement (ii). Let $\varphi, \psi \in \mathrm{IBr}_k(G)$, and let j be a primitive idempotent in $\mathcal{O}G$ such that $\mathcal{O}Gj$ is a projective cover of a simple kG-module with Brauer character ψ. By 5.14.5 we have $\chi(j) = d_\psi^\chi$, for any $\chi \in \mathrm{Irr}_K(G)$. Thus we have $\hat{\varphi}(j) = \sum_{\chi \in \mathrm{Irr}_K(G)} a_\varphi^\chi \chi(j) = \sum_{\chi \in \mathrm{Irr}_K(G)} a_\varphi^\chi d_\psi^\chi$. By (ii) this expression is 1 if $\psi = \varphi$ and 0 otherwise, proving (iii). □

Proof of Theorem 5.14.8 Since $\mathrm{Irr}_K(G)$ is a K-basis of $\mathrm{Cl}_K(G)$ there is a unique K-linear combination $\Psi = \sum_{\chi \in \mathrm{Irr}_K(G)} a_\chi \chi$ such that Ψ coincides with φ_T on $G_{p'}$ and such that Ψ vanishes on all p-singular elements. Thus $\langle \Phi_S, \varphi_T \rangle'_G = \langle \Phi_S, \Psi \rangle_G = \sum_{\chi \in \mathrm{Irr}_K(G)} a_\chi \langle \Phi_S, \chi \rangle_G$. Using 5.14.4 yields $\langle \Phi_S, \chi \rangle = d_{\varphi_S}^\chi$, and now the result follows from 5.14.9. □

If we restrict the character of a projective indecomposable $\mathcal{O}G$-modules to $G_{p'}$ and express this as a linear combination of the irreducible Brauer characters, we get back the Cartan matrix of kG.

Theorem 5.14.10 *Let G be a finite group. Suppose that K and k are splitting fields for G. For any $\varphi \in \mathrm{IBr}_k(G)$ and any $\Phi \in \mathrm{IPr}_{\mathcal{O}}(G)$ denote by $c(\varphi, \Phi)$ the unique integer satisfying*

$$\mathrm{Res}^G_{G_{p'}}(\Phi) = \sum_{\varphi \in \mathrm{IBr}_k(G)} c(\varphi, \Phi) \cdot \varphi.$$

Then the matrix $(c(\varphi, \Phi))_{\varphi \in \mathrm{IBr}_k(G), \Phi \in \mathrm{IPr}_{\mathcal{O}}(G)}$ is the Cartan matrix of kG.

Proof Let Y be a projective indecomposable $\mathcal{O}G$-module with character Φ. Then $k \otimes_{\mathcal{O}} Y \cong kGi$ for some primitive idempotent i in kG. Let S be a simple kG-module such that $\varphi = \varphi_S$ and let i be a primitive idempotent in kG such that kGi is a projective cover of S. By 5.13.10, the number $c(\varphi, \Phi)$ is equal to the number of composition factors isomorphic to S in a composition series of kGi, which is equal to the entry $c(i, j)$ of the Cartan matrix of kG, with notation as in the definition 4.10.1. □

Corollary 5.14.11 *Let G be a finite group. Suppose that K and k are splitting fields for G. The matrix $(\langle \varphi, \psi \rangle')$, where $\varphi, \psi \in \mathrm{IBr}_k(G)$, is the inverse of the Cartan matrix of kG. In particular, this matrix has coefficients in $\mathbb{Q}$.*

Proof Let $\varphi \in \mathrm{IBr}_k(G)$. Since the Cartan matrix of kG is nonsingular, it follows from 5.14.10, that $\varphi = \sum_{\Phi \in \mathrm{IPr}_{\mathcal{O}}(G)} c'(\varphi, \Phi)\mathrm{Res}^G_{G_{p'}}(\Phi)$, where the matrix $(c'(\varphi, \Phi))$ is the inverse of the Cartan matrix of kG. Brauer's reciprocity 5.14.8 implies that $\langle \varphi, \psi \rangle' = c'(\varphi, \Phi)$, where Φ is the character of a projective cover of a simple module with Brauer character ψ. □

5.15 Generalised decomposition maps for group algebras

Suppose that $\mathrm{char}(K) = 0$. Decomposition maps for group algebras admit generalisations to maps relating class functions generated by the sets $\mathrm{Irr}_K(G)$ and $\mathrm{IBr}_k(C_G(u))$, with u running over the p-elements in G.

Definition 5.15.1 Let G be a finite group and $u \in G$ a p-element. We define the *generalised decomposition map*

$$d^u_G : \mathrm{Cl}_K(G) \to \mathrm{Cl}_K(C_G(u)_{p'})$$

by setting $d_G^u(\chi)(s) = \chi(us)$ for any K-valued class function χ on G and any p'-element s in $C_G(u)$.

Clearly $d_G^1 = d_G$. If $G = C_G(u)$, then d_G and d_G^u are both maps from $\mathrm{Cl}_K(G)$ to $\mathrm{Cl}_K(G_{p'})$, but they do not coincide in general. Since $d_G^u(\chi)$ is a K-valued class function on $C_G(u)_{p'}$, it is a K-linear combination of elements of $\mathrm{IBr}_k(C_G(u))$, by 5.13.13. In other words, there are unique numbers $d^u_{\chi,\varphi} \in K$ satisfying

$$d_G^u(\chi)(s) = \sum_{\varphi\in \mathrm{IBr}_k(C_G(u))} d^u_{\chi,\varphi} \cdot \varphi(s)$$

for all $\chi \in \mathrm{Irr}_K(G)$, any p-element u in G and any p'-element s in $C_G(u)$. The coefficients $d^u_{\chi,\varphi}$ arising in this way are called *generalised decomposition numbers*. For $u = 1$, these coincide with the ordinary decomposition numbers d^χ_φ in 5.14.4, 5.14.5, but for $u \neq 1$, they need not be rational integers. We will see that the generalised decomposition numbers are algebraic integers in a cyclotomic field generated by a root of unity of p-power order. This is due to Brauer; our approach using local pointed elements on group algebras follows Puig. Let G be a finite group and u_ϵ a local pointed group on $\mathcal{O}G$; that is, u is a p-element in G and ϵ is a conjugacy class of primitive idempotents in $(\mathcal{O}G)^{\langle u\rangle}$ such that $\mathrm{Br}_{\langle u\rangle}(\epsilon) \neq 0$. Any two elements $j, j' \in \epsilon$ are conjugate by an invertible element in $(\mathcal{O}G)^{\langle u\rangle}$, and hence, for any class function χ on $\mathcal{O}G$ we have $\chi(uj) = \chi(uj')$. Moreover, $\mathrm{Br}_{\langle u\rangle}(\epsilon)$ is a conjugacy class of primitive idempotents in $kC_G(u)$, hence determines a unique isomorphism class of projective indecomposable $kC_G(u)$-modules, hence an isomorphism class of simple $kC_G(u)$-modules, and thus an irreducible Brauer character. We introduce the following notation.

Definition 5.15.2 Let G be a finite group and u_ϵ a local pointed element on $\mathcal{O}G$. For any K-valued class function χ on $\mathcal{O}G$ we set $\chi(u_\epsilon) = \chi(uj)$, where $j \in \epsilon$. We denote by φ_ϵ the irreducible Brauer character in $\mathrm{IBr}_k(C_G(u))$ of the simple quotient of the projective indecomposable $kC_G(u)$-module $kC_G(u)\bar{j}$, where $\bar{j} = \mathrm{Br}_{\langle u\rangle}(j)$.

If u_ϵ, v_τ are G-conjugate local pointed elements on $\mathcal{O}G$, then clearly $\chi(u_\epsilon) = \chi(v_\tau)$ for any K-valued class function χ on $\mathcal{O}G$. For u a p-element in G, the standard lifting theorems for idempotents imply that the map sending a local point ϵ of $\langle u\rangle$ on $\mathcal{O}G$ to φ_ϵ is a bijection between the set of local points of $\langle u\rangle$ on $\mathcal{O}G$ and the set $\mathrm{IBr}_k(C_G(u))$. Rather than indexing a sum over $\mathrm{IBr}_k(C_G(u))$, it is sometimes convenient to choose the set of local points of $\langle u\rangle$ on $\mathcal{O}G$ as indexing set.

Theorem 5.15.3 *Let G be a finite group. Suppose that K, k are splitting fields for all subgroups of G. Let $\chi \in \mathrm{Irr}_K(G)$ and u a p-element in G. For all*

$s \in C_G(u)_{p'}$ we have

$$d_G^u(\chi)(s) = \sum_\epsilon \chi(u_\epsilon) \cdot \varphi_\epsilon(s),$$

where ϵ runs over the set of local points of $\langle u \rangle$ on $\mathcal{O}G$. Equivalently, $d^u_{\chi,\varphi} = \chi(u_\epsilon)$, where $\varphi \in \mathrm{IBr}_k(C_G(u))$ and ϵ is the local point of $\langle u \rangle$ on $\mathcal{O}G$ such that $\varphi = \varphi_\epsilon$. Moreover, we have $\chi(u_\epsilon) \in \mathbb{Z}[\zeta]$, where ζ is a root of unity of order equal to the order of u.

Proof For any $\varphi \in \mathrm{IBr}_k(G)$ choose a primitive idempotent i_φ in $(\mathcal{O}G)^{\langle u \rangle}$ such that $\mathrm{Br}_u(i_\varphi) \neq 0$ and $kC_G(u)\mathrm{Br}_u(i_\varphi)$ is a projective cover of a simple module with Brauer character φ. In other words, if ϵ is the local point of $\langle u \rangle$ containing i_φ then $\varphi = \varphi_\epsilon$. In the sums that follow the summation index η runs over the set $\mathrm{Irr}_K(C_G(u))$ and φ runs over the set $\mathrm{IBr}_k(C_G(u))$. Restricting χ to $C_G(u)$ yields that $\chi(y) = \sum_\eta d^\chi_\eta \cdot \eta(y)$ for all $y \in C_G(u)$ for some non negative integers d^χ_η. Since $u \in Z(C_G(u))$, for any $s \in C_G(u)_{p'}$ we have $\eta(us) = \frac{\eta(u)}{\eta(1)}\eta(s)$. Applying the decomposition map to the group $C_G(u)$ and character η yields integers d^η_φ such that $\eta' = \sum_\varphi d^\eta_\varphi \cdot \varphi$, where η' is the restriction of $\eta \in \mathrm{Irr}_K(C_G(u))$ to $C_G(u)_{p'}$. Combining these equations yields therefore

$$\chi(us) = \sum_\eta d^\chi_\eta \frac{\eta(u)}{\eta(1)}\eta(s) = \sum_{\eta,\varphi} d^\chi_\eta \frac{\eta(u)}{\eta(1)} d^\eta_\varphi \varphi(s)$$

for all $s \in C_G(u)_{p'}$. By 5.14.5 we have $d^\eta_\varphi = \eta(j_\varphi)$, where j_φ is a primitive idempotent in $\mathcal{O}C_G(u)$ such that $\mathcal{O}C_G(u)j_\varphi$ is a projective cover of a simple $kC_G(u)$-module with Brauer character φ. Thus

$$\chi(us) = \sum_{\eta,\varphi} d^\chi_\eta \frac{\eta(u)}{\eta(1)}\eta(j_\varphi)\varphi(s) = \sum_{\eta,\varphi} d^\chi_\eta \eta(uj_\varphi)\varphi(s) = \sum_{\varphi,\eta} d^\chi_\eta \eta(uj_\varphi)\varphi(s)$$
$$= \sum_\varphi \chi(uj_\varphi)\varphi(s).$$

Let i_φ be a primitive idempotent in $(\mathcal{O}G)^{\langle u \rangle}$ such that $\mathrm{Br}_u(i_\varphi) = \bar{j}_\varphi$, where $\bar{j}_\varphi$ is the canonical image of j_φ in $kC_G(u)$. Then $i_\varphi - j_\varphi \in \ker(Br_u)$. By 5.12.16 we have $\chi(uj_\varphi) = \chi(ui_\varphi)$, and hence we get the formula

$$\chi(us) = \sum_\varphi \chi(ui_\varphi)\varphi(s)$$

for all $s \in C_G(u)_{p'}$ as required. For the last statement, let Y be an $\mathcal{O}$-free $\mathcal{O}G$-module with character χ. Then $\chi(ui_\varphi)$ is the value at u of the character of the $\mathcal{O}\langle u \rangle$-module $i_\varphi Y$, hence in $\mathbb{Z}[\zeta]$. □

Proposition 5.15.4 *Let G be a finite group, U a cyclic p-subgroup of G, and ϵ a local point of U on $\mathcal{O}G$. Let χ be the character of a finitely generated $\mathcal{O}$-free $\mathcal{O}G$-module. Then the product $\prod_u \chi(u_\epsilon)$, with u running over the set of elements in U that generate U, is a rational integer.*

Proof Let X be an $\mathcal{O}$-free $\mathcal{O}G$-module with χ as its character, and let $j \in \epsilon$. If $u \in U$ generates U, then $\chi(u_\epsilon) = \chi(uj)$ is the character of the $\mathcal{O}U$-module jX evaluated at u. Thus 5.15.4 is a special case of 3.5.4 (i). □

Generalised decomposition maps are related to the Brauer construction as follows.

Proposition 5.15.5 *Let G be a finite group and $u \in G$ a p-element. Set $Q = \langle u \rangle$ and $\Delta Q = \{(v, v) | v \in Q\}$. Suppose that K and k are splitting fields for the subgroups of G. Let U be a finitely generated p-permutation $\mathcal{O}G$-$\mathcal{O}Q$-bimodule. Denote by χ_U the character of U as an $\mathcal{O}(G \times Q)$-module and define a class function τ on G by $\tau(x) = \chi_U(x, u)$ for all $x \in G$. Then $d_G^u(\tau)$ is the Brauer character of $U(\Delta Q)$, viewed as a left $kC_G(Q)$-module.*

Proof Let $s \in C_G(Q)_{p'}$. We have $d_G^u(\tau)(s) = \tau(us) = \chi_U(us, u)$. This is the trace of the element $(us, u) \in C_G(Q) \times Q$ acting on U as simultaneous left multiplication by us and right multiplication by u^{-1}. The p-part of (us, u) is (u, u). It follows from 5.13.7 that $\chi_U(us, u)$ is equal to the Brauer character of $U(\Delta Q)$ evaluated at $(s, 1)$. □

We noted in 4.17.4, that decomposition maps commute with the maps induced by tensoring with bimodules that are finitely generated projective on both sides. There are analogous statements for generalised decomposition maps and p-permutation bimodules. We first show that the matrix of generalised decomposition numbers $\chi(u_\epsilon)$ is a nondegenerate square matrix. The proof we present is based on the following result of Puig.

Theorem 5.15.6 ([75, 3.4]) *Let G be a finite group. Suppose that K and k are splitting fields for all subgroups of G. Denote by $\mathcal{R}$ a set of representatives of the G-conjugacy classes of local pointed elements on $\mathcal{O}G$. For any local pointed element u_ϵ on $\mathcal{O}G$ denote by $\bar{u}_\epsilon$ the image of uj in $\mathcal{O}G/[\mathcal{O}G, \mathcal{O}G]$ for some $j \in \epsilon$. Then $\bar{u}_\epsilon$ is independent of the choice of j in ϵ, and the set $\{\bar{u}_\epsilon | u_\epsilon \in \mathcal{R}\}$ is an $\mathcal{O}$-basis of $\mathcal{O}G/[\mathcal{O}G, \mathcal{O}G]$ that is independent of the choice of $\mathcal{R}$.*

Proof Let u_ϵ be a pointed element on $\mathcal{O}G$ and let $j, j' \in \epsilon$. Then $j' = cjc^{-1}$ for some invertible element c in $(\mathcal{O}G)^{\langle u \rangle}$, hence $uj' = c(uj)c^{-1}$. Thus $uj - uj' = [ujc^{-1}, c] \in [\mathcal{O}G, \mathcal{O}G]$, which shows that $\bar{u}_\epsilon$ is independent of the choice of j in ϵ. The same argument with c replaced by a group element shows that $\bar{u}_\epsilon$

depends only on the G-conjugacy class of u_ϵ. For u a p-element in G, every local point ϵ of $\langle u\rangle$ on $\mathcal{O}G$ corresponds to a unique point of $kC_G(u)$, namely $\mathrm{Br}_u(\epsilon)$. Every such point of $kC_G(u)$ corresponds to an isomorphism class of simple $kC_G(u)$-modules. Denoting by $\ell(kC_G(u))$ the number of isomorphism classes of simple $kC_G(u)$-modules, we therefore get $|\mathcal{R}| = \sum_u \ell(C_G(u))$, where u runs over a set of representatives of the conjugacy classes of p-elements in G. Thus, by 1.15.11, we have $|\mathcal{R}| = |\mathrm{Irr}_K(G)|$. This in turn is equal to the $\mathcal{O}$-rank of $Z(\mathcal{O}G)$, hence of $\mathcal{O}G/[\mathcal{O}G, \mathcal{O}G]$, by 3.3.6 and 2.11.7. It suffices therefore to show that the elements of the form $\bar{u}_\epsilon$, with u_ϵ running over the local pointed elements on $\mathcal{O}G$, generate $\mathcal{O}G/[\mathcal{O}G, \mathcal{O}G]$ as an $\mathcal{O}$-module. Clearly the image of G in this quotient generates this $\mathcal{O}$-module. Let $x \in G$. Write $x = us = su$, with u a p-element and s a p'-element. Then s belongs to the commutative $\mathcal{O}$-algebra $\mathcal{O}\langle s\rangle$, which by 4.7.13 is a direct product of copies of $\mathcal{O}$, and hence s is an $\mathcal{O}$-linear combination of idempotents. Thus the image in $\mathcal{O}G/[\mathcal{O}G, \mathcal{O}G]$ of the set of elements of the form uj, with u a p-element in G and j an idempotent in $(\mathcal{O}G)^{\langle u\rangle}$, generates this $\mathcal{O}$-module. By decomposing j into primitive idempotents it follows that it suffices to take the images of uj with j a primitive idempotent in $(\mathcal{O}G)^{\langle u\rangle}$. If j belongs to a nonlocal point of $\langle u\rangle$, then $\chi(uj) = 0$ for any $\chi \in \mathrm{Irr}_K(G)$ by 5.12.16, hence $uj \in [\mathcal{O}G, \mathcal{O}G]$ by 4.12.21. Thus the images of uj, with u a p-element in G and j a primitive idempotent belonging to a local point of $\langle u\rangle$ on $\mathcal{O}G$ generate the $\mathcal{O}$-module $\mathcal{O}G/[\mathcal{O}G, \mathcal{O}G]$, whence the result. □

By passing to the dual basis in $(\mathcal{O}G/[\mathcal{O}G, \mathcal{O}G])^*$ and using the isomorphism $(\mathcal{O}G/[\mathcal{O}G, \mathcal{O}G])^* \cong Z(\mathcal{O}G)$ from 2.11.7 the above theorem yields a basis in $\mathbb{Z}(\mathcal{O}G)$, which can be used to describe the general decomposition numbers in terms of the ideal in $Z(\mathcal{O}G)$ generated by the elements $\chi^0 = \sum_{x\in G} \chi(x^{-1})x$, where $\chi \in \mathrm{Irr}_K(G)$; see [75]. We will describe this basis of $Z(\mathcal{O}G)$ in 5.15.18 and 5.15.19 below.

Theorem 5.15.7 (Brauer) *Let G be a finite group. Suppose that K and k are splitting fields for all subgroups of G. Denote by $\mathcal{R}$ a set of representatives of the G-conjugacy classes of local pointed elements on $\mathcal{O}G$. The matrix of generalised decomposition numbers $D = (\chi(u_\epsilon))$, with χ running over $\mathrm{Irr}_K(G)$ and u_ϵ over $\mathcal{R}$, is a nondegenerate square matrix with entries in $\mathbb{Z}[\zeta]$, for some root of unity ζ of p-power order.*

Proof The fact that the generalised decomposition numbers are in $\mathbb{Z}[\zeta]$ for some root of unity of p-power order was already observed in 5.15.3. Since $|\mathrm{Irr}_K(G)| = |\mathcal{R}|$, the generalised decomposition matrix D is indeed a square matrix. Denote by $\mathcal{C}$ a set of representatives of the conjugacy classes in G. The character table of G is the matrix $X = (\chi(x))$, with χ running over $\mathrm{Irr}_K(G)$ and $x \in \mathcal{C}$. This matrix is nondegenerate by 3.4.1. We noted in 1.5.5 that the image of $\mathcal{C}$ in $\mathcal{O}G/[\mathcal{O}G, \mathcal{O}G]$ is an $\mathcal{O}$-basis. By 5.15.6, the set $\{\bar{u}_\epsilon | u_\epsilon \in \mathcal{R}\}$ is another basis. Denote by $\Delta = (\delta_{u_\epsilon,x})$ the – necessarily nondegenerate – transition matrix with entries in $\mathcal{O}$ defined by $\bar{u}_\epsilon = \sum_{x\in\mathcal{C}} \delta_{u_\epsilon,x} x$ for any $u_\epsilon \in \mathcal{R}$ and any $x \in \mathcal{C}$. Then $\chi(u_\epsilon) = \sum_{x\in\mathcal{C}} \delta_{u_\epsilon,x}\chi(x)$. Thus $D = X \cdot {}^t\Delta$ is nondegenerate as claimed. □

Remark 5.15.8 The fact that the generalised decomposition matrix D is a square matrix follows also from 1.15.11, since a local pointed element u_ϵ determines a unique irreducible Brauer character of $C_G(u)$ as in 5.15.2.

If u_ϵ is a local pointed element on $\mathcal{O}G$, then so is u_ϵ^{-1}, since ϵ is a local point of $\langle u\rangle = \langle u^{-1}\rangle$ on $\mathcal{O}G$. Thus if u_ϵ runs over a set of representatives of the G-conjugacy classes of local pointed elements on $\mathcal{O}G$, then so does u_ϵ^{-1}.

Corollary 5.15.9 *Let G be a finite group. Suppose that K and k are splitting fields for all subgroups of G. Denote by $\mathcal{R}$ a set of representatives of the G-conjugacy classes of local pointed elements on $\mathcal{O}G$. For any local pointed element u_ϵ on $\mathcal{O}G$ set $\theta_{u_\epsilon} = \sum_{\chi\in\mathrm{Irr}_K(G)} \chi(u_\epsilon^{-1})\chi$. The set $\{\theta_{u_\epsilon} | u_\epsilon \in \mathcal{R}\}$ is a K-basis of the space $\mathrm{Cl}_K(G)$ of K-valued class functions on G.*

Proof This follows from the fact that $\mathrm{Irr}_K(G)$ is a basis of $\mathrm{Cl}_K(G)$, together with the previous Theorem. □

Unlike the elements in $\mathrm{Irr}_K(G)$, the θ_{u_ϵ} need not be pairwise orthogonal, but we will see that they satisfy orthogonality relations that are well suited for dealing with the block decomposition of $\mathcal{O}G$ in 6.13.8. If $u = 1$ then θ_{u_ϵ} is the character of the projective indecomposable $\mathcal{O}G$-module determined by ϵ; see 5.14.6. Thus the θ_{u_ϵ} with $u = 1$ span $K \otimes_{\mathbb{Z}} \mathrm{Proj}_{\mathcal{O}}(G)$; we will see in section 6.13 that the the θ_{u_ϵ} with $u \neq 1$ span $K \otimes_{\mathbb{Z}} L^0(G)$. Although the class functions θ_{u_ϵ} are not generalised characters in general, they have another interpretation relating them to characters of p-permutation bimodules:

Proposition 5.15.10 *Let G be a finite group. Suppose that K and k are splitting fields for all subgroups of G. Let u_ϵ be a local pointed element on $\mathcal{O}G$ and set $\theta_{u_\epsilon} = \sum_{\chi\in\mathrm{Irr}_K(G)} \chi(u_\epsilon^{-1})\chi$. Let $j \in \epsilon$ and denote by χ_j the character of the $\mathcal{O}(G \times \langle u\rangle)$-module $\mathcal{O}Gj$, with $(x, v) \in G \times \langle u\rangle$ acting on $\mathcal{O}Gj$ by left multiplication with x and right multiplication wth v^{-1}. Then $\theta_{u_\epsilon}(x) = \chi_j(x, u)$.*

Proof Consider $\mathcal{O}G$ as an $\mathcal{O}(G \times G)$-module, with $(x, y) \in G \times G$ acting as left multiplication by x and right multiplication by y^{-1}. By 3.3.12, the character of this module is given by $\psi(x, y) = \sum_{\chi\in\mathrm{Irr}_K(G)} \chi(x)\chi(y^{-1})$. Since j commutes with u, the $\mathcal{O}(G \times \langle u \rangle)$-module $\mathcal{O}Gj$ has as character the function $\chi_j(x, v) = \sum_{\chi \in \mathrm{Irr}_K(G)} \chi(x)\chi(v^{-1}j)$, where $x \in G$ and $v \in \langle u \rangle$. When specialised to $v = u$ this yields θ_{u_ϵ} as claimed. □

Proposition 5.15.11 *Let G be a finite group. Suppose that K and k are splitting fields for all subgroups of G. Let u_ϵ be a local pointed element on $\mathcal{O}G$ and set $\theta_{u_\epsilon} = \sum_{\chi \in \mathrm{Irr}_K(G)} \chi(u_\epsilon^{-1})\chi$. Let $j \in \epsilon$ and set $\bar{j} = \mathrm{Br}_u(j)$. Then $d_G^u(\theta_{u_\epsilon})$ is the Brauer character of the projective indecomposable $kC_G(u)$-module $kC_G(u)\bar{j}$. If v is a p-element in G that is not conjugate to u then $d_G^v(\theta_{u_\epsilon}) = 0$.*

Proof The first statement follows from 5.15.10 and 5.15.5 applied to the module $\mathcal{O}Gj$. For s a p'-element in $C_G(v)$, we have $d_G^v(\theta_{u_\epsilon})(s) = \theta_{u_\epsilon}(vs) = \sum_{\chi \in \mathrm{Irr}_K(G)} \chi(vs)\chi(u^{-1}j)$. By 5.15.10, this is the value of the character of $\mathcal{O}Gj$ at (vs, u), with vs acting by left multiplication and u by right multiplication with u^{-1}. The p-part of (vs, u) is (v, u). Thus, denoting by R the cyclic subgroup of $G \times \langle u \rangle$ generated by (v, u), the previous value is by 5.15.5 equal to the Brauer character of $(\mathcal{O}Gj)(R)$ evaluated at s. Since u and v are not conjugate we have $(\mathcal{O}G)(R) = \{0\}$ by 5.8.9, hence $(\mathcal{O}Gj)(R) = \{0\}$, which implies the second statement. □

This Proposition can be interpreted as saying that the sum of the generalised decomposition maps d_G^u, with u running over a set of representatives of the conjugacy classes of p-elements in G, induces a K-linear isomorphism

$$\mathrm{Cl}_K(G) \cong \oplus_u K \otimes_{\mathbb{Z}} \mathrm{Pr}_k(C_G(u))$$

where $\mathrm{Pr}_k(C_G(u))$ is the group of $\mathbb{Z}$-linear combinations of Brauer characters of projective indecomposable $kC_G(u)$-modules. We turn now to the commutation of the Brauer construction and generalised decomposition maps.

Theorem 5.15.12 *Let G, H be finite groups. Suppose that K and k are splitting fields for all subgroups of $G \times H$. Let u be a common p-element of G and H, and denote by ζ a primitive $|\langle u \rangle|$-th root of unity in K. Let M be a p-permutation $\mathcal{O}H$-$\mathcal{O}G$-bimodule that is finitely generated projective as a left $\mathcal{O}H$-module and as a right $\mathcal{O}G$-module. Denote by*

$$\Phi : \mathbb{Z}\mathrm{Irr}_K(G) \to \mathbb{Z}\mathrm{Irr}_K(H),$$

$$\bar{\Phi} : \mathbb{Z}[\zeta]\mathrm{IBr}_k(C_G(u)) \to \mathbb{Z}[\zeta]\mathrm{IBr}_k(C_H(u))$$

the maps induced by the functors $M \otimes_{\mathcal{O}G} -$ and $M(\Delta\langle u \rangle) \otimes_{kC_G(u)} -$, respectively. Suppose that whenever v is a p-element in G that is not G-conjugate to u we have $M(\langle (u, v) \rangle) = \{0\}$. Then we have

$$\bar{\Phi} \circ d_G^u = d_H^u \circ \Phi.$$

Proof We need to show the equality $\bar{\Phi}(d_G^u(\chi)) = d_H^u(\Phi(\chi))$ for any χ in $\mathbb{Z}\mathrm{Irr}_K(G)$. Both sides in this equation are K-linear in χ, and hence it suffices to show this equality for χ running over a K-basis of $\mathrm{Cl}_K(G)$. We choose the basis consisting of the θ_{v_ϵ} considered in 5.15.9, with v_ϵ running over a set of representatives of the G-conjugacy classes of local pointed elements on $\mathcal{O}G$. Fix a local pointed element v_ϵ on $\mathcal{O}G$ and set $\chi = \theta_{v_\epsilon}$. Consider first the case where v is not conjugate to u. Then the left side in the above equation is zero, by 5.15.11. We need to show that $d_H^u(\Phi(\chi))$ is zero as well. Let $i \in \epsilon$. The class function χ evaluated at $x \in G$ is by 5.15.10 equal to the character of $\mathcal{O}Gi$ evaluated at (x, v), with x acting by left multiplication and v acting by right multiplication with v^{-1}. Thus $\Phi(\chi)$ evaluated at $y \in H$ is the character of $M \otimes_{\mathcal{O}G} \mathcal{O}Gi \cong Mi$ evaluated at (y, v). In order to apply d_H^u to this class function, we need to consider $y \in H$ of the form $y = ut = tu$, with $t \in C_H(u)_{p'}$. By 5.15.5, the value of $d_H^u(\Phi(\chi))$ at any such t is equal to the Brauer character of $(Mi)(\langle(u, v)\rangle)$ at t. But $M(\langle(u, v)\rangle)$ is zero by the assumptions on M, and hence so is $d_H^u(\Phi(\chi))$. It remains to consider the case where v and u are conjugate in G. We may assume $v = u$. In that case, $d_G^u(\chi)$ is the Brauer character of $kC_G(u)j$, by 5.15.11, where $j = \mathrm{Br}_u(i)$. This is mapped under $\bar{\Phi}$ to the Brauer character of the $kC_H(u)$-module $M(\Delta\langle u\rangle) \otimes_{kC_G(u)} kC_G(u)j \cong (Mi)(\Delta\langle u\rangle)$. We need to show that this is equal to $d_H^u(\Phi(\chi))$. As before, $\Phi(\chi)$ evaluated at $y \in H$ is the character of Mi evaluated at (y, u). The same argument as before shows that applying d_H^u to this class function yields again the Brauer character of $(Mi)(\Delta\langle u\rangle)$. □

Following [12], the generalised decomposition map d_G^u has an adjoint.

Definition 5.15.13 Let G be a finite group and $u \in G$ a p-element. We define a linear map

$$t_G^u : \mathrm{Cl}_K(C_G(u)_{p'}) \to \mathrm{Cl}_K(G)$$

as follows. For $\varphi \in \mathrm{Cl}_K(C_G(u)_{p'})$ we denote by $t_G^u(\varphi)$ the unique class function on G satisfying $t_G^u(\varphi)(us) = \varphi(s)$ for all $s \in C_G(u)_{p'}$ and $t_G^u(\varphi)(x) = 0$ for all $x \in G$ such that the p-part of x is not conjugate to u in G.

The map t_G^1 is the map extending a class function defined on $G_{p'}$ to G by setting it to be zero on $G \setminus G_{p'}$.

Proposition 5.15.14 ([12, A.1.3]) *Let G be a finite group and $u \in G$ a p-element. Let $\chi \in \mathrm{Cl}_K(G)$ and $\varphi \in \mathrm{Cl}_K(C_G(u)_{p'})$. We have*

$$\langle d_G^u(\chi), \varphi\rangle'_{C_G(u)} = \langle \chi, t_G^{u^{-1}}(\varphi)\rangle_G.$$

Proof The right side is equal to $\frac{1}{|G|}\sum_{x\in G} \chi(x)t_G^{u^{-1}}(\varphi)(x^{-1})$. If the p-part of an element $x \in G$ is not conjugate to u, then the corresponding summand vanishes.

Suppose that the p-part of x is equal to u. The number of conjugates of u is $|G : C_G(u)|$, and hence the previous sum is equal to $\frac{1}{|C_G(u)|}\sum_{s\in C_G(u)_{p'}} \chi(us)\varphi(s^{-1})$, which is equal to the left side in the statement. □

Remark 5.15.15 Let P be a finite p-group. It is sometimes convenient to extend the notation for decomposition numbers to interior P-algebras over $\mathcal{O}$. Let A be an interior P-algebra over $\mathcal{O}$, and let $\chi : A \to K$ be an $\mathcal{O}$-linear symmetric function on A with values in K. Let u_ϵ be a local pointed element on A. We set $\chi(u_\epsilon) = \chi(uj)$, where $j \in \epsilon$. If $K \otimes_{\mathcal{O}} A$ is split semisimple and χ the character of a simple $K \otimes_{\mathcal{O}} A$-module X, then $\chi(1_\epsilon) = \chi(j) = \dim_K(jX)$, where $j \in \epsilon$, so in that case, the numbers $\chi(1_\epsilon)$, with ϵ running over the points of A, are the decomposition numbers considered in 4.16.1. In that situation, the numbers $\chi(u_\epsilon)$, with χ a character of a simple $K \otimes_{\mathcal{O}} A$-modules and u_ϵ a local pointed element on A are called *generalised decomposition numbers of the interior P-algebra A*.

Proposition 5.15.16 *Let G be a finite group. Denote by X the $\mathcal{O}$-submodule of $\mathcal{O}G$ spanned by all idempotents in $\mathcal{O}G$, and denote by $\mathcal{O}G_{p'}$ the $\mathcal{O}$-submodule spanned by all p'-elements in G. We have*

$$\mathcal{O}G_{p'} \subseteq X \subseteq \mathcal{O}G_{p'} + [\mathcal{O}G, \mathcal{O}G].$$

Proof Let y be a p'-element in G. Then $\mathcal{O}\langle y\rangle$ is a direct product of copies of $\mathcal{O}$, and hence every element in $\mathcal{O}\langle y\rangle$ is an $\mathcal{O}$-linear combination of idempotents. This shows the inclusion $\mathcal{O}G_{p'} \subseteq X$. The space X is spanned by the set of primitive idempotents. Conjugate idempotents have the same image in $\mathcal{O}G/[\mathcal{O}G, \mathcal{O}G]$. Thus the image of X in $\mathcal{O}G/[\mathcal{O}G, \mathcal{O}G]$ is an $\mathcal{O}$-submodule of rank at most the number ℓ of isomorphism classes of simple kG-modules. By 1.15.3, this is also the number of conjugacy classes of p'-elements in G. By 1.5.5, the image of $\mathcal{O}G_{p'}$ in $\mathcal{O}G/[\mathcal{O}G, \mathcal{O}G]$ is an $\mathcal{O}$-pure $\mathcal{O}$-submodule of rank ℓ. Since this $\mathcal{O}$-submodule is contained in the image of X, it follows that $\mathcal{O}G_{p'}$ and X have the same image in $\mathcal{O}G/[\mathcal{O}G, \mathcal{O}G]$. This implies the second inclusion. □

Corollary 5.15.17 *Let G be a finite group, let y be a p'-element in G, and b an idempotent in $Z(\mathcal{O}G)$. Then $yb \in \mathcal{O}G_{p'} + [\mathcal{O}G, \mathcal{O}G]$.*

Proof The first inclusion of 5.15.16 implies that y is a linear combination of idempotents in $\mathcal{O}G$. Multiplying an idempotent by b yields either 0 or again an idempotent, so yb is again a linear combination of idempotents, hence contained in $\mathcal{O}G_{p'} + [\mathcal{O}G, \mathcal{O}G]$ by the second inclusion of 5.15.16. □

For u_ϵ a local pointed element on $\mathcal{O}G$, we denote by φ_ϵ the associated irreducible Brauer character of $C_G(u)$ defined in 5.15.2.

Theorem 5.15.18 *Let G be a finite group and let u_ϵ be a local pointed element on $\mathcal{O}G$.*

(i) We have $t_G^u(\varphi_\epsilon)(u_\epsilon) = 1$ and $t_G^u(\varphi_\epsilon)(v_\tau) = 0$ for any local pointed element v_τ on $\mathcal{O}G$ that is not conjugate to u_ϵ.

(ii) The set $\{t_G^u(\varphi)\}$, with u running over a set of representatives of the G-conjugacy classes of p-elements in G and with φ running over $\mathrm{IBr}_k(C_G(u))$, is the dual basis in $(\mathcal{O}G/[\mathcal{O}G, \mathcal{O}G])^*$ of the basis $\{\bar{u}_\epsilon\}_{u_\epsilon \in \mathcal{R}}$ of $\mathcal{O}G/[\mathcal{O}G, \mathcal{O}G]$ from 5.15.6.

Proof Set $\varphi = \varphi_\epsilon$. Let v_τ be a local pointed element on $\mathcal{O}G$, and let $j \in \tau$. We have $(t_G^u(\varphi))(v_\tau) = (t_G^u(\varphi))(vj)$. By 5.15.16, applied to $C_G(v)$ instead of G, we have $j = j' + c$, where $j' \in \mathcal{O}C_G(v)_{p'}$ and $c \in [\mathcal{O}C_G(v), \mathcal{O}C_G(v)]$. Note that then also $vc \in [\mathcal{O}C_G(v), \mathcal{O}C_G(v)]$. Since $t_G^u(\varphi)$ is a class function on $\mathcal{O}G$, it follows that this function vanishes on additive commutators, and hence $(t_G^u(\varphi))(vj) = (t_G^u(\varphi))(vj')$. This is zero if v is not conjugate to u. If $u = v$, this is equal to $(t_G^u(\varphi))(uj') = \varphi(j')$, hence equal to $\hat{\varphi}(j)$, where $\hat{\varphi}$ coincides with φ on $C_G(u)_{p'}$ and vanishes on $C_G(u) \setminus C_G(u)_{p'}$. By 5.14.9, this is 1 if $j \in \epsilon$, and 0, otherwise. This proves (i), and (ii) is an equivalent reformulation of (i), with the maps $t_G^u(\varphi) : \mathcal{O} \to \mathcal{O}$ considered as maps from $\mathcal{O}G/[\mathcal{O}G, \mathcal{O}G]$ to $\mathcal{O}$. □

Using the isomorphism $(\mathcal{O}G/[\mathcal{O}G, \mathcal{O}G])^* \cong Z(\mathcal{O}G)$ from 2.11.7 the above theorem yields the following basis of $Z(\mathcal{O}G)$ which appears in work of Cliff, Plesken and Weiss [20]. We use the fact that a symmetrising form on $\mathcal{O}G$ induces a map $\mathcal{O}G/[\mathcal{O}G, \mathcal{O}G] \to \mathcal{O}$, hence is well-defined on local pointed elements.

Corollary 5.15.19 *Let G be a finite group. Denote by $\sigma : \mathcal{O}G \to \mathcal{O}$ the canonical symmetrising form of $\mathcal{O}G$. For any local pointed element u_ϵ on $\mathcal{O}G$ denote by $z(u_\epsilon)$ the element in $Z(\mathcal{O}G)$ corresponding to $t_G^u(\varphi_\epsilon)$ under the canonical isomorphism $(\mathcal{O}G/[\mathcal{O}G, \mathcal{O}G])^* \cong Z(\mathcal{O}G)$ from 2.11.7.*

(i) The set $\{z(u_\epsilon)\}$, with u_ϵ running over a set of representatives of the G-conjugacy classes of local pointed elements on $\mathcal{O}G$ is an $\mathcal{O}$-basis of $Z(\mathcal{O}G)$.

(ii) For any two local pointed elements u_ϵ, v_τ on $\mathcal{O}G$ we have $\sigma(z(u_\epsilon)v_\tau) = 1$, if u_ϵ and v_τ are G-conjugate, and $\sigma(z(u_\epsilon)v_\tau) = 0$, otherwise.

(iii) For any local pointed element u_ϵ on $\mathcal{O}G$ we have $z(u_\epsilon) = \sum_s \varphi_\epsilon(s^{-1})c(u^{-1}s)$, where s runs over a set of representatives of the

p'-conjugacy classes of $C_G(u)$, and where $c(u^{-1}s)$ is the sum of all G-conjugates of $u^{-1}s$.

Proof Statement (i) follows from 5.15.18. The inverse of the isomorphism $(\mathcal{O}G/[\mathcal{O}G,\mathcal{O}G])^* \cong Z(\mathcal{O}G)$ from 2.11.7 maps $z \in Z(\mathcal{O}G)$ to the map $\mathcal{O}G/[\mathcal{O}G,\mathcal{O}G]$ induced by the central function s_z defined by $s_z(a) = s(az)$, for all $a \in \mathcal{O}G$. Note that since s_z is a central function, the value $s_z(v_\tau) = s_z(vj)$, where $j \in \tau$, does not depend on the choice of j. Thus (ii) follows from 5.15.18. The isomorphism $(\mathcal{O}G/[\mathcal{O}G,\mathcal{O}G])^* \cong Z(\mathcal{O}G)$ from 1.5.5 maps $t_G^u(\varphi_\epsilon)$ to $z(u_\epsilon) = \sum_{x \in G} t_G^u(\varphi_\epsilon)(x^{-1})x$. Thus the coefficient of $z(u_\epsilon)$ at a group element x vanishes if the p-part of x is not conjugate to u^{-1}. If the p-part of x is equal to u^{-1}, then the coefficient is equal to $\varphi(s^{-1})$, where s is the p'-part of x. Statement (iii) follows. □

Bibliography

[1] M. Auslander and J. F. Carlson, *Almost split sequences and group rings*, J. Algebra **103** (1986), 122–140.

[2] M. Auslander and O. Goldman, *The Brauer group of a commutative ring*, Trans. Amer. Math. Soc. **97** (1960), 367–409.

[3] M. Auslander and I. Reiten, *Representation theory of Artin algebras VI: a functorial approach to almost split sequences*, Comm. Algebra **6**(3) (1978) 257–300.

[4] M. Auslander, I. Reiten, and S. O. Smalø, *Representation theory of Artin algebras*, Cambridge studies in advanced mathematics **36**, Cambridge University Press (1995).

[5] D. J. Benson, *Representations and cohomology, Vol. I: Cohomology of groups and modules*, Cambridge studies in advanced mathematics **30**, Cambridge University Press (1991).

[6] D. J. Benson, *Representations and cohomology, Vol. II: Cohomology of groups and modules*, Cambridge studies in advanced mathematics **31**, Cambridge University Press (1991).

[7] D. J. Benson and J. F. Carlson, *Nilpotent elements in the Green ring*, J. Algebra **104** (1986), 329–350.

[8] S. D. Berman, *On certain properties of integral group rings*, Dokl. Akad. Nauk. SSSR **91** (1953), 7–9.

[9] R. Boltje, *A canonical Brauer induction formula*, Astérisque **181–182** (1990), 31–59.

[10] S. Bouc, *Résolutions de foncteurs de Mackey*; in: "Groups, representations: cohomology, group actions and topology" (eds.: A. Adem, J. Carlson, S. Priddy, P. Webb), Proc. Symp. Pure Math. **63** (1998), 31–83.

[11] S. Bouc, *Bisets as categories and tensor product of induced bimodules*. Appl. Categor. Struct. **18** (2010), 517–521.

[12] M. Broué, *Radical, hauteurs, p-sections et blocs*, Ann. Math. **107** (1978), 89–107.

[13] M. Broué, *On Scott modules and p-permutation modules: an approach through the Brauer homomorphism*, Proc. Amer. Math. Soc. **93** (1985), 401–408.

[14] M. Broué, *Equivalences of blocks of group algebras*, in: Finite dimensional algebras and related topics, Kluwer (1994), 1–26.

[15] M. Broué and L. Puig, *Characters and local structure in G-algebras*, J. Algebra **63** (1980), 306–317.

[16] R.-O. Buchweitz, *Maximal Cohen-Macaulay modules and Tate-cohomology over Gorenstein rings*, unpublished manuscript (1987)

[17] D. W. Burry and J. F. Carlson, *Restrictions of modules to local subgroups*, Proc. Ams **84** (1982), 181–184.

[18] M. Cabanes and M. Enguehard, *The representation theory of finite reductive groups*. New Mathematical Monographs, Cambridge University Press (2004).

[19] L. G. Chouinard, *Transfer maps*, Comm. Alg. **8** (1980), 1519–1537.

[20] G. Cliff, W. Plesken, and A. Weiss, *Order-theoretic properties of the center of a block*, in: The Arcata Conference on Representations of Finite Groups (editor: P. Fong), Proc. Sympos. Pure Math. **47**, Amer. Math. Soc, Providence RI (1987), 413–420.

[21] D. B. Coleman, *On the modular group ring of a p-group*, Proc. AMS **15** (1964), 511–514.

[22] C. W. Curtis and I. Reiner, *Methods of representation theory* Vol. I, John Wiley and Sons, New York, London, Sydney (1981).

[23] E. Dade, *Deux groupes finis ayant la même algèbre de groupe sur tout corps*, Math. Z. **119** (1964), 345–348.

[24] E. C. Dade, *Block extensions*, Illinois J. Math. **17** (1973), 198–272.

[25] E. C. Dade, *The Green correspondents of simple group modules*, J. Algebra **78** (1982), 357–371.

[26] W. E. Deskins, *Finite Abelian groups with isomorphic group algebras*, Duke Math. J. **23** (1956), 35–40.

[27] A. Dugas and R. Martinez Villa, *A note on stable equivalences of Morita type*, J. Pure Appl. Algebra **208** (2007), 421–433.

[28] A. Facchini, *Module theory.* Modern Birkhäuser Classics. Springer Basel (1998).

[29] W. Feit, *The representation theory of finite groups*, North-Holland Mathematical Library **25**, North-Holland Publishing Company, Amsterdam (1982).

[30] W. Gaschütz, *Über den Fundamentalsatz von Maschke zur Darstellungstheorie der endlichen Gruppen.* Math. Z. **56** (1952) 376–387.

[31] M. Geck and G. Pfeiffer, *Characters of finite Coxeter groups and Iwahori algebras.* Oxford University Press (2000).

[32] D. Gorenstein, *Finite groups*, Chelsea Publishing Company, New York (1980).

[33] J. A. Green, *Some remarks on defect groups*, Math. Z. **107** (1968), 133–150.

[34] J. A. Green, *Polynomial representations of GL_n*, Lecture Notes in Math. **830** (2007), second edition, Springer Verlag, Berlin Heidelberg.

[35] M. Grime, *Adjoint functors and triangulated categories*, Comm. Algebra **36** (2008), 3589–3607.

[36] M. Hertweck and M. Soriano, *On the modular isomorphism problem: groups of order* 2^6, in: Groups, Rings, Algebras (W. Chin, J. Osterburg, D. Quinn, eds.), Contemp. Math **420**, Amer. Math. Soc., Providence, RI (2006), 141–161.

[37] G. Higman, *Units in group rings*, D. Phil. Thesis, Oxford Univ. (1939).

[38] T. Holm, R. Kessar, and M. Linckelmann, *Blocks with a quaternion defect group over a 2-adic ring: the case* $\tilde{A}_4$, Glasg. Math. J. **49** (2007), 29–43.

[39] B. Huppert, *Bemerkungen zur modularen Darstellungstheorie 1. Absolut unzerlegbare Moduln.* Archiv Math. (Basel) **26** (1975), 242–249.

[40] M. Ikeda, *On a theorem of Gaschütz.* Osaka Math. J. **4** (1952), 53–58.

[41] I. M. Isaacs, *Character theory of finite groups*, Dover (1994).

[42] D. A. Jackson, *The group of units of the integral group rings of finite metabelian and finite nilpotent groups*, Quart. J. Math. Oxford **20** (1969), 313–319.

[43] S. A. Jennings, *The structure of the group ring of a p-group over a modular field*, Trans. Amer. Math. Soc. **50** (1941), 175–185.

[44] L. Kadison, *On split, separable subalgebras with counitality condition*. Hokkaido Math. J. **24** (1995), 527–549.

[45] B. Keller, *On the construction of triangle equivalences*, in: Derived Equivalences for Group Rings (S. König, A. Zimmermann), Lecture Notes in Math. **1685**, Springer Verlag, Berlin-Heidelberg (1998), 155–176.

[46] R. Kessar, *A remark on Donovan's conjecture*, Archiv Math. (Basel) **82** (2004), 391–394.

[47] R. Knörr, *On the vertices of irreducible modules*, Ann. Math. **110** (1979), 487–499.

[48] S. König and A. Zimmermann, *Derived equivalences for group rings*, Lecture Notes in Mathematics **1685**, Springer Verlag Berlin Heidelberg (1998) pp. X+246.

[49] B. Külshammer, *Bemerkungen über die Gruppenalgebra als symmetrische Algebra*, J. Algebra **72** (1981), 1–7.

[50] B. Külshammer, T. Okuyama, and A. Watanabe, *A lifting theorem with applications to blocks and source algebras*, J. Algebra **232** (2000), 299–309.

[51] B. Külshammer and L. Puig, *Extensions of nilpotent blocks*, unpublished manuscript.

[52] B. Külshammer and L. Puig, *Extensions of nilpotent blocks*, Invent. Math. **102** (1990), 17–71.

[53] B. Külshammer and G. R. Robinson, *Characters of relatively projective modules II*, J. London Math. Soc. **36** (1987), 59–67.

[54] S. Lang, *Algebraic groups over finite fields*, Amer. J. Math. **78** (1956), 555–563.

[55] M. Linckelmann, *Stable equivalences of Morita type for self-injective algebras and p-groups*, Math. Z. **223** (1996) 87–100.

[56] M. Linckelmann, *On stable equivalences of Morita type*, in: S. König and A. Zimmermann, *Derived equivalences for group rings*, Lecture Notes in Mathematics **1685**, Springer Verlag Berlin Heidelberg (1998), 221–232.

[57] M. Linckelmann, *On splendid derived and stable equivalences between blocks of finite groups*, J. Algebra **242** (2001), 819–843.

[58] M. Linckelmann, *Induction for interior algebras*, Quart. J. Math. **53** (2002), 195–200.

[59] M. Linckelmann, *Finite generation of Hochschild cohomology of Hecke algebras of finite classical type in characteristic zero*, Bull. London Math. Soc. **43** (2011), 871–885.

[60] M. Linckelmann, *Tate duality and transfer in Hochschild cohomology*. J. Pure Appl. Algebra **217** (2013), 2387–2399.

[61] M. Linckelmann, *On stable equivalences with endopermutation source*, J. Algebra **434** (2015), 27–45.

[62] Y. Liu and Ch. Xi, *Constructions of stable equivalences of Morita type for finite-dimensional algebras. III*. J. Lond. Math. Soc. **76** (2007), 567–585.

[63] G. Lusztig, *Characters of reductive groups over finite fields*, Princeton University Press (1984).

[64] A. Marcuş, *Representation Theory of Group-graded Algebras*, Nova Science Publishers, Inc., Commack NY (1999).

[65] H. Matsumura, *Commutative ring theory.* Cambridge studies in advanced mathematics **8**, Cambridge University Press (1986).

[66] B. Mitchell, *Some applications of module theory to functor categories.* Bull. Amer. Math. Soc. **84** (1978) 867–885.

[67] T. Okuyama, *Module correspondences in finite groups*, Hokkaido Math. J. **10** (1981), 299–318.

[68] T. Okuyama and Y. Tsushima, *Local properties of p-block algebras of finite groups*, Osaka J. Math. **20** (1983), 33–41.

[69] D. S. Passman, *Isomorphic groups and group rings*, Pacific J. Math. **35** (1965), 561–583.

[70] L. Puig, *Sur un théorème de Green*, Math. Z. **166** (1979), 117–129.

[71] L. Puig, *Pointed groups and construction of characters.* Math. Z. **176** (1981), 265–292.

[72] L. Puig, *Local fusion in block source algebras*, J. Algebra **104** (1986), 358–369.

[73] L. Puig, *Pointed groups and construction of modules*, J. Algebra **116** (1988), 7–129.

[74] L. Puig, *On Thévenaz' parametrization of interior G-algebras*, Math. Z. **215** (1994), 321–335.

[75] L. Puig, *The center of a block*, in: Finite Reductive Groups (M. Cabanes, ed.), Progress in Math. **141** (1997), 361–372.

[76] I. Reiner, *Topics in integral representation theory*, Lecture Notes in Math. **744**, Springer Verlag, Berlin Heidelberg (1979).

[77] J. Rickard, *Morita theory for derived categories*, J. London Math. Soc. **39** (1989), 436–456.

[78] J. Rickard, *Derived equivalences as derived functors*, J. London Math. Soc. **43** (1991), 37–48.

[79] J. Rickard, *Splendid equivalence: derived categories and permutation modules*, Proc. London Math. Soc. **72** (1996), 331–358.

[80] J. Rickard, *Equivalences of derived categories for symmetric algebras*, J. Algebra **257** (2002), 460–481.

[81] K. W. Roggenkamp and L. L. Scott, *Isomorphisms of p-adic group rings*, Annals Math. **126** (1987), 593–647.

[82] R. Rouquier, *The derived category of blocks with cyclic defect groups*, in: Derived Equivalences for Group Rings (S. König, A. Zimmermann), Lecture Notes in Math. **1685**, Springer Verlag, Berlin-Heidelberg, 1998, 199–220.

[83] A. I. Saksononv, *On the group ring of finite groups*, Publ. Math. Debrecen **18** (1971), 187–209.

[84] J.-P. Serre, *Corps locaux*, Hermann, Paris (1968).

[85] J.-P. Serre, *Linear representations of finite groups*, Graduate Texts in Mathematics **42**, Springer-Verlag, New York (1982).

[86] V. Snaith, *Explicit Brauer induction*, Invent. Math. **94** (1988), 455–478.

[87] R. Steinberg, *Endomorphisms of linear algebraic groups.* Mem. Amer. Math. Soc. **80** (1968), Providence.

[88] J. Thévenaz, *Extensions of group representations from a normal subgroup*, Comm. Algebra **11** (1983), 391–425.

[89] J. Thévenaz, *The parametrization of interior algebras*, Math. Z. **212** (1993), 411–454.

[90] J. Thévenaz, *G-algebras and modular representation theory*, Oxford Science Publications, Clarendon, Oxford (1995).
[91] J. Thompson, *Vertices and sources*, J. Algebra **6** (1967), 1–6.
[92] P. J. Webb, *An introduction to the representations and cohomology of categories*, in: (eds: M. Geck, D. Testermann, J. Thévenaz) *Group representation theory*, EPFL Press, Lausanne (2007).
[93] C. A. Weibel, *An introduction to homological algebra*, Cambridge studies in advanced mathematics **38**, Cambridge University Press (1994).
[94] Ch. Xi, *Stable equivalences of adjoint type.* Forum Math. **20** (2008), 81–97.
[95] F. Xu, *Representations of categories and their applications.* J. Algebra **317** (2007), 153–183.

Index

Printed in the United States
By Bookmasters